DEMANA ▪ WAITS ▪ CLEMENS ▪ GREENE

# Intermediate Algebra
# A Graphing Approach

Franklin
**DEMANA** ▪
The Ohio State
University

Bert K.
**WAITS** ▪
The Ohio State
University

Stanley R.
**CLEMENS** ▪
Bluffton College

Margaret
**GREENE**
Florida Community
College at
Jacksonville

# Intermediate Algebra
# A Graphing Approach

**ADDISON-WESLEY PUBLISHING COMPANY, INC.**

Reading, Massachusetts ▪ Menlo Park, California
New York ▪ Don Mills, Ontario ▪ Wokingham, England ▪ Amsterdam
Bonn ▪ Sydney ▪ Singapore ▪ Tokyo ▪ Madrid ▪ San Juan ▪ Milan ▪ Paris

**Executive Editor:** Ann Heath
**Managing Editor:** Karen M. Guardino
**Production Supervisor:** Peggy McMahon
**Marketing Manager:** Brenda Bravener
**Manufacturing Manager:** Roy Logan
**Cover Design:** Marshall Henrichs
**Technical Art Supervisor:** Joseph K. Vetere
**Technical Art Coordination:** Alena Konecny
                                Loretta Bailey
**Technical Art Illustration:** Tech Graphics
                                Scientific Illustrators
**Copyeditor:** Laura K. Michaels
**Photo Research:** Carlene Coen-Murray
**Photo Credits:**    Focus On Sports, pp. 1; The Image Bank, pp. 74 and 294; Superstock, pp. 143, 369, 442, 505, 614, and 666; The Boston Red Sox Baseball Club, pp. 232; Edmond van Hoorick, Farbdia-Archiv, pp. 569; Black Star, pp. 714.

Library of Congress Cataloging-in-Publication Data


Intermediate algebra : a graphing approach / Franklin Demana . . . [et
   al.].
      p.   cm.
   Includes bibliographical references and index.
   ISBN  0-201-65001-0
   1. Algebra.    2. Algebra—Graphic methods—Data processing.
I. Demana, Franklin D., 1938-
QA152.2.I563    1994
512.9—dc20                                                      93-45357
                                                                   CIP

1 2 3 4 5 6 7 8 9 10  CRW  97969594

# PREFACE

In the past two decades we have seen powerful computational and graphical capabilities first available on mainframe computers, then on desktop computers, and recently on hand-held graphers. Technology is now convenient and affordable. The availability of technology causes mathematics teachers to rethink both the teaching and learning of mathematics.

Our goal in writing *Intermediate Algebra: A Graphing Approach* was to present a friendly, convenient, easy-to-use book which *fully* integrates graphing technology. **We assume that each student has access to a graphing calculator for both class work and for completing assignments.** Graphers are used throughout for concept development, for discovery learning, and for problem solving. At the same time, all of the essential problem-solving concepts and skills of the more traditional symbolic algebra course have been preserved.

Our own combined teaching experience of over one hundred years at The Ohio State University, Bluffton College, and Florida Community College at Jacksonville has given us many insights into the characteristics and needs of intermediate algebra students. There has always been a rich interplay between visualizing a concept and representing it symbolically. However, until recently, our ability to create visualizations has been limited to long, laborious, and often imprecise paper-and-pencil methods. Because of this past limitation, we generally have not fully utilized visualization as a teaching and learning vehicle. Many persons are visual thinkers and visual problem solvers. Modern hand-held graphers now make visualization methods accessible, precise, and time efficient.

There are a variety of pedagogical benefits to fully integrating a grapher into the classroom. We find that using a grapher helps students feel that mathematics is concrete and tangible. We have discovered that our classrooms are less "teacher centered" and more "student centered." As students interact with their grapher and with each other in the spirit of cooperative learning, they talk about mathematics with each other. And as students verbalize mathematics, they learn mathematics.

The content of this book will match a standard course syllabi for intermediate algebra as it has typically been taught. It includes all of the standing topics including paper-and-pencil problem-solving skills that intermediate algebra students are expected to learn. Students who successfully complete a course from this text should be successful in any subsequent college algebra and trigonometry course—even if that subsequent course does not integrate a grapher.

## Features

A number of features have been included in this text to make it a more valuable learning and teaching tool.

### ART

Throughout the book, pictures of grapher screens are provided that show true grapher output. Some of these screens show the home screen on which numerical calculations are completed. On occasion, grapher menus are shown to help students learn grapher technique. There are also instances in which grapher-produced tables of numerical data are shown. And of course, many figures show graphs. Adequate information is provided so that students are able to duplicate any of these graphs on their own grapher.

### MARGIN BOXES

Three different types of margin boxes appear in the text. The purpose of these boxes is to provide information and helpful hints about both the mathematics and about the grapher and its use. The three types of boxes are:

- **Reminder.** The information in these boxes often reviews mathematical convention, language, or notation. In some cases, a prerequisite concept is reviewed briefly. In other cases, warnings about common pitfalls are given.
- **Grapher Note.** These notes clarify information about the grapher and its use. They sometimes refer students to the lab manual to learn some specific grapher skill. Others make suggestions about grapher use.

- **Try This.** These boxes ask students to engage in a brief activity. Sometimes it is an activity involving paper-and-pencil skills. More often it is a grapher activity.

REFERENCES TO THE LAB MANUAL

A Lab Manual is available free to users of this text. Whenever a new grapher skill is introduced, that skill is referenced and taught in the lab manual. All of the popular brands of grapher models are included in this lab manual. This manual has a convenient organization that makes it easy for students to find the particular skill needed for identifying the necessary key strokes on the model of grapher they are using.

EXPLORE WITH A GRAPHER

Many sections of the book begin with a feature by this name. These feature activities use the grapher to introduce a new concept. Usually the pedagogical approach is "discovery learning" and often they conclude by asking the student to state the generalization that they have discovered. These activities are consistent with the pedagogical philosophy

I hear and I forget,

I see and I remember,

I do and I understand.

PROBLEM SITUATIONS

Throughout the text real-life problem situations are discussed. These problem situations allow both student and teacher to refer to various problem types by a name so that they can more easily be classified and revisited later in the course. We find that a graphing-calculator-based approach to problem solving dramatically changes results in the classroom. Instead of being bored and discouraged by conventional, contrived problems, students suddenly grow excited by their ability to explore problems that arise from real-world situations.

We have a three-strand philosophy regarding the interplay between algebraic and graphical methods for solving problems. These three strands can be stated as follows.

1. **Solve Algebraically, and Support Graphically or Numerically.**

   This means that the primary solution method is paper-and-pencil algebra. After an algebraic solution is found, the grapher is used to provide numerical or visual understanding of the problem. A numerical table or a graph can add credibility to the solution.

   We follow this strand when an algebraic solution is reasonably easy and manageable.

2. **Solve Graphically and/or Numerically, and Confirm Algebraically.**

This means that the primary solution method is graphical or numerical. After a graphical or numerical solution has been found, which is probably an approximate solution, algebraic methods are used to confirm an exact solution.

We follow this strand when the graphical visualization of the problem closely matches the physical description of the problem. In some cases, this graphical solution may even be a simulation of the problem.

3. **Solve Graphically** (because algebraic methods are not available, are too difficult, or are beyond the scope of this course.)

This means that, for all practical purposes, a graphical solution is the only one available.

Throughout the text you will find the *Solve Algebraically, Support Graphically or Numerically,* and *Solve Graphically, and Confirm Algebraically* labels used in solutions to examples. These solutions model these various approaches to problem solving.

EXERCISE SETS

There are nearly 5,000 exercises in this text. Special care and attention has been given to crafting exercise sets that give students a balance between drill exercises, which help gain mastery of key skills, and extension exercises, which encourage creative problem solving.

Each exercise set begins with paired exercises that reflect the types of problems that have been solved in the development of the section. These exercises, which mirror the examples in the development, also include application problems similar to those discussed in the Problem Situations.

Besides the basic list of exercises there are several additional categories of exercises—each with its own objectives:

- **Translating Words to Symbols.** These exercises focus on the important task of translating word phrases to mathematical symbols. These exercises are designed to help students with the process of translating a problem situation to its mathematical formulation.
- **Extending the Ideas.** Each exercise set includes several exercises in this category. Their purpose is to encourage students to move beyond the specific skills taught in the lesson. These exercises require a higher level of creativity as students synthesize combinations of skills in solving problems.
- **Looking Back—Making Connections.** These exercises review previously learned concepts and skills. They also require that students apply mathematical skills to solve problems which come from other disciplines or from other concentrations within mathematics.

END-OF-CHAPTER MATERIAL

**Chapter Summary.**    Each chapter ends with a Chapter Summary. This summary classifies the concepts, skills, and solution techniques that have been learned in the chapter using a three-column format. The first column names the concept or skill, the second column defines or describes the named item, and the third column gives an example of it. The material in this summary is organized according to natural classification schemes and is not listed in the order of the chapter development. This organization helps students comprehend a more global view of the chapter as they study for examinations.

## Supplements

*Instructor's Solutions Manual.* The Instructor's Solutions Manual contains worked-out solutions to every problem in the text.

*Student's Solutions Manual.* The Student's Solutions Manual contains worked-out solutions to every odd-numbered problem in the text.

*Graphing Calculator Lab Manual.* This manual combines instructions for using the latest Casio, Sharp, and Texas Instruments calculators with the text.

*Instructor's Manual and Printed Test Bank.* Included in this manual and printed test bank are three alternate test forms that correspond to every chapter, a sample syllabus, and an essay discussing the implementation of technology in the classroom.

**OmniTest II.** A unique algorithm-based test generator for Demana, Waits, Clemens, Greene, *Intermediate Algebra: A Graphing Approach* produces virtually unlimited versions of problems appropriate for the graphing calculator. These printed problems include graphic representation of the images which appear on the graphing calculator. This provides a permanent record of the calculator image with which instructors and students may annotate instructional notes and illustrative comments. These printed problems are also useful in a cooperative learning environment.

## Acknowledgments

We wish to thank the following reviewers for their thoughtful comments and suggestions: Elizabeth Betzel, *Columbus State Community College*; Eunice Everett, *Seminole Community College*; Beverly K. Michael, *University of Pittsburgh-Main Campus*; Stuart Moskowitz, *Butte College*; JoAnne Thomasson, *Pellissippi State Technical Community College*; Charles Vonder Embse, *Central Michigan University*; Tom Williams, *Rowan-Cabarrus Community College*; Elizabeth Wade, *Hiwasee College*; Barry Weber, *Florida Community College-Jacksonville*; Cora S. West, *Florida Community College-Jacksonville*.

We also wish to thank our class testers for their contributions: Susan Berman, *Rowan-Cabarrus Community College*; June W. Hundley, *Rowan-Cabarrus Community College*; David L. Sherer, *Rowan-Cabarrus Commu-*

*nity College*; Mary Margaret Shoaf-Grubbs, *College of New Rochelle*; Bonnie Simon, *Naugatuck Valley Technical Community College*; Cora S. West, *Florida Community College-Jacksonville*; Barry Weber, *Florida Community College-Jacksonville*; Harriet Wood, *Rowan-Cabarrus Community College*.

Our appreciation and thanks to our team of accuracy checkers for their dedication and precision: Cindy Bernlohr, *The Ohio State University*; Rachel Clemens, Penny Dunham, *Muhlenburg College*; Amy Edwards, *The Ohio State University*; Kevin Greene; Glenn Parrish, *Florida Community College-Jacksonville*; Sally Thomas, *Orange Coast College*; Cora S. West, *Florida Community College-Jacksonville*. In addition, we thank Cindy Bernlohr for her assistance throughout this project.

We would also like to thank Melissa Acuña, Ann Heath, Kathleen Manley, Peggy McMahon, Carlene Coen Murray, Laurie Petrycki, and Joe Vetere at Addison-Wesley for their hard work in making this first edition of *Intermediate Algebra: A Graphing Approach* a reality.

# CONTENTS

# Numerical Mathematics and the Graphing Calculator

### AN APPLICATION

When a baseball is thrown straight up into the air with an initial velocity of 88 ft/sec, the equation

$$h = -16t^2 + 88t$$

describes the ball's height $t$ seconds later. Many questions relating height and time can be answered using this equation.

## 1.1   Real Numbers and a Graphing Calculator

Repeating and Terminating Decimals   ▪   Comparing Real Numbers and the Real Number Line   ▪   Inequalities and Intervals on the Number Line   ▪   Mathematical Notation and Exponents   ▪   Order of Operations

Numbers are used to represent quantity. Number concepts as they are applied to real-world problems become the motivation for many topics in algebra.

The **natural numbers**, or **counting numbers**, are the numbers in the infinite set

$$N = \{1, 2, 3, \cdots\}.$$

Many applications require only the natural numbers. For example, the natural numbers are used to report data on the number of students enrolled in each class of a university. When the number 0 is included, the set then is called the set of **whole numbers.**

For some applications, we need to extend our number set to include the set called **integers**. Integers are the numbers in the set

$$I = \{\cdots, -3, -2, -1, 0, 1, 2, 3, \cdots\}.$$

For example, to communicate gain versus loss or profit versus debt, the numbers are selected from the integers.

In still other settings, we need to use numbers from the set of **rational numbers**. The rational numbers are the set of numbers of the form $\frac{m}{n}$, where $m$ and $n$ are integers, $n \neq 0$. For example, the numbers used to represent the three shaded regions of the chart in Figure 1.1 are rational numbers

$$\frac{98}{802} = 0.1221\cdots, \qquad \frac{385}{802} = 0.4800\cdots, \qquad \frac{319}{802} = 0.3977\cdots.$$

Rational numbers can be reported in any one of three representations—as a decimal, as shown in the above three examples; as a fraction, and as a percent, as shown in the figure. For example,

$$\frac{2}{5} \qquad 0.4 \qquad 40\%$$

are three different representations of the same rational number.

And note that in the pie chart, the decimal form of each rational number is represented as a percentage, rounded to the nearest tenth.

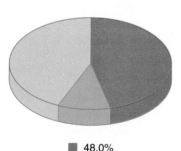

■ 48.0%
■ 12.2%
■ 39.8%

**FIGURE 1.1** A chart like this is called a Pie Chart.

## Repeating and Terminating Decimals

When a rational number $\frac{m}{n}$ is written in its decimal representation by finding $m \div n$, the resulting decimal may have an infinite number of digits. Each rational number can be represented either as a repeating or terminating decimal.

The following are examples of both types of rational numbers:

| **Repeating Decimals** | **Terminating Decimals** |
|---|---|
| $\dfrac{1}{3} = 0.333\cdots$ | $\dfrac{11}{25} = 0.44$ |
| $\dfrac{2}{7} = 0.285714285714\cdots$ | $\dfrac{43}{20} = 2.15$ |

### EXPLORE WITH A GRAPHER

Use a grapher to find the decimal representation of each of the following rational numbers. Decide which you think are **terminating** and which are **repeating**.

1. $\dfrac{5}{8}$    2. $\dfrac{1}{33}$    3. $\dfrac{5}{11}$

4. $\dfrac{9}{20}$    5. $\dfrac{11}{5}$    6. $\dfrac{5}{24}$

**Formulate a Definition**    Write several sentences that explain your definition of the words repeating and terminating.

A **repeating decimal** has to the right of the decimal point a block of digits that continues repeating. When the repeating digit is 0, however, the zeros are not written or displayed on the grapher screen and such a decimal is called a **terminating decimal**.

In a repeating decimal, the block of digits that repeat are often indicated by using a "bar" above the repeating block, as shown next.

$$\frac{1}{3} = 0.333\cdots \qquad \frac{2}{7} = 0.285714285714\cdots$$

$$= 0.\overline{3} \qquad\qquad = 0.\overline{285714}$$

For some repeating decimals, the block of digits that repeats is longer than the number of digits on your calculator display. For example, when you find $1 \div 19$, the repeating block, while it exists, is not evident on the calculator display.

**EXAMPLE 1** Finding Repeating and Terminating Decimals

Determine with a calculator, if possible, whether the decimal representation of each of the following rational numbers is repeating or terminating:

a) $\dfrac{143}{999}$

b) $\dfrac{723}{250}$

c) $\dfrac{9}{13}$

SOLUTION

a) $\dfrac{143}{999} = 143 \div 999$

$= 0.143143\cdots$

$= 0.\overline{143}$

The decimal representation of $\dfrac{143}{999}$ is repeating.

b) $\dfrac{723}{250} = 2.892$

The decimal representation of $\dfrac{723}{250}$ is terminating.

c) $\dfrac{9}{13} = 0.6923076923\cdots$

$= 0.\overline{692307}$

The decimal representation of $\dfrac{9}{13}$ is repeating. ■

**GRAPHER NOTE**

Because a calculator can display (or remember internally) only a finite number of digits, calculator output is in effect a terminating decimal. However, it is a close approximation to the actual repeating decimal.

**TRY THIS**

Show that on your calculator $123456 \times 0.33333$ and $123456 \times \dfrac{1}{3}$ are not equal. Explain why.

Some numbers are neither repeating nor terminating. Any number whose decimal representation is neither repeating nor terminating is called an **irrational** number. The **radical symbol**, $\sqrt{\ }$, is used to describe the positive square root of a number. For example, $\sqrt{25} = 5$ since $5 \cdot 5 = 25$. Although $\sqrt{25} = 5$ is a rational number, most radicals are irrational. Both of the following are irrational numbers. Confirm these two decimal values on your grapher.

$$\sqrt{2} = 1.414213562\cdots \qquad \pi = 3.141592654\cdots$$

We are now able to define the set of real numbers.

---

**DEFINITION 1.1** Real Numbers

The **real numbers** consist of all numbers that are either rational or irrational. The points on a number line are a model for these numbers.

---

Some of the important subsets of real numbers are shown in Figure 1.2. Notice that letters of the alphabet—$m$, $n$ and $x$—are used to reserve a place for a number. Such a letter is called a **variable.**

Real numbers ($R$)

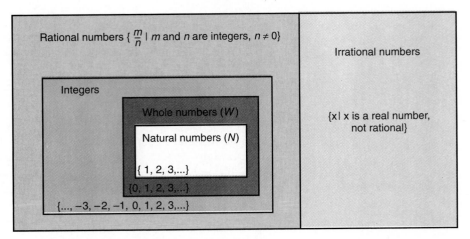

FIGURE 1.2 Important subsets of real numbers.

**REMINDER**

The brackets { and } are used when a set or a list is being identified. A set might be described by listing the elements of the set; for example, $A = \{2, 3, 5\}$. The set $Q = \{\frac{m}{n} \mid m$ and $n$ are integers, $n \neq 0\}$ is an alternate way of defining a set. The vertical bar | is read 'such that.' This latter method is called the *set builder* notation.

The relationships among these subsets of real numbers are also indicated by the tree diagram in Figure 1.3. All natural numbers are whole numbers. All whole numbers are integers. All integers are rational numbers. And all rational numbers are real numbers. A real number is *either* rational or irrational but not both.

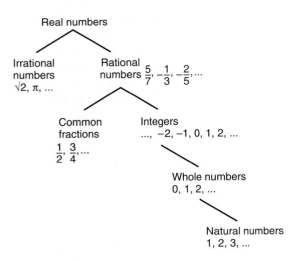

**FIGURE 1.3** The most important subsets of the set of real numbers.

## Comparing Real Numbers and the Real Number Line

When two numbers represent identical values, we call them **equal** and we use the symbol "=". For example, we write

$$\frac{3}{4} = 0.75$$

because the symbols on the right and left represent the same value.

There are three basic properties of equality used frequently.

---

### Properties of Equality

For all real numbers $a$, $b$, and $c$:

1. $a = a$ — Reflexive Property
2. If $a = b$, then $b = a$. — Symmetric Property
3. If $a = b$ and $b = c$, then $a = c$. — Transitive Property

---

We often use these properties without being fully aware of it. For example, when we reason,

$$x = 72 - 31$$

$$72 - 31 = 41 \text{ and therefore}$$

$$x = 41,$$

we are using the Transitive Property of Equality. We will use these properties in our equation-solving experiences.

It is often helpful to picture numbers as points on a line. A **real number line** consists of a line with a point—the **origin**—labeled as zero. Points to the left of zero are associated with the **negative** numbers and points to the right of zero are associated with the **positive** numbers (see Figure 1.4). Each real number is

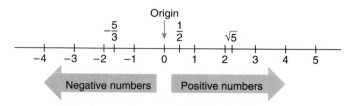

**FIGURE 1.4** The real number line is divided into positive numbers, negative numbers, and zero.

associated with a point on the line, and each point on the line is associated with a real number.

The real number line can be used to compare two real numbers. If any two distinct real numbers $a$ and $b$ are selected, one is to the left of the other on the real number line. If $a$ is to the left of $b$, we say that $a$ **is less than** $b$ and we write $a < b$. If $a$ is to the right of $b$, we say that $a$ **is greater than** $b$ and we write $a > b$.

The symbols $<$, $>$, $\leq$, and $\geq$ are called **inequality** symbols and are read as "less than" ($<$), "greater than" ($>$), "less than or equal to" ($\leq$), and "greater than or equal to" ($\geq$). Following is a summary of these symbols and their meanings.

| Symbol | Meaning | Example |
|--------|---------|---------|
| $a < b$ | $a$ is less than $b$. | $-5 < 2$ |
| $a > b$ | $a$ is greater than $b$. | $-3 > -5$ |
| $a \leq b$ | $a$ is less than or equal to $b$. | $2 \leq 5$ |
| $a \geq b$ | $a$ is greater than or equal to $b$. | $4 \geq 4$ |

**EXAMPLE 2**    Writing Inequality Statements

For the following, write two true statements by replacing the $\bigcirc$ symbol with two symbols selected from among $<$, $>$, $\leq$, and $\geq$:

a) $-3 \bigcirc 5$     b) $4 \bigcirc -1$     c) $-2 \bigcirc -2$

SOLUTION

a) $-3 < 5$     or     $-3 \leq 5$
b)  $4 > -1$     or      $4 \geq -1$
c) $-2 \leq -2$     or     $-2 \geq -2$     ■

Two important properties of inequalities that are assumed throughout are as follows.

---

### Properties of Inequality

For all real numbers $a, b,$ and $c,$

1. either $a < b, a = b,$ or $a > b$.     Trichotomy Property
2. if $a < b$ and $b < c$, then $a < c$.     Transitive Property

## Inequalities and Intervals on the Number Line

Infinite collections of numbers can be described by using a variable in conjunction with an inequality symbol. The shading of the line in Figure 1.5 represents the set of all real numbers that are less than 3; that is, it represents all real number solutions of the inequality $x < 3$. This set of numbers is called an **interval** on the real number line. The **interval notation** for this set is $(-\infty, 3)$.

**FIGURE 1.5** The shaded section represents all real numbers $x$ that satisfy the inequality $x < 3$. It is the interval $(-\infty, 3)$.

> **REMINDER**
>
> The symbol $\infty$ is called the **infinity symbol** and is used to communicate all numbers less than 3 in an interval like $(-\infty, 3)$ and all numbers greater than $-4$ in an interval like $(-4, \infty)$.

In Figure 1.5, an open circle is used to show that the number 3 is *not* included in the interval. When the endpoint of an interval is included, a closed circle is used on the line and a bracket "]" is used in the interval notation. For example, the interval $(-\infty, 3]$ represents all solutions to the inequality $x \le 3$.

**EXAMPLE 3**    Drawing Intervals on the Real Number Line

Draw on a real number line the interval $[-4, \infty)$, which represents the solution to the inequality $x \ge -4$.

SOLUTION
See Figure 1.6.

**FIGURE 1.6** This shaded section represents all real numbers $x$ that satisfy the inequality $x \ge -4$. It is the interval $[-4, \infty)$. ■

The intervals discussed in Example 3 and the paragraph preceding it are called **infinite intervals** because they extend without bound in either the positive or the negative direction.

The shaded section of the line in Figure 1.7 represents a **finite interval**. This interval consists of all real numbers greater than $-3$ and less than or equal to 2. This interval is represented using the interval notation $(-3, 2]$.

**FIGURE 1.7** The shaded section represents all real numbers greater than −3 and less than or equal to 2. It is the interval (−3, 2].

**EXAMPLE 4**    Drawing Intervals on the Real Number Line

Draw on a real number line the interval [−1, 3), which is the interval that represents the solution to the inequality −1 ≤ x < 3.

SOLUTION
See Figure 1.8.

**FIGURE 1.8** The shaded section represents all real numbers greater than or equal to −1 and less than 3. It is the interval [−1, 3). ■

## Mathematical Notation and Exponents

Notational conventions are used in mathematics as abbreviations for word phrases or mathematical symbols. It is important for reading and understanding mathematics to learn what these notational conventions mean. For example, the familiar symbols +, −, ×, and ÷ represent the operations of addition, subtraction, multiplication, and division, respectively. Sometimes · is used instead of × for multiplication and sometimes / is used instead of ÷ for division. For example, $5 \times 4 = 5 \cdot 4$ and $8/2 = 8 \div 2$.

In another example, an expression like

$$6 \cdot 6 \cdot 6 \cdot 6 \text{ is abbreviated } 6^4.$$

This notation communicates that four factors of 6 have been multiplied together. $6^4$ is read "six to the fourth power."

In general, for any number $b$ and any positive whole number $n$, $b$ to the $n$th power, written $b^n$, means

$$b^n = \underbrace{b \cdot b \cdot \; \cdots \; \cdot b}_{n \text{ factors of } b}.$$

In this expression, $b$ is called the **base** and $n$ is the **exponent**, or **power**.

This definition of "$b$ to the $n$th power" is used to evaluate numerical expressions in the next example.

**EXAMPLE 5**  Using Exponents

Evaluate each of the following expressions with paper and pencil and support your results for parts (a) and (b) with a grapher:

a) $3^5$  b) $\left(\dfrac{2}{5}\right)^3$  c) $\dfrac{3^4 \cdot 2^5}{6^2}$

SOLUTION

Figure 1.9 shows the evaluation of (a) and (b) on a grapher.

a) $3^5 = 3 \cdot 3 \cdot 3 \cdot 3 \cdot 3 = 243$

b) $\left(\dfrac{2}{5}\right)^3 = \dfrac{2}{5} \cdot \dfrac{2}{5} \cdot \dfrac{2}{5} = \dfrac{8}{125}$

c) $\dfrac{3^4 \cdot 2^5}{6^2} = \dfrac{(3 \cdot 3 \cdot 3 \cdot 3) \cdot (2 \cdot 2 \cdot 2 \cdot 2 \cdot 2)}{6 \cdot 6}$

$= \dfrac{(3 \cdot 3 \cdot 2 \cdot 2) \cdot (3 \cdot 3 \cdot 2 \cdot 2 \cdot 2)}{6 \cdot 6}$

$= 3 \cdot 3 \cdot 2 \cdot 2 \cdot 2$

$= 9 \cdot 8 = 72$   ■

```
3^5
                       243
(2/5)^3
                      .064
8/125
                      .064
```

**FIGURE 1.9**  The symbol $^\wedge$ represents exponentiation. Notice that parentheses are needed in line 2. Also notice that fractions are expressed in their decimal representation. The answer to line 2 is expressed as .064, although on some graphers the FRAC or an equivalent command can be used.

## Order of Operations

We begin this section with an exploration.

### EXPLORE WITH A GRAPHER

Evaluate each of the following on a grapher. Does the grapher give equal outputs?
1. $2 + 3 \times 4$ and $(2 + 3) \times 4$
2. $22 + 38 \div 2 \times 3$, $22 + 38 \div (2 \times 3)$, and $(22 + 38) \div 2 \times 3$
3. $2 \times 3^2$ and $(2 \times 3)^2$

**TRY THIS**

Evaluate on a grapher:
(a) $3^2$, (b) $(-3)^2$, and (c) $-3^2$. Explain why the last answer is negative.

You probably discovered in this exploration that the outcome of a calculation depends on which order the operations are executed. The expression

$$2 + 3 \times 4^2 - 6 \div 2$$

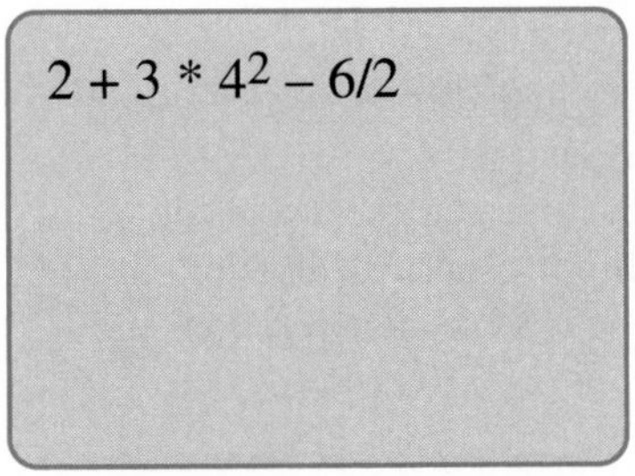

$$2 + 3 * 4^2 - 6/2$$

**FIGURE 1.10** The grapher display of the expression $2 + 3 \times 4^2 - 6 \div 2$. The symbol $*$ represents multiplication and $/$ represents division.

is written without using any parentheses ( ) or brackets [ ]. When it is keyed into a calculator without using either of these, the logic of the calculator must determine the order in which the operations are to be performed.

Here is how your grapher does the calculation in Figure 1.10. As shown in the following, first the $4^2$ is evaluated. Then the multiplication and division are done, and finally the addition and subtraction.

$$2 + 3 \times 4^2 - 6 \div 2 =$$
$$= 2 + 3 \times 16 - 6 \div 2$$
$$= 2 + 48 - 3$$
$$= 47$$

Throughout this text, the following agreements about order of operations are used. These agreements are consistent with the logic used in graphers.

---

## Order of Operations

If no parentheses ( ) or brackets [ ] are present, do the following:
1. Evaluate all powers, working from left to right.
2. Do any multiplications and divisions in the order in which they occur, reading from left to right.
3. Do any additions or subtractions in the order in which they occur, reading from left to right.

If parentheses or brackets are present, do the following:

1. First use the above steps within each pair of parentheses or brackets.
2. If the expression includes nested parentheses, evaluate the expression in the innermost set of parentheses first.

---

$$1/2 + 3 * 5$$
$$15.5$$
$$3 * 2^3 + 12/4 - 2$$
$$25$$

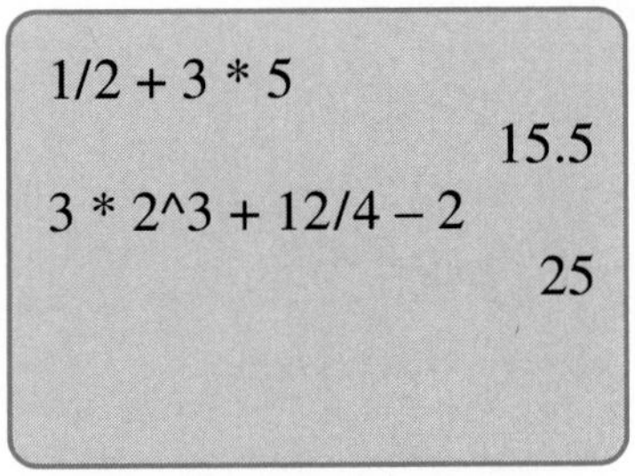

**FIGURE 1.11** The grapher evaluation of the expressions in Example 6.

This order of operations agreement is applied in Example 6.

**EXAMPLE 6**   *Applying the Order of Operations*

Figure 1.11 shows that the following expressions have been evaluated. In what order are each of these operations performed?

a) $1/2 + 3 * 5$
b) $3 * 2^3 + 12/4 - 2$

SOLUTION

a) In $1/2 + 3 * 5$, the $1 \div 2$ and $3 \times 5$ are performed first. So the expression is converted to $0.5 + 15$. The addition is performed last.

b) In $3 * 2^3 + 12/4 - 2$, evaluating $2^3$ is performed first and the expression is converted to $3 * 8 + 12/4 - 2$. Then the multiplication and division are performed and the expression becomes $24 + 3 - 2$. Next the operations are performed from left to right. ∎

## Exercises for Section 1.1

In Exercises 1–8, use a calculator to find a decimal representation for each of the rational numbers. Identify, if possible, whether the decimals are repeating or terminating. (Write the repeating decimals with a bar over the repeating block.)

1. $\dfrac{5}{7}$

2. $\dfrac{7}{11}$

3. $\dfrac{17}{20}$

4. $\dfrac{11}{25}$

5. $\dfrac{11}{6}$

6. $\dfrac{37}{36}$

7. $\dfrac{726}{250}$

8. $\dfrac{92}{40}$

In Exercises 9–17, identify each number as a member of as many of these sets as apply: natural number, whole number, integer, rational number, and irrational number.

9. $-5$

10. $129$

11. $0$

12. $\dfrac{1}{2}$

13. $\dfrac{-5}{8}$

14. $\dfrac{-9}{26}$

15. $\sqrt{3}$

16. $\sqrt{5}$

17. $\dfrac{-3}{-5}$

In Exercises 18–24, state whether the given statement is true or false.

18. Every whole number is also an integer.

19. Every integer is also a whole number.

20. All whole numbers are natural numbers.

21. Some rational numbers are also irrational.

22. A decimal representation of a real number is either a repeating decimal or a terminating decimal.

23. A decimal representation of an irrational number is neither repeating nor terminating.

24. Each integer $b$ is equivalent to the rational number $\dfrac{b}{1}$.

In Exercises 25–34, draw the given interval on a real-number line.

25. $(-\infty, 4)$

26. $(-\infty, -3]$

27. $[-3, \infty)$

28. $(1, \infty)$

29. $[-2, 1)$

30. $(-3, 4)$

31. $[-4, 2]$

32. $(1, 4]$

33. $(-4, -1]$

34. $[-3, 5]$

In Exercises 35–40, write two true statements by replacing the ○ symbol with two of the symbols selected from $<$, $>$, $\le$, and $\ge$.

35. $-3 \bigcirc -1$

36. $-2 \bigcirc 3.5$

37. $5.09 \bigcirc 5.009$

38. $-3.1 \bigcirc -3$

39. $4.38 \bigcirc 4.388$

40. $22.8 \bigcirc 21.6$

In Exercises 41–46, evaluate each expression.

41. $5^2 \cdot 2^3$

42. $3^2 \cdot 2^4$

43. $7^2 \cdot 6 \cdot 2^3$

44. $(3 \cdot 2)^3$

45. $\left(\dfrac{2}{3}\right)^2$

46. $\left(\dfrac{1}{2}\right)^3$

In Exercises 47–56, evaluate these expressions with paper and pencil. Support your answers with a grapher.

**47.** $7 + 3 \times 5 \div 2$      **48.** $6 \div 2 + 5 \times 8 - 2$

**49.** $2^3 + 7 - 3^2$      **50.** $18 - 2 \times 3 \div 4$

**51.** $2 + 48 \div 2 \times 3 \div 6$

**52.** $17 + 3^3 \times 2 - 5 + 24 \div 6$

**53.** $256 \div 4 \div 2^2 + 19 \times 2$

**54.** $19 - 5^3 \times 2 + 5^2 \times 3^3$

**55.** $2^3 \times 4^2 + 3 \times 5^3$

**56.** $18 \div 2 + 51 \div 3 - 15 \div 3$

In Exercises 57–64, evaluate the expressions with paper and pencil. Support your answers with a grapher.

**57.** $(2 \times 3 + 4 \times 7) \div (8 - 6)$

**58.** $(2 \times 5 + 4) - (6 \times 4) \div 2$

**59.** $[(2 + 3^2) + 3] \div 4 + 6$

**60.** $[6 + 3^2 \cdot (5 - 3)] \times (2^3 - 3)$

**61.** $2 + 3 \cdot [24 - (7 + 2 \cdot (6 - 2))]$

**62.** $123 - 2 \cdot [18 + 2 \cdot (19 - 2 \cdot (7 + 1))] - 4$

**63.** $[(3^2 + 2)^2 + 4]^2$

**64.** $\left[ [(19 - 2^3) + 3] \div 2 \right]^3$

TRANSLATING WORDS TO SYMBOLS

**65.** Is the expression "five squared plus three squared" ambiguous or is there only one symbolic interpretation of this phrase?

**66.** Explain two different interpretations that could be given the phrase "five plus three squared."

**67.** Use number symbols to write a unique expression for each of the phrases "five squared plus three squared" and "five plus three, squared." Do these expressions have the same numerical value?

**68.** Draw on a number line the interval of numbers that are greater than $-2$ and less than or equal to 3.

**69.** Draw on a number line the interval of numbers that are less than 5 and greater than or equal to $-3$.

**70.** Use pauses and voice inflection to state orally the phrase "two plus three times four" so that in one case it means $(2 + 3) \times 4$ and in the other case it means $2 + (3 \times 4)$.

**71.** Use pauses and voice inflection to state orally the phrase "two plus three squared" so that in one case it means $(2 + 3)^2$ and in the other case it means $2 + 3^2$.

EXTENDING THE IDEAS

Exercises 72 and 73 refer to the pie chart below. This chart shows the percentage of total company expenses allotted to the departments of Production, Finance, Sales, and Marketing of the company Technology Unlimited.

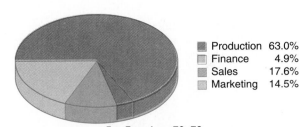

| | |
|---|---|
| ■ Production | 63.0% |
| ▨ Finance | 4.9% |
| ■ Sales | 17.6% |
| ▧ Marketing | 14.5% |

For Exercises 72–73.

**72.** *Finding Total Expenses.* If the total expenses of the company are $123,000, what are the expenses for the Marketing Department?

**73.** *Comparing Expenses.* How many more dollars of expenses does the Sales Department have than does the Marketing Department?

**74.** Some graphers allow comparisons between numbers. When a comparison is true, the calculator returns the value 1 and when a comparison is false, the calculator returns the value 0. For example, if you key in $3 < 5$ and press ENTER, the value 1 shows on the screen. If you key in $-3 < -5$ and press ENTER, the value 0 shows on the screen. What values will each of the following expressions return? Test your answers with your calculator.

a) $-3 < 5$

b) $-1 \geq -0.5$

c) $2 + 3 - 6 \leq 5$

d) $5 + 3 \cdot 2 \geq 15$

**75. Writing to Learn.** Write several sentences that explain the pattern in the decimal number $x = 0.10110111011110\cdots$. Discuss why this number $x$ is an irrational number, assuming the pattern continues.

In Exercises 76 and 77, use the bar chart below, which shows the school attendance for a certain week at Garfield School.

**76.** *Comparing Data.*  Is the change in attendance from Monday to Tuesday described as 14 or $-14$? Assign an integer that describes the change in attendance from Tuesday to Wednesday, from Wednesday to Thursday, and from Thursday to Friday.

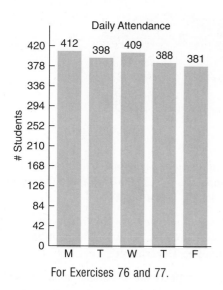

For Exercises 76 and 77.

**77.** *Numerical Information from a Bar Graph.* The attendance on Friday is what percentage of the attendance on Monday?

Exercises 78–80 refer to the following graph, which is intended to reflect the same data as in the graph for Exercises 76 and 77.

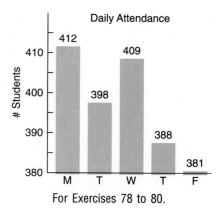

For Exercises 78 to 80.

**78.** *Visual Interpretation of Percentage.* Is the length of the bar for Friday approximately (a) 5%, (b) 25%, (c) 78%, or (d) 92% the length of the bar for Monday?

**79.** *Numerical Interpretation of Percentage.* Would you estimate Friday's attendance of 381 to be (a) 5%, (b) 25%, (c) 78%, or (d) 92% of Monday's attendance of 412?

**80. Writing to Learn.** Write a paragraph that discusses how Exercises 78 and 79 are related to the fact that the bar graph for Exercises 76 and 77 more accurately reflects the attendance data. What is wrong with the graph shown for Exercises 78 to 80?

LOOKING BACK—MAKING CONNECTIONS

**81.** *Estimating the Cost of Carpet.* Recall from your previous experiences in mathematics that the area $A$ of a rectangle of length $L$ and width $W$ can be found by using the formula $A = LW$. Find the area of the room in the following picture. If carpet costs approximately $2.85 per square foot, approximately how much will carpet for this room cost?

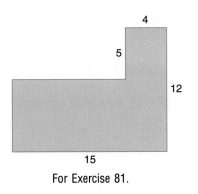

For Exercise 81.

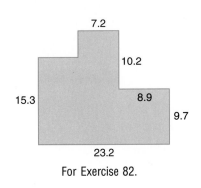

For Exercise 82.

**82.** *Estimating the Cost of Carpet.* Find the area of the room pictured. If carpet costs approximately $2.25 per square foot, approximately how much will carpet for this room cost?

**83.** Recall that the perimeter of a polygon is the distance around the polygon. Find the perimeters of the polygons in Exercises 81 and 82.

| 1.2 | Properties of Real Numbers and the Basic Rules of Operations |

Basic Operations ▪ Properties of Opposites ▪ Absolute Value ▪ Summary of Basic Properties ▪ Distance on a Number Line

The four basic operations on real numbers are **addition, subtraction, multiplication,** and **division,** denoted $+$, $-$, $\times$, and $\div$, respectively. These four operations can easily be located on a scientific calculator or a grapher.

When electrical engineers design the circuits for a calculator or computer (see Figure 1.12), they must ensure these circuits follow the order of operations outlined in Section 1.1 as well as all other properties satisfied by the real number operations.

### Basic Operations

Figure 1.13 on the following page shows that the real numbers are either **negative, zero,** or **positive.** Each positive number has an opposite that is negative, and each negative number has an opposite that is positive; for example,

$$3 \text{ is the opposite of } {}^{-}3$$

and

$${}^{-}3 \text{ is the opposite of } 3.$$

**FIGURE 1.12** Electrical engineers must design calculator circuit boards so that they obey all properties of the real numbers that we will study in this section.

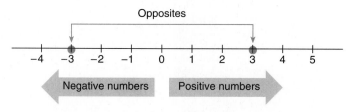

Opposites

Negative numbers    Positive numbers

**FIGURE 1.13**  The real number line.

Numbers that are written with a positive or negative sign, such as $^+4$ and $^-7$, are called **signed numbers.** The **absolute value** of a signed number is its distance from zero on the number line. For example, the absolute value of 3 and $-3$ are both equal to 3. When adding signed numbers, the following rules apply.

---

### Adding Signed Real Numbers

1. Add two numbers of the *same* sign by adding their absolute values. The sign of the answer ($+$ or $-$) is the same as the sign of the two numbers.

$$^-3.1 + {}^-5.2 = {}^-(3.1 + 5.2) = {}^-8.3 \qquad 7.25 + 1.23 = 8.48$$

2. Add two numbers with *different* signs by finding the difference between the absolute values of the numbers. The answer is positive if the positive number has the larger absolute value. The answer is negative if the negative number has the larger absolute value.

$$^-3 + 5 = (5 - 3) = 2 \qquad 4 + {}^-7 = {}^-(7 - 4) = {}^-3$$

---

The operations of addition ($+$) and multiplication ($\times$ or $\cdot$) are considered **primary operations**. Subtraction and division are defined as the **inverse operations** of addition and multiplication, respectively. That is,

| Subtraction | Division |
|---|---|
| $a - b = a + (^-b)$ | $a \div b = a\left(\dfrac{1}{b}\right) = \dfrac{a}{b}, \text{ provided } b \neq 0$ |

Example 1 illustrates addition and subtraction of real numbers. The properties of signed numbers are used.

**REMINDER**

Notice the difference between the symbol − and the symbol ⁻. The first signifies "subtraction" and the second signifies "the opposite of."

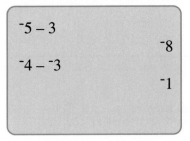

**FIGURE 1.14** Grapher support for parts (c) and (e).

**EXAMPLE 1**    Finding Sums and Differences

Apply the rules of addition and subtraction to simplify the following. Support parts (c) and (e) on a grapher.

a) $12 + {}^{-}2$
b) $^{-}15 + 8$
c) $^{-}5 - 3$
d) $4 - {}^{-}3$
e) $^{-}4 - {}^{-}3$.

SOLUTION

a) $12 + {}^{-}2 = {}^{+}(12 - 2) = 10$
b) $^{-}15 + 8 = {}^{-}(15 - 8) = {}^{-}7$
c) $^{-}5 - 3 = {}^{-}5 + {}^{-}3 = {}^{-}8$        **Recall that** $a - b = a + {}^{-}b.$
d) $4 - {}^{-}3 = 4 + 3 = 7$
e) $^{-}4 - {}^{-}3 = {}^{-}4 + 3 = {}^{-}1$

See Figure 1.14 for grapher support of parts (c) and (e). ■

When multiplying and dividing signed numbers, first determine the sign of the product or the quotient. Then multiply or divide the absolute values of the numbers. The following rules apply when finding the sign of the product or quotient.

---

### Multiplying and Dividing Signed Real Numbers

1. The product or quotient of two numbers with the *same* sign is positive.

$$^{-}6 \div {}^{-}2 = 3 \qquad 18 \div 3 = 6$$

2. The product or quotient of two numbers with *opposite* signs is negative.

$$^{-}14 \div 2 = {}^{-}7 \qquad 21 \div {}^{-}7 = {}^{-}3$$

---

## Properties of Opposites

In algebra it is correct to read the symbol $^{-}a$ as "the opposite of $a$." This statement is true no matter which real number $a$ represents. For example,

$$^{-}({}^{-}7) \text{ is the opposite of } {}^{-}7$$

and

$$^{-}7 \text{ is the opposite of } 7.$$

Consequently, $^-(^-7) = 7$, and this equation is an example of the following general rule.

---

### Double Negative Property

If $a$ is any real number, $^-(^-a) = a$.

---

**GRAPHER NOTE**

Find the key on your grapher that is used for "opposites." Notice that it is not the key used for subtraction.

The second rule that relates opposites is the **additive inverse property**, as follows:

---

### Additive Inverse Property

For any real number $a$, the numbers $a$ and $^-a$ are additive inverses. This means that

$$a + (^-a) = (^-a) + a = 0.$$

---

**EXAMPLE 2**   Using Properties of Opposites

State what property of opposites is used for each of the following simplifications:
  a) $^-(^-\sqrt{3}) = \sqrt{3}$
  b) $^-(5 + 2) = ^-7$
  c) $^-(^-3) + 2 = 3 + 2 = 5$
  d) $14 + ^-(6 + 8) = 14 + ^-14 = 0$

SOLUTION

  a) The double negative property was used.
  b) The definition of $5 + 2$ was used but no property of opposites was used.
  c) The double negative property was used.
  d) The definition of $6 + 8$ and the additive inverse property were used.  ■

We end this subsection with a summary of the properties of opposites as they are used in multiplication and addition. As you read these properties, realize that the $a$ and $b$ represent real numbers that could be either positive or negative.

**Properties of Opposites**

Let $a$ and $b$ be any real numbers.

| Property | Example |
|---|---|
| $(^-1)a = {}^-a$ | $(^-1)7 = {}^-7$ |
| $(^-a)b = {}^-ab$ | $(^-8)3 = {}^-8 \cdot 3 = {}^-24$ |
| $(^-a)(^-b) = ab$ | $(^-3)(^-5) = 15$ |
| ${}^-a + {}^-b = {}^-(a+b)$ | ${}^-7 + {}^-4 = {}^-(7+4) = {}^-11$ |
| $-\dfrac{a}{b} = \dfrac{{}^-a}{b} = \dfrac{a}{{}^-b}$ | $-\dfrac{2}{5} = \dfrac{{}^-2}{5} = \dfrac{2}{{}^-5}$ |

## Absolute Value

The **absolute value** of a real number $a$ represents its distance from the origin on the number line and is denoted $|a|$. Because distance is positive, the absolute value of any real number is positive, as shown in Figure 1.15; for example,

$$|4| = 4 \qquad |{}^-2.6| = 2.6 \qquad |{}^-\sqrt{2}| = \sqrt{2}.$$

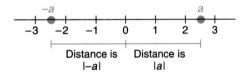

**FIGURE 1.15** On a number line, $a$ and ${}^-a$ are the same distance from 0. Therefore $|{}^-a| = |a|$.

Sometimes absolute value is used in an algebraic expression such as $|x^2 - 4|$. The following definition of absolute value applies to these algebraic settings as well.

**DEFINITION 1.2**   Absolute Value

If $a$ is any real number, the **absolute value** of $a$, denoted $|a|$, is defined by

$$|a| = \begin{cases} a, & \text{if } a \text{ is positive or zero} \\ {}^-a, & \text{if } a \text{ is negative.} \end{cases}$$

Notice that $^-|a|$ and $|^-a|$ are not the same numbers. For example, $|^-3|$ is 3 and $^-|3|$ is the opposite of 3. Pay attention to this subtle but important difference.

**EXAMPLE 3**   Finding Absolute Value

Apply the definition of absolute value to simplify the following expressions:

a) $^-|^-7|$
b) $^-|13.2|$
c) $|^-21| + 21$
d) $^-|21| + 21$

SOLUTION

a) $^-|^-7| = ^-(7) = ^-7$
b) $^-|13.2| = ^-13.2$
c) $|^-21| + 21 = 21 + 21 = 42$
d) $^-|21| + 21 = ^-21 + 21 = 0$

Example 4 asks you to use a grapher. For some graphers, if you use the subtraction key $\boxed{-}$ when the opposite key $\boxed{(-)}$ is needed or vice versa, an error message will appear. So even when using a grapher, you must interpret the notation correctly.

**EXAMPLE 4**   Using a Grapher to Calculate Absolute Value

Verify each of the following calculations with a grapher. Which ones are not correct?

a) $|73.2 - ^-14.8| = 88$
b) $|^-27.6 - ^-14.3| = 13.3$
c) $|17.9| - |^-26.7| = 8.8$
d) $|14.3| - |^-36.9| = 22.6$

SOLUTION
Parts (c) and (d) are not correct.

## Summary of Basic Properties

The following basic properties of algebra will be used throughout the remainder of the text.

### Basic Properties of Algebra

Let $a, b, c$, and $x$ be any real numbers. Then the following properties are true:

| Property | Example |
|---|---|
| Commutative Property of Addition | |
| $a + b = b + a$ | $2 + x = x + 2$ |
| Commutative Property of Multiplication | |
| $ab = ba$ | $(2 + x)x^2 = x^2(2 + x)$ |
| | |
| Associative Property of Addition | |
| $a + (b + c) = (a + b) + c$ | $2 + (x^2 + x) = (2 + x^2) + x$ |
| Associative Property of Multiplication | |
| $a(bc) = (ab)c$ | $3[x(x + 2)] = (3x)(x + 2)$ |
| | |
| Distributive Property | |
| $a(b + c) = ab + ac$ | $2(x + 3) = 2x + 6$ |
| $(a + b)c = ac + bc$ | $(x + 3)x = x^2 + 3x$ |
| | |
| Additive Identity Property | |
| $a + 0 = a$ | $3x^2 + 0 = 3x^2$ |
| Multiplicative Identity Property | |
| $a \cdot 1 = 1 \cdot a = a$ | $3y^3 \cdot 1 = 1 \cdot 3y^3 = 3y^3$ |

**EXAMPLE 5**    Recognizing the Basic Properties

Name the basic property that is being illustrated in each part of the following:

a) $3(x + 2) = (x + 2)3$
b) $3x + 3 = 3x + 3 \cdot 1$
c) $5x - 35 = 5(x - 7)$

SOLUTION

a) The Commutative Property of Multiplication
b) The Multiplicative Identity Property
c) The Distributive Property                                   ■

The next example illustrates that these properties are often used in combination.

**EXAMPLE 6**    Using the Basic Properties

Use the basic properties to show that for any value of $x$, the following is true:

a) $2 + (x + 7) = 9 + x$
b) $4x - 2 = 2(2x - 1)$

SOLUTION

a) $2 + (x + 7) = 2 + (7 + x)$    **Using the Commutative Property of Addition**
$\qquad\qquad\quad = (2 + 7) + x$    **Using the Associative Property of Addition**
$\qquad\qquad\quad = 9 + x$

b) $4x - 2 = 2 \cdot 2x - 2 \cdot 1$    **Using the Multiplicative Identity Property**
$\qquad\quad = 2(2x - 1)$    **Using the Distributive Property**    ■

## Distance on a Number Line

The distance between two points on a number line can be found by combining absolute value and subtraction. The method is summarized as follows.

---

### Distance on a Number Line

The **distance** between two points on a number line is the absolute value of the difference of the two numbers. If point $A$ has coordinate $a$ and point $B$ has coordinate $b$, then the distance between $A$ and $B$, denoted $d(A, B)$, is as follows:

$$d(A, B) = |a - b|$$

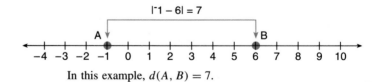

$|{}^-1 - 6| = 7$

In this example, $d(A, B) = 7$.

---

Often it is helpful to use a number line in the analysis of a problem situation. You can tell from the context of the probem where the zero of the number line should be.

**EXAMPLE 7**    APPLICATION: Finding Distance

A hot-air balloon captain drops ballast when the balloon is 860 ft above the ocean. The ballast lands on the bottom of the ocean at a depth of 212 ft below the surface. How far does the ballast fall?

SOLUTION

Figure 1.16 shows a balloon with a number line drawn. The zero of the number line corresponds to sea level.

$$d(\text{balloon, ballast}) = |860 - (-212)| = 1072$$

The ballast falls 1072 ft.                                                                                ■

A thermometer is a model of a number line that is used to record temperature. You can find temperature changes the same way you find distance on a number line.

**EXAMPLE 8**    APPLICATION: Determining Record Temperature Differences

Alaska has record high and low temperatures of 100°F and −80°F, respectively. Montana has records of 117°F and −70°F, while North Dakota has records of 121°F and −60°F. Which of these three states has the record temperature difference?

SOLUTION

$$\text{Alaska:} \quad d(\text{High, Low}) = |100 - (-80)| = 180$$
$$\text{Montana:} \quad d(\text{High, Low}) = |117 - (-70)| = 187$$
$$\text{North Dakota:} \quad d(\text{High, Low}) = |121 - (-60)| = 181$$

Montana has the greatest temperature difference.                                        ■

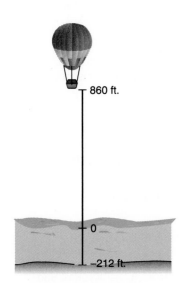

**FIGURE 1.16** The hot-air balloon problem.

## Exercises for Section  1.2

In Exercises 1–12, find the indicated sums and differences by applying the rules of addition and subtraction. Solve these problems mentally.

1. $18 + {}^-3 = ?$

2. ${}^-12 + {}^-4 = ?$

3. ${}^-16 + 8 = ?$

4. ${}^-9 + 15 = ?$

5. ${}^-8 + {}^-5 = ?$

6. ${}^-25 + 20 = ?$

7. $13 - {}^-3 = ?$

8. $3 - {}^-7 = ?$

9. ${}^-8 - {}^-5 = ?$

10. ${}^-7 - {}^-4 = ?$

11. $21 - {}^-2 = ?$

12. ${}^-18 - 3 = ?$

In Exercises 13–20, use properties of opposites to complete each problem mentally.

**13.** $^-(^-7) = ?$     **14.** $^-(^-7) + 3 = ?$

**15.** $6 + (^-6) = ?$     **16.** $^-17 + ^-(^-17) = ?$

**17.** $3 + (7 + ^-3) = ?$     **18.** $^-12 + (7 + 5) = ?$

**19.** $^-8 + |^-8| = ?$     **20.** $3 + (^-|^-3|) = ?$

In Exercises 21–28, find the indicated sums, differences, and products using a grapher.

**21.** $|126.7 - ^-18.2| = ?$

**22.** $|^-18.7 - 29.3| = ?$

**23.** $|^-35.7| - |83.9| = ?$

**24.** $|42.1| - |^-18.6| = ?$

**25.** $|27.5 - |91.4| \,| = ?$

**26.** $^-2.8 + 17.6 - ^-7.3 = ?$

**27.** $^-9.6(4.7 - 17.8) = ?$

**28.** $^-7.2(6.2 - 21.9) = ?$

In Exercises 29–38, state which basic properties are used to show that the given equation is true.

**29.** $x + 3 = 3 + x$

**30.** $a + (2 + x) = (a + 2) + x$

**31.** $3(xy) = (3x)y$

**32.** $5(x + 3) = (x + 3)5$

**33.** $3 + (x + 17) = x + 20$

**34.** $7(x + 3) = 7x + 21$

**35.** $x(x + 1) = x^2 + x$

**36.** $3(7x) = 21x$

**37.** $2(8x) = 16x$

**38.** $a + (3 + a) = 2a + 3$

In Exercises 39–46, explain how to use the Distributive Property to do the given multiplications mentally. For example, $5 \cdot 21 = 5 \cdot (20 + 1) = 5 \cdot 20 + 5 \cdot 1 = 105$.

**39.** $6 \cdot 21$     **40.** $4 \cdot 51$

**41.** $7 \cdot 12$     **42.** $41 \cdot 7$

**43.** $5 \cdot 62$     **44.** $3 \cdot 31$

**45.** $3 \cdot 91$     **46.** $2 \cdot 61$

In Exercises 47–50, decide whether the statement is true or false.

**47.** Given any real number $a$ except zero, $|a|$ is positive.

**48.** Given any nonzero real number $a$, $-a$ is a negative number.

**49.** Given any real number $a$, $|^-a| + a = 0$.

**50.** Given any nonzero real number $a$, $|\,|a| - |^-a|\,| = 0$.

In Exercises 51–58, consider each "−" symbol and state whether it represents subtraction or the opposite of. Then complete the calculation mentally.

**51.** $-(5 - 3) = ?$

**52.** $-5 - (-7) = ?$

**53.** $-|5 - (-2)| = ?$

**54.** $|-2 + 7| - (-8) = ?$

**55.** $-(-8 - 3) = ?$

**56.** $-2(7 + (-5)) = ?$

**57.** $(-3)(-5) - (-1) = ?$

**58.** $12 - (-2)(-3) = ?$

**59.** *Finding Net Gain.* On a certain series of plays, the Browns football team gained 3 yards on the first down, lost 7 yards on the second down, and gained 12 yards on the third down. How many yards have they gained so far?

**60.** *Finding Net Gain.* The XYZ corporation had two quarters of losses of $23 and $41 million, respectively. In the third quarter, they had a gain of $93 million and in the fourth quarter, they had a loss of $32 million. Did they have a gain for the entire year?

**61.** *Comparing Temperature Scales.* Chemists use both the Celsius (C) temperature scale and the Kelvin (K) temperature scale. The unit of 1 degree is the same amount in both scales. However, $0°K$ is equal to $-273.16°C$. What temperature in K corresponds to $22.3°C$?

**62.** *Comparing Temperature Scales.* If a liquid has a temperature of $23°C$, does its temperature need to drop (a) $23 - (-273.16)$ or (b) $23 - 273.16$ degrees in order to reach $0°K$?

**63.** *Determining Submarine Depth.* A submarine is cruising 32 ft below the surface of the ocean when it is commanded to dive to a depth of 128 ft. Fifteen minutes later, it rose 35 ft. The submarine is in radio communication with an airplane that is flying at an altitude of 22,000 ft.

a) Draw a sketch of this problem, including a vertical number line to represent elevations, with the zero representing the surface of the ocean.

b) What is the distance of the submarine's first dive?

c) What is the submarine's final depth?

d) What is the greatest distance (or elevation) between the submarine and the airplane?

**64.** *Determining Altitude Changes.* A hot-air balloon is flying at an altitude of 815 ft above the Intercoastal Waterway. When the balloon's captain drops ballast, the balloon rises 218 ft and the ballast falls to 15 ft below the surface of the waterway.

a) Draw a sketch of this problem and superimpose on it a vertical number line to represent elevations, with the zero representing the surface of the waterway.

b) What is the elevation of the balloon after the ballast is dropped?

c) How far does the ballast drop?

### TRANSLATING WORDS TO SYMBOLS

In Exercises 65–68, write the word description in symbols.

**65.** The absolute value of the sum of 12.3 and $^-5.8$

**66.** The sum of the absolute values of 12.3 and $^-5.8$

**67.** The difference between the absolute values of $^-13.2$ and $^-12.6$

**68.** The absolute value of the difference of $^-13.2$ and $^-12.6$

### EXTENDING THE IDEAS

**69.** Complete the following pattern:

$$3 \cdot x = {}^-6$$
$$2 \cdot x = {}^-4$$
$$1 \cdot x = {}^-2$$
$$0 \cdot x = \phantom{^-}0$$
$${}^-1 \cdot x = \,?$$
$${}^-2 \cdot x = \,?$$

**70.** In exercise 69, what value of $x$ makes each of the equations true statements?

**71.** Explain how Exercises 69 and 70 can be used to teach that the product of two negative integers is a positive integer.

### LOOKING BACK—MAKING CONNECTIONS

For familiar numbers, you should be able to recognize equivalences between percent, decimal, and fraction representations of rational numbers. For example, 20%, 0.2 and $\frac{1}{5}$ are all equivalent. In Exercises 72–79, write the other two forms of the number given.

**72.** $\dfrac{3}{4}$

**73.** 60%

**74.** 0.25

**75.** $\dfrac{1}{3}$

**76.** 1.5

**77.** 37.5%

**78.** 62.5%

**79.** $\dfrac{4}{5}$

| 1.3 | Algebraic Expressions and Problem Situations |

Introducing Problem Situations ▪ Using a Grapher to Evaluate an Expression ▪ Determining the Computational Method to Use

A **variable** is a symbol or a letter that reserves a place for a single number or many numbers. Each of the following expressions includes one or more variables:

$$x + 3 \qquad \bigcirc - 5 \qquad z - 6x$$

An **algebraic expression** is a collection of variables and real numbers—called **constants**—that are combined using the operations of addition, subtraction, multiplication, division, and exponentiation.

It is common to use the letters $x$ and $y$ as the variables in an algebraic expression. Following that convention, here are several other examples of algebraic expressions:

$$3x \qquad x^2 - 5 \qquad 2x - y^3 \qquad \frac{5 - x}{x^2 + 3}$$

The **terms** of an algebraic expression are those parts that are separated by a plus or minus sign. For example, in the expression $x^2 - 5$ the $x^2$ is called the **variable term** and the 5 is called the **constant term**.

To evaluate an algebraic expression, we replace the variables with specific numerical values.

**EXAMPLE 1**    Evaluating an Algebraic Expression

Evaluate the following expressions when the variables $x$ and $y$ are replaced with $x = 4$ and $y = -7$:

a) $2x + 3y$
b) $3x^2 - y$

SOLUTION
When $x$ is replaced by 4 and $y$ is replaced by $-7$,

a) $2x + 3y$ becomes $2(4) + 3(-7) = 8 + (-21)$

$$= -13$$

b) $3x^2 - y$ becomes $3(4)^2 - (-7) = 3(16) + 7$

$$= 55$$

---

**REMINDER**

In an algebraic expression, when a constant and a variable are written next to each other, multiplication is implied. For example, $2x$ means "$2 \times x$."

Algebraic expressions are often used in formulas and equations in scientific, mathematical, and business applications. For example, algebraic expressions are used in area and perimeter formulas for rectangles, as follows:

$$\text{Area} = L \times W \qquad \text{(where } L \text{ represents length and } W \text{ represents width)}$$

$$\text{Perimeter} = 2L + 2W \qquad \text{(where } L \text{ represents length and } W \text{ represents width)}$$

Another use of an algebraic expression is in the familiar formula for converting Fahrenheit temperature to Celsius temperature. We can find degrees on the Celsius scale using the formula

$$\text{Celsius degrees} = \frac{5}{9}(^\circ F - 32).$$

**GRAPHER NOTE**

When evaluating an expression like $\frac{5}{9}(41 - 32)$ on a grapher, it is necessary to use parentheses around the 5/9. The grapher display will look like $(5/9)(41 - 32)$.

Some graphers have an "editing" or a "replay" capability that allows you to quickly recall the previous expression for editing. For example, after evaluating $(5/9)(41 - 32)$, recall, replace the 41 with 50, and then evaluate again.

**EXAMPLE 2**    APPLICATION: Changing Fahrenheit to Celsius

Complete a table of values that shows the Celsius degrees that correspond to the Fahrenheit degrees of 32, 41, 50, 65, 85, 98.6, and $x$.

SOLUTION

Evaluate the expression $\frac{5}{9}(^\circ F - 32)$ for each given value of $F$ and place this value in the third column of the table below.

| Fahrenheit Temperature ($^\circ F$) | Expression $\frac{5}{9}(^\circ F - 32)$ | Corresponding Celsius Temperature ($^\circ C$) |
|:---:|:---:|:---:|
| 32 | $\frac{5}{9}(32 - 32)$ | 0 |
| 41 | $\frac{5}{9}(41 - 32)$ | 5 |
| 50 | $\frac{5}{9}(50 - 32)$ | 10 |
| 65 | $\frac{5}{9}(65 - 32)$ | 18.33 |
| 85 | $\frac{5}{9}(85 - 32)$ | 29.44 |
| 98.6 | $\frac{5}{9}(98.6 - 32)$ | 37 |
| $x$ | $\frac{5}{9}(x - 32)$ | $\frac{5}{9}(x - 32)$ |

## Introducing Problem Situations

In most areas of our lives, we are faced with mathematical problems, whether it's estimating the cost of carpeting a room, planning a family or personal budget, or planning for retirement. In each case, there is a set of facts or assumptions that set the parameters of the situation for us. Those facts or assumptions often allow us to write an algebraic expression that describes some quantity in our situation. Consequently, we call this situation a **problem situation**.

Throughout this text, a variety of problem situations will be described and studied. Many of these are given names so we can easily refer to them later in the text.

---

### CAR DEALER PROBLEM SITUATION

A certain car and truck dealer makes an average profit of $512 for each car sold and $1258 for each truck sold.

---

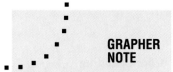

**GRAPHER NOTE**

Notice that the grapher displays in Figures 1.17 and 1.19(a) use the key $\boxed{:}$. This key is used to separate statements, a process called **concatenation**. If your grapher does not have concatenation, press ENTER after each statement as in Figure 1.19(b).

A problem situation is simply a setting in which mathematical problems occur. Once someone asks a specific quantitative question, a mathematical problem has been posed. Example 3 poses a mathematical problem about the **Car Dealer Problem Situation.**

**EXAMPLE 3**     CAR DEALER PROBLEM SITUATION: Finding the Profit

Complete a table that shows the profit for selling the following:

a) 12 cars and 5 trucks
b) 21 cars and 8 trucks
c) 29 cars and 7 trucks
d) $C$ cars and $T$ trucks

SOLUTION

Let $C$ represent the number of cars sold and $T$ represent the number of trucks sold.

| $C$ (no. cars) | $T$ (no. trucks) | Dollars of Profit |
|:---:|:---:|:---:|
| 12 | 5 | $512(12) + 1258(5) = 12,434$ |
| 21 | 8 | $512(21) + 1258(8) = 20,816$ |
| 29 | 7 | $512(29) + 1258(7) = 23,654$ |
| $C$ | $T$ | $512C + 1258T$ |

## Using a Grapher to Evaluate an Expression

Graphers permit the use of variables. Consequently, a grapher can be used to complete the table below by evaluating the expression $x^3 + 17$ for values $x = 25$, $x = 39$, and $x = 52$.

| $x$ | $x^3 + 17$ |
|-----|-----------|
| 25 | 15,642 |
| 39 | ? |
| 52 | ? |

$$25 \rightarrow X: X^3 + 17$$
$$15642$$

**FIGURE 1.17**
Evaluating the expression
$x^3 + 17$ if $x = 25$.

Begin by storing values for $x$, one at a time, in a memory location called $X$. Your grapher lab manual shows which key strokes to use to store a value in a memory location. Figure 1.17 shows a grapher screen after storing 25 in the memory location called $X$ and evaluating the expression $x^3 + 17$. Figure 1.18 shows a grapher screen after using the recall and editing features of the grapher to evaluate the expression for the other entries in the table. In Example 4, we evaluate an expression with two variables.

**EXAMPLE 4**    Evaluating an Expression  Using a Grapher

Use a grapher to evaluate the expression $2x^3 - 3y^2$ for $x = -23$ and $y = 19.3$.

$$25 \rightarrow X: X^3 + 17$$
$$15642$$
$$39 \rightarrow X: X^3 + 17$$
$$59336$$
$$52 \rightarrow X: X^3 + 17$$
$$140625$$

**FIGURE 1.18**  You can use the recall and editing feature of the grapher to evaluate the expression $x^3 + 17$.

SOLUTION

1. Store $-23$ for $x$ and 19.3 for $y$.
2. Key in $2x^3 - 3y^2$ and press $\boxed{\text{ENTER}}$. The grapher display should look like that in either Figure 1.19 (a) or (b).

$$^-23 \rightarrow X: 19.3 \rightarrow Y:$$
$$2X^3 - 3Y^2$$
$$^-25451.47$$

(a)

$$^-23 \rightarrow X$$
$$^-23$$
$$19.3 \rightarrow Y$$
$$19.3$$
$$2X^3 - 3Y^2$$
$$^-25451.47$$

(b)

**FIGURE 1.19**  Grapher display for evaluating $2x^3 - 3y^2$ (a) using concatenation and (b) without using concatenation.

Once particular values have been assigned to variables, you can easily evaluate many different expressions using those variables.

**EXAMPLE 5**   Evaluating an Expression Using a Grapher

If the variables $r, s$, and $t$ have values $r = 13.5$, $s = -23.7$, and $t = 3.28$, evaluate the following:

a) $r^2 - 3t$

b) $5s - 6t^2 + 17r$

c) $\dfrac{4t - 3s^3}{5r}$

SOLUTION

Store the given values in variables $r, s$, and $t$ as follows (on most graphers, these variables will appear as capital letters):

$$13.5 \rightarrow r : \quad -23.7 \rightarrow s : \quad 3.28 \rightarrow t$$

Then key in each of the given expressions and press $\boxed{\text{ENTER}}$. The expressions will appear on your grapher screen as shown in Figure 1.20.

```
R² – 3T
                        172.41
5S – 6T² + 17R
                       46.4496
(4T – 3S³)/ 5R
                    591.8411704
```

**FIGURE 1.20** Grapher display for evaluating three expressions.

Notice that the values of these three expressions are 172.41, 46.4496, and 591.8411704, respectively.  ∎

### GRAPHER NOTE

In the grapher display in Figure 1.20, the 5 and $R$ are multiplied. This is called an **implied** multiplication because their positions next to one another implies multiplication.

Most graphers carry out an implied multiplication as a higher priority than they do an explicit multiplication between numbers such as $5 \times 3$. Consequently, a set of parentheses is not needed around the $5R$.

## Determining the Computational Method to Use

When faced with a computation, we should decide whether to use a mental computation or a paper-and-pencil computation or to use a grapher. You might be tempted to reach for a grapher even when a mental calculation is faster. Develop instead the habit of asking yourself, "Which computation method should I use—mental, paper and pencil, or grapher?"

Example 6 illustrates the three methods of computation from which to choose when evaluating an expression, depending on the values of the variable.

**EXAMPLE 6**    Deciding on a Computational Method

Evaluate the expression $x^2 + x$ for the following values of $x$. In each case, choose an appropriate computational method.

a) $x = 13$    b) $x = 5$    c) $x = 19.3$

SOLUTION

a) When $x = 13$, the paper-and-pencil method may be appropriate.

$$x^2 + x \text{ becomes } 13^2 + 13 = 169 + 13$$
$$= 182$$

b) When $x = 5$, the expression can be evaluated mentally.

$$x^2 + x \text{ becomes } 30 \quad \textbf{Think: } \mathbf{25 + 5 = 30}$$

c) When $x = 19.3$, a grapher is certainly appropriate. The grapher screen will appear as in Figure 1.21.    ■

$19.3^2 + 19.3$

$391.79$

**FIGURE 1.21** One method of evaluating $x^2 + x$ for $x = 19.3$.

**Exercises for Section**  **1.3**

In Exercises 1–8, name the variable terms in the given expression.

1. $x + 5$    2. $7 - y$

3. $3x - 8$

4. $x^2 + 5x - 3$

5. $3 + \triangle$

6. $5x - 2\diamondsuit$

7. $3x - 2y^2$

8. $17x^2 - 5y^2$

In Exercises 9–16, use the paper and pencil method to evaluate the given expression for $x = 3$ and $y = 8$. Support your answer with a grapher.

9. $x + 2$

10. $3x - 1$

11. $x^2 - 3x$

12. $2y - 6x$

13. $2y - 3x$

14. $y^2 - 6y$

15. $5y - x^2$

16. $3y + x^2 - x$

In Exercises 17–20, use the paper-and-pencil method to evaluate the expression $3x^2 - 5x + 2$ for the given value of $x$.

17. $x = 2$

18. $x = 3$

19. $x = 4$

20. $x = 5$

In Exercises 21–24, evaluate the expression $2x + 3y^2$ for the given values of $x$ and $y$. Decide whether to complete the evaluation mentally, using paper and pencil, or with a grapher.

21. $x = 2$ and $y = 3$

22. $x = 1$ and $y = 2$

23. $x = -8$ and $y = 3$

24. $x = 5$ and $y = -2$

25. Evaluate the expression $3h - 5k + j$ for $h = 2$, $k = 1$, and $j = 12$.

26. Evaluate the expression $T^2 + 5M$ for $T = 7$ and $M = 6$.

In Exercises 27–30, use the grapher method of Examples 4 and 5 to evaluate the expression $12.3x^3 - 7x^2 + 3.89$ for the given value of $x$.

27. $x = 19.3$

28. $x = -8.6$

29. $x = -24.94$

30. $x = 13.62$

In Exercises 31–36, evaluate the given expression for $r = 5.31$, $s = 3.9$, and $t = -9.2$. Use either the paper-and-pencil method or a grapher.

31. $6r^2 - 5s$

32. $\dfrac{2r + 7t}{s^2}$

33. $\dfrac{2t}{4s^2}$

34. $6.2s^3 + 5.1t^2$

35. $(1.07r - 6.89s) + 9t^2$

36. $r(s^2 - 7.3t^2)$

In Exercises 37–42, decide which computational method to use—mental, paper-and-pencil, or grapher—and evaluate each expression for the indicated value of the variable.

37. $3x - 5$ for $x = 7$

38. $17.3x^2 - 17$ for $x = 5.9$

39. $x + 3y$ for $x = 19$ and $y = 9$

40. $3R + 7S$ for $R = 1$ and $S = 3$

41. $S^3 - 5S - 6$ for $S = 17$

42. $7Y - Y^2$ for $Y = 8$

43. *Finding Temperature.* The expression $\frac{5}{9}(F - 32)$ is equal to the Celsius temperature corresponding to Fahrenheit degrees. Complete a table showing the Celsius temperatures corresponding to the Fahrenheit temperatures of $-12°$, $25°$, $48°$, $68°$, and $105°$.

44. *Finding Temperature.* The expression $\frac{9}{5}C + 32$ is equal to the Fahrenheit temperature corresponding to Celsius degrees. Complete a table showing the Fahrenheit temperatures corresponding to the Celsius temperatures of $-5°$, $12°$, $15°$, $25°$, and $32°$.

45. *Finding Profit.* Sarah makes an average profit of $485 for each car sold and $695 for each truck sold at her dealership. If $C$ represents the number of cars sold and $T$ the number of trucks sold, the expression $485C + 695T$ represents the profit from these sales.

   a) What is Sarah's profit if she sells 32 cars and 6 trucks in January?

   b) February car sales go up by 3 cars and truck sales go down by 2 trucks. Does overall profit go up or down compared to January? By how much?

   c) One March Sarah sold 27 cars and 5 trucks and in April she sold 29 cars and 4 trucks. In which month does she make more profit? How much more?

46. *Finding Profit.* The "More for Less" car rental agency rents compact cars for $29.95/day and intermediate-sized cars for $34.95/day. If $C$ is the

number of compact cars and $I$ is the number of intermediate-sized cars rented in a day, then the expression $29.95C + 34.95I$ represents the total dollars of revenue generated in a day.

a) How many dollars are generated by the agency if it rents 85 compact cars and 93 intermediate-sized cars in one day?

b) How many dollars are generated by the agency if it rents 121 compact cars and 84 intermediate-sized cars in one day?

c) The agency rented on three successive days (a) 142 compacts and 119 intermediates, (b) 137 compacts and 124 intermediates, and (c) 153 compacts and 91 intermediates, respectively. On which day did it generate the most revenue?

## TRANSLATING WORDS TO SYMBOLS

In Exercises 47–49, write the word descriptions in symbols.

**47.** Three times $x$, plus four

**48.** Three times the quantity $x$ plus four

**49.** Three and seven-tenths times $y$, minus five times $x$

## EXTENDING THE IDEAS

**50.** Verify that if the ? is replaced with $x + 1$, then the pattern has been described. Each entry in the second row is found by evaluating this expression.

| Value of the variable | 1 | 2 | 3 | 4 | $\cdots$ | $x$ |
|---|---|---|---|---|---|---|
| Value of the expression | 2 | 3 | 4 | 5 | $\cdots$ | ? |

**51.** Discover the pattern. Replace the ? with an expression that describes this pattern.

| Value of the variable | 1 | 2 | 3 | 4 | $\cdots$ | $x$ |
|---|---|---|---|---|---|---|
| Value of the expression | 2 | 4 | 6 | 8 | $\cdots$ | ? |

**52.** *Modeling Data.* This bar graph records total revenue of a company in millions of dollars for the

past 8 yr. The company accountant uses the formula $(1.03^n) \cdot 112$ to predict revenue in the future, where $n = 1$ corresponds to 1987. How close does the expression $(1.03^n) \cdot 112$ match the total revenue for 1989? For 1991?

**53.** *Modeling Data.* Use the expression $(1.03^n) \cdot 112$ to predict the total revenue for 1995.

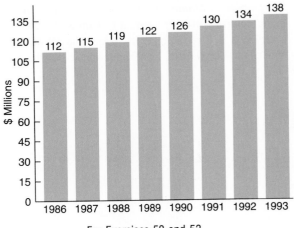

For Exercises 52 and 53.

## LOOKING BACK—MAKING CONNECTIONS

**54.** In the expression

$$\frac{7 + 3 \cdot 18(18 - 4)}{5},$$

what is the order of operations?

In Exercises 55–58, use information from the table below, which reviews the algebraic expressions that can be used to find the perimeter, circumference, or area of circles and rectangles.

|  | Perimeter/Circumference | Area |
|---|---|---|
| (rectangle) $W$, $L$ | $2L \times 2W$ | $L \times W$ |
| (circle) $R$ | $2\pi R$ | $\pi R^2$ |

**55.** Find the area and circumference of a circle with radius 2.3 in.

**56.** *Comparing Field Dimensions.* The official size for an American football field (including end zones) is 360 ft long and 160 ft wide and the official size for a soccer field is 330 ft long and 240 ft wide. Which field has the larger area and by how much?

**57.** *Container Manufacturing.* An engineer estimating manufacturing costs of a cereal box finds that a certain cereal box is 8 in. long, 3 in. wide, and 10 in. high. What is the area of the cardboard surface of the box?

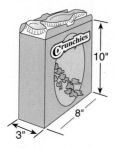

For Exercise 57.

**58.** *Agricultural Economics.* A large circular irrigation system irrigates a circular region 300 ft in radius. How many acres does this system irrigate? (There are 43,560 sq ft/acre.)

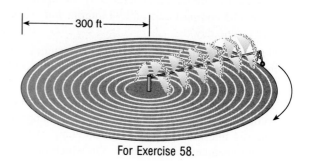

For Exercise 58.

**59.** Explain what the following display means:

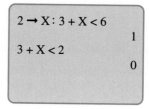

$$2 \rightarrow X : 3 + X < 6$$
$$\qquad\qquad\qquad\qquad 1$$
$$3 + X < 2$$
$$\qquad\qquad\qquad\qquad 0$$

---

**1.4** **Solving Equations and Inequalities**

> Solving an Equation by Guess-and-check ▪ Solving a Linear Equation Using Equivalent Equations ▪ Simplifying Expressions in Equation Solving ▪ Solving a Linear Inequality Using Equivalent Inequalities

An **equation** is a statement that two mathematical expressions are equal. Recall that the symbol for equality is =. In an equation, one expression is on the left side of the = symbol and the other expression is on the right side of the = symbol.

Use your grapher to answer each of the following questions yes or no. If the answer is no, how would you change the equation so that your answer is yes?

1. If $x = 4$, does $7x + 3 = 31$?
2. If $x = 7.2$, does $x^2 - 3x = 30.14$?
3. If $x = 4.3$, does $x^3 + 3x - 2 = 90.407$?

Explain how you used a grapher to answer this question.

While completing this exploration, you can see that if $x$ is replaced by 4 the statement $7x + 3 = 31$ is true. We call $x = 4$ a solution of the equation $7x + 3 = 31$. The **solution of an equation** is the collection of *all* numbers that make the equation true.

## Solving an Equation by Guess-and-check

The Guess-and-check strategy can be used to find a solution to an equation. First, guess a value for $x$ and check whether the guessed value of $x$ makes the equation a true statement. If not, modify the guess and try again.

For example, recall the equation for converting from degrees Celsius to degrees Fahrenheit:

$$C = \frac{5}{9}(F - 32)$$

To find $F$ when $C = 12.5$, we need to solve the equation

$$12.5 = \frac{5}{9}(F - 32).$$

Example 1 illustrates the Guess-and-check strategy for finding a solution to this equation.

**EXAMPLE 1**     Using Guess-and-check

Use the Guess-and-check strategy to find a solution to the equation $12.5 = \frac{5}{9}(F - 32)$.

(5/9)(40 − 32)
                4.4444444444
Ans ► FRAC
                    40/9

**FIGURE 1.22**
Evaluating the expression
$(5/9)(F - 32)$ for $F = 40$.

SOLUTION

The entries in this table record the results of using the Guess-and-check strategy. Figure 1.22 shows a grapher evaluation for $x = 40$.

| Guess for $F$ | Calculated Value of $\frac{5}{9}(F - 32)$ | Conclusion |
|---|---|---|
| 40 | 4.44 | Guess is too small. |
| 80 | 26.67 | Guess is too large. |
| 60 | 15.56 | Guess is too large. |
| 50 | 10 | Guess is too small. |
| 55 | 12.78 | Guess is too large. |
| 54 | 12.22 | Guess is too small. |
| 54.5 | 12.5 | Guess is correct. |

The solution to the equation is $F = 54.5$, that is, $54.5°F$ is equal to $12.5°C$.

It is useful to know that nearly any equation can be solved using the Guess-and-check strategy illustrated in Example 1. Although it is usually not a practical strategy in a paper-and-pencil environment because of the time it takes, it is a valid method in the age of computers.

There are other methods—algebraic and graphical—that might be superior to Guess-and-check. They might be superior because they might be quicker or because they might provide additional information about the equation such as whether there is a unique solution or more than one solution.

## Solving a Linear Equation Using Equivalent Equations

Throughout this section, we will study linear equations and algebraic methods of solving them.

---

**DEFINITION 1.3**   Linear Equation in One Variable

An equation is a **linear equation in one variable** $x$ if it can be written in the form

$$ax + b = c,$$

where $a$, $b$, and $c$ are any real numbers with $a \neq 0$.

---

One of the following equations is not a linear equation. Which one? The second one, because of the $x^2$ term.

$$3x - 4 = 9 \qquad x^2 + 8 = 32 \qquad 2x + 8 = 4$$

Linear equations, like many others, can be solved using the concept of equivalent equations. Two or more equations are **equivalent** if they have the same solution set. To illustrate with a simple example,

$$x + 3 = 4 \qquad \text{and} \qquad x + 1 = 2 \qquad \text{and} \qquad x + 5 = 6$$

are equivalent because the number 1 is the only solution for each equation.

Certain algebraic steps can be performed on an equation so that the resulting equation is equivalent to the original equation. For example, you can add any real number to each side of an equation to obtain an equation that is equivalent to the original equation. This Addition Property is stated next.

---

### Addition Property of Equality

For any real numbers $a, b$, and $c$, the equations

$$a = b$$

and

$$a + c = b + c$$

are equivalent equations.

---

Example 2 illustrates how this Addition Property of Equality can be used to solve equations.

### EXAMPLE 2    Using the Addition Property of Equality

Find the solution to the linear equation $x + 48 = 71$. Support numerically.

SOLUTION

Begin with the given equation and add the additive inverse of 48 to both sides of the equation, resulting in an equivalent equation.

$$x + 48 = 71 \tag{1}$$

$$(x + 48) + (-48) = 71 + (-48) \qquad \text{Use the Associative Property of Addition to regroup the left side of the equation.}$$

$$x + (48 + (-48)) = 71 + (-48) \qquad \text{Use the Additive Inverse Property.}$$

$$x + 0 = 23 \qquad \text{Use the Additive Identity Property.}$$

$$x = 23$$

**Support Numerically.** Figure 1.23 shows numerical support for this solution. ∎

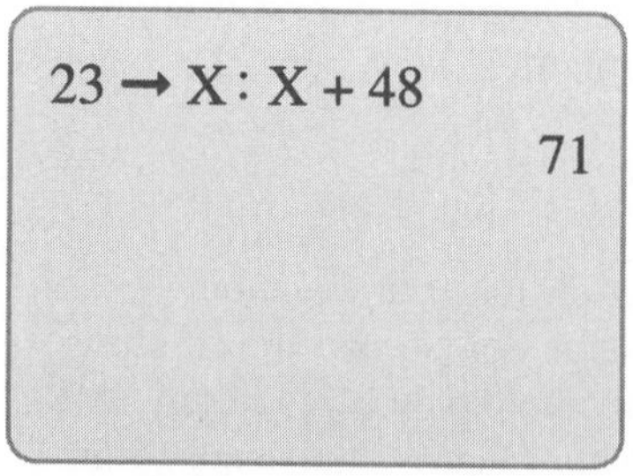

$23 \rightarrow X : X + 48$

$71$

**FIGURE 1.23**
Numerical support on a grapher that $x = 23$ is a solution of $x + 48 = 71$.

The following Multiplication Property of Equality says you can multiply both sides of an equation by any real number except zero to obtain an equation equivalent to the original equation.

---

### Multiplication Property of Equality

For any real numbers $a, b$, and $c$ with $c \neq 0$, the equations

$$a = b$$

and

$$a \cdot c = b \cdot c$$

are equivalent equations.

---

When using the Multiplication Property you will usually multiply both sides of an equation by the multiplicative inverse of a number. Recall that the **multiplicative inverse** of any nonzero number $a$ is the number $\frac{1}{a}$. Example 3 illustrates how to use a multiplicative inverse together with the Multiplication Property of Equality.

**EXAMPLE 3**    Using the Multiplication Property of Equality
Find the solution to the equation $4x + 17 = 48$. Support numerically.

SOLUTION
Begin with the given equation and add the additive inverse of 17 to both sides of the equation.

$$4x + 17 = 48$$

$$(4x + 17) + (-17) = 48 + (-17)$$

$$4x = 31$$

Next multiply both sides of this equation by $\frac{1}{4}$. (This is equivalent to saying divide both sides of the equation by 4.)

$$\frac{1}{4} \cdot 4x = \frac{1}{4} \cdot 31$$

$$x = 7.75$$

**Support Numerically.**  Figure 1.24 shows numerical support for this solution.  ■

## Simplifying Expressions in Equation Solving

Terms of an expression with the same variable and the same exponent are called **like terms**. In an expression, like terms should be combined to form a single term. For example, the terms of $5x - 2x$ can be combined to obtain the equivalent expression $3x$. Here is the rationale.

$$5x - 2x = (5 - 2)x \qquad \text{This equation is a form of the}$$
$$\text{Distributive Property.}$$

$$= 3x$$

Example 4 shows that to solve the equation $5(x + 4) - 2x = 16$, we first use the Distributive Property and combine like terms.

**EXAMPLE 4**    Simplifying and Solving an Equation

Solve the equation $5(x + 4) - 2x = 16$. Support numerically.

SOLUTION

First, remove parentheses and combine like terms.

$$5(x + 4) - 2x = 16 \qquad \text{Use Distributive Property.}$$

$$(5x + 20) - 2x = 16 \qquad \text{Combine like terms.}$$

$$(5x - 2x) + 20 = 16 \qquad \text{Simplify.}$$

$$3x + 20 = 16$$

Next, use the Addition and Multiplication Properties of Equality to complete the solution.

$$3x + 20 + (-20) = 16 + (-20)$$

$$3x = -4 \qquad \begin{array}{l} \text{Multiply each side of the} \\ \text{equation by } \frac{1}{3}. \end{array}$$

$$\frac{1}{3} \cdot 3x = \frac{1}{3}(-4)$$

$$x = -4/3$$

---

$$7.75 \rightarrow X : 4X + 17$$
$$48$$

**FIGURE 1.24**  Grapher support that $x = 7.75$ is a solution of $4x + 17 = 48$.

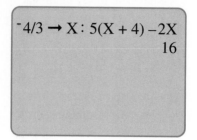

$^-4/3 \rightarrow$ X: 5(X + 4) −2X

16

**FIGURE 1.25** Grapher support that $x = -4/3$ is a solution to $5(x + 4) - 2x = 16$.

**Support Numerically.** Figure 1.25 shows numerical support of this solution. ■

If an equation includes fractions, multiply both sides of the equation by a whole number that is a multiple of all the denominators. This step will eliminate all the denominators and is illustrated in Example 5.

**EXAMPLE 5**　Simplifying and Solving

Solve the equation $\dfrac{z}{6} - \dfrac{z + 3}{15} = 7$. Support numerically.

SOLUTION

The number 30 is the smallest number that is a multiple of both 6 and 15, so multiply both sides of the equation by 30 in order to eliminate the denominators.

$$\frac{z}{6} - \frac{z + 3}{15} = 7$$

$$30 \cdot \left( \frac{z}{6} - \frac{z + 3}{15} \right) = 7 \cdot 30 \qquad \text{Note the use of commutativity of multiplication in this step.}$$

$$5z - 2(z + 3) = 210 \qquad \text{Use the Distributive Property.}$$

$$5z - 2z - 6 = 210$$

$$(5z - 2z) - 6 = 210 \qquad \text{Combine terms.}$$

$$3z - 6 = 210 \qquad \text{Add 6 to each side of the equation.}$$

$$3z - 6 + 6 = 210 + 6$$

$$3z = 216 \qquad \text{Divide both sides of the equation by 3.}$$

$$\frac{3z}{3} = \frac{216}{3}$$

$$z = 72$$

72 $\rightarrow$ Z: Z/6 − (Z + 3)/1

5

7

**FIGURE 1.26** Grapher support that $z = 72$ is a solution to the equation $\dfrac{z}{6} - \dfrac{z + 3}{15} = 7$.

**Support Numerically.** Figure 1.26 shows grapher support of this solution. ■

## Solving a Linear Inequality Using Equivalent Inequalities

Linear inequalities are similar to linear equations except that the $=$ is replaced with one of the inequality symbols $<$, $\leq$, $>$, or $\geq$.

**DEFINITION 1.4**   Linear Inequality in One Variable

A **linear inequality in one variable** $x$ can be written in the form

$$ax + b < c, \quad ax + b \leq c, \quad ax + b > c, \quad \text{or} \quad ax + b \geq c,$$

where $x$ is the variable and $a, b$, and $c$ are real numbers, with $a \neq 0$.

Following are some examples of linear inequalities in one variable:

$$2x + 4 < 7 \qquad y + 2 > 6 \qquad 3x \leq 9 \qquad 5y - 2 \geq 8$$

A number **belongs to the solution of an inequality** if replacing the variable with the number results in a true statement.

There is an Addition Property for Inequalities that is similar to the Addition Property for Equality.

## Addition Property of Inequality

For all real numbers $a, b$, and $c$, if

$$a < b, \quad \text{then} \quad a + c < b + c.$$

This Addition Property of Inequality remains true when $<$ is replaced by any of the other three inequality symbols $\leq$, $>$, or $\geq$.

The following Multiplication Property of Inequality is much like the Multiplication Property of Equality. However, it is necessary to notice whether the number used to multiply both sides of the inequality is positive or negative.

## Multiplication Property of Inequality

For all real numbers $a, b$, and $c$, if $a < b$ and

a) if $c > 0$, then $ac < bc$, and
b) if $c < 0$, then $ac > bc$.

This Multiplication Property remains true when the $<$ in $a < b$ is replaced by any of the other three inequality symbols $\leq$, $>$, or $\geq$.

Because division by $c$ is equivalent to multiplying by the reciprocal $\frac{1}{c}$, there is also a division property like this multiplication property. We will use this division property without stating it.

Example 6 shows the similarity between equation solving and inequality solving.

**EXAMPLE 6**   Using the Addition Property of Inequality

Find the solution to the linear inequality $3x - 19 < 32$.

SOLUTION
Begin with the given inequality and add 19 to both sides of the inequality.

$$3x - 19 < 32$$

$$3x - 19 + 19 < 32 + 19$$

$$3x < 51$$

$$\frac{3x}{3} < \frac{51}{3}$$

$$x < 17$$

All real numbers less than 17 are solutions to the given inequality. This solution is the infinite interval $(-\infty, 17)$.   ■

You will study how to solve inequalities algebraically to a greater extent in Chapter 4.

## Exercises for Section 1.4

In Exercises 1–8, use a grapher to determine whether the given value is a solution to the given equation. Record the grapher screen image.

**1.** Is $x = 29$ a solution to $5x - 13 = 132$?

**2.** Is $x = 17$ a solution to $15x + 29 = 285$?

**3.** Is $x = 9.5$ a solution to $x^2 + 17x = 251$?

**4.** Is $x = 12.1$ a solution to $5x^2 - 8x = 635$?

**5.** Is $x = 3$ a solution to $\dfrac{5}{x} + \dfrac{x}{3} = \dfrac{8}{3}$?

**6.** Is $x = 5$ a solution to $\dfrac{1}{x} + \dfrac{x}{4} = 1.45$?

**7.** Is $x = 5.2$ a solution to $x^2 + 1.5 = 28.54$?

**8.** Is $x = 2.8$ a solution to $21x + x^2 = 66.64$?

In Exercises 9–22, solve the given equation by using

the *Guess-and-check* strategy. Record your guesses and conclusions on a table as illustrated in Example 1.

**9.** $x - 6 = 19$      **10.** $x + 3 = -4$

**11.** $x + 4 = -11$      **12.** $x - 8 = 17$

**13.** $17 - x = 12$      **14.** $23 - x = -8$

**15.** $1.8x = 14.4$      **16.** $2.3x = 13.8$

**17.** $17x = 1870$      **18.** $14x = 44.8$

**19.** $3x + 5 = 9.5$      **20.** $4x - 7 = 3$

**21.** $3x^2 = 588$      **22.** $2x^2 - 5 = 877$

In Exercises 23–30, solve the given equation algebraically using the Addition and Multiplication Properties of Equality.

**23.** $2x + 5 = 23$      **24.** $4x + 2 = -5$

**25.** $2x - 5 = 14$      **26.** $3x + 5 = 7$

**27.** $5 = 3x + 73$      **28.** $16 = 6x - 92$

**29.** $11 - 2x = 14$      **30.** $-2x + 5 = 13$

In Exercises 31–40, simplify each side of the equation and use the Addition and Multiplication Properties to find an exact solution to the equation. Support your solution with a grapher.

**31.** $5(x - 3) = 17 - 3x$

**32.** $7(2 - x) + 3x = 11x - 8$

**33.** $2.3x + 9 = 5 - 4.8x$

**34.** $\dfrac{x - 3}{4} + \dfrac{7x}{3} = 3x - 2$

**35.** $\dfrac{x + 1}{3} + \dfrac{x - 1}{5} = \dfrac{x}{6}$

**36.** $(5 - 2.8x) + (7 + 1.2x) = 17.3$

**37.** $-2(x - 3) - (3x - 5) = 23 - 4x$

**38.** $6(x - 1) - (3x + 5) = 28 + 14x$

**39.** $2x - 1 = -2[2 - (3x - 4) + 7] + 2x$

**40.** $4[6 - (3x + 8) + 2] = 9 - 3x$

In Exercises 41–48, solve the given equation algebraically.

**41.** $21x - 8 = 34$      **42.** $7 - 3x = 14$

**43.** $9 - 4q = 5$      **44.** $5p - 4 = 17$

**45.** $3t - 23 = 19$      **46.** $4y + 21 = 83$

**47.** $17 = 3k + 46$      **48.** $8 = 4r + 21$

In Exercises 49–62, solve the given inequality. Describe the solution using interval notation.

**49.** $x + 12 < 28$      **50.** $x - 15 \geq 3$

**51.** $x - 8 \leq 5$      **52.** $x + 29 > 15$

**53.** $x + 14 \geq 33$      **54.** $x - 11 \leq 19$

**55.** $3x - 7 > 11$      **56.** $8x + 9 < 6$

**57.** $2x - \frac{1}{3} < 5$      **58.** $2x - 3 \geq \frac{5}{3}$

**59.** $3x - 17 \leq 12$      **60.** $5x + 2 > 12$

**61.** $x - \sqrt{2} < 5$      **62.** $2x + \sqrt{3} \leq 3$

In Exercises 63–66, decide whether you will use the *Guess-and-check* strategy or an algebraic method to find a solution. Then solve the equation.

**63.** $15 - 9x = 3(x - 2)$      **64.** $3x - 17 = 6 - 7x$

**65.** $4x^2 + 115 = 259$      **66.** $2x^2 - 23 = 75$

TRANSLATING WORDS TO SYMBOLS

Let $x$ represent a number and write an equation that represents each statement.

**67.** Three $x$ minus two is equal to seventy-five.

**68.** A number tripled is two greater than seventy-five.

**69.** Seventy is two less than the triple of a number.

**70.** Twice a number is 25.

**71.** A number tripled is two more than the number doubled.

**72.** If a number is increased by 14, the result is three times the number.

73. Six less than a number is its double.

74. Six more than a number is its triple.

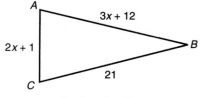

For Exercise 81.

EXTENDING THE IDEAS

An equation that has no solution is called a **contradiction**. For example, $x + 1 = x + 2$ is a contradiction. Why? In Exercises 75–78, simplify each equation to determine whether it is a contradiction.

75. $3(x + 2) - 2x + 3 = x + 6$

76. $5(x + 3) - 2(2x + 1) = x - 3$

77. $x(x + 3) = x(x - 2)$

78. $5 + (x + 2) = -[2 - (9 + x)]$

79. *Create Your Own Contradiction.* Show how to begin with the contradiction $x + 4 = x + 5$ and work backward to obtain $2(2x + 1) - (3x - 2) = (10x - 2) - (9x - 7)$, an equation that is not obviously a contradiction.

80. *Create Your Own Contradiction.* Use the method of Exercise 79 to create your own contradiction too complicated to be obvious.

LOOKING BACK—MAKING CONNECTIONS

81. *Using Algebra in Geometry.* An **isosceles triangle** has two sides that are equal in length. Find the lengths of each side of the following isosceles triangle:

82. *Using Algebra in Geometry.* Find the measure of each angle of the following right triangle:

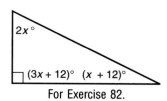

For Exercise 82.

83. *Manufacturing Storage Bins.* The volume of a cylinder with a base radius $r$ and height $h$ is $V = \pi r^2 h$. Find the volume of a storage bin in the shape of a cylinder with a radius of $r = 8$ ft and a height of $h = 12$ ft.

84. *Manufacturing Storage Bins.* Find the volume of a storage bin shaped like a cylinder with a radius $r = 7$ ft and height $h = 15$ ft.

85. *Manufacturing Satellites.* The volume of a sphere of radius $r$ is $V = \frac{4}{3}\pi r^3$. Find the volume of a spherical satellite of radius $r = 5$ in.

86. *Manufacturing Satellites.* Find the volume of a spherical satellite of radius $r = 9$ in.

**1.5** Numerical Representations of Problem Situations

Pólya's Four-step Process ▪ Solving a Problem Using Numerical Methods ▪ Using the Strategies—Complete a Table and Use a Variable ▪ Using the Strategy—Use a Formula

Mathematics is a powerful tool for solving economic, scientific, and consumer problems. Solving mathematical problems involves a process that often requires a variety of approaches, and good problem solvers learn how to analyze a problem and to consider a variety of approaches to the solution.

## Pólya's Four-step Process

George Pólya is sometimes called the father of modern problem-solving because of his significant writing and analysis of the mathematical problem-solving process. Born in Hungary in 1887, Pólya completed a Ph.D. at the University of Budapest. He joined the faculty of Stanford University in 1942 and lived until the age of 97, dying in 1985. His four-step problem-solving process continues to be helpful in this age of technology.

1. UNDERSTAND THE PROBLEM. The first step is to understand the problem. This involves doing such things as the following:

   - Read and reread the problem.
   - Restate the problem in your own words.
   - Learn the definitions of any unknown words.
   - Identify the data pertinent to the problem.
   - Consider specific numerical instances of the problem.

2. DEVISE A PLAN. The second step involves devising a plan. This involves choosing a strategy or a combination of strategies that you think will be helpful in solving the problem. Following are nine strategies that will be used throughout this text:

   - Draw a picture
   - Consider special cases
   - Guess-and-check
   - Complete a table
   - Find a graph
   - Look for a pattern
   - Use a variable
   - Use a formula
   - Write an equation

3. CARRY OUT THE PLAN. After formulating a set of strategies, you need to carry out the plan. This might involve doing the needed computation to complete a table. Or it might mean writing and solving an equation.

4. LOOK BACK. Finally, it is important to look back and review your steps to see if they have led to a reasonable conclusion. Decide whether your solution needs

**REMINDER**

Pólya identified four steps in problem solving.

1. Understand the problem.
2. Devise a plan.
3. Carry out the plan.
4. Look back.

to be refined. Reflect on your solution techniques and what characteristics of the problem led you to use the chosen strategy. What can you learn about solving problems in the future?

## Solving a Problem Using Numerical Methods

Consider this problem situation.

---

### THE GARDEN PROBLEM SITUATION

Maria has 24 ft of fence to put around her garden. She plans one edge of the garden to be along the side of her house, with the fence enclosing the other three sides of a rectangle. She wonders what rectangular shape to use to enclose the largest garden.

---

Pólya's four–step process will be illustrated as a solution to the **Garden Problem Situation** is found.

STEP 1: UNDERSTAND THE PROBLEM

After reading this problem and stating it in our own words, we consider several cases of the problem. Figure 1.27 shows two specific numerical cases that demonstrate that the area varies with different rectangular shapes. It is not obvious which shape will have the largest area.

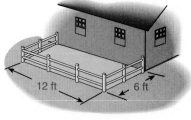
Area = 72 ft$^2$

Area = 64 ft$^2$

**F I G U R E  1 . 2 7**  Two possible gardens using 24 ft of fence.

STEP 2: DEVISE A PLAN

The two special cases suggest that we might find a solution to the problem by considering a larger number of special cases and arranging the resulting data in a table. We decide to use the strategy *Complete a Table*.

STEP 3: CARRY OUT THE PLAN

Complete a table arranging the data in an organized fashion. Begin with a width of 2 ft and increase the width by 1 ft for each case. Use all 24 ft of fence in each case.

| Width(ft) | Length(ft) | Area(ft$^2$) |
|-----------|------------|--------------|
| 2 | $24 - 2(2) = 20$ | 40 |
| 3 | 18 | 54 |
| 4 | 16 | 64 |
| 5 | 14 | 70 |
| 6 | 12 | 72 |
| 7 | 10 | 70 |
| 8 | 8 | 64 |

STEP 4: LOOK BACK

It appears that we have the largest area when the width is 6 ft and the length is 12 ft. In the next table, we refine the computation by investigating values near $W = 6$.

| Width(ft) | Length(ft) | Area(ft$^2$) |
|-----------|------------|--------------|
| 5.9 | 12.2 | 71.98 |
| 5.95 | 12.1 | 71.995 |
| 6 | 12 | 72 |
| 6.05 | 11.9 | 71.995 |
| 6.1 | 11.8 | 71.98 |

From this, we now feel quite confident that if $W \neq 6$, the area is probably less than 72 ft$^2$. We will show in Chapter 8 that this is the largest area, which occurs when the width is 6 ft and the length is 12 ft.

We call the table of data generated in Step 3 of the solution a **numerical representation** of the problem. Numerical representations of a problem are used in the next two examples.

**REMINDER**

Study the patterns in this table.

- As the width increases by 1, the length decreases by 2.
- As the width increases, the area first increases and then decreases.
- We use the words width and length to mean the length of two adjacent sides of a rectangle. Notice that the width is not necessarily shorter than the length.

**GRAPHER NOTE**

It is possible to generate numerical tables on some graphers. Consult your grapher lab manual to see whether it is possible.

**EXAMPLE 1**  THE GARDEN PROBLEM SITUATION: Finding the Largest Area

Suppose Maria uses the 24 ft to build a rectangular fenced region that includes a section dividing the rectangle into two identical parts (see Figure 1.28). Find the width that results in the garden with the maximum area.

SOLUTION

Mentally walk through the four steps of Pólya's process and use the strategy *Complete a Table*.

| Width(ft) | Length(ft) | Area(ft$^2$) |
|:---:|:---:|:---:|
| 1 | $24 - 3(1) = 21$ | 21 |
| 2 | $24 - 3(2) = 18$ | 36 |
| 3 | 15 | 45 |
| 4 | 12 | 48 |
| 5 | 9 | 45 |
| 6 | 6 | 36 |
| 7 | 3 | 21 |

**FIGURE 1.28** There are two identical parts to the garden.

It appears that the largest area occurs when the width is 4 ft. The data in this second table refine the computation.

| Width(ft) | Length(ft) | Area(ft$^2$) |
|:---:|:---:|:---:|
| 3.9 | 12.3 | 47.97 |
| 4 | 12 | 48 |
| 4.1 | 11.7 | 47.97 |

From this refinement, we feel confident that the maximum area is 48 ft$^2$. In Chapter 8, we will use algebraic and graphical methods to find the maximum area of a similar problem.  ■

The next problem situation deals with mixing acid and water.

**MIXTURE PROBLEM SITUATION**

John adds distilled water to a 30% solution of acid in order to dilute the strength of the acid. How does the amount of water added to the 30% solution affect the percent solution of the acid?

We must clarify what is meant by a 30% acid solution. Suppose we begin with 1.5 liters of pure acid and we add distilled water until 5 liters of solution have been obtained. We say that the resulting solution is a 30% acid solution because

$$\frac{1.5 \quad \text{liters of pure acid}}{5 \quad \text{liters of solution}} = 0.30 = 30\% \text{ acid solution.}$$

In general,

$$\frac{\text{liters of pure acid in the solution}}{\text{liters of solution}} \times 100 = \% \text{ acid solution.} \qquad (1)$$

**EXAMPLE 2**   THE MIXTURE PROBLEM SITUATION: Finding the % Solution

Suppose John adds distilled water to 5 liters of a 30% acid solution in increments of 1 liter. Complete a table showing the percent solution after each addition. Estimate how many liters of water must be added to obtain an 18% acid solution.

SOLUTION

30% of 5 is 1.5. Therefore if there are 1.5 liters of acid in a 5-liter acid solution, the solution is a 30% solution. Consider Figure 1.29.

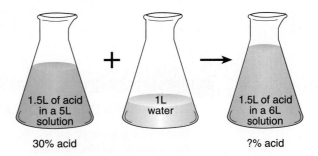

**FIGURE 1.29** One liter of water is added to a 30% solution of acid.

When 1 liter of water is added to the 5 liters of the 30% acid, the mixture has 1.5 liters of acid in 6 liters of solution. Equation (1) becomes $\frac{1.5}{5+1} = 0.25 = 25\%$, and the resulting solution is a 25% solution. This table shows the result of adding 1, 2, 3, 4, 5, or 6 liters of water.

| # Liters of Water Added | Ratio of Liters of Acid to Liters of Solution | Percent Acid |
|:---:|:---:|:---:|
| 1 | $\dfrac{1.5}{5+1} = 0.25$ | 25% |
| 2 | $\dfrac{1.5}{5+2} = 0.214$ | 21.4% |
| 3 | $\dfrac{1.5}{5+3} = 0.188$ | 18.8% |
| 4 | $\dfrac{1.5}{5+4} = 0.167$ | 16.7% |
| 5 | $\dfrac{1.5}{5+5} = 0.15$ | 15% |
| 6 | $\dfrac{1.5}{5+6} = 0.136$ | 13.6% |

We see from this table that slightly more than 3 liters must be added to form an 18% solution. Later we also will apply analytic and graphical methods to solve this problem with greater accuracy. ∎

**REMINDER**

Increasing the amount of distilled water causes the percentage of acid to decrease.

## Using the Strategies—Complete a Table and Use a Variable

Examples 1 and 2 and the discussion preceding them illustrate how to use the problem-solving strategies of *Draw a Picture, Consider Special Cases,* and *Complete a Table.* We next want to illustrate how to combine these strategies with the *Using a Variable* strategy.

---

### THE BOOKSTORE PROBLEM SITUATION

Bernard, a bookstore owner, buys his books from several wholesalers. He prices his books by using a 20% markup so that the retail price is 20% higher than the wholesale price.

---

**EXAMPLE 3**   THE BOOKSTORE PROBLEM SITUATION: Finding Markup and Retail Price

Show how Bernard can use the *Complete a Table* and *Find a Pattern* strategies to discover formulas for markup ($m$) and the retail price ($r$).

SOLUTION

Complete a table showing the wholesale price, the markup, and the retail price. Recall that $20\% = 0.2$.

| Wholesale Price($) | Markup($) | Retail Price($) |
|---|---|---|
| 10 | $(0.2)10 = 2.00$ | $10 + (0.2)10 = 12.00$ |
| 11 | $(0.2)11 = 2.20$ | $11 + (0.2)11 = 13.20$ |
| 12 | $(0.2)12 = 2.40$ | $12 + (0.2)12 = 14.40$ |
| 13 | $(0.2)13 = 2.60$ | $13 + (0.2)13 = 15.60$ |
| 14 | $(0.2)14 = 2.80$ | $14 + (0.2)14 = 16.80$ |
| 15 | $(0.2)15 = 3.00$ | $15 + (0.2)15 = 18.00$ |
| $\vdots$ | $\vdots$ | $\vdots$ |
| $x$ | $(0.2)x$ | $x + (0.2)x = x(1 + 0.2)$<br>$= 1.2x$ |

The formulas are $m = (0.2)x$ and $r = 1.2x$. We can use these formulas to generate a table on a grapher as shown in Figure 1.30.  ■

| X | Y1 | Y2 |
|---|---|---|
| 10 | 2.00 | 12.00 |
| 11 | 2.20 | 13.20 |
| 12 | 2.40 | 14.40 |
| 13 | 2.60 | 15.60 |
| 14 | 2.80 | 16.80 |
| 15 | 3.00 | 18.00 |
| 16 | 3.20 | 19.20 |

X = 10

**FIGURE 1.30** A table generated by one model grapher where $y_1 = 0.2x$ and $y_2 = 1.2x$.

## Using the Strategy—Use a Formula

We illustrate in this section how formulas can be used to solve problems. Later in this text, you will be expected to develop some formulas on your own. However, in this section the formulas needed to solve the problems will be provided.

**FIGURE 1.31** A baseball is thrown up into the air.

---

### THE BASEBALL PROBLEM SITUATION

A baseball is thrown straight up into the air with an initial velocity of 88 ft/sec. It can be shown that $t$ seconds after the ball is thrown, its height ($h$) above the ground is described by the formula

$$h = -16t^2 + 88t.$$

This formula is called the **algebraic representation** of this problem.

---

Many quantitative questions can be asked about this problem situation, each resulting in a unique problem. We explore two of these in the next two examples.

**EXAMPLE 4** THE BASEBALL PROBLEM SITUATION: Finding the Height of a Ball's Trajectory

If a baseball is thrown straight up into the air with a velocity of 88 ft/sec, find its height above the ground at each second for the first 5 sec. Is the ball falling 4 sec after it is thrown?

$3 \rightarrow$ T: $^-16$T$^2 + 88$T

120

**FIGURE 1.32** The formula is evaluated for $t = 3$ with a grapher.

SOLUTION

The formula to use is $h = -16t^2 + 88t$. Complete a table by evaluating this formula for $t = 1, 2, 3, 4,$ and 5. Figure 1.32 shows the result of evaluating $-16t^2 + 88t$ for $t = 3$.

| Seconds($t$) | Height($h = -16t^2 + 88t$) |
|:---:|:---:|
| 1 | 72 |
| 2 | 112 |
| 3 | 120 |
| 4 | 96 |
| 5 | 40 |

The height of the ball 4 sec after it is thrown is lower than its height at 3 sec. Therefore the ball is already falling 4 sec after it is thrown. ∎

It appears from the data in the table in Example 4 that the ball reaches its maximum height between 2 and 4 sec after it is thrown. In Example 5, we pursue finding its maximum height.

**EXAMPLE 5** THE BASEBALL PROBLEM SITUATION: Finding the Maximum Height

Find the height of the ball's trajectory 2.5 sec after it is thrown and every 0.1 sec after that until you can determine the ball's maximum height.

$2.6 \rightarrow$ T: $^-16$T$^2 + 88$T

120.64

$2.7 \rightarrow$ T: $^-16$T$^2 + 88$T

120.96

**FIGURE 1.33** A grapher evaluation when the formula, $-16t^2 + 88t$, is evaluated for $t = 2.6$ and $t = 2.7$.

SOLUTION

Use a grapher as shown in Figure 1.33 to find the data for the following table:

| Seconds($t$) | Height($h = -16t^2 + 88t$) |
|:---:|:---:|
| 2.5 | 120 |
| 2.6 | 120.64 |
| 2.7 | 120.96 |
| 2.8 | 120.96 |
| 2.9 | 120.64 |
| 3 | 120 |

Observe that the first three heights are repeated in reverse order in the next three heights, indicating that the ball is traveling up and then down. This pattern also suggests that the ball will be at its peak midway between 2.7 and 2.8 sec after it is thrown. Therefore, evaluate $-16t^2 + 88t$ when $t = 2.75$.

$$h = 16(2.75)^2 + 88(2.75) = 121$$

The maximum height of the ball's trajectory appears to be 121 ft. ■

In later sections, we will confirm both algebraically and graphically that this height is exactly 121 ft.

## Exercises for Section 1.5

TRANSLATING WORDS TO SYMBOLS

1. Write an algebraic expression for the phrase "17% of a number $x$."

2. Write an algebraic equation for the phrase "length $L$ is 3 more than width $W$."

3. Write an algebraic equation for the phrase "the area $A$ is width $W$ times 3 units greater than the width."

4. Write an algebraic expression for the phrase "$x$ less 17% of $x$."

5. Suppose Maria has only 18 ft of fence to enclose 3 sides of her garden. (See the **Garden Problem Situation**.)

   a) Draw and label pictures of two possible rectangular shapes for the garden in addition to the one following. (This figure shows one possible rectangular shape for the garden.) Find the area of each.

   b) Complete the following table to show all possible whole number widths for the garden and the corresponding lengths and areas.

   c) Which two whole number widths have the largest area? What width do you think would result in an even larger area? Confirm your guess.

| Width | Length | Area |
|-------|--------|------|
| 1 | 16 | 16 |
| 2 | ? | ? |
| ⋮ | ⋮ | ⋮ |

6. Joel is going to use 24 ft of fence to enclose all four sides of a rectangular garden. He wants to make the garden as large as possible.

   a) Draw and label pictures of two possible rectangular shapes for the garden. Find the area of each.

   b) Complete the following table to show all possible whole number widths for the garden and the corresponding lengths and areas:

| Width | Length | Area |
|-------|--------|------|
| 1 | 11 | 11 |
| 2 | ? | ? |
| ⋮ | ⋮ | ⋮ |

c) Find the width and length of the garden of maximum area.

7.  Suppose Joel uses 30 ft of fence to enclose all four sides of his rectangular garden.

a)  Complete the following table to show all possible whole number widths for the garden and the corresponding lengths and areas:

| Width | Length | Area |
|-------|--------|------|
| 1 | 14 | 14 |
| 2 | ? | ? |
| ⋮ | ⋮ | ⋮ |

b)  Which two whole number widths have the largest area? What width do you think would result in an even larger area? Confirm your guess.

c)  After reviewing the solutions to Exercises 6 (c) and 7 (b), what pattern do you see to these solutions? What generalization do you conjecture will always be true?

8.  *Diluting an Acid.* A 30% solution of acid is to be diluted by adding distilled water. (See the **Mixture Problem Situation**.)

a)  How many liters of pure acid are contained in 4 liters of this 30% acid solution?

b)  Suppose distilled water is added to 4 liters of this 30% acid solution. Complete a table that shows the percentage of acid in each of the solutions resulting from adding 1, 2, 3, 4, 5, 6, and 7 liters of distilled water to 4 liters of a 30% acid solution.

c)  Approximately how much water should be added to 4 liters of this 30% acid solution to obtain a 10% solution?

9.  *Diluting an Acid.* Suppose the acid in the **Mixture Problem Situation** is a 20% acid solution.

a)  How many liters of pure acid are in 5 liters of a 20% acid solution?

b)  Suppose distilled water is added to this solution. Complete a table that is a numerical representation of this problem situation.

c)  Estimate how many liters of water should be added to this 20% acid solution to obtain a 5% solution.

d)  Suppose 4 liters of distilled water are mixed with 5 liters of a 20% solution of acid. What percentage acid is the mixture?

e)  Use the *Make a Table* strategy to determine how many liters of distilled water should be added to 8 liters of a 25% solution of acid to result in a 15% solution.

10.  *Finding Wholesale Prices.* Suppose Bernard uses a 35% markup instead of a 20% markup. (See the **Bookstore Problem Situation.**)

a)  Complete a table that shows Bernard's markup and retail prices for books with wholesale prices of $5, $10, $15, $20, and $25.

b)  Use the pattern that you observe in the table completed in Exercise 10 (a) to find Bernard's markup and the retail price for a book with a wholesale price of $x$.

c)  What is the wholesale price of a book that has a retail price of $21.60?

11.  *Speed, Distance, and Time.* A vehicle is traveling at a constant speed of 66 ft/sec.

a)  Using the formula *distance* = *rate* · *time*, find the distance that the vehicle travels in each of the first 5 sec and complete the following table:

| Seconds | Distance Traveled |
|---------|-------------------|
| 1 | ? |
| 2 | ? |
| 3 | ? |
| 4 | ? |
| 5 | ? |

b) Observe the number pattern in the table you completed in Exercise 11 (a). How far does the vehicle travel in $t$ seconds?

c) How long does it take the vehicle to travel 165 ft?

**12.** *Acceleration, Distance, and Time.* A car accelerates as it leaves a traffic light. The car's distance, $d$, in feet, from the intersection during the first 8 sec is given by the formula $d = 12t^2$, where $t$ represents time in seconds.

a) Complete the following table:

| Time($t$) | Distance Traveled($d$) |
|-----------|------------------------|
| 1 | ? |
| 2 | ? |
| 3 | ? |
| 4 | ? |
| 5 | ? |
| ⋮ | ⋮ |
| $t$ | $12t^2$ |

b) How long does it take the car to travel 432 feet?

c) Use *Guess-and-check* to see how many seconds it takes the car to travel 165 ft. (Report the answer to the nearest tenth of a second.)

**13.** *Projectile Motion.* A pitching machine shoots a baseball straight up into the air at a speed of 112 ft/sec. It is $h$ feet above the ground $t$ seconds later. The algebraic representation for height in terms of time is $h = -16t^2 + 112t$.

a) Use the *Make a Table* strategy to find how many seconds pass until the thrown ball hits the ground.

b) What is the maximum height for this ball's trajectory and how many seconds after it is shot up does it reach this maximum height?

c) Would this ball hit the top of the Houston Astrodome? (Research the height of the Astrodome.)

EXTENDING THE IDEAS

**14.** *Finding Maximum Area.* Sarah uses 24 ft of fence to surround a rectangular garden around the corner of her house. She placed a post for the end of the fence 2 ft from the corner of the house in each direction from the corner, as shown in the figure. She would like her garden to be as large as possible.

For Exercise 14.

a) Draw and label pictures of two possible shapes for the garden. Find the area of each. Note that the garden is a rectangle with a 2-ft by-2-ft square removed from one corner.

b) Complete the following table to show all possible whole number widths for the garden and the corresponding lengths and areas:

| Width | Length | Area |
|-------|--------|------|
| 3 | 11 | $33 - 4 = 29$ |
| 4 | 10 | ? |
| ⋮ | ⋮ | ⋮ |

c) Find the length and width of the largest garden meeting the conditions of this problem.

15. In Sarah's garden problem the width is $W$, the length is $L$, and a third side of the garden is $W - 2$.

a) Find the length of the fourth side in terms of the variable $L$.

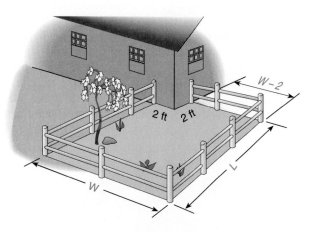

For Exercise 15.

b) If Sarah uses 30 ft of fence for her garden, write an equation using the variables $W$ and $L$ for the length of the fence.

c) Write an equation that describes the area of Sarah's garden in terms of the variables $W$ and $L$.

16. *Projectile Motion.* Suppose a baseball is thrown straight up into the air at a velocity of 60 ft/sec. The height($h$) of the ball's trajectory in feet $t$ seconds after the ball is thrown is described by the algebraic representation $h = -16t^2 + 60t$.

a) Complete a table that records the thrown ball's height above the ground at intervals of 0.5 sec.

b) Refine the table in Exercise 16 (a) to find when the thrown ball reaches its maximum height. What is this maximum height?

c) After how many seconds does the ball hit the ground?

17. State from memory the names of the problem-solving strategies that have been discussed.

LOOKING BACK—MAKING CONNECTIONS

18. If $m\angle 1 = 12x - 17$ and $m\angle 2 = 8x - 9$, what is the measure of $\angle 2$ if lines $p$ and $q$ are parallel. (If lines $p$ and $q$ are parallel, $m\angle 1 = m\angle 2$.)

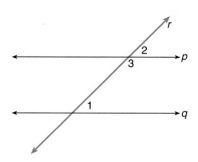

For Exercise 18.

19. Find the block of digits that repeat in the decimal representation of $\frac{1}{7}$.

20. Evaluate the expression $-2(7 - (-4)) + 3(-5 - (-8))$.

21. Find the decimal representation of

$$\frac{r^2 + 8}{r^3 - 5s^2 + 17}$$

for $r = -3$ and $s = 7$. Represent this number as a repeating decimal.

## 1.6 Algebraic Representations of Problem Situations

Translating from Words to Algebraic Expressions ▪ Using the Strategies—Write and Solve an Equation ▪ Investigating Additional Problem Situations

In Section 1.5, problems were analyzed by forming a table of numerical values that described the problem situation. We called it a numerical representation of the problem.

For some problems, it is helpful to use a variable as the value of a quantity instead of using specific numerical values. This variable is then used in algebraic expressions to represent other related quantities. An algebraic expression that uses a variable(s) is known as an **algebraic representation** of the problem situation.

In the solution to the **Bookstore Problem Situation** in Example 3 of Section 1.5, we used the variable $x$ to describe algebraically the pattern of the numerical data in the table. If the wholesale price is $\$x$, the markup is $\$0.2x$ and the retail price is $\$(x + 0.2x)$, or $\$1.2x$. This algebraic expression $1.2x$ is the algebraic representation of the problem situation.

---

### An Algebraic Representation of the Bookstore Problem Situation

If $R$ represents the retail price and $x$ represents the wholesale price, we call the equation

$$R = 1.2x$$

the **algebraic representation of the retail price $R$** in terms of the wholesale price $x$ when prices are increased by 20%.

This equation is also called the algebraic representation of the **Bookstore Problem Situation**.

---

Knowing the algebraic representation of a problem situation is often an important ingredient in finding a solution to a problem. We illustrate with the **Bookstore Problem Situation**.

---

**A Bookstore Problem and Its Reformulation**

1. **Problem:** What is the wholesale price of a book whose retail price is $17.10?
2. **Reformulation as an equation:** For what value of $x$ is the retail price $R$ equal to 17.10? That is, what value of $x$ is a solution of the equation $1.2x = 17.10$?

---

## Translating from Words to Algebraic Expressions

In the process of finding an algebraic representation of a problem situation, we must change word phrases into algebraic symbols. Learning to translate word phrases into algebraic expressions and sentences into equations is an important part of problem solving. Before continuing to analyze problem situations, we summarize several of the more frequently encountered word phrases and sentences.

---

**Common Word Phrases Converted to Algebraic Expressions**

| Word Phrase | Algebraic Expression |
|---|---|
| 3 less than a number $x$ | $x - 3$ |
| a number $x$ increased by 5 | $x + 5$ |
| 8 times a number $x$ | $8x$ |
| 18 more than a number $x$ | $x + 18$ |
| $x$ divided by 4 | $\dfrac{x}{4}$ |
| the difference between two numbers, $x$ and $y$ | $x - y$ |
| one-half of a number $x$ | $\dfrac{1}{2}x$ |
| the ratio of two numbers, $x$ and $y$ | $\dfrac{x}{y}$ |
| twice the sum of a number $x$ and 5 | $2(x + 5)$ |

---

**EXAMPLE 1**    Translating Words to Algebraic Expressions

Write an algebraic expression for each of the following word phrases:

a) 5 more than a number
b) a number decreased by 3.15
c) 7% of a number

SOLUTION

a) $x + 5$ or $5 + x$
b) $x - 3.15$
c) $0.07x$ ∎

When translating a sentence into an equation, the verb "is" in the sentence is often translated to $=$, as the following examples illustrate.

### Common Sentences Converted to Equations

| Sentence | Equation |
|---|---|
| Five is two less than a number. | $5 = x - 2$ |
| Seventeen is 1 more than 3 times a number. | $17 = 3x + 1$ |
| A number is one less than its double. | $x = 2x - 1$ |
| A number is one more than its triple. | $x = 3x + 1$ |

**EXAMPLE 2**    Translating Sentences to Equations

Write an equation for each of the following sentences:

a) 15 is one more than three times a number.
b) Tripling a number results in the double of eight less than the number.

SOLUTION

a) $3x + 1 = 15$
b) $3x = 2(x - 8)$ ∎

## Using the Strategies—Write and Solve an Equation

Often a solution to a problem situation can be achieved by writing and solving an equation. We call this strategy *Writing and Solving an Equation*. In Section 1.5, the **Bookstore Problem Situation** was studied by using the strategies *Complete a Table, Find a Pattern,* and *Use a Variable*. Now we will solve a related problem by using the strategy *Write and Solve an Equation*.

**REMINDER**

Following is a summary of the problem-solving strategies used so far:

- Draw a Picture
- Consider Special Cases
- Complete a Table
- Look for a Pattern
- Use a Variable
- Write and Solve an Equation
- Use Guess and Check
- Use Equivalent Equations

**EXAMPLE 3**  THE BOOKSTORE PROBLEM SITUATION: Finding the Wholesale Price

If Bernard's markup is 38%, what is the wholesale price for a book that sells in his store for $27.60?

SOLUTION

Let $x$ = the wholesale price.

**Markup:**  $0.38x$

**Retail Price:**  $x + 0.38x = 1.38x$

**Equation:** $1.38x = 27.60$   This equation is the algebraic representation of the problem.

Solve this equation.

$$1.38x = 27.60$$

$$\frac{1.38}{1.38} \cdot x = \frac{27.60}{1.38}$$   Divide each side of the equation by 1.38.

$$x = 20$$

The wholesale price of the book is $20.00. ∎

**REMINDER**

Throughout the remainder of this text, when a computation is completed on a grapher and reported as a decimal, we round to the nearest hundredth. So in Example 4, the solution will be reported as $x = 2.17$ if solved using the grapher. If an equation is solved algebraically, an exact solution is reported. If $\frac{0.26}{0.12} = \frac{13}{6}$ is evaluated by hand, the solution will be reported as $x = \frac{13}{6}$.

**EXAMPLE 4**  THE MIXTURE PROBLEM SITUATION: Finding the Amount of Acid Added

A chemist mixes some unknown amount of 30% acid solution with 2 liters of a 5% acid solution to obtain an 18% solution. How many liters of the 30% are added?

SOLUTION

Let $x$ = the number of liters of 30% needed. The information from the problem is summarized in the following table:

| Strength | Liters of Solution | Liters of Pure acid |
|----------|--------------------|---------------------|
| 5% solution | 2 | $0.05(2) = 0.1$ |
| 30% solution | $x$ | $0.30x$ |
| mixture of 18% solution | $x + 2$ | $0.18(x + 2)$ |

The pure acid comes from two sources: the 5% acid solution and the 30% acid solution. Therefore

$$\left\{\begin{array}{c}\text{Pure acid from} \\ \text{5\% solution}\end{array}\right\} + \left\{\begin{array}{c}\text{Pure acid from} \\ \text{the 30\% solution}\end{array}\right\} = \left\{\begin{array}{c}\text{Total liters} \\ \text{of pure acid}\end{array}\right\}$$

$$0.1 + 0.30x = 0.18(x + 2)$$

$$0.30x + 0.1 = 0.18x + 0.36$$

$$0.12x = 0.26$$

$$x = \frac{0.26}{0.12} = 2.167$$

Add 2.17 liters of 30% acid solution. ■

## Investigating Additional Problem Situations

----

□                                                                    □

### THE INVESTMENT PROBLEM SITUATION

Melissa has $15,000 to invest. She knows that usually investments with higher interest rates carry greater risk and those with lower interest rates carry less risk. So she decides to spread her investment among several risk levels.

   She invests a portion in an account that pays 7% a year and the remainder into an account that pays 9% a year.

□                                                                    □

----

**EXAMPLE 5**    THE INVESTMENT PROBLEM SITUATION: Finding the Amount to Invest

How much should Melissa invest in each account in order to earn $1170 interest in the first year?

SOLUTION

Let $x =$ the amount invested at 7% and $15000 - x =$ the amount invested at 9%.

   **Interest earned in 7% account:** $0.07x$

   **Interest earned in 9% account:** $0.09(15,000 - x)$

   **Total interest:** $0.07x + 0.09(15,000 - x) = 0.07x + 1350 - 0.09x$

$$= -0.02x + 1350$$

   **Equation:** $-0.02x + 1350 = 1170$   **This equation is the algebraic representation of the problem.**

Solve this equation.

$$-0.02x + 1350 = 1170$$

$$-0.02x = 1170 - 1350$$

$$-0.02x = -180$$

$$x = \frac{180}{0.02} = 9000$$

$$.07\,(9000) + 0.09\,(6000)$$

$$1170$$

**FIGURE 1.34** Grapher support for the solution to the investment problem.

Melissa should invest \$9000 in the account that pays 7% and \$6000 in the account that pays 9%. Figure 1.34 shows grapher support for this solution. ∎

Here is another problem situation.

---

### THE FOOT RACE PROBLEM SITUATION

At one instant in a foot race, Juan is running at a speed of 30 ft/sec and Carlos is running at 28 ft/sec. The variables in this problem situation are distance(*d*), rate(*r*), and time(*t*) and are related by the equation

$$\text{distance} = \text{rate} \times \text{time}$$

$$d = r \cdot t.$$

---

Consider the following aspect of the foot race problem situation.

**EXAMPLE 6**   THE FOOT RACE PROBLEM SITUATION: Finding the Catch-up Time

Suppose that at the instant Juan is running at 30 ft/sec, he is 10 ft behind Carlos, who is running at 28 ft/sec. How long will it take for Juan to catch Carlos?

SOLUTION
Figure 1.35 shows a picture of this problem.

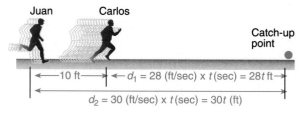

**FIGURE 1.35** How long will it take Juan to catch Carlos?

Let *t* be the time it takes for Juan to catch Carlos. The equation $d = r \times t$ is used to find algebraic representations for the distances $d_1$ and $d_2$ that Carlos and

Juan ran respectively. We conclude the following:

**Algebraic representation for distance Carlos ran:** $d_1 = 28t$
**Algebraic representation for distance Juan ran:** $d_2 = 30t$

We see from Figure 1.35 that

$$d_2 = d_1 + 10$$
$$30t = 28t + 10. \tag{1}$$

The solution to Eq. (1) is the solution to the problem.

$$30t = 28t + 10 \quad \text{Add the additive inverse of } 28t \text{ to both sides of the equation.}$$

$$30t + (-28t) = 28t + (-28t) + 10$$

$$(30 - 28)t = 10$$

$$2t = 10$$

$$t = 5$$

It will take Juan 5 sec to catch Carlos.

**Numerical Support.** When $t = 5$, $d_1 = 28 \cdot 5 = 140$ and $d_2 = 30 \cdot 5 = 150$. Juan runs 10 ft farther than Carlos in 5 sec. ∎

**TRY THIS**

To simulate this foot race set a grapher as follows. (Check your grapher lab manual to see if your grapher has parametric mode.)

- parametric mode,
- simultaneous mode,
- $X_{1T} = 30T$, $Y_{1T} = 5$,
- $X_{2T} = 28T + 10$, $Y_{2T} = 2$.
- WINDOW
  $T\text{min} = 0$, $T\text{max} = 5.5$,
  $T\text{step} = 0.01$,
  $X\text{min} = 0$, $X\text{max} = 175$,
  $X\text{scl} = 10$, $Y\text{min} = 0$,
  $Y\text{max} = 8$, $Y\text{scl} = 1$

Notice how you can use TRACE to check the distance between runners each 0.01 second along the way.

## Exercises for Section 1.6

In Exercises 1–12 **Translate Words to Symbols** by writing an algebraic expression that is the translation of the given word phrase.

1. A number is increased by 12

2. 4 times a number

3. Double two less than a number

4. The sum of 12.3 and a number

5. The difference of two numbers

6. A number is decreased by 14

7. 5 is increased by twice a number

8. The ratio of a number and 2.3

9. 15% of the sum of a number and 5

10. 7% of two less than a number

11. The quotient of a number and 5

12. 4 more than the triple of a number

In Exercises 13–20, write an equation that is equivalent to the given sentence.

13. Twice a number is 7.

14. A number plus 4 is eighteen.

15. Two times a number is four less than the number.

16. Three less than a number is 12.

17. The double of the sum of a number and 3 is 125.

18. A number is decreased by 4 and the result is doubled. The result is 36.

19. Thirty more than a number is halved to obtain 25.

20. The triple of the difference of a number and 2 is twice the sum of a number and one.

21. *Comparing Wholesale and Retail Prices.* Suppose that Bernard conducts a special sale in which his markup is reduced to 18%. (See the **Bookstore Problem Situation**.)

   a) Suppose $x$ represents the wholesale price of a book whose sale price is $35.50. Write an equation that is an algebraic representation of this problem situation.

   b) What is the wholesale price (to the nearest cent) of a book that has a sale price of $35.50?

   c) Bernard buys from his supplier 100 copies of this book with its sale price of $35.50. Write and solve an equation to find how much he must pay for the 100 books.

22. *Mixing Solutions.* Suppose a chemist has 5 liters of a 10% acid solution. A 20% acid solution is to be added to obtain a 12% solution. (See the **Mixture Problem Situation**.)

   a) If $x$ represents the number of liters of 20% solution added, write algebraic expressions that describe the number of liters of pure acid in the 20% solution and the number of liters of pure acid in the mixed solution.

   b) Write an equation that is an algebraic representation of this problem situation.

   c) Find how many liters of 20% solution must be added to the 10% solution to obtain a 12% solution.

In Exercises 23–38, solve each of the stated problems.

23. *Wholesale Price.* A new suit retails at $129.00. If the store owner has a 55% markup, how much did the owner pay wholesale for the suit?

24. *Wholesale Price.* Better Buy Appliance has a 30% markup. How much does the store owner pay for an appliance that retails for $560?

25. *Computer Sales.* In a recent year, IBM microcomputer sales were $123,000,000 more than Apple Computer's. If the two companies together had sales of $560,000,000, what were the sales for each?

26. *Investment Analysis.* John made a single investment into a savings account that pays 8% a year. At the end of 1 yr, John had $13,273.84 in the account. How much was his original investment?

27. *Diluting a Chemical.* Water is added to 5 gal of a 30% solution of sodium hypochloride to obtain a 22% solution. How much water was added?

28. *Diluting a Chemical.* A bug spray manufacturer sells a certain chemical as a 2% concentration. The chemical is manufactured at an 80% concentration. If 500 liters of the concentrate is diluted to the 2% concentration, how many liters of diluted chemical will result?

29. *Mixing a Solution.* A chemist wants to add a 25% acid solution to 4 liters of a 5% acid solution to obtain a 10% acid solution. How much 25% acid solution should be added?

**30.** *Mixing a Solution.* How many liters of a 30% alcohol solution must be added to 3 liters of a 12% alcohol solution to obtain a 20% solution?

**31.** *Investment Analysis.* Suppose that Reese invests $12,000—part in an account that earns 6.5% and the rest in an account that earns 8%. (See the **Investment Problem Situation.**)

a) Reese invests $x in the account that pays 8%. Write expressions for the amount of interest earned at the end of 1 yr in each of the two accounts.

b) Write an algebraic representation for the total amount of interest earned from the $12,000 investment.

c) How much money is invested in each account if the total interest at the end of 1 yr is $885?

**32.** *Comparing Running Speeds.* Suppose that while running at a pace of 28.5 ft/sec, Juan is 8 ft behind Carlos. Carlos is running at a speed of 28 ft/sec. (See the **Foot Race Problem Situation.**)

a) Suppose *t* represents the time needed for Juan to catch Carlos. Write an equation that is an algebraic representation of this problem situation.

b) If Juan and Carlos both continue running at their current pace, find the number of seconds until Juan catches Carlos.

c) If Carlos is 300 ft from the finish line at the instant that Juan is 8 ft behind him, who wins the race?

**33.** *Investment Analysis.* Wes won a $15,000 sales bonus from his company. He invests part of it in an account that pays 7.5% and the rest in an account that pays 9%. How much must he invest in the 9% account in order to earn $1312.50 in interest the first year?

**34.** *Investment Analysis.* Jody's bank will pay her 9% for every dollar invested in a certificate of deposit if she invests three times as many dollars into an account paying 6%. How much should she invest in each account to earn $3240 interest in the first year?

**35.** *Investment Analysis.* Sally invests $4500 into an account that pays 6.8%. How much should she invest in an account that pays 8.5% if she wants to earn $560 interest in the first year?

**36.** *Travel Time.* Juan averages 55 mph for a 340-mi trip. Write an equation whose solution is the time the trip takes Juan and solve for the time.

**37.** *Catch-up Time.* A horse named Frequent Winner is clocked at 45 ft/sec coming down the homestretch. Swift A-foot was trailing at 48 ft/sec. How long will it take for Swift A-foot to take the lead if he starts out 700 ft behind?

**38.** *Separation Time.* Two planes leave Denver at the same time. One travels east at 450 mph and the other travels west at 520 mph. How long is it until these planes are 1500 mi apart?

EXTENDING THE IDEAS

**39.** *Investigating Rectangular Dimensions.* A farmer has 738 ft of fence. He uses all of it to make a rectangular corral that is twice as long as it is wide. (Assume the width of the barn is one-third the length of the corral.)

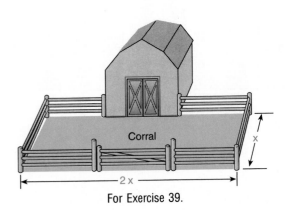

For Exercise 39.

a) Write an equation that is an algebraic representation of this problem situation.

b) Solve this equation to find the dimensions of the corral.

c) Find the area of the resulting corral. What fraction of an acre will the resulting corral be?

**40.** *Investment Analysis.* A certain company has a record of increasing its dividend each quarter. The dividends of the recent past, shown in the following bar graph, can be approximated by $1.14 + 0.20t$, where $t = 0$ corresponds to Jan. 1, 1990, $t = 1$ corresponds to Jan. 1, 1991, $t = 2$ corresponds to Jan. 1, 1992, etc.

a) What was the dividend paid in the third quarter of 1992?

b) Predict the dividend paid in the first quarter of 1994.

LOOKING BACK—MAKING CONNECTIONS

**41.** *Using Algebra in Geometry.* The sum of the measures of the angles of a triangle is 180°. If two angles of the triangle measure 12° and 38°, write an equation in $x$ whose solution represents the measure of the third angle.

**42.** *Using Algebra in Geometry.* The sum of the measures of the angles of a quadrilateral is 360°. If two of the angles have equal measures and the other two angles measure 37° and 71°, write an equation in $x$ whose solution represents the measure of each of the two equal angles.

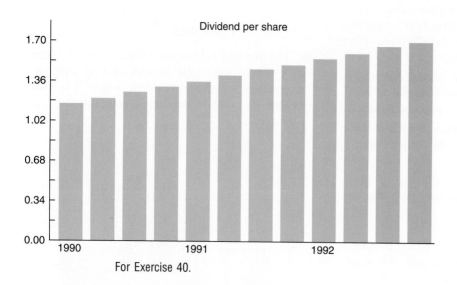

Dividend per share

For Exercise 40.

## Chapter 1 Summary

| | Sets of Numbers | Examples |
|---|---|---|
| Natural Numbers (Counting Numbers) | $\{1, 2, \ldots, \}$ | |
| Whole Numbers | $\{0, 1, 2, \ldots\}$ | |
| Integers | $\{\ldots, -2, -1, 0, 1, 2, \ldots\}$ | |
| Rational Numbers | $\left\{\dfrac{m}{n} \middle| m \text{ and } n \text{ are integers}, n \neq 0\right\}$ | $\dfrac{-3}{4}, \dfrac{2}{1}$ |
| Irrational Numbers | $\{x \mid x \text{ is a real number, not rational}\}$ | $\sqrt{2}, \pi$ |
| Real Numbers | The set of all numbers modeled by points on the number line. These numbers are either rational or irrational. | |

| | Decimal Representations | Examples |
|---|---|---|
| Repeating Decimals | Decimals with a block of digits that repeat to the right of the decimal. The repeating block is indicated by a bar over the digits that repeat. These numbers are rational. | $2.\overline{31}, \ 0.4\overline{35}$ |
| Terminating Decimals | These numbers are equivalent to repeating decimals whose repeating block is $\overline{0}$. These numbers are rational. | $-0.75, \ 4.875$ |
| Nonrepeating/ Nonterminating Decimal | These decimals do not terminate and do not have a repeating block of digits. These numbers are irrational. | $0.181181118\ldots$ |

| | Properties of Equality | Examples |
|---|---|---|
| Basic Properties | 1. $a = a$ (Reflexive Property) <br> 2. If $a = b$, then $b = a$ (Symmetric Property) <br> 3. If $a = b$ and $b = c$, then $a = c$ (Transitive Property) | $5 = 5$ <br> If $2 = x$, then $x = 2$. <br><br> If $x = 2y$ and $2y = 9$, then $x = 9$. |

| | | |
|---|---|---|
| Addition Property | If $a = b$, then $a + c = b + c$ | $x - 2 = 5$ implies that $(x - 2) + 2 = 5 + 2$. |
| Multiplication Property | If $a = b$, then $ac = bc$ with $c \neq 0$. | If $2x = 6$, then $\frac{1}{2}(2x) = \frac{1}{2}(6)$. |

| | Inequalities and Their Properties | Examples |
|---|---|---|
| Less than: $a < b$ | $a$ is left of $b$ on the number line. | $-5 < -1$ |
| Less than or equal to: $a \leq b$ | $a$ is left of $b$ on the number line or equal to $b$. | $2 \leq 5$;   $3 \leq 3$ |
| Greater than: $a > b$ | $a$ is right of $b$ on the number line. | $6 > -2$ |
| Greater than or equal to: $a \geq b$ | $a$ is right of $b$ or equal to $b$ on the number line. | $-1 \geq -3$, $-5 \geq -5$ |
| Trichotomy Property | Either $a < b, a = b,$ or $a > b$. | |
| Transitive Property | If $a < b$ and $b < c$, then $a < c$. | $2 < 5$ and $5 < 8$, therefore $2 < 8$ |
| Addition Property | If $a < b$, then $a + c < b + c$. | If $x - 4 < 6$, then $x - 4 + 4 < 6 + 4$. |
| Multiplication Property | If $a < b$, then $ac < bc$ if $c > 0$ and $ac > bc$ if $c < 0$. | If $-0.5x < 6$, then $2(-0.5x) < 2 \cdot 6$ and $-2(-0.5x) > -2 \cdot 6$. |

| | Interval Notation | Examples |
|---|---|---|
| Infinite Interval | $(-\infty, a], (-\infty, a), [a, \infty), (a, \infty),$ or $(-\infty, \infty)$ | $(-\infty, -3)$ |
| Finite Interval | $(a, b),$   $[a, b),$   $(a, b],$   or $[a, b]$ | $[-2, 4)$    $(3, 5]$ |

| | Absolute Value | Examples |
|---|---|---|
| Definition of Absolute Value | $|a| = \begin{cases} a & \text{if } a \geq 0 \\ -a & \text{if } a < 0 \end{cases}$ | $|-3| = -(-3) = 3$ |
| Distance between Two Points on a Number Line | $d(A, B) = |a - b|$, where $a$ is the coordinate of point $A$ and $b$ is the coordinate of point $B$. | $d(-3, 7) = |-3 - 7| = 10$ |

| | Operations of Real Numbers | Examples |
|---|---|---|
| Addition of Two Negative Numbers | The opposite of the sum of the absolute values of the numbers | $-3 + -5 = -(3 + 5) = -8$ |

| | | |
|---|---|---|
| Addition of Two Numbers with Opposite Signs | Find the difference in the absolute values of the numbers. The sign of the answer agrees with the sign of the number with larger absolute value. | $2 + -5 = -(5 - 2) = -3$ |
| Definition of Subtraction | $a - b = a + (^-b)$ | $4 - 6 = 4 + (-6) = -2$ |
| Definition of Division | $a \div b = a \cdot \dfrac{1}{b} = \dfrac{a}{b}$, provided $b \neq 0$ | $-3 \div 8 = -3 \cdot \dfrac{1}{8} = \dfrac{-3}{8}$ |
| Multiplication and Division of Signed Numbers | The product or quotient of two numbers with the same sign is positive. The product or quotient of two numbers with opposite signs is negative. | $(-5)(-8) = 40$<br>$(7)(-5) = -35$ |
| Definition of Exponents | $a^n = \underbrace{a \cdot a \cdot \; \cdots \; \cdot a}_{n \text{ factors of } a}$ | $4^3 = 4 \cdot 4 \cdot 4$ |
| Order of Operations | Within each set of parentheses, starting with the innermost set, perform the following steps:<br>1. Evaluate all powers.<br>2. Do any multiplications and divisions in the order in which they occur.<br>3. Do any additions or subtractions in the order in which they occur.<br>4. Continue until all parentheses are removed and all operations are performed. | $4 + 3^2 = 13$<br>$25 \div 5 \times 2 = 10$<br><br>$12 - 5 + 3 = 10$ |

| | Properties of Real Numbers | Examples |
|---|---|---|
| Properties of Negation | 1. $(-1)(a) = -a$<br>2. $(-a)(b) = -(ab)$<br>3. $(-a)(-b) = (ab)$<br>4. $-(a + b) = (-a) + (-b)$<br>5. $-\dfrac{a}{b} = \dfrac{-a}{b} = \dfrac{a}{-b}$ | $-1(4) = -4$<br>$-3(4) = -12$<br>$(-5)(-3) = 15$<br>$-(3 + 6) = (-3) + (-6) = -9$<br>$-\dfrac{5}{8} = \dfrac{-5}{8} = \dfrac{5}{-8}$ |
| Double Negative Property | $-(-a) = a$ | $-(-5) = 5$ |
| Additive Inverse Property | $a + (-a) = 0$ | $7 + (-7) = 0$ and<br>$-2 + (-(-2)) = 0$ |

| | **Algebra Concepts** | **Examples** |
|---|---|---|
| Numerical Expression | A collection of real numbers that are combined using operations | $2 - 5^3 + \sqrt{3}$ |
| Algebraic Expression | A collection of variables and real numbers that are combined using operations | $5x + 3b^2$ |
| Variable | A symbol or a letter that reserves the place for a number | $x$, $y$ or $\bigcirc$ |
| Term | A part of an expression that is separated by addition or subtraction | $5x$ and $3b^2$ are the terms of $5x + 3b^2$. |
| Like Terms | Terms of an expression with the same variable and same exponent on that variable | $3x^2$ and $x^2$ are like terms of $3x^2 + x + x^2 + 2x^2 y$. |
| Equation | A statement that two mathematical expressions are equal | $2(x - 4) = 2x - 8$ |

| | **Basic Properties of Algebra** | **Examples** |
|---|---|---|
| Commutative Properties of Addition and Multiplication | $a + b = b + a$ and $ab = ba$ | $2 + 5 = 5 + 2$ and $-3(4) = 4(-3)$ |
| Associative Properties of Addition and Multiplication | $a + (b + c) = (a + b) + c$ and $a(bc) = (ab)c$ | $2 + (5 + 3) = (2 + 5) + 3$ and $2(4x) = (2 \cdot 4)x = 8x$ |
| Distributive Property | $a(b + c) = ab + ac$ and $(a + b)c = ac + bc$ | $3(x + 2) = 3x + 6$ and $(x + 2)x = x^2 + 2x$ |
| Additive Identity | $a + 0 = 0 + a = a$ | $-7 + 0 = 0 + (-7) = -7$ |
| Multiplicative Identity | $a \cdot 1 = 1 \cdot a = a$ | $2 \cdot 1 = 1 \cdot 2 = 2$ |

| | **Equation Solving and Inequality Solving** | **Examples** |
|---|---|---|
| Linear Equation in One Variable | An equation that can be written in the form $ax + b = c$ | $2x - 4 = 7$ |
| Linear Inequality in One Variable | An inequality that can be written in the form $ax + b \bigcirc c$, where $\bigcirc$ can be replaced with $<, \leq, >,$ or $\geq$ | $3x - 2 < 7$ |

| A Solution to an Equation or Inequality | A number that makes the equation or inequality true when substituted for the variable in the equation or inequality | $x = 4$ is the solution to $2x - 7 = 1$. <br><br> $x = 2$ is a solution to the inequality $2x < 7$. |
|---|---|---|
| Using Guess-and-check to Solve an Equation | Make an initial guess; modify the guess based on information from the previous guess; repeat this process until a solution is found. | Guess: $x = 12$ is a solution to $x^2 = 169$. Check: $12^2 = 144 < 169$. The guess must be larger. Guess: $x = 13$. Check: $13^2 = 169$. The solution is $x = 13$. |
| Solving an Equation or Inequality Algebraically | Use properties of equality (or inequality) and the basic properties of algebra to solve. | $$2x + 8 = 21$$ $$2x + 8 + (-8) = 21 + (-8)$$ $$2x = 13$$ $$x = 6.5$$ |

| | **Problem Solving** | **Examples** |
|---|---|---|
| Pólya's Four-step Process in Problem Solving | 1. Understand the problem. <br> 2. Devise a plan. <br> 3. Carry out the plan. <br> 4. Look back. | |
| Problem Situation | A real-life setting in which mathematical problems occur. | The **Car Dealer Problem Situation** |

| | **Grapher Skills** | **Examples** |
|---|---|---|
| Evaluating Numerical Expressions | Evaluate $3 - 4^9 \div 7$. | $3 - 4 \wedge 9 \div 7$ ENTER |
| Evaluating Algebraic Expressions | Evaluate $3x^2 - 17$ for $x = 5$. | $3 \times 5^2 - 17$ ENTER or $5 \rightarrow X$ : $3X^2 - 17$ ENTER |

## Review Exercises for Chapter 1

In Exercises 1–2, use a calculator to find a decimal representation for each rational number. Identify, if possible, whether the decimals are repeating or terminating. If repeating, write the decimal with a bar over the block of numbers that repeat.

**1.** $\dfrac{125}{999}$

**2.** $\dfrac{225}{64}$

In Exercises 3–6, draw the given interval on a real number line.

**3.** $[-2, 2)$

**4.** $(2, 3]$

**5.** $(-\infty, 0)$

**6.** $[-1, \infty)$

In Exercises 7–10, identify each number as a member of as many of these sets as apply: natural number, whole number, integer, rational number, and irrational number.

**7.** $1.5$

**8.** $-7$

**9.** $0$

**10.** $\dfrac{2}{3}$

In Exercises 11–16, evaluate the expressions with paper and pencil. Support your answers with a grapher.

**11.** $28 - 4 \times 2^3 + 63 \div 3^2 \times 5$

**12.** $-(-17 - 3) - 5(14 - 6)$

**13.** $|12.7 - 16.2| - |-7.6|$

**14.** $[-3(16 - 17)]^3 + (-6 + 26) \div 2^2$

**15.** $-3 \times 7 - (-5)[9 + 2(-1)]$

**16.** $|31.4 - |-1.6| \,|$

In Exercises 17 and 18, state what properties are used to show that the given equation is true.

**17.** $2x - 12 = 2(x - 6)$   **18.** $(3x)(2x) = 6x^2$

**19.** Evaluate the expression $4x^2 - x + 3$ for $x = -1$.

**20.** Evaluate the expression $\dfrac{3P}{Q + 2}$ for $Q = 6$ and $P = -2$.

**21.** Evaluate the expression $j^2 - jk$ for $j = 5$ and $k = 3$.

**22.** The formula for the circumference of a circle is $C = 2\pi R$, where $R$ is the radius. Find the circumference of a circle with radius 13 in.

In Exercises 23–28, solve the given equation algebraically. Find an exact solution. Support the solution with a grapher.

**23.** $3y + 18 = 14$   **24.** $7k + 30 = 12$

**25.** $18.1 - 0.6x = 1.9x - 2.4$

**26.** $3(x - 3) - (6x - 1) = 1 - 4x$

**27.** $5(x + 1) - 2(x + 4) = 2x + 1$

**28.** $11.85 + 2.6x = 5.7x - 8.3$

In Exercises 29–32, solve the given inequality algebraically. Describe the solution using interval notation.

**29.** $x + 27 > 13$   **30.** $x + 8 \le 5$

**31.** $2x - 7 \ge 25$   **32.** $4x - 9 < 15$

For Exercises 33–35, write an equation that represents each of the statements.

**33.** The sum of the absolute values of $-3$ and 6 is $x$.

**34.** A number doubled is three less than the number squared.

**35.** The retail price $R$ of an item is the wholesale price $x$ marked up 35%.

**36.** Use the *Guess-and-check* strategy to solve $3x^2 + 168 = 600$.

**37.** *Projectile Motion.* A baseball is thrown straight up into the air with a velocity of 56 ft/sec. It is $h$ feet above the ground $t$ seconds later. The algebraic representation for height in terms of time is $h = -16t^2 + 56t$. Use the *Make a Table* strategy to find the maximum height for this ball. How many seconds after it is thrown does the ball reach this maximum?

**38.** *Cash Flow Management.* Stacie has responsibility for cash flow management at her company. On Monday morning the balance in her account was $36,280. On Monday afternoon, she paid $7,250 for materials that were delivered to the manufacturing plant, and on Tuesday, a total of $5,000 was received from customers. Stacie put half of this amount into the bank account and used the balance to repay an old debt. On Wednesday, Stacie purchased a company car for one of the salesman. The purchase price was $2,650 below the list price of $10,000. Employee salaries due to be paid on Friday totaled $33,500.

How much did Stacie need to borrow to cover the cost of Friday's salary payment?

**39.** *Analysis of Motion.* Tanya and James are jogging side-by-side at a pace of 12.5 ft/sec. If Tanya increases her pace to 13.3 ft/sec and James continues at the original pace, how many seconds will it take until James is 150 ft behind Tanya?

**40.** *Mixture.* How many milliliters of a 65% acid solution must be added to 15 milliliters of a 40% acid solution to get a 60% solution?

## Chapter 1 Test

In Questions 1 and 2, identify each number as a member of as many of these sets as apply: natural number, whole number, integer, rational number, and irrational number.

**1.** $-1.\overline{6}$          **2.** $\sqrt{7}$

**3.** Draw the interval $[-5, 2)$ on a real number line.

**4.** Draw the interval $(-2, \infty)$ on a real number line.

**5.** Evaluate with paper and pencil: $2 + 5 \times (3 - 2^2)$.

**6.** Evaluate with paper and pencil: $|(-6 + 8)^2 - (-4)(-3 - 5)|$.

**7.** Evaluate the expression $\dfrac{3a - 4b}{2c}$ for $a = -4$, $b = 1$, and $c = -5$.

**8.** Evaluate the expression $0.2x^3 + 0.01x - 4$ for $x = -2$.

In Questions 9 and 10, state which basic properties of algebra are used to show that the given equation is true.

**9.** $(2x + 4) - 7 = 2x + (4 - 7)$

**10.** $11 \cdot 14 = 10 \cdot 14 + 1 \cdot 14$

**11.** Solve algebraically: $\dfrac{x+4}{3} + \dfrac{x-5}{2} = 2x$.

**12.** Solve algebraically: $3(x - 1) + 6 = 5 - 2x$.

**13.** Solve algebraically: $2x - 5 \geq 17$. Describe the solution using interval notation.

**14.** Write in symbols: The sum of the absolute values of a number tripled and $-7.5$ is fifty-five.

**15.** *Rental Business.* A car rental agency rents compact cars for $27.95/day and luxury cars for $42.95/day.

a) Write an expression that represents the total dollars of revenue generated in 1 day if $C$ compact cars and $L$ luxury cars are rented.

b) On Monday the agency rented 38 compact cars and 35 luxury cars. How many dollars of revenue were generated?

c) On Thursday the agency rented 55 compact cars and 23 luxury cars. How many dollars of revenue were generated?

d) Was more revenue generated on Monday or Thursday? How much more?

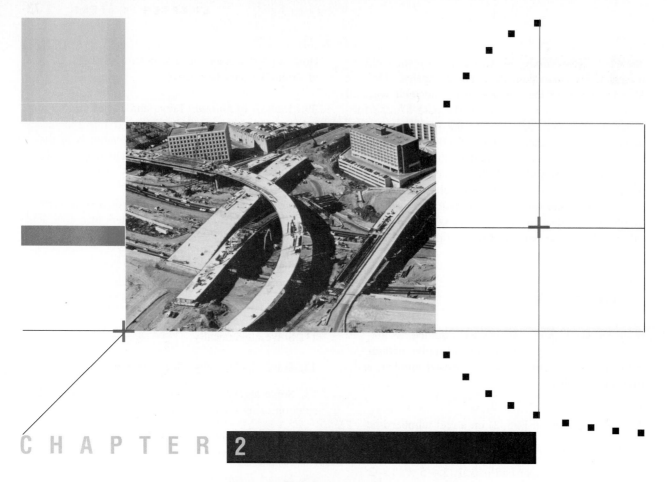

# Graphing Equations and Functions

## AN APPLICATION

When an interstate highway is built, the slope of the grades do not exceed a 6% hill. If one location on a highway with a 6% uphill grade is $K$ ft above sea level, the height ($h$) above sea level of a point $x$ ft along the highway is approximated by the equation

$$h = 0.06x + K$$

A **rectangular coordinate plane** is formed by a pair of real-number lines perpendicular to one another intersecting each other at the point labeled zero. The horizontal line is usually called the **x-axis** and the vertical line is usually called the **y-axis**. These two axes divide the plane into four regions called **quadrants**. The point of intersection of the two axes is called the **origin**.

The rectangular coordinate system in Figure 2.1 shows the origin, $O$, and points $P$ and $Q$ identified by the ordered pairs $(4, 2)$ and $(-6, -4)$. It is common practice to write both the name and the ordered pair of the point, for example, $P(4, 2)$ and $Q(-6, -4)$. A point such as $P(4, 2)$ is said to be **graphed**, or **plotted**, when it is marked on a rectangular coordinate system.

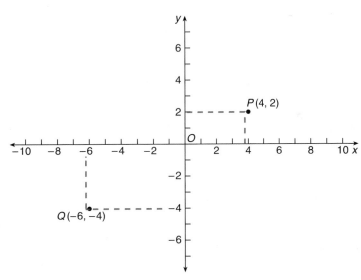

**FIGURE 2.1**  The rectangular coordinate system with two ordered pairs.

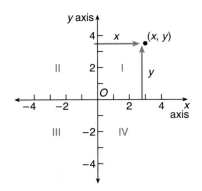

**FIGURE 2.2**  A rectangular coordinate system has four quadrants.

In general, each point of a rectangular coordinate plane corresponds to a unique ordered pair $(x, y)$ of real numbers called the **coordinates** of the point. The absolute value of the $x$-coordinate tells the perpendicular distance from the $y$-axis to the point, and the absolute value of the $y$-coordinate tells the perpendicular distance from the $x$-axis to the point (see Figure 2.2).

**EXAMPLE 1**    Graphing Points

Use paper and pencil to plot the points $P(2, 3)$, $Q(-3, 1)$, and $R(-4, -3)$ on a rectangular coordinate system.

SOLUTION

Plot point $P$ by beginning at the origin, then moving either two units to the right and three units up or three units up and two units right. Plot points $Q$ and $R$ in a similar fashion. See Figure 2.3.

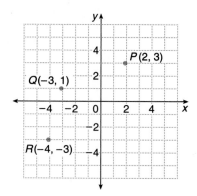

**FIGURE 2.3** A rectangular coordinate system with the points $P$, $Q$, and $R$ plotted.

**EXAMPLE 2**    Finding Coordinates of Points

Find the coordinates of the points $R$, $S$, $T$, and $U$ in Figure 2.4.

SOLUTION

The coordinates of these four points are as follows:

| Point | Coordinates |
| --- | --- |
| $R$ | $(-3, -3)$ |
| $S$ | $(4, 2)$ |
| $T$ | $(-4, 5)$ |
| $U$ | $(3, -4)$ |

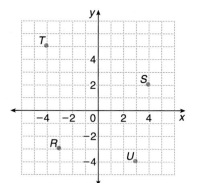

**FIGURE 2.4** A rectangular coordinate system with points $R, S, T$, and $U$ plotted.

Although the points $R, S, T$, and $U$ in Example 2 have integer coordinates, there are many points whose coordinates are non-integer real numbers. For example, where is the point $W(1/2, \sqrt{3})$ located?

## Representing a Coordinate Plane on a Grapher

One characteristic that distinguishes graphing calculators from other scientific calculators is their ability to model the coordinate plane. However, only a portion of the plane can be shown at any one time. This rectangular region is called a **window** (see Figure 2.5). You as operator of the grapher determine what window is to be displayed by using the WINDOW (RANGE) settings. For example, when the WINDOW (RANGE) settings are those shown in Figure 2.6, the grapher screen window represents points whose $x$-coordinates are between $-40$ and $80$ and whose $y$-coordinates are between $-30$ and $50$, precisely those points that fall in the rectangle in Figure 2.5. We say that this is the window $[-40, 80]$ by $[-30, 50]$. Throughout the remainder of the text this menu will simply be referred to as the WINDOW menu.

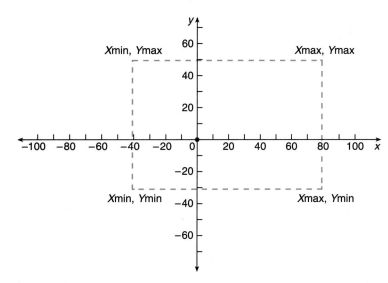

**FIGURE 2.5** The window $[-40, 80]$ by $[-30, 50]$.

```
WINDOW
Xmin = ⁻40
Xmax = 80
Xscl = 10
Ymin = ⁻30
Ymax = 50
Yscl = 10
```

**FIGURE 2.6** A grapher WINDOW (RANGE) setting for the window in Figure 2.5.

The $X$scl and $Y$scl shown in Figure 2.6 determine that each "tick" mark represents 10 units. By changing the scale you can vary the number of "tick" marks shown on the grapher screen. Try changing the scale to 5 or 15 and see how the appearance of the grapher screen changes.

**EXAMPLE 3** Setting the WINDOW Menu

Identify the WINDOW settings for the window $[-20, 10]$ by $[-15, 30]$ with each $x$-axis tick mark representing 5 units and each $y$-axis tick mark representing 10 units.

SOLUTION

**WINDOW**

$X\min = -20$
$X\max = 10$
$X\mathrm{scl} = 5$
$Y\min = -15$
$Y\max = 30$
$Y\mathrm{scl} = 10$   ■

After setting the WINDOW menu as shown in Example 3, the grapher representation of the plane should look approximately as shown in Figure 2.7. The window $[-10, 10]$ by $[-10, 10]$ is frequently used and is called the **Standard window**.

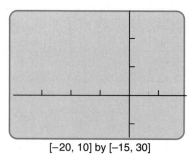

[−20, 10] by [−15, 30]

**FIGURE 2.7**   The window for Example 3.

Notice as you move the cursor around the screen, the coordinates of points of the plane are shown at the bottom of the screen. Also notice that there are many points in the region defined by the window that cannot be identified by the graph cursor.

**GRAPHER NOTE**

Check your grapher lab manual to learn how to move a cursor around the screen to display the coordinates of points in the window you have selected.

## Integer Windows

The coordinates of the points shown on the screen depend in part on the particular window you have selected. An **Integer window** is one in which the coordinates displayed as the cursor is moved around the grapher screen are integers.

When the origin is centered in the middle of the screen of an Integer window, we will call it the **Integer Window.** The particular window that turns out to be the Integer window varies from one grapher model to another. Complete the following exploration and find the Integer window for your grapher.

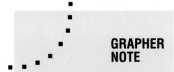

**GRAPHER NOTE**

Note the distinction between saying "*an* integer window" and "*the* integer window."

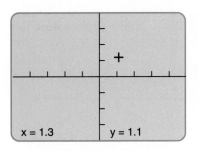

x = 1.3          y = 1.1

**FIGURE 2.8** The coordinates in this window are in tenths.

**EXPLORE WITH A GRAPHER**

Set the WINDOW of your grapher for each of these windows. Check the *x*- and *y*-coordinates as you move the cursor around the screen.

1. [−19, 19] by [−11, 11]
2. [−48, 47] by [−32, 31]
3. [−47, 47] by [−31, 31]
4. [−63, 63] by [−31, 31]

A. Which window is the Integer window for your grapher? (The answer depends upon which model grapher you are using.)
B. Experiment! Find an Integer window that is not centered at (0, 0).
C. Experiment! Find a window whose cursor coordinates are "tenths" (see Figure 2.8). We call this window **the 0.1 window** when the origin is centered in the window.

Throughout the remainder of this text, we will alternate back and forth between sketching graphs with paper and pencil and using a grapher to produce them.

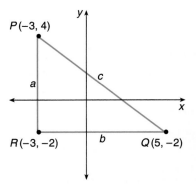

**FIGURE 2.9** The distance between points $P$ and $Q$ is equal to the length of the segment $c$. This length can be found using the Pythagorean theorem.

## Finding the Distance Between Points in a Coordinate Plane

The distance between any two points in a coordinate plane can be calculated if the coordinates of the points are known. For example, to find the distance between points $P(-3, 4)$ and $Q(5, -2)$, complete a right triangle as shown in Figure 2.9.

The legs of $\triangle PQR$ have lengths $a = 4 - (-2) = 6$ and $b = 5 - (-3) = 8$. By the Pythagorean theorem, we conclude that:

$$c^2 = a^2 + b^2$$

$$c^2 = 6^2 + 8^2$$

$$c^2 = 36 + 64 = 100$$

$$c = 10$$

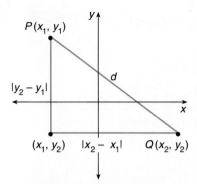

**FIGURE 2.10** Finding the distance $d$ between two points.

To find the distance between two general points $P(x_1, y_1)$ and $Q(x_2, y_2)$, complete a triangle as shown in Figure 2.10.

Apply the Pythagorean theorem to the triangle in Figure 2.10 to obtain the equation

$$d^2 = |x_2 - x_1|^2 + |y_2 - y_1|^2.$$

Solve for $d$ by taking the square root of both sides of the equation.

---

### Distance Formula

The distance $d$ between the points $P(x_1, y_1)$ and $Q(x_2, y_2)$ is

$$d = \sqrt{(x_2 - x_1)^2 + (y_2 - y_1)^2}.$$

This equation is known as the **distance formula**.

---

**REMINDER**

Subscripts are used on the coordinates of points to communicate the fact that they are different points.

This distance formula is applied in Example 4.

**EXAMPLE 4**    Finding Distance Between Points

Find the distance between the points $P(-4, 2)$ and $Q(3, 5)$.

SOLUTION

Figure 2.11 shows the grapher screen for this calculation.

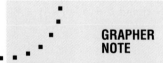

**GRAPHER NOTE**

If the expression under a radical includes several terms, parentheses are needed around the expression. See Figure 2.11.

$$\sqrt{((^-4 - 3)\wedge 2 + (2 - 5)\wedge 2)}$$

$$7.615773106$$

**FIGURE 2.11** Use a grapher to calculate the distance between $P(-4, 2)$ and $Q(3, 5)$.

$$d = \sqrt{(-4-3)^2 + (2-5)^2}$$
$$= \sqrt{(-7)^2 + (-3)^2}$$
$$= \sqrt{49+9}$$
$$= \sqrt{58}$$
$$= 7.62$$

The distance between points $P(-4, 2)$ and $Q(3, 5)$ is approximately 7.62. ■

Recall that a triangle is called **equilateral** if all sides are equal in length and it is called **isosceles** if at least two sides are equal in length. Example 5 uses the distance formula to determine whether a given triangle is equilateral or isosceles.

**EXAMPLE 5**  Studying a Triangle

Given points $P(1, 2)$, $Q(3, 7)$, and $R(6, 4)$, determine whether $\triangle PQR$ is equilateral or isosceles.

SOLUTION
Figure 2.12 shows $\triangle PQR$ and Figure 2.13 shows the grapher screen for the calculations of the lengths $PQ$, $PR$, and $QR$.

$$PQ = \sqrt{(1-3)^2 + (2-7)^2} = \sqrt{4+25}$$
$$= 5.39$$
$$PR = \sqrt{(1-6)^2 + (2-4)^2} = \sqrt{25+4}$$
$$= 5.39$$
$$QR = \sqrt{(3-6)^2 + (7-4)^2} = \sqrt{9+9}$$
$$= 4.24$$

Because $PQ = PR$ and $PR \neq QR$, we conclude that $\triangle PQR$ is isosceles but not equilateral. ■

**FIGURE 2.12** $\triangle PQR$.

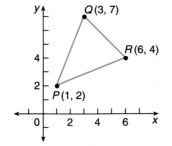

$$\sqrt{((1-3)^2 + (2-7)^2)}$$
$$5.385164807$$
$$\sqrt{((1-6)^2 + (2-4)^2)}$$
$$5.385164807$$
$$\sqrt{((3-6)^2 + (7-4)^2)}$$
$$4.242640687$$

**FIGURE 2.13**
Calculating the lengths $PQ$, $PR$, and $QR$.

Three points are **collinear** if they all lie on the same line. In particular, three points $P$, $Q$, and $R$ all lie on the same line if and only if the sum of two of the distances $PQ$, $PR$, and $QR$ is equal to the third distance.

**EXAMPLE 6**    Determining Collinear Points

Determine whether points $P(-1, -2)$, $Q(3, 1)$, and $R(11, 7)$ are collinear.

SOLUTION

$$PQ = \sqrt{(-1-3)^2 + (-2-1)^2} = \sqrt{(-4)^2 + (-3)^2} = \sqrt{16+9} = 5$$

$$QR = \sqrt{(3-11)^2 + (1-7)^2} = \sqrt{(-8)^2 + (-6)^2} = \sqrt{64+36} = 10$$

$$PR = \sqrt{(-1-11)^2 + (-2-7)^2} = \sqrt{(-12)^2 + (-9)^2} = \sqrt{144+81}$$

$$= \sqrt{225}$$

$$= 15$$

Because $PQ + QR = PR$, we see that the three points are collinear.    ■

## Exercises for Section  2.1

For Exercises 1–6, plot the given points on a piece of graph paper. Do the points appear to lie on a straight line?

**1.** $(-1, 2)$, $(1, 4)$, and $(3, 6)$

**2.** $(-3, -2)$, $(0, -1)$, and $(3, 5)$

**3.** $(-4, 3)$, $(-2, 2)$, and $(0, -1)$

**4.** $(-4, 2)$, $(0, -1)$, and $(8, -7)$

**5.** $(-8, -4)$, $(-2, -1)$, and $(-4, 1)$

**6.** $(2, 3)$, $(4, 1)$, and $(-2, 7)$

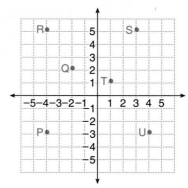

For Exercises 7 to 12.

For Exercises 7–10, name the coordinates of the indicated points in the figure below.

**7.** Point $R$              **8.** Point $S$

**9.** Point $P$              **10.** Point $Q$

**11.** Which of the following coordinates are the most likely for point $T$: (a) $(2, 1)$, (b) $(1, 2.5)$, or (c) $(1, 1)$.

**12.** Which of the following coordinates are the most likely for point $U$: (a) $(4.3, -3.1)$, (b) $(4, -3)$, or (c) $(5, -3.2)$.

For Exercises 13–16, write down the WINDOW menu settings for the given window.

**13.** $[-10, 15]$ by $[-8, 10]$ with each tick mark on the $x$-axis representing 5 units and each tick mark on the $y$-axis representing 1 unit

**14.** $[-20, 40]$ by $[-30, 30]$ with each tick mark on the $x$-axis representing 10 units and each tick mark on the $y$-axis representing 5 units

**15.** $[-300, 450]$ by $[30, 50]$ with each tick mark on the $x$-axis representing 100 units and each tick mark on the $y$-axis representing 5 units

**16.** [$-1000, 400$] by [$-100, 300$] with each tick mark on the $x$-axis representing 100 units and each tick mark on the $y$-axis representing 50 units

For Exercises 17–24, determine whether the given window is an Integer window on your grapher.

**17.** [$-47, 48$] by [$-31, 32$]

**18.** [$-46, 48$] by [$-30, 32$]

**19.** [$-14, 24$] by [$-11, 11$]

**20.** [$-28, 66$] by [$-22, 40$]

**21.** [$-37, 57$] by [$-41, 21$]

**22.** [$-94, 94$] by [$-62, 62$]

**23.** [$0, 95$] by [$0, 63$]

**24.** [$0, 38$] by [$0, 22$]

For Exercises 25–27, several WINDOW menu settings are given. Consider each window before answering the following question: for which window are the tick marks closest together? Make a mental decision. Then check with a grapher.

**25.**

| Window #1 | Window #2 | Window #3 |
|---|---|---|
| $X$min $= -10$ | $X$min $= -15$ | $X$min $= -20$ |
| $X$max $= 10$ | $X$max $= 15$ | $X$max $= 20$ |
| $X$scl $= 1$ | $X$scl $= 1$ | $X$scl $= 1$ |
| $Y$min $= -10$ | $Y$min $= -15$ | $Y$min $= -20$ |
| $Y$max $= 10$ | $Y$max $= 15$ | $Y$max $= 20$ |
| $Y$scl $= 1$ | $Y$scl $= 1$ | $Y$scl $= 1$ |

**26.**

| Window #1 | Window #2 | Window #3 |
|---|---|---|
| $X$min $= -20$ | $X$min $= -20$ | $X$min $= -20$ |
| $X$max $= 20$ | $X$max $= 20$ | $X$max $= 20$ |
| $X$scl $= 1$ | $X$scl $= 5$ | $X$scl $= 10$ |
| $Y$min $= -20$ | $Y$min $= -20$ | $Y$min $= -20$ |
| $Y$max $= 20$ | $Y$max $= 20$ | $Y$max $= 20$ |
| $Y$scl $= 1$ | $Y$scl $= 5$ | $Y$scl $= 10$ |

**27.**

| Window #1 | Window #2 | Window #3 |
|---|---|---|
| $X$min $= -20$ | $X$min $= -30$ | $X$min $= -40$ |
| $X$max $= 20$ | $X$max $= 30$ | $X$max $= 40$ |
| $X$scl $= 1$ | $X$scl $= 1$ | $X$scl $= 1$ |
| $Y$min $= -20$ | $Y$min $= -30$ | $Y$min $= -40$ |
| $Y$max $= 20$ | $Y$max $= 30$ | $Y$max $= 40$ |
| $Y$scl $= 1$ | $Y$scl $= 1$ | $Y$scl $= 1$ |

For Exercises 28–37, find the distance between the given points.

**28.** $(-3, 4)$ and $(3, 4)$       **29.** $(0, 0)$ and $(-4, 3)$

**30.** $(5, 3)$ and $(-2, 1)$       **31.** $(-3, -2)$ and $(4, 2)$

**32.** $(-3, -2)$ and $(-5, -1)$

**33.** $(7, -3)$ and $(1, 2)$

**34.** $(2, 0)$ and $(0, -5)$       **35.** $(1, 5)$ and $(-3, -2)$

**36.** $(0, 3b)$ and $(4b, 0)$       **37.** $(a, b)$ and $(b, a)$

For Exercises 38–41, determine whether the three given points all lie on the same line.

**38.** $(-1, -2)$, $(1, 1)$, and $(3, 4)$

**39.** $(-2, 1)$, $(1, 4)$, and $(4, 9)$

**40.** $(-8, 3)$, $(4, -2)$, and $(0, 0)$

**41.** $(-3, 4)$, $(1, -4)$, and $(-2, 2)$

**42. Writing to Learn.** After completing Exercises 25–27, write a paragraph explaining why the coordinate axes appear to be "fat" on the grapher screen with the following WINDOW setting.

| Window |
|---|
| $X$min $= -50$ |
| $X$max $= 50$ |
| $X$scl $= 1$ |
| $Y$min $= -50$ |
| $Y$max $= 50$ |
| $Y$scl $= 1$ |

For Exercises 43–46, state the coordinates of the points in each corner of the grapher screen for the window.

**43.** $[-10, 10]$ by $[-10, 10]$

**44.** $[-9.6, 9.4]$ by $[-6.4, 6.2]$

**45.** $[-5, 5]$ by $[-5, 5]$

**46.** $[-15, 15]$ by $[-10, 10]$

### TRANSLATING WORDS TO SYMBOLS

Exercises 47–50 refer to points $P(x_1, y_1)$ and $Q(x_2, y_2)$. For each, write the given algebraic expression.

**47.** The square of the difference of the $x$-coordinates

**48.** The square of the difference of the $y$-coordinates

**49.** The sum of the squares of the difference in $x$- and $y$-coordinates

**50.** The square root of the sum of the squares of the difference in $x$- and $y$-coordinates

### EXTENDING THE IDEAS

**51.** *Using Algebra in Geometry.* Use the distance formula to show that the triangle with vertices $P(2, 5)$, $Q(5, 0)$, and $R(1, 1)$ is an isosceles right triangle.

**52.** *Using Algebra in Geometry.* If a trapezoid $ABCD$ is isosceles, the diagonals $AC$ and $BD$ are equal in length. Use the distance formula to show that trapezoid $ABCD$ is isosceles when $A(-2, 1)$, $B(-1, 6)$, $C(4, 7)$, and $D(5, 2)$.

### LOOKING BACK—MAKING CONNECTIONS

For Exercises 53–55, decide whether the triangle whose vertices are the given points is equilateral, isosceles, or scalene. (Recall that a triangle is equilateral if all three sides are equal in length, isosceles if at least two sides are equal in length, and scalene if the three sides have different lengths.)

**53.** $(-2, -1)$, $(3, 0)$, and $(3, -2)$

**54.** $(-2, 3)$, $(5, 2)$, and $(3, -1)$

**55.** $(-1.5, 1)$, $(1, 5\sqrt{3}/2)$, and $(3.5, 1)$

**56.** Find the perimeter of the triangle with vertices $(-2, 1)$, $(0, 5)$, and $(3, -1)$

---

| 2.2 | Equations in Two Variables and Their Graphs |
|-----|---------------------------------------------|

> Finding Solution Pairs to Equations in Two Variables  ▪  Finding a Graph of an Equation in Two Variables  ▪  Finding the Graph of an Equation Using a Grapher  ▪  Finding Solution Pairs with a Grapher

In Chapter 1, we studied linear equations in one variable. These are equations of the form $ax + b = c$, where $x$ is the variable and $a$, $b$, and $c$ represent real numbers. Recall that the solution set of such an equation is the set of all numbers that make the equation true. In fact a linear equation in one variable has a **unique** solution.

A solution to an equation in two variables is an ordered pair of numbers that make the equation true. An equation with two variables has many solutions, as shown in Example 1.

**EXAMPLE 1** Verifying Several Solutions

Show that the ordered pairs $(2, 5)$, $(4, 4)$, and $(8, 2)$ are all solutions of the equation $14x + 28y = 168$. Support numerically.

SOLUTION
**Solve Algebraically.**

$$14x + 28y = 168$$

Substitute the first number of the pair for $x$ and the second number of the pair for $y$.

$$14(2) + 28(5) =$$
$$= 28 + 140 = 168$$

$$14(4) + 28(4) =$$
$$= 56 + 112 = 168$$

$$14(8) + 28(2) =$$
$$= 112 + 56 = 168$$

**Support Numerically.** Figure 2.14 shows the same three expressions evaluated on a grapher screen. This screen provides numerical support that the given ordered pairs are solution pairs.

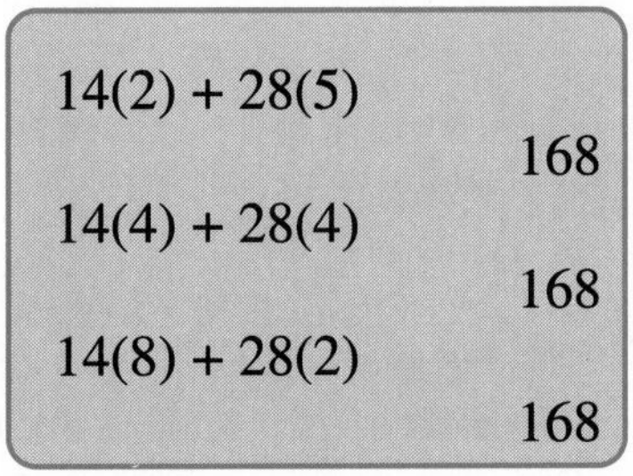

$$14(2) + 28(5)$$
$$168$$
$$14(4) + 28(4)$$
$$168$$
$$14(8) + 28(2)$$
$$168$$

**FIGURE 2.14** Numerical support on a grapher that $(2, 5)$, $(4, 4)$, and $(8, 2)$ are solution pairs. Notice the use of the REPLAY feature of the grapher. ■

### Finding Solution Pairs to Equations in Two Variables

Example 1 verifies three solution pairs to equation $14x + 28y = 168$. But there are many more. In fact for every real number $x$, you can find a real number $y$ such that $(x, y)$ is a solution pair of this equation.

To find solution pairs for an equation in two variables, substitute any specific value for one variable and solve the resulting equation for the second variable. Example 2 illustrates how this is done with $y = x - 2$, an equation that is simple enough that computations can be done mentally.

**EXAMPLE 2**   Finding a Table of Solution Pairs

Complete a table of solution pairs for the equation $y = x - 2$. Use the values $x = -3, -2, -1, 0, 1, 2,$ and $3$.

SOLUTION

Substitute each value of $x$ into the equation and find the corresponding value of $y$.

| $x$ | $-3$ | $-2$ | $-1$ | $0$ | $1$ | $2$ | $3$ |
|---|---|---|---|---|---|---|---|
| $y = x - 2$ | $-5$ | $-4$ | $-3$ | $-2$ | $-1$ | $0$ | $1$ |   ■

When equations are more complex, it is efficient to use a grapher. If an equation is not written in the form $y =$, prepare it for use with a grapher by first solving the equation for $y$. For example, for equation $14x + 28y = 168$, do the following:

$$14x + 28y = 168$$
$$28y = -14x + 168$$
$$y = -\frac{14}{28}x + \frac{168}{28}$$
$$y = -0.5x + 6$$

Figure 2.15 shows how to use the REPLAY feature of a grapher to evaluate the expression $-0.5x + 6$ for $x = -15$, $x = -10$, and $x = -5$ to find the solution pairs $(-15, 13.5)$, $(-10, 11)$, and $(-5, 8.5)$.

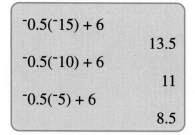

```
⁻0.5(⁻15) + 6
                13.5
⁻0.5(⁻10) + 6
                  11
⁻0.5(⁻5) + 6
                 8.5
```

**FIGURE 2.15**  Using the REPLAY feature of a grapher to solve for $y$.

**EXAMPLE 3**   Finding a Table of Solution Pairs

Complete a table of solution pairs for the equation $14x + 28y = 168$. Use the values $x = -15, -10, -5, 0, 5, 10,$ and $15$.

SOLUTION

Rewrite the equation $14x + 28y = 168$ into the equivalent form $y = -0.5x + 6$ and use a grapher as illustrated in Figure 2.15 to complete this table.

| $x$ | $-15$ | $-10$ | $-5$ | $0$ | $5$ | $10$ | $15$ |
|---|---|---|---|---|---|---|---|
| $y = -0.5x + 6$ | 13.5 | 11 | 8.5 | 6 | 3.5 | 1 | $-1.5$ |

∎

## Finding a Graph of an Equation in Two Variables

The **graph of an equation in two variables** is the graph of all solution pairs of the equation. For example, to sketch a graph of $14x + 28y = 168$ begin by completing a table of a few of the solution pairs, as shown in Example 3. Then graph each of these solution pairs.

### EXAMPLE 4   Graphing Solution Pairs

Use the paper-and-pencil method to graph the solution pairs of $14x + 28y = 168$ shown in Example 3.

SOLUTION

See Figure 2.16.

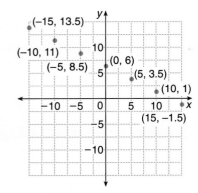

**FIGURE 2.16** The graph of the ordered pairs found in Example 3.

∎

**TRY THIS**

You may want to learn how to use the STAT or equivalent menu on your grapher to graph solution pairs like those shown on the graph for Example 4. Consult your grapher lab manual for details.

There are an infinite number of solution pairs that exist between each of the points on the graph in Example 4. To find a complete graph of the equation $14x + 28y = 168$, graph several solution pairs and then connect their points with a continuous line.

**EXAMPLE 5**   Finding a Graph of an Equation in Two Variables

Use the paper-and-pencil method to sketch a graph of the equation $14x + 28y = 168$.

SOLUTION

Graph the set of solution pairs found in Example 3. Observe that the solution points appear to lie on a "straight" line. Then connect the points with a continuous line, extending the line beyond the graphed seven points (see Figure 2.17).

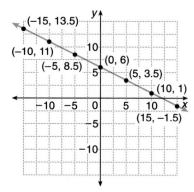

**FIGURE 2.17**   A graph of $14x + 28y = 168$.

Examples 3, 4, and 5 illustrate the **point plotting method of graphing.**

## Finding the Graph of an Equation Using a Grapher

**GRAPHER NOTE**

Consult the grapher lab manual that accompanies this text to learn how to find the graph of $y = 2x - 8$ in the Integer window.

In Section 2.1 you learned that the coordinate plane can be represented on a grapher. As one would expect from the name "graphing calculator" or "grapher," it is possible to produce graphs of equations on a grapher. This ability is probably the most powerful use of a graphing calculator.

Refer to your grapher manual to learn how to produce the graph of an equation on your grapher. Figure 2.18 shows the Equation-defining menu and the graph of the equation $y = 2x - 8$ in the Integer window.

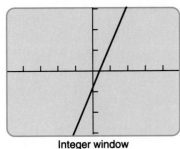

Equation-defining menu             Integer window

**FIGURE 2.18** The Equation-defining menu and the resulting graph. The tick marks are 10 units apart.

Example 6 shows another grapher-produced graph.

**EXAMPLE 6**    Finding a Graph with a Grapher

Use a grapher to find the graph of the equation $y = 1.8x + 7.2$ in the Integer window.

SOLUTION

Define $y = 1.8x + 7.2$ in a grapher as shown in Figure 2.19. Figure 2.20 shows the graph in the Integer window.

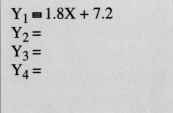

**FIGURE 2.19** Define $y = 1.8x + 7.2$ in your grapher.

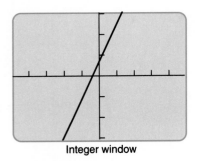

Integer window

**FIGURE 2.20** Graph of $y = 1.8x + 7.2$. The tick marks are 10 units apart.

**GRAPHER NOTE**

We suggest that throughout this section you use the Integer window. (See Explore with a Grapher on page 79.)

On the grapher, we use the key named  Y =  to select the Equation-defining menu. This key name suggests that the equation to be graphed must be written in the form

$$y = \text{some algebraic expression in variable } x.$$

To graph the equation $7x + 12y = 78$, rewrite it as follows:

$$7x + 12y = 78$$
$$12y = -7x + 78$$
$$y = -\frac{7}{12}x + \frac{78}{12}$$

**EXAMPLE 7**  Finding a Graph with a Grapher

Use a grapher to find the graph of the equation $7x + 12y = 78$ in the Integer window.

SOLUTION

Change $7x + 12y = 78$ to the form $y = -\dfrac{7}{12}x + \dfrac{78}{12}$. Key in the equation as shown in Figure 2.21. Figure 2.22 shows the graph in the Integer window.

$Y_1 = \ ^-(7/12)X + 78/12$
$Y_2$
$Y_3$
$Y_4$

**FIGURE 2.21** Enter $y = -\dfrac{7}{12}x + \dfrac{78}{12}$ on the appropriate grapher menu. Notice that ( ) are needed in (7/12).

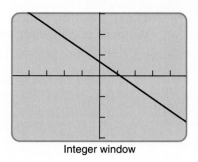

Integer window

**FIGURE 2.22**  A graph of $7x + 12y = 78$.

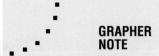

**GRAPHER NOTE**

Consult your grapher lab manual to learn how to move the trace cursor along the graph to show the solution pairs to the equation after its graph has been found.

### Finding Solution Pairs with a Grapher

Examples 3 and 4 illustrated how a table of solution pairs for an equation are found by hand calculation and plotted on a coordinate plane. In Example 5, the resulting points were then connected to complete a hand sketch of the graph.

When a graph is found on a grapher as illustrated in Examples 6 and 7, the grapher automatically calculates many solution pairs and draws an accurate graph. After the graph appears on the grapher screen, these solution pairs can be recovered by using the TRACE feature on the grapher. For example, Figure 2.23 shows the

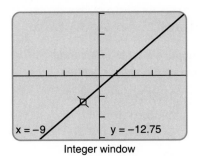

**FIGURE 2.23** The graph of $y = (3/4)x - 6$ with the trace cursor showing the solution pair $(-9, -12.75)$ to the equation $3x - 4y = 24$.

REMINDER

Notice that $-(5/4)x$ and $-5x/4$ are equivalent expressions. Using the form $-5x/4$ saves several key strokes on the grapher.

graph of the equation $3x - 4y = 24$ or $y = (3/4)x - 6$ in the Integer window. The solution pair $(-9, -12.75)$ is shown at the bottom of the screen.

By moving the left and right arrow keys, you can find many of the other solution pairs to $3x - 4y = 24$, such as the following:

$$(-7,-11.25) \quad (-6,-10.5) \quad (-5,-9.75) \quad (-4,-9) \quad (-3,-8.25)$$
$$(-2, -7.5) \quad (-1, -6.75) \quad (0, -6) \quad (1, -5.25) \quad (2, -4.5)$$

**EXAMPLE 8**    Finding Solution Pairs

Find the graph of $5x + 4y = 20$ in the Integer window and all of the solution pairs whose $x$-coordinate is an integer between $x = -9$ and $x = 7$.

SOLUTION

Begin by writing $5x + 4y = 20$ in the equivalent expression $y = -(5/4)x + 5$. Figure 2.24 shows the graph of $y = -(5/4)x + 5$ in the Integer window and also shows the solution pair $x = -9$, $y = 16.25$.

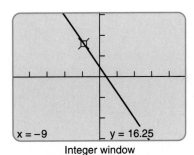

**FIGURE 2.24** The graph of $5x + 4y = 20$ with the trace cursor at the solution pair $(-9, 16.25)$.

Other solution pairs to this equation include the following:

$$(-8, 15) \quad (-7, 13.75) \quad (-6, 12.5) \quad (-5, 11.25)$$
$$(-4, 10) \quad (-3, 8.75) \quad (-2, 7.5) \quad (-1, 6.25)$$
$$(0, 5) \quad (1, 3.75) \quad (2, 2.5) \quad (3, 1.25)$$
$$(4, 0) \quad (5, -1.25) \quad (6, -2.5) \quad (7, -3.75) \quad \blacksquare$$

In all, there are an infinite number of solution pairs whose $x$-coordinate is between $x = -9$ and $x = 8$.

## Exercises for Section 2.2

For Exercises 1–6, select the ordered pairs that are solution pairs for the given equation in two variables.

1. $2x + y = 6$;  $(0, 6), (1, 5), (3, 0)$

2. $3x - y = 5$;  $(0, -5), (2, 1), (1, 2)$

3. $y = -2x + 5$;  $(2, 1), (-3, 11), (5, 15)$

4. $3x - 4y = 12$;  $(0, -3), (2, 1), (4, 0)$

5. $17x - 6y = 12$;  $(3, 6.5), (18, 23.5), (-12, -17)$

6. $3x - y = 21$;  $(3, 13), (2, -15), (-2, -15)$

For Exercises 7–14, find the coordinate that makes the ordered pair a solution pair for the given equation.

7. $y = 5x - 3$;  $(7, ?)$

8. $y = -14x + 23$;  $(-8, ?)$

9. $y = 48 - 16x$;  $(5, ?)$

10. $y = 0.5x + 17.5$;  $(4, ?)$

11. $6x + 2y = 24$;  $(?, -15)$

12. $5x - 2y = 165$;  $(?, -75)$

13. $7.6x - 3.2y = 8$;  $(35, ?)$

14. $1.2x + 0.48y = 48$;  $(92, ?)$

For Exercises 15–24, complete a table of solution pairs for $x = -3, -2, -1, 0, 1, 2,$ and 3 and sketch a graph of the given equation.

15. $y = 5x - 3$

16. $y = 4 - 3x$

17. $y = 5x + 1$

18. $y = x - 3$

19. $3x + y = 5$

20. $2x + y = 15$

21. $y - 5x = 12$

22. $y + 7x = 15$

23. $6x - 3y = 5$

24. $12x + 4y = 7$

For Exercises 25–34, complete a table of solution pairs and sketch the graph of the given equation.

25. $3x + 6y = 15$

26. $12x - 3y = 112$

27. $x - 4y = 84$

28. $6x + y = 78$

29. $y - x + 4 = 0$

30. $y + x - 6 = 0$

31. $y = x - 1 + 3x$

32. $y = x + 3 - 2x$

33. $3x = 2y - 12$

34. $5x = 2y + 3$

For Exercises 35–44, use a grapher to find the graph of each of the given equations in the Integer window. Sketch the graph produced by your grapher on a copy of the screen like the one shown below.

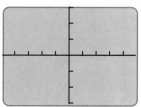

Integer window

35. $y = 5x + 13$

36. $y = -0.5x + 13$

37. $y = 17 - 2x$

38. $y = 8 - 0.25x$

39. $y = (2/5)x + 12$

40. $y = -4x - 9$

41. $2x + 4y = 17$

42. $3x - 9y = 88$

43. $8x - 5y = 21$

44. $15x + 12y = 225$

For Exercises 45–52, find the graph of the given equation in the Integer window. Record the solution pairs for $x = -30$, $x = -15$, $x = 15$, and $x = 30$.

45. $y = 3x - 6$

46. $y = (1/2)x + 2$

47. $2x - 4y = 7$

48. $3x + 2y = 9$

49. $2x + 5y = 30$

50. $3x - 8y = 48$

51. $10x + 16y = 144$

52. $9x - 4y = 27$

TRANSLATING WORDS TO SYMBOLS

In Exercises 53–55, write the equation for the given word description.

53. Two $x$ plus nine $y$ is equal to two hundred forty-three.

54. $y$ is equal to the quantity $x$ minus 3, quantity squared, plus two.

55. $y$ is equal to three minus $x$ squared.

EXTENDING THE IDEAS

For Exercises 56–59, use a grapher to find the graph of each equation in the window $[-10, 10]$ by $[-10, 10]$. Match the equation with its graph.

56. $y = |0.5x - 2|$

57. $y = x^3 - 5x^2 + 7x - 5$

58. $y = 2\sqrt{x + 8}$

59. $y = -0.5x^2 - 3x + 5$

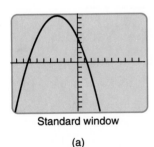

Standard window

(a)

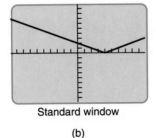

Standard window

(b)

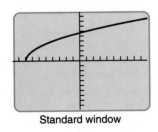

Standard window

(c)

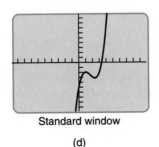

Standard window

(d)

For Exercises 60 and 61, graph each equation and press ⬛TRACE⬛ to complete its table of solution pairs. Use the 0.1 window.

60. $y = -0.25x + 2$

| $x$ | $-2.2$ | $-1.7$ | $-1.2$ | $-0.7$ | $0.8$ | $1.3$ | $2.5$ |
|---|---|---|---|---|---|---|---|
| $y$ | ? | ? | ? | ? | ? | ? | ? |

61. $y = x^2 - 2x - 1$

| $x$ | $-1.2$ | $-0.9$ | $-0.4$ | $0.3$ | $0.8$ | $1.7$ | $2.5$ |
|---|---|---|---|---|---|---|---|
| $y$ | ? | ? | ? | ? | ? | ? | ? |

62. Find the graphs of both $4x - 2y = 2$ and $x + 2y = 13$ in the Integer window. Press ⬛TRACE⬛ to find the coordinates of the point at which these graphs intersect each other. (Experiment to see what happens when you use the up and down arrow keys as the trace cursor moves along the graph.)

63. **Writing to Learn.** Write a paragraph explaining what you did to solve the problem posed in Exercise 62.

LOOKING BACK—MAKING CONNECTIONS

64. a) Solve the equation $2.5x - 22.5 = 0$.
    b) Find the graph of $y = 2.5x - 22.5$ in the Integer window. What is the relationship between the solution in part (a) and the point at which this graph crosses the $x$-axis?

65. a) Solve the equation $-2x + 9 = 0$.
    b) Find the graph of $y = -2x + 9$ in the Integer window. What is the relationship between the solution in part (a) and the point at which this graph crosses the $x$-axis?

# Linear Equations and the Slope of a Line

Sketching Graphs of Linear Equations in Variables $x$ and $y$  ▪  Finding the Intercepts Algebraically  ▪  Graphing Horizontal and Vertical Lines  ▪  Finding the Slope of a Line

In this section, you will learn several of the basic concepts related to graphs of linear equations. In particular, you will study about $x$- and $y$-intercepts and the concept of the slope of a line.

## Sketching Graphs of Linear Equations in Variables $x$ and $y$

In Section 2.2, you graphed the equations $14x + 28y = 168$ (Figure 2.17) and $7x + 12y = 78$ (Figure 2.22). In both cases, the graph is a straight line. Consequently these equations are each called **a linear equation in variables $x$ and $y$**. These two linear equations are written in standard form, as follows.

---

### Standard Form of a Linear Equation in Two Variables

The **standard form of a linear equation in variables $x$ and $y$** is an equation in the form

$$Ax + By = C,$$

where $A$, $B$, and $C$ are real numbers and at least one of $A$ or $B$ is not zero.

---

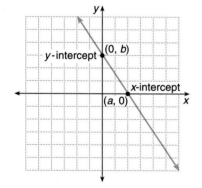

**FIGURE 2.25** The graph of the line with $x$-intercept $(a, 0)$ and $y$-intercept $(0, b)$.

It can be shown that the graph of a linear equation in two variables is always a straight line. Consequently, it is sufficient to graph only two solution pairs in order to sketch the graph. Two of the easiest points to find are the $x$-intercept and the $y$-intercept. The **$x$-intercept** (if any) is the point at which the graph crosses the $x$-axis. The **$y$-intercept** (if any) is the point at which the graph crosses the $y$-axis. For example, see Figure 2.25.

## Finding the Intercepts Algebraically

We see in Figure 2.25 that the $y$-intercept is the point at which the graph crosses the $y$-axis; consequently, it is the point on the graph whose $x$-coordinate is 0. The $x$-intercept is the point at which the graph crosses the $x$-axis; therefore it is the point on the graph whose $y$-coordinate is zero. So the two intercepts of a linear equation in $x$ and $y$ are easy to find since one coordinate of each point is zero.

---

### Finding the Intercepts Algebraically

1. To find the $y$-intercept, let $x = 0$ and solve for $y$.
2. To find the $x$-intercept, let $y = 0$ and solve for $x$.

---

This method of finding the intercepts is the basis of one of the easiest ways to sketch a linear equation, particularly when the linear equation is in standard form. This intercept method depends on the fact that a line has been determined once two points on that line are selected.

**TRY THIS**

Some lines have only one intercept. Sometimes the $x$- and $y$-intercept are the same. Give some examples.

---

### The Intercept Method of Sketching a Linear Equation

1. Find both the $x$-intercept and the $y$-intercept of the line.
2. Graph the two intercepts.
3. Draw the line determined by the two intercepts.

---

Example 1 shows how to use the intercept method to sketch a graph.

**EXAMPLE 1**    Finding Intercepts and Drawing a Graph

Find the $x$- and $y$-intercepts of $5x + 4y = 20$ and graph the equation.

SOLUTION

$$5(0) + 4y = 20 \quad \text{Let } x = 0 \text{ to find the } y\text{-intercept.}$$

$$y = 5$$

$$5x + 4(0) = 20 \quad \text{Let } y = 0 \text{ to find the } x\text{-intercept.}$$

$$x = 4$$

The $y$-intercept is $(0, 5)$ and the $x$-intercept is $(4, 0)$ (see Figure 2.26). ∎

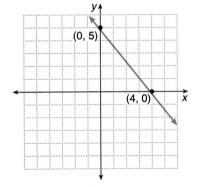

**FIGURE 2.26** The graph of $5x + 4y = 20$. This is the same graph as that shown in Figure 2.24 in Section 2.2.

If a line goes through the origin, then the point $(0, 0)$ is both the $x$- and $y$-intercept. Consequently, the intercept method of sketching does not apply.

## Graphing Horizontal and Vertical Lines

A horizontal line has no $x$-intercept and a vertical line has no $y$-intercept. Consequently, the method of Example 1 cannot be used when graphing horizontal and vertical lines.

**EXAMPLE 2**   Graphing a Vertical Line

Find the graph of the equation $x = 3$.

SOLUTION

The graph consists of all ordered pairs in the coordinate plane whose $x$-coordinate is 3. Consequently, it is the vertical line with $x$-intercept $(3, 0)$ (see Figure 2.27).

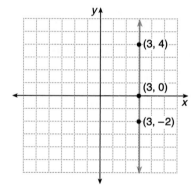

**FIGURE 2.27** Graph of the equation $x = 3$.

**EXAMPLE 3**   Graphing a Horizontal Line

Find the graph of the equation $y = -2$.

SOLUTION

The graph consists of all ordered pairs in the coordinate plane whose $y$-coordinate is $-2$. Consequently, it is the horizontal line with $y$-intercept $(0, -2)$ (see Figure 2.28).

**REMINDER**

The equation $x = 3$ is equivalent to the equation $1 \cdot x + 0 \cdot y = 3$. Consequently, $x = 3$ can be thought of as a linear equation in standard form.

**GRAPHER NOTE**

Consult your grapher lab manual to learn how to graph a vertical line on a grapher.

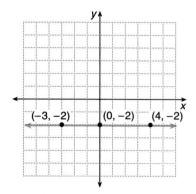

**FIGURE 2.28** Graph of the equation $y = -2$.

Examples 2 and 3 show graphs that are vertical and horizontal lines, respectively. The equations that determine these lines, although in standard form, seem like special cases. We summarize these two cases next.

---

### Graphing Horizontal and Vertical Lines

1. The vertical line with $x$-intercept $(a, 0)$ is the graph of the equation $x = a$.
2. The horizontal line with $y$-intercept $(0, b)$ is the graph of the equation $y = b$.

---

## Finding the Slope of a Line

The direction and measure of "steepness" of a line can be found by examining the **slope** of the line. Slope is the number of units of vertical "rise" or "fall" for each unit of horizontal change, moving from left to right. It is the ratio of "the change in $y$" divided by "the change in $x$." We sometimes say that it is the ratio of "rise to run." For example, consider the line that passes through points $(2, 1)$ and $(6, 9)$ (see Figure 2.29). There are 8 units of rise to 4 units of run. In other words, the slope of the line is equal to $8/4 = 2$.

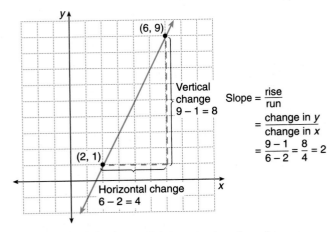

$$\text{Slope} = \frac{\text{rise}}{\text{run}}$$
$$= \frac{\text{change in } y}{\text{change in } x}$$
$$= \frac{9 - 1}{6 - 2} = \frac{8}{4} = 2$$

**FIGURE 2.29** The slope of the line through points $(2, 1)$ and $(6, 9)$ is 2.

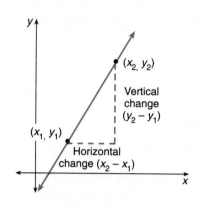

**FIGURE 2.30** Since the line is not vertical, $x_2 - x_1$ is not zero and the slope of the line is defined.

---

**DEFINITION 2.1**   Slope of a Line

The **slope** of a nonvertical line passing through points $(x_1, y_1)$ and $(x_2, y_2)$ is the number $m$ defined by

$$m = \frac{y_2 - y_1}{x_2 - x_1} = \frac{\text{change in } y}{\text{change in } x}.$$

---

The definition of slope assumes that the two points $(x_1, y_1)$ and $(x_2, y_2)$ are not on a vertical line (see Figure 2.30). If the line is a vertical line the denominator is 0, and the fraction is undefined. Therefore the slope of a vertical line is undefined.

**EXAMPLE 4**   Finding the Slope of the Line Through Two Points

Find the slope of the line determined by the following points: a) $(-2, 1)$ and $(5, 3)$  b) $(-3, 5)$ and $(3, -2)$

SOLUTION
See Figure 2.31.

a) $m = \dfrac{y_2 - y_1}{x_2 - x_1} = \dfrac{3 - 1}{5 - (-2)}$       b) $m = \dfrac{y_2 - y_1}{x_2 - x_1} = \dfrac{-2 - 5}{3 - (-3)}$

$\quad = \dfrac{2}{7}$                   $\quad = \dfrac{-7}{6}$

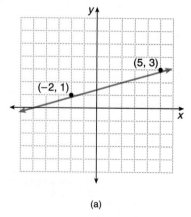

(a)

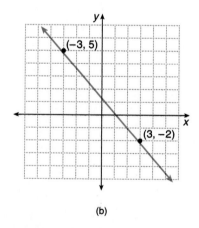

(b)

**FIGURE 2.31**   (a) A line with positive slope rises from left to right. (b) A line with negative slope falls from left to right.

Example 4 shows how to find the slope determined by two points in a rectangular coordinate plane. This approach is the basis of an algebraic method for finding the slope of the graph of a linear equation, as illustrated in Example 5.

### EXAMPLE 5    Finding the Slope Algebraically

Find the slope of the line $12x - 21y = 1260$.

SOLUTION

Find the intercepts and use those two points to find the slope.

$$12(0) - 21y = 1260$$

$$y = \frac{1260}{-21} = -60$$

$$12x - 21(0) = 1260$$

$$x = \frac{1260}{12} = 105$$

The $y$-intercept is $(0, -60)$ and the $x$-intercept is $(105, 0)$. Therefore the slope $m$ of the line is as follows:

$$m = \frac{y_2 - y_1}{x_2 - x_1}$$

$$= \frac{0 - (-60)}{105 - 0} = \frac{60}{105} = \frac{4}{7}$$    ∎

## Using a Grapher to Find the Slope

The graph of the line $-3x + 5y = 15$ is shown in the Integer window in Figure 2.32. Notice that Figure 2.32 (a) shows the solution pair $(0, 3)$ found using the TRACE feature. A *single move to the right* by moving the trace cursor to the right once shows the solution pair $(1, 3.6)$ (see Figure 2.32 (b)).

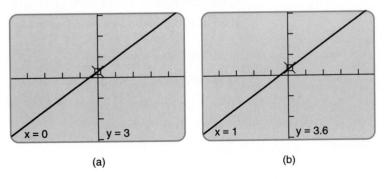

(a)                                              (b)

**FIGURE 2.32**  Adjacent trace coordinates on the line $-3x + 5y = 15$.

Applying the definition of slope to the two points $(0, 3)$ and $(1, 3.6)$ shown in Figure 2.32, we conclude the following:

$$m = \frac{y_2 - y_1}{x_2 - x_1}$$

$$= \frac{3.6 - 3}{1 - 0}$$

$$= 3.6 - 3 = 0.6$$

Consequently you can easily find the slope following this procedure.

---

### Finding the Slope of a Line on a Grapher

1. Find the graph of the line in the Integer window.
2. Position the trace cursor at a point on the line. Let $y_1$ be its $y$-coordinate.
3. Move the trace cursor one unit right. Let $y_2$ be its $y$-coordinate and calculate $y_2 - y_1$.

---

Example 6 demonstrates this grapher method. It is a particularly attractive method when the numbers are decimals or fractions.

### EXAMPLE 6    Finding the Slope Graphically

Use the trace coordinates on a grapher to find the slope of the line $3x + 4y = 32$.

SOLUTION

1. First change the form of $3x + 4y = 32$ to prepare it for input to the grapher.

$$3x + 4y = 32$$

$$4y = -3x + 32$$

$$y = -\frac{3}{4}x + \frac{32}{4}$$

2. Find the graph of $y = -\frac{3}{4}x + 8$ in the Integer window.

3. Figure 2.33 (a) and (b) show that $(0, 8)$ and $(1, 7.25)$ are points on the line. Therefore

$$m = \frac{y_2 - y_1}{x_2 - x_1} = \frac{7.25 - 8}{1 - 0} = 7.25 - 8 = -0.75.$$

The slope of line $3x - 4y = 32$ is $-0.75$.

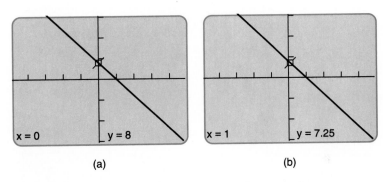

(a)          (b)

**FIGURE 2.33** Two adjacent trace coordinates on line $3x + 4y = 32$.

Example 7 illustrates that when the slope of a line and one point on the line are known, the coordinates of other points on the line can be found.

**EXAMPLE 7**    Solving an Equation Involving Slope

A line with slope 3 passes through the points $(-1, -3)$ and $(x, 15)$. Find the value of the unknown coordinate.

SOLUTION

Using the definition of slope, $m = \dfrac{y_2 - y_1}{x_2 - x_1}$, we obtain the following equation:

$$3 = \frac{15 - (-3)}{x - (-1)}$$

$$3 = \frac{18}{x + 1}$$

$$3(x + 1) = 18$$

$$3x + 3 = 18$$

$$3x = 15$$

$$x = 5$$

Therefore, the line passes through the points $(-1, -3)$ and $(5, 15)$.

We end this section examining what happens to the graph of a linear equation when it includes absolute value.

### EXAMPLE 8   Finding a Graph

Find the graph of the equation $y = |x - 14|$.

SOLUTION

Figure 2.34 shows this equation on the grapher, while Figure 2.35 shows the graph of $y = |x - 14|$ in the Integer window.

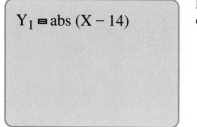

$Y_1 = \text{abs} (X - 14)$

**FIGURE 2.34** Entering the equation $y = |x - 14|$ in the grapher.

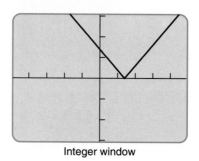

Integer window

**FIGURE 2.35** The graph of $y = |x - 14|$.

Observe that the graph in Figure 2.35 does not go below the $x$-axis. This characteristic of the graph is consistent with the fact that the absolute value of a number is always positive or zero.

## Exercises for Section  2.3

In Exercises 1–8, use an algebraic method to find the $x$- and $y$-intercepts and graph each linear equation.

1. $6x - 2y = 36$
2. $3x + 12y = 144$
3. $12x + 15y = 60$
4. $-9x + 6y = 51$
5. $5x - 3y = 15$
6. $6x + 15y = 105$
7. $-3x + 7y = 84$
8. $32x - 24y = 96$

In Exercises 9–16, use the intercept method to sketch the graph of the given linear equation.

9. $2x + 4y = 15$
10. $5x + 4y = 20$
11. $3x - 5y = 15$
12. $2x - 7y = 28$
13. $6x - 5y = 120$
14. $5x + 2y = 25$
15. $8x + 3y = 12$
16. $7x - 8y = 56$

In Exercises 17–22, sketch the graph of each horizontal or vertical line.

**17.** $y = 3$   **18.** $x = 2$

**19.** $x = -3.5$   **20.** $y = -5$

**21.** $x = 8$   **22.** $y = 7$

**23.** Find an equation of the vertical line whose $x$-intercept is $(3.2, 0)$.

**24.** Find an equation of the horizontal line whose $y$-intercept is $(0, -4.8)$.

In Exercises 25–32, find the slope of the line passing through the given pair of points.

**25.** $(3, -1)$ and $(5, 7)$

**26.** $(2, 5)$ and $(11, 8)$

**27.** $(1, 3)$ and $(4, 12)$

**28.** $(-4, 5)$ and $(3, -9)$

**29.** $(2.3, 5.1)$ and $(-3.6, 4.2)$

**30.** $(9.2, -3)$ and $(7.6, 1.4)$

**31.** $(23, 14)$ and $(-12, -11)$

**32.** $(12, 183)$ and $(13, 119)$

In Exercises 33–40, use an algebraic method to find the slope of the line that is the graph of the given linear equation.

**33.** $3x - 2y = 17$   **34.** $5x + 3y = 48$

**35.** $8x + 2y = 34$   **36.** $7x - 5y = 25$

**37.** $3x + 18y - 42 = 0$   **38.** $4x - 7y + 28 = 0$

**39.** $19x - 43y = 817$   **40.** $51x + 17y = 1734$

In Exercises 41–46, use a grapher to find the slope of the line that is the graph of the given linear equation.

**41.** $12x - 4y = 40$

**42.** $18x - 12y = 36$

**43.** $-8x + 16y = 64$

**44.** $2y = 8x + 22$

**45.** $6x = 48 - 12y$

**46.** $4y = 56x + 28$

In Exercises 47–52, find the unknown coordinate so that the slope through the two points is equal to the given value.

**47.** $(2, 3)$ and $(x, 5)$; $m = 2$

**48.** $(-1, 5)$ and $(7, y)$; $m = -\dfrac{1}{3}$

**49.** $(-3, 6)$ and $(1, y)$; $m = \dfrac{1}{2}$

**50.** $(x, -5)$ and $(7, 4)$; $m = -1$

**51.** $(0, 9.2)$ and $(x, 7.5)$; $m = 3.4$

**52.** $(x, 3.5)$ and $(5.2, 9.4)$; $m = 0.5$

In Exercises 53–58, find the graph of the given equation.

**53.** $y = |x - 1|$   **54.** $y = |x + 3|$

**55.** $y = |x + 2|$   **56.** $y = |2x - 3|$

**57.** $y = |3x - 8|$   **58.** $y = |5 - x|$

## TRANSLATING WORDS TO SYMBOLS

Exercises 59–63 refer to points $P(x_1, y_1)$ and $Q(x_2, y_2)$. For each, write an algebraic expression for the given phrase.

**59.** The difference in the coordinates $y_2$ and $y_1$

**60.** The difference in the coordinates $x_2$ and $x_1$

**61.** The difference in $y$-coordinates divided by the difference in $x$-coordinates

**62.** The quotient of the difference in $y$-coordinates and the difference in $x$-coordinates

**63.** The ratio of the differences in the $y$-coordinates by the difference in $x$-coordinates

**64.** *Highway Engineering.* Interstate 70 west of Denver, Colorado, has a section that is posted as a 6% grade. That means that for a horizontal change of 100 ft, there is a 6-ft vertical change.

6% grade.

For Exercise 64.

a) Find the slope of this highway.

b) On a highway with a 6% grade, what is the horizontal distance required to climb 450 ft?

c) In a 3-mi horizontal distance along the highway, how much vertical climb is there? (There are 5280 ft in each mile.)

**65.** *Investment Analysis.* Sarah is studying the impact on her financial condition of investing $$x$ into an account that earns 8% a year.

a) If $I$ represents the amount of interest earned in the first year, express $I$ in terms of $x$.

b) If $y$ represents the balance in her account (amount invested + interest earned) at the end of 1 yr, express $y$ in terms of $x$.

c) What is the slope of the linear equation found in part (b) ?

**66.** *Financial Planning.* Rachel has $7000 to invest. She plans to invest part of it in an account that pays 5% and the rest in an account that pays 8%. (There is a penalty for early withdrawal in this 8% account.)

a) Suppose Rachel invests $$x$ in the 5% account and the rest in the 8% account. If $I$ is the sum of the interest earned in each account in the first year, express $I$ in terms of $x$.

b) If $y$ represents the balance in her two accounts (amount invested + interest earned) at the end of 1 yr, express $y$ in terms of $x$.

c) What is the slope of the linear equation found in part (b) ?

EXTENDING THE IDEAS

In Exercises 67–70, suppose L is a line through point $(a, b)$ with slope 2.

**67.** Which of these points are also on line L: (a) $(a+2, b+1)$, (b) $(a+1, b+2)$, or (c) $(a+2, b+2)$?

**68.** Which two of these points are on line L: (a) $(a+2, b+4)$, (b) $(a-1, b+2)$, or (c) $(a-1, b-2)$?

**69.** How many of these points are on line L: (a) $(a + 0.1, b + 0.2)$, (b) $(a + 0.05, b + 0.1)$, and (c) $(a - 27, b - 54)$?

**70.** If $(3, 2)$ is a point on L, use the definition of slope to write a linear equation in $a$ and $b$.

In Exercises 71–74, find the slope of the line shown on the grapher screen. Assume the line passes through the points in the bottom left-hand and top right-hand corners of the screen.

**71.**

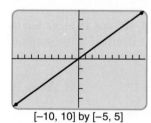

$[-10, 10]$ by $[-5, 5]$

**72.**

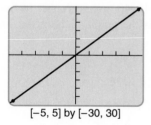

$[-5, 5]$ by $[-30, 30]$

**73.**

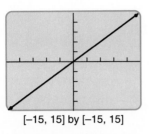

[−15, 15] by [−15, 15]

**74.**

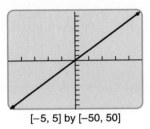

[−5, 5] by [−50, 50]

**75. Writing to Learn.** Estimate the slope of each line in the following grapher screens and write a paragraph explaining which of the two shows the line with the greater slope.

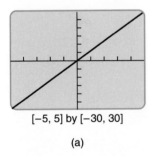

[−5, 5] by [−30, 30]

(a)

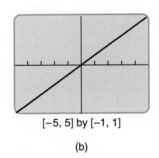

[−5, 5] by [−1, 1]

(b)

**76. Writing to Learn.** Estimate the slope of each line in the following grapher screens and write a paragraph explaining which of the two shows the line with the greater slope.

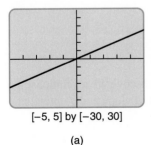

[−5, 5] by [−30, 30]

(a)

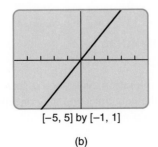

[−5, 5] by [−1, 1]

(b)

**77. Writing to Learn.** Find graphs of both $y = x + 8$ and $y = |x + 8|$ in the Integer window. Write a paragraph that describes the relationship between these two graphs.

LOOKING BACK—MAKING CONNECTIONS

In Exercises 78–87, sketch the graph of each linear equation.

**78.** $2x - y = 13$

**79.** $3x + 2y = 12$

**80.** $x = 1.7$

**81.** $y = -4$

**82.** $5x + 2y = 10$

**83.** $3x - 7y = 42$

**84.** $y - 13 = 7.3$

**85.** $x + 3 = 5$

**86.** $14y - 2x = 5.6$

**87.** $24 - 3x = 12y$

**88.** Find the slope of the line through the points $(a, 0)$ and $(0, b)$. If $a$ and $b$ are both positive numbers, is this slope positive or negative? Explain.

**89.** Rewrite the equation $4x - 17 = 2x - (5 - x)$ so that it is in the form $ax + b = 0$.

**90.** Find an algebraic solution to $17x - 23 = 4x + 11$. Express the answer as a fraction in reduced form.

## 2.4 The Slope–Intercept Form and Point–Slope Form of Linear Equations

The Slope–Intercept Form of a Linear Equation ▪ Using a Point and the Slope to Graph a Line ▪ Finding an Equation Whose Graph is a Given Line ▪ Point–Slope Form of a Linear Equation ▪ Determining Parallel and Perpendicular Lines

Before reading further, complete the following exploration with a grapher.

### EXPLORE WITH A GRAPHER

For each of the equations

$$y = 2x + 5, \quad y = -3x + 7, \quad y = 0.5x - 17, \quad y = -4x + 12$$

find the following:

a) The graph of the equation in the Integer window.
b) The slope of the graph in (a) by using the method given in Example 6 in Section 2.3. How does this slope compare to the coefficient of $x$ in the equation being graphed?
c) The $y$-intercept of each line. How is it related to the constant in the equation being graphed?

**Generalization:** State a generalization based on this exploration.

## The Slope–Intercept Form of a Linear Equation

In completing the above exploration, you should have discovered a relationship between the coefficients in the equations and the concepts of slope and $y$-intercept. We summarize this relationship next.

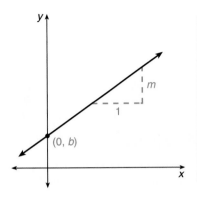

**FIGURE 2.36** The line whose equation is $y = mx + b$ has $y$-intercept of $(0, b)$ and slope $m$.

---

### Slope–Intercept Form of a Linear Equation

The graph of the equation

$$y = mx + b \qquad (1)$$

is a line whose slope is $m$ and whose $y$-intercept is $(0, b)$. (See Figure 2.36.)

The form of Equation (1) is called the **slope–intercept** form.

---

**EXAMPLE 1**  Finding the Slope and $y$-intercept

Find the slope and $y$-intercept of the graph of $y = -3x + 2$.

SOLUTION

The equation $y = -3x + 2$ is in the slope–intercept form. Therefore the following holds:

$$\text{slope} = -3 \quad \textbf{The slope is equal to the coefficient of x.}$$

$$y\text{-intercept} = (0, 2) \quad \textbf{The y-intercept is determined by the constant.} \quad ■$$

To find the slope and $y$-intercept of a line whose equation is not written in the slope–intercept form, change the equation to that form.

**EXAMPLE 2**  Finding the Slope and $y$-intercept of a Line

Find the slope and $y$-intercept of the following lines:
   a) $4x - 2y = -9$
   b) $3x - 12 = y - 7$

SOLUTION

Write each equation in the slope–intercept form.

a)
$$4x - 2y = -9$$
$$-2y = -4x - 9$$
$$y = 2x + \frac{9}{2}$$

The slope of the line $4x - 2y = -9$ is $m = 2$ and the $y$-intercept is $\left(0, \dfrac{9}{2}\right)$.

b)
$$3x - 12 = y - 7$$
$$3x - 5 = y$$
$$y = 3x - 5$$

The slope of the line $3x - 12 = y - 7$ is $m = 3$ and the $y$-intercept is $(0, -5)$. ∎

## Using a Point and the Slope to Graph a Line

If you know the coordinates of one point on a line and you also know the slope of the line, it is possible to sketch the line by hand. Consider for example the line with slope $m = \dfrac{2}{3}$ that passes through a point $P$. Recall that

$$m = \frac{\text{change in } y}{\text{change in } x} = \frac{2}{3}.$$

So from the point $P$, you can obtain another point on line L by a vertical move of $+2$ followed by a horizontal move of $+3$ (see Figure 2.37).

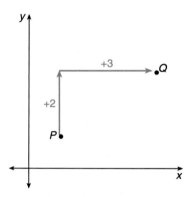

**FIGURE 2.37** If the slope is 2/3, from $P$ move 2 units up and 3 units right.

Example 3 illustrates that this observation can be used to find the graph of a line.

**EXAMPLE 3**    Sketching the Graph of a Line

Sketch the graph of the line with slope $m = 2/3$ that passes through $P(1, 2)$.

SOLUTION

On a piece of graph paper, first graph $P(1, 2)$. Then locate a second point $Q$ from point $P$ by moving 2 units up and 3 units right (see Figure 2.38 (a) on the following page). Then draw the line $PQ$ (see Figure 2.38 (b)).

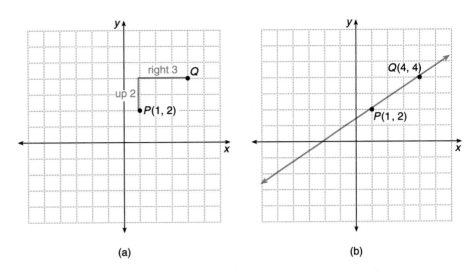

**FIGURE 2.38** Graph of the line through $P(1, 2)$ with slope $m = 2/3$. ■

If the slope is negative, locate $Q$ by moving down rather than up from $P$.

**EXAMPLE 4**    Graphing the Line Through $P$ with Slope $m$

Find the graph of the line through $P(-4, 2)$ with slope $m = -5/8$.

SOLUTION
Locate $Q$ by moving from $P$ down 5 units and right 8 units (see Figure 2.39 (a)). Figure 2.39 (b) shows the completed graph.

**TRY THIS**

In Figure 2.39 down 5 units (the direction $-5$) and right 8 units (the direction 8) is the slope $m = \dfrac{-5}{8}$. However, from $Q$ to $P$, up 5 units (the direction 5) and left 8 units (the direction $-8$) is the slope $m = \dfrac{5}{-8}$. This is visual support that $\dfrac{-5}{8} = \dfrac{5}{-8} = -\dfrac{5}{8}$. Experiment! Illustrate this on your grapher.

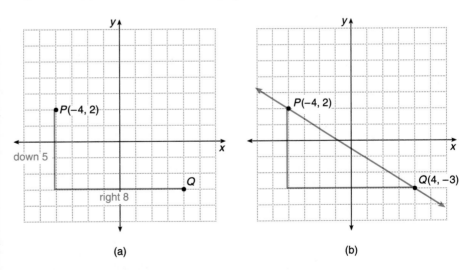

**FIGURE 2.39** The line through $P(-4, 2)$ with slope $m = -5/8$. ■

## Finding an Equation Whose Graph Is a Given Line

So far in this section, you have learned how to

a) find the slope and $y$-intercept of a line if you know an equation of the line, and

b) find the graph of a line if you know one point and its slope.

Next you will learn how to find an equation of a line if you know its slope and the coordinates of one point on the line. Example 5 assumes that the known point is the $y$-intercept of the line.

**EXAMPLE 5**     Finding the Slope–Intercept Form of an Equation

Write the slope–intercept form of the equation whose graph is the line L with $y$-intercept $(0, -3)$ and slope $m = 5$.

SOLUTION

Substitute the values $m = 5$ and $b = -3$ into the slope–intercept form.

$$y = mx + b \quad \textbf{Let } \boldsymbol{m = 3} \textbf{ and } \boldsymbol{b = -3.}$$

$$y = 5x - 3$$

The equation in slope–intercept form whose graph is L is $y = 5x - 3$ (see Figure 2.40).

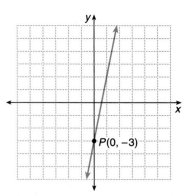

**FIGURE 2.40** The graph of the equation $y = 5x - 3$. The slope of this line is $m = 5$. What is its $x$-intercept?

## Point–Slope Form of a Linear Equation

Suppose the known point of a line is not the $y$-intercept. For example, consider the line L through $P(-3, 1)$ with slope $m = 2/3$. Select an arbitrary point $Q(x, y)$ so that the slope of line $PQ$ is 2/3 (see Figure 2.41). This point $Q$ must be a point of L.

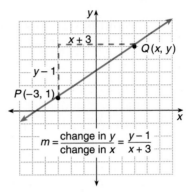

**FIGURE 2.41**  Point $Q$ is chosen so that line $PQ$ has slope $\dfrac{4}{6} = \dfrac{2}{3}$.

We know that an arbitrary point on line L has coordinates $(x, y)$. Using the two points $(x, y)$ and $(-3, 1)$ and our definition of slope, we obtain the equation

$$m = \frac{\text{change in } y}{\text{change in } x} = \frac{y - 1}{x + 3}.$$

Because the slope of the line is $m = 2/3$, we conclude that

$$\frac{y - 1}{x + 3} = \frac{2}{3}.$$

Multiplying both sides of the equation by $3(x + 3)$ results in the following equivalent equations:

$$3(y - 1) = 2(x + 3)$$

$$3y - 3 = 2x + 6$$

$$-2x + 3y = 9$$

So $-2x + 3y = 9$ is a linear equation in standard form whose graph is line L.

**EXAMPLE 6**     Finding an Equation of a Line

Find an equation in standard form of the line that passes through $P(3, -2)$ with slope $m = -\dfrac{1}{2}$.

SOLUTION

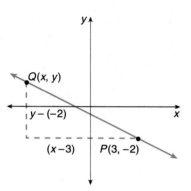

**FIGURE 2.42** Graph of the line $x + 2y = -1$.

Consider the line through $(3, -2)$ with slope $m = -\dfrac{1}{2}$ and let $Q(x, y)$ be a point on this line (see Figure 2.42). Apply the definition of slope to obtain

$$\frac{y - (-2)}{x - 3} = -\frac{1}{2}.$$

Multiply both sides of the equation by $2(x - 3)$ to obtain the following equivalent equations:

$$2(y + 2) = -1(x - 3)$$
$$2y + 4 = -x + 3$$
$$x + 2y = -1$$

An equation in standard form of this line is $x + 2y = -1$.  ∎

Example 6 shows how to find the equation of a line given its slope and one point on the line. Example 7 establishes the general equation for this situation, where the known point is $P(a, b)$ and the slope is $m$.

**EXAMPLE 7**     Finding an Equation of a Line

Find an equation of the line with slope $m$ that passes through point $P(a, b)$.

SOLUTION

Let $Q(x, y)$ be any point such that line $PQ$ has slope $m$. Apply the definition of slope to obtain the following:

$$\frac{y - b}{x - a} = m$$  **Multiply both sides of the equation by $(x - a)$.**

$$y - b = m(x - a)$$  ∎

The equation $y - b = m(x - a)$ is called the **point–slope form** of the equation of the line.

---

### Point–Slope Form

The **point–slope form** of the equation of the line with slope $m$ passing through the point $(a, b)$ is

$$y - b = m(x - a).$$

---

The *standard form, slope–intercept form*, and *point–slope form* are the three forms for an equation of a line L that you have learned. It is often helpful to change from one form to another, as the next example demonstrates.

**EXAMPLE 8**  Changing from Point–Slope Form to Slope–Intercept Form and Standard Form

Find the $y$-intercept of line L with slope $m = 2$ that passes through point $P(4, 1)$.

SOLUTION
The point–slope form of the equation of line L is

$$y - 1 = 2(x - 4).$$

**Change to Slope–Intercept Form:** Change the equation to slope–intercept form.

$$y - 1 = 2(x - 4)$$
$$y - 1 = 2x - 8$$
$$y = 2x - 8 + 1$$
$$y = 2x - 7$$

The equation $y = 2x - 7$ is the slope–intercept form of an equation for L. The $y$-intercept is $(0, -7)$ (see Figure 2.43).

**Change to Standard Form:** Change the equation to the standard form.

$$y = 2x - 7 \qquad \text{Begin with the slope–intercept form, and add } -2x \text{ to both sides of the equation.}$$

$$-2x + y = -7 \qquad \text{Multiply both sides of the equation by } -1 \text{ to obtain an equivalent equation.}$$

$$2x - y = 7$$

The equation $2x - y = 7$ is a linear equation in standard form. ∎

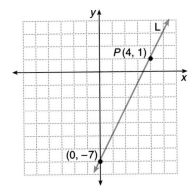

**FIGURE 2.43** Graph of the line with slope $m = 2$ that passes through point $P(4, 1)$.

## Determining Parallel and Perpendicular Lines

The slope of a line can be used to determine whether two lines are parallel or perpendicular to each other. Before reading further, complete the following exploration with a grapher.

### EXPLORE WITH A GRAPHER

Graph each of the following pairs of equations in the Integer window. Decide if the graphs appear to be parallel lines, perpendicular lines, or neither.

1. $y = 3x - 2$ and $y = 3x + 4$
2. $y = 0.5x + 2$ and $y = -2x - 3$
3. $y = 0.7x + 2$ and $y = 0.9x - 3$
4. $y = 7 - 2x$ and $y = -2x - 3$
5. $y = x - 3$ and $y = -x + 5$
6. $y = -\frac{1}{3}x - 4$ and $y = 3x + 1$
7. $y = 0.6x - 4$ and $y = 0.6x + 7$

**Generalize** How are the slopes of the two lines related when the lines are parallel? When they are perpendicular?

The generalization that emerges from this exploration is the following.

**GRAPHER NOTE**

Lines whose slopes are negative reciprocals of each other will not appear to be perpendicular when graphed on a grapher unless the window is a "Square" window. Read in your grapher lab manual about Square windows.

### Parallel and Perpendicular Lines

Let $\ell_1$ be a line with equation $y = m_1x + b$ and $\ell_2$ be a line with equation $y = m_2x + c$. Then

- $\ell_1$ and $\ell_2$ are parallel if and only if their slopes are equal; that is, if and only if $m_1 = m_2$, and
- $\ell_1$ and $\ell_2$ are perpendicular if and only if their slopes are negative reciprocals of each other, that is, if and only if $m_1 \cdot m_2 = -1$.

Example 9 uses information about the slope of a given line to find an equation of lines parallel to and perpendicular to the given line.

**EXAMPLE 9**   Finding Parallel and Perpendicular Lines

Find equations of the lines $\ell_1$ and $\ell_2$ passing through $(0, 3)$ that are respectively parallel and perpendicular to the line $\ell_3$: $y = -\dfrac{1}{4}x - 3$.

SOLUTION

The $y$-intercept for both lines is $(0, 3)$.

**Equation of $\ell_1$:** Because $\ell_1$ is parallel to $\ell_3$, the slope of $\ell_1$ is also $m = -1/4$ and the equation is

$$y_1 = -0.25x + 3.$$

**Equation of $\ell_2$:** Because $\ell_2$ is perpendicular to $\ell_3$, the slope of $\ell_2$ is $m = 4$ and the equation is

$$y_2 = 4x + 3.$$

Notice that lines $\ell_1$ and $\ell_2$ do not appear to be perpendicular in Figure 2.44 since the standard window is not a "square" window.   ■

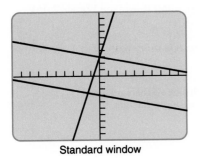

Standard window

**FIGURE 2.44**  Graphs of $y_3 = -\frac{1}{4}x - 3$, $y_1 = -\frac{1}{4}x + 3$, and $y_2 = 4x + 3$.

**EXAMPLE 10**   Finding Equations of Parallel and Perpendicular Lines

Find equations in standard form of the lines $\ell_1$ and $\ell_2$ passing through the point $(-2, 3)$ such that (a) $\ell_1$ is parallel to the line $3x - 5y = 17$ and (b) $\ell_2$ is perpendicular to $3x - 5y = 17$.

SOLUTION

Change the equation $3x - 5y = 17$ to the slope–intercept form to find the slope of the line.

$$3x - 5y = 17$$
$$-5y = -3x + 17$$
$$y = \frac{3}{5}x - \frac{17}{5}$$

Therefore, the slope of $\ell_1$ is $m = \dfrac{3}{5}$ and the slope of $\ell_2$ is $m = -\dfrac{5}{3}$.

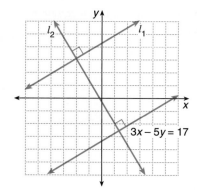

**FIGURE 2.45** Graphs of $\ell_1$ and $\ell_2$.

**Equation of $\ell_1$:** Using the point–slope form of the equation, the following holds with $m = \frac{3}{5}$ and point $(-2, 3)$:

$$y - 3 = \frac{3}{5}(x - (-2))$$ **Multiply each side of the equation by 5 to eliminate the fraction.**

$$5(y - 3) = 3(x + 2)$$

$$5y - 15 = 3x + 6$$

$$-3x + 5y = 21$$

$$3x - 5y = -21$$

The graphs of $3x - 5y = 17$ and $3x - 5y = -21$ are parallel lines (see Figure 2.45).

**Equation of $\ell_2$:** Using the point–slope form of the equation with $m = \dfrac{-5}{3}$ and point $(-2, 3)$, the following holds:

$$(y - 3) = -\frac{5}{3}(x - (-2))$$ **Multiply each side of the equation by 3 to eliminate the fraction.**

$$3(y - 3) = -5(x + 2)$$

$$3y - 9 = -5x - 10$$

$$5x + 3y = -1$$

The graphs of $3x - 5y = 17$ and $5x + 3y = -1$ are perpendicular (see Figure 2.45). ∎

## Exercises for Section 2.4

In Exercises 1–10, find the slope and the $y$-intercept of the graph of each equation.

1. $y = 3x - 4$

2. $y = -5x + 3$

3. $3x + 4y = 7$

4. $2x - 5y + 12 = 0$

5. $5x - 2y = 15$

6. $4x + 3y = 24$

7. $7x - 4y = 3x - 2$

8. $2y + 7x = 12 - 5x$

9. $3x = 7y - 2$

10. $5x + y = 6$

11. To sketch the line with slope $m = \dfrac{3}{4}$ that passes through the point $P(2, 1)$, which of the following graphs would you use?

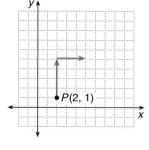

(a)

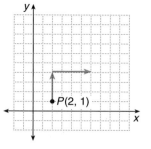

(b)

For Exercise 11.

In Exercises 12–21, sketch the line with the given slope $m$ that passes through the given point $P(a, b)$.

**12.** $m = 2$ through $P(2, 3)$

**13.** $m = -3$ through $P(-1, 4)$

**14.** $m = -1$ through $P(-3, 7)$

**15.** $m = \dfrac{1}{2}$ through $P(-4, -2)$

**16.** $m = \dfrac{3}{4}$ through $P(2, 1)$

**17.** $m = 1.25$ through $P(-2, 3)$

**18.** $m = -\dfrac{2}{3}$ through $P(-1, 5)$

**19.** $m = -3$ through $P(0, 5)$

**20.** $m = -\dfrac{5}{4}$ through $P(-3, 4)$

**21.** $m = \dfrac{2}{5}$ through $P(-1, 0)$

In Exercises 22–27, find an equation of the line with the given slope and $y$-intercept. Write the equation in slope–intercept form.

**22.** $m = 2$ and $y$-intercept $(0, -3)$

**23.** $m = 5$ and $y$-intercept $(0, 3)$

**24.** $m = -2$ and $y$-intercept $(0, 4)$

**25.** $m = \dfrac{3}{4}$ and $y$-intercept $\left(0, \dfrac{1}{2}\right)$

**26.** $m = 3.5$ and $y$-intercept $(0, 2.1)$

**27.** $m = -2.5$ and $y$-intercept $(0, -5.4)$

In Exercises 28–35, find the slope and $y$-intercept of each line.

**28.** $(y - 4) = 2(x + 3)$    **29.** $2(y - 3) = 3(x - 5)$

**30.** $3(y + 4) = 2(x - 1)$    **31.** $-2(y + 3) = 5(x - 6)$

**32.** $3x - 4y = 8$    **33.** $5x + 3y = 14$

**34.** $-3x + 12y = 21$    **35.** $-5x - 3y = -12$

In Exercises 36–41, find an equation of the line with the given slope $m$ that passes through the given point $P(a, b)$. Write the equation in slope–intercept form.

**36.** $m = 1$ through $P(2, 3)$

**37.** $m = 3$ through $P(-1, -4)$

**38.** $m = -2$ through $P(-3, 4)$

**39.** $m = -2.5$ through $P(-2, 5)$

**40.** $m = 5$ through $P(-4, -7)$

**41.** $m = \dfrac{1}{2}$ through $P(-1, 2)$

In Exercises 42–47, determine algebraically which pairs of lines are parallel, perpendicular, or neither. Use the Integer window to support the solutions graphically.

**42.** $y = 2x - 3$ and $2y = 4x + 3$

**43.** $2x + 3y = 7$ and $3x - 2y = 6$

**44.** $4x + 12y = 14$ and $3x - y = 12$

**45.** $2x + 3y = 7$ and $3x - 4y = 6$

**46.** $x + 3y = 14$ and $6y = 12 - 2x$

**47.** $y = 2x + 1$ and $y - \dfrac{1}{2}x = 4$

In Exercises 48–53, find an equation in standard form for the line through the given point that is parallel to the given line.

**48.** $P(-3, 2)$; $y = 2x - 4$

**49.** $P(1, 2)$; $y = -3x + 7$

**50.** $P(-3, -2)$; $2x + 3y = 18$

**51.** $P(-1, 2)$; $x - 3y = 14$

**52.** $P(-4, -3)$; $2x - 5y = 12$

**53.** $P(5, 4)$; $-4x - 3y = 21$

In Exercises 54–59, find an equation in standard form for the line through the given point that is perpendicular to the given line.

**54.** $P(-1, 2)$; $y = -3x - 4$

**55.** $P(3, -2)$; $y = 2x + 3$

**56.** $P(-1, 4)$; $2x - 3y = 18$

**57.** $P(1, 3)$; $2x + 3y = 11$

**58.** $P(4, -2)$; $2x + 7y = 15$

**59.** $P(3, 2)$; $4x - 3y = 12$

In Exercises 60–61, we will use the fact that the concept of slope can be defined by the quotient

$$\frac{\text{change in } y}{\text{change in } x}.$$

This quotient is also described as the **average rate of change of quantity $y$ with respect to quantity $x$.**

**60.** A graph shows that the average rate of change of distance $d$ with respect to time $t$ is 5. Would you label the horizontal axis distance or time?

**61.** From any point on the graph described in Exercise 60, move right 4 units. How many units must you move up to reach the line?

EXTENDING THE IDEAS

**62.** Find $b$ if the line $y = 3x + b$ passes through the point $P(-2, 3)$.

**63.** Find $b$ if the line $y = -2x + b$ passes through the point $P(1, 4)$.

**64.** Find $m$ if the line $y = mx + 5$ passes through the point $P(-2, -4)$.

**65.** Find $m$ if the line $y = mx - 3$ passes through the point $P(4, -1)$.

**66.** If the equation $(y - b) = m(x - a)$ in point–slope form is transformed to the slope–intercept form $y = mx + k$, express $k$ in terms of $a, b$, and $m$.

The slope–intercept form of a straight line allows you to find the equation of a line if you know its slope and $y$-intercept. The point–slope form allows you to find the equation of a line if you know its slope and one point on the line. In Exercises 67–69, find the equation in standard form of the line that passes through a given set of two points.

**67.** $P(1, 2)$ and $Q(4, 1)$ (*Hint*: Find the slope of the line and use the point–slope form.)

**68.** $P(-2, 3)$ and $Q(5, -2)$

**69.** $(x_1, y_1)$ and $(x_2, y_2)$

**70.** *Tool-and-die Design.* A tool-and-die designer is completing the drawings for a new engine part. He draws a line with slope 3 through the point $P(-2, -1)$. The designer needs to find the $x$-intercept of this line. What is it?

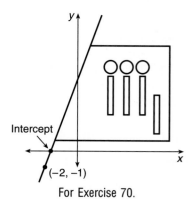

For Exercise 70.

**71.** *Navigation.* A navigator is plotting a course beginning at the point $P(-3, 2)$ of a coordinate system. His final destination is a location that is 200 miles east and 400 miles south of his starting point. If L is the line from $P$ to the final destination, find the point on L with coordinates $(x, 0)$.

**72.** *Using Algebra in Geometry.* Consider $\triangle ABC$ where $A(1, 7)$, $B(2, -2)$, and $C(-3, 2)$.

a) Use slope to show that $\triangle ABC$ is a right triangle.

b) Use the distance formula to show that $\triangle ABC$ is a right triangle.

LOOKING BACK—MAKING CONNECTIONS

**73.** Find a graph of the line through the points $(2, 5)$ and $(19, 23)$ in the Integer window. Use the TRACE feature to confirm that the two given points are on the line.

**74.** *Depreciation.* A piece of equipment with an initial value of $\$35,000$ depreciates over 15 yr and has a salvage value of $\$2500$. In determining the company's tax obligation, an accountant uses **straight line depreciation**. The line that passes through the points $(0, 35000)$ and $(15, 2500)$ describes this depreciation. Find the equation of this line in standard form.

**75.** Find the graph in the Integer window of a line with slope $m = -3/4$. Move the cursor from a point on the line down 3 units and right 4 units to another point on the line. Also move the cursor from a point on the line up 3 units and left 4 units to another point on the line. Explain why this illustrates that $\dfrac{-3}{4} = \dfrac{3}{-4} = -\dfrac{3}{4}$.

## 2.5   Functions and Graphs of Functions

Functions in Table Form ▪ Functions on a Grapher ▪ Function Notation ▪ Linear Functions ▪ Graph of a Function ▪ Finding a Complete Graph of a Functihon ▪ Finding the Domain and Range of a Function ▪ Equations That Do Not Determine Functions ▪ Looking Ahead—The Power of the Grapher

In the **Bookstore Problem Situation** discussed in Chapter 1, the retail price $R$ could be found by multiplying the wholesale price $x$ by 1.2. That is,

$$R = 1.2x.$$

Given any wholesale value $x$, there is a unique retail value $1.2x$. Think of the $x$ as an "input" value and the $R = 1.2x$ as the "output" value.

| input | $x$ | 5 | 10 | 15 | 20 | 25 | ... | $x$ |
|---|---|---|---|---|---|---|---|---|
| output | $R$ | 6 | 12 | 18 | 24 | 30 | ... | $1.2x$ |

A significant property of the algebraic representation $R = 1.2x$ is that each input value $x$ has *only one* output value $1.2x$. When that property is true of an algebraic representation, we call the representation a **function**.

---

**DEFINITION 2.2** Function

A **function** $f(x)$, read "$f$ of $x$," is a pairing of each number $x$ (or input) with a unique number $f(x)$ (or output). A function is determined by the ordered pairs $(x, f(x))$. We call $f(x)$ **the image of $x$** under $f$, or the **value of $f(x)$**.

The **domain** of the function consists of all input values and the **range** of the function consists of all output values.

---

## Functions in Table Form

Functions are encountered routinely in our daily lives in tabular form: height and weight charts, time and temperature charts, price and sales tax charts, income tax tables, and so forth. In each case for each input value, there is a unique output value. For example, the Fahrenheit to Celsius temperature conversion table shown next represents a function whose equation is $C = \dfrac{5}{9}(F - 32)$.

| Input <br> °F | Output <br> (5/9)(°F − 32) |
|---|---|
| −50 | −45.6 |
| −40 | −40 |
| −20 | −28.9 |
| 0 | −17.8 |
| 32 | 0 |
| 50 | 10 |
| 70 | 21.1 |
| 90 | 32.2 |

For each input temperature value, there is one and only one output temperature value. The entries in the first column of the table are the *domain* and the entries in the second column are the *range* of the function.

## Functions on a Grapher

Functions also are found on a grapher or a scientific calculator. For example, the calculator keys

$$\boxed{x^2} \quad \boxed{\sqrt{\phantom{x}}} \quad \boxed{x^{-1}}$$

$\sqrt{2}$

1.414213562

$\sqrt{4}$

2

$\sqrt{5}$

2.236067977

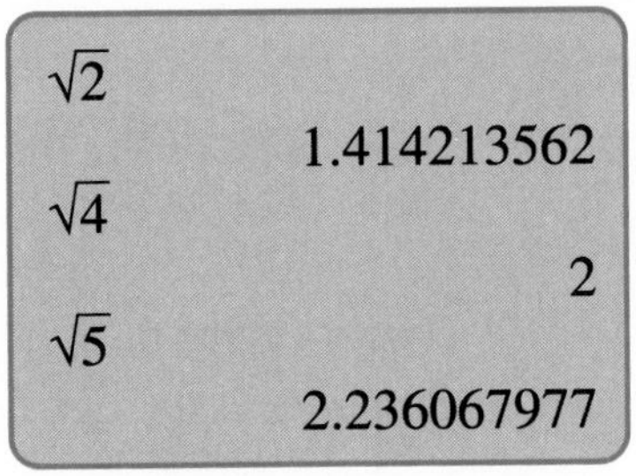

**FIGURE 2.46** The input value 2 has the output value $\sqrt{2}$, which is approximately 1.414213562.

each represent a function. For each *input* value $x$, there is only function key has been pressed.

For example, Figure 2.46 shows that for the function $y = \sqrt{x}$, the input value 2 has the output value 1.414213562. We will study many of the functions on a grapher more thoroughly later in this book.

## Function Notation

Many equations define functions; for example, the following:

$$y = x^2 \qquad y = \sqrt{x} \qquad y = 4x^2 + x - 2$$

For each of these equations, each value of $x$ results in one and only one value of $y$—the condition required of a function.

If $x$ is an input value of a function, the output value that corresponds to $x$ is denoted by $f(x)$. The notation and terminology used for the function $f(x) = 4x^2 + x - 2$ is summarized as follows:

| Input | Output | Function Rule |
|-------|--------|---------------|
| $x$ | $f(x)$ | $f(x) = 4x^2 + x - 2$ |

The symbol $f(x)$ is read "$f$ of $x$" and is called the **value of $f$ at $x$**. This value $f(x)$ is the $y$-value that corresponds to the value of $x$, so $y = f(x)$. To find the values of $f$ at $x = -1$, substitute $-1$ for $x$; for example,

$$f(-1) = 4(-1)^2 + (-1) - 2$$
$$= 4(1) - 1 - 2$$
$$= 4 - 3$$
$$= 1$$

In a similar fashion, we can find that $f(3) = 37$.

Note that we can use our grapher to evaluate $f(x)$ if $x$ is a specific constant value.

**REMINDER**

We have been studying functions in many examples in Chapters 1 and 2, so they are more familiar than you might expect. For example, the function $f(x) = 4x^2 + x - 2$ could have been written as $y = 4x^2 + x - 2$.

**EXAMPLE 1**    Finding Functional Values

Let $f$ be the function defined by the equation $f(x) = x^2 + 1$. Find the following:
  a) $f(-2)$
  b) $f(3)$
  c) $f(a)$
  d) $f(x - 3)$

Support parts (a) and (b) with a grapher.

SOLUTION

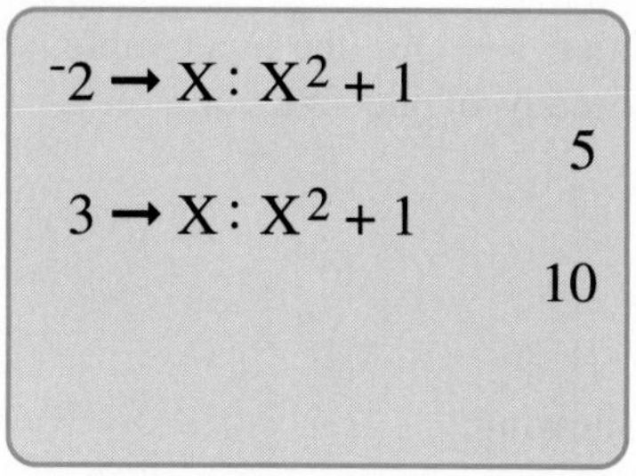

$$-2 \rightarrow X : X^2 + 1$$
$$5$$
$$3 \rightarrow X : X^2 + 1$$
$$10$$

a)    $f(-2) = (-2)^2 + 1 = 5$    **Replace x with −2.**

b)    $f(3) = (3)^2 + 1 = 10$

c)    $f(a) = a^2 + 1$    **Replace x with a.**

d)    $f(x - 3) = (x - 3)^2 + 1 = x^2 + 2x(-3) + (-3)^2 + 1$    **Replace x with x − 3 and simplify.**

$$= x^2 - 6x + 10$$

Figure 2.47 supports the answers in parts (a) and (b).    ■

**FIGURE 2.47**   Finding $f(-2)$ and $f(3)$ for $f(x) = x^2 + 1$.

Figure 2.47 supports our algebraic work and shows grapher output for $f(-2)$ and $f(3)$. It is also important to observe that most graphers do only numerical calculations. Usually they cannot find the algebraic expression $f(a)$.

## Linear Functions

In Section 2.4, we studied the slope–intercept form $y = mx + b$. This is the form that defines a linear function.

---

■                                                                              ■

**DEFINITION 2.3**   Linear Function

A function $y = f(x)$ is a **linear function** if it can be written in the form

$$f(x) = mx + b,$$

where $m \neq 0$ and $m$ and $b$ are real-number constants.

■                                                                              ■

---

We can tell from this definition that any linear equation of the form $y = mx + b$ determines a unique linear function. Its graph is a line with slope $m$ and $y$-intercept of $b$.

The variable $x$ can be any value, so the domain of a linear function is all real numbers. Also, the set of output values $f(x)$ covers all real-number values, so the range of a linear function also is all real numbers.

---

### Domain and Range of a Linear Function

If $f(x) = mx + b$, where $m$ and $b$ are real numbers, $m \neq 0$, then

$$\text{domain of } f = (-\infty, \infty) \quad \text{and} \quad \text{range of } f = (-\infty, \infty).$$

---

Example 2 illustrates that a linear equation in standard form can be changed to an equivalent form that defines a linear function.

**EXAMPLE 2**    Finding a Linear Function

Show that the equation $9x - 6y - 14 = 0$ determines a linear function by showing that it is equivalent to an equation in the form $y = mx + b$.

SOLUTION

$$9x - 6y - 14 = 0$$

$$-6y = -9x + 14$$

$$y = \frac{-9}{-6}x + \frac{14}{-6}$$

$$y = \frac{3}{2}x - \frac{7}{3}$$

Therefore the equation $9x - 6y - 14 = 0$ determines the function $f(x) = \frac{3}{2}x - \frac{7}{3}$. ∎

## Graph of a Function

The **graph of a function** consists of all points in the rectangular coordinate plane whose $x$-coordinate represents an input value and whose $y$-coordinate represents the corresponding output value of the function. In other words,

$$\text{graph of } f = \{(x, y) \quad \text{such that} \quad y = f(x)\}.$$

A hand-sketch of the graph of the function $f(x) = x^2 - 3x$ can be found using the point-plotting method described in Section 2.2. Example 3 illustrates how to find such a hand-sketch.

**EXAMPLE 3** Sketching a Graph of a Function in Two Variables

Find a table of solution pairs and sketch by hand a graph of the function $f(x) = x^2 - 3x$.

SOLUTION

1. Complete a table of solution pairs.

| $x$ | $-2$ | $-1$ | 0 | 1 | 2 | 3 | 4 | 5 |
|---|---|---|---|---|---|---|---|---|
| $f(x) = x^2 - 3x$ | 10 | 4 | 0 | $-2$ | $-2$ | 0 | 4 | 10 |

2. Plot these solution pairs on a coordinate plane as shown in Figure 2.48 (a).

3. It can be shown that a complete graph is the smooth curve suggested by Figure 2.48 (b).

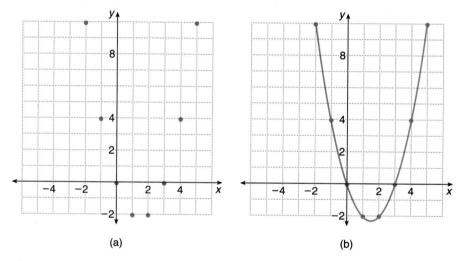

(a)                    (b)

**FIGURE 2.48** The point-plotting method for hand-sketching the graph of the function $f(x) = x^2 - 3x$.

This same function can be graphed on a grapher. The method is identical to that used to graph linear functions. Figure 2.49 (a) shows the function defining menu and Figure 2.49 (b) shows the completed graph. The Standard window is often a convenient one to use when the function is not linear.

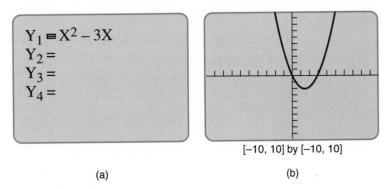

(a)

[−10, 10] by [−10, 10]

(b)

**FIGURE 2.49**  Graph of $f(x) = x^2 - 3x$ in the Standard window.

## Finding a Complete Graph of a Function

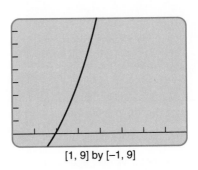

[1, 9] by [−1, 9]

**FIGURE 2.50**  This is a part of the graph of $f(x) = x^2 - 3x$.

When finding the graph of a function on a grapher, ensure the window selected shows all the important characteristics of the graph. Notice that if the graph of $f(x) = x^2 - 3x$ is found in the window [1, 9] by [−1, 9], as shown in Figures 2.50 and 2.51, important characteristics of the graph will not be seen.

A graph of a function is called a **complete graph** when all the important characteristics of the graph can be seen. In particular a complete graph will show all the $x$-intercepts of the function. With experience you will learn how to select windows that show a complete graph of a given function. Example 4 asks that a complete graph be selected.

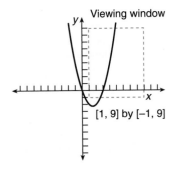

**FIGURE 2.51**  The window [1, 9] by [−1, 9] does not show all the important characteristics of the graph of $f(x) = x^2 - 3x$.

**EXAMPLE 4**    Identifying a Complete Graph with a Grapher

Graph the function $f(x) = x^2 + 10$ in each of the following windows:

a) $[-5, 5]$ by $[-5, 5]$
b) $[-10, 10]$ by $[-10, 10]$
c) $[-10, 10]$ by $[-50, 50]$

Which window shows the most likely complete graph?

SOLUTION

The window $[-5, 5]$ by $[-5, 5]$ includes no points of a graph, and the $[-10, 10]$ by $[-10, 10]$ window includes only one point of a graph. The best window is $[-10, 10]$ by $[-50, 50]$ (see Figure 2.52).

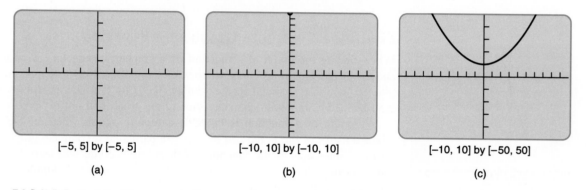

| $[-5, 5]$ by $[-5, 5]$ | $[-10, 10]$ by $[-10, 10]$ | $[-10, 10]$ by $[-50, 50]$ |
| :---: | :---: | :---: |
| (a) | (b) | (c) |

**FIGURE 2.52**  Three views of $f(x) = x^2 + 10$: (a) no points visible; (b) one point visible; and (c) a complete graph.

Complete graphs of functions like $f(x) = x^2 + 10$ will be discussed again in Chapter 8. We ask you to accept on faith now that Figure 2.52 (c) shows a complete graph of $f(x) = x^2 + 10$; that is, the graph has no other important behavior. Finding and confirming complete graphs for many functions is the role of precalculus and calculus.

## Finding the Domain and Range of a Function

Recall that the domain of the function $f$ consists of the set of real numbers for which the value $f(x)$ is defined. In the case of $f(x) = x^2 - 3x$, the variable $x$ can represent any real number, so the domain of $f$ consists of all real numbers. On the other hand, in the case of $g(x) = \sqrt{x}$ the variable $x$ cannot be a negative number, so the domain of $g$ consists of all nonnegative real numbers.

The range of a function consists of all output values; that is, all $f(x)$ values. The range can be determined from a complete graph of the function to the extent that the graph suggests all the $y$-coordinates of points on the graph.

Example 5 illustrates how to find the domain and range of a function.

**EXAMPLE 5**   Finding the Domain and Range of a Function

Use algebraic methods to find the domain and range of the function $f(x) = \sqrt{x - 3}$. Express each in interval notation and provide graphical support.

SOLUTION

**Solve Algebraically:** It is not possible to find the square root of a negative number. Therefore the expression under the radical must be greater than or equal to zero.

$$x - 3 \geq 0 \quad \text{The expression under a square root must be nonnegative.}$$

$$x \geq 3$$

The domain of $f$ is the interval $[3, \infty)$.

The expression $\sqrt{x - 3}$ can assume any positive value or zero. Therefore the range of $f$ is all positive real numbers or zero, which is the interval $[0, \infty)$.

**Support Graphically:** Figure 2.53 shows a complete graph of $f(x) = \sqrt{x - 3}$ in the Integer window. Notice that as you move the trace cursor along the curve, there is no $y$-coordinate defined when $x < 3$. Furthermore, the graph lies entirely on or above the $x$-axis. As you trace, the $y$-values continue increasing as you move to the right. These two observations support that the domain and range are as stated above, as follows:

$$\text{domain of } f = [3, \infty) \qquad \text{range of } f = [0, \infty)$$

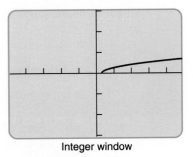

Integer window

**FIGURE 2.53** Graph of $f(x) = \sqrt{x - 3}$.

Example 6 provides an example of finding the range of a function graphically.

### EXAMPLE 6    Finding the Range

Find complete graphs of $f(x) = x^2 - 4$ and $g(x) = |x^2 - 4|$. Use the graphs to find the range of each function. Confirm the solution algebraically.

SOLUTION

**Solve Graphically:** Figure 2.54 shows complete graphs of $f$ and $g$ in the Standard window.

**REMINDER**

You can graph functions $f$ and $g$ in other windows and never find behavior other than that suggested in Figure 2.54.

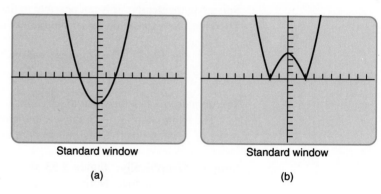

Standard window

(a)

Standard window

(b)

**FIGURE 2.54**   (a) $f(x) = x^2 - 4$ and (b) $g(x) = |x^2 - 4|$.

All numbers $y$ such that $y \geq -4$ appear as $y$-coordinates of points on the graph of $f$. Therefore

$$\text{range of } f = [-4, \infty).$$

All points on the graph of $g$ lie on or above the $x$-axis. Therefore

$$\text{range of } g = [0, \infty).$$

**Confirm Algebraically:**

$$x^2 \geq 0 \qquad \textbf{The square of any number is nonnegative.}$$

$$x^2 - 4 \geq 0 - 4$$

$$x^2 - 4 \geq -4$$

The range of $f = [-4, \infty)$. Because $g(x) = |x^2 - 4| \geq 0$, we see that the range of $g = [0, \infty)$.   ■

Notice that the absolute value of any expression is positive or zero; the graph of $g(x) = |x^2 - 4|$ is always on or above the $x$-axis. The absolute value symbols in $g$ have the effect of reflecting the part of the graph of $f$ that lies below the $x$-axis to above the $x$-axis. Example 6 leads to the following general principle.

---

### Graph of Absolute Value

To obtain a complete graph of $y = |f(x)|$ from a complete graph of $y = f(x)$, reflect through the $x$-axis the portion of the graph of $y = f(x)$ that lies below the $x$-axis.

---

## Equations That Do Not Determine Functions

Not all equations in variables $x$ and $y$ determine a function of $x$. For example, the equation in Example 7 does not determine a function $y = f(x)$.

**EXAMPLE 7**  Determining When an Equation Is Not a Function of $x$

Show that $y^2 = x$ does not determine a function of $x$.

SOLUTION
By substituting values for $x$, in the equation $y^2 = x$, we can find values for $y$ as indicated in this table.

| $x$ | 0 | 0.5 | 1 | 1.5 | 2 | 2.5 | 3 | 4 | 5 |
|-----|---|-----|---|-----|---|-----|---|---|---|
| $y$ | 0 | $\pm 0.7$ | $\pm 1$ | $\pm 1.2$ | $\pm 1.4$ | $\pm 1.6$ | $\pm 1.7$ | $\pm 2$ | $\pm 2.2$ |

Because there are two $y$-values corresponding to all nonzero $x$-values, this equation does not meet the condition for being a function of $x$. The graph of the equation $y^2 = x$ is shown in Figure 2.55. ∎

It is clear visually that both values $y = 1$ and $y = -1$ are associated with the value $x = 1$. This is because the vertical line $x = 1$ intersects the graph of $y^2 = x$ at the two points $(1, 1)$ and $(1, -1)$. Consequently, the equation $y^2 = x$ does not determine a function(see Figure 2.55). In general, an equation does not determine a function of $x$ if there exists at least one vertical line that intersects the graph of the equation in more than one point. This observation leads to the **vertical line test** for functions.

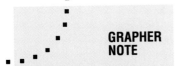

**GRAPHER NOTE**

Consult your grapher lab manual to learn how to produce the graph of $y^2 = x$ on a grapher.

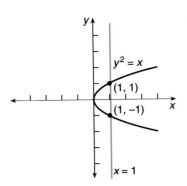

**FIGURE 2.55** Testing whether $y^2 = x$ is a function using $x = 1$.

### Vertical Line Test for a Function

If each vertical line intersects the graph of an equation in no points or in exactly one point, then the equation determines a function of $x$.

**EXAMPLE 8**    Using the Vertical Line Test

Which of the following are graphs of functions?

SOLUTION

All four graphs in Figure 2.56 satisfy the vertical line test except (b). Therefore only (b) is not the graph of a function of $x$.

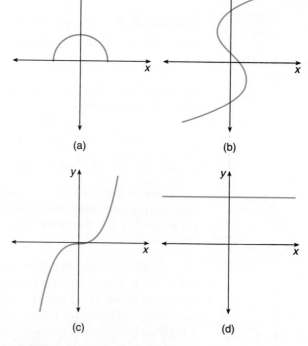

(a)

(b)

(c)

(d)

**FIGURE 2.56**  Which graphs satisfy the vertical line test?

## Looking Ahead—The Power of the Grapher

You will learn throughout this textbook that graphers can be used to find complete graphs of functions that would be impossible or impractical to find by paper-and-pencil point-plotting methods. Example 9 gives us a glimpse of this power and shows a complete graph of the function $f(x) = x^4 - 3x^2 + 5x - 6$ in the window $[-5, 5]$ by $[-20, 20]$.

**EXAMPLE 9** Finding the Range with a Grapher

A complete graph of the function $f(x) = x^4 - 3x^2 + 5x - 6$ can be seen in the window $[-5, 5]$ by $[-20, 20]$. Use this graph to estimate the range of the function $f$.

SOLUTION

Figure 2.57 shows a complete graph of the function $f(x) = x^4 - 3x^2 + 5x - 6$ in the window $[-5, 5]$ by $[-10, 20]$. By moving the trace cursor to the lowest point of the graph, we estimate that the $y$-coordinates of the points of the graph include all values greater than approximately $y = -15.2$. We estimate that the range of $f$ is the interval $[-15.2, \infty)$.

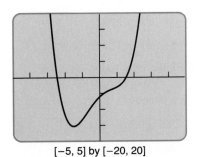

$[-5, 5]$ by $[-20, 20]$

**FIGURE 2.57** A complete graph of $f(x) = x^4 - 3x^2 + 5x - 6$. ■

We might guess that the graph in Figure 2.57 is a complete graph. The algebraic confirmation of this is indeed the role of calculus. All we can do with a grapher is make a reasonable conjecture that we have found a complete graph.

## Exercises for Section **2.5**

The functions $f(x) = 2x + 3$, $g(x) = 1/(x - 1)$, and $h(x) = x^2 - 5$ apply to Exercises 1–12. Find:

**1.** $f(0)$ and $f(1)$ 

**2.** $g(-2)$ and $g(-1)$

**3.** $f(3)$, $g(3)$ and $h(3)$ 

**4.** $f(2)$, $g(2)$ and $h(2)$

**5.** $h(5) + h(17)$ 

**6.** $g(5) + g(1/5)$

**7.** $f(-2) - f(3)$ 

**8.** $f(2) \cdot g(2) - h(1/5)$

**9.** $f(a)$ 

**10.** $g(b)$

**11.** $f(a + 2)$ and $f(a) + 2$

**12.** $g(x + 3)$ and $g(x) + 3$

In Exercises 13–18, change the linear equation in standard form to the form $f(x) = mx + b$ and find the graph of the linear equation in the Integer window. Move the trace cursor until the graph scrolls off the screen to convince yourself that the domain and range of the linear function are all real numbers.

**13.** $3x - 5y = 12$ 

**14.** $2x - y = 17$

**15.** $4x + 12y = 144$ 

**16.** $9x - 18y - 130 = 0$

**17.** $6y - 2x = 17$ 

**18.** $5y + 12x - 32 = 0$

In Exercises 19–26, find a table of solution pairs for $x = -3, -2, -1, 0, 1, 2,$ and 3 and sketch by hand a complete graph of the given function.

**19.** $f(x) = 3x - 7$ 

**20.** $g(x) = -2x + 11$

**21.** $g(x) = 14 - 3x$ 

**22.** $f(x) = 9 - 2x$

**23.** $h(x) = x^2 - 3$ 

**24.** $f(x) = 6 - 2x^2$

**25.** $g(x) = 7 - x + x^2$ 

**26.** $h(x) = -x^2 + 5$

In Exercises 27–32, find the graph of the given function in each of the specified windows. Choose the window that shows what is most likely a complete graph of the given function.

**27.** Graph $f(x) = 0.5x - 13$ in (a) $[-10, 10]$ by $[-10, 10]$, (b) $[-100, 100]$ by $[-100, 100]$, and (c) $[-5, 5]$ by $[-10, 10]$. Which window shows a complete graph of $f$?

**28.** Graph $f(x) = 11 - 0.3x$ in (a) $[-10, 10]$ by $[-10, 10]$, (b) $[-100, 100]$ by $[-100, 100]$, and (c) $[-5, 5]$ by $[-20, 10]$. Which window shows a complete graph of $f$?

**29.** Graph $f(x) = x^2 - 13$ in (a) $[-250, 250]$ by $[-100, 100]$, (b) $[-100, 100]$ by $[-100, 100]$, and (c) $[-5, 5]$ by $[-10, 10]$. Which window shows a complete graph of $f$?

**30.** Graph $f(x) = (x - 8)^2 - 4$ in (a) $[-10, 10]$ by $[-10, 10]$, (b) $[-5, 5]$ by $[-10, 100]$, and (c) $[-5, 20]$ by $[-10, 100]$. Which window shows a complete graph of $f$?

**31.** Graph $f(x) = \sqrt{8 - x}$ in (a) $[-5, 5]$ by $[-5, 5]$, (b) $[-30, 30]$ by $[-10, 10]$, and (c) $[-100, 100]$ by $[-5, 5]$. Which window shows a complete graph of $f$?

**32.** Graph $f(x) = 15\sqrt{x + 12}$ in (a) $[-10, 10]$ by $[-10, 10]$, (b) $[-10, 10]$ by $[-10, 100]$, and (c) $[-15, 15]$ by $[-10, 100]$. Which window shows a complete graph of $f$?

In Exercises 33–36, consider the graph of the function $y = f(x)$ given in the accompanying figure. Estimate the indicated values. (Each tick mark represents a value of 1.)

**33.** $f(0)$, $f(-1)$, and $f(4)$

**34.** $x$ if $f(x) = 0$

**35.** $x$ if $f(x) = 2$ 

**36.** $x$ if $f(x) = -8$

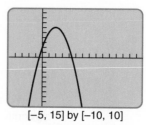

$[-5, 15]$ by $[-10, 10]$

For Exercises 33–36.

In Exercises 37 and 38, determine the domain and the range by studying the graphs.

**37.**

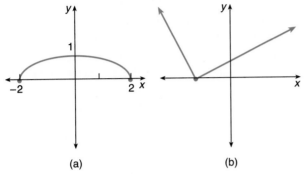

(a)                                    (b)

For Exercise 37.

**38.**

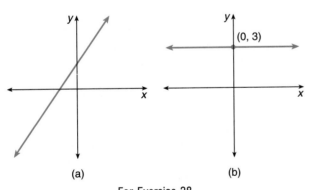

(a)                                    (b)

For Exercise 38.

In Exercises 39–44, determine the domain of the given function.

**39.** $f(x) = x^2 + 1$         **40.** $g(x) = 2 - x$

**41.** $f(x) = 3x - 17$       **42.** $g(x) = 9 - 2x^2$

**43.** $f(x) = 5\sqrt{x}$          **44.** $g(x) = \sqrt{x - 1}$

In Exercises 45–52, find the graph of the given function in the Integer window and use the graph and the TRACE feature to find the function's range.

**45.** $f(x) = x^2 - 5$         **46.** $f(x) = x^2 + 8$

**47.** $g(x) = \sqrt{x}$          **48.** $g(x) = \sqrt{x + 3}$

**49.** $h(x) = 5x - 2$        **50.** $h(x) = -13 + 4x$

**51.** $f(x) = |2x - 4|$      **52.** $h(x) = |3 - 0.5x|$

In Exercises 53–60, find both the domain and range of the given function. (Use either algebraic or graphical methods.)

**53.** $f(x) = |x - 3|$        **54.** $g(x) = \sqrt{x} - 4$

**55.** $h(x) = \sqrt{x + 3}$     **56.** $f(x) = x^2 + 4$

**57.** $f(x) = 3x^2 - 1$       **58.** $g(x) = x^2 - 7$

**59.** $f(x) = |x^2 - 5|$        **60.** $g(x) = |4 - x^2|$

In Exercises 61–66, complete a table of solution pairs for the given equation for $x = 0, 1, 2, 3,$ and 4. Decide whether the equation determines a function of $x$.

**61.** $y^2 = 2x$              **62.** $x^2 = y$

**63.** $x^2 = 2y$             **64.** $3y^2 = x$

**65.** $x = y^2 - 3$          **66.** $x + 2 = y^2$

In Exercises 67–69, use the vertical line test to decide which are graphs of functions of $x$.

**67.**

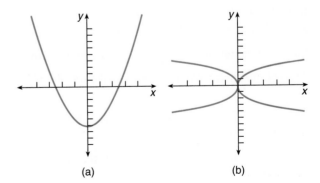

(a)                                    (b)

**68.**

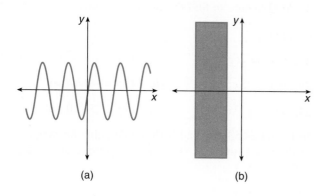

(a)                              (b)

**69.** .

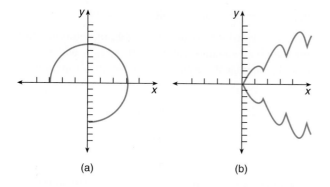

(a)                              (b)

TRANSLATING WORDS TO SYMBOLS

In Exercises 70–73, describe a function $f(x)$ as an expression in $x$ under the given condition.

**70.** The function adds three to the square of its input $x$.

**71.** The function subtracts five from the double of its input $x$.

**72.** The function's output is the input $x$ times two more than its input $x$.

**73.** The function's output is five more than the square of the input $x$.

**74. Writing to Learn.** Explain the difference between writing "the absolute value of a difference" and writing "the difference of the absolute values."

**75.** *Analyzing a Family of Rectangles.* A rectangle is 20 ft longer than it is wide. Let $x$ represent the width and $P(x)$ the perimeter of the rectangle.

a) Write an algebraic expression for the length of the rectangle if $x$ represents its width.

b) Write an algebraic expression for the perimeter $P(x)$ as a function of $x$.

c) Find a complete graph of the algebraic representation $P(x)$ for the perimeter of the rectangle.

d) Find the width of the rectangle if the perimeter is 150 ft.

EXTENDING THE IDEAS

In Exercises 76–81, determine the domain of the given function.

**76.** $f(x) = \dfrac{5}{x-3}$

**77.** $g(x) = \dfrac{3}{x^2-1}$

**78.** $f(x) = |x+2|$

**79.** $g(x) = \dfrac{x+3}{|x+2|}$

**80.** $f(x) = \dfrac{\sqrt{x+2}}{x-1}$

**81.** $g(x) = \dfrac{x+3}{x^2-3x}$

In Exercises 82–87, find the graph of the given function in the window $[-5, 5]$ by $[-20, 20]$ and estimate the range of the function.

**82.** $f(x) = 3x^2 - 5x - 12$

**83.** $f(x) = 2x^2 + 3x - 9$

**84.** $f(x) = -2x^2 + 3x + 7$

**85.** $f(x) = x^4 - 5x^2 + 3x - 7$

**86.** $f(x) = x^4 - 6x^2 + 2$

**87.** $f(x) = -x^4 + 6x^2 + 3x + 2$

**88.** One plumber charges a flat fee of $35 plus $20/hr. A second plumber charges $25/hr.

a) Describe the algebraic representation of the cost $C_1(t)$ of hiring plumber number one as a function of $t$.

b) Describe the algebraic representation of the cost $C_2(t)$ of hiring plumber number two as a function of $t$.

c) Find a complete graph of both functions $C_1(t)$ and $C_2(t)$ in the same window. What do the coordinates of the point of intersection of these two graphs tell us about this problem?

**89. Writing to Learn.**

a) The range of the function $f(x) = 3x - 7$ is all real numbers. Suppose $a$ is any real number. Find a real number $x$ such that $f(x) = a$.

b) Write a paragraph explaining why part (a) is a proof that the range is $(-\infty, \infty)$.

LOOKING BACK—MAKING CONNECTIONS

**90.** The base angles of an isosceles triangle have equal measures. If these base angles are twice the size of the third angle, how many degrees are there in each angle? (Recall that the sum of the degree measures of the angles of a triangle is $180°$.)

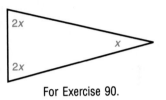

For Exercise 90.

---

| Chapter 2 Summary |
| --- |

|  | Rectangular Coordinate Plane | Examples |
| --- | --- | --- |
| Coordinate Plane<br><br>$x$-axis<br><br>$y$-axis<br><br>Origin<br><br>Quadrants I to IV | 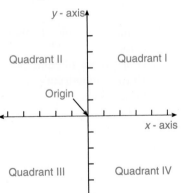 | |

| | **Rectangular Coordinate Plane** | **Examples** |
|---|---|---|
| Coordinates of a Point | The $x$-coordinate of point $P(a, b)$ is $a$ and the $y$-coordinate is $b$. | The $x$-coordinate is 3 and the $y$-coordinate is $-4$ for $P(3, -4)$. |
| Distance Formula | $d = \sqrt{(x_1 - x_2)^2 + (y_1 - y_2)^2}$ | Given $P(-1, 3)$ and $Q(2, 5)$, $$d = \sqrt{(-1-2)^2 + (3-5)^2}$$ $$= \sqrt{13}.$$ |

| | **Graphs of Equations in Two Variables** | **Examples** |
|---|---|---|
| Ordered Pair | A pair of real numbers in which the order of the pair is taken into account. | $(1, -3)$ |
| $x$-intercept | Point at which the graph crosses the $x$-axis. This occurs when $y = 0$. | $2x - 3y = 6$ has $x$-intercept $(3, 0)$ and $y$-intercept $(0, -2)$. |
| $y$-intercept | Point at which the graph crosses the $y$-axis. This occurs when $x = 0$. | |
| A Solution of an Equation in Two Variables | An ordered pair $(a, b)$ that completes a true statement when substituted into the equation. | $(1, 2)$ is a solution of $3x - 4y = -5$. |
| Finding Solution Pairs | Substitute a value for one variable and calculate to find the other. | Substitute $x = 5$ into $y = 3x - 4$ and calculate $y = 3(5) - 4 = 11$. $(5, 11)$ is a solution pair. |
| Point Plotting Method of Graphing | 1. Complete a table of solution pairs. 2. Graph these solution pairs. 3. Draw a smooth curve through these points. | 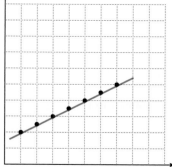 |

| | **Graphs of Equations in Two Variables** | **Examples** |
|---|---|---|
| Grapher Method of Graphing | Enter the appropriate equation into the grapher and choose an appropriate window. For linear equations, the Integer window is usually used. | Integer window |

| | **Linear Equations** | **Examples** |
|---|---|---|
| Slope of a Line | $m = \dfrac{y_2 - y_1}{x_2 - x_1} = \dfrac{\text{change in } y}{\text{change in } x}$ | |
| Standard Form | $Ax + By = C$, where at least one of $A$ or $B$ is not zero. | $5x - 2y = 8$ |
| Slope–Intercept Form | $y = mx + b$, where $m$ is the slope and the $y$-intercept is $(0, b)$. | $y = -2x + 6$, given slope $m = -2$ and $y$-intercept $(0, 6)$. |
| Point–Slope Form | $y - b_1 = m(x - a_1)$, where $m$ is slope and $(a_1, b_1)$ is a given point on the line | $y - 2 = 6(x + 3)$, slope $m = 6$ and $P(-3, 2)$ is on the line. |
| Intercept Method of Graphing a Linear Equation | 1. Find both $x$- and $y$-intercepts. 2. Graph the intercepts. 3. Draw a line through the two points. | |
| Horizontal line | A line with no $x$-intercept | $y = 3$ |
| Vertical line | A line with no $y$-intercept | $x = 5$ |

| | **Functions** | **Examples** |
|---|---|---|
| Definition of Function | A pairing $(x, y)$ that associates each $x$ with exactly one $y$ | $f(x) = 3x^2 - 4$ |
| Domain | All input values; all possible values of $x$ | The domain of $f(x) = \sqrt{x + 2}$ is the interval $[-2, \infty)$. |

|  | **Functions** | **Examples** |
|---|---|---|
| Range | All output values; all possible values of $f(x)$ | The range of $f(x) = |x - 3|$ is the interval $[0, \infty)$. |
| Vertical Line Test | If every vertical line intersects the graph of an equation in at most one point, then the equation determines a function of $x$. | |
| Linear Function | A function that can be written in the form $f(x) = mx + b$, where $m$ and $b$ are real numbers | $f(x) = 5x - 3$ |
| Graph of a Function | All pairs $(x, f(x))$ where $x$ is in the domain | |
| Complete Graph of a Function | A graph of the function in a window that shows all the important characteristics of the function | 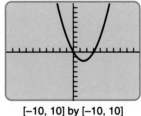 [−10, 10] by [−10, 10] |

| | **Grapher Skills** | **Examples** |
|---|---|---|
| Adjust the WINDOW values to find the window of your choice. | Select values for $X$min, $X$max, $Y$min, $Y$max, $X$scl, and $Y$scl | |
| Setting $X$scl and $Y$scl | Select the scale values so that tick marks show on the axes as desired. | |
| The Integer Window | The window centered at $(0, 0)$ with Integer screen coordinates. Window settings vary from one grapher model to another. | On the grapher we use, the window settings are $[-47, 47]$ by $[-31, 31]$. |
| The 0.1 Window | The window centered at $(0, 0)$ with screen coordinates in tenths. Window settings vary from one grapher model to another. | On the grapher we use, the window settings are $[-4.7, 4.7]$ by $[-3.1, 3.1]$. |

| | **Grapher Skills** | **Examples** |
|---|---|---|
| The Standard Window | The rectangular portion of the coordinate plane with both $x$- and $y$-coordinates between $-10$ and $10$ | Standard Window window settings: $[-10, 10]$ by $[-10, 10]$ |
| Finding the Slope of a Line Using a Grapher | 1. Find a graph of the line.<br><br>2. Use an Integer Window and the TRACE feature to find the $y$-coordinate of one pixel minus the $y$-coordinate one pixel to the left. | |
| Finding the Range of a Function | Find a complete graph and use the TRACE feature to identify the range. | |

## Review Exercises for Chapter 2

**1.** Use paper and pencil to plot the points $Q(5, -3)$, $R(-4, 0)$, and $S(-2, 1)$ on a rectangular coordinate plane.

**2.** State the coordinates of the points in each corner of the grapher screen for the window $[-5, 15]$ by $[0, 150]$.

**3.** Find the distance between the points $(2, -7)$ and $(-8, -3)$.

**4.** Find the distance between the points $(-3, 1)$ and $(0, 5)$.

In Exercises 5 and 6, find the coordinate that makes the ordered pair a solution pair for the given equation.

**5.** $5x + 4y = 16$;  $(3, ?)$   **6.** $y = 6x + 29$;  $(?, 11)$

**7.** Complete a table of solution pairs for the equation $84x + 105y = 840$, using the values $x = -10, -5, 0$,

5, and 10. Sketch the graph of the given equation on graph paper.

In Exercises 8 and 9, use a grapher to find the graph of each equation in the Integer window. Sketch the graph produced by your grapher on a copy of the screen like the one shown below.

**8.** $3x - 4y = 18$

**9.** $15x + 2y = 14$

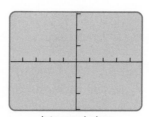

Integer window

For Exercises 8 and 9.

In Exercises 10 and 11, use an algebraic method to find the $x$- and $y$-intercepts and graph each linear equation.

10. $3x + 4y + 84 = 0$      11. $18x - 10y + 45 = 0$

12. Sketch the graph of $x = -5$.

13. Sketch the graph of $y = 2$.

14. Find the slope of the line $16x - 11y = 25$.

15. Find the slope of the line $2x + 9y = 14$.

16. Find the slope of the line determined by points $(-2, 4)$ and $(3, 1)$.

17. Find the slope of the line determined by points $(12, -1)$ and $(2, -8)$.

18. Find an equation of the horizontal line that passes through the point $(6, -6)$.

19. Sketch the line with slope $m = \dfrac{1}{4}$ that passes through the point $P(3, -2)$.

20. Sketch the line with slope $m = -\dfrac{2}{3}$ that passes through the point $P(-2, -3)$.

21. Determine whether $5x - 15y = 6$ and $6x + 2y = 5$ are perpendicular, parallel, or neither. Support your answer graphically.

22. Determine whether $3x + 6y = 10$ and $5x - 10y = 2$ are perpendicular, parallel, or neither. Support your answer graphically.

In Exercises 23–26, find an equation of the line with the given conditions. Write the equation in slope–intercept form.

23. Slope $m = \dfrac{2}{3}$ and $y$-intercept $(0, -2)$

24. Slope $m = -1.4$ and $y$-intercept $(0, 3)$

25. Slope $m = -3$ and passes through the point $(-4, 1)$

26. Slope $m = \dfrac{3}{2}$ and passes through the point $(6, -5)$

In Exercises 27–30, find an equation of the line with the given conditions. Write the equation in standard form.

27. The line passes through the points $(3, 5)$ and $(2, 7)$.

28. The line passes through the points $(-2, -1)$ and $(2, 1)$.

29. The line passes through the point $(5, 2)$ and is perpendicular to the line $x - 4y = 5$.

30. The line passes through the point $(-1, 7)$ and is parallel to the line $3x + 2y = 7$.

In Exercises 31–34, consider the graph of the function $y = f(x)$ given in the accompanying figure. Estimate the indicated values. (Each tick mark represents a value of 1.)

31. $f(0)$                    32. $f(2)$

33. $x$ if $f(x) = 7$          34. $x$ if $f(x) = 0$

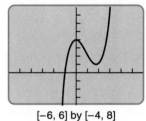

[−6, 6] by [−4, 8]
For Exercises 31–34.

35. Find $f(-2)$ and $f(a)$ if $f(x) = x^2 + 3x$.

36. Find $g(0)$ and $g(b + 1)$ if $g(x) = |2x - 1|$.

37. Graph $f(x) = 2x^3 - x^2 - 10$ in each of the specified windows. Choose the window that shows what is most likely a complete graph of $f$.

  a) $[-50, 50]$ by $[-20, 20]$

  b) $[-10, 10]$ by $[-10, 10]$

  c) $[-5, 5]$ by $[-20, 20]$

**38.** Graph $f(x) = |0.5x - 5|$ in each of the specified windows. Choose the window that shows what is most likely a complete graph of $f$.

   a) $[-10, 10]$ by $[-10, 10]$

   b) $[-15, 15]$ by $[-80, 80]$

   c) $[-20, 20]$ by $[-10, 10]$

**39.** Find both the domain and the range of $f(x) = \sqrt{x - 8}$.

**40.** Find both the domain and the range of $f(x) = x^2 + 2x - 15$.

**41.** Complete a table of solution pairs for the equation $x^2 = y^2$ for $x = 0, 1, 2, 3$, and 4. Decide whether the equation determines a function of $x$.

**42.** Determine whether the points $(-2, 7)$, $(0, 3)$, and $(3, -3)$ are collinear.

**43.** Find $m$ if the line $y = mx - 4$ passes through the point $(5, -7)$.

**44.** Find the slope of the line shown in each of the following grapher screens. Assume that the line passes through the points in the bottom left-hand and top right-hand corners of the screen. Which screen shows the line with the greater slope?

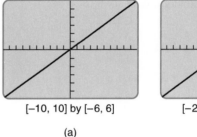

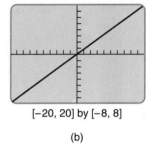

     $[-10, 10]$ by $[-6, 6]$         $[-20, 20]$ by $[-8, 8]$

          (a)                      (b)

## Chapter 2 Test

**1.** Find the distance between the points $(3, -3)$ and $(8, 1)$.

**2.** Find the value of $x$ that makes the ordered pair $(x, -5)$ a solution pair for $3x - 2y = 19$.

**3.** Sketch the graph of $2x - 5y = 20$.

**4.** Sketch the graph of $y = -1$.

**5.** Sketch the graph of the line with slope $m = -\dfrac{1}{3}$ that passes through the point $(1, 4)$.

**6.** Use an algebraic method to find the $x$- and $y$-intercepts of $2x - 5y = 35$.

**7.** Find the slope of the line that passes through the points $(-4, 5)$ and $(-1, 3)$.

**8.** Find the slope of the line that is parallel to the line $6x + 7y = 8$.

**9.** Find an equation of the vertical line that passes through the point $(-2, 1)$.

**10.** Find an equation of the line that passes through the points $(-1, 3)$ and $(0, 4)$. Write the equation in standard form.

**11.** Find an equation of the line with slope $m = 2$ and $x$-intercept $(-2, 0)$. Write the equation in slope–intercept form.

**12.** Find an equation of the line that passes through the point $(1, 2)$ and is perpendicular to the line $3x - y = 7$. Write the equation in standard form.

**13.** Using a grapher, find $f\left(\dfrac{5}{6}\right)$ if $f(x) = x^2 + 2x - 3$.

14. Find $f(a - 1)$ if $f(x) = 3x - 6$.

15. Determine both the domain and the range of the function by graphing $f(x) = -x^2 - 4x + 1$ in the Integer window.

16. Find $m$ if the line $y = mx + 6$ passes through the point $(-2, 5)$.

17. Graph the equation $15y = -25x + 75$ in the Integer window. Find the $x$- and $y$-intercepts.

18. Graph $f(x) = 3x - 40$ in each of the specified windows. Choose the window that shows what is most likely a complete graph of $f$.

a) Standard window

b) $[-10, 10]$ by $[-15, 15]$

c) $[-25, 25]$ by $[-50, 50]$

19. Graph $f(x) = x^2 - x - 12$ in each of the specified windows. Choose the window that shows what is most likely a complete graph of $f$.

a) $[-5, 5]$ by $[-5, 5]$

b) Standard window

c) Integer window

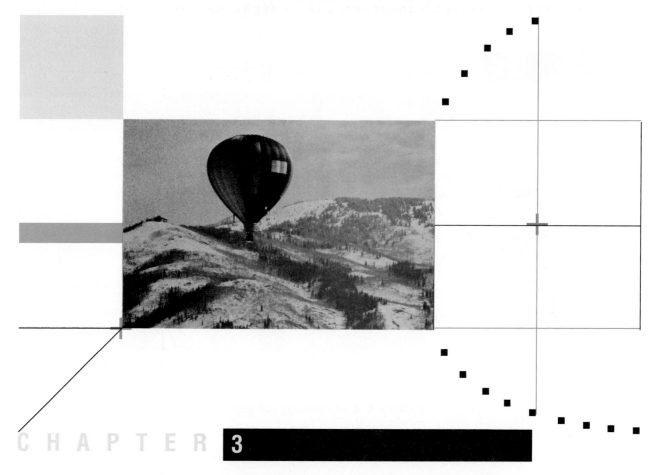

# Solving Equations and Systems of Equations

## AN APPLICATION

The captain of a hot-air balloon may drop ballast in order to increase the rate at which the balloon rises. The time $t$ that it takes for the ballast to hit the ground, if it is dropped from $d$ feet above the ground, is described by the equation

$$t = \frac{1}{4}\sqrt{d}.$$

| **3.1** | # Finding Graphical Solutions |

The Multigraph Method of Solving Linear Equations   ▪   The *x*-intercept Method of Solving Linear Equations

**GRAPHER NOTE**

Consult your grapher lab manual to see how to select and deselect graphs. The manual also explains that when two graphs are produced in the same window, the graphs can be generated either sequentially or simultaneously.

In Chapter 1, you solved equations and inequalities algebraically. In this chapter, you will use a grapher to find solutions. Figure 3.1 shows that when the equal sign is highlighted in both equations, both graphs appear in the same window.

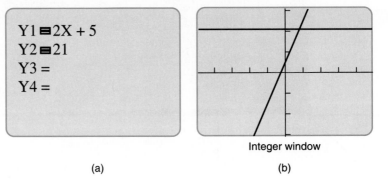

(a)                                                                          (b)

**FIGURE 3.1**   Defining and graphing both $y = 2x + 5$ and $y = 21$ simultaneously.

Use your grapher's trace feature as you complete this exploration.

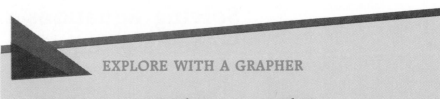

**EXPLORE WITH A GRAPHER**

| **Equation** | **Graph #1** | **Graph #2** |
| --- | --- | --- |
| $3x - 15 = 12$ | $y_1 = 3x - 15$ | $y_2 = 12$ |

1. Find the graphs of both $y_1$ and $y_2$ in the Integer window.
2. Use TRACE to find the coordinates of the point of intersection of the two graphs. How do you know that you have found the point of intersection? (Explore the impact of using the up and down arrow keys in conjunction with the right and left arrow keys.)
3. What is the relationship between the coordinates of the point of intersection and the solution to the related equation?

**Experiment**   Repeat these three steps for several equations that you write.

## The Multigraph Method of Solving Linear Equations

In the exploration you solved the equation $2x + 5 = 21$ by producing a graph of both $y_1 = 2x + 5$ and $y_2 = 21$ in the same window. In particular you identified the point where the graph of $y_1 = 2x + 5$ intersects the graph of $y_1 = 21$.

If $x_0$ is the $x$-coordinate of this point of intersection, then we can name this point in two ways. It is the point:

$(x_0, y_1)$ when viewed as a point on the graph of $y_1 = 2x + 5$,

and it is the point

$(x_0, y_2)$ when viewed as a point on the graph of $y_2 = 21$.

In other words $x_0$ is the value for which $y_1 = y_2$. Consequently $x_0$ is the solution to the equation.

Because this method requires that we find two graphs, it is called the **multigraph method**.

---

### Multigraph Method for Solving an Equation of the form $ax + b = c$

1. Find both the graphs of $y_1 = ax + b$ and $y_2 = c$ in the Integer window.
2. Use TRACE to move the cursor along the graph as close to the point of intersection $(x_0, y_0)$ as possible. (Note that the UP and DOWN keys move the cursor back and forth between the two graphs. The point of intersection has been found when the UP/DOWN arrows do not change the $y$-coordinate of the cursor.)
3. If the equation has an integer solution, the point of intersection can be found exactly using the TRACE cursor.

The value $x = x_0$ is the solution to the equation.

**GRAPHER NOTE**

In Section 3.2, grapher methods of approximation will be introduced that can be used if the solution to the equation is not an integer.

Example 1 illustrates this method.

**EXAMPLE 1**   Solve Graphically: Multigraph Method

Use the multigraph method to solve the equation $17 - 1.5x = -10$. Support this result numerically and confirm algebraically.

SOLUTION

**Solve Graphically.** Find the graph of $y_1 = 17 - 1.5x$ and $y_2 = -10$ in the Integer window. Move the trace cursor to the point of intersection of the two graphs (see Figure 3.2).

The point of intersection is $(18, -10)$ and the solution to the equation is $x = 18$.

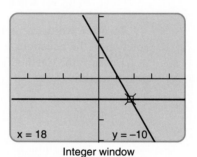

Integer window

**F I G U R E   3 . 2**   Graphs of $y = 17 - 1.5x$ and $y = -10$.

$$18 \rightarrow X : 17 - 1.5X$$
$$^-10$$

**F I G U R E   3 . 3**   $x = 18$ is a solution of the equation $17 - 1.5x = -10$.

**Support Numerically.** Store the value 18 for the variable $x$ and evaluate the expression $17 - 1.5x$, as shown in Figure 3.3.

**Confirm Algebraically.**

$$17 - 1.5x = -10$$
$$-1.5x = -10 - 17$$
$$-1.5x = -27$$
$$x = \frac{-27}{-1.5}$$
$$x = 18$$    ∎

Example 2 illustrates that the multigraph solution method can be applied to solve linear equations that have not been simplified algebraically to the form $ax + b = c$.

**EXAMPLE 2**    Solve Graphically: Multigraph Method

Use the multigraph method to solve the equation $(4x - 5) - (x + 11) = 23$. Support this result numerically and confirm algebraically.

SOLUTION

**Solve Graphically.** Figure 3.4 (a) shows the function-defining menu and Figure 3.4 (b) shows the graph of $y_1 = (4x - 5) - (x + 11)$ and $y_2 = 23$ in the Integer window. Move the trace cursor to the point of intersection of the two graphs.

The point of intersection is $(13, 23)$ and the solution to the equation is $x = 13$.

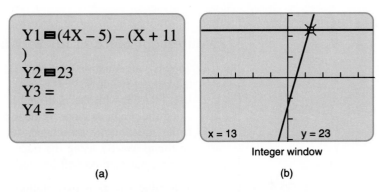

(a)

Integer window

(b)

**FIGURE 3.4**  Notice the use of parentheses on the function-defining menu in (a). In (b) we see the graphs of the two equations.

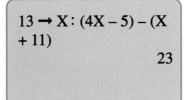

**FIGURE 3.5**  $x = 13$ is a solution of the equation $(4x - 5) - (x + 11) = 23$.

**Support Numerically.** Figure 3.5 shows numerical support of this solution on the grapher.

**Confirm Algebraically.**

$$(4x - 5) - (x + 11) = 23$$
$$4x - 5 - x - 11 = 23$$
$$3x - 16 = 23$$
$$3x = 39$$
$$x = 13$$

## The $x$-intercept Method of Solving Linear Equations

Suppose you want to solve an equation like $(5x - 12) - (3x - 4) = 0$, which has a zero on one side of the equation. To solve this equation by the multigraph method, you would graph $y_1 = (5x - 12) - (3x - 4)$ and $y_2 = 0$. However, the graph of $y_2 = 0$ is the $x$-axis. So we are interested in finding the intersection of the graph of $y_1$ and the $x$-axis. This point is the $x$-intercept of the line $y_1 = (5x - 12) - (3x - 4)$. This method of solving equations is called the **$x$-intercept method**, which we summarize next.

---

### $x$-intercept Method for Solving a Linear Equation

1. To solve an equation $ax + b = 0$, find the graph of $y = ax + b$ in the Integer window.
2. Use TRACE with the right and left arrow keys to move the cursor along the graph as close to the $x$-intercept as possible.
3. If the equation has an integer solution, the $x$-intercept $(x_0, 0)$ can be found exactly using the trace cursor. (If the solution is not an integer, our method will be modified later.)

   The value $x = x_0$ is the solution to the equation.

---

Examples 3 and 4 illustrate the $x$-intercept method.

**EXAMPLE 3**    Solve Graphically: $x$-intercept Method

Use the $x$-intercept method to solve the equation $(5x - 12) - (3x - 4) = 0$. Confirm this solution algebraically and support it numerically.

SOLUTION

**Solve Graphically.** Find the graph of the function $y = (5x - 12) - (3x - 4)$ in the Integer window. Figure 3.6 shows the trace cursor positioned at the $x$-intercept $(4, 0)$. The $x$-intercept is $x = 4$, and the solution to the equation is $x = 4$.

**FIGURE 3.6** Graph of $y = (5x - 12) - (3x - 4)$ in the Integer window.

**Support Numerically.** Figure 3.7 shows the numerical support for this solution.

$$4 \rightarrow X : (5X - 12) - (3X - 4)$$
$$0$$

**FIGURE 3.7** $x = 4$ is the solution of the equation $(5x - 12) - (3x - 4) = 0$.

**Confirm Algebraically.**

$$(5x - 12) - (3x - 4) = 0$$
$$5x - 12 - 3x + 4 = 0$$
$$2x - 8 = 0$$
$$2x = 8$$
$$x = 4 \qquad \blacksquare$$

The $x$-intercept method applies only to cases in which one side of the equation to be solved is zero. Sometimes, if this method is used, it is necessary to change the given equation to an equivalent form to obtain a zero on one side of the equation. Example 4 illustrates this method.

□ **REMINDER**

In Examples 1 to 4, the equations were solved by a graphical method, supported numerically, and confirmed algebraically. Throughout the remainder of the text, a graphical solution method will not ordinarily both be supported numerically and confirmed algebraically.

**EXAMPLE 4**    Solve Graphically: $x$-intercept Method

Use the $x$-intercept method to solve the equation $7x - 2(2x + 11) = 5$. Support the solution numerically and confirm it algebraically.

SOLUTION

Begin by subtracting 5 from both sides of the equation.

$$7x - 2(2x + 11) = 5$$
$$7x - 2(2x + 11) - 5 = 0$$

**Solve Graphically.** Figure 3.8 shows the graph of $y = 7x - 2(2x + 11) - 5$ in the Integer window and shows that $(9, 0)$ is the $x$-intercept. The solution to the equation is $x = 9$.

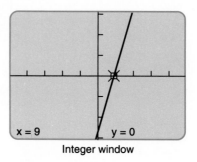

Integer window

**FIGURE 3.8** The $x$-intercept of the graph of $y = 7x - 2(2x + 11) - 5$ is $(9, 0)$.

**Support Numerically.** Figure 3.9 shows numerical support for this solution.

$$9 \rightarrow X : 7X - 2(2X + 11)$$
$$5$$

**FIGURE 3.9** $x = 9$ is a solution to $7x - 2(2x + 11) = 5$.

**Confirm Algebraically.**

$$7x - 2(2x + 11) = 5$$
$$7x - 4x - 22 = 5$$
$$3x = 27$$
$$x = 9$$

## Exercises for Section 3.1

In Exercises 1–12, use the multigraph method to solve the given equation. (*Hint*: Use the Integer window.) Support the solution numerically and confirm algebraically.

1. $2x - 24 = 18$
2. $3x - 39 = -18$
3. $16 - 2x = 24$
4. $42 - 3x = 6$
5. $2x - 19 = 9$
6. $3x + 9 = 24$
7. $0.4x - 3.2 = -10$
8. $0.7x - 6.3 = 14$
9. $18 - 2x = 4$
10. $14 - 3x = 5$
11. $1.6x - 7.6 = 2$
12. $1.3x - 15 = 7.1$

In Exercises 13–24, use the $x$-intercept method to solve the given equation. (*Hint*: Use the Integer window.) Support the solution numerically and confirm algebraically.

13. $2x - 18 = 0$
14. $3x - 21 = 0$
15. $21 - 3x = 0$
16. $51 - 3x = 0$
17. $(2x - 5) + (x - 7) = 0$
18. $(5x - 2) - (3x - 10) = 0$
19. $(6x + 3) - (5x - 14) = 0$
20. $(3x - 4) - (7 + 4x) = 0$
21. $0.5x - 17 = -9$
22. $0.3x - 6 = -9$
23. $0.4x - 7 = -15$
24. $1.3x + 2 = 15$

In Exercises 25–32, solve the given equation by a graphical method. Choose between the multigraph and the $x$-intercept method. (Use the Integer window.)

25. $2x + 1 = 16 - x$
26. $3x - 19 = x + 13$
27. $3x - 1 = x - 7$
28. $7x - 3 = 6x + 5$
29. $9x + 15 = 7x - 13$
30. $15 - 6x = 4 - 5x$
31. $(x + 5) + (2x - 4) = x - 5$
32. $(17 + x) - (13 - 2x) = 4x$

### TRANSLATING WORDS TO SYMBOLS

In Exercises 33–40, translate each sentence into an equation or an inequality.

33. Five more than twice a number is twenty-one.

34. The triple of a number is four more than eight.

35. Doubling a number results in five less than seventeen.

36. The double of a number is twenty-eight.

37. Three times a number is seventeen.

38. Four times a number is less than or equal to six.

39. Two more than three times a number is less than twenty-one.

40. Five less than two times a number is greater than or equal to forty-two.

In Exercises 41–47, write and solve an equation that represents the given problem. Use an algebraic or graphical method to solve the equation.

41. When 1/2 of a number is subtracted from 15, the answer is 8. What is the number?

42. When 2 times a number is added to 12, the answer is $-17$. What is the number?

43. *Hotel Tax.* A certain city has a 10% room tax for hotel rooms. At a discount hotel, the room rent plus tax was $28.60. How much was the room rent?

44. *State Sales Tax.* A certain state has a 5% sales tax. A book cost $27.20 including tax. What was the list price of the book?

45. *Salary Increase Calculations.* Betty received a 5% salary increase. Her salary after the increase was $29,400. What was her salary before the increase?

46. *Rental Company Charges.* A 32-ft extension ladder can be rented at Sam's Rent-All for $6/day and $0.25/hr. John's total bill for renting this ladder is $8.75. For how many hours did he rent the ladder?

47. *Company Driving Expenses*. Jack's company has an expense account that pays him $12/day plus $0.25/mi for each mile driven. His payment for a certain day was $21.75. How many miles did he drive that day?

paragraphs to explain how to use each of these approaches to solve the equation

$$(3x - 7) - (2x + 6) = 28$$

EXTENDING THE IDEAS

48. Devise a graphical solution method that can be used to find a value of $x$ such that

$$4x + 16 = -x - 19.$$

*Hint*: Model your method after the multigraph method of solving equations.)

49. Devise a graphical solution method that can be used to find a value of $x$ such that

$$3x + 15 = -2x - 20.$$

50. **Writing to Learn**. So far in this text we have studied several methods for solving equations: guess-and-check, algebraic, and graphical. Write several

LOOKING BACK—MAKING CONNECTIONS

51. If the equation $(3x - 4) - (x + 4) = 6$ is solved graphically by the $x$-intercept method, the graph you obtain will be a line. What is its slope?

52. If the equation $(3x + 2) - (5x + 7) = 13$ is solved graphically by the $x$-intercept method, the graph you obtain will be a line. What is its slope?

53. Write an equation in slope-intercept form of the line perpendicular to the graph of $y = 3x - 2$ at the point $(1, 2)$.

54. Write an equation in slope-intercept form of the line perpendicular to the graph of $y = -2x + 1$ at the point $(-3, 7)$.

---

## 3.2 Solving Equations: Using Zoom-In

Finding a Graphical Solution Using a 0.1 Window ▪ Approximating a Solution to an Equation ▪ Increasing the Precision of the Approximation ▪ Using the Grapher ZOOM-IN Menu ▪ Using the Grapher in Problem Situations

The solutions to the equations in Section 3.1 could all be found using the Integer window. As we will see in this section, many times a graphical solution cannot be found unless we zoom in or magnify on a smaller portion of the plane.

## Finding a Graphical Solution Using a 0.1 Window

The equation $2x - 7 = 0$ can be easily solved using the algebraic methods discussed in Section 1.5. However, in the interest of illustrating the benefits of using windows other than an Integer window, we will solve $2x - 7 = 0$ graphically. More-refined graphical methods are worth learning because many equations can be solved most easily by graphical methods; in some cases, solving by algebraic methods is impossible.

**REMINDER**

Here is the algebraic solution.

$$2x - 7 = 0$$

$$2x = 7$$

$$x = 7/2 = 3.5$$

USING THE INTEGER WINDOW TO SOLVE 2X − 7 = 0

Figure 3.10 shows that $(3, -1)$ and $(4, 1)$ are both points on the graph of the equation $y = 2x - 7$.

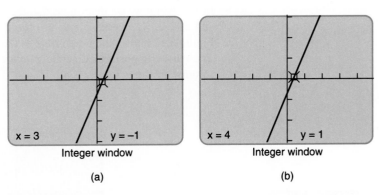

**(a)**                    **(b)**

**FIGURE 3.10** Graphs of $y = 2x - 7$ showing a point slightly below the $x$-axis in (a) and another point slightly above the $x$-axis in (b) in the Integer window. Notice when $x = 3$ the $y$ value is negative, and when $x = 4$ the $y$ value is positive. This indicates $y = 0$ is between $x = 3$ and $x = 4$.

**GRAPHER NOTE**

The WINDOW settings for the 0.1 window depends on which model grapher you are using. Refer to your grapher lab manual to learn what WINDOW settings should be used to obtain the 0.1 window on your grapher.

So the graph crosses the $x$-axis somewhere between $x = 3$ and $x = 4$. However, the exact $x$-intercept cannot be shown in the Integer window. So we investigate another window.

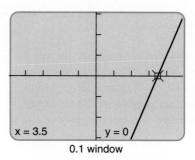

**FIGURE 3.11** A graph of $y = 2x - 7$ in the 0.1 window.

USING THE 0.1 WINDOW TO SOLVE $2X - 7 = 0$

The 0.1 window is one in which the origin is centered in the window and the coordinates of the points in the window all end evenly in tenths. Figure 3.11 shows the graph of $y = 2x - 7$ in this window. We say that we have **zoomed in** at the origin.

The coordinates shown in Figure 3.11 indicate that the $x$-intercept is $(3.5, 0)$. The solution to the equation $2x - 7 = 0$ is $x = 3.5$.

**EXAMPLE 1**    Using the 0.1 Window

Use the $x$-intercept method to find a graphical solution to the equation $(3x + 7) - (5 - 2x) = 4$. Support your solution numerically.

SOLUTION
**Solve Graphically.**

1. These equations are equivalent.

$$(3x + 7) - (5 - 2x) = 4$$

$$(3x + 7) - (5 - 2x) - 4 = 0$$

We begin by graphing $y = (3x + 7) - (5 - 2x) - 4$ in the Standard window. In this window, we determine that the graph crosses the $x$-axis between $x = 0$ and $x = 1$. So, we decide to use the 0.1 window.

2. Figure 3.12 shows the graph of $y = (3x + 7) - (5 - 2x) - 4$ in the 0.1 window. The cursor is positioned at the $x$-intercept, which is $(0.4, 0)$. Therefore the solution to the equation is $x = 0.4$.

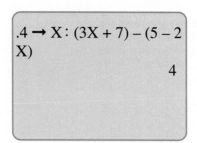

**FIGURE 3.13**
Numerical support that $x = 0.4$ is a solution to the equation $(3x + 7) - (5 - 2x) = 4$.

**FIGURE 3.12** A graph of $y = (3x + 7) - (5 - 2x) - 4$.

**Support Numerically.** Figure 3.13 shows a numerical support of this solution. ∎

### Approximating a Solution to an Equation

There are some equations whose exact solutions cannot be found using a graphical method. For example, the equation $3x - 2 = 0$ has the solution $x = 2/3$. The decimal representation of $2/3$ is $0.666\ldots$, an infinite decimal that does not terminate. Therefore the exact solution cannot be found graphically. In cases like this, the solution can be approximated with whatever degree of accuracy is desired. Example 2 illustrates how to find an approximate solution.

**EXAMPLE 2**    Approximating a Solution Graphically

In the 0.1 window, find the two trace cursor positions on the graph of $y = 3x - 2$ that are closest to the $x$-axis. Use the coordinates of these points to approximate a solution to the equation $3x - 2 = 0$.

SOLUTION

Figure 3.14 shows the graph of $y = 3x - 2$ in the 0.1 window with the trace cursor positioned at two points $(0.6, -0.2)$ and $(0.7, 0.1)$, the two points closest to the $x$-axis.

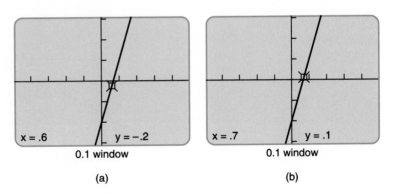

x = .6    y = -.2    0.1 window    (a)

x = .7    y = .1    0.1 window    (b)

**FIGURE 3.14**    The graph of $y = 3x - 2$ and two different trace coordinates.

We conclude that the solution to the equation $3x - 2 = 0$ is between $x = 0.6$ and $x = 0.7$. So $x = 0.65$ is a reasonable approximation of the solution.    ∎

## Increasing the Precision of the Approximation

To increase the precision of the approximation found in Example 2, we use a window that magnifies the interval [0.6, 0.7] by using the WINDOW settings shown in Figure 3.15. Notice that the $X$scl setting tells us that the "tick" marks along the $x$-axis in Figure 3.16 are 0.01 apart.

Window
Xmin = .6
Xmax = .7
Xscl = .01
Ymin = ⁻.2
Ymax = .2
Yscl = .1

**FIGURE 3.15** Use this WINDOW setting to approximate the solution with an error of at most 0.01.

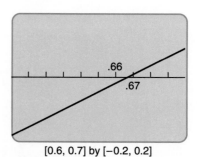

[0.6, 0.7] by [−0.2, 0.2]

**FIGURE 3.16** The graph of $y = 3x - 2$ showing that the solution to the equation $3x - 2 = 0$ is between $x = 0.66$ and $x = 0.67$.

Figure 3.17 shows a hand-sketch that magnifies a portion of Figure 3.16. The $x$-intercept of $y = 3x - 2$ is $x_0$ and is the exact solution of the equation $3x - 2 = 0$. So the difference between 0.66 and $x_0$ is less than 0.01 and the difference between 0.67 and $x_0$ is also less than 0.01.

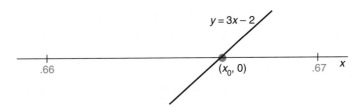

**FIGURE 3.17** The graph of $y = 3x - 2$ crosses the $x$-axis between $x = 0.66$ and $x = 0.67$.

This process of considering smaller and smaller intervals around a solution is referred to as the **zoom-in** process. We could continue this process further if desired to obtain a solution to whatever precision is needed. However, throughout the remainder of this text we shall use the following agreement.

### Accuracy Agreement When Using a Grapher

The phrase "find a solution of an equation" means "find an approximate solution with an error of at most 0.01."

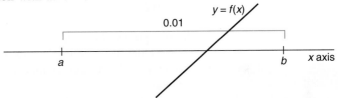

If $y = f(x)$ crosses the $x$-axis in the interval $[a, b]$, an interval of width 0.01, it is correct to report either $a$ or $b$, or any number between the two, as **a solution of $f(x) = 0$ with an error of at most 0.01**.

**Convention.**   We agree to estimate the third digit (by visual inspection of the graph) and report a three-digit solution. Because the third digit is a visual estimate, your third digit may differ.

Applying this convention to Figure 3.16, we inspect the graph and estimate that the graph of $y = 3x - 2$ crosses about 7/10 of the distance from 0.66 to 0.67. We report the solution to the equation $3x - 2 = 0$ as $x = 0.667$.

Although the equation $3x - 2 = 0$ can be solved algebraically to obtain the solution $x = \dfrac{2}{3}$, you will not see the fraction $\dfrac{2}{3}$ on the grapher. You will see instead its approximation. However, using our accuracy agreement, 0.667 or 0.66 or 0.67 or 0.665 are all correct as solutions.

**EXAMPLE 3** Finding a Graphical Solution with an Error of At Most 0.01

Find a graphical solution to the equation $0.7x - 0.5 = 0$.

SOLUTION

1. The graph of $y = 0.7x - 0.5$ in the 0.1 window is shown in Figure 3.18.

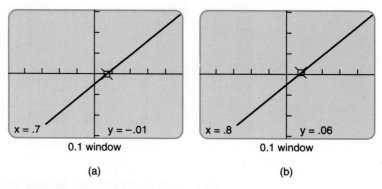

| | |
|---|---|
| x = .7    y = −.01 | x = .8    y = .06 |
| 0.1 window | 0.1 window |
| (a) | (b) |

**FIGURE 3.18** The graph of $y = 0.7x - 0.5$. Notice that in (a) when $x = 0.7$ the $y$ value is negative, and in (b) when $x = 0.8$ the $y$ value is positive. This indicates the solution is between $x = 0.7$ and $x = 0.8$.

2. The two points closest to the $x$-intercept are $(0.7, -0.01)$ and $(0.8, 0.06)$, so we know that the solution is between $x = 0.7$ and $x = 0.8$.
3. Choose WINDOW settings that will magnify the interval $[0.7, 0.8]$ as shown in Figure 3.19.
4. Figure 3.20 shows that the graph of $y = 0.7x - 0.5$ crosses the $x$-axis between $x = 0.71$ and $x = 0.72$.

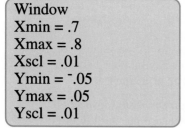

Window
Xmin = .7
Xmax = .8
Xscl = .01
Ymin = ⁻.05
Ymax = .05
Yscl = .01

**FIGURE 3.19**
WINDOW settings to show a zoom-in of the solution.

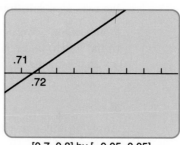

[0.7, 0.8] by [−0.05, 0.05]

**FIGURE 3.20** Graph of $y = 0.7x - 0.5$ after zooming in.

The solution to the equation $0.7x - 0.5 = 0$ is $x = 0.715$ with an error of at most 0.01.  ■

## Using the Grapher ZOOM-IN Menu

In Example 3, the solution to the equation $0.7x - 0.5 = 0$ was found by manually changing the WINDOW settings as shown in Figure 3.19. In this section, we demonstrate how to use the grapher's ZOOM-IN capability to improve the precision of a solution.

SOLVING $3.6X - 12.92 = 0$

We illustrate the zoom-in process as we solve the equation $3.6x - 12.92 = 0$. (We use decimal constants to illustrate that the process is no harder using decimal numbers than integers.) Follow these steps.

1. Find the graph of $y = 3.6x - 12.92$ in an appropriate viewing window. (Either the Integer window or the Standard window is appropriate in this case.)
2. Because we want to find the solution with an error of at most 0.01, set $X$scl= 0.01 on the WINDOW menu. The $x$-axis appears "fat" since the tick marks are squeezed so close together (see Figure 3.21).

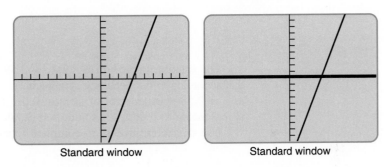

Standard window          Standard window

**FIGURE 3.21** The graph before and after setting $X$scl= 0.01.

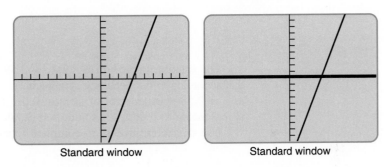

ZOOM
1: Box
2: Zoom In
3: Zoom Out
4: Set·Factors
5: Square
6: Standard

**FIGURE 3.22** ZOOM menu on a grapher.

**GRAPHER NOTE**

Consult your grapher lab manual to learn how to use the ZOOM menu of your grapher. Figure 3.22 shows the ZOOM menu on the grapher we use. For your work in this section, we encourage you to set the ZOOM factors at $X$Fact= 10 and $Y$Fact= 10 and to use the ZOOM menu item #2 (ZOOM-IN).

3. ZOOM factors should be set at $X$Fact= 10 and $Y$Fact= 10. Position the cursor on the $x$-intercept and zoom in several times. It is clear that the graph crosses the axis between $x = 3.58$ and $x = 3.59$ (see Figures 3.22 and 3.23).

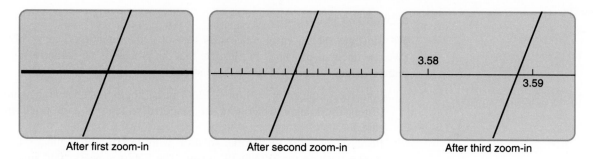

After first zoom-in    After second zoom-in    After third zoom-in

**F I G U R E   3 . 2 3**   A three-step zoom process for finding a solution with an error of at most 0.01.

The solution to the equation is $x = 3.59$ with an error of at most 0.01. Notice that according to our accuracy agreement, it is also acceptable to report the solution as $x = 3.58$ with an error of at most 0.01. However, following our convention it is reasonable to report this solution as 3.589.

Use this zoom process to complete Examples 4 and 5.

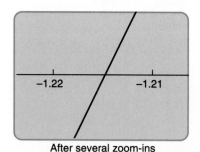

After several zoom-ins

**F I G U R E   3 . 2 4**   Graph of $y = 2.7x + 3.28$ after zooming in.

**EXAMPLE 4**   Solving an Equation Using Zoom-in

Find a solution to the equation $2.7x + 3.28 = 0$.

SOLUTION

1. Find a graph of $y = 2.7x + 3.28$ in the Standard window.
2. Set $X$scl= 0.01 in the WINDOW menu.
3. Zoom in until the graph is easily seen crossing the $x$-axis between two tick marks (see Figure 3.24).

The solution to the equation $2.7x + 3.28 = 0$ is $x = -1.215$ with an error of at most 0.01. ∎

**EXAMPLE 5** Solving an Equation Using Zoom-in

Find a solution to the equation $0.43x - 7 = 4.71$.

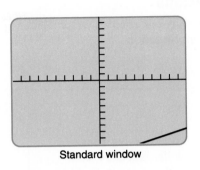

**FIGURE 3.25** Graph of $y = 0.43x - 11.71$.

SOLUTION

Change $0.43x - 7 = 4.71$ to the equivalent form $0.43x - 11.71 = 0$.

1. Figure 3.25 shows the graph of $y = 0.43x - 11.71$ in the Standard window. To see a larger portion of the graph, we next use the Integer window.

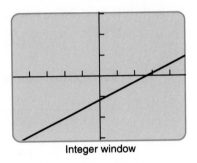

**FIGURE 3.26** Graph of $y = 0.43x - 11.71$.

2. Set $X$scl$= 0.01$ on the WINDOW menu. Position the cursor on $x$-intercept and zoom in. (See Figure 3.26.)
3. Repeat the zoom-in process until you arrive at a graph like the one in Figure 3.27.

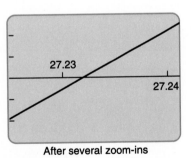

**FIGURE 3.27** Graph of $y = 0.43x - 11.71$.

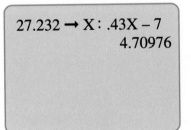

**FIGURE 3.28**
Numerical support of the "approximate solution" $x = 27.232$ is shown. The result is "close to" 4.71, as expected.

The solution to the equation $0.43x - 7 = 4.71$ is $x = 27.232$ with an error of at most 0.01.

Figure 3.28 shows numerical support of the "approximate solution" $x = 27.232$. The value of $0.43x - 7$ is "close to" 4.71, as expected. ∎

## Using the Grapher in Problem Situations

Both algebraic and graphical methods can be used in our analysis of problem situations. Consider the following **Inflation Problem Situation**.

---

### INFLATION PROBLEM SITUATION

Inflation is an economic situation in which the cost of living rises. Suppose the cost of an item increases the same dollar amount $I$ each year. If the cost of the item in a given year is $C_0$, then its cost $C$ after $t$ years is

$$C = C_0 + It.$$

This equation is the algebraic representation of this problem situation.

---

The sketch in Figure 3.29 shows a graphical representation for this problem situation. The cost of the product in the given year is the $y$-intercept, $C_0$.

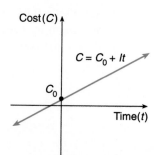

**FIGURE 3.29** A graphical representation of the **Inflation Problem Situation.** The slope $I$ is the constant dollar increase each year.

For example, suppose a product that costs $23 today (represented by $t = 0$) increases $0.69/yr. When $t$ is positive, $C = 23 + 0.69t$ represents the cost $t$ years after the $t = 0$ year. When $t$ is negative, $C = 23 + 0.69t$ represents the cost $t$ years *before* the $t = 0$ year.

To find the graphical representation of a problem situation, you must choose a window consistent with the magnitude of the numbers in the problem, as illustrated in Example 6.

**EXAMPLE 6**  THE INFLATION PROBLEM SITUATION: Finding a Graphical Representation of Lifestyle Cost

Suppose the cost of a certain lifestyle requires $32,000 in 1993. If the cost of this lifestyle increases $960/yr, find a graphical representation of the cost of this lifestyle from 1988 to 1998.

SOLUTION

1. An algebraic representation of lifestyle cost is $C = 32,000 + 960t$, where $t = 0$ represents 1993 and $t = 5$ represents 1998.
2. The grapher works in variables $x$ and $y$, therefore we change the equation $C = 32,000 + 960t$ to $y = 32,000 + 960x$. To represent the years 1988 to 1998, we must let $X\min = -5$ and $X\max = 5$. The vertical dimension must include values 32,000 and some larger. So we choose the window $[-5, 5]$ by $[-5000, 45000]$.

Figure 3.30 shows a graphical representation of this lifestyle cost.

■
∴
■
**GRAPHER NOTE**

When graphing equations from a problem situation, the graphs may not appear in the Standard or Integer windows. If you have this situation, use TRACE in your first window to get an idea of how large or small the $y$-values should be on your next window.

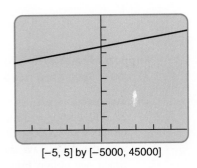

$[-5, 5]$ by $[-5000, 45000]$

**FIGURE 3.30**  A graphical representation of the cost of living.

**EXAMPLE 7**  THE INFLATION PROBLEM SITUATION: Finding Future Costs

Suppose a $32,000 lifestyle at the beginning of 1993 increases $960/yr.
  a) How much will this lifestyle cost at the beginning of 1997?
  b) How long after the beginning of 1993 will this lifestyle cost $35,000?

SOLUTION

**Solve Graphically.** The algebraic representation of this problem is
$C = 32,000 + 960t$.

a) Because the grapher uses variables $x$ and $y$, we change the equation
$C = 32,000 + 960t$ to $y = 32,000 + 960x$. We must find the cost $C$ when
$t = 1997 - 1993 = 4$. So we choose the window in which $X$min and
$X$max are the values used for the 0.1 window, and $Y$min $= -5000$ and
$Y$max $= 45,000$. Figure 3.31 shows that the point (4, 35840) is on this line.
The $32,000 lifestyle in 1993 will cost $35,840 at the beginning of 1997.
(Why?)

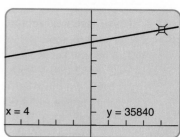

x = 4    y = 35840

**FIGURE 3.31** A graph of
$y = 32,000 + 960x$ in the 0.1
window with vertical dimension
changed to $[-5000, 45000]$.

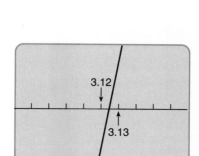

**GRAPHER NOTE**

Throughout the remainder
of the text, we shall describe
the window in Figure 3.31
as the 0.1 window with the
vertical dimension changed
to $[-5000, 45000]$.

**Support Numerically.** To find the cost at the beginning of 1997, evaluate
$32,000 + 960t$ when $t = 4$. So $C = 32,000 + 960(4) = 35,840$.

b) To solve this part, we need to solve the equation
$32,000 + 960t = 35,000$. To do so graphically, use the $x$-intercept method.
Graph $y = 32000 + 960x - 35000$ in the window $[-5, 5]$ by $[-5000, 45000]$
with the WINDOW menu setting $X$scl $= 0.01$. After using ZOOM-IN several
times, we obtain Figure 3.32. The answer appears to be $x = 3.125$. So in
3.125 yr from 1993, $35,000 will maintain a lifestyle that in 1993 required
$32,000. (This is about 46 days into 1996, or sometime in February of that
year.)

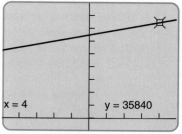

3.12

3.13

**FIGURE 3.32**
A graph of
$y = 32000 + 960x - 35000$.

**Confirm Algebraically.** Solve $C = 32,000 + 960t$ for $t$ when $C = 35,000$.

$$35,000 = 32,000 + 960t$$

$$3000 = 960t$$

$$\frac{3000}{960} = t$$

$$3.125 = t$$

Consider the following **Depreciation Problem Situation.**

---

☐                                                                                        ☐

## DEPRECIATION PROBLEM SITUATION

Any vehicle, building, or major piece of equipment, has a certain
useful life. It **depreciates**, or loses value, over time. The **straight-line
depreciation method** assumes that the item depreciates the same amount
each year.

☐                                                                                        ☐

---

Example 8 studies the **Depreciation Problem Situation** for a specific case.

**EXAMPLE 8**     THE DEPRECIATION PROBLEM SITUATION: Finding When
the Value is Zero

Suppose a punch press costs $28,000 new and a constant 4.5% of its original
value depreciates each year. Find an algebraic representation for value $V$ after $t$
years. When will the value of the press be zero? Solve algebraically and support
graphically.

SOLUTION

Let $t$ = the age of the press (in years).

> **Depreciation after $t$ years:** Each year the press loses a constant
> $0.045(28) = 1.26$ thousand dollars in value. After $t$ years, the value lost is
> $1.26t$ thousand dollars.
> **Value after $t$ years:** $V = 28 - 1.26t$ (in thousands of dollars)

To solve the problem we need to solve the equation $28 - 1.26t = 0$.

**Solve Algebraically.**

$$28 - 1.26t = 0$$
$$-1.26t = -28$$
$$t = \frac{-28}{-1.26}$$
$$t = 22.222$$

So in about 22.222 years, according to the straight-line depeciation method,
the press will have zero value.

**Support Graphically.** Because the grapher uses variables $x$ and $y$, we change the equation $V = 28 - 1.26t$ to $y = 28 - 1.26x$.

1. Find the graph of $y = 28 - 1.26x$ in the Integer window.
2. Zoom in to find the $x$-intercept. Figure 3.33 shows the graph after the Zoom in.

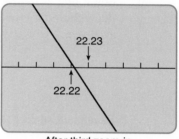

After third zoom-in

**FIGURE 3.33** A graph of $y = 28 - 1.26x$ after zooming in three times with $X$Fact= 10 and $Y$Fact= 10.

The solution to $28 - 1.26x = 0$ is $x = 22.222$ with an error of at most 0.01. ∎

## Exercises for Section 3.2

In Exercises 1–10, solve the given equation using either the Integer window or the 0.1 window, whichever is appropriate for finding an exact solution. Use either the multigraph or the $x$-intercept method.

1. $2x + 18 = 26$
2. $14 - 3x = -4$
3. $2x - 1 = 2$
4. $5x - 3 = 1$
5. $1.2x - 1 = 2$
6. $1.5x - 2 = -0.2$
7. $1.2 - 0.8x = 1.6$
8. $0.4x - 2.4 = -1$
9. $(4x + 3) - (x + 12) = 15$
10. $(x + 2) - (0.5x + 3) = 12$

In Exercises 11–18, use the 0.1 window. Graph the equation that you would use for a graphical solution of the given equation. Find the two trace cursor positions on the graph closest to the $x$-axis. Use the coordinates of these points to find an approximate solution for the associated equation. Confirm each solution algebraically.

11. $3x - 7 = 0$
12. $1.3x - 2 = 0$
13. $1.4x + 3.2 = 0$
14. $1.05x - 2.5 = 0$
15. $1.08x - 1.9 = 0$
16. $2.1 - 1.45x = 0$
17. $1.09x - 1.7 = 2$
18. $2.1x - 1.75 = 1.5$

In Exercises 19–26, find a graphical solution to the given equation. (This means find a solution with an error of at most 0.01.) *(Hint:* Begin with the Integer window and set $X$scl= 0.01. Set $X$Fact= 10 and $Y$Fact= 10 for ZOOM.)

19. $3x - 14 = 0$         20. $2x + 1.05 = 0$

21. $(2x - 18) + (x - 2) = 0$

22. $(4x - 15) + (2x + 1) = 0$

23. $18 - 2.1x = 7$       24. $21 - 1.4x = 12$

25. $1.35x - 12 = 14$     26. $2.6x + 12 = 3$

### TRANSLATING WORDS TO SYMBOLS

In Exercises 27–30, write an algebraic expression in terms of $x$ for each expression.

27. 4% of a number is subtracted from 5000.

28. 3% of a number is added to 7000.

29. 7% of a number is subtracted from the number.

30. 5% of a number is added to the number.

31. *Projecting Consumer Prices.* A suit of clothes that costs $125 in 1993 increases an average of $8/yr for several years before and after 1993.

   a) Find an algebraic representation of this **Inflation Problem Situation.**

   b) Find a graphical representation of this problem situation in a window that represents the time from the beginning of 1988 to the beginning of 1998.

   c) What does the graphical model suggest about when the price of the suit of clothes will be $142.50?

32. *Projecting Future Real Estate Costs.* One certain style house that sells for $88,000 in a certain city in 1993 has been averaging a $2000 increase each year for the past 5 yr. Assume that this pattern continues for the next 5 yr.

   a) Find an algebraic representation of this **Inflation Problem Situation.**

   b) Find a graphical representation of this problem situation in a window that represents the time from the beginning of 1988 to the beginning of 1998.

   c) Using this graphical representation, find during which year the price of this style house will reach $95,400. Confirm algebraically.

33. *Depreciation of a Truck.* A certain model truck cost $65,000 new in 1994. The company accountant uses straight-line depreciation and assumes that the truck loses 12.5% of this new purchase value each year.

   a) Find an algebraic representation of this **Depreciation Problem Situation.**

   b) Find a graphical representation of this problem situation in a window that represents the time from the beginning of 1989 to the beginning of 1999.

   c) Using this graphical representation, find during what year the depreciated value of this truck will be $48,500. Confirm algebraically.

34. *Real Estate Depreciation.* A new apartment building costs $225,000 on Jan. 1, 1994. For tax purposes, an accountant assumes that the building depreciates an equal amount, $9,000, each year for 25 yr, ending with a value of $0.

   a) Find an algebraic representation of this **Depreciation Problem Situation.**

   b) Find a graphical representation of this problem situation that reflects the entire 25-yr life of the building. Does this graphical representation assume that the building depreciates a little each day or that it depreciates once each year?

   c) The apartment building owner decides to sell the building when its depreciated value is $110,000. During what year should the sale take place? Solve graphically and confirm numerically.

EXTENDING THE IDEAS

**35.** *Comparing Depreciated Values.* Suppose two trucks were purchased in the same year. The $55,000 truck depreciates $8000/yr and the $32,000 truck depreciates $2500/yr. Find an algebraic representation that describes the value of each truck $t$ years after the truck is purchased. Graph both representations in the same window. Do the two trucks ever have the same depreciated value? If so, when?

**36.** *Comparing Total Sales.* One store whose total sales in 1990 were $200,000 has been experiencing an increase in sales of $22,000/yr. Another store that had total sales in 1990 of $325,000 has been experiencing an increase of $8000/yr. Find an algebraic representation that describes the total sales of each store $t$ years later. Graph both representations in the same window. Do the two stores ever have the same amount of total sales? If so, in what year?

LOOKING BACK—MAKING CONNECTIONS

Solve each of the following equations algebraically:

**37.** $5x - 3(2x - 3) = x + 4$

**38.** $7x - (5 + 2x) = 3x - 2$

**39.** $(5x - 6) + (9 - 3x) = 7x + 19$

**40.** $-3x = (4x - 2) - 8x$

---

## 3.3   Some Nonlinear Equations

Selecting a Window  ▪  Solving Nonlinear Equations  ▪  Determining the Number of Solutions to an Equation  ▪  Finding an Appropriate Window in a Graphical Solution  ▪  Solving Equations When Problem Solving

In Sections 3.1 and 3.2, you learned how to solve linear equations graphically using the multigraph method and the $x$-intercept method. You also found that in some cases, exact solutions can be found while in many cases, finding approximate solutions was the best or only easy alternative.

Throughout these sections, functions were linear. In this section, we will solve equations in which the associated functions and equations are not linear. In particular we shall consider nonlinear functions of the following two general types:

$$f(x) = ax^2 + bx + c \quad \text{and} \quad f(x) = \pm a\sqrt{x + b},$$

where $a, b$, and $c$ are any real numbers and $\sqrt{x}$ represents the square root of $x$.

## Selecting a Window

To solve equations graphically, we must first find the graph of the associated functions in some reasonable window. When graphing nonlinear functions, it is often convenient to choose a window with different dimensions on the vertical scale than on the horizontal scale. The following exploration will provide insight regarding how to choose windows for nonlinear functions.

### EXPLORE WITH A GRAPHER

1. Find the graph of (a) $f(x) = x^2$, (b) $f(x) = 25x^2$, and
   (c) $f(x) = 150x^2$ in each of the following windows:
   a) $[-5, 5]$ by $[-5, 10]$
   b) $[-5, 5]$ by $[-5, 100]$
   c) $[-5, 5]$ by $[-5, 1000]$

   For each function, which window seems most appropriate?

2. Find the graph of (a) $f(x) = \sqrt{x}$, (b) $f(x) = 10\sqrt{x}$, and
   (c) $f(x) = 0.15\sqrt{x}$ in each of the following windows:
   a) $[-5, 10]$ by $[-3, 5]$
   b) $[-5, 10]$ by $[-3, 35]$
   c) $[-5, 10]$ by $[-1, 1]$

   For each function, which window seems most appropriate?

3. Find an appropriate window for: (a) $f(x) = \sqrt{x - 3}$ and
   (b) $f(x) = -18(x - 6)^2 + 5$.

## Solving Nonlinear Equations

The two methods for solving linear equations studied in Sections 3.1 and 3.2—the multigraph method and the $x$-intercept method—also apply to nonlinear equations. Paper-and-pencil algebraic methods become more difficult as the equations become more complex. In fact the equations discussed in this section cannot be solved using the paper-and-pencil algebraic methods developed so far in this book. A powerful feature of graphical methods is that they do *not* become more difficult as the equations become more complex. Example 1 illustrates this point.

**EXAMPLE 1** Solve Graphically: $x$-intercept Method

Use the $x$-intercept method to solve the equation $4\sqrt{x+5} = 3.5$.

SOLUTION

To use the $x$-intercept method, we first rewrite the equation $4\sqrt{x+5} = 3.5$ in the equivalent form $4\sqrt{x+5} - 3.5 = 0$.

1. Find the graph of the equation $y = 4\sqrt{x+5} - 3.5$ in the Standard window (see Figures 3.34 and 3.35 (a)).
2. Set $X$scl$= 0.01$ on the WINDOW menu.
3. Move the cursor close to the $x$-intercept and zoom in. Repeat this process until the graph looks like that in Fig. 3.35 (b).

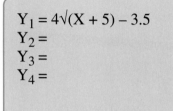

$Y_1 = 4\sqrt{(X+5)} - 3.5$
$Y_2 =$
$Y_3 =$
$Y_4 =$

**FIGURE 3.34**
The Function-defining menu for a graph of $y = 4\sqrt{x+5} - 3.5$. Note that the parentheses in the above expression are needed to indicate the square root of a binomial.

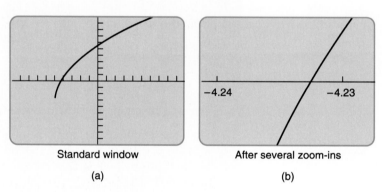

Standard window
(a)

After several zoom-ins
(b)

**FIGURE 3.35** Graphs of $y = 4\sqrt{x+5} - 3.5$ before and after zooming in.

The solution is $x = -4.234$ with an error of at most 0.01.  ■

**REMINDER**

Be sure you can explain why 4 is reported as the last digit in the solution of Example 1.

Example 2 is similar to Example 1 except that the multigraph method is used rather than the $x$-intercept method.

**EXAMPLE 2** Solve Graphically: Multigraph Method

Use a graphical method to solve the equation $-\sqrt{x-1} = -2.5$. Support numerically.

SOLUTION

**Solve Graphically.** To use the multigraph method, we need to graph both $y_1 = -\sqrt{x-1}$ and $y_2 = -2.5$. Figure 3.36 (a) shows these two graphs in the Standard window. The $x$-coordinate of the point of intersection is the solution to the equation.

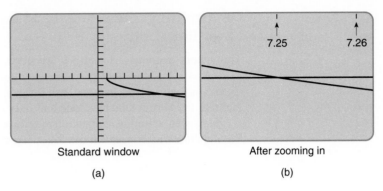

Standard window

(a)

After zooming in

(b)

**FIGURE 3.36**  Graphs of $y_1 = -\sqrt{x-1}$ and $y_2 = -2.5$

Set $X$scl= 0.01 and zoom in until you see the tick marks. We see from Figure 3.36 (b) that the solution to the equation $-\sqrt{x-1} = -2.5$ is $x = 7.250$ with an error of at most 0.01. Actually, it turns out that the solution is exactly 7.25.

**Support Numerically.** For numerical support, see Figure 3.37.

$$7.25 \rightarrow X: \ ^-\sqrt(X-1)$$
$$^-2.5$$

**FIGURE 3.37**  Numerical support of the solution for Example 2.  ■

When using the multigraph method, the solution to the equation is always the $x$-coordinate of any points of intersection of the two graphs. Notice that the horizontal line in Fig. 3.36 (b) is not the $x$-axis. In that case, the tick marks 7.25 and 7.26 are shown along the top edge of the window.

## Determining the Number of Solutions to an Equation

Throughout Sections 3.1 and 3.2, we observed that a linear equation of the form $ax + b = c$ has one, and only one, solution whenever $a \neq 0$. Each equation in Examples 1 and 2 also had one, and only one, solution. That was evident from the fact that the two graphs have one and only one point of intersection.

However, a nonlinear equation may have no solution or more than one solution. We illustrate this concept in the following example.

**EXAMPLE 3**    Finding the Number of Solutions

Find the number of solutions to the equation $2\sqrt{x + 4} + 3 = 1.5$.

SOLUTION

We use the multigraph method. Figure 3.38 shows the graphs of $y_1 = 2\sqrt{x + 4} + 3$ and $y_2 = 1.5$ in the Standard window.

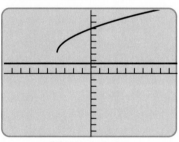

Standard window

**FIGURE 3.38** The graph
of $y_1 = 2\sqrt{x + 4} + 3$ and $y_2 = 1.5$.

Assuming that the graph of $y_1 = 2\sqrt{x + 4} + 3$ continues to increase as it moves off the screen to the right, we conclude that the graphs of $y_1$ and $y_2$ do not intersect and that there is no solution to the equation $2\sqrt{x + 4} + 3 = 1.5$. ■

So Figure 3.38 shows that not all equations have a solution. The next example demonstrates that equations can have more than one solution.

**EXAMPLE 4**    Solve Graphically: Algebraic Methods Are
Not Yet Available

Find all the solutions to the equation $8 + 4x - 3x^2 = -3$.

SOLUTION

We use the multigraph method. Figure 3.39 shows complete graphs of $y_1 = 8 + 4x - 3x^2$ and $y_2 = -3$ in the window $[-5, 5]$ by $[-10, 10]$. It is visually evident that there are two solutions to the equation $8 + 4x - 3x^2 = -3$.

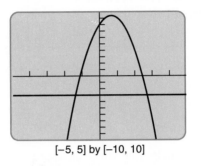

$[-5, 5]$ by $[-10, 10]$

**FIGURE 3.39** Graphs of $y_1 = 8 + 4x - 3x^2$ and $y_2 = -3$.

1. Set $X$scl$= 0.01$ on the WINDOW menu and zoom in on the point of intersection in the third quadrant until you can identify the tick marks. Figure 3.40 (a) shows the result. One solution is $x = -1.361$ with an error of at most 0.01.

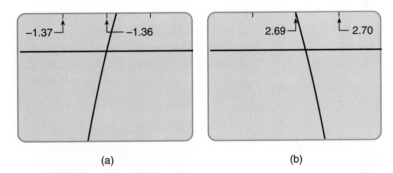

(a)                                (b)

**FIGURE 3.40** The graphs of $y_1 = 8 + 4x - 3x^2$ and $y_2 = -3$ after using zoom-in on each of the two points of intersection.

2. Return to the original window $[-5, 5]$ by $[-10, 10]$ shown in Figure 3.39 and proceed to zoom in at the point of intersection in the fourth quadrant. Figure 3.40 (b) shows the result. The second solution is $x = 2.694$ with an error of at most 0.01. ∎

Examples 3 and 4 suggest the following generalization about the number of solutions.

---

### Number of Solutions to Certain Equations

1. An equation of the form $a\sqrt{x+b} = k$, where $a, b,$ and $k$ are real-number constants, has either zero or one solution.
2. An equation of the form $ax^2 + bx + c = k$, where $a, b, c,$ and $k$ are real-number constants, has either zero, one, or two solutions.

---

## Finding an Appropriate Window in a Graphical Solution

Knowing the number of solutions to search for can help you select an appropriate viewing window. For example, suppose we want to solve the equation $2x^2 - 26x - 3 = 9$. We have not yet studied algebraic methods that can be used in this case, so the only alternative is a graphical method.

When solving nonlinear equations, we recommend beginning with the Standard window and changing to other windows as needed. Figure 3.41(a) shows a graph of $y_1 = 2x^2 - 26x - 3$ and $y_2 = 9$ in the Standard window.

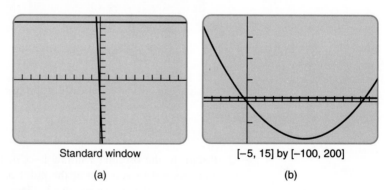

|  |  |
|:---:|:---:|
| Standard window | [−5, 15] by [−100, 200] |
| (a) | (b) |

**FIGURE 3.41** A graph of $y_1 = 2x^2 - 26x - 3$ and $y_2 = 9$ in two different windows.

In Figure 3.41 (a), notice that the graphs of $y_1 = 2x^2 - 26x - 3$ and $y_2 = 9$ intersect. The graph on the left shows one intersection, but it does not show enough for us to decide if there is only one intersection. However, with the TRACE feature we can find $x$- and $y$-values that can suggest a more appropriate viewing window. As a result of using a different window, the graph on the right indicates that there are two solutions.

Example 5 shows how finding an appropriate window can be used when solving an equation graphically.

**EXAMPLE 5**   Solving an Equation Graphically

Solve the equation $-x^2 + 16x - 50 = 6$.

SOLUTION

Begin by finding the graphs of $y_1 = -x^2 + 16x - 50$ and $y_2 = 6$ in the Standard window, as shown in Figure 3.42.

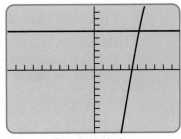

Standard window

**FIGURE 3.42**   The graphs of $y = -x^2 + 16x - 50$ and $y = 6$ intersect.

Notice that the graph of $y = -x^2 + 16x - 50$ seems to be curving to the right as it rises. We suspect that the window needs to be extended up and right to see the second point of intersection, so we try the window $[-5, 15]$ by $[-10, 15]$ (see Figure 3.43). Finally, set $X\text{scl} = 0.01$ and zoom in at each point of intersection to find the solutions.

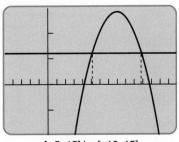

$[-5, 15]$ by $[-10, 15]$

**FIGURE 3.43**   Graph of $y_1 = -x^2 + 16x - 50$ and $y_2 = 6$ showing both points of intersection.

The two solutions to the equation $-x^2 + 16x - 50 = 6$ are $x = 5.172$ and $x = 10.828$ with an error of at most 0.01.   ■

We will learn algebraic methods for solving an equation like the one in the previous examples in Chapters 5 and 8.

## Solving Equations When Problem Solving

Consider the following problem situation.

---

### THE HOT-AIR BALLOON PROBLEM SITUATION

The captain of a hot-air balloon often drops ballast as the balloon rises. The algebraic representation for determining how long, in seconds $t$, it takes for the ballast to reach the ground (ignoring air resistance) is

$$t = \frac{1}{4}\sqrt{d},$$

where $d$ is the distance, in feet, above the ground when the ballast is dropped.

---

**EXAMPLE 6**   THE HOT-AIR BALLOON PROBLEM SITUATION:
Finding the Altitude of a Hot-air Balloon

If ballast hits the ground 5.3 sec after it is dropped, from how high above the ground is it dropped? (See Figure 3.44.)

**FIGURE 3.44**  The Hot-Air Balloon Problem Situation

SOLUTION

Because the algebraic representation of this problem is $t = \frac{1}{4}\sqrt{d}$, we need to solve the equation $5.3 = \frac{1}{4}\sqrt{d}$.

We know from the context of the problem that $x$ (representing distance $d$) is always positive and that $y$ (representing time $t$) is also positive. So we graph both $y_1 = 0.25\sqrt{x}$ and $y_2 = 5.3$ in the window [0, 10] by [0, 10] (see Figure 3.45). We see from this figure that the graph of $y = 0.25\sqrt{x}$ is increasing slowly enough that we must increase the horizontal dimension of the window a lot in order to see the intersection of the two graphs. Figure 3.46 (a) shows the intersecting graphs in the window [0, 1000] by [0, 10].

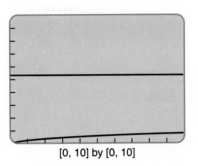

[0, 10] by [0, 10]

**FIGURE 3.45** Graphs of $y_1 = 0.25\sqrt{x}$ and $y_2 = 5.3$ with the point of intersection not visible.

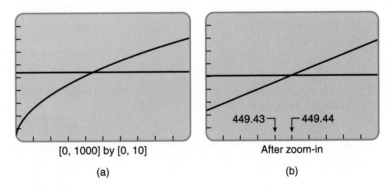

[0, 1000] by [0, 10]

(a)

449.43 ⌐    ⌐ 449.44

After zoom-in

(b)

**FIGURE 3.46** Graphs of $y_1 = 0.25\sqrt{x}$ and $y_2 = 5.3$: (a) after increasing the horizonal dimension and (b) after zooming in on the point of intersection.

Figure 3.46 (b) shows the graph after zoom-in, with the tick mark values indicated. The balloon is approximately 449.440 ft above the ground when the ballast is dropped.    ■

## Exercises for Section   3.3

In Exercises 1–6, graph each function in the given windows and determine which window seems most appropriate for each function.

**1.** Find the graph of $f(x) = 5\sqrt{x + 8}$.

 a) [−10, 10] by [−10, 10]

 b) [−10, 10] by [−10, 25]

 c) [−5, 5] by [−15, 15]

**2.** Find the graph of $f(x) = 0.3\sqrt{x} - 2$.

 a) [−10, 10] by [−10, 10]

 b) [0, 10] by [−1, 10]

 c) [0, 15] by [0, 2]

**3.** Find the graph of $f(x) = 125\sqrt{x + 4}$.

 a) [−10, 10] by [−10, 10]

 b) [−10, 10] by [−10, 200]

 c) [−10, 10] by [−10, 700]

4. Find the graph of $f(x) = 3x^2$.

a) $[-10, 10]$ by $[10, 10]$

b) $[-5, 5]$ by $[10, 200]$

c) $[-10, 10]$ by $[-1, 1]$

5. Find the graph of $f(x) = 35x^2$.

a) $[-10, 10]$ by $[-10, 10]$

b) $[-10, 10]$ by $[-500, 2000]$

c) $[-10, 10]$ by $[-10000, 10000]$

6. Find the graph of $f(x) = 2x^2 - 12x - 15$.

a) $[-10, 10]$ by $[-10, 10]$

b) $[-5, 5]$ by $[-25, 25]$

c) $[-10, 10]$ by $[-35, 25]$

In Exercises 7–12, use the $x$-intercept method to solve each equation. Set $X$scl$= 0.01$ and use ZOOM-IN.

7. $2\sqrt{x + 3} = 1.2$

8. $3\sqrt{x + 2} = 3.1$

9. $5\sqrt{x - 3} = 7.6$

10. $-\sqrt{x - 7} = -3$

11. $-2\sqrt{x + 1} = -3$

12. $12\sqrt{x + 5} = 32$

In Exercises 13–18, use the multigraph method to solve each equation. Set $X$scl$= 0.01$ and use ZOOM-IN.

13. $\sqrt{2x - 3} = 5$

14. $\sqrt{3x - 1} = 8$

15. $-2\sqrt{4x + 3} = -5$

16. $-5\sqrt{2x + 4} = -6$

17. $0.1\sqrt{x + 5} = 0.4$

18. $0.2\sqrt{x + 3} = 0.7$

In Exercises 19–24, use a graphical method to determine the number of solutions to each equation.

19. $\sqrt{x + 5} + 2 = 1$

20. $2\sqrt{x + 3} = 5$

21. $x^2 + 3x = 2$

22. $x^2 + 5x = -6$

23. $x^2 - 4x = -7$

24. $x^2 + 5x - 12 = 6$

In Exercises 25–32, find an appropriate window and solve the given equation graphically. Use either the $x$-intercept or the multigraph method.

25. $x^2 + 5 = 12$

26. $3x^2 + 7 = 15$

27. $x^2 - 11x + 2 = -6$

28. $x^2 - 7x - 2 = -11$

29. $0.7\sqrt{x - 3} = 0.3$

30. $125\sqrt{x} = 55$

31. $88\sqrt{x} = 33$

32. $18\sqrt{x + 7} = 55$

TRANSLATING WORDS TO SYMBOLS

Write an equation that is described by each sentence in Exercises 33–36.

33. Twenty-three times the square root of a number is forty-eight.

34. Five times the square root of twice a number is twelve.

35. Twice a number squared is six less than fourteen.

36. Four times a number squared is five more than twenty-three.

37. *Hot-Air Balloon Navigation.* Refer to the **Hot-Air Balloon Problem Situation** to answer the following questions:

a) The captain of a hot-air balloon drops ballast when the balloon is 800 ft above the ground. How long does it take for the ballast to hit the ground?

b) The captain drops more ballast, which takes 12 sec to hit the ground. How high above the ground is the captain when this ballast is dropped?

c) The captain drops ballast when the balloon is 900 ft above the ground. Six seconds after the ballast hits the ground, more ballast hits it. How high above the ground is the balloon when the second ballast is dropped?

d) The captain drops ballast when the balloon is 1600 ft above the ground. Eight seconds after the first ballast hits the ground, more ballast hits the ground. How high above the ground is the balloon when the second ballast is dropped?

EXTENDING THE IDEAS

38. Solve the equation $8x + 3.1 = 6x + 3$ twice using two different graphical methods. First use the

multigraph method and then the $x$-intercept method. Which method do you prefer and why?

39. Solve the equation $0.11x - 0.3 = 0$ twice using the $x$-intercept graphical method both times. First use ZOOM setting $X$fact= 5 and $Y$fact= 5. Then use ZOOM setting $X$fact= 5 and $Y$fact= 10. Which method do you prefer?

40. Use a grapher to show visually that $\sqrt{x+5} \neq \sqrt{x} + \sqrt{5}$.

41. Use a grapher to show visually that $(x-3)^2 \neq x^2 - 3^2$.

LOOKING BACK—MAKING CONNECTIONS

Solve each of the following equations algebraically. Support the solutions numerically.

42. $3(x+5) = 2x - 5$     43. $2000 - 4x = 800 - x$

44. $300 + 1.05x = 50 + 1.1x$

45. $500 - 15x = 225$

46. Write a **Depreciation Problem** whose algebraic representation is the equation in Exercise 45.

---

| 3.4 | Graphical Representations of Problem Situations |

Revisiting the Garden Problem Situation ▪ Using a Graphical Representation to Solve Problems ▪ Solving Problems Both Graphically and Algebraically

In Section 1.4 you completed data tables in search of a solution to a problem. The completed set of data was called a numerical representation of the problem situation.

In Chapter 1 and 2 problems were solved using a numerical or alegebraic representation of the problem situation. We shall review these two representations and introduce a third type, a **graphical representation of the problem situation.** This type of representation is important because of the visual information provided. The numerical and algebraic representations are used in finding a graphical representation.

**FIGURE 3.47** A particular fence arrangement for the **Garden Problem Situation.**

## Revisiting the Garden Problem Situation

Recall that in the **Garden Problem Situation,** 24 ft of fence is used to surround a rectangular garden on three sides (see Figure 3.47). There are three representations of this problem.

**Numerical Representation.**

| Width(ft) | Length(ft) | Area(ft$^2$) |
|-----------|------------|--------------|
| 2 | $24 - 2(2) = 20$ | 40 |
| 4 | 16 | 64 |
| 6 | 12 | 72 |
| 8 | 8 | 64 |
| 10 | 4 | 40 |
| 11 | 2 | 22 |

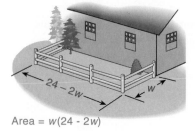

Area = $w(24 - 2w)$

**FIGURE 3.48**  A fence arrangement with width $w$ and length $24 - 2w$.

**Algebraic Representation.** If the width of the garden is represented by the variable $w$, its length is $24 - 2w$ (see Figure 3.48). Consequently, the area of the garden is represented by the equation

$$A = w(24 - 2w),$$

which is an algebraic representation of the problem situation.

**A Graphical Representation.** Solution pairs of the equation $A = w(24 - 2w)$ can be graphed to give a visual sense of how the area ($A$) changes for varying values of $w$. Figure 3.49 shows both a hand-sketch and the graph of $y = x(24 - 2x)$ produced by a grapher. Either of these graphs is a graphical representation of the problem situation.

The graph in Figure 3.49 (a) can be completed by graphing the data in a numerical representation of the problem. However, one important use of the algebraic

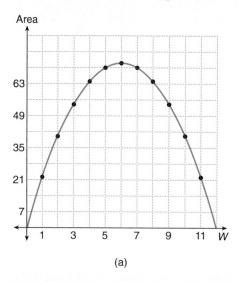

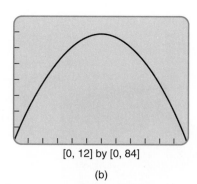

[0, 12] by [0, 84]

(a)

(b)

**FIGURE 3.49**  The width of a rectangle is some positive number, so $w$ cannot represent a negative number. If $w = 12$, the length $24 - 2w = 0$ and no rectangle is formed. Accordingly we choose the viewing window in Figure 3.49 (b) with $X$min$= 0$ and $X$max$= 12$.

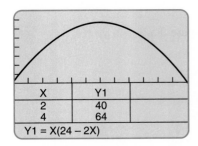

**FIGURE 3.50** A grapher screen of the numerical, algebraic, and graphical representation of the garden problem.

representation is to generate the graph in Figure 3.49 (b) on your grapher. Each of these three representations are shown on the grapher screen in Figure 3.50.

## Using a Graphical Representation to Solve Problems

To illustrate how a graphical representation can be used to solve problems, we continue analyzing the **Garden Problem Situation**. We ask, What is the width of the garden if its area is 63 ft²?

Figure 3.51 shows that to answer this question graphically, begin at 63 on the vertical axis (Area) and move horizontally to the point $P$ on the graph. Move down the axis to the number $w_0$. The real number $w_0$ is a solution to the problem. It appears that $w_0 = 4$.

A second solution is the value $w_1$ in Figure 3.51 (b). It appears that $w_1 = 8$. Notice that $w_0$ is also the $x$-coordinate of point $P$ and $w_1$ is the $x$-coordinate of point $Q$.

The next example illustrates how to use a grapher to find approximations to these values $w_0$ and $w_1$.

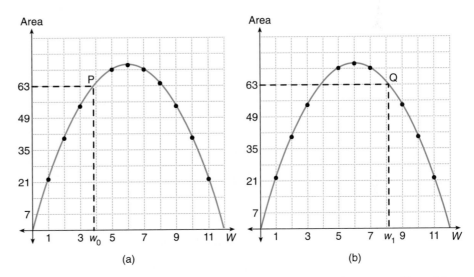

**FIGURE 3.51** Graph of the garden problem situation with two possible widths indicated.

**EXAMPLE 1**     THE GARDEN PROBLEM SITUATION: Finding the Width for a Given Area

Use a graphical representation to find the width (to the nearest tenth) of the garden when its area is 63 ft². That is, find the values of $w_0$ and $w_1$ in Figure 3.51.

SOLUTION

Enter $y_1 = 63$ and $y_2 = x(24 - 2x)$, as shown in Figure 3.52.

$$Y_1 = 63$$
$$Y_2 = X(24 - 2X)$$

**FIGURE 3.52** Enter $y_1 = 63$ and $y_2 = x(24 - 2x)$.

Because only 24 ft of fence is available, the largest that $w$ can be is 12. That is, $x$ must be between 0 and 12, so choose the viewing window [0, 12] by [0, 84]. Figure 3.53 shows that there are two solutions to this problem. Using ZOOM-IN, we find that the solutions are $w = 3.879$ and $w = 8.121$ with an error of at most 0.01. The width of the garden (to the nearest tenth) is either approximately 3.9 ft or 8.1 ft when its area is 63 ft$^2$.

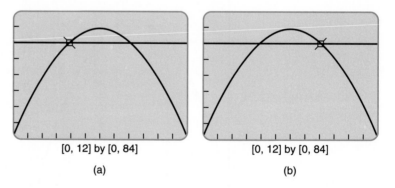

|  [0, 12] by [0, 84]  |  [0, 12] by [0, 84]  |
| :---: | :---: |
| (a) | (b) |

**FIGURE 3.53** Use ZOOM-IN to find the solutions.

Figure 3.54 shows that the shapes of these two solution rectangles are quite different.

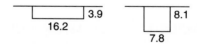

**FIGURE 3.54** Sketch of the two rectangles with width 3.9 and 8.1 and both with area about 63 ft$^2$.

The next two examples revisit the **Baseball Problem Situation.** In both examples, a grapher is used.

### EXAMPLE 2    THE BASEBALL PROBLEM SITUATION: Finding the Time

Recall that if a ball is thrown straight up in the air with an initial velocity of 88 ft/sec, its height after $t$ seconds is given by the algebraic representation $h = -16t^2 + 88t$. Find the time (to the nearest tenth of a second) at which the ball will be 110 ft above the ground on its way down.

SOLUTION

Find the graph of $y = 110$ and the graph of $y = -16x^2 + 88x$ in the window $[0, 5.5]$ by $[0, 150]$.

Figure 3.55 shows that the ball passes through the 110-ft height both on its way up and its way down. Using ZOOM-IN, we can determine that the time on the way down is $t = 3.579$ with an error of at most 0.01.

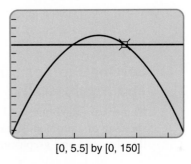

[0, 5.5] by [0, 150]

**FIGURE 3.55**  The graph of $y = 110$ and $y = -16x^2 + 88x$.

Rounding to the nearest tenth of a second, we conclude that the ball has a height of 110 ft on the way down when the time is approximately $t = 3.6$ sec.    ∎

### EXAMPLE 3    THE BASEBALL PROBLEM SITUATION: Finding the Time

Using the algebraic representation $h = -16t^2 + 88t$ from Example 2, how many seconds after the ball is thrown straight up into the air does it reach a height of 130 ft?

**GRAPHER NOTE**

To graph $y = -16t^2 + 88t$ on a grapher, we must use variables $x$ and $y$, so we enter the equation $y = -16x^2 + 88x$. Because $t$ represents the time after the ball has been thrown, the window is chosen to represent those positive values of $t$ (or $x$) when the ball is in the air.

**TRY THIS**

Set the mode of your grapher to Parametric. Then enter $X_{1T} = 3$ and $Y_{1T} = -16T^2 + 88T$. On the WINDOW menu set $T$min= 0, $T$max= 5.5, and $T$step= 0.1. Graph and trace. Hold down either the right or left arrow key for a simulation of the ball. For what values of $T$ is the ball on the rise and when is it falling?

SOLUTION

Find the graph of $y = 130$ and $y = -16x^2 + 88x$ in the window $[0, 5.5]$ by $[0, 150]$ (see Figure 3.56).

It is evident from Figure 3.56 that the ball never reaches a height of 130 ft.

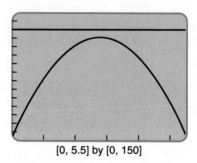

[0, 5.5] by [0, 150]

**FIGURE 3.56** The graph of $y = 130$ and $y = -16x^2 + 88x$.  ■

## Solving Problems Both Graphically and Algebraically

Numerical, algebraic, and graphical representations all contribute to our understanding of a problem situation. In this section, we explore the interaction between the algebraic and graphical representations.

□                                                                                          □

### THE 500-MI TRIP PROBLEM SITUATION

Birds, insects, walkers, joggers, trucks, and spacecraft all travel at different speeds. Describe all the combinations of average speed and time of travel for a 500-mi trip.

□                                                                                          □

**EXAMPLE 4**   THE 500-MI TRIP PROBLEM SITUATION:
Finding Representations

Find a numerical, algebraic, and graphical representation of this problem situation.

**REMINDER**

The last row of the numerical representation table is found by observing the pattern throughout the table.

SOLUTION
**Numerical Representation.**

| Average Speed(mph) | Time of Trip(hr) | Rate·Time |
|---|---|---|
| 10 | $500/10 = 50$ | 500 |
| 15 | $500/15 = 33.33$ | 500 |
| 20 | 25 | 500 |
| 25 | 20 | 500 |
| 30 | 16.67 | 500 |
| 35 | 14.29 | 500 |
| 40 | 12.5 | 500 |
| 45 | 11.11 | 500 |
| 50 | 10 | 500 |
| $r$ | $500/r = t$ | $r \cdot t = 500$ |

**Algebraic Representation.** Because distance = rate × time, the equation $d = r \cdot t$ describes this situation, where $d$ represents distance, $r$ represents rate, and $t$ represents time. Therefore the algebraic representation for this problem situation is

$$r \cdot t = 500, \qquad \text{or} \qquad t = \frac{500}{r}.$$

**Graphical Representation.** Two graphical representations are shown in Figure 3.57. One is drawn by plotting points and the other by a grapher. Remember to rewrite the algebraic representation in terms of $x$ and $y$ for use on your grapher.

**TRY THIS**

Use your grapher to plot the points in the table. Refer to your grapher lab manual.

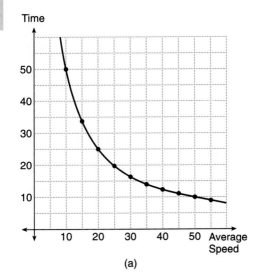

(a)

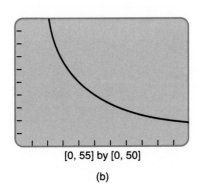

[0, 55] by [0, 50]

(b)

**FIGURE 3.57** (a) Points of the table in Example 4 plotted with the graph of $y = 500/x$. (b) The graph of $y = 500/x$.

Notice that the numerical and algebraic representations of the problem situation are an integral part of finding the graphical representation, which is the most visual. Example 3 illustrates that many problems can be solved simply by a visual inspection of the graph. In the past, a graphical representation might have been used to estimate a solution to a problem. With a modern grapher, a graphical representation can be used to find solutions to a specified accuracy.

**EXAMPLE 5**   THE 500-MI TRIP PROBLEM SITUATION: Finding the Average Speed

Find the average speed (to the nearest tenth) for a trip of 500 miles that takes 27 hours. Use both a graphical and an algebraic method and show numerical support.

SOLUTION

**Solve Graphically.** Draw the graph of $y = 27$ and $y = 500/x$ in the $[0, 55]$ by $[0, 50]$ window. Using ZOOM-IN, we find that $x = 18.519$ is a solution with an error of at most $0.01$ (see Figure 3.58).

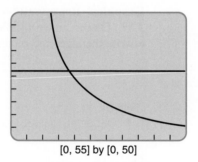

$[0, 55]$ by $[0, 50]$

**FIGURE 3.58**   Graphs of $y = 27$ and $y = \frac{500}{x}$.

If the average speed (to the nearest tenth mile per hour) is 18.5 mph, the trip will take 27 hours.

**Confirm Algebraically.** The algebraic representation of the problem situation is $t = 500/r$. We want to find $r$ when $t = 27$.

Substitute $t = 27$ into the equation $t = 500/r$ and solve for $r$.

$$t = \frac{500}{r} \qquad \text{Substitute } t = 27 \text{ into this equation.}$$

$$27 = \frac{500}{r} \qquad \text{Solve this equation for } r.$$

$$27r = 500$$

$$r = \frac{500}{27} = 18.\overline{518}$$

If the average speed is 18.5 mph, the trip will take approximately 27 hours.

**Support Numerically.** The table in Figure 3.59 shows the average speed is about 18.5 when $x = 27$ and $y_2 = 500/x$.  ■

| X | Y$_2$ | |
|---|---|---|
| 25 | 20 | |
| 26 | 19.231 | |
| 27 | 18.519 | |
| 28 | 17.857 | |
| 29 | 17.241 | |
| 30 | 16.667 | |
| 31 | 16.129 | |
| Y2 = 18.5185185185 | | |

**FIGURE 3.59** A table where $x = $ time and $y_2 = 500/x$. The entries in the first column represent time and the second column speed.

In both the graphical and algebraic solutions to Example 5, an algebraic representation was used. The next example illustrates how to find an algebraic representation of a problem situation.

---

☐                                                                            ☐

### RENT-ALL STORE PROBLEM SITUATION

Rent-All Appliance rents carpet cleaners for a flat amount of $8 for the entire rental period plus $1.75 for each hour the carpet cleaner is rented. What is the total cost of the rental?

☐                                                                            ☐

---

**EXAMPLE 6**     THE RENT-ALL STORE PROBLEM SITUATION: Finding Representations

Find an algebraic and a graphical representation of the **Rent-All Store Problem Situation**.

SOLUTION

**Algebraic Representation.** Let $t = $ the number of hours the carpet cleaner is used and let $R = $ the total rent.

> **Fixed rent** (in dollars): 8
>
> **Hourly rate rent** (in dollars): $1.75t$
>
> **Total rent(R)** (in dollars): $8 + 1.75t$

The algebraic representation for this problem is $R = 1.75t + 8$.

**Graphical Representation.** Figure 3.60 shows the graph of $y = 1.75x + 8$ in the window $[0, 5]$ by $[0, 15]$.

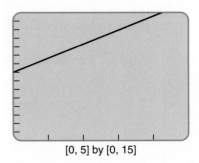

[0, 5] by [0, 15]

**FIGURE 3.60**   A graph of $y = 1.75x + 8$.

**EXAMPLE 7**   THE RENT-ALL STORE PROBLEM SITUATION: Finding Rental Time

Using the **Rent-All Store Problem Situation,** determine how many hours the carpet cleaner was rented if the total rent is $11.50.

SOLUTION

**Solve Graphically.** Graph $y = 1.75x + 8$ and $y = 11.5$ in the window $[0, 5]$ by $[0, 15]$ and zoom in to find the point of intersection of the two graphs.

The $x$-coordinate of this point of intersection is 2.00 with an error of at most 0.01. Therefore we conclude that the carpet cleaner was rented for 2 hours (see Figure 3.61).

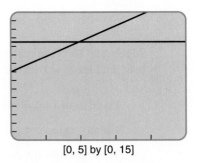

[0, 5] by [0, 15]

**FIGURE 3.61**   The graph of $y = 11.5$ and $y = 1.75x + 8$.

| X | Y1 | |
|-----|--------|---|
| 1.5 | 10.625 | |
| 2 | 11.5 | |
| 2.5 | 12.375 | |
| 3 | 13.25 | |
| 3.5 | 14.125 | |
| 4 | 15 | |
| 4.5 | 15.875 | |
| Y1 = 1.75X + 8 | | |

**FIGURE 3.62**
Numerical support for Example 7.

**Confirm Algebraically.** The algebraic representation of the problem situation is $R = 1.75t + 8$. We want to find $t$ when $R = 11.5$.

$$1.75t + 8 = 11.5$$

$$1.75t = 3.5$$

$$t = \frac{3.5}{1.75} = 2$$

The carpet cleaner was rented for 2 hours.

The table in Figure 3.62 supports the cost of the rental for 2 hours to be $11.50. The remaining entries in the $y_1$ column correspond to the cost of renting the carpet cleaner for a time ranging from 1.5 to 4.5 hours. ∎

## Exercises for Section 3.4

In Exercises 1–9, use algebraic and graphical representations to solve problems. You may find it helpful to first develop a numerical representation as an aid in finding an algebraic representation.

1. *Garden Problem Situation.* Sally has 30 ft of fence to surround a rectangular garden on three sides.

 a) The following figure was used to analyze the garden problem when 24 ft of fence is used. How does this figure change when 30 ft of fence is to be used to surround the garden?

b) Find an algebraic representation of this problem situation using 30 ft of fence.

c) Sketch a graphical representation of the problem situation using 30 ft of fence. Label both the horizontal and vertical axes using words "width" and "area."

d) Find (on a grapher) a graphical representation of this problem situation using 30 ft of fence. Trace the grapher window shown here and sketch the graph you found.

[0, 15] by [0, 150]
For Exercise 1 (d).

e) Repeat the graph using the window [0, 25] by [−100, 200]. A portion of this graph does not refer to the problem situation. Which values of $x$ on this graph do not represent this problem situation? Why?

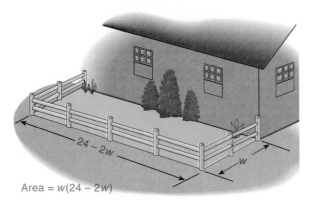

Area = $w(24 - 2w)$

24 − 2w

w

For Exercise 1 (a).

f) Use a graphical representation to find the width (to the nearest tenth of a foot) of the garden if its area is 102 ft$^2$.

g) Use trace coordinates to estimate the width of the garden when its area is as great as possible.

**2.** *Garden Problem Situation.* Scott has 72 ft of fence to surround his rectangular garden on three sides.

a) Find an algebraic representation of this problem situation.

b) Find (on a grapher) a graphical representation of this problem situation. Trace the grapher window shown here and sketch the graph you found.

[0, 15] by [0, 150]

For Exercise 2 (b).

c) Repeat the graph using the window [0, 50] by [−100, 750]. A portion of this graph does not refer to this problem situation. What value of $X$max should be used if window [0, $X$max] by [0, 750] shows only the part of the graph that represents the problem situation?

d) Use a graphical representation to find the width (to the nearest tenth of a foot) of the garden if its area is 398 ft$^2$.

e) Use trace coordinates to estimate the width of the garden when its area is as great as possible.

**3.** *Corral Problem Situation.* Suppose 96 ft of fence is to be used to enclose two pens against a building, as shown in the following figure. Find the areas of the two pens.

For Exercises 3 (a) and 3 (b).

a) Recall that a numerical representation of a problem often helps in finding an algebraic representation. Consequently find at least 6 rows of data in a numerical representation of the problem.

b) Find an algebraic representation of the problem situation using $x$ to represent the width, as shown in the figure.

c) Find a graphical representatation of the problem situation and sketch the graph you found, indicating the scale. Trace the grapher window shown next and sketch the graph you found.

[0, 32] by [0, 800]

For Exercise 3 (c).

d) Use a graphical method to find the width of each pen (to the nearest tenth of a foot) if the total area of the two pens is 679 ft.

e) Use trace coordinates to estimate the maximum area possible for the two pens using 96 ft of fence.

4. *Profits of a Bookstore.* A bookstore has a 40% markup from the wholesale price to the retail price.

a) Find an algebraic representation of this problem situation.

b) Find a graphical representation of this problem situation. Use an Integer window $[0, X\text{max}]$ by $[0, Y\text{max}]$. (See your grapher lab manual to learn how to find this window.)

c) What are the wholesale prices for books that sell at retail for $57.40, $40.60, $30.80, and $19.60?

5. *Projectile Motion.* It can be shown that when a baseball is thrown straight up into the air with a velocity of 50 ft/sec, $h = -16t^2 + 50t$ is an algebraic representation (ignoring air resistance) for the ball's height $t$ seconds after being thrown.

For Exercise 5.

a) Find a graph of the algebraic representation in the window $[0, 10]$ by $[0, 100]$. Use the trace process to determine a window $[0, X\text{max}]$ by $[0, Y\text{max}]$ so that the graph of the problem situation looks much like Fig. 3.53.

b) How many seconds (to the nearest tenth) after the ball is thrown is the ball 24 ft high?

c) Estimate the maximum height of the ball (to the nearest tenth). How many seconds after it is thrown into the air does it reach this height?

d) If the ball is thrown into the air with an initial velocity of $v_0$ feet per sec, the algebraic representation for the height of the ball $t$ seconds

after it is thrown is $h = -16t^2 + v_0 t$. Find what initial velocity (to the nearest whole number) will result in the ball reaching a maximum of 100 feet. (*Hint:* Begin by graphing $h = -16t^2 + 60t$ in the window $[0, 5.5]$ by $[0, 100]$ and adjust the value of $v_0$ by trial and error until achieving a maximum height of 100 ft.)

6. *Comparing Speed and Time for a Trip.* Consider all combinations of time and average speed for a 300-mile trip.

a) Find numerical and algebraic representations of this problem situation.

b) Find a graphical representation of this problem situation, both by hand and using a grapher. Graph average speed on the horizontal axis and time on the vertical axis.

c) What is the average speed (to the nearest tenth) if the time for the trip is 47 hr?

d) What is the average speed (to the nearest tenth) if the time for the trip is 7 hr?

e) Did you use an algebraic or a graphical method to solve (c) and (d)? Solve each again, using whichever method you did not use first. Which method do you prefer and why?

7. *Rental Company Charges.* Suppose an appliance rents for a flat fee of $12 for the rental period plus $1.80/hr. What is the total cost of the rental in terms of time $t$?

a) Find an algebraic representation for this problem situation.

b) Find a graphical representation for this problem situation.

c) How much would it cost to rent the appliance for a 24-hr period?

d) How long did a customer rent this appliance if the total rental was $33.15? Solve this problem using both graphical and algebraic methods.

8. *Car Rental Costs.* A car rental company will rent to its customers a full-sized car for $39.95/day plus $0.25/mi.

a) How much would it cost to rent this car for (a) 2 days and 125 mi? (b) 2 days and 148 mi? (c) 2 days and 224 mi?

b) Suppose the car was rented for one day and driven $x$ miles. Find an algebraic representation for the total rent $R$ in terms of miles $x$ traveled.

c) Find a graph of this algebraic representation in the window $[0, 285]$ by $[0, 100]$. (On the grapher we use, this is an Integer window in which each $x$-coordinate is a multiple of 3. You might need to use a slightly different window.)

d) If the total rent was $71.45, how many miles was the car driven? Find the solution both graphically and algebraically.

e) If you plan to rent a car for 1 day and drive it for 120 mi, would you prefer paying $39.95/day plus $0.25/mi or $54.95/day plus $0.15/mi. Explain your answer using both algebraic and graphical methods.

9. *Investment Analysis.* One account that carries slightly greater risk pays 9% interest and another account with less risk pays 7% interest. Morris has $13,000 to invest.

a) Find an algebraic representation for the total interest at the end of 1 yr if Morris invests $x$ in the 9% account and the remaining money in the 7% account.

b) To find a graphical representation of this problem situation in the window $[0, X\text{max}]$ by $[0, Y\text{max}]$, find the values for $X\text{max}$ and $Y\text{max}$ and explain why you chose them.

c) Find a graphical representation for the problem situation using this window.

### TRANSLATING WORDS TO SYMBOLS

Write an equation for each of the following sentences.

10. $y$ is 25% of $x$.

11. $y$ is 25% more than $x$.

12. $y$ is less than $x$ by 25% of $x$.

### EXTENDING THE IDEAS

Exercises 14–15 are related to Exercise 13.

13. *Catch-up Time.* A truck traveling 40 mph begins 12 mi ahead of a car traveling 60 mph. How far will the truck have traveled when the car catches up to it? Explain how the graphs $y_1 = 60x$ and $y_2 = 40x + 12$, both produced in the viewing window $[0, 1]$ by $[0, 50]$, can be used to answer this question.

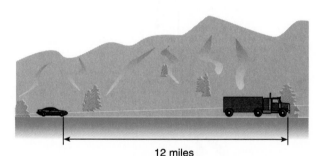

12 miles

For Exercise 13.

14. *Motion Simulation.* If your grapher has Parametric mode, enter MODE, select Simultaneous mode and Parametric mode and enter $x_{1T} = 60t$, $y_{1T} = 2$, $x_{2T} = 40t + 12$, and $y_{2T} = 5$ as shown on the screen below.

X1T = 60T
Y1T = 2
X2T = 40T + 12
Y2T = 5

For Exercise 14.

Choose the following WINDOW settings: Tmin= 0, Tmax= 2, Tstep= .01 and the window [0, 120] by [0, 6]. After completing this setup, press ENTER . Explain how this image is related to Exercise 13.

15. *Analysis of Motion Simulation.* Experiment and discover how the output of Exercise 14 together with the TRACE feature can be used to find when the car passes the truck in Exercise 13.

LOOKING BACK—MAKING CONNECTIONS

16. *Temperature Comparisons.* The formulas for converting Celsius ($C$) to Fahrenheit ($F$) and conversely are $F = \frac{9}{5}C + 32$ and $C = \frac{5}{9}(F - 32)$, respectively.

Explain how the trace process can be used on the following graphs of the equations to find conversions from Celsius to Fahrenheit and back?

$$Y1 = (9/5)X + 32$$
$$Y2 = (5/9)(X - 32)$$

For Exercise 16.

17. What is the meaning of the point of intersection of the two graphs found in Exercise 16?

---

**3.5  Solving Systems of Equations by Graphing**

Verifying the Solution of a System of Equations ▪ Solving a System with Non-integer Solutions ▪ Solving Problems Using a System of Equations ▪ Types of Systems of Equations

Whenever an equation like $3x - 2 = 5$ is solved by the multigraph method, graphs of both $y = 3x - 2$ and $y = 5$ are found in the same viewing window. The solution is the $x$-coordinate of the point of intersection of the two graphs. In this section, we alter that common situation only slightly. Instead of one of the two graphs being a horizontal line, we will consider any two linear equations taken as a pair. We begin this discussion with an exploration.

## EXPLORE WITH A GRAPHER

**GRAPHER NOTE**

It is often convenient to use zoom factors of $X$Fact= 10 and $Y$Fact= 10 when you are using the Integer window.

Graph each pair of the following linear equations on the same coordinate system. Begin with the Integer window and zoom out if necessary, until you see both lines. Then use zoom-in to find the $x$- and $y$- coordinates of the point of intersection of the lines.

1. $y = x - 30$ and $y = -x + 6$
2. $y = 2x + 50$ and $y = x + 150$
3. $y = -2x + 180$ and $y = 3x - 320$

Each pair of equations in this exploration is an example of a **system of linear equations**. For example, the first pair in the exploration is the following system of equations:

$$y = x - 30$$
$$y = -x + 6$$

You have found that the $x$- and $y$-coordinates of the point of intersection of the two lines are $x = 18$ and $y = -12$. We call the ordered pair $(18, -12)$ **the solution of the system of equations**.

## Verifying the Solution of a System of Equations

**REMINDER**

"Confirm" means finding an exact solution by a pencil-and-paper method or with algebra. "Support numerically" indicates we are finding an approximation, usually on a grapher, that supports the exact solution.

A specific point $(a, b)$ is a solution of a system of two equations in $x$ and $y$ if the substitution $x = a$ and $y = b$ satisfies both equations of the system.

**EXAMPLE 1**    Confirming the Solution of a System of Equations

Confirm that $(3, 5)$ is the solution of the following system of equations:

$$6x - 2y = 8$$
$$-4x + 6y = 18$$

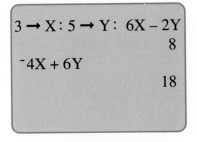

**FIGURE 3.63** Grapher support of the solution of Example 1.

**SOLUTION**

**Confirm Numerically Using a Paper-and-Pencil Method.** Replace $x$ with 3 and $y$ with 5 in each equation.

$$6(3) - 2(5) = 8 \qquad \text{True Statement}$$

$$-4(3) + 6(5) = 18 \qquad \text{True Statement}$$

Because both equations are true, the ordered pair $(3, 5)$ is the solution to the system.

**Support Numerically.** Store the value 3 as the variable $x$ and 5 as $y$. Find the value of the expressions $6x - 2y$ and $-4x + 6y$, as shown in Figure 3.63.  ■

Example 1 shows how to verify a solution once you have found a candidate for a solution. But how do you find a candidate for a solution? Use the technique of the exploration. Once this candidate has been found, use either the paper-and-pencil or the grapher method of Example 1 to verify that it is a solution.

**EXAMPLE 2**    Solving a System of Equations Graphically

Use a grapher to solve the following system of equations:

$$y = 3x - 2$$

$$y = -0.5x + 5$$

**SOLUTION**

**Solve Graphically.** Find the graph of the system of equations in the Integer window (see Figure 3.64).

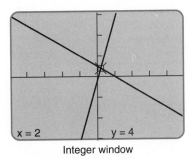

Integer window

**FIGURE 3.64** A graph of $y = 3x - 2$ and $y = -0.5x + 5$.

Use the TRACE feature to move the cursor to the point of intersection, which is the ordered pair $(2, 4)$. The solution is the pair $x = 2$ and $y = 4$.

**Support Numerically.** Figure 3.65 shows numerical support that $(2, 4)$ is the solution in (a) by storing the variable and (b) using a table.

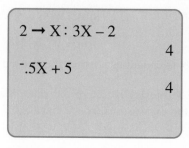

| X | Y1 | Y2 |
|---|-----|-----|
| 1 | 1 | 4.5 |
| 2 | 4 | 4 |
| 3 | 7 | 3.5 |
| 4 | 10 | 3 |
| 5 | 13 | 2.5 |
| 6 | 16 | 2 |
| 7 | 19 | 1.5 |
| X = 2 | | |

**FIGURE 3.65** Numerical support for the solution $(2, 4)$ to the system of equations.

When both equations in the system are given in slope–intercept form as they were in Example 2, the equations are already in the form needed to enter into the grapher. Example 3 illustrates that it is often necessary to first write each equation in slope–intercept form.

**EXAMPLE 3**  Solving a System of Equations Graphically

Use a grapher to solve the followingsystem of equations:

$$x + y = 9$$
$$2x - y = 15$$

SOLUTION

**Solve Graphically.** Change each equation to the slope–intercept form, as follows:

Change the system $\begin{cases} x + y = 9 \\ 2x - y = 15 \end{cases}$ to the equivalent form $\begin{cases} y = 9 - x \\ y = 2x - 15 \end{cases}$.

Graph the system $y = 9 - x$ and $y = 2x - 15$ in the Integer window (see Figure 3.66).

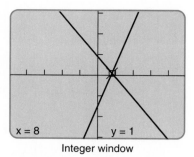

Integer window

**FIGURE 3.66** The graph of the system $x + y = 9$ and $2x - y = 15$ in the Integer window.

Moving the cursor to the point of intersection, we see that the coordinates of this point are $(8, 1)$. The solution is $x = 8$ and $y = 1$.

**Support Numerically.** Figure 3.67 shows numerical support that $(8, 1)$ is the point of intersection of the two graphs.

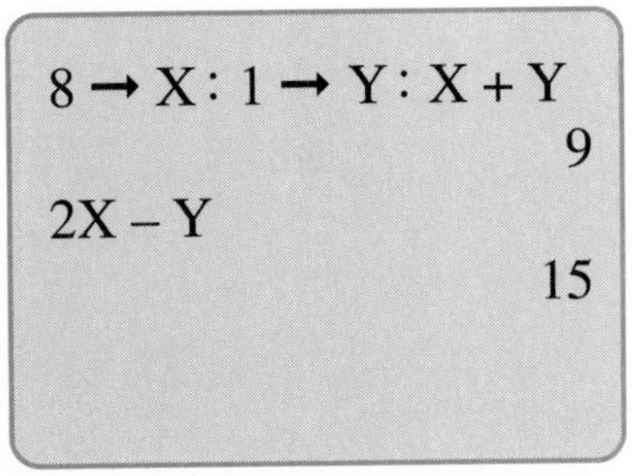

```
8 → X : 1 → Y : X + Y
                      9
2X – Y
                     15
```

**FIGURE 3.67** Numerical support of the solution for Example 3. ▪

In Examples 1, 2, and 3, the solution of the given system was a pair of integers. Consequently it was convenient to use the Integer window so that the trace coordinates were integers. We recommend that for systems of linear equations, you always begin with the Integer window, even when it is likely that the solution is not a pair of integers.

## Solving a System with Non-integer Solutions

A grapher is particularly handy when the coefficients are not integers. In such cases, the solution is usually not a pair of integers, and ZOOM-IN probably is needed. Example 4 illustrates such a system.

**EXAMPLE 4**    Solving a System with Non-integer Solutions

Use a grapher to solve the following system of equations:

$$3.8x - 2y = -4.3$$

$$0.3x - 2.5y = -7.6$$

SOLUTION

**Solve Graphically.** Change each equation to the slope–intercept form.

$$\begin{cases} 3.8x - 2y = -4.3 \\ 0.3x - 2.5y = -7.6 \end{cases} \text{ becomes } \begin{cases} y = 1.9x + 2.15 \\ y = 0.12x + 3.04 \end{cases}$$

Figure 3.68 shows the graph of this system in the Integer window.

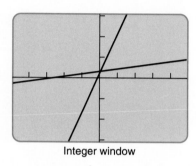

Integer window

**FIGURE 3.68** A graph of the system $3.8x - 2y = -4.3$ and $0.3x - 2.5y = -7.6$ in the Integer window before zooming in.

Using zoom-in with $X$Fact= 10 and $Y$Fact= 10, we see that the point of intersection is $(0.50, 3.10)$. Consequently the solution to the system is $x = 0.50$ and $y = 3.10$.

**Support Numerically.** Figure 3.69 shows numerical support of this solution. ■

```
0.5 → X: 3.10 → Y:
3.8X – 2Y
                ⁻4.3
0.3X – 2.5Y
                ⁻7.6
```

**FIGURE 3.69** Grapher support of the solution for Example 4.

## Solving Problems Using a System of Equations

Sometimes the algebraic representation of a problem situation is a system of linear equations. Here is an example.

---

**THEATER ATTENDANCE PROBLEM SITUATION**

The drama class sold 924 tickets to the class play. Adult tickets cost $4.00/each and student tickets cost $2.50/each. The total amount of money taken in by the performance was $2796.

---

Example 5 illustrates how to find an algebraic representation of this problem situation.

**EXAMPLE 5**    THE THEATER ATTENDANCE PROBLEM SITUATION:
Finding a Representation

Find an algebraic representation of the **Theatre Attendance Problem Situation.**

SOLUTION

Let $x$ be the number of student tickets sold and $y$ the number of adult tickets sold.

**Dollars from student tickets:** $2.50x$

**Dollars from adult tickets:** $4.00y$

**Total number of tickets sold:** $924$

**Total amount of money taken in:** $2796.

This information leads to this system of equations.

$$x + y = 924$$

$$2.50x + 4.00y = 2796$$

This system of two equations is the algebraic representation of the problem situation. ■

In Example 6 we now answer a question about the problem situation.

**EXAMPLE 6**    THE THEATER ATTENDANCE PROBLEM SITUATION:
Finding the Attendance

How many students attended the theater given the data in the attendance problem?

SOLUTION

We need to find the point of intersection of the two graphs of the equations in this system.

$$x + y = 924$$

$$2.50x + 4.00y = 2796$$

**Solve Graphically.** Change each of the following equations to the slope–intercept form.

$$\begin{cases} x + y = 924 \\ 2.50x + 4.00y = 2796 \end{cases} \rightarrow \begin{cases} y = -x + 924 \\ y = -0.625x + 699 \end{cases}$$

Begin with the Integer window and zoom out (with $X$Fact= 10 and $Y$Fact= 10) until you see the graphs. Then zoom in at the point of intersection and use TRACE to determine that $(600, 324)$ is the point of intersection (see Figure 3.70).

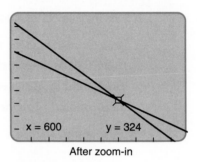

After zoom-in

**FIGURE 3.70** The graphs of $x + y = 924$ and $2.50x + 4.00y = 2796$.

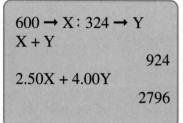

$600 \rightarrow X : 324 \rightarrow Y$
$X + Y$
$\qquad\qquad 924$
$2.50X + 4.00Y$
$\qquad\qquad 2796$

**FIGURE 3.71**
Numerical support of the solution for Example 6.

**Support Numerically.** Figure 3.71 shows numerical support of this solution. There were 600 students in attendance.  ■

In business, the total **cost** ($C$) of producing an item is typically the initial cost plus the cost per unit. The amount of money generated by selling the units is called **revenue** ($R$). The point where the cost and the revenue are the same is the **break-even point**. The units sold after this point produce a profit. By finding an algebraic representation for both the cost and the revenue and graphing each of these equations, we can find the point of intersection, that is, the break-even point.

☐ ☐

## BREAK-EVEN PROBLEM SITUATION

Fotobadge, Inc. invests $5000 in equipment to produce identification badges for employees of other companies. Each badge costs $0.65 to produce and each is sold for $1.35.

☐ ☐

In Example 7, focuses on this problem situation.

**EXAMPLE 7**  THE BREAK-EVEN PROBLEM SITUTATION:
Finding the Break-even Point

How many badges must be sold before Fotobadge, Inc. begins making a profit?

SOLUTION
**Solve Graphically.**

1. Begin by finding the system of equations that is an algebraic representation for the problem situation. Let

$$x = \text{the number of badges sold}$$

$$C = \text{the total cost to produce}$$

$$R = \text{the total revenue}$$

$$C = 5000 + 0.65x \qquad \text{This equation represents the total cost.}$$

$$R = 1.35x \qquad \text{This equation represents the total revenue.}$$

2. Find the graph of the algebraic representation of the problem situation in an appropriate window. Variables $x$ and $y$ must both be positive and sales of 10,000 badges result in costs under \$20,000, so we choose the window [0, 10000] by [0, 20000] (see Figure 3.72).

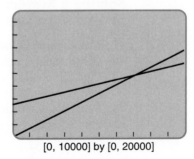

[0, 10000] by [0, 20000]

**FIGURE 3.72**  The graphs of $y = 5000 + 0.65x$ and $y = 1.35x$.

3. Use zoom-in to find the $x$-coordinate of the point of intersection, with an error of at most 0.01, to be $x = 7142.857$. (Note that only integer values make sense as a solution in this problem.)
   The company must sell 7143 badges to show its first profit.

**Support Numerically.** Figure 3.73 shows a grapher-generated table with $y_1$ representing cost and $y_2$ representing revenue. The 7143 entry for $x$ shows the cost to be 9643 and the revenue as 9643.1. This is the first entry to show a profit.

| X | Y1 | Y2 |
|------|--------|--------|
| 7140 | 9641 | 9639 |
| 7141 | 9641.7 | 9640.4 |
| 7142 | 9642.3 | 9641.7 |
| 7143 | 9643 | 9643.1 |
| 7144 | 9643.6 | 9644.4 |
| 7145 | 9644.3 | 9645.8 |
| 7146 | 9644.9 | 9647.1 |

X = 7143

**FIGURE 3.73** A grapher-generated table. Compare the $y_1$ (cost) entry with the $y_2$ (revenue) entry for each $x$ (number of badges).

Here is a summary of the procedure for solving a system of linear equations using a grapher.

---

### Solving a System of Two Linear Equations in Two Variables Using the Graphing Method

1. Change each equation to slope–intercept form.
2. Graph each equation in the same window.
3. Find the point of intersection (if it exists). If necessary, zoom in and find the solution with an error of at most 0.01.
4. Support numerically on your grapher that this ordered pair makes both equations in the system true.

---

## Types of Systems of Equations

The graph of each system of equations we have considered so far has been a pair of intersecting straight lines. The point of intersection of the two lines was the unique solution to the system. However, there are two other possible situations. The graphs could be either a pair of distinct parallel lines or the same line. In the first case, there is no solution to the two equations, and in the second case, there is an infinite set of possible solutions.

When a system of equations has a unique solution (a pair of intersecting lines), we say that the equations in the system are **independent**. When a system has no solution (a pair of parallel lines), we say that the equations in the system are **inconsistent**. And when the system has an infinite number of solutions (the same line), we say that the equations are **dependent** (see Figure 3.74).

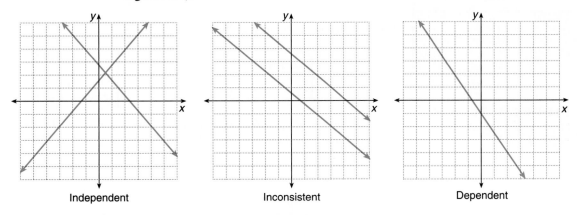

Independent          Inconsistent          Dependent

**FIGURE 3.74**  Three types of systems of linear equations.

The next examples illustrate how to detect the inconsistent and dependent cases.

**EXAMPLE 8**    Solving an Inconsistent System

Solve the following system of equations:

$$3x - 6 = y$$

$$-6x + 2y = 9$$

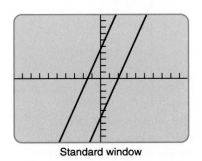

Standard window

**FIGURE 3.75**  Graph of $y_1 = 3x - 6$ and $y_2 = 3x + 4.5$.

SOLUTION
Change each equation to the slope–intercept form.

$$\begin{cases} 3x - 6 = y \\ -6x + 2y = 9 \end{cases} \quad \text{becomes} \quad \begin{cases} y = 3x - 6 \\ y = 3x + 4.5 \end{cases}$$

The slopes of the lines are equal and the $y$-intercepts are not equal; therefore the lines are parallel. The grapher screen in Figure 3.75 provides supporting visual evidence.

There is no solution to the given system of equations.    ∎

**EXAMPLE 9** Solving a Dependent System

Solve the following system of equations:

$$3x + 4y = 12$$
$$8y = -6x + 24$$

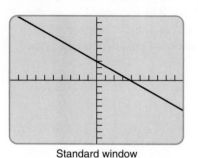

Standard window

**FIGURE 3.76** Graph of $3x + 4y = 12$ and $8y = -6x + 24$.

SOLUTION

Change each equation to the slope–intercept form.

$$\begin{cases} 3x + 4y = 12 \\ 8y = -6x + 24 \end{cases} \quad \rightarrow \quad \begin{cases} y = \dfrac{-3}{4}x + 3 \\ y = \dfrac{-3}{4}x + 3 \end{cases}$$

The slopes are equal, the $y$-intercepts are equal, and the two graphs are the same line (see Figure 3.76). Consequently there are an infinite number of solutions. ■

## Exercises for Section 3.5

In Exercises 1–4, use your grapher to determine if the given ordered pair is the solution to the system of equations.

**1.** $(1, 2)$: $\begin{cases} 2x + 3y = 8 \\ 3x - 8y = -13 \end{cases}$

**2.** $(-1, 9)$: $\begin{cases} 2x - y = -11 \\ -x + y = 8 \end{cases}$

**3.** $(0, 5)$: $\begin{cases} 4x - 2y = -10 \\ x + y = 6 \end{cases}$

**4.** $(7, 0)$: $\begin{cases} x - y = 7 \\ -2x - y = -14 \end{cases}$

In Exercises 5–16, find a graphical solution to the system of equations in the Integer window. Then find numerical support of the solution.

**5.** $\begin{cases} y = -x + 2 \\ y = 2x - 34 \end{cases}$

**6.** $\begin{cases} y = -2x + 50 \\ y = 5x - 55 \end{cases}$

**7.** $\begin{cases} -3x + y = 35 \\ y + x = -5 \end{cases}$

**8.** $\begin{cases} -2x + y = -32 \\ -2x + 64 = y \end{cases}$

**9.** $\begin{cases} y = -x + 10 \\ -25 = 2x - y \end{cases}$

**10.** $\begin{cases} y = 3x + 33 \\ y = -5x - 87 \end{cases}$

**11.** $\begin{cases} 5x + 3y = 80 \\ x + 2y = -5 \end{cases}$

**12.** $\begin{cases} 3x + 2y = -22 \\ 5x + y = -60 \end{cases}$

**13.** $\begin{cases} x + 2y = 0 \\ 3x - 5y = 99 \end{cases}$

**14.** $\begin{cases} 6x - 2y = -30 \\ 5x - y = -5 \end{cases}$

**15.** $\begin{cases} x + y = 27 \\ 3x - y = 41 \end{cases}$

**16.** $\begin{cases} x - y = 8 \\ 5x - 3y = 68 \end{cases}$

In Exercises 17–26, find the solution, with an error of at most 0.01, to the system of equations by graphing the system. Begin with the Integer window centered at $(0, 0)$ and use zoom-in with $X$Fact= 10 and $Y$Fact= 10.

**17.** $\begin{cases} 5x - 4y = 26 \\ 4x + 2y = 53.3 \end{cases}$

**18.** $\begin{cases} x + 2y = 29.4 \\ 5x - 2y = 45 \end{cases}$

19. $\begin{cases} y = -x + 5 \\ y - 3x = 67 \end{cases}$

20. $\begin{cases} x + y = 26.2 \\ -2x + y = 3.7 \end{cases}$

21. $\begin{cases} x + y = 22 \\ 5x + y = 36 \end{cases}$

22. $\begin{cases} 5x + 2y = -4 \\ x - y = -26.7 \end{cases}$

23. $\begin{cases} x + y = 5.35 \\ 3x - y = 3.65 \end{cases}$

24. $\begin{cases} 5x - y = 21.01 \\ 2x + y = 1.04 \end{cases}$

25. $\begin{cases} x + y = -10.74 \\ 2x - y = 57.27 \end{cases}$

26. $\begin{cases} -x + y = -23.75 \\ 2x - y = 58 \end{cases}$

In Exercises 27–32, find the solution to the system of equations by graphing the system and using zoom-in as necessary to find a solution with an error of at most 0.01.

27. $\begin{cases} 3x + y = 20 \\ x - 2y = 10 \end{cases}$

28. $\begin{cases} 5x - y = 7 \\ 3x + y = 2 \end{cases}$

29. $\begin{cases} 9x + y = 8 \\ x - y = 0 \end{cases}$

30. $\begin{cases} 2x - 7y = 14 \\ 4x - y = 9 \end{cases}$

31. $\begin{cases} y = -x + 7 \\ 8 = 2x - y \end{cases}$

32. $\begin{cases} y = 3x + 8 \\ 5x + y = 24 \end{cases}$

In Exercises 33–38, determine whether the system of equations has 0, 1, or an infinite number of solutions. Support your answer on the grapher.

33. $\begin{cases} 3x - y = 5 \\ -3x + y = 6 \end{cases}$

34. $\begin{cases} 2x - y = 1 \\ 2x + 7 = y \end{cases}$

35. $\begin{cases} 3x - 1 = y \\ 2x + y = 4 \end{cases}$

36. $\begin{cases} 4x - 7 = y \\ 12x - 3y = 21 \end{cases}$

37. $\begin{cases} \dfrac{-3}{4}x - 2 = y \\ 3x + 4y = -8 \end{cases}$

38. $\begin{cases} -4x + 10 = y \\ 4x + y = -15 \end{cases}$

39. *Ticket Sale Analysis.* Suppose the Drama Club found out in advance that they would have only 750 people attending the play it was presenting. They

decided to raise the adult ticket price to $5. The student ticket price remained $2.50. They raised a total of $3250. How many adults attended?

a) Let $x$ represent the number of adults attending and $y$ the number of students attending. Write an algebraic representation for the total number of tickets sold.

b) Write an algebraic representation for the total amount of money raised.

c) Solve the system of equations obtained in Exercises 39 (a) and (b) to obtain the number of adults attending the performance.

40. *Analyzing Production Costs.* Acme Corp. has an initial startup cost of $15,000$ to produce glass cutters. The cost per cutter is $0.75 and each sells for $2.25.

a) Write an algebraic representation for the total cost of producing $x$ glass cutters.

b) Write an algebraic representation for the total amount of revenue if $x$ glass cutters are sold.

c) Solve the system of equations obtained in Exercises 40 (a) and (b) to find the break-even point.

41. *Analyzing Production Costs.* Recapture Precious Moments Video Co. finds that it can buy blank video tapes for $1.25/tape. The average cost involved in copying a tape is $5.25. The equipment cost is $8000. The company charges $25 for making one copy.

a) Write an algebraic representation for the total cost of copying $x$ tapes.

b) Write an algebraic representation for the total amount of revenue if $x$ tapes are sold.

c) Solve the system of equations obtained in (a) and (b) to find how many copies must be made and sold before the business breaks even.

42. *Analyzing Garden Dimensions.* Linda has a rectangular garden whose perimeter is 92 ft. The length is 5 ft more than the width.

a) Write an algebraic representation in the form of an equation for the length $l$ in terms of the width $w$.

b) Write an algebraic representation in the form of an equation for the perimeter of the garden in terms of the length $l$ and width $w$.

c) Find the length and the width of Linda's garden by solving the system of equations obtained in (a) and (b).

d) Linda needs to apply liquid fertilizer to the garden. One bottle covers 450 sq. ft. How many bottles must she buy?

## TRANSLATING WORDS TO SYMBOLS

**43.** Write an algebraic expression that means 25% of $x$.

**44.** Write an algebraic expression that means 7.5% of $x$.

**45.** Write an algebraic expression that represents the total cost of producing $x$ bicycles if the initial cost is $12,000$ and the cost per bike is $125.50$.

## EXTENDING THE IDEAS

**46.** Find a system of linear equations in two variables that has $(2, -3)$ as its solution. (Answers will vary.)

**47.** The first equation of a system of linear equations is $2x + 5y = 9$. Find a second equation that would make the system have no solution. (Answers will vary.)

**48.** The first equation of a system of linear equations is $3x - 7y = 10$. Write another equation for the system that would make the system have an infinite number of solutions. (Answers will vary.)

**49.** Can a system of linear equations have exactly two solutions? Why or why not?

**50. Writing to Learn.** Write in your own words how you can determine the number of solutions to a system of equations by graphing the equations.

**51. Writing to Learn.** Write in your own words how you can determine the number of solutions to a system of equations without graphing the equations.

## LOOKING BACK—MAKING CONNECTIONS

**52.** Complete the following in the Integer window.

a) Graph the system of equations $y_1 = -3x + 16$ and $y_2 = 5x - 24$.

b) Next graph the equation $y_3 = y_1 - y_2$.

c) How is the zero of $y_3$ related to the solution of the system?

**53.** Find an equation of the line that has $y$-intercept $(0, 2)$ and slope $m = \frac{2}{3}$ and an equation of the line with $y$-intercept $(0, -6)$ and slope $m = \frac{1}{2}$. Find their point of intersection by solving this system of equations.

**54.** Find an equation of the line that passes through the point $(1, 8)$ with slope $m = -1$ and an equation of the line that passes through the points $(0, -3)$ and $(5, 0)$. Find the point of intersection of these two lines by solving the system of equations.

**3.6**

# Solving Systems
# of Equations Algebraically

The Substitution Method ▪ The Elimination Method ▪ Eliminating
Variables Whose Coefficients Are Not Opposites ▪ Elimination of
Variables Whose Coefficients Are Fractions ▪ Solving Problem
Situations using a System of Equations

In Section 3.5, graphical techniques were used to solve a system of linear equations. The focus of Section 3.6 is on algebraic techniques, that is, the more traditional paper-and-pencil methods. The grapher will be used to support conclusions determined by algebraic techniques.

## The Substitution Method

There are basically two methods for solving a system of two linear equations in two variables. In the first of these, the **substitution method**, you solve one of the equations for one variable, say $y$, and substitute that expression for $y$ in the second equation. The result is one equation in one variable. By solving for this one variable, one of the coordinates of the solution point has been found.

**EXAMPLE 1**    Finding the $x$-coordinate of the Point
of Intersection

Find the $x$-coordinate of the solution to the following system of equations:

$$2x + 3y = 18$$

$$y = 5x - 11$$

Support graphically and numerically.

SOLUTION

**Solve Algebraically.** The second equation has already been solved for $y$. Substitute $5x - 11$ for $y$ in the first equation.

$$2x + 3y = 18 \quad \textbf{Replace } y \textbf{ with } 5x - 11.$$

$$2x + 3(5x - 11) = 18$$

Next, solve the resulting equation for $x$.

$$2x + 3(5x - 11) = 18$$

$$17x - 33 = 18$$

$$17x = 51$$

$$x = 3$$

The $x$-coordinate of the point of intersection is $x = 3$.

**Support Graphically and Numerically.** Figure 3.77 provides grapher support for the algebraic solution.

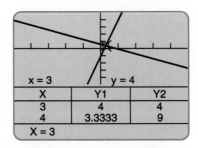

**FIGURE 3.77** Grapher support that the $x$-coordinate of the point of intersection is $x = 3$.

Example 1 illustrates half of the substitution process for solving a system of two linear equations in two variables. One coordinate of the point of intersection was found. Take this value and substitute it into one of the original equations and solve for the second variable. Example 2 illustrates the complete process.

**EXAMPLE 2**    Solving a System of Equations By Substitution

Find the solution to the following system of equations:

$$x - 2y = 12$$

$$x + y = 6$$

Support numerically and graphically.

SOLUTION

**Solve Algebraically.** Solve the second equation for $y$.

$$x + y = 6$$

$$y = 6 - x$$

Substitute this expression for $y$ into the first equation.

$$x - 2y = 12$$

$$x - 2(6 - x) = 12$$

$$x - 12 + 2x = 12$$

$$3x - 12 = 12 \quad \textbf{Add 12 to both sides.}$$

$$3x = 24 \quad \textbf{Divide each side by 3.}$$

$$x = 8$$

Next, find the $y$-coordinate by substituting $x = 8$ into either of the original equations and solve for $y$. We choose the second equation.

$$x + y = 6$$

$$8 + y = 6 \quad \textbf{Replace x with 8.}$$

$$y = -2$$

The solution to the system is $x = 8$ and $y = -2$.

**Support Graphically and Numerically.** Figure 3.78 provides graphical support that $(8, -2)$ is the solution to the system as well as numerical support in the form of a table.

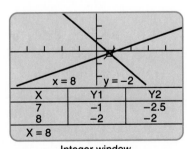

Integer window

**FIGURE 3.78** Grapher support for the solution $(8, -2)$ to Example 2 with $y_1 = 6 - x$ and $y_2 = (1/2)x - 6$.

In Example 2, we first solved for $y$ and substituted that $y$-value into an equation to solve for $x$. It is just as correct to first solve for $x$, and substitute into an equation to solve for $y$. Make whatever choice seems to result in the easiest algebra.

**EXAMPLE 3**   Solving a System of Equations by Substitution

Find the solution to the following system of equations:

$$3x + 6y = 11.5$$
$$x - 3y = 0.5$$

Support numerically and graphically.

SOLUTION

**Solve Algebraically.** Solve one equation for $x$. (It is easier to solve the second equation for $x$ than the first equation for $x$ because the coefficient of $x$ is 1.)

$$x - 3y = 0.5$$
$$x = 0.5 + 3y$$

Substitute this expression into the first equation and solve for $y$.

$$3x + 6y = 11.5 \qquad \text{Replace } x \text{ with } 0.5 + 3y:$$
$$3(0.5 + 3y) + 6y = 11.5$$
$$1.5 + 9y + 6y = 11.5$$
$$1.5 + 15y = 11.5 \qquad \text{Subtract 1.5 from each side.}$$
$$15y = 10 \qquad \text{Divide each side by 15.}$$
$$y = \frac{2}{3}$$

Replace $y$ with $\frac{2}{3}$ in either equation and solve for $x$. We choose to use the second equation.

$$x - 3y = 0.5$$
$$x - 3\left(\frac{2}{3}\right) = 0.5$$
$$x - 2 = 0.5$$
$$x = 2.5 \quad \text{or} \quad \frac{5}{2}$$

The solution to the system of equations is the pair $x = \frac{5}{2}, y = \frac{2}{3}$.

**Support Graphically.** Figure 3.79 shows graphical support of this solution.

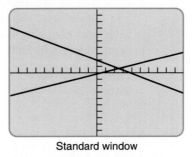

Standard window

**FIGURE 3.79** Graphical support for the solution to Example 3.

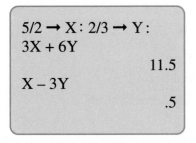

$5/2 \rightarrow X : 2/3 \rightarrow Y :$
$3X + 6Y$
$\qquad\qquad 11.5$
$X - 3Y$
$\qquad\qquad .5$

**FIGURE 3.80**
Numerical support of the solution to Example 3.

**Support Numerically.** Figure 3.80 shows numerical support of this solution.  ■

Here is a summary of the substitution method of solving a system of two linear equations in two variables.

---

### Substitution Method: A Procedure for Solving A System

To solve a system of two linear equations in two variables, do the following:

1. Solve for one of the two variables in terms of the other. (Use either equation.)
2. Substitute this expression into the other equation, resulting in an equation in one variable.
3. Solve this equation to obtain a numerical value for one variable.
4. Substitute this value into either of the original equations and solve for the other variable.

   The resulting ordered pair is a solution of the system.

---

When finding an algebraic solution, it is always a good idea to find graphical or numerical support for the solution of a system of equations.

## The Elimination Method

The second algebraic method for solving systems of two linear equations with two unknowns is the **elimination method**.

In this method, the two equations are added with a goal of eliminating one variable and obtaining a single equation in the other variable. For this method to work, the coefficients of the variable to be eliminated must differ only in sign.

**EXAMPLE 4**   Solving a System of Equations by Elimination

Solve the following system of equations using the elimination method:

$$3x - 5y = 24$$
$$-3x - 2y = -10$$

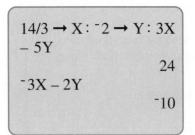

**Standard window**

**FIGURE 3.81** Graphical support for the solution to Example 4.

SOLUTION

**Solve Algebraically.** The coefficients of $x$ differ only in sign. Add the two equations.

$$3x - 5y = 24$$
$$\underline{-3x - 2y = -10}$$
$$-7y = 14 \qquad \textbf{Divide both sides by } -\textbf{7.}$$
$$y = -2$$

Therefore $y = -2$ is the $y$-coordinate of the solution.

Substitute this value into either equation to obtain the $x$-coordinate. We choose the first equation.

$$3x - 5y = 24 \qquad \textbf{Replace } y \textbf{ with } -\textbf{2.}$$
$$3x - 5(-2) = 24$$
$$3x + 10 = 24$$
$$3x = 14$$
$$x = \frac{14}{3}$$

The solution to the system is the pair $x = \dfrac{14}{3}$, and $y = -2$.

$14/3 \rightarrow X : {}^-2 \rightarrow Y : 3X - 5Y$

$\qquad\qquad\qquad 24$

${}^-3X - 2Y$

$\qquad\qquad\qquad {}^-10$

**FIGURE 3.82**
Numerical support of the solution to Example 4.

**Graphical and Numerical Support.** Figure 3.81 shows graphical support of the solution and Figure 3.82 shows numerical support on a grapher. ∎

## Eliminating Variables Whose Coefficients Are Not Opposites

In Example 4, the coefficients of $x$ were opposites. When we added the two equations together, the $x$-terms were eliminated. However, in the following system, the coefficients of neither $x$ nor $y$ are opposites. Before adding the equations, we need to prepare the system for the elimination method.

$$2x + 3y = 4 \tag{1}$$

$$5x - 2y = 29 \tag{2}$$

Here are two options.

a) Multiply Eq. (1) by 5 and Eq. (2) by $-2$ so the coefficients of $x$ become 10 and $-10$. Then

$$\begin{cases} 2x + 3y = 4 \\ 5x - 2y = 29 \end{cases} \quad \text{becomes} \quad \begin{cases} 10x + 15y = 20 \\ -10x + 4y = -58. \end{cases}$$

b) Multiply Eq. (1) by 2 and Eq. (2) by 3 so the coefficients of $y$ become 6 and $-6$. Then

$$\begin{cases} 2x + 3y = 4 \\ 5x - 2y = 29 \end{cases} \quad \text{becomes} \quad \begin{cases} 4x + 6y = 8 \\ 15x - 6y = 87. \end{cases}$$

Only one of these options needs to be followed. After completing either one, a final solution can be found by adding the equations and proceeding as in Example 4. Example 5 illustrates the entire method.

**EXAMPLE 5**    Solving a System of Equations by Elimination

Solve the following system of equations:

$$2x + 3y = 4$$

$$5x - 2y = 29$$

SOLUTION

**REMINDER**

Remember, multiply *both* sides of an equation by a number in order to obtain an equivalent equation.

**Solve Algebraically.** Prepare for eliminating the $x$-terms by multiplying the first equation by 5 and the second by $-2$.

$$\begin{cases} 2x + 3y = 4 \\ 5x - 2y = 29 \end{cases} \quad \text{becomes} \quad \begin{cases} 10x + 15y = 20 \\ -10x + 4y = -58 \end{cases}$$

Add the resulting equations together to obtain the following:

$$0x + 19y = -38$$
$$19y = -38$$
$$y = -2$$

Substitute $y = -2$ into the first equation.

$$2x + 3y = 4 \quad \textbf{Replace } y \textbf{ with } -2.$$
$$2x + 3(-2) = 4$$
$$2x - 6 = 4$$
$$2x = 10$$
$$x = 5$$

The solution to the system is the pair $x = 5$ and $y = -2$.

**Support Numerically.** Figure 3.83 shows numerical support of the solution. ■

---

5 → X: ‾2 → Y: 2X + 3Y

　　　　　　　4

5X − 2Y

　　　　　　　29

**FIGURE 3.83**
Numerical support on a grapher of the solution to Example 5.

## Elimination of Variables Whose Coefficients Are Fractions

Example 6 shows how to use the elimination method when the system contains fractions. It is usually advisable to eliminate the fractions before eliminating one of the variables. Multiply each side of each equation by the least common multiple (LCM) of all the denominators of the equation. Figure 3.84 shows an estimate of the solution.

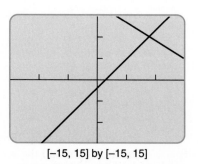

[−15, 15] by [−15, 15]

**FIGURE 3.84** This graph suggests (10, 10) as a preliminary estimate of the solution in Example 6.

**EXAMPLE 6**     Solving a System Containing Fractions

Solve the following system of equations:

$$\frac{1}{3}x + \frac{2}{5}y = 7$$

$$\frac{-2}{3}x + \frac{1}{2}y = -1$$

**REMINDER**

When adding or subtracting fractions with unlike denominators using paper-and-pencil methods it is necessary to rewrite the fractions as equivalent fractions with common denominators. The least common denominator (LCD) or the least common multiple (LCM) is the smallest integer that is a multiple of all the denominators under consideration.

SOLUTION

**Solve Algebraically.** Multiply both sides of the first equation by the LCM $(3, 5) = 15$ and the second equation by LCM $(2, 3) = 6$.

$$\frac{15}{1} \cdot \frac{1}{3}x + \frac{15}{1} \cdot \frac{2}{5}y = (15)(7)$$

$$\frac{6}{1} \cdot \frac{-2}{3}x + \frac{6}{1} \cdot \frac{1}{2}y = (6)(-1)$$

Simplify to obtain the resulting system of equations that is free of fractions.

$$5x + 6y = 105$$

$$-4x + 3y = -6$$

The variable $y$ seems to be the easier to eliminate.

Multiply the second equation by $-2$, resulting in the following system:

$$5x + 6y = 105$$

$$8x - 6y = 12$$

Add the equations.

$$13x = 117$$

$$x = 9$$

Substitute $x = 9$ into one of the equations that is free of fractions, for example, $5x + 6y = 105$.

$$5x + 6y = 105 \quad \text{Replace } x \text{ with 9.}$$

$$5(9) + 6y = 105$$

$$45 + 6y = 105 \quad \text{Subtract 45 from each side.}$$

$$6y = 60 \quad \text{Divide each side by 6.}$$

$$y = 10$$

The solution is the pair $x = 9$ and $y = 10$.

**Support Numerically.** Figure 3.85 shows numerical support using a grapher.

$9 \rightarrow X: 10 \rightarrow Y: (1/3)X + (2/5)Y$

7

$(^-2/3)X + (1/2)Y$

$^-1$

**FIGURE 3.85**
Numerical support with a grapher of the solution to Example 6.

Here is a summary of the elimination method of solving a system of two linear equations in two variables.

---

### Elimination Method: A Procedure for Solving a System

To solve a system of two linear equations in two variables, do the following:

1. Write the pair of equations so that each equation is in the form $ax + by = c$, with $a$, $b$, and $c$ being integers.
2. Multiply both sides of the equations as needed so that the coefficients of one of the variables differ only in sign.
3. Add the two equations together resulting in one equation in one variable.
4. Solve the equation to obtain a numerical value for this variable.
5. Substitute this value into one of the equations and solve for the other variable.

---

It is always a good idea to find numerical or graphical support for the algebraic solution of a system of equations.

## Solving Problem Situations Using a System of Equations

Section 3.5 introduced problems whose algebraic representation was a system of equations. In that section, systems of equations were solved using a graphical method. In contrast this section illustrates that such problems can also be solved by

- the substitution method or
- the elimination method.

You will recall that in Section 1.6, we studied a situation in which Melissa invested $15,000$—a portion at 7% and the remaining at 9%. In Example 7, that problem situation is reconsidered with different interest rates.

**EXAMPLE 7**    THE INVESTMENT PROBLEM SITUATION:
Finding an Algebraic Representation

Bill invests $5000 and earns $272.50 in annual interest from two investments—one earning 6.5% and the other 5%. Find an algebraic representation of this situation.

SOLUTION
Let

$$x = \text{the amount invested at } 6.5\%,$$

and

$$y = \text{the amount invested at } 5\%.$$

**Interest earned in 6.5% account:** $0.065x$

**Interest earned in 5% account:** $0.05y$

**Total amount of money invested:** $5000

**Total interest earned:** $0.065x + 0.05y$

Combine these expressions to write a system of equations that is an algebraic expression of the problem situation.

$$x + y = 5000$$

This equation represents.
the total amount
of money invested.

$$0.065x + 0.05y = 272.50$$

This equation represents
the total amount
of interest received. ∎

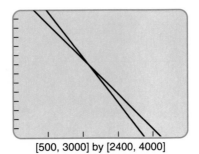

[500, 3000] by [2400, 4000]

**FIGURE 3.86** This graph can be obtained by beginning with the Integer window with $X$Fact $= 10$ and $Y$Fact $= 10$. Then zoom out three times. After placing the cursor at the point of intersection, zoom in once.

In Example 8, we use a substitution method to solve this system of equations whose solution answers the problem.

**EXAMPLE 8**    THE INVESTMENT PROBLEM SITUATION:
Finding the Amounts Invested

Using the data in Example 7, how much did Bill invest at each rate?

SOLUTION
The algebraic representation of this problem situation found in Example 7 is

$$x + y = 5000$$
$$0.065x + 0.05y = 272.50.$$

Figure 3.86 shows a graph that can be used to find an estimate of a solution to this system.

1. Solve the first equation for $x$.

$$x = 5000 - y$$

2. Substitute this expression for $x$ in the second equation.

$$0.065x + 0.05y = 272.50 \quad \textbf{Replace } x \textbf{ with 5000} - y.$$

$$0.065(5000 - y) + 0.05y = 272.50$$

$$325 - 0.065y + 0.05y = 272.50$$

$$325 - 0.015y = 272.50$$

$$-0.015y = -52.50$$

$$y = 3500$$

3. Substitute the value 3500 for $y$ in the first equation.

$$x + 3500 = 5000$$

$$x = 1500$$

Bill invests \$1500 at 6.5% and \$3500 at 5% interest. Figure 3.87 shows numerical support of the solution. ■

$1500 \rightarrow X : 3500 \rightarrow$
$Y : 0.065X + 0.05Y$
$\qquad\qquad 272.5$

**FIGURE 3.87** Grapher support of the solution to Example 8.

---

**REMINDER**

Make certain units are all the same. If using dollars, one dime is \$.10 and one nickel is \$.05. If using cents, change \$5.60 to 560 cents.

---

The following problem situation is the type that many persons are likely to analyze by trial and error. It is instructive to verify that it also can be solved using a system of linear equations. Which method do you prefer?

> ### COIN PROBLEM SITUATION
>
> Joshua has the face amount of \$5.60 in his antique coin collection, all in dimes and nickels. He knows he has 80 coins altogether.

Example 9 uses the elimination method to solve a system of equations.

**EXAMPLE 9** THE COIN PROBLEM SITUATION:
Finding the Coin Distribution

How many dimes and nickels are in Joshua's coin collection in the **Coin Problem Situation?**

SOLUTION

Let

$$x = \text{the number of dimes}$$

and

$$y = \text{the number of nickels.}$$

**The total number of coins:** $x + y$

**The total number of cents in dimes:** $10x$

**The total number of cents in nickels:** $5y$

$$x + y = 80$$     **This equation represents the total number of coins.**

$$10x + 5y = 560$$     **This equation represents the total value of the coins in cents.**

Figure 3.88 shows a graphical representation of the problem situation. It indicates that there should be a unique solution.

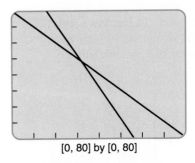

[0, 80] by [0, 80]

**FIGURE 3.88** This graph can be obtained by beginning with a [0, 80] by [0, 80] window. Position the cursor at the point of intersection and access the Integer window. This is the graphical support for the (32, 48) solution.

To solve this system algebraically, eliminate one of the variables by adding the equations.

$$-5x - 5y = -400$$     **Multiply the first equation by $-5$.**

$$10x + 5y = 560$$     **Add these equations together.**

$$5x = 160$$

$$x = 32$$

Substitute for $x$ into either equation, say the first equation.

$$32 + y = 80$$

$$y = 48 \quad \textbf{Add } -32 \textbf{ to each side.}$$

There are 32 dimes and 48 nickels in Joshua's collection. Figure 3.89 show numerical support of this solution. ∎

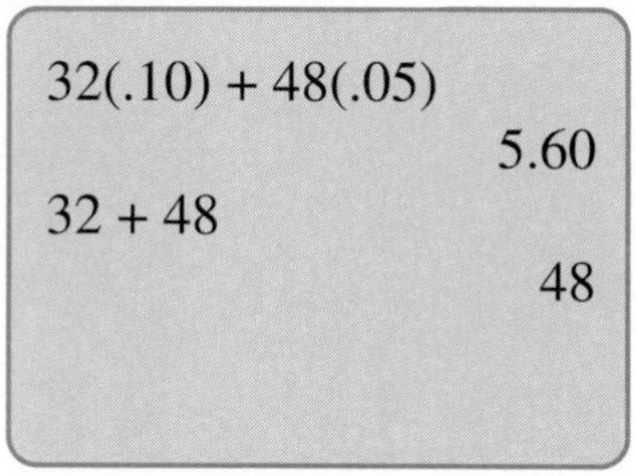

$32(.10) + 48(.05)$

$\qquad\qquad 5.60$

$32 + 48$

$\qquad\qquad 48$

**FIGURE 3.89**
Numerical support on a grapher
of the solution to Example 9.

## Exercises for Section **3.6**

The first step in solving a system of equations by the substitution method is to solve for one of the two variables in terms of the other. In Exercises 1–4, tell which variable you would choose and which equation you would pick to solve for that variable initially.

1. $\begin{cases} 3x + 5y = 9 \\ 2x + y = 8 \end{cases}$
   2. $\begin{cases} x + y = 72 \\ 2x - 3y = 19 \end{cases}$

3. $\begin{cases} 2x + 8y = 25 \\ x - y = 9 \end{cases}$
   4. $\begin{cases} 7x - 4y = 15 \\ 3x + y = 18 \end{cases}$

In Exercises 5–10, write the resulting equation after substituting the expression for $x$ in the second equation into the first equation.

5. $\begin{cases} 2x + y = 10 \\ x = 4 \end{cases}$
   6. $\begin{cases} 4x - y = 20 \\ x = 4 \end{cases}$

7. $\begin{cases} 3x - 2y = 12 \\ -y + 9 = x \end{cases}$
   8. $\begin{cases} x + y = 15 \\ 1 - 2y = x \end{cases}$

9. $\begin{cases} 6x - y = -5 \\ y - 5 = x \end{cases}$
   10. $\begin{cases} 4x - 7y = 35 \\ 2y + 15 = x \end{cases}$

In Exercises 11–22, use the substitution method to solve each system of equations. Use a grapher to support your solution.

11. $\begin{cases} 3x + y = 15 \\ y = 3 \end{cases}$
    12. $\begin{cases} x - y = 9 \\ x = 6 \end{cases}$

13. $\begin{cases} 2x - 3y = -9 \\ 3x - 4 = y \end{cases}$
    14. $\begin{cases} 3x - 2y = 12 \\ 9 - x = y \end{cases}$

15. $\begin{cases} x + y = 15 \\ 1 - 2y = x \end{cases}$
    16. $\begin{cases} 6x - y = -5 \\ y - 5 = x \end{cases}$

17. $\begin{cases} 2x + y = 0 \\ 3x + 2y = 1 \end{cases}$
    18. $\begin{cases} x + y = -6 \\ 3x - 2y = -18 \end{cases}$

19. $\begin{cases} x + 2y = 13 \\ 2x - 4y = 8 \end{cases}$
    20. $\begin{cases} 4x - y = 8 \\ 2x + y = -1 \end{cases}$

21. $\begin{cases} x + y = 3 \\ 4x + y = 10 \end{cases}$
    22. $\begin{cases} 6x + y = 3 \\ 3x + y = 8 \end{cases}$

In Exercises 23–26, tell what each side of each equation must be multiplied by in order to have the $x$-terms eliminated. (*Hint*: there is more than one correct answer.)

23. $\begin{cases} 2x - 3y = -9 \\ 3x + 4y = 29 \end{cases}$
    24. $\begin{cases} 5x - 3y = -1 \\ 2x + y = -7 \end{cases}$

25. $\begin{cases} x - y = -7 \\ x + y = 9 \end{cases}$    26. $\begin{cases} 2x + y = 5 \\ x + 3y = 20 \end{cases}$

In Exercises 27–30, tell what each side of each equation must be multiplied by in order to have the $y$-terms eliminated. (*Hint*: there is more than one correct answer.)

27. $\begin{cases} 2x - 3y = 8 \\ 6x - 9y = 27 \end{cases}$    28. $\begin{cases} x + 2y = 15 \\ 3x + 6y = 19 \end{cases}$

29. $\begin{cases} x + y = 9 \\ 2x + 5y = 27 \end{cases}$    30. $\begin{cases} x + y = 15 \\ 2x + 5y = 54 \end{cases}$

In Exercises 31–42, use the elimination method to solve each system of equations. Use a grapher to support your solution.

31. $\begin{cases} 2x - 3y = -9 \\ 3x + 4y = 29 \end{cases}$    32. $\begin{cases} 5x - 3y = -1 \\ 2x + y = -7 \end{cases}$

33. $\begin{cases} x - y = -7 \\ x + y = 9 \end{cases}$    34. $\begin{cases} 2x + y = 5 \\ x + 3y = 20 \end{cases}$

35. $\begin{cases} 2x - 3y = 8 \\ 6x - 9y = 27 \end{cases}$    36. $\begin{cases} x + 2y = 15 \\ 3x + 6y = 19 \end{cases}$

37. $\begin{cases} x + y = 9 \\ 2x + 5y = 27 \end{cases}$    38. $\begin{cases} x + y = 15 \\ 2x + 5y = 54 \end{cases}$

39. $\begin{cases} x + y = 925 \\ 2x + y = 1150 \end{cases}$    40. $\begin{cases} -3x + y = -7 \\ 2x + y = 8 \end{cases}$

41. $\begin{cases} 3x + 2y = 4 \\ 5x + 3y = 7 \end{cases}$    42. $\begin{cases} 3x + 2y = 78 \\ 8x + 5y = 200 \end{cases}$

In Exercises 43–54, choose either the elimination or the substitution method to solve the system of equations. Use a grapher to support your solution.

43. $\begin{cases} 2x - y = 1 \\ x + y = 2 \end{cases}$    44. $\begin{cases} 3x - y = -2 \\ 6 - y = x \end{cases}$

45. $\begin{cases} 2x + 4y = 2 \\ -x + y = 8 \end{cases}$    46. $\begin{cases} 13 - 3y = x \\ 2x - y = 5 \end{cases}$

47. $\begin{cases} x + y = -8 \\ 2x - 3y = 14 \end{cases}$    48. $\begin{cases} 3x + 2y = 6 \\ 3x - 8y = 12 \end{cases}$

49. $\begin{cases} 5x - 2y = -17 \\ 2x + 2y = 3 \end{cases}$    50. $\begin{cases} 2x - y = 12 \\ 3x = y - 3 \end{cases}$

51. $\begin{cases} 3x + 6y = -42 \\ 3y = -2x + 6 \end{cases}$

52. $\begin{cases} x - 5y = 11 \\ 2x + 3y = -4 \end{cases}$

53. $\begin{cases} 4x - 2y = 6 \\ 2x - 3 = y \end{cases}$    54. $\begin{cases} 2x - 2y = 2 \\ x - y = 1 \end{cases}$

In Exercises 55–62, eliminate the fractions in the system by multiplying each side of each equation by the LCD. Give the resulting system of equations.

55. $\begin{cases} \dfrac{1}{4}x + \dfrac{1}{3}y = 12 \\ \dfrac{1}{2}x - \dfrac{2}{3}y = -8 \end{cases}$    56. $\begin{cases} \dfrac{1}{5}x - \dfrac{1}{3}y = -7 \\ \dfrac{1}{2}x + \dfrac{2}{5}y = 1 \end{cases}$

57. $\begin{cases} \dfrac{1}{3}x + \dfrac{2}{3}y = -4 \\ \dfrac{1}{2}x - \dfrac{1}{3}y = 6 \end{cases}$    58. $\begin{cases} \dfrac{2}{3}x + \dfrac{1}{2}y = 2 \\ \dfrac{1}{6}x - \dfrac{1}{2}y = -12 \end{cases}$

59. $\begin{cases} \dfrac{3}{2}x + \dfrac{2}{5}y = \dfrac{9}{10} \\ \dfrac{1}{2}x + \dfrac{6}{5}y = \dfrac{3}{10} \end{cases}$    60. $\begin{cases} \dfrac{1}{2}x + \dfrac{1}{3}y = 4 \\ \dfrac{1}{4}x - \dfrac{1}{2}y = -2 \end{cases}$

61. $\begin{cases} \dfrac{1}{5}x + 9 = y \\ \dfrac{3}{5}x + \dfrac{1}{2}y = 8 \end{cases}$

62. $\begin{cases} \dfrac{1}{2}x - \dfrac{2}{3}y = -9 \\ x = \dfrac{1}{3}y \end{cases}$

In Exercises 63–68, eliminate the fractions in the system and then solve the resulting system by choosing either the graphing, substitution, or elimination method.

63. $\begin{cases} \dfrac{1}{2}x + \dfrac{1}{2}y = 1 \\ \dfrac{1}{4}x - \dfrac{1}{3}y = 4 \end{cases}$

64. $\begin{cases} \dfrac{1}{4}x + \dfrac{1}{8}y = -1 \\ \dfrac{3}{5}x - \dfrac{1}{2}y = -4 \end{cases}$

65. $\begin{cases} \dfrac{1}{2}x + \dfrac{1}{4}y = 3 \\ x - \dfrac{1}{2}y = -2 \end{cases}$

66. $\begin{cases} \dfrac{1}{2}x + \dfrac{1}{3}y = 13 \\ \dfrac{1}{5}x + \dfrac{1}{8}y = 5 \end{cases}$

67. $\begin{cases} \dfrac{1}{5}x + \dfrac{1}{2}y = -4 \\ -\dfrac{1}{2}y = x \end{cases}$

68. $\begin{cases} \dfrac{2}{3}x - \dfrac{1}{5}y = -\dfrac{34}{5} \\ \dfrac{3}{4}x + \dfrac{1}{6}y = -10 \end{cases}$

**69. Investment Planning.** Anne invests $9500 in two investments. One pays 5% interest annually and the other pays 7.5% interest annually. Her combined annual income from the two investments is $582.50.

a) Let $x$ be the amount invested at 5% and $y$ be the amount invested at 7.5%. Find an algebraic representation in the form of a system of equations for the problem situation.

b) How much did she invest at each rate?

**70. Investment Planning.** Barry invests $7900 in two different investments. One pays 6.5% interest annually and the other pays 5% interest annually. He makes $462.50 income from the two investments.

a) Let $x$ stand for the amount invested at 6.5% and $y$ be the amount invested at 5%. Find an algebraic representation in the form of a system of equations for the problem situation.

b) How much does he invest at each rate?

**71. Comparing Car Rental Plans.** Geoff went to Quality Rent-a-Car to rent a car. He paid a total of $175 for the car rental, which he rented for 5 days and drove 250 miles. Sammy Jo also rented the same type of car from Quality. Her bill was $108 after she drove 180 miles in 3 days. Let $x$ be the fixed charge per day and $y$ be the charge per mile.

a) Find a system of equations for the algebraic representation of the problem situation.

b) Find the charge per day and the charge per mile by solving the system of equations.

**72. Comparing Car Rental Plans.** Corey rented a car for 6 days and drove 325 miles. His bill was $215. Sara rented the same type of car for 4 days and drove 250 miles. Her bill was $150. Let $x$ be the charge per day and $y$ be the charge per mile.

a) Find a system of equations for the algebraic representation of the problem situation.

b) Find the charge per day and the charge per mile by solving the system of equations.

c) How much would Corey have to pay if he had kept the car only 5 days and had driven 200 miles?

**73. Coin Distribution.** Pedro had 23 coins consisting of quarters and nickels. The value of the coins is $2.95. How many of each denomination does he have?

**74. Coin Distribution.** Marmaretha has 33 coins in her purse. All are either dimes or quarters. She has a total of $5.55. How many dimes does she have?

**75. Consumer Purchases.** Kevin bought some jeans that cost $29.95/pair and socks at $4.50/pair. He bought four more pairs of socks than jeans and spent $86.90 altogether. How many pairs of jeans and socks did he buy?

**76. Consumer Purchases.** Wendy and Mike go to Hamburger Haven for lunch. Mike spends $11.05, not including tax, for 5 burgers, 3 fries, and 1 drink. Wendy spends $7.15, not including tax, for 3 burgers, 2 fries, and 1 drink. If the drinks are $.95 cents each, what is the cost of each burger and each order of fries?

**77. Using Algebra in Geometry.** If the difference of the measures of two acute angles of a right triangle is 40°, what are the measures of each of the three angles in the triangle?

**78.** The sum of two numbers is 96. One number is twice the other. Find the numbers.

**79.** The difference of two numbers is 43. Their sum is 93. Find the numbers.

EXTENDING THE IDEAS.

**80.** Solve the following system of equations:

$$\begin{cases} \dfrac{1}{x} + \dfrac{1}{y} = 4 \\[2mm] \dfrac{1}{x} - \dfrac{1}{y} = -6 \end{cases}$$

**81.** Solve the following system of equations:

$$\begin{cases} \dfrac{2}{x} + \dfrac{3}{y} = 2 \\[2mm] \dfrac{1}{x} - \dfrac{2}{y} = 8 \end{cases}$$

**82.** One equation of a system of equations is $3x + 4y = 9$. What does $A$ have to be in the other equation, $Ax - 8y = 15$, if there is no solution to the system? Explain your answer.

**83.** One equation of a system of equations is $x - y = 8$. What does $A$ have to be in the other equation, $3x - 3y = A$, if there is no solution to the system? Explain your answer.

**84.** One equation of a system of equations is $2x + 3y = 9$. What does $A$ have to be in the other equation, $Ax + 6y = 18$, so that there is an infinite number of solutions to the system? Explain your answer.

LOOKING BACK—MAKING CONNECTIONS

**85.** *Using Algebra in Geometry.* Two angles are said to be **complementary** if the sum of their measures is 90°. Find the two complementary angles if the measure of one is 15 more than twice the measure of the other.

**86.** *Using Algebra in Geometry.* Two angles are **supplementary** if the sum of their measures is 180°. Find the angles if the measure of one is five times the other.

**87.** *Using Algebra in Geometry.* The perimeter of a rectangle is 54 ft. The length is 3 ft more than the width. Find the dimensions of the rectangle.

**88.** The perimeter of a rectangle is 96 in. The length is twice the width. Find the dimensions of the rectangle.

**89.** *Finding Discounted Prices.* The Beehive Apparel Shoppe runs a 40%-off sale over the Labor Day weekend.

a) Let $x$ represent the original price of the garment and $y$ represent the sale price. Write an algebraic representation, in the form of an equation, for the sale price in terms of the original price.

b) Graph this equation and use the TRACE feature to find the sale price of a garment with an original price of $89.95.

c) Use the TRACE feature to find the original price of a suit, if Eleanor bought it on sale for $68.21.

## Chapter 3 Summary

| | Solving Linear Equations in One Variable with a Graphical Method | Examples |
|---|---|---|

**Multigraph Method**

1. Graph each side of the equation in the Integer window.
2. Use TRACE to determine the $x$-value of the point of intersection.
3. If the equation has an integer solution, the intersection point can be found exactly. If not, the intersection point can be approximated by zooming in and tracing.

For $3x - 7 = 5(x - 3)$:

1. Graph $y_1 = 3x - 7$ and $y_2 = 5(x - 3)$ in the Integer window.
2. Trace to the point of intersection.

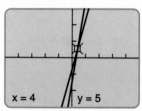

x = 4       y = 5

The solution is $x = 4$.

**$x$-intercept Method**

1. Rewrite the equation in the form $ax + b = 0$.
2. Graph $y = ax + b$ in the Integer window.
3. Use TRACE to determine the value of the $x$-intercept. Zoom in if necessary.

For $3x - 7 = 5(x - 3)$:

1. Rewrite the equation as $-2x + 8 = 0$.
2. Graph $y_1 = -2x + 8$ in the Integer window.

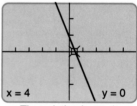

x = 4       y = 0

The solution is $x = 4$.

| | Solving Nonlinear Equations | Examples |
|---|---|---|

**Equations of the Form $a\sqrt{x + b} = k$**

Use a graphical method to solve the equation. There will be 0 or 1 solution.

$2\sqrt{x + 3} = 5$

**Equations of the Form $ax^2 + bx + c = 0$**

Use a graphical method to solve the equation. There will be 0, 1, or 2 solutions.

$3x^2 - 5x - 2 = 0$

| | Types of Systems of Two Linear Equations in Two Variables | Examples |
|---|---|---|
| Independent System | A system that has a unique solution. On a graph, the two lines have one point of intersection. | $y = 2 - 7x$ and $y = 5x + 8$ intersect at $(-0.5, 5.5)$. |
| Dependent System | A system that has an infinite number of solutions. When graphed, the two equations are the same line. | The graphs of $x - y = 5$ and $2x - 2y + 4 = 14$ are the same line. |
| Inconsistent System | A system that has no solution. On a graph, the two lines are parallel. | $9x + 3y = 7$ and $3x + y = 10$ each have slope $m = -3$ and different $y$-intercepts so there are no intersection points. |

| | Finding Solutions to a System of Equations | Examples |
|---|---|---|
| Graphing Method for Solving a System | 1. Change each equation to slope–intercept form.<br>2. Graph each equation in the same window.<br>3. Find the point of intersection (if it exists). Zoom in, if necessary, approximating the solution with an error of at most 0.01.<br>4. Support the solution numerically. | <br>$x = 5$    $y = 2$<br>(5, 2) is the solution to $x - y = 3$ and $0.2x + y = 3$. |
| Substitution Method for Solving a System | 1. Solve one of the equations for one variable in terms of the other.<br>2. Replace the variable with its equivalent expression in the equation not yet used.<br>3. Solve this equation for the variable that remains. | To solve the system,<br>$$x + y = 4$$<br>$$y - 2x = 7$$<br>complete the substitution $y = 2x + 7$ into<br>$$x + y = 4$$ |

| | **Finding Solutions to a System of Equations** | **Examples** |
|---|---|---|
| | 4. Substitute the variable value found into either of the original equations and solve for the other variable. | Solve $x + (2x + 7) = 4$ for $x$, then find $y$. Solution: $(-1, 5)$. |
| Elimination Method for Solving a System | 1. Write each equation in the form $ax + by = c$. | To solve the system |
| | 2. Multiply both sides of each equation so that the coefficients of one variable differ only in sign. | $$2x - 3y = 6$$ $$-2x + 4y = -2$$ |
| | 3. Add the equations together. | add the two equations to eliminate $x$. Solve for $y$, then find $x$. |
| | 4. Solve the resulting equation for the variable that remains. | Solution: $(9, 4)$. |
| | 5. Substitute this value into either of the original equations and solve for the other variable. | |

| | **Additional Concepts on Solving Equations** | **Examples** |
|---|---|---|
| Zoom-In Process | The process of considering progressively smaller windows around a solution for greater precision | |
| Finding a Graphical Solution That Is Not Exact: Accuracy Agreement | Zoom in on the solution point until there is precision to an error of at most 0.01. | The graphical solution to $7x = 4$ is 0.571. |
| Convention for Reporting a Solution | All approximated solutions should have three-digit decimal solutions. This assures an error of at most 0.01. | |
| Algebraic Confirmation for a Solution | Find an exact solution with algebra or by pencil-and-paper methods. | Solve $2(2x - 3) = 2 - 7x$: $$4x - 6 = 2 - 7x$$ $$11x = 8$$ $$x = \frac{8}{11}$$ |

|  | **Additional Concepts on Solving Equations** | **Examples** |
|--|--|--|
| Numerical Support for a Solution | Verify the solution either by pencil-and-paper method, or by using a grapher to store the variable, or by using a grapher-generated table. | To support that 3 is a solution to $2(x - 7) + 8 = 0$: |

1. With pencil and paper: Substitute 3 for $x$ and simplify:

$$2(3 - 7) + 8 = 0$$
$$2(-4) + 8 = 0$$
$$-8 + 8 = 0$$
$$0 = 0$$

2. On a grapher:

```
3 → X
                    3
2(X – 7) + 8
                    0
```

---

**Grapher Skills**

| Zoom-In and Zoom-Out | A grapher adjustment of the window dimensions to get either a closer or more distant view of the current screen. |
|--|--|
| Zoom Factors | Factors that adjust how quickly the grapher zooms in or out. |
| $X$scl, $Y$scl | The distance between the tick marks on the grapher screen. The $y$-axis values can be adjusted separately from the $x$-axis values. |

## Review Exercises for Chapter 3

In Exercises 1 and 2, use the $x$-intercept method to solve graphically the given equation. Support each solution both algebraically and numerically.

1. $x = 20 + 5x$          2. $24 - x = 2x$

In Exercises 3 and 4, use the multigraph method to solve the given equation. Support each solution both algebraically and numerically.

3. $(5x - 7) - (2x + 3) = x + 2$

4. $5(2x + 1) - (5x - 3) = 2 + 3x$

In Exercises 5 and 6, use the 0.1 window to solve the given equation. Use either the multigraph method or the $x$-intercept method and confirm each solution algebraically.

5. $1 - 0.4x = 0.2$          6. $1.5x + 4 = 0.55$

In Exercises 7–10, find a graphical solution to the given equation. Set $X$Scl= 0.01 and use zoom-in.

7. $3.25x + 12 = 0$          8. $5.6x + 2 = 0.9$

9. $\sqrt{x - 2} = 2.1$          10. $2\sqrt{x + 4} = 5.6$

In Exercises 11 and 12, graph each function in the given windows and determine which window seems most appropriate for each function.

11. Find the graph of $f(x) = -16x^2 + 80x + 25$.
   a) $[-10, 10]$ by $[-10, 10]$
   b) $[-10, 10]$ by $[-200, 200]$
   c) $[0, 10]$ by $[0, 150]$

12. Find the graph of $f(x) = -0.04\sqrt{x + 7}$.
   a) $[-10, 10]$ by $[-10, 10]$
   b) $[-10, 10]$ by $[-2, 2]$
   c) $[-10, 10]$ by $[-0.5, 0.5]$

13. Use a graphical method to determine the number of solutions to $\sqrt{3x - 5} = 12$.

In Exercises 14 and 15, find an appropriate window and solve the given equation graphically.

14. $x^2 + x = 29$          15. $5\sqrt{2x - 7} = 1$

In Exercises 16 and 17, use your grapher to check if the given ordered pair is the solution to the following system of equations.

16. $(2, -1)$ $\begin{cases} 3x - 2y = 8 \\ 4x + 3y = 5 \end{cases}$

17. $(2, 5)$ $\begin{cases} 8x - 3y = 1 \\ 4x - y = 2 \end{cases}$

In Exercises 18 and 19, determine whether the system of equations have 0, 1, or an infinite number of solutions. Support your answer on your grapher.

18. $\begin{cases} 13 - 6x = 2y \\ 3x + y = 5 \end{cases}$          19. $\begin{cases} \dfrac{3}{2}x - 6y = 48 \\ \dfrac{1}{4}x - 8 = y \end{cases}$

In Exercises 20 and 21, find the solution to the system of equations by graphing the system and using zoom-in as necessary to find a solution with an error of at most 0.01.

20. $\begin{cases} 2x + 2y = 15 \\ 3x - y = 8 \end{cases}$          21. $\begin{cases} 2x - y = 16 \\ x + 4y = 11 \end{cases}$

In Exercises 22–24, use the substitution method to solve each system of equations. Use a grapher to support your solution.

22. $\begin{cases} 3x - 4y = 12 \\ x + 3y = 56 \end{cases}$          23. $\begin{cases} 7x - 2y = 8 \\ 3x - y = 1 \end{cases}$

24. $\begin{cases} 9x - y = 7 \\ 3x + 2y = 28 \end{cases}$

In Exercises 25–27, use the elimination method to solve each system of equations. Use a grapher to support your solution.

25. $\begin{cases} -2x + 3y = 13 \\ 3x + 5y = 9 \end{cases}$

26. $\begin{cases} 4x + 7y = 9 \\ 5x - 6y = 26 \end{cases}$

27. $\begin{cases} 4x - 3y = 5 \\ 7x - 5y = 10 \end{cases}$

In Exercises 28 and 29, eliminate the fractions in the system of equations and then solve the resulting system by using either the graphing, substitution, or elimination method.

28. $\begin{cases} \dfrac{1}{2}x + \dfrac{2}{3}y = \dfrac{4}{3} \\ \dfrac{1}{5}x - \dfrac{3}{2}y = -3 \end{cases}$

29. $\begin{cases} -\dfrac{3}{4}x + \dfrac{1}{2}y = 6 \\ \dfrac{2}{3}x - \dfrac{1}{6}y = 3 \end{cases}$

In Exercises 30–32, write an equation that is described by each sentence.

30. Doubling the square root of a number is eight.

31. The value of $x$ pennies and $y$ dimes is $8.24.

32. Sixty-five percent of a number is three more than half the number.

33. When three times a certain number is added to $-5$, the answer is the same as that if that number is doubled. What is the number?

34. *Car Rental Costs.* A car rental company rents a compact car for $25/day plus $0.30/mi. Suppose a car was rented for 2 days and driven $x$ miles.

a) Find an algebraic representation for the total rent $R$ in terms of $x$.

b) Find a graph of the algebraic representation in an Integer window that shows the graph in the first quadrant.

c) From this graph, determine the total rent if the car was driven 220 miles in the two-day period. Confirm algebraically.

d) From the same graph, determine how many miles were driven if the total cost for two days was $140. Confirm algebraically.

35. *Analyzing Cost.* The "Conqueror" bicycle came out in 1990 and was priced at $175. Since then the price of the bicycle has increased $12/yr due to inflation. Asssume the price continues to increase at the same pace.

a) Write an algebraic representation of the cost of the bicycle.

b) Find a graph of the algebraic representation in an Integer window that shows the graph in the first quadrant.

c) From this graph, determine the price of the bicycle in 1997.

d) In what year will the price of the bicycle be $355?

36. *Investment Analysis.* David invested $20,000 in two different investments. One pays 8% interest annually and the other pays 6% interest annually. In one year, the combined income from the two investments is $1431.00.

a) Let $x$ represent the amount of money invested at 8% and $y$ represent the amount invested at 6%. Find an algebraic representation in the form of a system of equations for the problem situation.

b) Solve the system to determine how much David invested at each rate.

## Chapter 3 Test

1. What function(s) would you graph if you were asked to solve $3x - 8 = 7x + 5$ (a) using the multigraph method? (b) using the $x$-intercept method?

2. Graph $f(x) = 0.25x^2 - 2x - 15$ in the given windows and determine which window seems most appropriate for $f$.

   a) $[-10, 10]$ by $[-10, 10]$

   b) $[-10, 20]$ by $[-10, 20]$

   c) $[-20, 20]$ by $[-20, 20]$

3. Use a graphical method to determine the number of solutions to $5x^2 + 33x + 55 = 0$.

In Questions 4 and 5, find an appropriate window for the following equations and solve graphically.

4. $\sqrt{2x + 11} = 4.62$      5. $18x^2 + 189x + 61 = 0$

6. Solve $11(x - 1) + x = 10(2 - x) + 3$ using a graphical method. Confirm algebraically.

7. Determine whether the following system of equations has 0, 1, or an infinite number of solutions. Support your answer on your grapher.

$$\begin{cases} 3 - 0.6y = 0.2x \\ 7x + 21y = 105 \end{cases}$$

In Questions 8 and 9, use the substitution method to solve the system of equations. Use a grapher to support your solution.

8. $\begin{cases} 7x + y = 4 \\ 4x - 3y = 3 \end{cases}$      9. $\begin{cases} 3x + 2y = 21 \\ x - 5y = 10 \end{cases}$

In Questions 10 and 11, use the elimination method to solve the system of equations. Use a grapher to support your solution.

10. $\begin{cases} 6x + 5y = 4 \\ 2x + 3y = 1 \end{cases}$      11. $\begin{cases} 8x + 5y = 17 \\ -7x + 6y = 37 \end{cases}$

12. Eliminate the fractions in the following system of equations and then solve the resulting system by choosing either the graphing, substitution, or elimination method.

$$\begin{cases} \dfrac{4}{7}x + \dfrac{1}{2}y = \dfrac{16}{7} \\ 3x + \dfrac{6}{5}y = \dfrac{3}{5} \end{cases}$$

13. Write as an equation: Six less than twice a number is 40% of the number.

14. The sum of two numbers is 29 and their difference is 42.

   a) Find an algebraic representation of this problem in the form of a system of equations.

   b) Solve for the two numbers.

15. Forty feet of fence is used to surround a rectangular garden on three sides, as shown in the following figure:

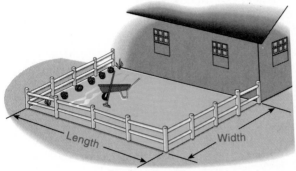

For Question 15.

a) Find an algebraic representation for the length of the garden when the width is $x$ feet.

b) Find an algebraic representation for the area of the garden when the width is $x$ feet.

c) Use a graphical representation to find the dimensions (to the nearest tenth of a foot) of the garden if its area is 187.5 ft$^2$.

d) Use trace coordinates to estimate the width of the garden when its area is as great as possible.

# Linear Inequalities and Systems of Linear Inequalities

### AN APPLICATION

A certain outfielder discovered that whenever a fly ball goes more than 180 ft above ground level, he tends to lose track of the path of the ball because he is temporarily blinded by the sun. If a hit ball has a vertical velocity of 120 ft/sec, its height $t$ seconds later is described by

$$h = -16t^2 + 120t.$$

# Solving Linear Inequalities Algebraically

Linear Inequality in One Variable ▪ Solving a Linear Inequality Algebraically ▪ Graphical Methods to Support Algebraic Solutions ▪ Two Special Cases ▪ Inequalities as Algebraic Representations

In Chapters 1–3, we examined linear equations. In addition, Chapter 1 introduced the symbols $<$, $\leq$, $>$, and $\geq$. That is, given any real number $a$, there are four basic inequalities associated with it, as follows:

$$x < a \qquad x \leq a \qquad x > a \qquad x \geq a$$

Each of these inequalities determines an infinite set of numbers, illustrated in Figure 4.1 either as a shaded portion of a number line or in interval notation.

**REMINDER**

$5 > x$ and $x < 5$ are equivalent statements. It is easier to graph and write the solution in interval notation if the variable is on the left.

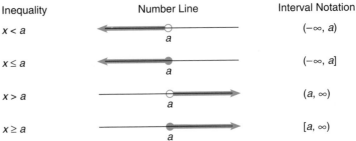

**FIGURE 4.1**    Number lines and Interval notation.

On a number line, if the endpoint of the interval is not included we use an open circle. When the endpoint of the interval is included, we use a closed circle. In interval notation, a parenthesis is always used with $-\infty$ or $\infty$.

## Linear Inequality in One Variable

Definition 1.4 introduced the concept of linear inequality in one variable, for example,

$$2x + 4 < 7, \qquad y + 2 > 6, \qquad 3x \leq 9, \qquad 5y - 2 \geq 8.$$

A number belongs to the **solution of an inequality** if replacing the variable with the number results in a true statement. Generally, there are infinitely many numbers that are solutions to a linear inequality. Example 1 shows how to confirm whether a particular number is a solution.

**EXAMPLE 1**     Confirming a Solution of an Inequality

Determine which of the following values of $x$ is a solution to the inequality $2x - 3.75 < 5$.

a. $x = -21.3$
b. $x = 8$

SOLUTION

**Confirm Algebraically.**

a)
$$2x - 3.75 < 5$$
$(*)$     $2(-21.3) - 3.75 < 5$     **Replace $x$ with $-21.3$ and simplify.**
$$-42.6 - 3.75 < 5$$
$$-46.35 < 5$$     **True statement**

This confirms that $-21.3$ is a solution to the inequality.

b)
$$2x - 3.75 < 5$$
$(*)$     $2(8) - 3.75 < 5$     **Replace $x$ with 8 and simplify.**
$$16 - 3.75 < 5$$
$$12.25 < 5$$     **False statement**

This confirms that 8 is *not* a solution to the inequality.

**Support Numerically.** Figure 4.2 shows $2x - 3.75$ evaluated for each $x$-value.
a.   $-46.35$ less than 5 means that $-21.3$ is a solution.
b.   12.25 not less than 5 means that 8 is not a solution.     ■

**GRAPHER NOTE**

Notice the similarity between the lines prefixed with $(*)$ and the grapher expressions in Figure 4.2. Use the grapher feature of recalling and editing the last answer.

```
2(⁻21.3) – 3.75
                     ⁻46.35
2(8) – 3.75
                      12.25
```

**FIGURE 4.2** The grapher evaluation of $2x - 3.75$ for $x = -21.3$ and $x = 8$.

## Solving a Linear Inequality Algebraically

Properties that we use to solve inequalities are much like those we use to solve equations. For example, the Addition Property of Inequality stated in Section 1.4

tells us that any real number can be added to both sides of an inequality and the inequality is preserved. Example 2 illustrates the use of this property.

**EXAMPLE 2**    Using the Addition Property of Inequality

Solve the inequality $2x - 5 < x + 7$ and write the solution in interval notation.

SOLUTION

$$2x - 5 < x + 7$$    **Subtract $x$ from each side and simplify.**

$$2x - x - 5 < x - x + 7$$

$$x - 5 < 7$$    **Add 5 to each side and simplify.**

$$x - 5 + 5 < 7 + 5$$

$$x < 12$$

The solution of the inequality is all real numbers in the interval $(-\infty, 12)$. Figure 4.3 shows a picture of this solution using a number line.  ■

**REMINDER**

Because subtraction can be written in terms of addition, $a - b = a + (-b)$, the Addition Property of Inequality holds true for subtraction as well.

**FIGURE 4.3**  A number line graph of the solution to Example 2.

The inequality in Example 3 is solved using the Division Property. Recall that dividing both sides of an inequality by $-3$ is equivalent to multiplying both sides by $-\dfrac{1}{3}$.

**REMINDER**

Remember that when dividing both sides of an inequality by a negative number, the inequality sign is reversed.

**EXAMPLE 3**    Solving Inequalities Using the Division Property

Solve $-3x \geq 18$. Graph the solution on a number line and write the solution in interval notation.

SOLUTION

$$-3x \geq 18$$    **Divide each side of the inequality by $-3$ and reverse the inequality symbol.**

$$\frac{-3x}{-3} \leq \frac{18}{-3}$$

$$x \leq -6$$

The solution is the set of numbers in the interval $(-\infty, -6]$. A graph of the solution is shown in Figure 4.4.  ■

**FIGURE 4.4**  Number line graph of $x \leq -6$. Since the endpoint $-6$ is included in the solution, a bracket is used in the interval notation $(-\infty, -6]$.

Examples 2 and 3 required the use of only one property, either the Addition Property or the Division Property. Often, solving an inequality algebraically requires using properties such as the Commutative, Associative, and Distributive Properties as well as those of inequalities. Example 4 illustrates such a problem.

### EXAMPLE 4   Using More Than One Property to Solve an Inequality

Solve $3(x-5)+2 > 8$, graph the solution on a number line, and write the solution in interval notation.

**FIGURE 4.5** The number line graph of $x > 7$. Because the endpoint 7 is not included in the solution, a parenthesis is used in the interval notation $(7, \infty)$.

SOLUTION

$$3(x-5)+2 > 8 \qquad \text{Use the Distributive Property and simplify.}$$

$$3x - 15 + 2 > 8$$

$$3x - 13 > 8 \qquad \text{Add 13 to each side of the inequality and simplify.}$$

$$3x > 21 \qquad \text{Divide each side by 3 and simplify.}$$

$$x > 7$$

The solution is the set of numbers in the interval $(7, \infty)$. A graph of the solution is shown in Figure 4.5.  ■

## Graphical Methods to Support Algebraic Solutions

Examples 2–4 demonstrated how to solve linear inequalities using algebraic methods. Next, we use the inequality $4x - 5 \geq 2x + 7$ to illustrate the *solve algebraically* and *support graphically* methods that we have used frequently when solving equations. We begin with an algebraic solution.

$$4x - 5 \geq 2x + 7 \qquad \text{Add } -2x \text{ to each side.}$$

$$2x - 5 \geq 7 \qquad \text{Add 5 to each side.}$$

$$2x \geq 12 \qquad \text{Divide each side by 2.}$$

$$x \geq 6$$

The solution interval is $[6, \infty)$, which contains all real numbers greater than or equal to 6.

Graphical support for this solution can be obtained by following these steps:

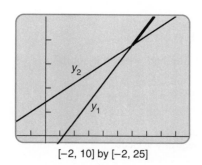

[−2, 10] by [−2, 25]

**FIGURE 4.6** A graph of $y_1 = 4x - 5$ and $y_2 = 2x + 7$. The bold portion of the $y_1$ line designates where $y_1 \geq y_2$, which is $[6, \infty)$.

1. Find the graphs of the left side $y_1 = 4x - 5$ and the right side $y_2 = 2x + 7$ in the same window.
2. Identify the portion of the $y_1$ graph that lies on or above the $y_2$ graph (see Figure 4.6).

3. Use TRACE to see that the $x$-coordinates of the points identifed in step 2 are all greater than or equal to 6.

Example 5 illustrates the solve algebraically and support graphically method.

## REMINDER

Remember to reverse the inequality symbol when multiplying or dividing by a negative number. At each step, you might find it helpful to first decide on the inequality direction and place the inequality symbol on the paper before applying the properties.

### EXAMPLE 5    Solve Algebraically and Support Graphically

Solve $3(x - 2) + 4 < 8(x + 1)$. Write the solution in interval notation and show grapher support for the solution.

SOLUTION

**Solve Algebraically.**

$$3(x - 2) + 4 < 8(x + 1) \qquad \text{Use the Distributive Property and simplify.}$$

$$3x - 6 + 4 < 8x + 8$$

$$3x - 2 < 8x + 8 \qquad \text{Add } -8x \text{ to each side.}$$

$$-5x - 2 < 8 \qquad \text{Add 2 to each side.}$$

$$-5x < 10 \qquad \begin{array}{l}\text{Divide each side by } -5.\\ \text{Remember to reverse the inequality sign.}\end{array}$$

$$x > -2$$

The solution interval is $(-2, \infty)$, which contains all real numbers greater than $-2$.

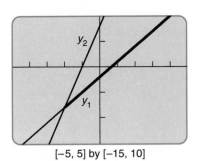

[−5, 5] by [−15, 10]

**FIGURE 4.7** A graph of $y_1 = 3(x - 2) + 4$ and $y_2 = 8(x + 1)$. The bold portion of the $y_1$ line is where $y_1 < y_2$, which is $(-2, \infty)$.

**Support Graphically.** Figure 4.7 shows the graph of $y_1 = 3(x - 2) + 4$ and $y_2 = 8(x + 1)$. The bold portion of the $y_1$ line shows where $y_1$ is below $y_2$, that is, where $y_1 < y_2$. It appears that the graph of $y_1$ lies below the graph of $y_2$ whenever $x > -2$.    ■

We showed in Section 3.6 that we could eliminate fractions from an equation when solving algebraically by multiplying both sides of the equation by the least common multiple (LCM) of all the denominators present. The same technique works when solving inequalities, as Example 6 illustrates.

### EXAMPLE 6    Solve Algebraically and Support Graphically

Solve the inequality $\dfrac{2x}{3} + \dfrac{1}{2} > \dfrac{5}{6}$. Write the solution in interval notation and support graphically.

SOLUTION

**Solve Algebraically.** We will first eliminate the fractions by multiplying each side by 6.

$$\frac{2x}{3} + \frac{1}{2} > \frac{5}{6} \qquad \text{Multiply each side by the LCM, which is 6.}$$

$$6\left(\frac{2x}{3} + \frac{1}{2}\right) > 6\left(\frac{5}{6}\right) \qquad \text{Use the Distributive Property and simplify.}$$

$$6\left(\frac{2x}{3}\right) + 6\left(\frac{1}{2}\right) > 6\left(\frac{5}{6}\right)$$

$$4x + 3 > 5 \qquad \text{Add } -3 \text{ to each side of the inequality.}$$

$$4x > 2 \qquad \text{Divide each side by 4.}$$

$$x > 0.5$$

The solution is $(0.5, \infty)$, which is all numbers greater than 0.5.

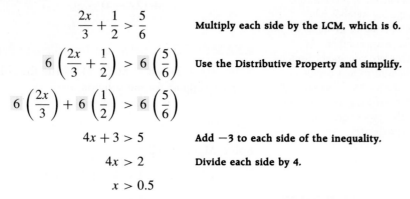

[−5, 5] by [−5, 5]

**FIGURE 4.8** The graph of $y_1 = 2x/3 + 1/2$ and $y_2 = 5/6$. The bold portion of the $y_1$ line is where $y_1 > y_2$ which is $(0.5, \infty)$.

**Support Graphically.** Figure 4.8 shows the graph of $y_1 = 2x/3 + 1/2$ and $y_2 = 5/6$. The bold portion of the $y_1$ line shows where $y_1 > y_2$. It appears that the graph of $y_1$ is above the graph of $y_2$ when $x > 0.5$. ■

## Two Special Cases

Sometimes when we try to isolate the variable on one side of the inequality, we find that the variable drops out of the problem entirely. This indicates that the truth or falsity of the statement is independent of the value of the variable. If the resulting statement, after the variable is eliminated, is true, the statement is always true no matter what value is given to the variable. The solution then is *all real numbers*. If the statement is false, it is always false and therefore there is *no solution*. Example 7 illustrates each of these cases.

**EXAMPLE 7**  Solve Algebraically and Support Graphically

Solve each of the following inequalities:

a) $3x + 15 + (2x - 4) > 5(x - 6)$
b) $2(x - 1) + 14 < 4x + 7 - 2(x + 2)$

SOLUTION

**Solve Algebraically.**

a)    $3x + 15 + (2x - 4) > 5(x - 6)$    **Remove the parentheses and combine like terms.**

$5x + 11 > 5x - 30$    **Subtract 5x from each side.**

$11 > -30$    **True statement**

The solution is all real numbers.

b)    $2(x - 1) + 14 < 4x + 7 - 2(x + 2)$    **Use the Distributive Property.**

$2x - 2 + 14 < 4x + 7 - 2x - 4$    **Combine like terms.**

$2x + 12 < 2x + 3$    **Subtract 2x from each side of the inequality.**

$12 < 3$    **False statement**

There is no solution.

**Support Graphically.**

a)    Figure 4.9 (a) shows that the graph of $y_1 = 3x + 15 + (2x - 4)$ appears to always be above $y_2 = 5(x - 6)$. This is graphical support that $y_1 > y_2$ for all values of $x$.

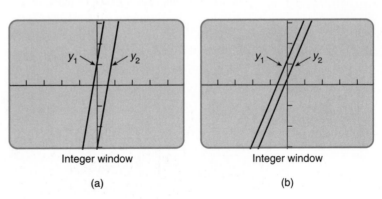

Integer window          Integer window

(a)          (b)

**F I G U R E  4 . 9**   (a) The graph of $y_1 = 3x + 15 + (2x - 4)$ and $y_2 = 5(x - 6)$ and (b) the graph of $y_1 = 2(x - 1) + 14$ and $y_2 = 4x + 7 - 2(x + 2)$.

b)    Figure 4.9 (b) shows that the graph of $y_1 = 2(x - 1) + 14$ appears to always be above the graph of $y_2 = 4x + 7 - 2(x + 2)$. This is graphical support that there are no solutions to the inequality.    ■

**REMINDER**

To say that the graph of $y_1$ is above the graph of $y_2$ means that for a given value of $x$, the corresponding $y_1$-value will be greater than the corresponding $y_2$-value.

## Inequalities as Algebraic Representations

Often the algebraic representations of problem situations include inequalities that need to be solved. Consider the following problem situation.

---

### AVERAGING TEST SCORES PROBLEM SITUATION

After his first three tests, Dan Patrick wants to determine what score he needs on the fourth exam in American history in order for his average to reach or exceed the minimum 92% he needs for an A.

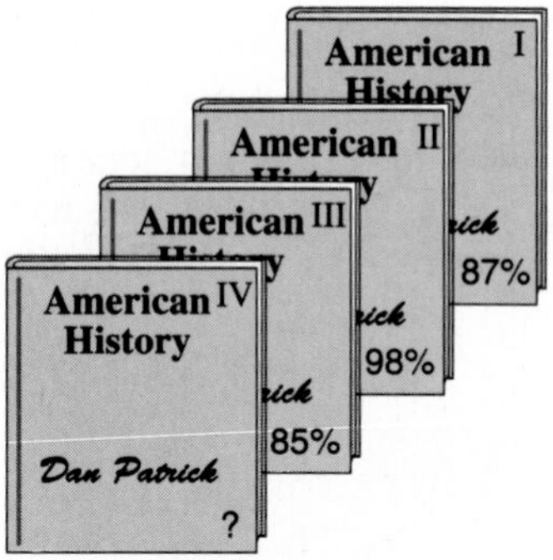

---

**FIGURE 4.10** Dan's exam scores in American History.

**EXAMPLE 8**   THE AVERAGING TEST SCORES PROBLEM SITUATION: Finding the Score on the Fourth Exam

If Dan's scores on the first three exams are 87%, 98%, and 85%, what score does he need on his fourth exam to reach or exceed an A average?

SOLUTION

Let $x =$ the grade on the fourth exam. The average of four tests needs to be greater than or equal to 92 (see Figure 4.10).

$$\frac{87 + 98 + 85 + x}{4} \qquad \text{The average of four test scores}$$

$$\frac{87 + 98 + 85 + x}{4} \geq 92 \qquad \text{An inequality that must be true to achieve an A average.}$$

$$\frac{270 + x}{4} \geq 92 \qquad \text{Multiply each side by 4.}$$

$$270 + x \geq 368 \qquad \text{Add } -270 \text{ to each side.}$$

$$x \geq 98$$

This means that Dan must score 98% or higher on his fourth exam to achieve an A average in the course.  ■

We next look at another problem situation that involves inequalities.

---

### CHECKING ACCOUNT FEES PROBLEM SITUATION

The Educational Community Credit Union gives college students a special rate on its checking accounts. They charge a monthly service fee of $1.25 plus $0.15 for each check processed.

---

**EXAMPLE 9**  THE CHECKING ACCOUNT PROBLEM SITUATION:
Finding the Number of Checks

Dwight needs to keep his monthly charges on his checking account to less than $3.50/mo. How many checks can Dwight write each month?

SOLUTION

**Solve Algebraically.** Write and solve an algebraic representation for the problem situation. Let $x$ be the number of checks Dwight writes.

  **cost for $x$ checks:** $0.15x$

$$0.15x + 1.25$$    **An expression representing the total checking account cost per month**

$$0.15x + 1.25 < 3.50$$    **An algebraic representation of the problem situation. Solve this inequality.**

$$0.15x + 1.25 < 3.50$$    **Subtract 1.25 from each side.**

$$0.15x < 2.25$$    **Divide each side by 0.15.**

$$x < 15$$

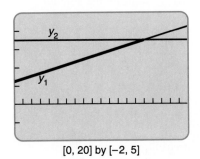

[0, 20] by [−2, 5]

**FIGURE 4.11** The graph of $y_1 = 0.15x + 1.25$ and $y_2 = 3.5$. The integers that are $x$-coordinates of points that lie on the bold portion of the $y_1$ line represent the solution to Example 9.

**Support Graphically.** Figure 4.11 shows graphical support for this solution. Dwight must write fewer than 15 checks to stay within his budget. ∎

---

## Exercises for Section 4.1

In Exercises 1–8, use numerical methods to determine if the given number is a member of the solution interval for the inequality.

**1.** $x = 3.56; x < 6$      **2.** $x = -3.56; 2x > 10$

**3.** $x = -2; 3x + 5 \geq 2$      **4.** $x = 5; -x > 8$

**5.** $x = 1.5; 2x - 3 < 6$    **6.** $x = -2.4; x - 5 \leq -3$

**7.** $x = -5.6; -2x \leq 3$    **8.** $x = -2.3; 2x + 6 > 5$

In Exercises 9–12, express the interval represented on the graph in interval notation.

**9.**

**10.**

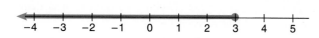

**11.**

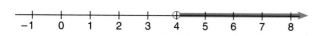

**12.**

In Exercises 13–20, write an inequality that characterizes the interval.

**13.** $(5, \infty)$    **14.** $(-\infty, 9]$

**15.** $(-\infty, 2]$    **16.** $(-5.18, \infty)$

**17.** $[0, \infty)$    **18.** $(-\infty, 0)$

**19.** $(-\infty, 7.25)$    **20.** $(3.8, \infty)$

In Exercises 21–28, use interval notation to represent the inequality.

**21.** $x < 6.35$    **22.** $x < -9.1$

**23.** $x \geq 2.5$    **24.** $x > -8$

**25.** $-2.3 < x$    **26.** $23.8 < x$

**27.** $x \leq -3$    **28.** $x < 5$

TRANSLATING WORDS TO SYMBOLS

In Exercises 29–34, write the phrase in interval notation and as an inequality.

**29.** The set of all numbers greater than or equal to 13

**30.** The set of all numbers less than 5.5

**31.** The set of all numbers at least $-2$

**32.** The set of all numbers at most 3.4

**33.** The set of all numbers less than or equal to $-55$

**34.** The set of all numbers greater than 13

In Exercises 35–42, use the Addition Property of Inequality to solve the inequality. Indicate the solution on a number line graph.

**35.** $x + 2 < 9$    **36.** $x - 5 \geq 12$

**37.** $5x + 2 \leq 4x + 17$    **38.** $2x + 3 \leq x + 9$

**39.** $2x + 4 < x - 7$    **40.** $3x - 5 > 2x + 2$

**41.** $5x - 7 \geq 4x + 28$    **42.** $5x + 3 > 4x + 10$

In Exercises 43–50, use the Multiplication or Division Property of Inequality to solve the inequality. Indicate the solution using interval notation.

**43.** $3x > 15$    **44.** $-5x < 25$

**45.** $-7x \geq 35$    **46.** $8x \geq 64$

**47.** $4x < 9$    **48.** $-2x < 17$

**49.** $-3.2x \leq 6.5$    **50.** $-5x > 2.5$

In Exercises 51–60, solve the inequality algebraically. Indicate the solution using interval notation and support it graphically.

**51.** $3(x + 5) - 7 \leq 2(x + 3)$

**52.** $-(x + 3) - 5 < 2(3 - x)$

**53.** $2x - 3 > 4x + 5$

**54.** $5(x - 1) < 3(x + 2)$

**55.** $2(x + 3) > 2(x - 1) - 5$

56. $6x + 7 < 3(2x - 5)$

57. $4x - 6 \geq 2(x + 3)$

58. $3(x - 4) + 2 \geq x + 3(x - 2)$

59. $7 - (x - 6) + 2x \leq 5x - 9$

60. $3x + 15 \leq 4(x + 7) - x + 2$

In Exercises 61–68, eliminate the fractions using the Multiplication Property of Inequality and then solve the equivalent inequality. Indicate the solution using interval notation and support it graphically.

61. $\dfrac{x}{3} + \dfrac{1}{4} > 2$

62. $\dfrac{2x}{5} + 4 < \dfrac{1}{2}$

63. $\dfrac{1}{2} + \dfrac{x}{4} \leq -3$

64. $\dfrac{x}{5} - \dfrac{3x}{2} \geq 6$

65. $\dfrac{2}{3}(x + 5) \leq \dfrac{1}{2}(x - 6)$

66. $\dfrac{1}{4}(4x - 24) \geq \dfrac{2}{3}(x + 3)$

67. $\dfrac{3(x + 2)}{5} - \dfrac{(2x - 3)}{2} \leq 4$

68. $-\dfrac{(x + 3)}{2} + \dfrac{(x - 5)}{3} > 0$

In Exercises 69–74, solve the inequality algebraically. Indicate the solution using interval notation and support it graphically.

69. $2.1 + x < 7.89$

70. $9.75 - x \geq 10.89$

71. $3.5x + 2.7 \geq 13.2$

72. $2.35 - 5.6x \leq 27.55$

73. $-4.96 - 7.85x \geq 12.7025$

74. $-3.28 + 8.96x > 33.456$

*Averaging Test Scores.* In Exercises 75–77, consider the **Averaging Test Scores Problem Situation.**

75. Angie earned a 93%, 85%, and 89% on her first three exams.

a) Let $x$ equal the score of the fourth exam. Write an algebraic representation of the average of the four exams.

b) What must Angie score on the fourth exam to have an average greater than or equal to 90%?

76. Kim scored 82% and 76% on her first two biology exams. What is the lowest score he can make on the third exam and have at least an 80% average?

77. In Horace's chemistry class, the quiz average counts as one exam grade. If he scored an average of 87% on the quizzes and 96%, 92%, and 85% on the exams, what must he score on the final exam to have at least a 92% average for the term?

78. *Personal Financial Planning.* In the **Checking Account Fees Problem Situtation,** John wants to keep his checking account fees under $8/mo. His bank charges a monthly fee of $2.50 plus $0.20 per check written.

a) Let $x$ equal the number of checks he can write in one month. Write an inequality that is an algebraic representation of his monthly charges as being under $8/mo for his checking account.

b) Solve this inequality in (a) and find the maximum number of checks John can write per month.

In Exercises 79–83, solve the stated problem.

79. *Car Rental.* Lian wants to visit San Francisco over Spring Break. He has only $255 to spend on car rental. He can rent a car from Budget/Wise for $25/day and $0.25/mi. He plans to use the car for 5 days. What is the maximum mileage he can afford to put on the car?

80. *Car Rental.* After visiting Disney World, Gabe plans to rent a car to tour Orlando for 3 days. He has $85 to spend on car rental. Hietz Rental will rent him a car for $15/day plus $0.10/mi. How much mileage can he afford to put on the car before returning it?

81. *Surfing Equipment Rental.* During his vacation in Hawaii, Jason wants to try surfing. Upon arriving he finds it will cost him $28/day to rent a board and $225 to buy a new one. After how many days would the

price to rent the board be more than the price to buy it?

**82.** *Bicycle Rental.* Christine wants to rent a bicycle for the day. It costs $7/hr or $25/day to rent it. After how many hours would it be more economical to rent by the day?

**83.** *Transportation Industry Planning.* Florida Wire needs to ship via trucks coils of wire that weigh 6000 lbs/each. The truck and trailer together weigh 34,000 lb and only whole coils can be shipped. In order to travel the interstate highways, the truck and its load cannot exceed 78,000 lbs. How many coils can be shipped at one time?

**84. Writing to Learn.** Write in your own words the numbers included in the interval $x \leq 9$.

**85. Writing to Learn.** Write in your own words how to tell whether to use a parenthesis or a bracket to name an interval.

**86. Writing to Learn.** Explain why the inequality symbol reverses when you multiply or divide each side of an inequality by the same negative number.

TRANSLATING WORDS TO SYMBOLS

In Exercises 87 and 88, write the sentence in symbols.

**87.** The sum of a number and 6.5 is more than 10.35.

**88.** The difference of 40.6 and $x$ is less than 25.

In Exercises 89–95, write the sentence as an inequality.

**89.** $x$ is positive.

**90.** $x$ is nonnegative.

**91.** $x$ is at least 3.

**92.** The maximum value that $x$ can be is 150.

**93.** The minimum value that $x$ can be is 65.

**94.** Half of the sum of a number and 6 is at least 10.

**95.** Half of the difference of a number and 5 is more than 12.

EXTENDING THE IDEAS

**96.** *Projectile Motion.* Matthew throws a baseball up into the air with the initial velocity of 64 ft/sec. The ball is 6 ft above the ground when it leaves his hand. When will the ball be 54 ft above the ground if the height is given by the equation $h = -16t^2 + vt + h_0$, where $t$ = time, $h$ = height, $v$ = velocity, and $h_0$ = initial height?

LOOKING BACK—MAKING CONNECTIONS

In Exercise 97 and 98, write an inequality for the problem situation.

**97.** The perimeter of a triangle whose sides are $a$, $b$, and $c$ is less than 52.

**98.** The area of a rectangle with width $x$ and length $2x$ is at least 81.

**99.** Find the solution to the following equation by using the $x$-intercept method: $3(x + 6) - 4x = 2(3x + 5) + 1$.

**100.** Find the solution to the following equation by using the multigraph method: $\dfrac{x + 3}{2} = \dfrac{5x - 6}{3}$.

**101.** *Mixing Chemicals.* A chemist needs a solution that is at least 6% silver nitrate. He has 5 liters of a solution that is 4% silver nitrate and wants to mix this with a 10% silver nitrate solution. How many liters of the 10% solution are needed to have a solution that is at least 6% silver nitrate?

**102.** *Investment Planning.* Bill needs to make at least $500 on his investments this year. He wants to put his money in two different banks. He invests $4000 in a bank paying 7% interest and wants to put the rest of his money in another bank that pays 8% interest. How much must he put in the bank paying 8% in order for his combined interest earned to be at least $500?

## 4.2    Solving Inequalities Graphically

The grapher provides a strong link between algebraic procedures and visual methods. With the grapher you can "see" what the solution is. This visual sense is particularly helpful when solving inequalities.

### Solving Inequalities with the *x*-intercept Method

The following exploration introduces you to the *x*-intercept method for solving $2x - 4 < 0$.

### EXPLORE WITH A GRAPHER

1. Find the graph of $y = 2x - 4$ in the Integer window.
2. Move the trace cursor along the line. What values of $x$ result in negative $y$-values? What values of $x$ result in positive $y$-values?
3. Which of these phrases describe the solution to the inequality $2x - 4 < 0$?
   a)  The $x$-coordinates of all points on the graph of $y = 2x - 4$ below the $x$-axis
   b)  All $x$-values less than 2
   c)  All numbers in the interval $(-\infty, 2)$

You should have decided that all three statements in the exploration described the solution to the inequality $2x - 4 < 0$.

We develop the $x$-intercept method of solving the inequality graphically by solving the inequality

$$0.8x - 4 < 0.$$

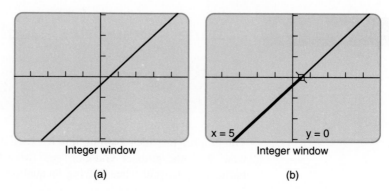

**FIGURE 4.12** (a) The graph of $y = 0.8x - 4$. (b) The points whose $x$-coordinates satisfy $0.8x - 4 < 0$ are included in the bold portion of the line.

Begin by finding the graph of $y = 0.8x - 4$. Figure 4.12 (b) identifies those points on the graph of $y = 0.8x - 4$ whose $y$-coordinates are negative. Note that if $y = 0.8x - 4$, the statements $y < 0$ and $0.8x - 4 < 0$ are equivalent.

The points on the bold portion of the line in Figure 4.12 (b) correspond to the solution of the inequality $0.8x - 4 < 0$. More precisely, the solution to the inequality consists of the $x$-coordinates of all the points on the graph below the $x$-axis. We say that the solution is the interval $(-\infty, 5)$, or $x < 5$.

The previous paragraph considers one of the four inequalities associated with the equation $0.8x - 4 = 0$. Here is a summary of the visual and interval representations of all four inequalities.

| Inequality | Visual Representation | Solution |
|---|---|---|
| $0.8x - 4 < 0$ | Points of the graph below the $x$-axis | $(-\infty, 5)$, or $x < 5$ |
| $0.8x - 4 \leq 0$ | Points of the graph on or below the $x$-axis | $(-\infty, 5]$, or $x \leq 5$ |
| $0.8x - 4 > 0$ | Points of the graph above the $x$-axis | $(5, \infty)$, or $x > 5$ |
| $0.8x - 4 \geq 0$ | Points of the graph on or above the $x$-axis | $[5, \infty)$, or $x \geq 5$ |

**EXAMPLE 1**   Solving an Inequality with the $x$-intercept Method

Use the $x$-intercept method to solve $-0.5x + 11 \geq 0$ graphically. Confirm the solution algebraically.

SOLUTION

**Solve Graphically.** Find the graph of the equation $y = -0.5x + 11$ in the Integer window (see Figure 4.13).

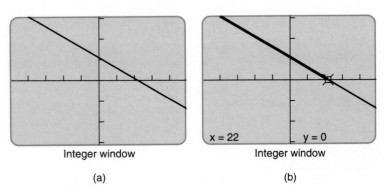

Integer window | Integer window
(a) | (b)

**FIGURE 4.13** (a) The graph of $y = -0.5x + 11$. (b) The points whose $x$-coordinates satisfy $-0.5x + 11 \geq 0$ are on the bold portion of the line.

**GRAPHER NOTE**

Observe that as the trace cursor is moved left from the point $(22, 0)$, the $y$-coordinates are all positive. This numerical observation reinforces the visual observation that we have found the solution to the inequality.

The point where the graph crosses the $x$-axis is $(22, 0)$. Therefore the solution to the inequality $-0.5x + 11 \geq 0$ is the interval $(-\infty, 22]$.

**Confirm Algebraically.**

$$-0.5x + 11 \geq 0$$

$$-0.5x \geq -11 \quad \textbf{Divide by } -0.5 \textbf{ and reverse the inequality symbol.}$$

$$x \leq 11/0.5$$

$$x \leq 22$$

Example 1 solves an inequality with a zero on one side. When the given inequality does not have a zero on one side, we first change it to an equivalent form with a zero on one side of it.

**EXAMPLE 2** Solving an Inequality with the $x$-intercept Method

Use the $x$-intercept method to solve the inequality $0.3x - 5 \leq 7$.

SOLUTION
Begin by subtracting 7 from both sides of the inequality.

$$0.3x - 5 \leq 7$$

$$0.3x - 5 - 7 \leq 7 - 7$$

$$0.3x - 12 \leq 0$$

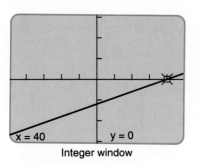

Integer window

**FIGURE 4.14** The graph of $y = 0.3x - 12$ with the $x$-intercept indicated.

Next find the graph of $y = 0.3x - 12$ in the Integer window and find the $x$-intercept (see Figure 4.14).

The solution is represented by the portion of the line on or below the $x$-axis. Because $(40, 0)$ is the $x$-intercept, we see that the solution to the inequality is $(-\infty, 40]$. ∎

**EXAMPLE 3**   Solving an Inequality with a Non-integer Endpoint

Solve the inequality $4.25(x - 6) + 2.75 < 5(x + 1.25)$ graphically.

SOLUTION

We use the $x$-intercept method.

1. Rewrite the inequality so that the right-hand side is 0.

$$4.25(x - 6) + 2.75 - 5(x + 1.25) < 0$$

Find the graph of $y_1 = 4.25(x - 6) + 2.75 - 5(x + 1.25)$ in the Integer window as seen in Figure 4.15 (a).

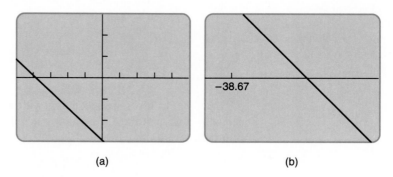

(a)                              (b)

**FIGURE 4.15**  Begin with the Integer window. Then zoom in to find the $x$-intercept with an error of at most 0.01.

After changing $X$scl to 0.01, position the cursor near the $x$-intercept and zoom in several times. Figure 4.15 (b) shows that the $x$-intercept is $x = -38.667$ with an error of at most 0.01. The graph is below the $x$-axis (or $<$) for all $x$-values greater than $-38.66$.

The solution of the inequality is $x > -38.66$ or $(-38.66, \infty)$. ∎

## Solving Problem Situations with Linear Inequalities

Often the graph of an inequality can be used to solve a problem situation. Consider the following problem situation.

---

□                                                                      □

### SALE MERCHANDISE PROBLEM SITUATION

The Executive Suit Co. has a 30%-off sale on all merchandise in stock. There is a 6% sales tax on clothing.

□                                                                      □

---

**EXAMPLE 4**   THE SALES MERCHANDISE PROBLEM SITUATION: Finding the Original Price

If Ruby has at most $125 to spend, find the original price of the suit she is able to afford. (Use a graphical method.)

SOLUTION

Begin by finding an algebraic representation for the problem situation. Note that with a 30% sale, the sale price is 70% of the original price.

1. Let $x =$ the original price of the suit. Then the following is true:

   **sale price of the suit:** $0.70x$

   **sales tax on the sale price:** $0.06(0.70x)$

   **total cost of purchase:** $0.70x + 0.06(0.70x)$

   Therefore an algebraic representation of the problem situation is

   $$0.70x + 0.06(0.70x) \leq 125.$$

2. Use the $x$-intercept method to solve this inequality graphically. Begin by graphing $y_1 = 0.70x + 0.06(0.70x) - 125$ in the Standard window and zoom out until the graph appears. Figure 4.16 (a) on the following page shows the result.

3. Set $X$scl$= 0.01$ and zoom in to find the $x$-intercept with an error of at most 0.01. Figure 4.16 (b) shows that the $x$-intercept is 168.464 with an error of at most 0.01.

   The solution to the inequality $0.70x + 0.06(0.70x) \leq 125$ is $x \leq 168.46$. Therefore Ruby can buy any suit priced less than or equal to $168.46.

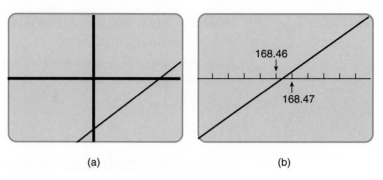

(a)                                    (b)

**FIGURE 4.16**  A graph of $y_1 = .70x + 0.06(0.70x) - 125$
(a) after using the Standard window and zooming out twice with
XFact= 5, YFact= 5 and (b) after zooming in to find the $x$-intercept.   ■

## Solving Problem Situations with Nonlinear Inequalities

In the next two examples, we revisit the **Garden Problem Situation** and the **Baseball Problem Situation** first introduced in Section 3.4. The particular problems considered result in algebraic representations that are inequalities. In both examples, a grapher is used.

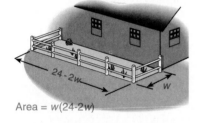

Area = $w(24\text{-}2w)$

**FIGURE 4.17**  The **Garden Problem Situation.**

**EXAMPLE 5**    THE GARDEN PROBLEM SITUATION: Finding the Width

Figure 4.17 reminds us that in the **Garden Problem Situation,** 24 ft of fence is used to surround a rectangular garden on three sides. The area was represented by the algebraic representation $A = w(24 - 2w)$. Find the width of the garden when its area is more than 50 ft$^2$.

SOLUTION
We shall use a graphical method.

**Solve Graphically.**

1. Graph $y_1 = 50$ and $y_2 = x(24 - 2x)$ in an appropriate window. Because $x$ must be between 0 and 12 and $y$ (representing the area) must be greater than 50, the window $[0, 12]$ by $[0, 85]$ seems reasonable.
2. Figure 4.18 shows that the graph of $y_2 = x(24 - 2x)$ intersects the graph of $y_1 = 50$ at the points $P$ and $Q$. Using TRACE and ZOOM-IN, we find that the $x$-coordinate of $P$ is $x = 2.683$ with an error of at most 0.01.
3. Zoom out to return to the original window and repeat the zoom-in process to find that the $x$-coordinate of $Q$ is $x = 9.317$ with an error of at most 0.01. So the $y = x(24 - 2x)$ values are greater than 50 when $2.683 < x < 9.317$.

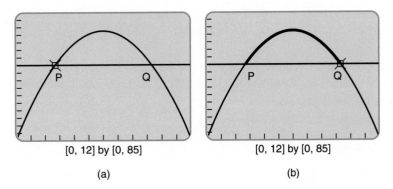

[0, 12] by [0, 85]

(a)

[0, 12] by [0, 85]

(b)

**FIGURE 4.18** The portion of the curve representing an area greater than 50 ft² are those points between points $P$ and $Q$. The solution is the $x$-values between the $x$-coordinates of points $P$ and $Q$.

Therefore when the width of the garden is between 2.683 and 9.317, the area is more than 50 ft². ∎

The next example revisits the baseball problem situation.

**EXAMPLE 6**    THE BASEBALL PROBLEM SITUATION: Finding the Time

Recall that if a ball is thrown straight up into the air with an initial velocity of 88 ft/sec, its height after $t$ seconds is given by the algebraic representation $h = -16t^2 + 88t$. Find the interval of time (with an error of at most 0.01) that the ball will be more than 90 ft above ground level.

SOLUTION

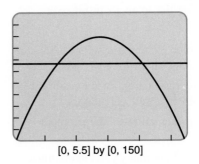

[0, 5.5] by [0, 150]

**FIGURE 4.19** The graph of $y_1 = 90$ and $y_2 = -16x^2 + 88x$.

1. We first must choose a reasonable window. Since $x$ represents time after the ball is thrown, $x$ must be greater than zero. Try $X$max= 5. The $y$-value represents height and we need to find when the height is greater than 90 ft. It is reasonable to try $Y$max= 150. So consider the window [0, 5] by [0, 150].
2. Find the graph of $y_1 = 90$ and the algebraic representation $y_2 = -16x^2 + 88x$ in the window [0, 5] by [0, 150]. After the first attempt, we see that a better window is [0, 5.5] by [0, 150] (see Figure 4.19).
3. Trace to the first point of intersection and zoom in to find that the $x$-coordinate of this point of intersection (with an error of at most 0.01) is $x = 1.358$.
4. Use ZOOM-OUT to return to the original window and Trace to the second point of intersection. Zoom in to find that the $x$-coordinate at this point is $x = 4.142$.

Therefore the ball is more than 90 ft off the ground from 1.358 to 4.142 sec. We can express this in interval notation as (1.358, 4.142). ∎

## Exercises for Section   4.2

In Exercises 1–8, write an equivalent inequality with the left-hand side simplified and the right-hand side equal to 0.

1. $2x - 5 < 6$

2. $5x - 4 > 20$

3. $3x - 9 < 5x + 2$

4. $4x + 3 \leq 2x - 8$

5. $3(x + 4) - 6 \geq 2x - 9$

6. $2x - (3x + 7) \geq 5(x - 4)$

7. $2(x - 1) - 3 \geq x - 5$

8. $x - (2x + 1) \geq 8(x - 3)$

In Exercises 9–18, solve the inequality using the $x$-intercept method.

9. $x + 3 < 5$

10. $2x - 6 > 8$

11. $3(x - 2) > 6$

12. $2(x + 3) - 5 > 6x + 7$

13. $4(x - 1) + 3 < 2x - 5$

14. $5x - 7 > 12$

15. $3(x - 7) + 2 \leq 6x + 2$

16. $\dfrac{3x}{5} - 4 \geq -1$

17. $\dfrac{2}{3} - \dfrac{x}{2} > 1$

18. $\dfrac{2x + 1}{3} + 5 < 8$

In Exercises 19–30, solve the inequality graphically. (*Hint:* Use the Integer window. Confirm the solution algebraically.)

19. $2x - 18 \geq 0$

20. $3x - 21 < 0$

21. $28 - 4x > 0$

22. $6 - 2x \leq 0$

23. $1.8x + 2 \leq 11$

24. $5 - 1.2x > 1.4$

25. $2.1x - 8.5 < 2$

26. $2.7x - 13.1 \geq 3.1$

27. $2x - 48 > 4$

28. $1.9x - 24.5 < 4$

29. $7.4 - 1.8x \leq 2$

30. $14.5 - 2.3x \geq 3$

In Exercises 31–34, use either the $x$-intercept or the multigraph method to solve the inequality. Explain why you chose the method you did.

31. $3(x + 2) - 4.6 > 7.9$

32. $5x + 2(x - 3) \geq 8.6$

33. $5.1x + 2.7 \leq 3.62x - 7.5$

34. $\dfrac{1}{2}(x + 3) - 4 > 2(x - 5) + 6$

In Exercises 35–41, solve the problem situation.

35. *Baseball Team Fund Raising.* The Padres Little League baseball team raises money by selling boxes of candy at $4/ea. The team purchases the candy for $2.75/box.

   a) Write an algebraic representation for the profit the baseball team will make when it sells $x$ boxes of candy. (Profit is the difference between revenue and expenses.)

   b) Find a graphical representation of the problem situation.

   c) Using this graph, how many boxes must the team sell to make a profit of at least $495?

   d) Using the same graph, determine what the sales would have to be for the team to profit over $1000?

36. *Analyzing Profit.* Office Warehouse sells calculators for $19.75/ea. They buy them for $14.50/ea.

   a) Write an algebraic representation for the profit the store will make when it sells $x$ calculators.

   b) Find a graphical representation of the problem situation.

   c) Using this graph, find how many calculators must be sold to make a profit of at least $150.

37. *Club Fund Raising.* The Drama Club needs to raise $1500 for a new stage curtain. Its members perform *Romeo and Juliet* and charge $6/ticket. How many tickets must be sold to raise the funds for the

curtain if their cost for staging the production is $525?

a) Write an algebraic representation for the profit the Drama Club will make if they sell $x$ tickets.

b) Find a graphical representation of the problem situation.

c) Using this graph, find how many tickets they must sell to raise the $1500 for the new stage curtain.

**38.** *Commission Sales.* Meera is paid $200 a week plus 4% commission on her sales.

a) Find an algebraic representation of her weekly gross pay (take-home pay before deductions).

b) Find a graphical representation of the problem situation.

c) Use this graphical representation to find how much she has to sell in order to have gross pay of at least $450 in one week?

d) Use the same graphical representation to find out her gross pay in one week if she sold $1052.63 worth of merchandise.

**39.** *Commission Sales.* Pam is paid $140 a week plus a commission of $20 on each order she places.

a) Write an algebraic representation for the amount of money Pam earned at the week's end if she placed $x$ orders that week.

b) Find a graphical representation of the problem situation.

c) Use this graph to find how many orders she must place to make at least $360 a week?

**40.** *Landscape Design.* Byron has 100 ft of fencing to enclose his rectangular garden, with one side of the rectangle being his house.

a) Find an algebraic representation of the problem situation describing the area of the garden if the width is $x$ feet.

b) For what values of the width is the area 200 ft$^2$?

c) For what values of the width is the area more than 200 ft$^2$?

**41.** *Projectile Motion.* It can be shown that when a baseball is thrown straight up into the air with a velocity of 70 ft/sec, $h = -16t^2 + 70t$ is an algebraic representation for the ball's height $t$ seconds after being thrown.

a) In what time interval (with an error of at most 0.01) after the ball is thrown is the ball more than 30 ft high?

b) Estimate the maximum height of the ball and at what time it reaches this height.

c) In what time interval (with an error of at most 0.01) is the ball descending?

d) How long after the ball is thrown does it hit the ground?

TRANSLATING WORDS TO SYMBOLS

In Exercises 42–44, write the sentence in symbols.

**42.** The sum of a number and 32 is less than half the difference of that number and 56.

**43.** Twice the sum of $x$ and 5.6 is at least 2.8.

**44.** The average of 87 and $x$ is at most 92.

**45.** *Grapher Activity.* Change your grapher to Dot mode and graph the inequality $y_1 = 0.8x - 4 < 0$. The following figure shows the inequality on the function-definition menu.

$$Y_1 = 0.8X - 4 < 0$$
$$Y_2 =$$
$$Y_3 =$$
$$Y_4 =$$

For Exercise 45.

Explain why this graph in the Integer window gives a "visual" picture of a solution to $x < 5$. (*Hint:* Check your grapher lab manual to see how the grapher handles inequalities.)

**46. Writing to Learn.** Describe in your own words the multigraph method of finding the solution to an inequality.

**47. Writing to Learn.** Describe in your own words the $x$-intercept method of finding the solution to an inequality.

**48. Writing to Learn.** Explain in a paragraph which graphing method you prefer and why you choose that method.

EXTENDING THE IDEAS

**49.** On a number line, indicate all numbers that are greater than 6 and at the same time less than 12.

**50.** On a number line, indicate all numbers that are either less than 2 or greater than 9.

LOOKING BACK—MAKING CONNECTIONS

**51.** Write an inequality for the area of a circle with radius $r$ whose area is more than 17 cm$^2$.

**52.** Write an inequality for the area of a triangle with base 6 ft, height $h$, and area less than 30 ft$^2$.

**53.** Write an inequality showing the perimeter of a regular octagon to be less than 96 ft.

**54.** The length of a rectangle is 3 ft more than twice the width. Write an inequality showing the perimeter to be more than 100 ft.

**55.** The Try This margin comment by Example 6 in Section 1.6 shows how Parametric mode can be used to simulate motion. Use this mode to simulate the motion in the **Baseball Problem Situation.**

---

## 4.3 Compound Inequalities and Double Inequalities

Solving Compound Inequalities with *And* ▪ Solving Compound Inequalities with *Or* ▪ Solving Double Inequalities ▪ Solving Problem Situations with Double Inequalities

A **compound inequality** consists of two inequalities joined with either the word *and* or the word *or*. Following are two examples of compound inequalities:

$$x > 3 \quad and \quad x < 7$$

$$x < 5 \quad or \quad x > 8$$

Each of these is associated with an interval. Using interval notation, we write the following:

$$(3, \infty) \quad and \quad (-\infty, 7)$$

$$(-\infty, 5) \quad or \quad (8, \infty)$$

Compound inequalities with the conjuction *and* can be combined into a single statement called a **double inequality**. For example,

$$x > -3 \quad and \quad x \le 2 \quad \text{is equivalent to} \quad -3 < x \le 2.$$

The double inequality $-3 < x \le 2$ can be represented as a finite interval, as shown in Figure 4.20.

**FIGURE 4.20** The interval $(-3, 2]$.

The four types of double inequalities and their representations as compound inequalities are shown in Figure 4.21.

| Double Inequality | Number Line | Interval Notation | Compound Inequality |
|---|---|---|---|
| $a < x < b$ | $\underset{a \qquad b}{\circ\!\!-\!\!-\!\!\circ}$ | $(a, b)$ | $x > a$ and $x < b$ |
| $a \le x < b$ | $\underset{a \qquad b}{\bullet\!\!-\!\!-\!\!\circ}$ | $[a, b)$ | $x \ge a$ and $x < b$ |
| $a < x \le b$ | $\underset{a \qquad b}{\circ\!\!-\!\!-\!\!\bullet}$ | $(a, b]$ | $x > a$ and $x \le b$ |
| $a \le x \le b$ | $\underset{a \qquad b}{\bullet\!\!-\!\!-\!\!\bullet}$ | $[a, b]$ | $x \ge a$ and $x \le b$ |

**FIGURE 4.21** Inequalities and their interval notations.

## Solving Compound Inequalities with *And*

Example 1 illustrates how to solve a compound inequality with *and* as the conjunction.

**EXAMPLE 1**    Solving a Compound Inequality Algebraically

Solve the compound inequality $3x - 7 < 8$ and $-2x + 8 < 20$.

SOLUTION
Solve each inequality individually and find all numbers that are common to both intervals.

$$3x - 7 < 8 \qquad\qquad -2x + 8 < 20$$
$$3x < 15 \qquad \text{and} \qquad -2x < 12$$
$$x < 5 \qquad\qquad\qquad x > -6$$

**REMINDER**

In a double inequality, all the inequality signs should go in the same direction. Consequently it is not correct to write $8 > x < 5$. Also it is not correct to write $8 < x < 5$, since no values of $x$ are both less than 5 and greater than 8.

The number-line sketch in Figure 4.22 shows two infinite intervals that represent the solutions of the individual inequalities. The solution of the compound inequality is the finite interval $(-6, 5)$, which is the intersection of the two infinite intervals, and is drawn as a shaded line segment. This solution can also be written as $-6 < x < 5$.

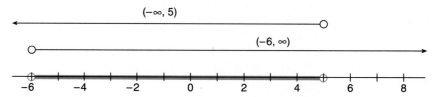

**FIGURE 4.22** The shaded portion of the number line represents the solution of the compound inequality $(-6, 5)$.

When solutions are represented using interval notation, it is correct to use the mathematical symbols $\cup$, which we read as *or* and $\cap$, which we read as *and*. For example,

$$(-\infty, 5) \cap (-6, \infty) \qquad \text{means} \qquad (-\infty, 5) \text{ and } (-6, \infty),$$

$$(-\infty, 5) \cup (-6, \infty) \qquad \text{means} \qquad (-\infty, 5) \text{ or } (-6, \infty).$$

With this notation, the solution to Example 1 also can be expressed using interval notation as $(-\infty, 5) \cap (-6, \infty)$.

## Solving Compound Inequalities with *Or*

The solution of a compound inequality like

$$x < 5 \qquad \text{or} \qquad x > 8$$

cannot be expressed as a single finite interval. Notice that a number is in the set of solutions if it is *either* less than 5 *or* greater than 8, but not both. Figure 4.23 shows that this solution set consists of two infinite intervals taken together.

**FIGURE 4.23** The solution to $x < 5$ or $x > 8$ consists of two infinite intervals.

In Example 2 we solve a compound inequality with *or* as the conjunction.

**REMINDER**

The symbol $\cup$ is read *union* and means *or*. The symbol $\cap$ is read *intersection* and means *and*.

**EXAMPLE 2**    Solving a Compound Inequality with *Or*

Solve the following compound inequality:

$$2x + 5 < 6 \quad \text{or} \quad 3(x - 4) + 2 > 2x - 5$$

SOLUTION

$$
\begin{array}{ccl}
 & & 3(x - 4) + 2 > 2x - 5 \\
2x + 5 < 6 & & 3x - 12 + 2 > 2x - 5 \\
2x < 1 & \text{or} & 3x - 10 > 2x - 5 \\
x < \dfrac{1}{2} & & x - 10 > -5 \\
 & & x > 5
\end{array}
$$

Figure 4.24 represents $(-\infty, 0.5) \cup (5, \infty)$, which is the solution to the compound inequality $2x + 5 < 6 \quad \text{or} \quad 3(x - 4) + 2 > 2x - 5$.

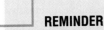

**FIGURE 4.24**   The bold portions of the number line represent $(-\infty, 0.5)$ $\cup (5, \infty)$, which is the union of the solution intervals of each inequality.  ∎

Next, we summarize the methods used in Examples 1 and 2.

**REMINDER**

When working with double inequalities, always check that the two inequality symbols are in the *same* direction. Sometimes we incorrectly mix greater than and less than inequalities in the same three-part sentence.

---

## Solving Compound Inequalities Algebraically

1. Solve individual inequalities in the compound inequality separately.
2. When the conjunction is *and*, find the *intersection* ($\cap$) of the two solution intervals.
3. When the conjunction is *or*, find the *union* ($\cup$) of the two solution intervals.

## Solving Double Inequalities

Example 3 illustrates that a double inequality like $-15 < 4x - 7 < 21$ can be solved graphically by modifying the multigraph method that has been used for solving single inequalities.

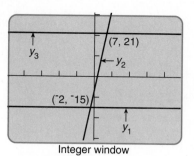

**FIGURE 4.25** The graph $y_1 = -15$, $y_2 = 4x - 7$, and $y_3 = 21$.

**EXAMPLE 3**    Solve Graphically and Confirm Algebraically

Solve $-15 < 4x - 7 \leq 21$ graphically. Confirm the solution algebraically.

SOLUTION

**Solve Graphically.** Find the graphs of $y_1 = -15$, $y_2 = 4x - 7$, and $y_3 = 21$, all in the Integer window. Use TRACE to determine that the graph of $y_2 = 4x - 7$ intersects the other two lines at points $(-2, -15)$ and $(7, 21)$. Observe that when $-2 < x \leq 7$, the graph of $y_2$ is above $y_1$ and on or below $y_3$. So the solution to $-15 < 4x - 7 \leq 21$ is $-2 < x \leq 7$ or in interval notation $(-2, 7]$ (see Figure 4.25).

**Confirm Algebraically.** Isolate the variable in the middle of the inequality.

$$-15 < 4x - 7 \leq 21 \quad \textbf{Add 7 to each of the three parts.}$$

$$-8 < 4x \leq 28 \quad \textbf{Divide each part by 4.}$$

$$-2 < x \leq 7$$

The algebra confirms that the solution is $(-2, 7]$.    ∎

As illustrated by the algebraic confirmation in Example 3, a double inequality can be solved algebraically by applying appropriate properties of inequalities. But in the case of a double inequality, we must apply the properties to all three parts of the inequality.

With double inequalities, we want to isolate the variable in the middle portion of the inequality. Example 4 illustrates solving algebraically and supporting graphically.

**EXAMPLE 4**    Solve Algebraically and Support Graphically

Solve the double inequality $\dfrac{1}{5} < \dfrac{-x + 5}{2} < \dfrac{7}{2}$ algebraically. Support the solution graphically.

SOLUTION
**Solve Algebraically.**

$$\frac{1}{5} < \frac{-x+5}{2} < \frac{7}{2}$$

Multiply each of the three parts by 10.

$$10\left(\frac{1}{5}\right) < 10\left(\frac{-x+5}{2}\right) < 10\left(\frac{7}{2}\right)$$

$$2 < 5(-x+5) < 5(7)$$

$$2 < -5x+25 < 35$$

Add $-25$ to each of the three parts.

$$-23 < -5x < 10$$

Divide each part by $-5$; which reverses inequality symbols.

$$\frac{23}{5} > x > -2$$

Write the inequality in equivalent form to correspond to interval notation.

$$-2 < x < \frac{23}{5}$$

The solution is the interval $(-2, 4.6)$.

**Support Graphically.** Figure 4.26 provides graphical support for this algebraic solution. ■

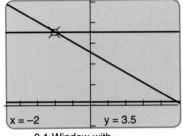

x = −2        y = 3.5

0.1 Window with vertical dimension [−1, 5]

**FIGURE 4.26** The graph $y_1 = 1/5$, $y_2 = (-x+5)/2$, and $y_3 = 7/2$.

## Solving Problem Situations with Double Inequalities

In the next example, we revisit the **Averaging Test Scores Problem Situation** introduced in Section 4.1. Recall that in this situation, Dan wanted to know what score he needed on the fourth exam to earn a grade of A in his American history course. In biology, Dan scored 83%, 79%, 93%, and 86% on his first four exams. The minimum percentages needed for an A and B are the same in both classes: 92% for an A and 84% for a B.

**EXAMPLE 5**    THE AVERAGING TEST SCORES PROBLEM SITUATION:
Finding a Particular Score

What must Dan score on the fifth exam to earn a B in his biology class? Write an algebraic representation for this problem situation and solve the inequality.

SOLUTION
Let $x = $ Dan's grade on the fifth exam. The algebraic representation for the average for the five grades is

$$\frac{83+79+93+86+x}{5}.$$

**FIGURE 4.27** Dan's exam scores in biology.

The average for five grades must be greater than or equal to 84 and less than 92 (see Figure 4.27).

$$84 \leq \frac{83 + 79 + 93 + 86 + x}{5} < 92$$   An inequality that must be true if a B average is achieved

$$84 \leq \frac{341 + x}{5} < 92$$   Simplify.

$$420 \leq 341 + x < 460$$   Multiply each of the three parts by 5.

$$79 \leq x < 119$$   Add $-341$ to each of the three parts.

This means that Dan must score at least or greater than a 79% on his fifth exam to have an B average in the course. Because 100% is the highest score that can be made on an exam, it is not possible for Dan to receive the 119 score needed to earn an A.   ∎

## Exercises for Section   **4.3**

In Exercises 1–8, graph the solution to the compound inequality as an interval on a number line.

**1.** $x > 3$ and $x < 9$

**2.** $x < 11$ and $x > -2$

**3.** $x < 2$ or $x > 6$

**4.** $x < -5$ or $x > 4$

**5.** $x \leq 10$ and $x > 2$

**6.** $x \geq 3$ and $x \leq 7$

**7.** $x \leq -9$ or $x \geq 0$

**8.** $x \geq -5$ or $x < -8$

In Exercises 9–18, use interval notation to write the solution of the compound inequality.

**9.** $x < 3$ and $x > -3$

**10.** $x < 7$ and $x > -5$

**11.** $5x - 2 > 3$ and $5x - 2 > -3$

**12.** $x + 7 \leq 3$ and $x + 7 \geq -3$

**13.** $x - 2 < 5$ and $x - 2 > -5$

**14.** $x + 2 < -5$ or $x + 2 > 5$

**15.** $2x - 6 \leq -2$ or $2x - 6 \geq 2$

**16.** $5x - 1 \leq -9$ or $5x - 1 \geq 9$

**17.** $x > 4$ and $x \leq 0$

**18.** $x < 3$ and $x > 9$

In Exercises 19–22, write a compound inequality that describes the set. (Answers can vary.)

**19.** $(-\infty, 1) \cup (5, \infty)$

**20.** $(-\infty, -5] \cup [10, \infty)$

**21.** $(-\infty, 2) \cap (-5, \infty)$

**22.** $(-\infty, 16) \cap (3, \infty)$

In Exercises 23–30, represent the inequality in interval notation.

**23.** $4 < x < 10$

**24.** $-2.5 < x < 7.5$

**25.** $-5.35 \leq x < 8.9$

**26.** $-5.1 < x \leq 3.4$

**27.** $2 < x \leq 7$

**28.** $-5.76 \leq x \leq 4.75$

**29.** $7.24 \leq x \leq 19.5$

**30.** $4.35 < x < 12.8$

In Exercises 31–38, write a double inequality that represents the interval.

**31.** $(5, 8)$

**32.** $(2, 9]$

**33.** $[-3, 2]$

**34.** $(-5.15, 18]$

**35.** $[23, 62.4)$

**36.** $(-45.3, -3]$

**37.** $(7.5, 9.99)$

**38.** $[-4.35, 9.75)$

In Exercises 39–46, write the interval as one interval.

**39.** $(-\infty, 4) \cap (2, 8)$    **40.** $(3, \infty) \cup (-1, 5)$

**41.** $(-\infty, 6) \cup (3, \infty)$    **42.** $(-3, 4] \cap (2, 8)$

**43.** $(-\infty, 4) \cap (-1, 8)$    **44.** $(-\infty, 0) \cap (-3, 5)$

**45.** $(-2, \infty) \cap (5, \infty)$    **46.** $(-5, \infty) \cap (3, \infty)$

In Exercises 47–50, determine if $-1.15$ is a member of the solution interval for the inequality.

**47.** $-2.3 < x < 6$    **48.** $-5 \le x \le -3$

**49.** $23.8 \ge x \ge 15.2$    **50.** $4 < x < 5$

In Exercises 51–62, solve the double inequality algebraically. Express the solution in interval notation and support it graphically.

**51.** $2 < x + 3 < 9$    **52.** $5 \le 2x - 7 < 8$

**53.** $-3.5 \le 3x - 1 \le 11$    **54.** $-5 < x + 5 < 10.5$

**55.** $2.1 < x - 7 < 5.8$    **56.** $7 \le 2x - 3 < 9$

**57.** $3 \le x + 7 < 5$    **58.** $-1 \le 2x + 3 < 5$

**59.** $4 \le 3(x - 6) < 15$    **60.** $-15 \le 5(x + 2) \le 20$

**61.** $\dfrac{7}{2} < 5x + 4 \le 14$    **62.** $\dfrac{8}{5} < 2x - 5 < \dfrac{9}{2}$

In Exercises 63–76, solve the double inequality graphically. Express the solution in interval notation and confirm it algebraically.

**63.** $-23.7 < 3x - 24 < -18$

**64.** $8.1 < 2x + 1.5 < 11$

**65.** $-3 < x + 2 < 8$    **66.** $6 \le -2x + 7 < 9$

**67.** $2.5 \le -x - 2 \le 5.5$    **68.** $5 < x - 10 < 10$

**69.** $12 < -0.5x - 7 < 15$

**70.** $-5 \le 2x + 3 < 4$

**71.** $-12 \le 3x + 6 < 3$    **72.** $-10 \le -2x + 1 < 6$

**73.** $\dfrac{1}{2} < x - 4 \le \dfrac{5}{3}$    **74.** $\dfrac{2}{5} < -2x + 5 < \dfrac{3}{2}$

**75.** $\dfrac{5}{2} < \dfrac{x + 4}{3} \le 7$    **76.** $\dfrac{1}{2} < \dfrac{-x + 5}{3} < \dfrac{7}{6}$

## TRANSLATING WORDS TO SYMBOLS

In Exercises 77–80, write the sentence in symbols.

**77.** The sum of a number and 32 is between 56 and 79.

**78.** The difference of $x$ and 5.6 is between 2.8 and 7.9.

**79.** The average of 87 and $x$ is between 92 and 100.

**80.** The average of 82, 75, 93, and $x$ is between 85 and 92.

**81.** *Averaging Test Scores.* Brenda scored 75%, 110%, and 94% on her first three economics exams.

  a) Let $x$ represent the score of the fourth exam and write an algebraic representation for the average of the four exams.

  b) Write an inequality representing the average of the four exams as being in the 85–92% range.

  c) Solve the inequality in part (b) by an algebraic method to find what score Brenda needs to receive a B in the course if the B range is 85–92.

**82.** *Averaging Temperatures.* Irv is studying meteorology and is finding average temperatures of different locations over a 5-day span. He found the weather in March in Florida varies from day to day. For example, in Ormond Beach, the average daily temperature for a 4-day span was $75°$, $62°$, $85°$, and $86°$.

  a) Let $x$ represent the temperature on the fifth day and write an algebraic representation for the average temperature over a 5-day span.

  b) Write an inequality representing the average temperature of the five days as being in the 80–85° range.

  c) Solve the inequality in part (b) by an algebraic method to find the temperature's range on the fifth day.

**83.** *Averaging Daily Sales.* The Beachy Boys Surf Shop finds that rental business is slow on weekdays but picks up on the weekend. Their average daily sales are either less than $1500 or more than $3500. If $x$ is the average amount of daily sales, write an algebraic expression in the form of a compound inequality that describes the problem situation.

**84. Writing to Learn.** In your own words, describe the process of solving a compound inequality with *and* as its conjunction.

**85. Writing to Learn.** In your own words, describe the process of solving a compound inequality with *or* as its conjunction.

**86. Writing to Learn.** Explain why the solution interval of a double inequality is a single interval.

LOOKING BACK—MAKING CONNECTIONS

**87.** Explain what we mean by the absolute value of 4.

**88.** Is $|-6.29| - |8.96|$ a positive or negative number?

**89.** Is $|5.5| - |3.1|$ a positive or negative number?

**90.** Which of the following statements is always true: $|x| > 0$ or $|x| \geq 0$?

**91.** Which of the following statements is always true: $|x| < 0$ or $|x|$ is nonnegative?

---

## 4.4 Solving Absolute Value Equations and Inequalities

Solving Absolute Value Equations Algebraically ▪ Introducing Basic Absolute Value Inequalities ▪ Solving Absolute Value Inequalities Algebraically

Sometimes equations and inequalities include absolute values. Translating back and forth among words, symbols, and pictures is important in understanding absolute value. Recall that $|x|$ is the distance that the number $x$ is from the origin. The following chart relates the concept of absolute value using a word phrase, an equation, and a picture.

**GRAPHER NOTE**

Most graphers use ABS to represent the absolute value function. For example, $|x - 7|$ is entered as *abs* $(x - 7)$. On the other hand, *abs* $x - 7$ is equal to $|x| - 7$.

**Word Phrase** $x$ is 4 units from the origin.

**Equation** $|x| = 4$

**Picture**

The equation $|x| = 4$ referred to above is a special case of the following general absolute value equation.

---

### A Basic Absolute Value Equation

If $a$ is a positive real number,

$$|x| = a \text{ if and only if } x = a \text{ or } x = -a.$$

---

Following are several specific instances of this basic equation:

$$|x| = 5 \text{ if and only if } x = 5 \text{ or } x = -5$$

$$|x| = 14 \text{ if and only if } x = 14 \text{ or } x = -14$$

Examples 1 and 2 illustrate that this basic absolute value equation is key to solving equations involving absolute value. Example 1 also shows that some equations involving absolute value can be solved mentally.

**EXAMPLE 1**  Solving Equations Mentally

Solve each of the following equations mentally:

a) $|x - 1| = 7$
b) $|x + 2| = 4$
c) $|2x| = 8$

SOLUTION

a) Observe: $x - 1$ must be 7 or $-7$.
    Answer: $x = 8$ and $x = -6$.
b) Observe: $x + 2$ must be 4 or $-4$.
    Answer: $x = 2$ and $x = -6$.
c) Observe: $2x$ must be 8 or $-8$.
    Answer: $x = 4$ and $x = -4$. ■

**REMINDER**

Some absolute value equations have no solutions. For example, $|x - 3| = -2$ has no solution since the absolute value of a number is always nonnegative.

The mental thought process used in Example 1 is the basis for the algebraic method used in Example 2 for solving absolute value equations.

**EXAMPLE 2**  Solving an Equation and Confirming its Solution

Solve $|x - 4| = 3$ algebraically. Confirm the solution numerically.

SOLUTION

**Solve Algebraically.** From the basic absolute value equation, we see that the expression $x - 4$ must be either 3 or $-3$. That is

$$|x - 4| = 3 \text{ if and only if} \quad \begin{array}{l} x - 4 = 3 \text{ or} \\ x - 4 = -3. \end{array}$$

Solve each of these equations.

$$\begin{array}{ccc} x - 4 = 3 & & x - 4 = -3 \\ & \text{or} & \\ x = 7 & & x = 1 \end{array}$$

**Confirm Numerically.** Here is a paper-and-pencil confirmation. Replace $x$ with 7 and 1 and show each result in a true statement.

$$\begin{array}{cc} |x - 4| = 3 & |x - 4| = 3 \\ |7 - 4| = 3 & |1 - 4| = 3 \\ |3| = 3 & |-3| = 3 \end{array}$$ ■

## Solving Absolute Value Equations Algebraically

The method used in Example 2 becomes the basis for the algebraic method of solving an equation that is used in Examples 3 and 4.

**EXAMPLE 3**    Solve Algebraically and Support Numerically

Solve $|3x - 5| = 9$ algebraically. Support numerically using a grapher.

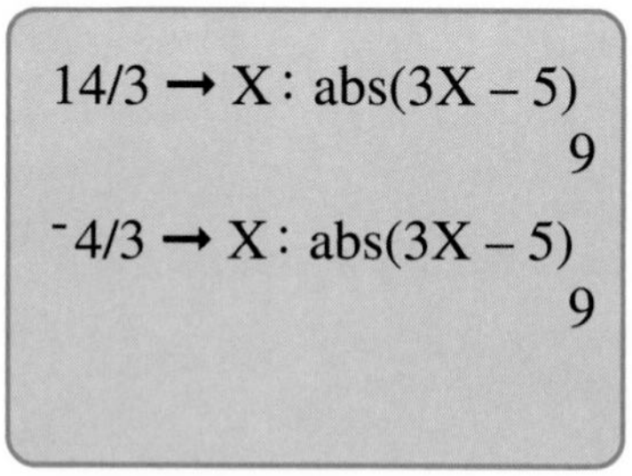

SOLUTION

**Solve Algebraically.** The expression inside the absolute value must be 9 or $-9$.

$$\begin{array}{ccc} 3x - 5 = 9 & & 3x - 5 = -9 \\ 3x = 14 & \text{or} & 3x = -4 \\ x = \dfrac{14}{3} & & x = -\dfrac{4}{3} \end{array}$$

The solutions to $|3x - 5| = 9$ are $x = \dfrac{14}{3}$ and $x = -\dfrac{4}{3}$.

**Support Numerically.** Figure 4.28 shows grapher support of the solutions.    ■

In Example 4, we solve the equation algebraically and use the graph to support the solution.

---

$14/3 \rightarrow X : \text{abs}(3X - 5)$
$\qquad\qquad\qquad\qquad 9$

$^-4/3 \rightarrow X : \text{abs}(3X - 5)$
$\qquad\qquad\qquad\qquad 9$

**FIGURE 4.28**
Numerical support of the solutions in Example 3.

**EXAMPLE 4**    Solving Absolute Value Equations That Contain Fractions

Solve the equation $|\frac{2}{3}x - 5| - 4 = 9$ algebraically. Support the solution by graphing the equation using the multigraph method.

SOLUTION

**Solve Algebraically.** Add 4 to both sides of the equation to *isolate* the absolute value.

$$\left|\frac{2}{3}x - 5\right| - 4 = 9$$

$$\left|\frac{2}{3}x - 5\right| = 13$$

Remove the absolute value symbols and solve the resulting equations.

$$\frac{2}{3}x - 5 = 13 \qquad\qquad \frac{2}{3}x - 5 = -13$$

$$2x - 15 = 39 \qquad \text{or} \qquad 2x - 15 = -39$$

$$2x = 54 \qquad\qquad\qquad 2x = -24$$

$$x = 27 \qquad\qquad\qquad x = -12$$

The solutions to $\left|\frac{2}{3}x - 5\right| - 4 = 9$ are $x = 27$ and $x = -12$.

**Support Graphically.** Figure 4.29 shows the graphs of $y_1 = \text{abs}\left(\frac{2}{3}x - 5\right) - 4$ and $y_2 = 9$ in the Integer window. The intersections of the two graphs is $(-12, 9)$ and $(27, 9)$. Therefore the solution to the equation $\left|\frac{2}{3}x - 5\right| - 4 = 9$ is $x = -12$ and $x = 27$.    ■

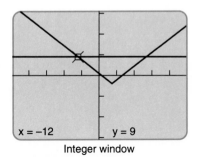

x = −12        y = 9

Integer window

**FIGURE 4.29**  Grapher support of the solutions to $\left|\frac{2}{3}x - 5\right| - 4 = 9$.

## Introducing Basic Absolute Value Inequalities

The solution of an inequality is closely related to the solution of the associated equation. For example, the solutions to the equation $|x| = 5$ are the endpoints of the solutions to the inequalities $|x| < 5$ and $|x| > 5$.

**Less-than Inequality.**

$|x| < 5$ if and only if $x$ is less than 5 units from the origin.

Figure 4.30 on the following page shows a number-line picture of this solution.

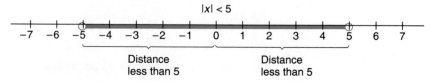

**FIGURE 4.30** The solution of $|x| < 5$ is $-5 < x < 5$, or using interval notation, $(-5, 5)$.

### Greater-than Inequality.

$|x| > 5$ if and only if $x$ is greater than 5 units from the origin.

Figure 4.31 shows a number-line picture of this solution.

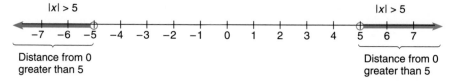

**FIGURE 4.31** The solution of $|x| > 5$ is $x < -5$ or $x > 5$, or, using interval notation, $(-\infty, -5) \cup (5, \infty)$.

The solution to an absolute value inequality is typically described either by a double inequality or in interval notation. The following chart shows each description.

---

## Equivalent Representations of Absolute Value Inequalities ($a > 0$)

| Absolute Value Inequality | Solution Described by an Inequality | Solution Described in Interval Notation |
|---|---|---|
| $|x| < a$ | $-a < x < a$ | $(-a, a)$ |
| $|x| \leq a$ | $-a \leq x \leq a$ | $[-a, a]$ |
| $|x| > a$ | $x < -a$ or $x > a$ | $(-\infty, -a) \cup (a, \infty)$ |
| $|x| \geq a$ | $x \leq -a$ or $x \geq a$ | $(-\infty, -a] \cup [a, \infty)$ |

Here are several examples of these equivalent representations.

| Absolute Value Inequality | Solution Described by an Inequality | Solution Described in Interval Notation |
|---|---|---|
| $|x| < 5$ | $-5 < x < 5$ | $(-5, 5)$ |
| $|x| \geq 8$ | $x \leq -8$ or $x \geq 8$ | $(-\infty, -8] \cup [8, \infty)$ |

## Solving Absolute Value Inequalities Algebraically

The transition from these basic absolute value inequalities to others requires facility in translating between symbols and words. Example 5 illustrates this process.

### EXAMPLE 5    Changing from Symbols to Words

Describe the solution to each of the following equations or inequalities in words:

a) $|x + 2| < 7$
b) $|x - 5| > 2$

SOLUTION

a) $x + 2$ is between $-7$ and 7.
b) $x - 5$ is greater than 2 or less than $-2$.    ■

Examples 6 and 7 illustrate how to solve an inequality algebraically and support its solution by using previous graphical methods.

### EXAMPLE 6    Solve Algebraically and Support Graphically

Solve the inequality $|x - 12| \leq 34$ algebraically. Express the solution as an interval and support it graphically.

SOLUTION
**Solve Algebraically.** Rewrite the inequality as a double inequality and solve.

$$|x - 12| \leq 34 \qquad \text{Write the inequality without the absolute value symbol.}$$

$$-34 \leq x - 12 \leq 34 \qquad \text{Isolate } x \text{ in the middle.}$$

$$-34 + 12 \leq x - 12 + 12 \leq 34 + 12$$

$$-22 \leq x \leq 46$$

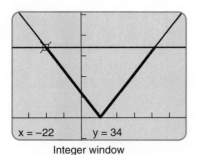

x = −22     y = 34

Integer window
centered at (12, 16)

**FIGURE 4.32** Grapher support for the solution to $|x - 12| \leq 34$ is represented by the bold portion of the graph.

The solution to $|x - 12| \leq 34$ is the interval $[-22, 46]$.

**Support Graphically.** Find the graphs of $y_1 = abs\ (x - 12)$ and $y_2 = 34$ in the Integer window centered at $(12, 16)$. Do this by positioning the cursor at $(12, 16)$ in an Integer window and accessing the Integer option from the ZOOM menu. Then use the trace cursor to verify that the points of intersection of $y_1$ and $y_2$ are $(-22, 34)$ and $(46, 34)$.

The solution to $|x - 12| \leq 34$ is all $x$-values such that the graph of $y_1$ is on or below $y_2$. Figure 4.32 shows that $y_1$ is on or below $y_2$ in the interval $[-22, 46]$. ∎

**EXAMPLE 7**    Solve Algebraically and Support Graphically

Solve $|x - 10| > 25$. Use an algebraic method and express the solution as an interval. Support the solution graphically.

SOLUTION

**Solve Algebraically.** Rewrite the inequality as a compound sentence.

$$|x - 10| > 25 \qquad \begin{array}{l}\textbf{Write the inequality}\\ \textbf{without the absolute value symbol.}\end{array}$$

$$x - 10 < -25 \text{ or } x - 10 > 25$$

$$x < -15 \text{ or } x > 35$$

The solution to $|x - 10| > 25$ is $(-\infty, -15) \cup (35, \infty)$.

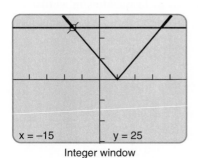

x = −15     y = 25

Integer window

**FIGURE 4.33** Grapher support of the solution of $|x - 10| > 25$ with one intersection at $x = -15$.

**Support Graphically.** Figure 4.33 shows the graph of $y_1 = abs\ (x - 10)$ and $y_2 = 25$ in the Integer window. These graphs intersect at $(-15, 25)$ and $(35, 25)$. The solution is all $x$-values such that the graph of $y_1$ is above the graph of $y_2$. The graph supports the solution $(-\infty, -15) \cup (35, \infty)$. ∎

The graphical support in Examples 6 and 7 is based on the multigraph method. This method also can be used as the basis for solving an absolute value inequality, as illustrated in Example 8.

**EXAMPLE 8**    Solve Graphically and Confirm Algebraically

Solve the inequality $15 - |2x + 9| > 13$. Use the multigraph method and confirm the solution algebraically.

SOLUTION

**Solve Graphically.** Graph $y_1 = 15 - abs\ (2x + 9)$ and $y_2 = 13$ in the Integer window. Zoom in to verify that the two graphs intersect when $x = -5.5$ and

$x = -3.5$ (see Figure 4.34). The graph of $y_1$ lies above the graph of $y_2$, between the two points of intersection. Thus the solution of $15 - |2x + 9| > 13$ is the interval $(-5.5, -3.5)$.

**Confirm Algebraically.** Begin by isolating the absolute value.

| | |
|---|---|
| $15 - |2x + 9| > 13$ | **Add $-15$ to each side of the inequality.** |
| $-|2x + 9| > -2$ | **Multiply each side by $-1$, which reverses the inequality symbol.** |
| $|2x + 9| < 2$ | **Eliminate the absolute value symbol.** |
| $-2 < 2x + 9 < 2$ | **Add $-9$ to each of the three parts and simplify.** |
| $-11 < 2x < -7$ | |
| $-5.5 < x < -3.5$ | |

The solution is the interval $(-5.5, -3.5)$.  ∎

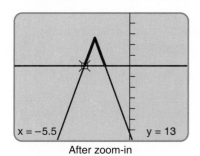

After zoom-in

**FIGURE 4.34** This graph can be obtained by beginning with the Integer window. Position the cursor at $(-4, 14)$ and zoom in with $X$Fact$= 10$, $Y$Fact$= 10$.

**TRY THIS**

If your grapher has table-generating capabilities, set the Tbl Min$= -6$ and $\triangle$Tbl$= 0.5$. Using the inequality in Example 8 with $y_1 = 15 - $ abs $(2x + 9)$, compare the $y_1$ values with 13. For which entries are the $y_1$ values greater than 13.

Example 9 solves an inequality using the $x$-intercept method. Notice that when solving by a graphing method, there is no need to isolate the absolute value term.

**EXAMPLE 9**   Solve Graphically—$x$-intercept Method

Solve $|2x + 3| + 2 \geq 10$ graphically. Confirm the solution algebraically.

SOLUTION

**Solve Graphically.** Using the $x$-intercept method we see that $|2x + 3| + 2 \geq 10$ is equivalent to $|2x + 3| - 8 \geq 0$. Find the graph of $y_1 = $ abs $(2x + 3) - 8$ in the Integer window. Using zoom-in with $X$Fact$= 10$ and $Y$Fact$= 10$, we see that the graph intersects the $x$-axis at $x = -5.5$ and $x = 2.5$ (see Figure 4.35). Therefore the solution is $(-\infty, -5.5] \cup [2.5, \infty)$.

**Confirm Algebraically.**

| | |
|---|---|
| $|2x + 3| + 2 \geq 10$ | **Isolate the absolute value term by adding $-2$ to each side of the inequality.** |
| $|2x + 3| \geq 8$ | **Remove the absolute value symbol.** |

$$2x + 3 \leq -8 \quad \text{or} \quad 2x + 3 \geq 8$$
$$2x \leq -11 \qquad\qquad 2x \geq 5$$
$$x \leq -5.5 \qquad\qquad x \geq 2.5$$

This confirms the solution is $(-\infty, -5.5] \cup [2.5, \infty)$.  ∎

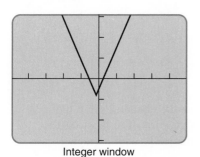

Integer window

**FIGURE 4.35** A graph of $y = $ abs $(2x + 3) - 8$ in the Integer window before zoom-in.

## Exercises for Section 4.4

In Exercises 1–6, solve the equation mentally.

**1.** $|x| = 10$

**2.** $|x| = 13$

**3.** $|x| + 5 = 9$

**4.** $|x| + 1 = 18$

**5.** $|x + 2| = 12$

**6.** $|x - 7| = 11$

In Exercises 7–12, determine whether the given values of the variable are solutions of the inequality.

**7.** $|x| < 5$     a) $x = 3$     b) $x = 7$

**8.** $|x| \leq 8$     a) $x = 10$     b) $x = 8$

**9.** $|x| > 2$     a) $x = -6$     b) $x = 5$

**10.** $|x| + 1 \geq 15$

    a) $x = 13$

    b) $x = 7$

**11.** $|x - 2| + 5 < 6$

    a) $x = 9$

    b) $x = 2$

**12.** $|2x + 1| < 12$

    a) $x = 3$

    b) $x = -1$

In Exercises 13–28, solve the equation algebraically and support the solution numerically.

**13.** $|x - 3| = 5$

**14.** $|5 - x| = 12$

**15.** $|3x - 4| = 8$

**16.** $|4x - 1| = \dfrac{5}{8}$

**17.** $|8x - 25| = 12$

**18.** $|7x + 14| = 15$

**19.** $\left|\dfrac{1}{3}x + 4\right| = 12$

**20.** $\left|8 - \dfrac{2}{3}x\right| = 24$

**21.** $|2x + 3| = 15$

**22.** $|x - 4| = 8$

**23.** $|3x - 8| = 13$

**24.** $|5x - 7| = -14$

**25.** $|2x + 3| = -5$

**26.** $|x + 4| - 5 = 11$

**27.** $|x - 7| + 2 = 9$

**28.** $|x - 4.1| - 2.5 = 8.5$

In Exercises 29–32, solve the equation algebraically and support the solution graphically.

**29.** a) $|x + 1| = 8$

    b) $|x + 1| < 8$

    c) $|x + 1| > 8$

**30.** a) $|x - 3| = 15$

    b) $|x - 3| \leq 15$

    c) $|x - 3| \geq 15$

**31.** a) $|x - 2| = 34$

    b) $|x - 2| < 34$

    c) $|x - 2| > 34$

**32.** a) $|x + 7| = 15$

    b) $|x + 7| < 15$

    c) $|x + 7| > 15$

In Exercises 33–38, state whether the solution interval is one interval or the union of two intervals.

**33.** $|x| < 5$

**34.** $|x| \leq 12$

**35.** $|x| > 4$

**36.** $|x - 2| \geq 15$

**37.** $|x + 3| < 8$

**38.** $|x - 2| + 3 > 9$

In Exercises 39–46, solve the equation using the multigraph method in the Integer window. Confirm the solution numerically using a grapher.

**39.** $|x + 1| = 11$

**40.** $|x - 3| = 25$

**41.** $|5 - x| + 2 = 5$

**42.** $|2x - 5| - 8 = -1$

**43.** $|5x - 1| = -9$

**44.** $|x + 2| + 8 = 12$

**45.** $|x + 5| + 2 = 9$

**46.** $|2x + 4| + 3 = 9$

In Exercises 47–58, solve the equation using the x-intercept method with an error of at most 0.01. Support the solution numerically with a grapher.

**47.** $|3x - 1| = 11$

**48.** $|x - 5| = 35$

**49.** $|2x + 3| = 17$

**50.** $|4x - 7| = -36$

**51.** $|2x - 7| = -10$

**52.** $|3x - 9| - 1 = 15.98$

**53.** $|x - 2| - 4 = 15$    **54.** $|3x - 4| + 1 = 16$

**55.** $|x + 2.1| = 10$    **56.** $|x + 3.1| = 9.6$

**57.** $|4x + 5.1| + 4.3 = 30.4$

**58.** $|3x + 6.5| + 8.7 = 24.8$

In Exercises 59–68, solve the inequality algebraically. Support the solution graphically.

**59.** $|x| < 7$    **60.** $|x| > 10$

**61.** $|x - 2| < 4$    **62.** $|2 - x| < 4$

**63.** $|x| + 5 \geq 14$    **64.** $|x| - 7 \geq 25$

**65.** $|2x - 3| > 18$

**66.** $|3x + 5| + 2 < 12$

**67.** $|3.3x + 2.1| + 2.6 \leq 10.8$

**68.** $|1.5x - 4.8| + 2.1 \leq 17.5$

In Exercises 69–76, solve the inequality by a graphing method.

**69.** $|x + 1| < 11$    **70.** $|x - 3| \leq 25$

**71.** $|2x - 5| \geq 18$    **72.** $|5x - 1| < -9$

**73.** $|4x + 7| > -24$    **74.** $|3x - 4| + 1 > 16$

**75.** $|4x + 8.1| + 5.3 > 32.4$

**76.** $|3x + 1.2| + 7.8 > 22.5$

## TRANSLATING WORDS TO SYMBOLS

In Exercises 77–82, use absolute value to write an inequality that represents each sentence.

**77.** $x$ is within 5 units of 0 on the number line.

**78.** $x$ is within 5 units of 2 on the number line.

**79.** $x$ is at least 8 units from 0 on the number line.

**80.** $x$ is less than 4 units from $-3$ on the number line.

**81.** $x$ is more than 2 units from 7 on the number line.

**82.** $x$ is at least 3 units from 10 on the number line.

**83. Writing to Learn.** Explain in your own words how you can use a grapher to determine that the solution to $|3x + 17| > -24$ is the interval $(-\infty, \infty)$.

**84. Writing to Learn.** Explain in your own words how you can use a grapher to determine that there is no solution to $|2x + 7| < -6$.

## EXTENDING THE IDEAS

In Exercises 85–89, write an inequality involving absolute values to represent the given interval.

**85.**

**86.**

**87.**

**88.**

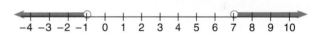

**89.**

**90.** *Projectile Motion.* A football is kicked from ground level with the initial velocity of 65 ft/sec. This situation can be expressed by the equation $h = -16t^2 + 65t$, where $h$ represents the height in feet and $t$ represents the time in seconds.

a) Graph the equation in an appropriate window.

b) Use this graph to find the time the football hits the ground.

c) Use the graph to find when the football is more than 50 ft above ground level.

d) Use the graph to find at what time the football reaches its maximum height.

For Exercise 90.

LOOKING BACK—MAKING CONNECTIONS

**91.** What is 35% of $325?

**92.** What is 1.2% of 3728?

**93.** 150 is what percent of 500?

**94.** What is 125% of 625?

**95.** *Projecting Price.* A textbook for a lifesaving course cost $22.50. Find the cost of the same book in 3 yr if the projected price increase is 3.5%/yr.

**96.** *Projecting Price.* A certain manufacturer wants to predict the cost involved in manufacturing bicycle tires. The cost presently is $5.35/tire. Estimate the cost in 2 yr if there is a predicted increase of 2.5% in each of the next 2 yr.

---

## 4.5   Linear Inequalities in Two Variables

Identifying All Ordered Pairs That Are Solutions to a Linear Inequality ▪ Graphing a Linear Inequality in Two Variables ▪ Two Special Cases of Linear Inequalities

In Chapter 2, you learned about linear equations and that the graph of $y = mx + b$ is a straight line. Recall that when the slope $m$ is positive, the line is rising as you read the graph from left to right and when $m$ is negative, the line is falling as you read the graph from left to right. The coordinates of every point on the line satisfy the equation $y = mx + b$.

The focus of this section is on linear inequalities with two variables, as defined next.

---

**DEFINITION 4.1**    A Linear Inequality in Two Variables

A **linear inequality in two variables** $x$ **and** $y$ is an inequality that can be written in one of the following forms:

$$y < mx + b \qquad y \leq mx + b \qquad y > mx + b \qquad y \geq mx + b$$

---

How are the graphs of the inequality and the equation related? The following exploration will help answer this question.

EXPLORE WITH A GRAPHER

Graph $y = -2x + 3$. Move the screen cursor (not the TRACE cursor) to any point on the graph. Now move the cursor up and down.

**Questions**

1. Does the $x$-coordinate of the moving point change as the cursor moves up and down?
2. What happens to the $y$-coordinate of the moving point?
3. When the moving point is above the line, do the $y$-coordinates satisfy $y > -2x + 3$ or $y < -2x + 3$?
4. When the moving point is below the line, do the $y$-coordinates satisfy $y > -2x + 3$ or $y < -2x + 3$?

**Generalize.** If you were to shade in all the points that satisfy the inequality $y > -2x + 3$, what points would be shaded?

This exploration suggests that for a given value of $x$, there are many values of $y$ such that $y < -2x + 3$ and there are many values of $y$ such that $y > -2x + 3$.

In Example 1, several $(x, y)$ pairs are given and you are asked to determine whether a given ordered pair is a solution to an inequality.

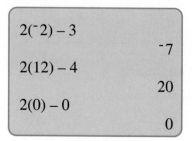

**FIGURE 4.36** Grapher support of Example 1.

**TRY THIS**

Consider the point $(0, 0)$. Does it satisfy the inequality $y \leq 4x - 5$ or $y \geq 4x - 5$? Explain how this observation can be used to get a quick check which region is the solution to a given linear inequality.

**EXAMPLE 1**   Confirming Solutions to a Linear Inequality

Use both paper-and-pencil confirmation and numerical support on a grapher to determine which of the following ordered pairs a) $(-2, 3)$, b) $(12, 4)$, or c) $(0, 0)$ are solutions of the inequality $2x - y > 7$.

SOLUTION

**Paper-and-Pencil Confirmation.** Substitute $x$ and $y$ in $2x - y > 7$ with the coordinates of each ordered pair and see if the inequality is true.

a)   $2(-2) - (3) > 7$   **False, because $-7 \not> 7$.**

b)   $2(12) - 4 > 7$   **True statement**

c)   $2(0) - 0 > 7$   **False, because $0 \not> 7$:**

**Numerical Support.** Figure 4.36 shows the numerical support of the solution.

We conclude by either method that the points $(-2, 3)$ and $(0, 0)$ are not in the solution and the point $(12, 4)$ is in the solution.   ∎

## Identifying All Ordered Pairs That Are Solutions to a Linear Inequality

Figure 4.37 shows that point $(1, -1)$ lies on the line $y = 4x - 5$. The points on the vertical line through $(1, -1)$ above the line $y = 4x - 5$ are those points $(1, y)$, where $y > 4(1) - 5$.

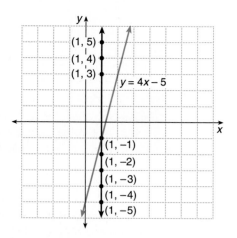

**FIGURE 4.37** The graph of the equation $y = 4x - 5$.

The points on the vertical line through $(1, -1)$ below the line $y = 4x - 5$ are those points $(1, y)$, where $y < 4(1) - 5$. Similar statements can be made about each vertical line. We summarize as follows:

1. The points whose coordinates satisfy $y > 4x - 5$ are those points that lie above the line $y = 4x - 5$. We call this the *boundary line* (see Figure 4.38a).
2. The points whose coordinates satisfy $y < 4x - 5$ are those points that lie below the boundary line $y = 4x - 5$ (see Figure 4.38b).

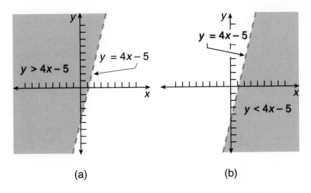

(a)                                           (b)

**FIGURE 4.38**  A sketch of the solution of $y >$ $4x - 5$ in (a) and of $y < 4x - 5$ in (b). Note that a dashed line is used to indicate $>$ and $<$.

## Graphing a Linear Inequality in Two Variables

As you saw in Figure 4.38, the graph of a linear inequality is really a picture or a sketch of the solution of the inequality.

---

### Sketching the Graph of a Linear Inequality in Two Variables

1. Rewrite the inequality, if necessary, so that it has the form

$$y \bigcirc mx + b,$$

   where one of the inequality symbols $<$, $\leq$, $>$, or $\geq$ appears in place of the symbol $\bigcirc$.
2. Sketch the graph of $y = mx + b$ using the fact that the $y$-intercept is $b$ and the slope is $m$. This line acts as a boundary line.
3. Shade above the line if the inequality symbol is $>$ or $\geq$. Shade below the line if the symbol is $<$ or $\leq$. Include the line if the inequality is $\leq$ or $\geq$. Draw a dashed line if the symbol is $<$ or $>$.

---

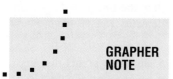

**GRAPHER NOTE**

Check your grapher lab manual to see if your grapher has a SHADE command. If it does, you can use SHADE to shade the region of the plane above or below a line.

Example 2 illustrates this method of sketching the graph of a linear inequality. It also illustrates how to use the SHADE command on a grapher to provide graphical support.

**EXAMPLE 2**    Graphing a Linear Inequality in Two Variables

Sketch the graph of the inequality $y < 3x - 7$. Support your sketch with a grapher.

SOLUTION

**Paper-and-Pencil Sketch.** Graph the line $y = 3x - 7$. The inequality is $<$, so that means that those points on any vertical line below the line $y = 3x - 7$ should be a part of the shaded region representing the solution. Therefore, shade the region below the line. Figure 4.39 shows the line $y = 3x - 7$ dashed to indicate that the boundary line is not included in the solution.

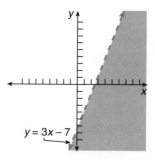

**FIGURE 4.39** A sketch of $y < 3x - 7$.

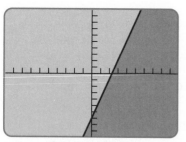

Standard window

**FIGURE 4.40** Grapher support for $y < 3x - 7$. Note the boundary line is dashed but the grapher cannot indicate this.

**Support Graphically.** Figure 4.40 shows the graph of $y = 3x - 7$ with the region below it shaded. Notice that on a grapher, you cannot distinguish whether the boundary line is included in the region, so we mentally note that the line should not be included (see Figure 4.40). ■

Example 3 illustrates a linear inequality that is not already written in the form

$$y \bigcirc mx + b.$$

Notice that in the process of rewriting the inequality so that it is equivalent to this form, you are at the same time finding the slope–intercept form needed for inputting the equation on the grapher.

**EXAMPLE 3**    Graphing an Inequality in Two Variables

Graph the inequality $2x - 3y < 9$ on a grapher.

SOLUTION

Begin by isolating $y$ to prepare the expression for a grapher.

$$2x - 3y < 9 \qquad \text{Add } -2x \text{ to each side of the inequality.}$$

$$-3y < -2x + 9 \qquad \text{Divide each side by } -3 \text{ and reverse the inequality.}$$

$$y > \frac{2}{3}x - 3$$

Graph $y = \frac{2}{3}x - 3$ in the Standard window. Then shade the graph above the line $y = \frac{2}{3}x - 3$.

Figure 4.41 shows a grapher image of this region. Recall that the points on the line are not included in the region, since the inequality is $>$ rather than $\geq$. ■

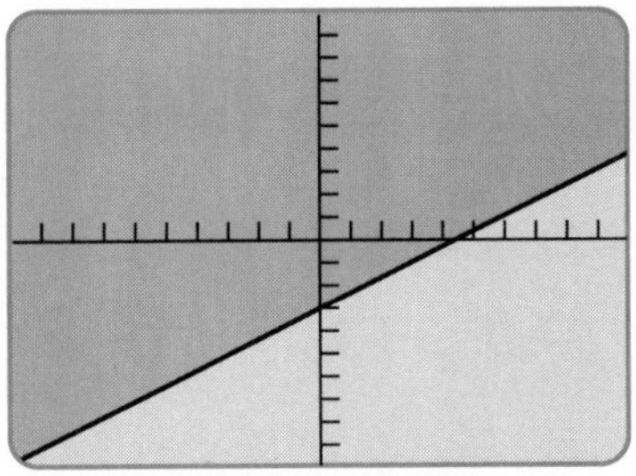

Standard window

**FIGURE 4.41** Grapher support for $y > \frac{2}{3}x - 3$.

## Two Special Cases of Linear Inequalities

Although an inequality like $y < 4$ has only one variable, it can be written $y < 0x + 4$ and graphed on a coordinate plane. The graph of $y = 0x + 4$ is a horizontal line with slope $m = 0$ and the graph of the solution to $y < 4$ is the region of the plane below that line.

Similarly, the graph of $x > -2$ is the region to the right of the vertical line $x = -2$.

**EXAMPLE 4**    Graphing an Inequality in One Variable

Graph the solution to the following

a) $y \leq 4$
b) $x > -2$

in a coordinate plane.

SOLUTION

Figure 4.42 (a) shows the graph of $y \leq 4$ and Figure 4.42 (b) shows the graph of $x > -2$, both sketched by hand. In fact it is not possible to graph $x = -2$ on a grapher in function mode.

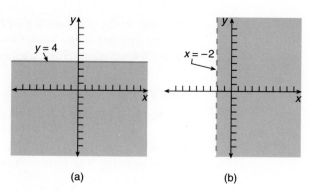

**FIGURE 4.42** A sketch of (a) $y \leq 4$ and (b) $x > -2$.

## Exercises for Section 4.5

In Exercises 1–6, determine if the indicated point is a member of the solution interval of the inequality.

**1.** $(2, 3)$; $y < 2x + 5$     **2.** $(-3, 1)$; $y > -x + 7$

**3.** $(2.5, 7)$; $y \geq 2x - 5$     **4.** $(3.1, 5.6)$; $x + y < 8$

**5.** $(4, 6.5)$; $x - y \geq 9$     **6.** $(0, 5)$; $y - 2x \leq 4$

In Exercises 7–16, state the slope of the line associated with the inequality. When sketching the graph of the inequality, should the line be drawn solid or dashed?

**7.** $y \leq 5$     **8.** $y < -6$

**9.** $y > 3x - 7$     **10.** $y \geq -6x + 2$

**11.** $y < -5x + 15$     **12.** $y \leq x - 10$

**13.** $y > \dfrac{2}{5}x - 9$     **14.** $y < \dfrac{4}{9}x - 12$

**15.** $x > 2y + 3$     **16.** $x \leq 3y - 4$

In Exercises 17–32, sketch the graph of the inequality.

**17.** $x > 5$     **18.** $x \leq -3$

**19.** $y \leq -2$     **20.** $y < -6$

**21.** $y > -1$     **22.** $y \leq -10$

**23.** $x > 0$     **24.** $y < 0$

**25.** $y \geq -2x + 9$     **26.** $y \geq 7x - 12$

**27.** $y \leq -\dfrac{3}{4}x - 6$     **28.** $y < \dfrac{1}{2}x - 4$

**29.** $y > -\dfrac{2}{3}x - 16$     **30.** $y > -\dfrac{1}{4}x + 3$

**31.** $y \leq \dfrac{1}{6}x + 3$     **32.** $y \leq \dfrac{2}{3}x - 5$

In Exercises 33–38, rewrite the inequality into the form $y \bigcirc mx + b$, where one of symbols $<$, $\leq$, $>$, or $\geq$ appears in $\bigcirc$.

**33.** $2y + 5x > 14$     **34.** $3x - 5y \leq 20$

**35.** $4y \geq -6x + 12$     **36.** $2x - y \geq 2$

**37.** $2x - 3y < 12$     **38.** $3x + 2y \leq 12$

In Exercises 39–42, use an Integer window to graph the boundary line of the region to be shaded. Tell whether the boundary line is included in the shaded region. Draw a sketch of this region on paper.

**39.** $18x - 12y < 234$

**40.** $12x + 8y \leq 120$

**41.** $22x + 14y \leq 418$

**42.** $-16x + 9y > 208$

**43.** *Personal Budgeting.* Dave and Wendy budgeted $2500/month for expenses. Their rent is $500, electricity $85, and food $300/month. Write an inequality for the remaining costs $C$ that enables them to keep within their budget. Solve the inequality for $C$.

**44.** *Long Distance Service.* Kyle has $8 for a long-distance phone call. The charges are $0.95 for the first minute and $0.75 for each additional minute. Write an inequality for the total amount of time $t$ he can talk. Solve the inequality.

**45.** *Ticket Sales.* Tickets for the Spring Fling Concert cost $3 for students and $5 for adults. Write an inequality indicating that the total receipts must be at least $2500.

**46.** *Club Fund Raising.* A club is selling fresh oranges and grapefruit as a fund raiser. If the club makes a profit of $4 for each box of oranges and $6.50 for each box of grapefruit, write an inequality that indicates the profit is at least $1200.

### TRANSLATING WORDS TO SYMBOLS

In Exercises 47–53, write an inequality expressing the word phrase.

**47.** $x$ is a positive number.

**48.** $x$ is a nonnegative number.

**49.** $y$ is more than 6.

**50.** $y$ is at least 10.

**51.** $y$ is no more than 325.

**52.** $x$ is at least 17 but no more than 56.

**53.** $x$ is between 25 and 75.

**54. Writing to Learn.** Explain to another student when to use a dashed line as the boundary line when graphing a linear inequality.

**55. Writing to Learn.** Explain to another student when to use a solid line as the boundary line when graphing a linear inequality.

### EXTENDING THE IDEAS

**56.** *Band Uniform Purchase.* A high school band is planning to buy new uniforms and some new flags for the flag corp. Each uniform costs $250 and each flag cost $120. Band members have $22,000 available to spend. They wonder how many of each item they can buy.

a) Let $x$ represent the number of flags bought and $y$ the number of uniforms bought. What does the expression $120x + 250y$ represent?

b) Is the algebraic representation of this problem situation $120x + 250y \leq 22,000$ or $120x + 250y \geq 22,000$?

c) If a pair $(x_0, y_0)$ is a solution of this problem situation, in which quadrant does it lie? Explain why.

d) Sketch a graph of this problem situation.

**57.** *Bakery Revenues.* A bakery bakes fresh whole wheat bread and rye bread each day. They lose $0.50 for each loaf of whole wheat not sold by the end of the day and $0.75 for each loaf of rye bread not sold by the end of the day.

a) Let $x$ represent the number of loaves of whole wheat not sold and $y$ the number of rye loaves not sold. What does the expression $0.50x + 0.75y$ represent?

b) Write an inequality that represents the situation of the bakery losing less than $25 in a day.

c) Find the graphical representation of this problem situation.

In Exercises 58–62, write an inequality whose solution is the indicated region.

**58.** The region above the line through the points $(2, 3)$ and $(3, 8)$.

**59.** The region below the line with slope $m = -3$ and $y$-intercept 4.

**60.** The region to the right of the line determined by $(3, 7)$ and $(3, -9)$.

**61.** The region above the line determined by $(-3, 5)$ and $(2, 5)$.

**62.** The region to the left of the line with no slope and $x$-intercept 6.

**LOOKING BACK—MAKING CONNECTIONS**

In Exercises 63–65, find the solution to the system of equations using an algebraic method.

**63.** $\begin{cases} y = 3x + 32 \\ y = -x - 4 \end{cases}$

**64.** $\begin{cases} y = -x + 8 \\ y = -5x + 36 \end{cases}$

**65.** $\begin{cases} x + 5y = -13 \\ x - y = 5 \end{cases}$

In Exercises 66–68, find the solution to the system of equations using a graphing method.

**66.** $\begin{cases} 5x - 6y = 50 \\ y = 5x - 25 \end{cases}$

**67.** $\begin{cases} y = -x \\ 2x - y = -15 \end{cases}$

**68.** $\begin{cases} y = 3x + 10 \\ y = -5x - 38 \end{cases}$

---

## 4.6   Solving Systems of Inequalities

Graphing Systems of Inequalities  ▪  Systems of Nonlinear Inequalities

An ordered pair of numbers belongs to the solution of a system of linear inequalities if it satisfies each inequality in the system. Example 1 illustrates how to determine whether a given ordered pair is a solution to such a system.

**EXAMPLE 1**    Confirm Points in the Solution

Confirm whether the following

a) $(1, 2)$
b) $(10, 3)$
c) $(7, -1)$

belong to the solution of the system

$$y < -x + 9$$
$$y > 3x - 15$$

SOLUTION

a) Replace $x$ with 1 and $y$ with 2 in each inequality.

$$2 < -1 + 9 \qquad \text{True statement}$$

$$2 > 3(1) - 15 \qquad \text{True statement}$$

The ordered pair $(1, 2)$ does belong to the solution of the system.

b) Replace $x$ with 10 and $y$ with 3 in the each inequality.

$$3 < -10 + 9 \qquad \text{False statement}$$

$$3 > 3(10) - 15 \qquad \text{False statement}$$

The pair $(10, 3)$ leads to two false statements; therefore it does not belong to the solution of the system.

c) Replace $x$ with 7 and $y$ with $-1$ in the each inequality.

$$-1 < -7 + 9 \qquad \text{True statement}$$

$$-1 > 3(7) - 15 \qquad \text{False statement}$$

Since $(7, -1)$ does not satisfy the second inequality, it is not in the solution of the system. ∎

## Graphing Systems of Inequalities

The solution of a system of inequalities is the set of *all* ordered pairs $(x, y)$ that satisfy both of the individual inequalities. So the solution to a system of two linear inequalities in two variables consists of all those points in common to both of the solutions of the individual inequalities.

Example 2 illustrates how to do a hand-sketch of such a solution.

**EXAMPLE 2**    Graphing the Solution to a System of Inequalities

Sketch a graph of the solution of the following system of inequalities using paper-and-pencil methods:

$$y < -x + 9$$

$$y > 3x - 15$$

Support the sketch with a grapher.

SOLUTION
**Hand-sketch.**

1. Graph $y < -x + 9$ and, because the inequality symbol is $<$, shade below the dashed line (see Figure 4.43a).

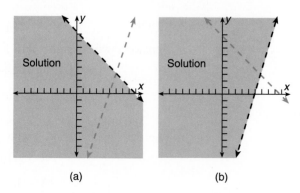

(a)          (b)

**FIGURE 4.43** (a) Shows the solution of $y < -x + 9$ and the dashed line $y = 3x - 15$. (b) Shows the solution of $y > 3x - 15$ and the dashed line $y = -x + 9$.

2. Graph $y > 3x - 15$ and, because the inequality symbol is $>$, shade above the dashed line (see Figure 4.43 b).

    The solution of the system is the region that is shaded in both graphs when we overlay Figure 4.43 (a) on Figure 4.43 (b) (see Figure 4.44).

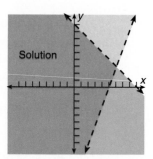

**FIGURE 4.44** The solution to the system in Example 2.

In other words, it is the intersection of the solution of both graphs.

**Support Graphically.** Figure 4.45 shows the solution of the inequality drawn on a grapher. The SHADE command was used to find this graph. ■

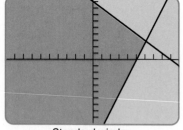

Standard window

**FIGURE 4.45** The graph of the intersection of the shaded regions in Example 2. Notice the grapher does not indicate a dashed line if the inequality symbol is $>$ or $<$.

In Example 3, we see a system in which the inequality of each system is rewritten into an equivalent form before the graphs are sketched.

**EXAMPLE 3**    Solving a System of Inequalities

Sketch the graph of the solution of the following system of inequalities:

$$3x - 4y < 12$$

$$2x + y \leq 10$$

SOLUTION

We begin by writing an equivalent system in slope–intercept form.

Change the system
$$3x - 4y < 12$$
$$2x + y \leq 10$$

to the equivalent form
$$y > \frac{3}{4}x - 3$$
$$y \leq -2x + 10$$

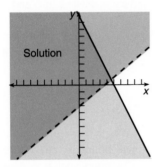

**FIGURE 4.46** A sketch of the solution of the system in Example 3.

1. Sketch the graphs of $y_1 = (3/4)x - 3$ and $y_2 = -2x + 10$ noting that $y_1$ should be a dashed line and $y_2$ a solid line.
2. Shade in the region above the line $y_1$ and below $y_2$. Figure 4.46 shows this sketch.  ■

For most systems of inequalities, especially if the system involves more than two inequalities, it is difficult or even impossible to find the shaded region using the SHADE feature of a grapher. We recommend using the grapher as an aid in determining the boundaries of the region and to complete the graph using a paper-and-pencil method. Example 4 illustrates a system with four inequalities.

**EXAMPLE 4**    Solving a System with Four Inequalities

Sketch the graph of the solution to the following system of inequalities:

$$x \geq 0$$

$$y > 0$$

$$y < x + 2$$

$$y > -3x + 5$$

SOLUTION

Notice that the first two inequalities restrict the intersection to the first quadrant. Therefore we need graph only the portions of the third and fourth inequalities that are in the first quadrant.

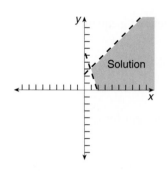

**FIGURE 4.47**  The sketch of the intersection of the shaded regions in Example 4.

1. Sketch the first quadrant with the $x$-axis as a solid line, since $x \geq 0$, and the $y$-axis as a dashed line, since $y > 0$.
2. Sketch the graph of $y < x + 2$, making the boundary a dashed line and shading below the line.
3. Sketch the graph of $y > -3x + 5$, making the boundary a dashed line and shading above the line.
4. Figure 4.47 shows that the solution to the system of inequalities is the region common to both shaded regions. ■

## Systems of Nonlinear Inequalities

It is possible to have a system of inequalities with a Nonlinear boundary curves are not linear, that is, the boundary is a parabola. Example 5 illustrates such a system of inequalities.

**EXAMPLE 5**    Solving a System with a Nonlinear Boundary

Sketch the graph of the solution to the following system of inequalities:

$$y > x^2$$
$$y \leq -x + 5$$

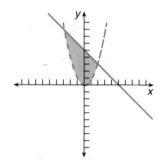

**FIGURE 4.48**  The sketch of the intersection of the shaded regions in Example 5.

SOLUTION

1. Sketch the graphs of $y_1 = x^2$ and $y_2 = -x + 5$.
2. Shade the region above $y_1$ and below $y_2$. Note that because $>$ is used instead of $\geq$, the points on the boundary $y = x^2$ are not included in the region. Figure 4.48 shows the graph of the solution. ■

In Example 6, we revisit the absolute value graph.

**EXAMPLE 6**    Solving a System in which the Boundary Is an Absolute Value

Sketch the solution to the following system of inequalities:

$$y \geq |x| - 6$$
$$y \leq 3$$

SOLUTION

1. Sketch the graphs of $y_1 = |x| - 6$ and $y_2 = 3$.
2. Sketch the region that is above $y_1$ and below $y_2$. Note that because the inequalities are $\leq$ and $\geq$, the points on both boundaries are included in the solutions. Figure 4.49 shows a sketch of the region. ∎

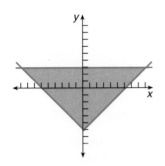

**FIGURE 4.49** A sketch of the solution to the system in Example 6.

## Exercises for Section 4.6

In Exercises 1–4, confirm algebraically and support numerically on your grapher that $(3, -5)$ is a solution of the inequalities.

1. $\begin{cases} 2x - 3y > 7 \\ x - y < 10 \end{cases}$

2. $\begin{cases} x + y < 5 \\ x - y > 7 \end{cases}$

3. $\begin{cases} 2y < x - 8 \\ 2x + y < 9 \end{cases}$

4. $\begin{cases} 3x + 2y < 12 \\ x - 2y < 15 \end{cases}$

In Exercises 5–10, sketch the solution to the first inequality in graph (a) and the solution to the second inequality in graph (b). Then in graph (c), sketch both of the graphs from (a) and (b) and identify the region common to both.

5. $\begin{cases} y < -x + 7 \\ y > x - 9 \end{cases}$

6. $\begin{cases} x - 2y > -1 \\ -x + 5y < 4 \end{cases}$

7. $\begin{cases} 2x - 5y \geq 13 \\ 3x + 4y < -15 \end{cases}$

8. $\begin{cases} 2x - 3y \leq 11 \\ 7x - 4y > 6 \end{cases}$

9. $\begin{cases} y - 2x \geq 0 \\ y < 5 \end{cases}$

10. $\begin{cases} -2x + y < -1 \\ x - 2y > 2 \end{cases}$

In Exercises 11–14, sketch the graph of the solution of the system of linear inequalities.

11. $\begin{cases} x + y > 3 \\ 2x + y > 5 \end{cases}$

12. $\begin{cases} 2x + 5y \geq 10 \\ 3x - y < -5 \end{cases}$

13. $\begin{cases} 0.2x - 0.3y \leq 1.5 \\ 0.5x + 4y < 2.0 \end{cases}$

14. $\begin{cases} 2.3x - 5.1y < 51.3 \\ 3.5x + 7.4y < -35.5 \end{cases}$

24. $\begin{cases} y \geq |x| - 8 \\ y \leq -|x| + 8 \end{cases}$

## EXTENDING THE IDEAS

In Exercises 15–20, identify the two functions whose graphs are boundaries of the solution of the inequalities. Graph these functions on a grapher in the Standard window. Then sketch on paper the region that is a solution to the system.

15. $\begin{cases} y \geq x^2 \\ y < x + 4 \end{cases}$

16. $\begin{cases} y < -x^2 + 4 \\ y \geq x \end{cases}$

17. $\begin{cases} y \leq |x + 3| \\ y > -8 \end{cases}$

18. $\begin{cases} y \leq -x^2 + 9 \\ y > x \end{cases}$

19. $\begin{cases} y \geq (x - 3)^2 + 2 \\ y < x + 4 \end{cases}$

20. $\begin{cases} y \geq |x + 5| - 4 \\ y < 2 \end{cases}$

In Exercises 21–24, sketch the graph of the solution of the system of inequalities. Use a grapher as needed.

21. $\begin{cases} y \leq -x + 10 \\ y \geq |x - 2| - 5 \end{cases}$

22. $\begin{cases} y \geq x^2 - 8 \\ y \leq -x^2 + 8 \end{cases}$

23. $\begin{cases} y < x^2 - 8 \\ y > -x^2 + 8 \end{cases}$

## LOOKING BACK—MAKING CONNECTIONS

25. *Business Profits.* Hawkins Electronics invests $22,000 in equipment to start its business of producing answering machines. Each machine costs $20.25 to produce and is sold for $39.95.

a) If $x$ equals the number of answering machines sold and $C(x)$ the total cost of producing $x$ machines, write an algebraic representation for $C(x)$.

b) If $R(x)$ represents the total revenue obtained by selling $x$ machines, write an algebraic representation for $R(x)$.

c) Does the expression $20.25x + 22,000$ represent total profit of producing $x$ machines or total cost required to produce $x$ machines?

d) What does the solution to the inequality $39.95x > 20.25x + 22,000$ tell about this problem situation?

e) How many answering machines must be produced (assuming they are all sold) in order for the company to realize a profit?

26. *Business Profits.* Smithers Electronics spends $525,000 on research and development for a new line of industrial machine. Each machine sells for $11,300 and costs $8,200 to manufacture.

a) If $x$ equals the number of machines sold and $C(x)$ represents the total cost of producing $x$ machines, write an algebraic representation for $C(x)$.

b) If $R(x)$ represents the total revenue obtained by selling $x$ machines, write an algebraic representation for $R(x)$.

c) Does the expression $8,200x + 525,000$ represent total profit of producing $x$ machines or total investment required to produce $x$ machines?

d) What does the solution to the inequality $11,300x > 8,200x + 525,000$ tell about this problem situation?

e) How many machines must be produced (assuming that they are all sold) in order for Smithers Electronics to realize a profit?

## Chapter 4 Summary

| | **Inequalities, Number Line, and Interval Notation** | **Examples** |
|---|---|---|
| Basic Inequalities | $x < a$, $x \le a$, $x > a$, and $x \ge a$ | $x \ge 5$ |
| Compound Inequalities | Two or more inequalities joined with *and* or *or* | $x > -1$ and $x < 3$<br>$x \le -2$ or $x > 5$ |
| Double Inequalities | A compound inequality with two statements joined by *and* and can be written as a single statement | $x > 4$ and $x < 9$ can be written as the double inequality $4 < x < 9$. |

| | **Solving Inequalities** | **Examples** |
|---|---|---|
| Solving Algebraically | Use algebra properties to find a solution in terms of $x \bigcirc a$, where $\bigcirc$ is either $<$, $\le$, $>$, or $\ge$. The solution also can be written as an interval. | Solve $4(x - 15) + x \ge 3$:<br>$4x - 60 + x \ge 3$<br><br>$5x - 60 \ge 3$<br><br>$5x \ge 63$<br>So $x \ge 12.6$, or $[12.6, \infty)$. |
| Support Algebraic Solution Graphically | 1. Graph the left side of the inequality ($y_1$) and the right side ($y_2$) in the same window.<br><br>2. Check that the $x$-value of the intersection is equal to the endpoint of the solution interval found algebraically and check where $y_1$ is above or below $y_2$, depending on which inequality sign is used. | 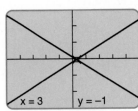<br>Integer window<br>To solve $x - 4 > 2 - x$;<br>graph $y = x - 4$ and $y = 2 - x$.<br>The solution is $(3; \infty)$. |

| | | |
|---|---|---|
| Solving Graphically with the $x$-intercept Method | 1. Rewrite the inequality with all terms on the left-hand side of the inequality sign and 0 on the right-hand side.<br>2. Graph the left-hand side to determine the $x$-intercept.<br>3. If $<$ is used, the solution is all values of $x$ where the line lies below the $x$-axis. If $>$ is used, the solution is all values of $x$ where the line lies above the $x$-axis. (Include the $x$-intercept if $\leq$ or $\geq$ is used.) | <br>To solve $2x + 7 < 3$, rewrite as $2x + 4 < 0$ and graph $y = 2x + 4$. The solution is $(-\infty, -2)$. |
| Special Cases | If the variable drops out after simplifying to solve an inequality, either the solution is $(-\infty, \infty)$ (a true statement) or there is no solution (a false statement). | $x + 1 < x + 2$<br>$x - x < 2 - 1$<br>$0 < 1$<br>This is always true, so the solution is $(-\infty, \infty)$.<br>$x - 3 > x + 2$<br>$x - x > 2 + 3$<br>$0 > 5$<br>This is always false, so there is no solution. |

| | **Solving Compound and Double Inequalities** | **Examples** |
|---|---|---|
| Solving Compound Inequalities Algebraically | 1. Solve each inequality individually.<br><br>2. If *and* is used, find the intersection of all the individual solutions. (List the values that are in common to all solutions.)<br>If *or* is used, find the union of all solutions. (List all values found for the individual solutions.) | Solve the pair $x + 2 > 3$ and $x + 3 < 8$. The solution of the individual inequalities are $x > 1$ and $x < 5$. The pair written together is $1 < x < 5$. The solution is the interval $(1, 5)$. |

| Solving Double Inequalities Algebraically | Use algebraic properties to isolate the variable in the middle portion of the inequality. | $2 < 3x + 5 \leq 14$ <br><br> $2 - 5 < 3x \leq 14 - 5$ <br><br> $-3 < 3x \leq 9$ <br><br> $-1 < x \leq 3$ |
|---|---|---|
| Solving Double Inequalities Graphically | 1. Graph the three parts of the inequality in the same window. For $a < bx + c < d$, graph $y_1 = a$, $y_2 = bx + c$, and $y_3 = d$. <br><br> 2. Determine where $y_2$ intersects $y_1$ and $y_3$. The solution is all values in between these intersection points. | 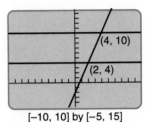 <br> $[-10, 10]$ by $[-5, 15]$ <br> To solve $4 < 3x - 2 < 10$, graph $y_1 = 4$, $y_2 = 3x - 2$, and $y_3 = 10$. The solution is $(2, 4)$. |

| | **Solving Absolute Value Equations and Inequalities** | **Examples** |
|---|---|---|
| Solving Absolute Value Equations Algebraically | To solve $|ax+b| = c$, solve each of the equations $ax + b = c$ and $ax + b = -c$ individually. | Solve $|4x - 5| = 3$. <br><br> The solutions of $4x - 5 = 3$ and $4x - 5 = -3$ are $x = 2$ and $x = \dfrac{1}{2}$, respectively. |
| Solving Absolute Value Inequalities with $<$ or $\leq$ | To solve an inequality $|ax + b| < c$, solve $-c < ax + b < c$. | Solve $|x - 2| < 1$. <br><br> $-1 < x - 2 < 1$ is equivalent to $1 < x < 3$. The solution is the interval $(1, 3)$. |
| Solving Absolute Value Inequalities with $>$ or $\geq$ | To solve $|ax + b| > c$, consider the compound inequality, $ax + b < -c$ *or* $ax + b > c$. Solve this compound inequality. | Solve $|2x + 1| > 5$. <br><br> Solve the compound inequality $2x + 1 < -5$ *or* $2x + 1 > 5$. This pair is equivalent to $x < -3$ *or* $x > 2$. The solution is the interval $(-\infty, -3) \cup (2, \infty)$. |

Solving Absolute Value Inequalities graphically.

To solve $|ax + b| \leq c$, find the graph of $y_1 = |ax + b|$ and $y_2 = c$ in the same window. The solution is all $x$-coordinates of points on the graph of $y_1$ that are on or below the graph of $y_2$.

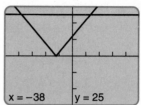

To solve $|x + 13| \leq 25$; graph $y_1 = |x + 13|$ and $y_2 = 25$. The solution is the interval $[-38, 12]$.

---

### Solving Linear Inequalities in Two Variables

**Examples**

Graphing Method of Solving

1. Rewrite the inequality in the form $y \bigcirc mx + b$, where $\bigcirc$ is either $<$, $\leq$, $>$, or $\geq$.
2. Sketch the line $y = mx + b$.
3. Draw a solid line if $\bigcirc$ is $\leq$ or $\geq$; draw a dashed line if $\bigcirc$ is $<$ or $>$.
4. Shade above the line if $\bigcirc$ is $>$ or $\geq$; shade below the line if $\bigcirc$ is $<$ or $\leq$.

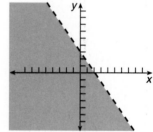

The solution to $3x + 2y \leq 6$.

Special Cases

1. For $y < a$, shade the area below the line $y = a$. If $y \leq a$, include the line in the solution.
2. For $y > a$, shade the area above the line $y = a$. If $y \geq a$, include the line in the solution.
3. For $x < a$, shade the area to the left of the line $x = a$. If $x \leq a$, include the line in the solution.
4. For $x > a$, shade the area to the right of the line $x = a$. If $x \geq a$, include the line in the solution.

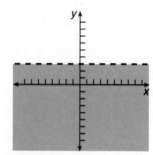

The solution to $y < 3$.

Graphing Method of Solving

**Solving Systems of Inequalities**

1. Graph each inequality and shade in each area of the solution.
2. The solution to the system is the portion that is shaded by all individual inequalities.

**Examples**

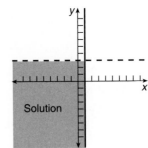

The solution to the system
$y < 3$ and $x \le 1$.

## Review Exercises for Chapter 4

In Exercises 1–4, use interval notation to represent the inequality.

**1.** $x > 5:5$

**2.** $x \le -2$

**3.** $x < 3$ or $x \ge 6$

**4.** $x \le 10$ and $x > 4$

**5.** Graph the solution to $x \le 3$ and $x > 0$ as an interval on a number line.

**6.** Write $.-\infty; -2] \cup .-3; 6/$ as one interval.

**7.** Write $[-2; 5/ \cap [-1; \infty/$ as one interval.

**8.** Is $x = -1:5$ a solution of $-3x \ge 2$?

**9.** Is $x = -5$ a solution of $|x + 2| + 3 < 6$?

**10.** Is $.2; -1/$ a solution of $\begin{cases} 4x - y \ge 8 \\ 2x + 3y \le 1 \end{cases}$?

In Exercises 11–14, solve the inequality graphically using either the x-intercept or the multigraph method. Express the solution in interval notation and confirm it algebraically.

**11.** $4 + 2.x + 1/ > 3x - 1$

**12.** $3.x - 1/ \le 2.2x - 3/ + x$

**13.** $4 \le 3x - 5 \le 7$

**14.** $10 < 5 - x < 13$

In Exercises 15–20, solve the inequality algebraically. Express the solution in interval notation and support it graphically.

**15.** $-4 < 2.x - 3/ < 5$

**16.** $-2 \le 2x + 1 \le 3$

**17.** $|x - 6| > 2$

**18.** $|x + 3| \le 8$

**19.** $6x \ge 4$ and $3x + 7 \le 16$

**20.** $6 - x < 2$ or $4x - 15 > 9$

In Exercises 21–24, solve the equation algebraically and support it numerically.

**21.** $|5x + 3| = 12$

**22.** $|2x - 7| = 25$

**23.** $\left|\frac{1}{2}x - 5\right| = 7$

**24.** $\left|\frac{1}{4}x - 6\right| = 1$

In Exercises 25–28, sketch the graph of the inequality.

**25.** $y \geq 3x + 5$

**26.** $y > 2x - 3$

**27.** $y < \dfrac{1}{4}x + 5$

**28.** $y \leq \dfrac{3}{5}x - 2$

In Exercises 29–32, sketch the graph of the solution of the system of inequalities.

**29.** $\begin{cases} x - 3y \geq 4 \\ 5x - 2y \leq 7 \end{cases}$

**30.** $\begin{cases} 2x + y < 2 \\ x - 3y < 15 \end{cases}$

**31.** $\begin{cases} y \leq 6 - x^2 \\ y \geq x + 4 \end{cases}$

**32.** $\begin{cases} y \geq |x - 2| \\ y < -\dfrac{1}{2}x + 3 \end{cases}$

**33.** Write an inequality involving absolute value that represents the interval pictured below.

**34.** Write as an inequality: The maximum value $x$ can be is $-5$.

**35.** Write in symbols: The sum of a number and 10 is between 8.1 and 16.6.

**36.** Write in symbols using absolute value notation: $x$ is at least 4 units from 7 on the number line.

**37.** The length of a rectangle is 5 in. longer than the width. Let $w$ represent the width. Write an inequality in terms of $w$ that shows the perimeter to be at least 40 in.

**38.** Denise scored 75% and 82% on her first two history exams.

a) Let $x$ represent the score of the third exam. Write an expression that represents the average of the three scores.

b) What must Denise score on the third exam to have an average of at least 80%?

**39.** A fund raiser for the Recycling Club is planned where tickets cost $2 for members and $3 for nonmembers. The auditorium that will be used has maximum seating capacity of 500 people. The club has a goal of selling at least $1250 in tickets. Let $x$ represent the number of members who buy tickets and $y$ the number of nonmembers who buy tickets.

a) Write an inequality that represents the total number of tickets sold. Graph this inequality.

b) Assume the club exceeds the goal of $1250 in ticket sales. Write an inequality that represents the total amount of ticket sales. Graph this inequality.

c) Is there a region that is common to both inequalities? If so, determine one solution in this region.

## Chapter 4 Test

**1.** Write the compound inequality $x \leq -2$ or $x > 5$ in interval notation.

**2.** Write $(-\infty, 3] \cap [0, 4)$ as one interval.

**3.** Solve $|3x + 8| = 19$ algebraically.

In Questions 4–6, solve the inequality algebraically. Express the solution in interval notation.

**4.** $8 + 8(x + 1) \geq 5(x + 2) - 2(x - 1)$

**5.** $2(3x + 4) + 1 > 2(4x + 1)$

**6.** $|0.1x - 8| < 0.6$

**7.** Solve the double inequality $-5 \leq \dfrac{2x + 9}{3} \leq 3$ algebraically. Express the solution in interval notation.

In Questions 8 and 9, solve each compound inequality algebraically. Express the solution in interval notation.

**8.** $2x \leq 3$ and $5x + 3 < 2$

**9.** $3x - 10 \leq 12$ or $6x + 11 \leq 7x + 4$

**10.** Graph $|x| > 8$ on a number line.

11. Sketch the graph of the inequality $x - 2y < 8$.

12. Sketch the graph of the inequality $y \geq 5$.

13. Sketch the graph of the solution of the following system of inequalities:

$$\begin{cases} y \geq \dfrac{1}{3}x + 4 \\ 4x - 2y < 3 \end{cases}$$

14. Write in symbols using absolute value notation: $x$ is less than 6 units from 3 on the number line.

15. Lynn has a checking account with Five Star Bank, which has a monthly service fee of $2.50 plus $0.12 for each check processed. Let $x$ equal the number of checks Lynn writes in 1 month.

a) Write an expression that is an algebraic representation of the monthly service fee on her checking account.

b) What is the total service fee if Lynn writes 18 checks in 1 month?

c) Lynn would like to keep her monthly service fees under $4/month. Write an inequality that represents this situation.

d) Solve the inequality found in Question 15(c) to find the maximum number of checks Lynn can write in 1 month and still keep her monthly service fees under $4.

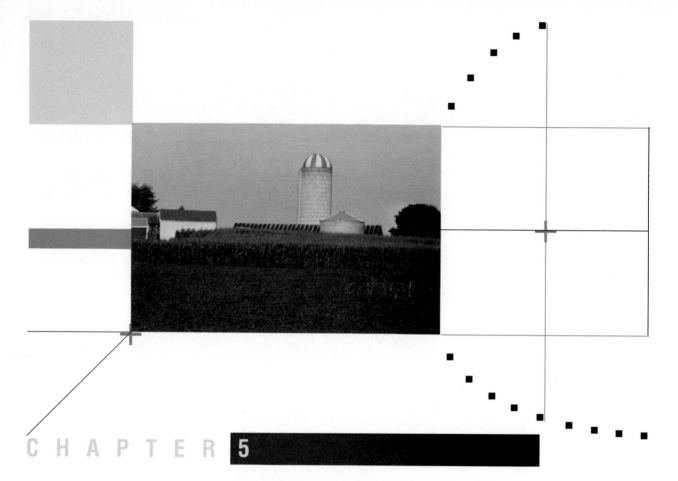

# CHAPTER 5

## Polynomials

AN APPLICATION

The volume of a storage tank varies depending on its diameter and height. The volume of a silo with radius $r$ ft and height $h$ ft can be found using the formula

$$V = \pi r^2 (h - r) + \frac{2}{3}\pi r^3.$$

| 5.1 | # Exponents and Their Properties |

In Section 1.1, the symbol $b^n$ was defined. We review its definition here.

---

**DEFINITION 5.1**    Positive Integer Exponent

Suppose $b$ represents any real number and $n$ is a positive integer. Then the symbol $b^n$ is defined as follows:

$$b^n = \underbrace{b \cdot b \cdot \,\cdots\, \cdot b}_{n \text{ factors of } b}$$

---

So $b^n$ is a symbol that is shorthand for a product of $n$ factors. The number $b$ is called the **base** and $n$ is called the **exponent**. In a numerical setting, the base is often a number; in an algebraic setting, it is often a variable. Example 1 demonstrates how to identify the base and exponent.

**EXAMPLE 1**    Identifying the Base and Exponent

Identify the base and exponent in the following:

a) $(-5)^4$
b) $7^{12}$
c) $3x^5$
d) $-9y^3$

**REMINDER**

In the expression $3x^5$, the base is $x$, whereas, in the expression $(3x)^5$, the base is $3x$.

SOLUTION

a) The base is $-5$ and the exponent is 4.
b) The base is 7 and the exponent is 12.
c) The base is $x$ (not $3x$) and the exponent is 5.
d) The base is $y$ (not $9y$ or $-9y$) and the exponent is 3.    ▪

Part (c) of Example 1 illustrates that in the expression $3x^5$, the constant 3 is not part of the base. Rather it is the coefficient in the expression $3x^5$. More generally, in the expression $ax^n$, $a$ is called the **coefficient** of the expression.

Notice that, in general,

$$(-b)^n \neq -b^n.$$

The base of $(-b)^n$ is $(-b)$ and the base of $-b^n$ is $b$. Figure 5.1 shows a comparison between $-3^2$ and $(-3)^2$. In the first two cases, the base is 3 and in the third, the base is $(-3)$.

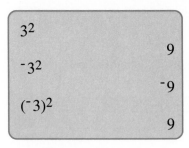

**FIGURE 5.1**  Notice the difference when 3 and $-3$ are the base.

**EXAMPLE 2**    Evaluating Exponential Expressions

Use a paper-and-pencil method to find the value of the following expressions:

a) $-3^4$
b) $(-5)^3$
c) $(2x)^5$
d) $2(3x)^2$

SOLUTION

a) $\quad -3^4 = -(3 \cdot 3 \cdot 3 \cdot 3) = -81$

b) $\quad (-5)^3 = (-5)(-5)(-5) = -125$

c) $\quad (2 \cdot x)^5 = (2 \cdot x) \cdot (2 \cdot x) \cdot (2 \cdot x) \cdot (2 \cdot x) \cdot (2 \cdot x)$      **Use the Associative and Commutative Properties of Multiplication.**

$\qquad = (2 \cdot 2 \cdot 2 \cdot 2 \cdot 2) \cdot (x \cdot x \cdot x \cdot x \cdot x)$

$\qquad = 32x^5$

d) $\quad 2(3x)^2 = 2(3x \cdot 3x)$

$\qquad = 2 \cdot 9x^2 = 18x^2$

Parts (b), (c), and (d) of Example 2 demonstrate the importance of parentheses in interpreting exponential expressions.

## Applying the Product Rule for Exponents

When multiplying two exponential expressions with the same base, we add the exponents. Consider the product of $b^3 \cdot b^4$.

$$b^3 \cdot b^4 = \underbrace{(b \cdot b \cdot b)}_{3 \text{ factors of } b} \cdot \underbrace{(b \cdot b \cdot b \cdot b)}_{4 \text{ factors of } b} = b^7$$

The exponent of the product can be found by adding the exponents of the two factors. Here is a complete statement of the Product Rule.

---

### Product Rule for Exponents

If $m$ and $n$ are positive integers and $b$ is any nonzero real number, then

$$b^n \cdot b^m = b^{m+n}.$$

---

To see that the product is true, we use the definition of exponent several times.

$$b^n \cdot b^m = \underbrace{(b \cdot b \cdot b \cdots b)}_{n \text{ factors of } b} \cdot \underbrace{(b \cdot b \cdot b \cdots b)}_{m \text{ factors of } b}$$

$$= \underbrace{(b \cdot b \cdot b \cdots b)}_{n + m \text{ factors of } b}$$

**GRAPHER NOTE**

When using TRACE on some graphers, a small number appears in the upper right-hand corner to indicate which function is being traced. The up and down arrow keys switch the cursor between the $Y_1$ and $Y_2$ graphs.

Example 3 illustrates how this Product Rule is used to evaluate and simplify expressions. When an expression involves only one variable, a graphical support can be used for the algebra. Graph the original expression as $y_1$ and the simplified result as $y_2$. When the two graphs appear to be identical, we have obtained visual support for the algebra. For further analysis, use the TRACE feature to compare the $y$ values of both graphs at the same $x$ value.

**EXAMPLE 3**    Using the Product Rule

Simplify the following expressions:

a) $a^8 \cdot a^2$

b) $ax^4 \cdot bx^3$

c) $(2x^3)(3x^2)$

Provide graphical support for part (c).

SOLUTION

**Simplify Algebraically.**

a) $a^8 \cdot a^2 = a^{(8+2)} = a^{10}$

b) $ax^4 \cdot bx^3 = a \cdot b \cdot x^4 \cdot x^3 = abx^7$

c) $2x^3 \cdot 3x^2 = 2 \cdot 3 \cdot x^3 \cdot x^2 = 6x^{3+2} = 6x^5$

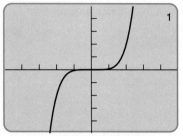

[−5, 5] by [−500, 500]

**FIGURE 5.2** The graph of both $y_1 = (2x^3)(3x^2)$ and $y_2 = 6x^5$ appear to be identical. This is visual support for Example 3, part (c).

**Support Graphically (part c).** Figure 5.2 shows the graphs of both $y_1 = (2x^3)(3x^2)$ and $y_2 = 6x^5$. Because the two graphs appear to be identical, we have visual support that the expressions $2x^3 \cdot 3x^2$ and $6x^5$ are equal. ■

## Extending the Definition of Exponent

Notice that in Definition 5.1, the exponent was restricted to a positive integer. In this section, we explain how we can extend the definition to include exponents that are negative integers or zero.

We would like expressions involving zero or negative exponents to continue to follow the Product Rule for Exponents. Consequently, we use that property in defining a meaning for zero and negative exponents.

Suppose $b$ is any positive number and $n$ is any positive integer and consider the meaning of $b^0$ and $b^{-n}$.

MEANING OF $b^0$

Apply the Product Rule to the following expression.

$$b^n \cdot b^0 = b^{n+0} = b^n.$$

Because $b^n \cdot b^0$ results in $b^n$, then $b^0$ must be 1.

MEANING OF $b^{-n}$

Apply the Product Rule to the following expression.

$$b^n \cdot b^{-n} = b^{n+(-n)} = b^0 = 1.$$

Because $b^n \cdot b^{-n} = 1$, we see that $b^{-n}$ must be the reciprocal of $b^n$.

---

**DEFINITION 5.2**   Extending the Definition of Exponent

If $b$ is any real number and $n$ is any positive integer, then

$$b^0 = 1 \qquad \text{and} \qquad b^{-n} = \frac{1}{b^n}.$$

---

**Throughout the remainder of this text, we assume that exponents can be either positive or negative integers and that the Product Rule for Exponents always applies.** Example 4 requires using the extended definition of exponent and the Product Rule for Exponents.

**EXAMPLE 4**    Simplifying Expressions

Write each of the following using only positive exponents with a single term in the denominator (if the denominator is not 1):

a) $2^{-3} \cdot x^{-2}$

b) $a^3 \cdot a^{-5}$

c) $\dfrac{7}{x^{-3}}$

d) $3^{-2} + x^{-2}$

Support part (d) graphically.

**REMINDER**

Notice that
$2^{-1} + 4^{-1} = 3/4$ and that
$(2+4)^{-1} = 1/6$. Therefore
$2^{-1} + 4^{-1} \neq (2+4)^{-1}$.

SOLUTION

**Simplify Algebraically.**

a) $2^{-3} \cdot x^{-2} = \dfrac{1}{2^3} \cdot \dfrac{1}{x^2} = \dfrac{1}{8x^2}$

b) $a^3 \cdot a^{-5} = a^{3+(-5)} = a^{-2} = \dfrac{1}{a^2}$

c) $\dfrac{7}{x^{-3}} = \dfrac{7}{1/x^3} = \dfrac{7}{1} \cdot \dfrac{x^3}{1} = 7x^3$

d) $3^{-2} + x^{-2} = \dfrac{1}{9} + \dfrac{1}{x^2} = \dfrac{x^2}{9x^2} + \dfrac{9}{9x^2} = \dfrac{x^2+9}{9x^2}$

**Support Graphically (part d).**   Figure 5.3 shows the graph of both $y_1 = 3^{-2} + x^{-2}$ and $y_2 = (x^2 + 9)/9x^2$. Because the two graphs appear to be identical, we have visual support that the expressions

$$3^{-2} + x^{-2} \text{ and } (x^2 + 9)/9x^2$$

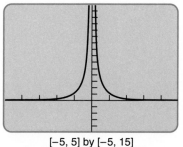

[−5, 5] by [−5, 15]

**FIGURE 5.3**  A graph of $y_1 = 3^{-2} + x^{-2}$, which consists of two parts—one part in quadrant I and the other in quadrant II. A graph of $y_2 = (x^2 + 9)/9x^2$ appears to be identical.

are equal.

For additional support, compare the trace coordinates of the two curves. Identical trace coordinates indicate an equality between expressions.  ■

## Applying the Quotient Rule for Exponents

The Product Rule can be combined with the extended definition of exponent to verify a quotient rule. Suppose that $b$ is any nonzero real number and $n$ and $m$ are integers, either positive or negative. Then

$$\frac{b^n}{b^m} = b^n \cdot \frac{1}{b^m} = b^n \cdot b^{-m} = b^{n-m}.$$

These steps are explained by applying the definition of multiplication of fractions, definition of exponents, and the Product Rule for Exponents. We summarize the procedure as the Quotient Rule for Exponents.

---

### Quotient Rule for Exponents

If $m$ and $n$ are integers and $b$ is any nonzero real number, then

$$\frac{b^n}{b^m} = b^{n-m} \qquad \text{or} \qquad \frac{b^n}{b^m} = \frac{1}{b^{m-n}}.$$

---

Example 5 illustrates the use of the Quotient Rule.

### EXAMPLE 5    Simplifying Expressions

Write the following quotients using only positive exponents. Show graphical support for part (d).

a) $\dfrac{2^6}{2^4}$

b) $\dfrac{x^8}{x^3}$

c) $\dfrac{3^{-4}}{3^{-2}}$

d) $\dfrac{x^{-4}}{x^3}$

SOLUTION

**Simplify Algebraically.**

a) $\dfrac{2^6}{2^4} = 2^{6-4} = 2^2 = 4$

b) $\dfrac{x^8}{x^3} = x^{8-3} = x^5$

c) $\dfrac{3^{-4}}{3^{-2}} = 3^{-4-(-2)} = 3^{-2} = \dfrac{1}{3^2} = \dfrac{1}{9}$

d) $\dfrac{x^{-4}}{x^3} = x^{-4-3} = x^{-7} = \dfrac{1}{x^7}$ or alternatively, $\dfrac{x^{-4}}{x^3} = \dfrac{1}{x^{3-(-4)}} = \dfrac{1}{x^7}$

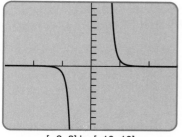

[−3, 3] by [−10, 10]

**FIGURE 5.4** The graph of $y_1 = x^{-4}/x^3$ consists of two parts—one part in quadrant I and the other in quadrant III. The graph of $y_2 = 1/x^7$ appears to be identical.

**Support Graphically.** Figure 5.4 shows that the graph of both $y_1 = x^{-4}/x^3$ and $y_2 = 1/x^7$ (part d). Because the two graphs appear to be identical, we have visual support that the expressions $x^{-4}/x^3$ and $1/x^7$ are identical. ∎

## Applying Exponent Rules with Different Bases

In both the Product and Quotient Rules, all factors of the expressions have the same base. Next we state several rules that apply when different bases are raised to a power.

---

**Power Rules**

If $a$ and $b$ are any nonzero real numbers and $n$ is any integer, then the following holds:

$$(ab)^n = a^n \cdot b^n$$

$$\left(\frac{a}{b}\right)^n = \frac{a^n}{b^n}$$

$$(a^n)^m = a^{nm}$$

---

**EXAMPLE 6**    Simplifying Expressions

Simplify the following expressions:

a) $(2a)^3 b^2$

b) $\left(\dfrac{9x^2}{3y^{-1}}\right)^3$

c) $\dfrac{x^{-3}}{y^{-2}} \cdot \left(\dfrac{x^2}{y^4}\right)^2$

d) $\left(\dfrac{3x^2}{2y}\right)^3 \cdot \dfrac{x^{-2}}{(2y)^{-3}}$

SOLUTION

a) $(2a)^3 b^2 = 2^3 a^3 b^2 = 8a^3 b^2$

b) $\left(\dfrac{9x^2}{3y^{-1}}\right)^3 = \left(\dfrac{3x^2}{y^{-1}}\right)^3 = \dfrac{(3x^2)^3}{(y^{-1})^3} = \dfrac{3^3 x^6}{y^{-3}} = 27x^6 y^3$

c) $\dfrac{x^{-3}}{y^{-2}} \cdot \left(\dfrac{x^2}{y^4}\right)^2 = \dfrac{x^{-3}}{y^{-2}} \cdot \dfrac{(x^2)^2}{(y^4)^2} = \dfrac{x^{-3}}{y^{-2}} \cdot \dfrac{x^4}{y^8} = \dfrac{x^{-3+4}}{y^{-2+8}} = \dfrac{x}{y^6}$

d) $\left(\dfrac{3x^2}{2y}\right)^3 \cdot \dfrac{x^{-2}}{(2y)^{-3}} = \dfrac{(3x^2)^3}{(2y)^3} \cdot \dfrac{x^{-2}}{2^{-3} y^{-3}} = \dfrac{3^3 x^6}{2^3 y^3} \cdot \dfrac{x^{-2}}{2^{-3} y^{-3}}$

$$= \dfrac{3^3 x^{6-2}}{2^{3-3} y^{3-3}} = 27x^4 \qquad\blacksquare$$

## Using Scientific Notation

Often numbers that occur in science or other applications are very large or very small numbers. For example, light travels about 30,970,000,000,000,000 feet in 1 year. The radius of a hydrogen atom in its natural state is 0.00000008 cm. So often it is more convenient to express numbers in **scientific notation** in which exponents represent very large or very small numbers. In scientific notation, a number is written as a factor between 1 and 10 times the appropriate power of 10.

---

**DEFINITION 5.3**   Scientific Notation

A number is written in **scientific notation** when it is expressed in the form

$$a \times 10^n,$$

where $1 \le |a| < 10$ and $n$ is an integer.

---

Here are several examples of numbers written in scientific notation:

$$1.23 \times 10^3 \qquad 2.5 \times 10^{-5} \qquad 4.625 \times 10^{12}$$

To express these numbers in standard notation, simply write the power of 10 in standard notation and multiply.

$$1.23 \times 10^3 = \underbrace{1.23 \times 1000}_{\substack{\text{Move the} \\ \text{decimal point} \\ \text{right 3 places.}}} = 1230$$

$$2.5 \times 10^{-5} = \underbrace{2.5 \times 0.00001}_{\substack{\text{Move the} \\ \text{decimal point} \\ \text{left 5 places.}}} = 0.000025$$

$$4.625 \times 10^{12} = \underbrace{4.625 \times 1{,}000{,}000{,}000{,}000}_{\substack{\text{Move the} \\ \text{decimal point} \\ \text{right 12 places.}}} = 4{,}625{,}000{,}000{,}000$$

**GRAPHER NOTE**

Read your grapher lab manual to learn how to use Sci or Eng mode and how to represent numbers in scientific notation on your grapher.

A number can be changed from scientific notation to standard notation by moving the decimal to the right when the exponent of 10 is positive and to the left when the exponent of 10 is negative. Likewise a number in standard notation can be changed to scientific notation by reversing the direction the decimal point moves.

---

### Changing a Number from Standard Notation to Scientific Notation

To change a number in standard notation to scientific notation, do the following:

1. Place an arrow ($\uparrow$) to the right of the first nonzero digit.
2. Count the number of digits from the arrow to the decimal point. This number is the absolute value of the exponent of 10.
3. The exponent is positive when the arrow is to the left of the decimal point and negative when the arrow is to the right of the decimal point.

---

**EXAMPLE 7**    Writing a Number in Scientific Notation

Write the following

 a) 12, 501
 b) 0.0028

in scientific notation.

SOLUTION

a) $1_\uparrow 2{,}501 = 1.2501 \times 10^4$   **Count 4 digits to the right from the arrow to the decimal.**

b) $0.002_\uparrow 8 = 2.8 \times 10^{-3}$   **Count 3 digits to the left from the arrow to the decimal.**   ■

Scientists often complete computations by first changing numbers from standard notation to scientific notation, completing the computation using properties of exponents, and then converting the answer back to standard notation.

### TRY THIS

Evaluate $2^{33}$ and $2^{34}$ on your grapher. Notice that it automatically expresses $2^{34}$ in scientific notation. Whenever a number is too large to be expressed in the usual decimal form, the grapher converts it to scientific notation. Now, if available on your grapher, change the MODE to Sci and again evaluate $2^{33}$ and $2^{34}$. Consult your grapher lab manual for further discussion of these two modes.

**EXAMPLE 8**   Using Scientific Notation

Use scientific notation to complete the following computation:

$$\frac{(851{,}000) \cdot (0.00000537)}{0.0000003}$$

Express the answer in scientific notation.

SOLUTION

$$\frac{(851{,}000) \cdot (0.00000537)}{0.0000003} = \frac{(8.51 \times 10^5) \cdot (5.37 \times 10^{-6})}{3 \times 10^{-7}}$$

$$= \frac{(8.51)(5.37)}{3} \cdot \frac{10^5 \cdot 10^{-6}}{10^{-7}}$$

**Use a grapher to find (8.51)(5.37)/3.**

$$= 15.2329 \times 10^{5-6+7}$$

**Change to scientific notation.**

$$= 1.52329 \times 10^1 \times 10^6 = 1.52329 \times 10^7$$

The factors in this product can be entered directly into a grapher in scientific notation.   ■

In Example 8, scientific notation reduces the grapher computation to $\frac{(8.51)(5.37)}{3}$. The power of 10 computation uses properties of exponents and often can be completed mentally. Even when using a grapher, scientific notation can save keystroke time.

## Exercises for Section 5.1

In Exercises 1–12, evaluate the expression either mentally or using paper and pencil. Support using your grapher.

1. $3^2 \cdot 2^3$

2. $(-2)^5$

3. $2 \cdot 5^3$

4. $(2 \cdot 5)^3$

5. $5^{-2}$

6. $3^2 \cdot 2^{-3}$

7. $4^{-1} + 2^{-1}$

8. $2^{-2} + 3^{-1}$

9. $2^{-3}(3^2 + 7^2)$

10. $3^4 \cdot 3^{-7}$

11. $\dfrac{4^3}{2^6}$

12. $\dfrac{3^3}{3^{-2}}$

In Exercises 13–18, use the Product Rule for Exponents to simplify the expression.

13. $a^2 \cdot a^4$

14. $2b^3 \cdot b^2$

15. $(3x)^2 \cdot x^4$

16. $6x^3 \cdot x$

17. $2^3 y^4 y^2$

18. $(3^2 a^3)(8a^2)$

In Exercises 19–24, use the Quotient Rule for Exponents and Definition 5.2 to simplify the expression.

19. $\dfrac{3^{-4}}{3^4}$

20. $\dfrac{x^5}{x^4}$

21. $\dfrac{a^2 b^5}{ab^3}$

22. $\dfrac{4x^3 y^2}{2xy}$

23. $\dfrac{3xy^2 z^4}{9x^2 yz^2}$

24. $\dfrac{17b^{-3} a^4 b^3}{a^2}$

In Exercises 25–38, simplify the expression by using properties of exponents. Write all the answers with positive exponents.

25. $2(a^2 b^3)^{-3}$

26. $(a^4 x^2)^2 (3a^{-2})$

27. $[(2a^2)^3 (4ab)^2]^3$

28. $(a^{-3} x^{-2})^2$

29. $\dfrac{a^2 x^3}{bx^2}$

30. $\dfrac{(a^2 b^5)^2}{a^3 b^4}$

31. $\dfrac{a^{-2} x^3}{a^{-3} x}$

32. $\dfrac{(xy)^3 x^{-2}}{x^2 y^3}$

33. $\left(\dfrac{x^2 y^2}{3xy^3}\right)^3$

34. $\left(\dfrac{x^{-2} y^3}{xy^{-2}}\right)^{-2}$

35. $\left(\dfrac{3ab}{9x^2 y}\right)^2 \cdot \left(\dfrac{xy^3}{6a^2 b^3}\right)^{-1}$

36. $\dfrac{ax^2}{c^3 d^2} \cdot \left(\dfrac{cd}{a^3 x}\right)^2$

37. $\left(\dfrac{p^{-2} q^{-3}}{4q^{-2}}\right)^2 \cdot 6p^2 q$

38. $(r^2 s^3)^2 \left(\dfrac{r^{-2} s^3}{r^2 s}\right)^{-3}$

In Exercises 39–48, write the number in scientific notation.

39. 1,587,000,000

40. 0.000023

41. 0.000000037

42. 1,580,000,000,000

43. 579,000

44. 0.0004829

45. $1,320 \times 10^{12}$

46. $45 \times 10^{-2}$

47. $0.00023 \times 10^5$

48. $280,000 \times 10^{-12}$

In Exercises 49–52, use paper-and-pencil methods together with scientific notation to complete the computation. Express your answer in scientific notation.

49. $(23,000,000) \times (0.00237)$

50. $(0.0000048) \times (45,000,000)$

51. $\dfrac{(25,800)(0.00031)}{0.002}$

52. $\dfrac{(0.00087)(0.0021)}{0.00009}$

In Exercises 53–57, enter the factor directly into a grapher in scientific notation and evaluate the expression.

53. $\dfrac{(0.0000034)(92,400,000)}{(120,000)(0.00004)}$

54. $\dfrac{(2.5 \times 10^{12})(3.8 \times 10^{-3})}{(5.2 \times 10^{-5})(9.1 \times 10^9)}$

**55.** *Applications to Astronomy.* Our sun is 93,000,000 mi away from Earth. Use scientific notation to express this distance in inches.

**56.** *Applications to Astronomy.* Light travels approximately $1.86 \times 10^5$ mi/sec. How many miles does it travel in 1 yr? (This distance is called a light-year.)

**57.** *Applications to Astronomy.* A certain star is 570 light-years away. How many miles from Earth is this star? (See Exercise 56.)

TRANSLATING WORDS TO SYMBOLS

The key to correct oral communication of algebraic expressions often depends on where a speaker pauses while speaking a phrase. In Exercises 58 and 59, the word (*pause*) indicates where a pause should occur when speaking the phrase.

**58.** Write an algebraic expression represented by each spoken phrase. Explain how the two expressions are different.

a) The square of $x$ (*pause*) times 2. That is, the quantity $x$ squared is multiplied by 2.

b) The square of (*pause*) $x$ times 2. That is, the square of quantity $x$ times two.

**59.** Explain symbolically the difference between the following:

a) The square of $a$ (*pause*) divided by $b$. That is, the quantity $a$ squared is divided by $b$.

b) The square of (*pause*) $a$ divided by $b$. That is, the quantity $a$ divided by $b$ is squared.

In Exercises 60–62, decide whether the expression is true or false. If true, write a symbolic expression that has been described by the statement.

**60.** The square of a quotient of two terms is equal to the quotient of the squares of the two terms.

**61.** A cube of a product of two terms is equal to the product of the cubes of the two terms.

**62.** The cube of a sum of two terms is equal to the sum of the cubes of the two terms.

EXTENDING THE IDEAS

In Exercises 63–64, perform the indicated operation as needed to simplify.

**63.** $\dfrac{x^3 \cdot x^{m+4}}{x^{m+6}}$

**64.** $\dfrac{(x^{m-5})^2}{(x^{-3})^{4m}}$

LOOKING BACK—MAKING CONNECTIONS

In Exercises 65–68, solve each of the following equations or inequalities. Express the solution to the inequalities in interval notation.

**65.** $5x - 3(x - 2) = 14 - 2x$

**66.** $(4x + 3) + 1 = 4(x - 5) + 5x$

**67.** $8x + 6 > 2(x + 3) - 24$

**68.** $5x - 7 \leq 7(x - 3) + 23$

**5.2**  **Polynomial Expressions**

> Evaluating Polynomial Expressions ▪ Other Ways of Evaluating a Polynomial with a Grapher ▪ Adding Polynomials ▪ Subtracting Polynomials ▪ Polynomials with Several Variables ▪ Applying Polynomials: Area and Volume Formulas

A **polynomial term in** $x$ is a real number or a product of a real number and $x$ raised to a whole number exponent. Examples of polynomial terms are

$$5x \qquad 3x^2 \qquad 8 \qquad 15x^3.$$

Each of the following are **not** polynomial terms:

$$\frac{1}{x+1} \qquad 2x^{1/2} \qquad 3x^{-1} \qquad \sqrt{x+1} \qquad \frac{3}{x^2}$$

A **polynomial expression in** $x$ is a polynomial term or a finite sum of polynomial terms in which the variable $x$ is raised to a nonnegative integer power in each term. Each of the following are polynomial expressions in $x$:

$$3x^4 - 5x^2 + 3 \qquad 2x^3 - 4x^2 + x - 2 \qquad 7x^5 - 5x^3 + 3x + 8$$

---

**DEFINITION 5.4**   Polynomial Expression

A **polynomial expression in** $x$ is an expression that can be written in the form

$$a_n x^n + a_{n-1} x^{n-1} + \cdots + a_1 x + a_0,$$

where each of the $a_n$'s are real number constants and the exponents are nonnegative integers.

---

**REMINDER**

Subscripts are used to identify with which power of $x$ the coefficient belongs. For example, $a_3$ is the coefficient of $x^3$ in Definition 5.4.

We usually refer to polynomial expressions simply as polynomials. Sometimes in a polynomial the variable is a letter other than $x$. For example, the following are polynomials with variables $y$, $s$, and $t$:

$$2y^3 - 4y^2 + 3 \qquad 5s^4 - 3s^2 + s - 2 \qquad t^2 - 6t + 2$$

The constants in a polynomial are called the **numerical coefficients** and the letters are called the **variables**. The **degree** of a term in a polynomial in $x$ is the exponent of $x$ in that term. For example, $4x$ in the polynomial $4x^3 - 3x^2 + 4x - 2$ is the **first degree** term and $-3x^2$ is the **second degree** term.

The **degree of a polynomial** in $x$ is equal to the degree of the term of highest degree.

**EXAMPLE 1**    Identifying a Polynomial

Identify whether the given expression is a polynomial. If a polynomial, state its degree.

a) $4x^3 - 3x^2 + 7x - 2$

b) $3x + 4\sqrt{x+2} - x^2$

c) $5x^4 - \sqrt{3}x^2 + 2x - 5$

d) $x^2 - 2x^{-1} + 3$

SOLUTION

a) $4x^3 - 3x^2 + 7x - 2$ is a polynomial of degree 3.

b) The expression $3x + 4\sqrt{x+2} - x^2$ is not a polynomial because the term $4\sqrt{x+2}$ includes a radical with the variable $x$. $\sqrt{x+2}$ can also be written as $(x+2)^{\frac{1}{2}}$. It is not the product of a number and a nonnegative *integer* power of a variable.

c) $5x^4 - \sqrt{3}x^2 + 2x - 5$ is a polynomial of degree 4.

d) $x^2 - 2x^{-1} + 3$ is not a polynomial. One exponent of $x$ is a negative integer.

          ■

A polynomial expression with three terms is sometimes called a **trinomial** and expressions with two or one term are called **binomials** and **monomials**, respectively.

| Monomials | Binomials | Trinomials |
|---|---|---|
| $x$ | $x^2 - 5$ | $3x^2 + x - 2$ |
| $5$ | $7 + y$ | $y^2 + 2x - 3$ |
| $3p$ | $p^3 - 7p$ | $4p^5 + 7p + 5p^2$ |

## Evaluating Polynomial Expressions

A polynomial expression like $x^2 - 4x + 5$ can be evaluated. When the variable $x$ is replaced by a numerical value, the expression represents a number. We say that the expression *has been evaluated.*

    A polynomial can be evaluated in your head, by a paper-and-pencil method, or by using a grapher.

**Paper-and-Pencil Method.**

| Value of variable $x$ | Value of the expression $x^2 - 4x + 5$ |
|---|---|
| $x = 3$ | $(3)^2 - 4(3) + 5 = 2$ |
| $x = -1$ | $(-1)^2 - 4(-1) + 5 = 10$ |

```
3 → X: X2 – 4X + 5
                  2
-1 → X: X2 – 4X + 5
                 10
```

**FIGURE 5.5** Evaluating $x^2 - 4x + 5$ for $x = 3$ and $x = -1$ on a grapher.

**Grapher Method.** Figure 5.5 shows a grapher screen after the same polynomial $x^2 - 4x + 5$ has been evaluated for the values $x = 3$ and $x = -1$.

In Example 2, a degree 3 polynomial is evaluated.

**EXAMPLE 2** Evaluating a Polynomial

Find the value of the polynomial $2x^3 - 4x^2 + 5x - 7$ for the following. Show both a paper-and-pencil method and a grapher method.

a) $x = -1$
b) $x = 2$

SOLUTION

**Paper-and-Pencil Method.** In the expression $2x^3 - 4x^2 + 5x - 7$, replace $x$ with (a) $-1$ and (b) 2.

$$\text{a)}\quad 2(-1)^3 - 4(-1)^2 + 5(-1) - 7 = -2 - 4 - 5 - 7$$
$$= -18$$

$$\text{b)}\quad 2(2)^3 - 4(2)^2 + 5(2) - 7 = 2(8) - 4(4) + 5(2) - 7$$
$$= 16 - 16 + 10 - 7$$
$$= 3$$

```
-1 → X: 2X3 – 4X2 + 5
X – 7
                  -18
2 → X: 2X3 – 4X2 + 5
X – 7
                    3
```

**FIGURE 5.6** Grapher screen when a grapher is used to evaluate the function $y = f(x)$.

**Grapher Method.** Figure 5.6 shows a grapher screen in which the polynomial $2x^3 - 4x^2 + 5x - 7$ has been evaluated for $x = -1$ and $x = 2$. ■

When we write the polynomial expression $2x^3 - 4x^2 + 5x - 7$ as $y = 2x^3 - 4x^2 + 5x - 7$ or $f(x) = 2x^3 - 4x^2 + 5x - 7$, we are thinking of the polynomial as a function. Notice Example 2 shows that for the polynomial function $f(x) = 2x^3 - 4x^2 + 5x - 7$, $f(-1) = -18$ and $f(2) = 3$.

## Other Ways of Evaluating a Polynomial with a Grapher

Figure 5.6 shows a grapher screen on which the polynomial $2x^3 - 4x^2 + 5x - 7$ was evaluated for $x = -1$ and $x = 2$. When a polynomial expression is to be evaluated for more than one value of $x$, the evaluation can be done more efficiently by using the function-defining menu. Figure 5.7 (a) on the following page shows the polynomial $2x^3 - 4x^2 + 5x - 7$ entered on the function-defining menu available on some graphers.

Figure 5.7 (b) shows the $Y_1$ polynomial evaluated on the Home screen for $x = 1$, $x = 2$, and $x = 3$.

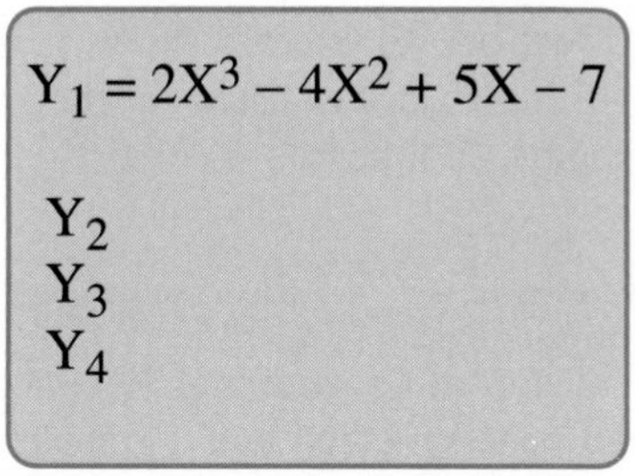

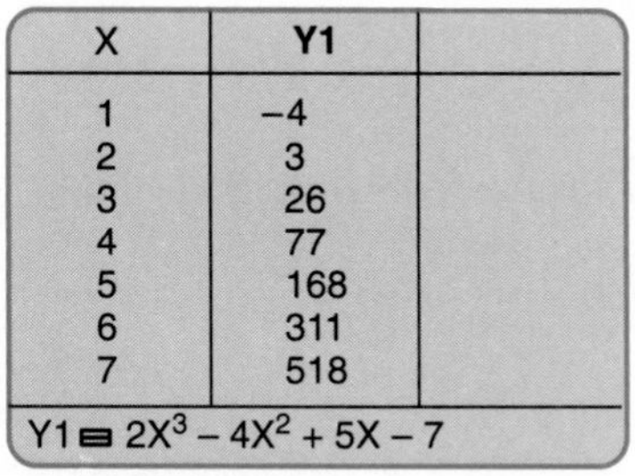

| X | Y1 | |
|---|-----|---|
| 1 | −4 | |
| 2 | 3 | |
| 3 | 26 | |
| 4 | 77 | |
| 5 | 168 | |
| 6 | 311 | |
| 7 | 518 | |

Y1 ▪ 2X³ − 4X² + 5X − 7

**FIGURE 5.8** Some graphers have the capability of generating a table like this. In this case, $y_1 = 2x^3 - 4x^2 + 5x - 7$.

$$3 \rightarrow X: 2X^3 - 4X^2 + 5X - 7$$
$$26$$
$$4 \rightarrow X: 2X^3 - 4X^2 + 5X - 7$$
$$77$$

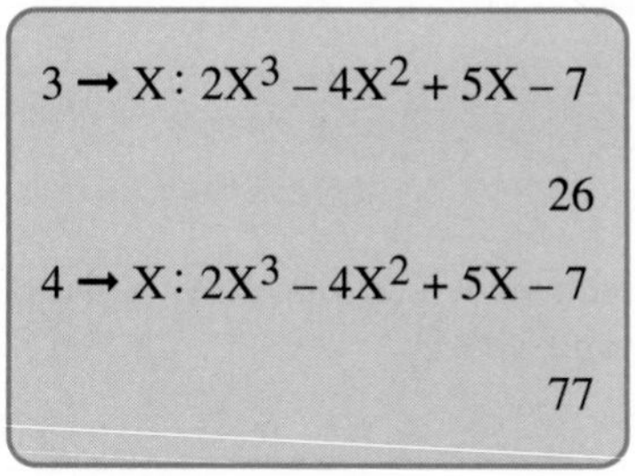

**FIGURE 5.9**
Constructing a table.

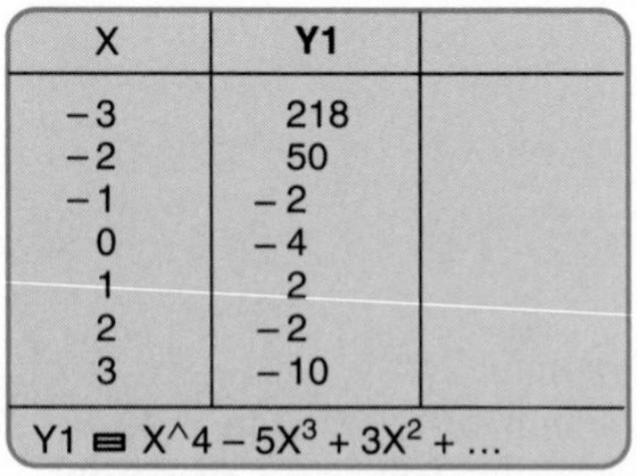

| X | Y1 | |
|-----|-------|---|
| −3 | 218 | |
| −2 | 50 | |
| −1 | −2 | |
| 0 | −4 | |
| 1 | 2 | |
| 2 | −2 | |
| 3 | −10 | |

Y1 ▪ X^4 − 5X³ + 3X² + ...

**FIGURE 5.10**
Evaluation of the polynomial $y_1 = x^4 - 5x^3 + 3x^2 + 7x - 4$.

$$Y_1 = 2X^3 - 4X^2 + 5X - 7$$
$$Y_2$$
$$Y_3$$
$$Y_4$$

Function-defining menu

(a)

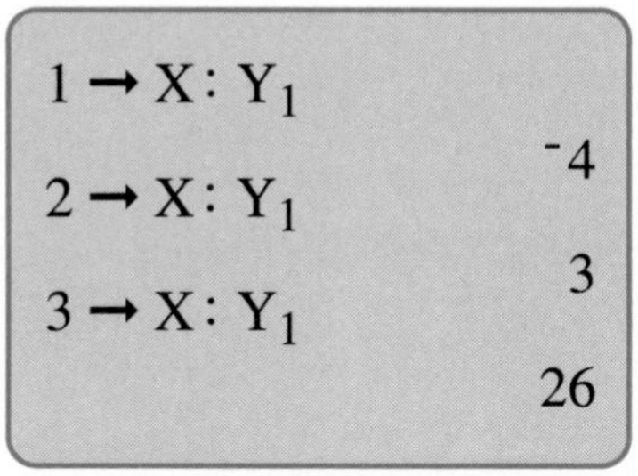

$$1 \rightarrow X: Y_1$$
$$2 \rightarrow X: Y_1$$
$$^-4$$
$$3 \rightarrow X: Y_1$$
$$3$$
$$26$$

Home screen evaluation

(b)

**FIGURE 5.7** Another technique of evaluating a polynomial using a grapher.

Some graphers have a built-in capability to generate a table of function values. Figure 5.8 shows a table of $(X, Y_1)$ values as it appears on the screen of a grapher with table-generating capabilities. The $Y_1$ polynomial $2x^3 - 4x^2 + 5x - 7$ is being evaluated for integers $x$ from 1 to 7.

If your grapher does not have this feature, you can make the table by replaying and editing $N \rightarrow X: 2x^3 - 4x^2 + 5x - 7$ on your grapher, where $N$ is the number being evaluated, as shown in Figure 5.9.

**EXAMPLE 3**  Evaluating a Polynomial with a Grapher

Evaluate the polynomial $x^4 - 5x^3 + 3x^2 + 7x - 4$ for all integer values between $x = -3$ and $x = 3$.

SOLUTION
Define $Y_1$ on your grapher as $y_1 = x^4 - 5x^3 + 3x^2 + 7x - 4$. Using any of the grapher methods discussed above, complete this table of values.

| $x$ | −3 | −2 | −1 | 0 | 1 | 2 | 3 |
|------|-----|-----|-----|-----|-----|-----|------|
| $y_1$ | 218 | 50 | −2 | −4 | 2 | −2 | −10 |

Figure 5.10 shows the output possible when a grapher has a table-generating capability. ■

## Adding Polynomials

The Distributive Property can be used to combine terms of a polynomial.  For example, the expression $4x^3 + 3x^2 + 2x^2 - 7$ can be simplified as follows:

$$4x^3 + 3x^2 + 2x^2 - 7 = 4x^3 + (3 + 2)x^2 - 7$$
$$= 4x^3 + 5x^2 - 7$$

When using the Distributive Property in this way, we say we have combined like terms.

In much the same way, we can add two polynomials. The Associative, Commutative, and Distributive Properties are used to rearrange the order of the terms and the Distributive Property is used to simplify the resulting expression. For example, the sum of $4x^2 + 3x - 2$ and $3x^2 - 5x + 1$ is as follows:

$$(4x^2 + 3x - 2) + (3x^2 - 5x + 1) = (4x^2 + 3x^2) + (3x - 5x) + (-2 + 1)$$
$$= (4 + 3)x^2 + (3 - 5)x + (-2 + 1)$$
$$= 7x^2 - 2x - 1$$

Notice that the sum of two polynomials can be found by adding the coefficients of like terms. Often it is convenient to write this addition vertically rather than horizontally. To illustrate, we find the sum of $3x^3 - 4x^2 + 5x - 2$ and $-4x^3 + 2x^2 - 3x + 4$.

$$
\begin{array}{r}
3x^3 - 4x^2 + 5x - 2 \\
-4x^3 + 2x^2 - 3x + 4 \\
\hline
-x^3 - 2x^2 + 2x + 2
\end{array}
$$

The sum of polynomials like those above can be supported both numerically and graphically, as illustrated in Example 4.

**GRAPHER NOTE**

If $Y_1$ and $Y_2$ are each polynomials, a graph of their sum $Y_1 + Y_2$ can be found on a grapher. Refer to your grapher lab manual.

**EXAMPLE 4**    Adding Polynomials

Find the sum of the polynomials $6x^4 + 2x^3 + 2x^2 + 10x + 1$ and $-5x^4 - 7x^3 + 3x^2 - 3x - 8$. Show both numerical and graphical support.

SOLUTION

We arrange the addition vertically.

**Solve Algebraically.**

$$
\begin{array}{r}
6x^4 + 2x^3 + 2x^2 + 10x + 1 \\
-5x^4 - 7x^3 + 3x^2 - 3x - 8 \\
\hline
x^4 - 5x^3 + 5x^2 + 7x - 7
\end{array}
$$

**Support Graphically.** On a grapher, define $y_1 = 6x^4 + 2x^3 + 2x^2 + 10x + 1$ and $y_2 = -5x^4 - 7x^3 + 3x^2 - 3x - 8$. Then turn off $y_1$ and $y_2$ and graph $y_3 = y_1 + y_2$ (the sum of the two given polynomials) and $y_4 = x^4 - 5x^3 + 5x^2 + 7x - 7$ (the result of the above algebra). Figure 5.11 (a) on the following page shows the function-defining menu, and Figure 5.11 (b) shows these graphs in the window $[-5, 5]$ by $[-10, 50]$. The fact that the two graphs appear to be identical supports the algebra.

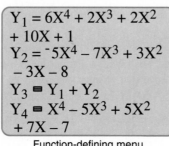

Function-defining menu

(a)

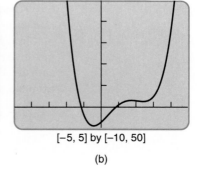

[−5, 5] by [−10, 50]

(b)

**FIGURE 5.11**  Graphs of $y_3 = y_1 + y_2$ and $y_4 = x^4 - 5x^3 + 5x^2 + 7x - 7$ appear to be identical.

| X | Y3 | Y4 |
|---|----|----|
| -3 | 233 | 233 |
| -2 | 55 | 55 |
| -1 | -3 | -3 |
| 0 | -7 | -7 |
| 1 | 1 | 1 |
| 2 | 3 | 3 |
| 3 | 5 | 5 |

X = −3

**FIGURE 5.12**  These numerical data support the algebra that polynomials $y_3 = y_1 + y_2$ and $y_4$ are identical.

**Support Numerically.** Figure 5.12 shows a table. We see that the values of $Y_3$ and $Y_4$ are identical for all the values of $X$ selected. This provides additional support that the addition was completed correctly.  ■

## Subtracting Polynomials

If $a$ and $b$ are two polynomials, we find the difference $a - b$ by using the rule $a - b = a + {}^-b$, where ${}^-b$ is obtained from $b$ by changing the sign of each term. Example 5 illustrates.

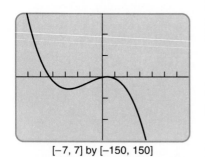

[−7, 7] by [−150, 150]

**FIGURE 5.13**  Graphs of $y = (5x^3 - 3x^2 + 4x + 2) - (7x^3 + 4x^2 - 2x + 3)$ and $y = -2x^3 - 7x^2 + 6x - 1$ appear to be identical.

**EXAMPLE 5**    Subtracting Polynomials

Find the difference $(5x^3 - 3x^2 + 4x + 2) - (7x^3 + 4x^2 - 2x + 3)$.

SOLUTION

**Solve Algebraically.** Let $a = 5x^3 - 3x^2 + 4x + 2$ and $b = 7x^3 + 4x^2 - 2x + 3$. Then we find ${}^-b$ by changing the sign of each term of $b$. So ${}^-b = -7x^3 - 4x^2 + 2x - 3$ and $a - b = a + {}^-b$ can be written vertically as follows:

$$\begin{array}{r} 5x^3 - 3x^2 + 4x + 2 \\ -7x^3 - 4x^2 + 2x - 3 \\ \hline -2x^3 - 7x^2 + 6x - 1 \end{array}$$

**Support Graphically.** Figure 5.13 shows the graph of both $(5x^3 - 3x^2 + 4x + 2) - (7x^3 + 4x^2 - 2x + 3)$ and $-2x^3 - 7x^2 + 6x - 1$. The fact that the graphs appear to be identical supports the algebra.  ■

## Polynomials with Several Variables

So far we have studied polynomials in one variable. Each of the following expressions are polynomials with more than one variable:

$$3xy + y^2 - x^3 \qquad s^2t + 4t - 5st \qquad 3r^2h$$

The degree of a term in a polynomial with more than one variable is the sum of the exponents on the variables of that term. For example, the degree of $3xy$ is two. As usual, the degree of a polynomial expression is equal to the degree of the highest term.

Polynomials with more than one variable can be added or subtracted just as polynomials with one variable. Simply use the principle of combining like terms. Example 6 illustrates.

**EXAMPLE 6**    Simplifying Expressions

Simplify the following expressions:

a)  $(3x^2 - 4xy + 3y^2) - (2y^2 + xy - 5x^2) + 7xy$

b)  $[2x - (3xy + 5x + [-xy + 4y] - 3y) + 5x]$

**REMINDER**

In Example 6, the term $-xy$ can be considered to be $-1xy$, which makes it easier to add like terms.

SOLUTION

Use properties of addition and subtraction to rearrange these expressions and combine like terms.

a)  Change subtraction to addition by changing the sign of each term in the parentheses.

$$(3x^2 - 4xy + 3y^2) - (2y^2 + xy - 5x^2) + 7xy$$

$$= (3x^2 - 4xy + 3y^2) + (-2y^2 - xy + 5x^2) + 7xy \qquad \text{Group together like terms.}$$

$$= (3x^2 + 5x^2) + (-4xy - xy + 7xy) + (3y^2 - 2y^2) \qquad \text{Combine like terms.}$$

$$= 8x^2 + 2xy + y^2$$

b)  Begin by removing innermost grouping symbols.

$$[2x - (3xy + 5x + [-xy + 4y] - 3y) + 5x]$$

$$= [2x - (3xy + 5x - xy + 4y - 3y) + 5x]$$

$$= [2x - 3xy - 5x + xy - 4y + 3y + 5x]$$

$$= 2x - 2xy - y$$

∎

## Applying Polynomials: Area and Volume Formulas

Many area and volume formulas are polynomial expressions with several variables. For example, given a closed box with length $\ell$, width $w$, and height $h$, the volume and surface area are found using the following formulas (see Figure 5.14):

**Volume of box:** $\ell \cdot w \cdot h$    **Surface Area of box:** $2\ell h + 2w\ell + 2hw$

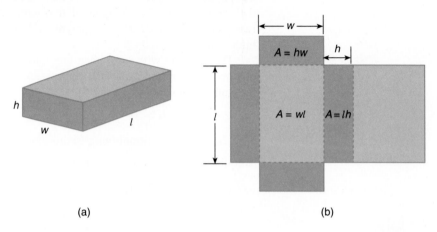

(a)                                              (b)

**FIGURE 5.14** A closed box (a) with dimensions $\ell$, $w$, and $h$ and (b) the box opened up and flattened out. The total surface area is the sum of the areas of each of the rectangles.

Notice that the right sides of both the area and volume formula are polynomials in variables $\ell, w$, and $h$. The same is true for the volume and surface area of a closed cylinder and sphere (see Figure 5.15).

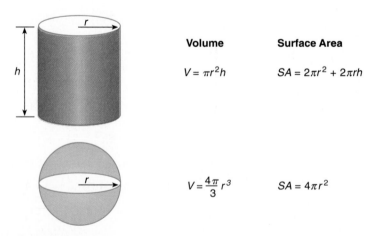

**Volume**          **Surface Area**

$V = \pi r^2 h$          $SA = 2\pi r^2 + 2\pi r h$

$V = \dfrac{4\pi}{3} r^3$          $SA = 4\pi r^2$

**FIGURE 5.15** These volume and surface area formulas are polynomial equations in variables $r$ and $h$.

**EXAMPLE 7**    Finding a Polynomial Formula for Volume

Find the volume and the surface area of the solid shown in Figure 5.16 as a polynomial in variables $x$ and $y$.

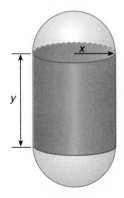

**FIGURE 5.16** A cylinder with radius $x$ and height $y$ with hemispherical caps on either end.

SOLUTION

**Volume.** This solid consists of a cylinder and two hemispherical caps.

> **Volume of cylinder:** $\pi x^2 y$
> **Volume of the two hemispheres:** $\frac{4\pi}{3}x^3$
> **Total Volume:** $\pi x^2 y + \frac{4\pi}{3}x^3$

The volume of this solid is represented by a polynomial in variables $x$ and $y$.

**Surface Area.** The surface of this solid consists of two hemispheres and the lateral surface of the cylinder (not the circular ends of the cylinder).

> **Surface Area of the two hemispheres:** $4\pi x^2$
> **Surface Area of the lateral surface of the cylinder:** $2\pi xy$
> **Total Surface Area:** $4\pi x^2 + 2\pi xy$

The surface area of this solid is represented by a polynomial in variables $x$ and $y$. ∎

**EXAMPLE 8**    Finding the Volume of a Package of Tennis Balls

Three tennis balls are packaged in a container shaped like the solid in Figure 5.17. If the diameter of the package is $d = 7.5$ cm and the total height of the package is $h = 26.5$ cm, find the volume of the container.

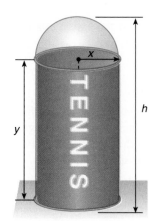

**FIGURE 5.17** Tennis ball container with radius of the base, $x$, height of the cylinder without the dome, $y$, and total height, $h$.

SOLUTION

An algebraic representation for the volume of a container of this shape is the polynomial

$$\pi x^2 y + \frac{1}{2}\cdot\frac{4\pi}{3}x^3 \quad \text{or} \quad \pi x^2 y + \frac{2\pi}{3}x^3$$

If $d = 7.5$, the radius $x = 3.75$. If the total height $h = 26.5$, then the height $y$ of the cylinder portion of the container is $y = 26.5 - 3.75 = 22.75$. Therefore we find the volume by evaluating the polynomial

$$\pi x^2 y + \frac{2\pi}{3} x^3$$

for $x = 3.75$ and $y = 22.75$. Figure 5.18 shows a grapher evaluation set to show 2 decimal places.

$3.75 \rightarrow X : 22.75 \rightarrow Y :$
$\pi X^2 Y + (2\pi/3)(X)^3$
$$1115.51$$

**FIGURE 5.18** The evaluation of the polynomial $\pi x^2 y + \frac{2\pi}{3} x^3$ for $x = 3.75$ and $y = 22.75$.

The volume of the container is 1115.51 cm$^3$. ∎

## Exercises for Section 5.2

In Exercises 1–6, decide whether each is a polynomial expression, a polynomial function, or neither.

1. $x^3 - 5x^2 + 3x - 2$
2. $x^2 y + 3x - 2y^2 - xy$
3. $5x + 3\sqrt{x + 1}$
4. $y = (\sqrt{2})x^2 - 3x + 4$
5. $f(x) = 2x^8 - 5x^3 - 3x + 1$
6. $f(x) = -5x^3 + 2x^{-2} + 3$

In Exercises 7–12, identify the expression as a monomial, binomial, or trinomial. State the degree of the expression and identify the coefficient of the term of highest degree.

7. $3x - 2$
8. $15x^2$
9. $3x^4 - 5x^2 + 3$
10. $8x + x^2 - 4$
11. $9x - 4x^3$
12. $8x + 3x^5$

In Exercises 13–16, polynomial expressions are given.

Find the values of $a, b$, and $c$ if the expression is written in the form $ax^2 + bx + c$.

13. $-3x^2 + 4x - 5$
14. $14x - 5x^2$
15. $3x - x^2 + 12$
16. $7 - 8x - 4x^2$

In Exercises 17–20, use a paper-and-pencil method to find the sum of the given polynomials. The graph of the sum appears below in the $[-5, 5]$ by $[-10, 10]$ window. Select the correct graph.

17. $f(x) = 4x^2 - 2x - 4$ and $g(x) = 3x - 2x^2 + 1$
18. $f(x) = x^3 - 4x^2 + 5x + 1$ and $g(x) = x^3 + x^2 - 7x - 3$
19. $f(x) = 2x^2 - 3x^3 + 7x + 1$ and $g(x) = 2x^3 - 4x^2 - 4x - 3$
20. $f(x) = x^3 - 4x^2 + 5x - 3$ and $g(x) = 5x^2 - x^3 - 8x + 1$

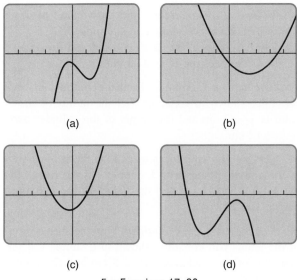

(a)   (b)

(c)   (d)

For Exercises 17–20.

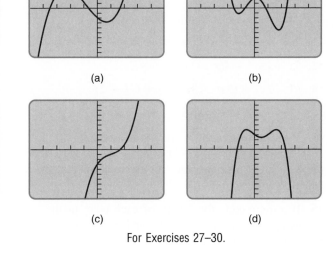

(a)   (b)

(c)   (d)

For Exercises 27–30.

In Exercises 21–24, evaluate the polynomial function for the given values of $x$. Choose between a paper-and-pencil method and a grapher method.

**21.** $f(x) = -2x^3 + 4x^2 + 5x - 15$ for $x = 1$ and $x = 13$

**22.** $f(x) = 3x^2 + 5x - 2$ for $x = -7$ and $x = 5$

**23.** $f(x) = x^4 - 5x^3 + 3x^2 + 2x - 7$ for $x = 0$ and $x = 3$

**24.** $f(x) = x^4 - 3x^2 + 5$ for $x = -3$ and $x = 4$

In Exercises 25 and 26, complete a table of $(x, y)$ values for $x = -4, -3, -2, -1, 0, 1, 2, 3, 4$.

**25.** $y = 7x^3 + 5x^4 - 3x^2 + x - 17$

**26.** $y = -4x^5 + 9x^2 + 3x - 2$

In Exercises 27–30, choose the graph (shown in $[-5, 5]$ by $[-10, 10]$) of the given polynomial.

**27.** $y = x^3 - 3x^2 + 4x - 3$

**28.** $y = x^3 + 2x^2 - 3x - 2$

**29.** $y = x^4 - x^3 - 4x^2 + x + 2$

**30.** $y = -x^4 + 2x^3 + x^2 - 2x + 3$

In Exercises 31–34, complete the given addition or subtraction algebraically and support graphically or numerically.

**31.** $(2x^2 - 4x + 5) - (-3x^2 + 2x - 3)$

**32.** $(-x^3 - 2x + 11) - (3x^2 + 5x - 8)$

**33.** $(y^2 - 3y^3 + 2) - (2y^3 + 7y - 3)$

**34.** $[(5x^2 + 3x) - (x^3 - 4x + 1)] + (3x^2 - 2x)$

In Exercises 35 and 36, polynomials with two variables are given. Complete the indicated addition.

**35.** Add
$$3x^2y + \ xy + 2y$$
$$\underline{-\ x^2y + 3xy + 5y}$$

**36.** Add
$$3px^3 - 2qx^2 + 3rx + 7$$
$$\underline{-1px^3 - \ qx^2 + 4rx - 2}$$

In Exercises 37–40, complete the additions and subtractions by writing the work horizontally.

**37.** $(3p^3 - 4p^2 + 7p - 2) + (2p^3 + 3p^2 - 2p + 3)$

**38.** $(3x^3 + 7x^2 - 3x + 8) - (-x^2 + 2x^3 - 5x + 1)$

**39.** $(7xy^2 - 4x^2y^2 + 2) - (3x^2y^2 - 3xy^2 + 4)$

**40.** $(-3p^2q + 2pq - 4q) - (4p^2q - 3pq + 5q)$

In Exercises 41–44, simplify each expression.

**41.** $2(x^2 + 3x - 2) + [x + (x^2 - 3x)]$

**42.** $5x + [3x^2 + 4(x - 3)] + x^2 - 7$

**43.** $(3 - 2x) - (5x^2 - 3x + 4) + (7x - 2)$

**44.** $(2x^3 - 4x + x^2) - [(5x - 3) - 7x^2 + x^3]$

**45. Writing to Learn.** A student insists that $(x + 2x^2 + 3) + (x^2 - x + 1) = 2x^2 + x + 3$. Describe three different methods you can use to show that the equation is false.

**46.** If $x$ represents the length of each edge in the following figure, what does $x^2$ and $x^3$ represent relative to the square and cube?

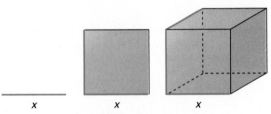

For Exercise 46.

**47.** If $x$ is the length of an edge of the larger box in the following figure, which polynomial comes closest to representing the volume of the two boxes?

a) $x^3 + x^2$

b) $x^3 + 0.5x^3$

c) $x^3 + (0.5x)^3$

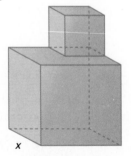

For Exercise 47.

**48.** *Volume of a Container.* Find the volume of the tennis ball container shown in Example 8 if the diameter of the package is $d = 8$ cm and the total height of the package is $h = 28$ cm.

**49.** *Volume of a Container.* Find the volume of the container shown in Example 7 if the radius of the solid is $x = 2$ cm and the height of the cylinder part is $y = 12$ cm.

**50.** *Surface Area of a Container.* Find the surface area of the container shown in Example 7 if the radius of the solid is $x = 2$ cm and the height of the cylinder part is $y = 12$ cm.

**51.** *Surface Area of a Container.* Find the surface area of the tennis ball container shown in Example 8 if the diameter of the package is $d = 8$ cm and the total height of the package is $h = 22$ cm.

**52.** *Volume of a Solid.* Write a polynomial in variables $x$ and $y$ that represents the volume of the following solid:

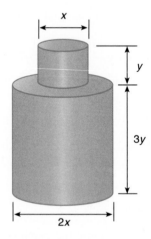

**53.** *Volume of a Solid.* Find the volume of the solid in Exercise 52 if $x = 1.2$ cm and $y = 1.8$ cm.

**54.** *Applications of Algebra to Geometry.* The following solid is built from 4 cubes, each with length $x$ on an edge. Write a polynomial that represents its volume (*V*) and surface area (*SA*).

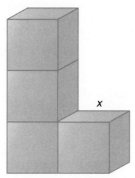

For Exercise 54.

**55.** *Applications of Algebra to Geometry.* Find the volume and surface area of the solid in Exercise 54 if the length of an edge is $x = 1.5$ cm.

TRANSLATING WORDS TO SYMBOLS

In Exercises 56–59, write an expression for the word descriptions.

**56.** Three times the square of a number

**57.** The sum of twice a number and its square

**58.** The sum of the square and twice the cube of a number

**59.** Three times a number less its square

EXTENDING THE IDEAS

In Exercises 60–62, simplify the expression.

**60.** $(x^{3n} + 2x^{2n} - x^n + 2) + (4x^{3n} - x^{2n} + 5x^n + 3)$

**61.** $(x^{2q} + 7x^q - 3) - (5x^{2q} - 3x^q + 4)$

**62.** $[3(p + q)^4 - 2(p + q)^2 - 3] - [3(p + q)^4 + (p + q)^2 + 5]$

**63.** Enter $y_1 = x^3 - 4x^2 + 3x - 2$ on the Y= screen of a grapher.

  a) Complete a table of $(x, y_1)$ values for $x = -4$, $-3, -2, -1, 0, 1, 2, 3, 4$. When do the $y$-values change from negative to positive?

  b) What does your answer in part (a) tell you about where the graph of $y = x^3 - 4x^2 + 3x - 2$ crosses the $x$-axis?

  c) Graph $y = x^3 - 4x^2 + 3x - 2$ in the Standard window. Does the graph support your conclusion?

**64.** Enter $y_2 = x^3 + 2x^2 - 7x - 5$ on the Y= screen of a grapher.

  a) Complete a table of $(x, y_2)$ values for $x = -4$, $-3, -2, -1, 0, 1, 2, 3, 4$. When do the $y$-values change from negative to positive or positive to negative?

  b) What does your answer in part (a) tell you about where the graph of $y_2 = x^3 + 2x^2 - 7x - 5$ crosses the $x$-axis?

  c) Graph $y_2 = x^3 + 2x^2 - 7x - 5$ in the Standard window. Does the graph support your conclusion?

**65.** *Free Fall vs. Propelled Fall.* One ball is dropped from a 180-ft tower at the same instant another ball is thrown down with an initial velocity of 42 ft/sec from a neighboring 280-ft tower. The distance above the ground $t$ seconds later for the dropped ball is $d_1 = -16t^2 + 180$ and for the thrown ball is $d_2 = -16t^2 - 42t + 280$. (See the figure on page 320.)

  a) Find the expression that represents the difference in elevation between the two balls at each instant.

  b) Find the graphs of $d_1$, $d_2$, and $|d_2 - d_1|$ in the same viewing rectangle [0, 5] by [0, 300]. Explain how these graphs are graphical representations of the problem situation.

  c) Are the two balls the same height off the ground at any instant? If so, when? What method did you use to answer the question?

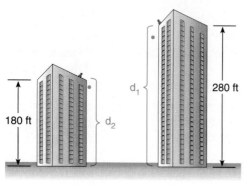

For Exercise 65.

and explain the meaning of the problem situation.

LOOKING BACK—MAKING CONNECTIONS

**68.** Write an expression in variable $x$ that describes the area of the shaded region in the following figure:

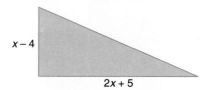

**66.** *Volume of a Container.* In Example 7, the volume of the container is given by the polynomial in variables $x$ and $y$: $V = \pi x^2 y + \frac{4\pi}{3} x^3$.

a) Let $y = 10$ and find the graph of $V$ in the window $[0, 5]$ by $[0, 150]$.

b) Use a grapher to solve the equation $V(x) = 20$ and explain the meaning of the problem situation.

**67.** *Volume of a Container.* In Example 8, the volume of the container is given by the polynomial in variables $x$ and $y$: $V = \pi x^2 y + \frac{2\pi}{3} x^3$.

a) Let $y = 8$ and find the graph of $V$ in the window $[0, 5]$ by $[0, 150]$.

b) Use a grapher to solve the equation $V(x) = 50$

**69.** Write an expression in variable $x$ that describes the area of the shaded region in the following figure:

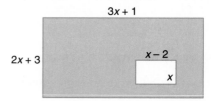

---

**5.3** **Multiplying Polynomials**

Multiplying Binomials ▪ Using the FOIL Method ▪ Identifying and Using Patterns ▪ Multiplying Trinomials

Recall that a *monomial* is an expression with one term. The Commutative and Associative Properties of Multiplication are used together with properties of exponents when two monomials such as $3x^4$ and $4x^3$ are multiplied.

$$(3x^4)(4x^3) = 3[x^4(4x^3)]$$

The Associative Property is used twice to regroup the factors.

$$= 3[(x^4 \cdot 4)x^3]$$

The Commutative Property is used to reverse the order of factors.

$$= 3[(4 \cdot x^4)x^3]$$

Use the Associative Property to regroup twice.

$$= 3[4(x^4 \cdot x^3)]$$

$$= (3 \cdot 4)x^4 x^3$$

Multiply 3 · 4 and use properties of exponents.

$$= 12x^7$$

The Distributive Property $a(b + c) = ab + ac$ can be generalized to more than two terms. The generalized Distributive Property is

$$a(b_1 + b_2 + \cdots + b_n) = ab_1 + ab_2 + \cdots + ab_n.$$

The Distributive Property extends to subtraction as well as to addition, as illustrated in Example 1.

### EXAMPLE 1  Multiplying a Monomial and a Polynomial

Find the product of $3x^2$ and $5x^3 - 3x^2 - 4x + 3$.

SOLUTION

$$3x^2(5x^3 - 3x^2 - 4x + 3) = (3x^2)(5x^3) - (3x^2)(3x^2) - (3x^2)(4x) + (3x^2)(3)$$

$$= 15x^5 - 9x^4 - 12x^3 + 9x^2 \qquad \blacksquare$$

## Multiplying Binomials

Example 1 illustrates that multiplying a polynomial by a monomial can be done by an application of the generalized Distributive Property. This property also is the key to multiplying two binomials together. For example, multiplying $(x+4)(2x+3)$ requires recognizing how the expression fits the pattern of the Distributive Property $a(b + c) = ab + ac$.

**Solve Algebraically.**

$$\underbrace{(x + 4)}_{a}\,(\,\underbrace{2x}_{b} + \underbrace{3}_{c}\,) = \underbrace{(x + 4)}_{a}\,\underbrace{(2x)}_{b} + \underbrace{(x + 4)}_{a}\,\underbrace{3}_{c}$$

Use the Distributive Property again.

$$= x(2x) + 4(2x) + x(3) + 4(3)$$

Multiply and combine like terms.

$$= 2x^2 + 8x + 3x + 12$$

$$= 2x^2 + 11x + 12$$

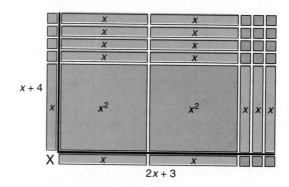

**FIGURE 5.19** Visual support of the algebra with tiles.

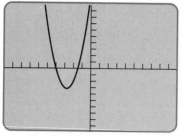

Standard window

**FIGURE 5.20**
The graphs of both
$y_1 = (x + 4)(2x + 3)$ and
$y_2 = 2x^2 + 11x + 12$ appear to
be identical.

**Support Visually.** The algebra tiles shown in Figure 5.19 represent a rectangle $2x + 3$ long and $x + 4$ wide.

The area of this rectangular region is $(x + 4)(2x + 3)$. On the other hand, we see that the rectangle consists of two $x^2$ tiles, eleven $x$ tiles, and twelve unit tiles. These tiles provide visual confirmation that

$$(x + 4)(2x + 3) = 2x^2 + 11x + 12.$$

**Support Graphically.** We can provide graphical support that $(x + 4)(2x + 3) = 2x^2 + 11x + 12$. Graph the left-hand side $(x + 4)(2x + 3)$ as $y_1$ and the right-hand side $2x^2 + 11x + 12$ as $y_2$. Figure 5.20 shows that the two graphs appear to be identical.

## Using the FOIL Method

Observe that the four terms $2x^2, 3x, 8x,$ and $12$ of the right-hand side of the equation

$$(x + 4)(2x + 3) = 2x^2 + 3x + 8x + 12$$

are the products of the first, the outer, the inner, and the last terms of $(x + 4)$ and $(2x + 3)$.

F: first
O: outer
I: inner
L: last

The word FOIL is a memory device for multiplying two binomials. Example 2 illustrates how to use the FOIL method.

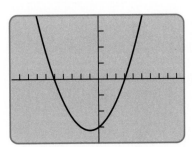

**GRAPHER NOTE**

While the emphasis in this section is on algebraic methods, use the grapher to support the algebra whenever possible.

**EXAMPLE 2**   Using the FOIL Method

Use the FOIL method to multiply the following binomials. Support part (a) graphically.

a) $(x - 3)(x + 5)$

b) $(2x + 1)(x + 2)$

c) $(u + v)(u - v)$

d) $(x^2 + 3)(2x^3 - 5x)$

SOLUTION
**Solve Algebraically.**

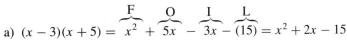

a) $(x - 3)(x + 5) = \overbrace{x^2}^{F} + \overbrace{5x}^{O} - \overbrace{3x}^{I} - \overbrace{(15)}^{L} = x^2 + 2x - 15$

b) $(2x + 1)(x + 2) = 2x^2 + 4x + x + 2 = 2x^2 + 5x + 2$

c) $(u + v)(u - v) = u^2 - uv + uv - v^2 = u^2 - v^2$

d) $(x^2 + 3)(2x^3 - 5x) = (x^2)(2x^3) - (x^2)(5x) + (3)(2x^3) - (3)(5x)$

$$= 2x^5 - 5x^3 + 6x^3 - 15x$$

$$= 2x^5 + x^3 - 15x$$

[−10, 10] by [−20, 20]

**FIGURE 5.21**
The graphs of both
$y_1 = (x - 3)(x + 5)$ and
$y_2 = x^2 + 2x - 15$ appear to be identical.

**Support Graphically (part a).** Figure 5.21 shows the graphs of both $y_1 = (x - 3)(x + 5)$ and $y_2 = x^2 + 2x - 15$. The fact that these two graphs seem to be identical supports part (a). ■

## Identifying and Using Patterns

The product of any two binomials can always be completed by using the FOIL method, which simply shortcuts using the Distributive Property. However, certain patterns of binomials occur frequently enough that it is helpful to learn these patterns. Pattern recognition is a frequently used problem-solving method.

SQUARE OF A BINOMIAL

The product of a binomial times itself is called the square of the binomial. For example, using the FOIL method we obtain

$$(2x + \underbrace{3)(2x}_{6x} + 3) = 4x^2 + 6x + 6x + 9 = 4x^2 + 12x + 9.$$

Because the outer and inner products are the same, this product can be diagrammed as follows:

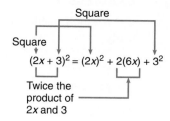

The pattern for the square of a binomial is summarized next.

---

### Squaring a Binomial

$$(a + b)^2 = a^2 + 2ab + b^2$$

$$(a - b)^2 = a^2 - 2ab + b^2$$

To square a binomial do the following:

1. Square the first term.
2. Add (or subtract) two times the product of the two terms.
3. Add the square of the last term.

---

Example 3 applies this squaring a binomial formula.

**EXAMPLE 3**    Squaring a Binomial

Square the following binomials. Support (a) graphically.

a) $(x + 5)^2$

b) $(3x - 4)^2$

c) $(2x + 3y)^2$

SOLUTION

a) $(x + 5)^2 = x^2 + 10x + 25$

b) $(3x - 4)^2 = (3x)^2 - 2(3x)(4) + (4)^2 = 9x^2 - 24x + 16$

c) $(2x + 3y)^2 = (2x)^2 + 2(2x)(3y) + (3y)^2 = 4x^2 + 12xy + 9y^2$

Figure 5.22 shows the graph of both $y_1 = (x + 5)^2$ and $y_2 = x^2 + 10x + 25$, thus providing grapher support of the algebra in part (a). ■

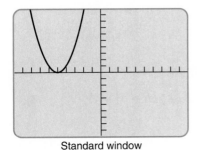

Standard window

**FIGURE 5.22** The graph of both $y_1 = (x + 5)^2$ and $y_2 = x^2 + 10x + 25$ appear to be identical.

PRODUCT OF A SUM AND DIFFERENCE

In Example 2 (c), the FOIL method was used to find the product of a sum and a difference, $(u + v)(u - v)$. The product of the outer terms and inner terms has a sum of zero, leaving only the difference of two squares.

$$(a + b)(a - b) = a^2 - ab + ab - b^2 = a^2 - b^2$$

$+ab$

$-ab$

This pattern is one of the easier ones to remember.

---

### The Product of a Sum and Difference

If $a$ and $b$ represent any algebraic expressions,

$$(a + b)(a - b) = a^2 - b^2.$$

---

This pattern is used in Example 4.

**EXAMPLE 4**    Product of the Sum and Difference of Two Quantities

Complete the following indicated operations:

a) $(y - 3)(y + 3)$

b) $(2x - 5)(2x + 5)$

c) $(x^2 + 4x)(x^2 - 4x)$

d) $[(a + b) - 7][(a + b) + 7]$

SOLUTION

a) $(y - 3)(y + 3) = y^2 - 9$

b) $(2x - 5)(2x + 5) = (2x)^2 - 5^2 = 4x^2 - 25$

c) $(x^2 + 4x)(x^2 - 4x) = (x^2)^2 - (4x)^2 = x^4 - 16x^2$

d) $[(a + b) - 7][(a + b) + 7] = (a + b)^2 - 49 = a^2 + 2ab + b^2 - 49$   ■

## Multiplying Trinomials

The FOIL method applies only when both factors in a product are binomials. When at least one factor has more than two terms, the basic method—using the generalized Distributive Property—again applies. Example 5 illustrates.

**EXAMPLE 5**   Using the Distributive Property

Find the product of $(2x^3 + 3)(x^2 - 4x + 5)$.

SOLUTION

Use the Distributive Property $a(b - c + d) = ab - ac + ad$, where $a = 2x^3 + 3$.

$$\overbrace{(2x^3 + 3)}^{a}\,(x^2 - 4x + 5) = \overbrace{(2x^3 + 3)}^{a}\,(x^2) - \overbrace{(2x^3 + 3)}^{a}\,(4x) + \overbrace{(2x^3 + 3)}^{a}\,(5)$$

$$= (2x^3)(x^2) + (3)(x^2) - (2x^3)(4x) - (3)(4x)$$
$$+ (2x^3)(5) + (3)(5)$$

$$= 2x^5 + 3x^2 - 8x^4 - 12x + 10x^3 + 15$$

**Arrange the terms in decreasing degrees.**

$$= 2x^5 - 8x^4 + 10x^3 + 3x^2 - 12x + 15$$   ■

It is often convenient to arrange multiplication vertically. Here is how the product $(x + 2)(x^2 - 4x + 3)$ would be written vertically.

$$x^2 - 4x + 3$$
$$\underline{\hspace{2.5cm} x + 2}$$
$$2x^2 - 8x + 6 \qquad \text{2 times each term of row 1.}$$
$$\underline{x^3 - 4x^2 + 3x \hspace{1.5cm} \text{x times each term of row 1.}}$$
$$x^3 - 2x^2 - 5x + 6$$

Each term of row 1 is multiplied by 2 and then each term of row 1 is multiplied by $x$. The terms are aligned vertically with like terms in the same column. The product is completed by adding like terms.

**EXAMPLE 6**     Arranging Multiplication Vertically

Find the product $(2x^3 - 4x^2 + 3x - 1)(x^2 + 2x + 4)$.

SOLUTION

$$2x^3 - \ 4x^2 + \ 3x - 1$$
$$\underline{\hspace{2cm} x^2 + \ 2x + 4}$$
$$8x^3 - 16x^2 + 12x - 4 \qquad \text{4 times each term of row 1.}$$
$$4x^4 - 8x^3 + \ 6x^2 - \ 2x \qquad \text{2x times each term of row 1.}$$
$$\underline{2x^5 - 4x^4 + 3x^3 - \ \ x^2 \hspace{1.7cm} \text{x}^2 \text{ times each term of row 1.}}$$
$$2x^5 + 0x^4 + 3x^3 - 11x^2 + 10x - 4$$

The product is $2x^5 + 3x^3 - 11x^2 + 10x - 4$.  ■

The formula used to find the volume of a box in Section 5.2 illustrates a situation where multiplication of polynomials is desired.

---

### THE BOX PROBLEM SITUATION—FINDING VOLUME

A 1.5 in. square is removed from all four corners of a piece of cardboard that is 2 inches longer than it is wide. The sides are folded up to form a box with no lid.

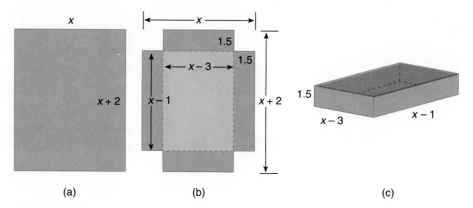

**FIGURE 5.23**   (a) The sheet of cardboard that measures $x$ by $x + 2$, (b) the 1.5 in. squares removed from the corners, and (c) the sides folded up to form a box with no lid.

Figure 5.23 illustrates the process of cutting out the square and folding the sides up to form a box with no lid.

**EXAMPLE 7**   THE BOX PROBLEM SITUATION: Representing the Volume of a Box

Write an algebraic representation as a polynomial in $x$ for the volume of the box with no lid in the **Box Problem Situation.**

SOLUTION

Let $x = $ the width of the cardboard

**length of cardboard $= x + 2$**

**side of each square to be removed $= 1.5$**

**width of cardboard with squares removed $= x - 3$**

**length of cardboard with squares removed $= x - 1$**

Apply the formula for the volume of a box: Volume $=$ length $\times$ width $\times$ height.

$$V = (x - 1)(x - 3)(1.5) \qquad \text{Use FOIL to multiply the binominals.}$$

$$= (x^2 - 4x + 3)(1.5) \qquad \text{Use the Distributive Property.}$$

$$= 1.5x^2 - 6x + 4.5 \qquad \blacksquare$$

## Exercises for Section 5.3

In Exercises 1–8, multiply the given monomial and polynomial.

1. $2x$ and $x^2 + 3$
2. $x^2$ and $3x^3 - 4x + 2$
3. $-x$ and $x^2 + 3x - 2$
4. $-2x$ and $3x^2 - 5x$
5. $2x^4$ and $5 - 3x$
6. $\frac{1}{2}x^3$ and $16 - 3x^2$
7. $5x^2$ and $3x^3 - 4x^2 + 2x - 3$
8. $7x$ and $-x^2 + 5x - 3$

In Exercises 9–22, use the FOIL method to multiply the two binomials.

9. $(x + 3)$ and $(x - 2)$
10. $(x - 5)$ and $(x + 7)$
11. $(2x + 1)$ and $(3x - 2)$
12. $(3x - 1)$ and $(2x - 3)$
13. $(5x + 3)$ and $(7x - 2)$
14. $(2x - 1)$ and $(3x + 2)$
15. $(3x^2 + 4)$ and $(x + 3)$
16. $(x + 4)$ and $(x^2 + 1)$
17. $(x - 2)$ and $(x^2 + 3)$
18. $(x - 1)$ and $(x^2 - 5)$
19. $(3x^2 + 5)$ and $(2x^3 + x)$
20. $(x^2 - 2)$ and $(5 - 3x)$
21. $(7x - x^2)$ and $(6 - x^3)$
22. $(x - x^3)$ and $(x^2 - x^3)$

In Exercises 23–26, choose the graph (shown in the window $[-10, 10]$ by $[-20, 20]$) of the given polynomial.

23. $y = (x - 2)(x - 5)$
24. $y = (x - 4)(x + 5)$
25. $y = (2x - 3)(x + 4)$
26. $y = (x + 1)(x + 6)$

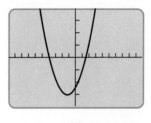
(a)

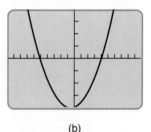
(b)

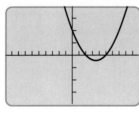
(c)

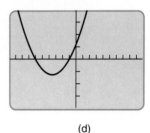
(d)

For Exercises 23–26.

In Exercises 27–34, use the FOIL method to find the product of the expression. Use a grapher to support the algebra when possible.

27. $(x + 3)(x + 5)$
28. $(x + 4)(x + 1)$
29. $(x - 2)(x + 4)$
30. $(x + 2)(x - 3)$
31. $(2x - 3)(x + 5)$
32. $(3x - 1)(2x + 3)$
33. $(3y - a)(y + 2a)$
34. $(a + 2b)(2a - b)$

In Exercises 35–38, graph the given function in the 0.1 window. Use TRACE to find the $x$-intercepts. How are the $x$-intercepts related to the expression that defines the function?

35. $y = 0.2(x - 4)(x + 3)$
36. $y = 0.3(x - 1)(x + 2)$
37. $y = 0.2(2x + 1)(x - 3)$
38. $y = 0.1(4x - 3)(5x + 3)$

39. **Writing to Learn.** Define $y_1 = (2x + 3)^2$ and $y_2 = 4x^2 + 9$. How does the following table show that $4x^2 + 9$ is not a square of a binomial? Discuss at least

two reasons why this table can be used to conclude that $(2x + 3)^2 \neq 4x^2 + 9$. Support your conclusion graphically.

| X | Y1 | Y2 |
|---|----|----|
| 0 | 9 | 9 |
| 1 | 25 | 13 |
| 2 | 49 | 25 |
| 3 | 81 | 45 |
| 4 | 121 | 73 |
| 5 | 169 | 109 |
| 6 | 225 | 153 |

X = 0

For Exercise 39.

In Exercises 40–51, express the function as a trinomial by squaring the given expression. Use a grapher to support the algebra.

**40.** $f(x) = (x + 5)^2$    **41.** $g(x) = (x + 7)^2$

**42.** $h(x) = (x - 4)^2$    **43.** $k(x) = (x - 9)^2$

**44.** $f(x) = (x + 12)^2$    **45.** $g(x) = (x - 8)^2$

**46.** $h(x) = \left(x + \frac{5}{2}\right)^2$    **47.** $f(x) = \left(x - \frac{3}{4}\right)^2$

**48.** $g(x) = \left(x - \frac{7}{2}\right)^2$    **49.** $h(x) = \left(x - \frac{8}{3}\right)^2$

**50.** $k(x) = (x - 3.5)^2$    **51.** $f(x) = (x - 0.25)^2$

**52. Writing to Learn.** A student insists that $(x + 3)^2 = x^2 + 3^2$. Describe three different methods you can use to show that the equation is false.

**53. Writing to Learn.** Write a paragraph that explains how you can use a grapher to support the claim that $(x + 3)(x - 2)(x + 5) = x^3 + 6x^2 - x - 30$.

In Exercises 54–59, use the Distributive Property to find the given product.

**54.** $(x + 2)(x^3 - 3x^2 + 4)$

**55.** $(x^3 + 5)(x^2 + 4x - 2)$

**56.** $(2x - 3)(x^2 + 5x - 3)$

**57.** $(x^2 - 3)(x^3 + 5x + 1)$

**58.** $(x^2 + 3)(x^2 - 4x + 4)$

**59.** $(5 - x^2)(x^3 - 3x^2 - 3)$

TRANSLATING WORDS TO SYMBOLS

In Exercises 60 and 61, write the word description as an equation.

**60.** The number 15 is the product of a number and nine more than its square.

**61.** The number 12 is the product of a number and four more than its square.

**62.** *Container Manufacturing.* Suppose the area of the bottom of a box is $x^2$ and that its height is $5x - 3$. Write the volume of the box in terms of a polynomial in $x$.

**63.** *Container Manufacturing.* Suppose the area of the bottom of a box is $2x$ and that its height is $3x + 2$. Write the volume of the box in terms of a polynomial in $x$.

EXTENDING THE IDEAS

**64.** If the polynomial $y = (x + 3)(x - 4)(x - 1)$ was written in the form $y = ax^3 + bx^2 + cx + d$, what is the value of $d$? What is the value of $b$?

**65.** If the polynomial $y = (2x - 1)(x + 3)(5x + 2)$ was written in the form $y = ax^3 + bx^2 + cx + d$, what is the value of $a$? What is the value of $c$?

**66. Writing to Learn.** Suppose the polynomial $y = (x + 3)(2x + 1)(x + 2)$ was written in the form $y = ax^3 + bx^2 + cx + d$. Explain how you can find the value of $b$ without multiplying the entire expression $(x + 3)(2x + 1)(x + 2)$.

In Exercises 67–70, use Associative, Commutative, and Distributive Properties or the FOIL method to complete the multiplication. Simplify the resulting expression.

**67.** $(x - 3)(x + 2 + x^2)$

**68.** $[(a + b) + 2][(a + b) - 2]$

**69.** $[3 - (x + y)][3 + (x + y)]$

**70.** $[4 - (x + y)][4 + (x + y)]$

**71.** The following figure is a pictorial representation for what product?

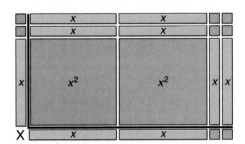

For Exercise 71.

In Exercises 72–74, verify the equation by drawing a pictorial representation similar to that for Exercise 71.

**72.** $(x + 1)(x + 2) = x^2 + 3x + 2$

**73.** $(x + 3)(x + 1) = x^2 + 4x + 3$

**74.** $(2x + 1)(x + 2) = 2x^2 + 5x + 2$

**75.** Complete these multiplications and use the strategy *find a pattern*.

$$(x + 1)(x^2 - x - 1)$$

$$(x + 1)(x^3 - x^2 - x - 1)$$

$$(x + 1)(x^4 - x^3 - x^2 - x - 1)$$

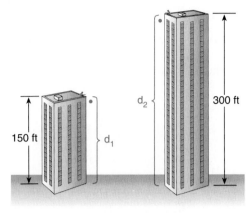

For Exercise 76.

**76.** *Free Fall vs. Propelled Fall.* One ball is dropped from a 150-ft tower at the same instant another ball is thrown down with an initial velocity of 85 ft/sec from a neighboring 300-ft tower. The distance above the ground $t$ seconds later for the dropped ball is $d_1 = -16t^2 + 150$ and for the thrown ball is $d_2 = -16t^2 - 85t + 300$.

a) Find the expression that represents the difference in height above the ground between the two balls at each instant.

b) Find the graphs of $d_1, d_2,$ and $|d_2 - d_1|$ in the same $[0, 5]$ by $[0, 300]$ window. Explain how these graphs are graphical representations of the problem situation.

c) Are the two balls the same height off the ground at any instant? If so, when? What method did you use to answer the question?

d) Which ball will hit the ground first?

In Exercises 77–80, use both algebraic and graphical information to determine the coordinates of the points where the graph of the polynomial function crosses the $x$-axis.

**77.** $f(x) = (x - 4)(x + 1)(x + 5)$

**78.** $f(x) = (3x - 2)(x + 3)(x - 5)$

**79.** $f(x) = (x - 4)(x^2 + 1)$

**80.** $f(x) = (x^2 + 5)(x^2 + 1)$

LOOKING BACK—MAKING CONNECTIONS

**81.** *Using Algebra in Geometry.* Write an expression in variable $x$ that describes the area of the shaded region in the following figure:

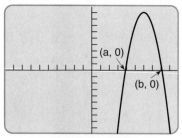

For Exercise 81.

**82.** *Using Algebra in Geometry.* Write an expression in variable $x$ that describes the area of the shaded region in the following figure:

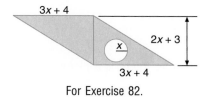

For Exercise 82.

## 5.4   Polynomial Equations and Factoring

Solving Polynomial Equations Written in Factored Form  ▪  Finding Common Factors  ▪  Factoring Out the Greatest Common Monomial Factor  ▪  Solving Equations Algebraically by Factoring  ▪  Factoring by Grouping

In Section 3.4, you studied the **Garden Problem Situation**. Recall that solving the problem depended on finding solutions to the equation $x(24 - 2x) = 63$, an equation that can be written in the following equivalent ways:

$$x(24 - 2x) = 63$$

$$24x - 2x^2 = 63$$

$$-2x^2 + 24x - 63 = 0$$

The third equation is easily recognized as a polynomial equation that can be solved using the $x$-intercept method. The $x$-intercept graphical method of finding a solution to this equation requires that we find the graph of the function

$$f(x) = -2x^2 + 24x - 63$$

(see Figure 5.24) and zoom in to locate the $x$-intercepts of this graph. The following four statements are logically equivalent; that is, they have the same meaning.

Standard window

**FIGURE 5.24** The graph of $y = -2x^2 + 24x - 63$.

---

### Equivalent Statements That Link the Visual and the Algebraic

1. The graph of $y = f(x)$ crosses the $x$-axis at the points $(a, 0)$ and $(b, 0)$.
2. $(a, 0)$ and $(b, 0)$ are $x$-intercepts of $y = f(x)$.
3. $x = a$ and $x = b$ are solutions to the equation $f(x) = 0$.
4. $f(a) = 0$ and $f(b) = 0$

---

Because $f(a) = 0$ and $f(b) = 0$, we call $x = a$ and $x = b$ **zeros** of the function $y = f(x)$. So another way of describing the solution to the **Garden Problem Situation** is to say that we need to find the **zeros** of the function $f(x) = -2x^2 + 24x - 63$. Another name for a zero is **root**. So finding the zeros of a polynomial is equivalent to finding the roots of that polynomial. Figure 5.24 provides visual support for the link between roots of $-2x^2 + 24x - 63 = 0$ and the $x$-intercept graphical method of solving the equation.

The graphical method will be applied to solving polynomial equations in Section 5.6. Let's begin, however, with algebraic methods.

## Solving Polynomial Equations Written in Factored Form

In arithmetic we say that 4 and 5 are **factors** of 20 since $4 \cdot 5 = 20$. In a similar fashion, we say that $3x$ and $5x$ are factors of $15x^2$ since $3x \cdot 5x = 15x^2$. In particular, if $a \cdot b = 15x^2$, both $a$ and $b$ are called **factors of $15x^2$**.

**EXAMPLE 1**    Finding Factors of $15x^2$

List all factors of $15x^2$ that include whole number factors of 15.

SOLUTION
Each of the following is a factor of $15x^2$:

$$1, 3, 5, 15, x, 3x, 5x, 15x, x^2, 3x^2, 5x^2, 15x^2 \qquad \blacksquare$$

---

> **REMINDER**
>
> A factor of any expression divides into the expression evenly. Sometimes when factoring large numbers, it is more convenient to think of what numbers go into that number evenly.

In Section 5.3, you learned how to multiply polynomials. Using these methods, we can complete the multiplication $(x + 3)(x - 1)(x - 4)$ to obtain

$$(x + 3)(x - 1)(x - 4) = x^3 - 2x^2 - 11x + 12.$$

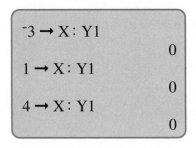

**FIGURE 5.25**
Numerical support
of the solutions to
$(x + 3)(x - 1)(x - 4) = 0$.

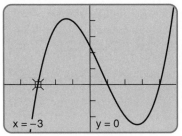

0.1 window with vertical
dimension adjusted to [−15, 25].

**FIGURE 5.26**
The graph of
$y = (x + 3)(x - 1)(x - 4)$.

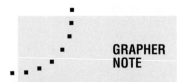

**GRAPHER NOTE**

Consult your lab manual
for how to change the
vertical dimension in the 0.1
window.

We say that $(x + 3)$, $(x - 1)$, and $(x - 4)$ are **factors** of the polynomial

$$x^3 - 2x^2 - 11x + 12.$$

**EXAMPLE 2**   Solving an Equation in Factored Form

Find three solutions of the equation $(x+3)(x-1)(x-4) = 0$. Support the solutions both numerically and graphically.

SOLUTION
**Solve Algebraically.** When $x$ is replaced with a value that makes one of the factors $(x + 3)$, $(x - 1)$, or $(x - 4)$ zero, then the product $(x + 3)(x - 1)(x - 4)$ is also zero. So we set each factor equal to zero and solve.

$$x + 3 = 0 \qquad x - 1 = 0 \qquad x - 4 = 0$$
$$x = -3 \qquad\quad x = 1 \qquad\quad x = 4$$

Three solutions of the equation $(x + 3)(x - 1)(x - 4) = 0$ are $x = -3$, $x = 1$, and $x = 4$.

**Support Numerically.** Define $y_1 = (x+3)(x-1)(x-4)$ on a grapher. Figure 5.25 shows numerical support of the three solutions.

**Support Graphically.** Figure 5.26 shows the graph of $y = (x + 3)(x - 1)(x - 4)$ in the 0.1 window in which the vertical dimension has been adjusted to $[-15, 25]$. Notice that $(-3, 0)$, $(1, 0)$, and $(4, 0)$ are $x$-intercepts, or equivalently, $x = -3$, $x = 1$, and $x = 4$ are solutions. ■

As you reflect on Example 2, here are several important observations.

A. The statement that $x = -3$, $x = 1$, and $x = 4$ are solutions to the equation $(x + 3)(x - 1)(x - 4) = 0$ is equivalent to saying that $x = -3$, $x = 1$, and $x = 4$ are zeros, or roots, of the function $f(x) = (x + 3)(x - 1)(x - 4)$. In fact, by substitution we find the following:

$$f(-3) = 0 \cdot (-4) \cdot (-7) = 0$$
$$f(1) = 4 \cdot 0 \cdot (-3) = 0$$
$$f(4) = 7 \cdot 3 \cdot 0 = 0$$

B. Because $(x + 3)(x - 1)(x - 4) = x^3 - 2x^2 - 11x + 12$, we see that $x = -3$, $x = 1$, and $x = 4$ are also solutions to the equation

$$x^3 - 2x^2 - 11x + 12 = 0.$$

C. We have shown that $x^3 - 2x^2 - 11x + 12 = 0$ has **at least** three solutions, but we have not shown that this equation has **only** three solutions.

The following zero-product property of real numbers can be used to conclude that there are **only** three solutions. This property says that a product of real numbers can be zero only if at least one of the factors is zero.

---

### Zero-product Property of Real Numbers

Suppose $a$ and $b$ are real numbers. Then

$$a \cdot b = 0 \text{ if and only if } a = 0 \text{ and/or } b = 0.$$

---

The impact of this property of real numbers is that the **only** solutions to the equation $x^3 - 2x^2 - 11x + 12 = 0$ are the three that were found in Example 2. This property is the basis for solving polynomial equations that have been written in factored form.

### EXAMPLE 3    Solving an Equation in Factored Form

Find all solutions of the equation $(4x - 3)(2x + 7)(2x - 3) = 0$.

SOLUTION
When $x$ is replaced with a value that makes one of the factors zero in the expression

$$(4x - 3)(2x + 7)(2x - 3),$$

the product of the three factors is also zero. So we set each factor equal to zero and solve.

$$4x - 3 = 0 \qquad 2x + 7 = 0 \qquad 2x - 3 = 0$$
$$4x = 3 \qquad 2x = -7 \qquad 2x = 3$$
$$x = \frac{3}{4} \qquad x = -\frac{7}{2} \qquad x = \frac{3}{2}$$

The three solutions of the equation $(4x - 3)(2x + 7)(2x - 3) = 0$ are $x = \frac{3}{4}$, $x = -\frac{7}{2}$, and $x = \frac{3}{2}$. ∎

Because $(4x - 3)(2x + 7)(2x - 3) = 16x^3 + 20x^2 - 108x + 63$, we have found solutions to the equation $16x^3 + 20x^2 - 108x + 63 = 0$ in Example 3.

In Examples 2 and 3, the factors of the given polynomials were already known. In most circumstances, the factors are not known but need to be found as a part of the algebraic solution. Next we turn our attention to learning the skill of finding these factors algebraically.

## Finding Common Factors

An expression that is a factor of two or more expressions is called a **common factor** of the set of expressions. For example,

$$3xy^2 \text{ is a common factor of } 6x^2y^2, \ 24x^2y^4, \ 3x^2y^3, \text{ and } 96x^3y^2$$

and

$$2x \text{ is a common factor of } 6x^2, \ 2x^3, \text{ and } 12x^3.$$

Although $2x$ is a common factor of $6x^2$, $2x^3$, and $12x^3$, it is not the greatest common factor. The **greatest common factor** of $6x^2$, $2x^3$, and $12x^3$ is $2x^2$, since 2 is the greatest common factor of $\{2, 6, 12\}$ and $x^2$ is the largest power of $x$ that is a factor of each term.

**REMINDER**

We shall use the direction **Factor** to mean find factors where coefficients are factored into whole numbers.

**EXAMPLE 4**    Finding the Greatest Common Factor

Find the greatest common factor of each of the following sets of terms:

a) $\{5x^3, 15x, 9x^4\}$
b) $\{6r^2s, 15r^2s^3, 21rs^4\}$

SOLUTION

a) The greatest common factor of $\{5, 15, 9\}$ is 1, so the greatest common factor of $\{5x^3, 15x, 9x^4\}$ is $x$.
b) The greatest common factor of $\{6, 15, 21\}$ is 3, so the greatest common factor of $\{6r^2s, 15r^2s^3, 21rs^4\}$ is $3rs$.    ■

## Factoring Out the Greatest Common Monomial Factor

The terms of the numerical expression $12 + 26$ have the common factor of 2. If we write the expression as $2 \cdot 6 + 2 \cdot 13$, we can use the Distributive Property to obtain the following equality:

$$12 + 26 = 2 \cdot 6 + 2 \cdot 13 \quad \textbf{Use the Distributive Property.}$$

$$= 2(6 + 13)$$

When using the Distributive Property in this way, we say that we have "factored out" a 2.

Similarly when each term of a polynomial has a common monomial factor, that monomial can be factored out. For example, $2x^2$ can be factored out from the polynomial $8x^4 - 6x^3 + 12x^2$ as follows:

$$8x^4 - 6x^3 + 12x^2 = 2x^2(4x^2 - 3x + 6)$$

Because the terms of $4x^2 - 3x + 6$ have no common factor, we see that $2x^2$ must be the greatest common factor of $8x^4 - 6x^3 + 12x^2$.

**EXAMPLE 5**    Factoring Out the Greatest Common Factor

Factor out the greatest common factor of the following. Show graphical support for part (c).

a) $12x^3 - 3x^2 + 9x$

b) $14x^2y - 21xy^3 + 35x^2y^2$

c) $3x^2(2x^3 + 6x) + x^3(x^3 + 3x)$

SOLUTION

a) The greatest common factor is $3x$.

$$12x^3 - 3x^2 + 9x = 3x(4x^2 - x + 3)$$

b) The greatest common factor is $7xy$.

$$14x^2y - 21xy^3 + 35x^2y^2 = 7xy(2x - 3y^2 + 5xy)$$

c) Notice that if 2 is factored out of $(2x^3 + 6x)$, then $x^3 + 3x$ can be identified as a common factor of $3x^2(2x^3 + 6x) + x^3(x^3 + 3x)$. Begin by rewriting the original expression.

$$3x^2(2x^3 + 6x) + x^3(x^3 + 3x) = 3x^2 \cdot 2(x^3 + 3x) + x^3(x^3 + 3x)$$

$$= (x^3 + 3x)(6x^2 + x^3) \qquad \begin{array}{l}\textbf{Factor out } x \textbf{ from the}\\ \textbf{first factor and } x^2 \\ \textbf{from the second factor.}\end{array}$$

$$= x^3(x^2 + 3)(6 + x)$$

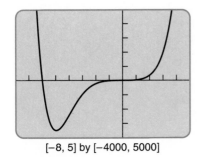

[−8, 5] by [−4000, 5000]

**FIGURE 5.27** Graphs of $y_1 = 3x^2(2x^3 + 6x) + x^3(x^3 + 3x)$ and $y_2 = x^3(x^2 + 3)(6 + x)$.

Figure 5.27 shows the graphs of both $y_1 = 3x^2(2x^3 + 6x) + x^3(x^3 + 3x)$ and $y_2 = x^3(x^2 + 3)(6 + x)$. Comparing the trace coordinates of both $y_1$ and $y_2$ adds additional support of the algebra.    ■

## Solving Equations Algebraically by Factoring

Examples 2 and 3 illustrate how to solve an equation algebraically if it has already been factored into a product of binomial factors. Often factoring must be included in the equation-solving process. An algebraic method for solving polynomial equations is summarized next.

---

### Solving Polynomial Equations Algebraically by Factoring

1. Rewrite the equation if necessary so that one side is 0.
2. Combine like terms and factor the resulting expression, if possible.
3. Set each factor equal to zero and use the Zero-product Property to find the solutions.

---

Notice how this method is applied in Example 6.

**EXAMPLE 6** Solve Algebraically and Support Graphically

Solve the equation $3x^2 + 11x = 2x$ algebraically. Support the solution graphically.

SOLUTION

**Solve Algebraically.** We write the equation $3x^2 + 11x = 2x$ in an equivalent form and factor it.

$$3x^2 + 11x = 2x$$

$$3x^2 + 11x - 2x = 0$$

$$3x^2 + 9x = 0$$

$$3x(x + 3) = 0$$

Set each factor equal to zero and solve.

$$3x = 0 \qquad x + 3 = 0$$

$$x = 0 \qquad x = -3$$

The equation $3x^2 + 11x = 2x$ has the two solutions $x = -3$ and $x = 0$.

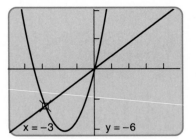

$x = -3$     $y = -6$

0.1 window with $y$-scale adjusted to $[-12, 10]$.

**FIGURE 5.28** Graphs of $y = 3x^2 + 11x$ and $y = 2x$.

**Support Graphically.** Figure 5.28 shows the graphs of $y = 3x^2 + 11x$ and $y = 2x$ in the 0.1 window with the vertical dimension adjusted to $[-12, 10]$. These two graphs appear to cross at $(0, 0)$ and $(-3, -6)$, which supports that $x = 0$ and $x = -3$ are solutions. ∎

## Factoring by Grouping

In Examples 5 and 6, the factoring was completed by factoring out the greatest common factor. Sometimes it is necessary to factor a common factor out of groups of terms before completing the final factorization. For example, to factor

$$ax - ay + 2x - 2y$$

group the terms as follows:

$$(ax - ay) + (2x - 2y)$$

Each of these two groups has a common factor: factor $a$ from the first group and factor 2 from the second group. The complete factorization is as follows:

$$ax - ay + 2x - 2y = (ax - ay) + (2x - 2y)$$

$$= a\,(x - y) + 2\,(x - y)$$

$$= (x - y)\,(a + 2)$$

**EXAMPLE 7**    Factoring by Grouping

Factor each of the following expressions by grouping:

a) $ax + bx + ay + by$

b) $x^3 + 3x^2 - 2x - 6$

SOLUTION

**REMINDER**

To factor an expression is to write it as a product. It is not in factored form until the final answer is a multiplication problem.

a) Group together the $x$ terms and the $y$ terms.

$$ax + bx + ay + by = (ax + bx) + (ay + by)$$

$$= x\,(a + b) + y\,(a + b) \quad \text{Factor out the common binomial.}$$

$$= (a + b)\,(x + y)$$

b) Group together the first two terms and the last two terms.

$$x^3 + 3x^2 - 2x - 6 = (x^3 + 3x^2) - (2x + 6)$$

$$= x^2\,(x + 3) - 2\,(x + 3) \quad \text{Factor out the common binomial.}$$

$$= (x + 3)\,(x^2 - 2) \qquad \blacksquare$$

**EXAMPLE 8**    Solve Algebraically and Support Numerically

Solve the equation $x^3 + x^2 - 4x = 4$ algebraically and support the solution numerically.

SOLUTION

**Solve Algebraically.**

$$x^3 + x^2 - 4x = 4$$ To use the Zero-product Property, we must set the equation equal to 0.

$$x^3 + x^2 - 4x - 4 = 0$$ Group terms and factor.

$$x^2(x + 1) - 4(x + 1) = 0$$ Factor again.

$$(x + 1)(x^2 - 4) = 0$$

$$(x + 1)(x - 2)(x + 2) = 0$$

Set each factor equal to zero and solve.

$$x + 1 = 0 \qquad x - 2 = 0 \qquad x + 2 = 0$$

$$x = -1 \qquad x = 2 \qquad x = -2$$

The three solutions to the equation are $x = -1$, $x = -2$, and $x = 2$.

**Support Numerically.** Figure 5.29 shows numerical support using a grapher.

■

```
-1 → X: X³ + X² – 4X
                     4
-2 → X: X³ + X² – 4X
                     4
 2 → X: X³ + X² – 4X
                     4
```

**FIGURE 5.29**
Evaluating the polynomial
$x^3 + x^2 - 4x$ for $x = -1$,
$x = -2$, and $x = 2$.

## Exercises for Section 5.4

In Exercises 1–4, list all factors of the term that are whole numbers times a power of $x$.

**1.** $8x$

**2.** $5x^2$

**3.** $6x^2$

**4.** $35x^3$

In Exercises 5–10, list all factors of the term or expression.

**5.** $x^2(x + 1)$

**6.** $x^3y^2$

**7.** $3xy^3$

**8.** $x^2(x - 3)^2$

**9.** $9x(2x - 5)^2$

**10.** $14x^2y$

In Exercises 11–14, solve each equation. Support the solution graphically.

**11.** $(5x - 3)(2x + 5)(x + 2) = 0$

**12.** $(2x - 1)(4x + 3)(7x - 2) = 0$

**13.** $(2x - 3)(7 - 3x)(5 - x) = 0$

**14.** $(x - \pi)(2x - 3)(4 - 6x) = 0$

In Exercises 15–18, write an equation whose solutions are the given values.

**15.** $x = -3$, $x = 1$, and $x = 5$

**16.** $x = -1$, $x = 2$, and $x = 3$

**17.** $x = 1$, $x = 3$, and $x = \frac{1}{2}$

**18.** $x = -3$, $x = -6$, and $x = \frac{3}{4}$

In Exercises 19–22, the graph of each function in either a $[-5, 5]$ by $[-3, 3]$ or a $[-5, 5]$ by $[-15, 15]$ window is shown below. Select the correct graph by visual inspection and support with a grapher.

**19.** $y = (x - 1)(x + 1)(x + 3)$

**20.** $y = (x + 1)(x + 2)(x + 3)$

**21.** $y = (x + 1)(x - 1)(x - 2)$

**22.** $y = (2x - 1)(2x - 3)(x - 3)$

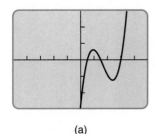

(a)

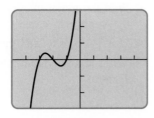

(b)

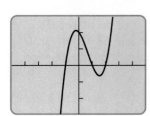

(c)

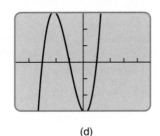

(d)

For Exercises 19–22.

In Exercises 23–30, find the greatest common factor of each set of terms.

**23.** $\{2x^2, 8x, 6x^3\}$ **24.** $\{6x^3, 3x, 12x^2\}$

**25.** $\{15x^2y, 5xy, 25xy^3\}$ **26.** $\{10r^2, 5s^3, 25rs\}$

**27.** $\{9xy, 27x^3, 15y^2\}$

**28.** $\{5(x + 3)^2, 17(x + 3)\}$

**29.** $\{3(x + 3)^2, 5(x + 3)^4\}$

**30.** $\{3xy^2z, 9x^2y^3z^2, 15xyz^2\}$

In Exercises 31–43, factor out the greatest common factor for the given expression.

**31.** $3x^4 - 5x^2 + 2x$ **32.** $9x^3 + 27x^2 - 12x$

**33.** $5x + 25x^3 - 10x^2$ **34.** $6x^2 - 3x + 15$

**35.** $(x + y)^2 - 5(x + y)$

**36.** $12x^2y^3 - 3xy + 20x^3y^2$

**37.** $3(x + 1)^3 - 4(x + 1)^2$

**38.** $2x^2(x + 3) - 4x(x + 3)$

**39.** $15x^2m^3 - 21xm^2 + 33x^4m^2$

**40.** $24p^2q - 16p^3q^2 + 8pq^2$

**41.** $p(x - y) - q(x - y) + r(x - y)$

**42.** $x^2(x - y) + x(x - y) + 3(x - y)$

**43.** $(x + y)x + (x + y)y$

In Exercises 44–51, solve the equation algebraically. Support the solution numerically and graphically.

**44.** $2x^2 - 6x = 0$ **45.** $10x^2 + 15x = 0$

**46.** $3x^2 - 18x = 3x$ **47.** $15x^2 - 15x = 5x$

**48.** $4x^3 - 5x^2 - x = 2x^2 - x$

**49.** $2x^4 - x^2 = 7x^2$

**50.** $(x - 1)x + 3(x - 1) = 0$

**51.** $2x(x - 2) + (x - 2) = 0$

In Exercises 52–59, factor the expression by grouping.

**52.** $xy + ay - bx - ab$

**53.** $2xy - ay - 6bx + 3ab$

**54.** $x^2 + 2x - 3x - 6$ **55.** $x^2 - 3x - 4x + 12$

**56.** $x^3 + 2x - 3x^2 - 6$ **57.** $2x^3 - 6x + x^2 - 3$

**58.** $x^3 + 3x^2 - 2x - 6$ **59.** $3x^3 - 6x^2 + x - 2$

TRANSLATING WORDS TO SYMBOLS

**60.** Two numbers differ by 3 and have a product of 0. Find one pair of numbers satisfying these conditions. Find a second pair.

**61.** Two numbers differ by 4 and have a product of 0. Find one pair of numbers satisfying these conditions. Find a second pair.

**62.** One number and three more than twice the number have a product of 0. What are the two numbers?

**63.** One number and two more than three times the number have a product of 0. What are the two numbers?

EXTENDING THE IDEAS

In Exercises 64–67, find a complete graph of these polynomials in $[-5, 5]$ by $[-10, 10]$. Recall that if $x - a$ is a factor of a polynomial expression, then $x = a$ is a zero. Consequently, when a polynomial expression has no real-number zeros, it will not have any linear factors $x - a$, with $a$ a real number. Determine visually whether each polynomial has a linear factor.

**64.** $x^2 + 5$          **65.** $3 - 4x^2$

**66.** $x^3 - 3x^2 + 4$

**67.** $x^4 - 2x^3 + 2x^2 - 3x + 4$

**68.** This exercise focuses on the relationship between $x$-intercepts and linear factors.

a) Find the graph of $y = x^3 + 2x^2 - x - 2$ in the 0.1 window.

b) Use TRACE to find the zeros.

c) Write this polynomial as a product of three linear factors.

**69.** This exercise focuses on the relationship between $x$-intercepts and linear factors.

a) Find the graph of $y = x^3 + \frac{5}{2}x^2 - \frac{1}{2}x - 3$ in the 0.1 window.

b) Use TRACE to find the zeros.

c) Write this polynomial as a product of three linear factors.

LOOKING BACK-MAKING CONNECTIONS

In Exercises 70–75, use a graph of the associated function in the Standard window to help find the solution to each of the inequalities. Write the solution in interval notation.

**70.** $(x - 3)(x - 8) < 0$          **71.** $(x + 2)(x - 4) < 0$

**72.** $(5 - x)(x - 3) > 0$          **73.** $(x - 6)(7 - x) > 0$

**74.** $(2x + 1)(3x + 7) < 0$

**75.** $(x - 8)(x + 3) > 0$

## 5.5   Factoring Trinomials

Factoring Perfect Square Trinomials   ▪   Factoring a Trinomial in $x^2 + bx + c$ Form   ▪   Using a Grapher as a Tool in Factoring   ▪   Factoring a Trinomial in $ax^2 + bx + c$ Form   ▪   Some Special Factoring

In Section 5.4, you learned that when a polynomial equation is expressed as a product of linear factors,

$$(x - a)(x - b) \cdots = 0,$$

it is possible to find all the solutions to the equation.

So one method of solving a polynomial equation algebraically involves looking for such linear factors. We focus in this section on factoring trinomial equations.

## Factoring Perfect Square Trinomials

Factoring a trinomial is in many respects a matter of pattern recognition. In Section 5.3, we noted the following patterns for finding the squares of binomials:

$$(a + b)^2 = a^2 + 2ab + b^2$$

$$(a - b)^2 = a^2 - 2ab + b^2$$

When $(a + b)^2$ is replaced by $a^2 + 2ab + b^2$, we say that a binomial has been squared. When $a^2 + 2ab + b^2$ is replaced by $(a + b)^2$, we say that a trinomial has been factored.

When a polynomial $x^2 + bx + c$ fits the correct pattern, it can be recognized as the square of a binomial. What is this correct pattern? The constant term must be the square of half the coefficient of the $x$-term. Figure 5.30 shows the pattern.

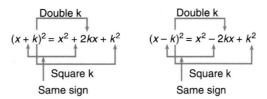

**FIGURE 5.30**  Notice the pattern of the signs.

When the coefficient of the $x$-term is an irrational number, then the constant $k$ may be an irrational number. Otherwise we shall allow only whole number constants in the factorizations.

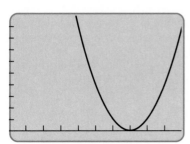

[0, 10] by [−2, 10]

**FIGURE 5.31**  A perfect square function like $y = (x - 7)^2$ has only one zero.

**EXAMPLE 1**    Factoring a Perfect Square

Determine that the following trinomials are perfect squares and factor them:

a) $x^2 - 14x + 49$

b) $x^2 + 2\sqrt{3}x + 3$

SOLUTION

a) The constant 49 is the square of 7, which is half of the coefficient 14. Therefore

$$x^2 - 14x + 49 = (x - 7)^2.$$

Figure 5.31 suggests that, in general, the graph of a perfect square polynomial $y = (x - k)^2$ has only one zero.

b) The constant 3 is the square of $\sqrt{3}$. Therefore

$$x^2 + 2x\sqrt{3} + 3 = (x + \sqrt{3})^2.$$

As a check, use the FOIL method to multiply the factors together.  ∎

## Factoring a Trinomial in $x^2 + bx + c$ Form

Most trinomials in $x^2 + bx + c$ form are not perfect squares. In the next several examples, we explore trinomials that can be factored into a product of two linear factors $(x + m)(x + n)$, where $m$ and $n$ are distinct integers. Study the following pattern:

$$(x + m)(x + n) = x^2 + nx + mx + mn = x^2 + (m + n)x + mn$$

$$= x^2 + bx + c$$

Here are some conclusions from this pattern.

---

### Factoring $x^2 + bx + c$ into a Product of Linear Factors: $(x + m)(x + n)$

1. $b$, the coefficient of $x$ in $x^2 + bx + c$, is the sum $m + n$.
2. $c$, the constant term in $x^2 + bx + c$, is the product $mn$.
3. $-m$ and $-n$ are zeros, or roots, of the equation $x^2 + bx + c = 0$.

---

So to factor $x^2 + bx + c$ into a product of linear factors $(x + m)(x + n)$, where $m$ and $n$ are integers, look for a pair of integers whose product is $c$ and whose sum is $b$. Example 2 shows that making a table is often helpful.

**EXAMPLE 2**    Factoring a Trinomial

Factor the following trinomials:

a) $x^2 + 10x + 21$

b) $x^2 - 9x + 20$

SOLUTION

a) To factor $x^2 + 10x + 21$, we look for integers $m$ and $n$ whose product is 21 and whose sum is 10. Values $m = 3$ and $n = 7$ will work. Then

$$x^2 + 10x + 21 = (x + 3)(x + 7).$$

b) To factor $x^2 - 9x + 20$, we look for integers $m$ and $n$ whose product is 20 and whose sum is $-9$. Because the product is positive and the sum is negative, we know that both integers must be negative. The following values all give a product of 20:

| $m$ | $n$ |
|---|---|
| $-1$ | $-20$ |
| $-2$ | $-10$ |
| $-4$ | $-5$ |

But only values $m = -4$ and $n = -5$ give a sum of $-9$.

$$x^2 - 9x + 20 = (x - 4)(x - 5)$$ ∎

## Using a Grapher as a Tool in Factoring

The grapher can be a useful tool in solving polynomial equations algebraically by means of factoring. Example 3 shows that a graph can provide quick visual estimates of the roots of a polynomial and consequently suggests which integer pairs are most likely to be used in finding a factorization.

**EXAMPLE 3**    Factoring a Trinomial

Factor the following trinomials. Use a graph to help determine which integer pairs are the most likely values to be used in the factorization.

a)  $x^2 - 2x - 8$,    b)  $x^2 + 3x - 28$.

SOLUTION

a) Figure 5.32 (a) shows a graph of $y = x^2 - 2x - 8$ with zeros that appear to be $x = -2$ and $x = 4$. This suggests that the factors are $x + 2$ and $x - 4$. We can confirm that these are the correct factors by using the FOIL method.

$$(x - 4)(x + 2) = x^2 - 2x - 8$$

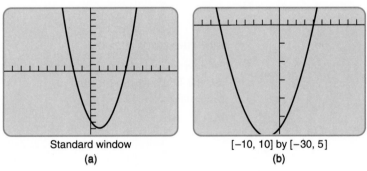

| Standard window | $[-10, 10]$ by $[-30, 5]$ |
|---|---|
| (a) | (b) |

**FIGURE 5.32**    A graph of (a) $y = x^2 - 2x - 8$ and (b) $y = x^2 + 3x - 28$.

b) Figure 5.32 (b) shows a graph of $y = x^2 + 3x - 28$ with zeros that appear to be $x = -7$ and $x = 4$. This suggests that the factorization of $x^2 + 3x - 28$ is $(x + 7)(x - 4)$. This can be confirmed algebraically using FOIL.

$$(x + 7)(x - 4) = x^2 + 3x - 28$$ ∎

## Factoring a Trinomial in $ax^2 + bx + c$ Form

The trinomials of the form $ax^2 + bx + c$ differ from those discussed in Examples 2 and 3 in that the coefficient of $x^2$ is not equal to 1. We again can use a graph as an aid in factoring, or we can use a trial and error method. We illustrate factoring $6x^2 + 11x + 3$ by using each of the methods.

A GRAPHING METHOD

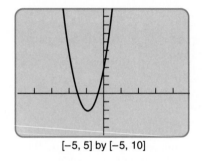

[−5, 5] by [−5, 10]

**FIGURE 5.33** A graph of $y = 6x^2 + 11x + 3$.

To factor the polynomial $6x^2 + 11x + 3$, graph $y = 6x^2 + 11x + 3$ and use the TRACE feature to find that the zeros are in the neighborhood of $x = -1\frac{1}{2}$ and $x = -1/3$. (See Figure 5.33.) This visual information can also be used to help decide what pairs of integers to use. If indeed $x = -1/3$ is a root, then $(3x + 1)$ is one of the factors. In the factorization $(3x + 1)(\underline{\quad} x + \underline{\quad})$, the product of the first terms must be $6x^2$. This means that the coefficient of $x$ in the second factor is 2. The product of the last terms must be 3.

$$(3x + 1)(\underline{\quad} x + \underline{\quad})$$

**Place factors in the blanks so that the product of the first terms is $6x^2$ and of the last terms is 3.**

$$(3x + 1)(2x + 3)$$

We confirm this solution algebraically using FOIL,

$$(3x + 1)(2x + 3) = 6x^2 + 11x + 3.$$

TRIAL-AND-ERROR METHOD

To gain insight into the factoring process, assume that $c, d, m,$ and $n$ are integers, use the FOIL method to find the product $(cx + m)(dx + n)$, and compare it to the trinomial $6x^2 + 11x + 3$.

$$(cx + m)(dx + n) = cdx^2 + cnx + dmx + mn$$

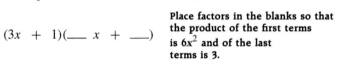

$$= (cd)\, x^2 + (cn + dm)\, x + mn$$

$$= 6x^2 + 11x + 3$$

We need to find *integer values* for $c, d, m$, and $n$ that meet these conditions.

$$cd = 6 \qquad mn = 3$$

$$cn + dm = 11$$

We complete a factorization by filling in the blanks in this diagram.

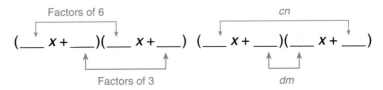

Factors of 6

$cn$

$(\underline{\quad} x + \underline{\quad})(\underline{\quad} x + \underline{\quad})$   $(\underline{\quad} x + \underline{\quad})(\underline{\quad} x + \underline{\quad})$

Factors of 3

$dm$

The sum of these products = 11.

Use trial and error to find terms $cn$ and $dm$ whose sum is 11. Notice that $c = 3, d = 2, m = 1$, and $n = 3$ satisfy this condition. The factorization of $6x^2 + 11x + 3$ is

$$6x^2 + 11x + 3 = (3x + 1)(2x + 3).$$

With experience and practice, you will gain facility in this process and will often be able to complete this trial-and-error method mentally.

Example 4 uses the graphing method, and Example 5 illustrates both methods.

**EXAMPLE 4**    Factoring a Trinomial

Factor the trinomial $5x^2 + 13x - 6$ using a graph to help determine which integer pairs are the most likely values to be used in the factorization.

SOLUTION

We begin by graphing $y = 5x^2 + 13x - 6$ in the $[-5, 5]$ by $[-15, 10]$ window. Figure 5.34 shows a possible zero at $x = -3$. Therefore the possible factor is $(x + 3)$. Consider $5x^2 + 13x - 6 = (x + 3)(? + ?)$.

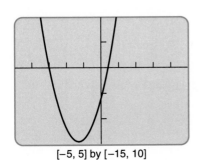

$[-5, 5]$ by $[-15, 10]$

**FIGURE 5.34** A graph of $y = 5x^2 + 13x - 6$.

$(x + 3)(\underline{\quad} x + \underline{\quad})$

**Place factors in the blanks so that the product of the first terms is $5x^2$ and the product of the last terms is $-6$.**

$(x + 3)(5x - 2)$

Therefore, $5x^2 + 13x - 6 = (x + 3)(5x - 2)$ which can be confirmed using FOIL.    ■

The two methods for factoring a trinomial in the form $ax^2 + bx + c$ can be summarized as follows.

---

### Factoring Trinomials of the Form $ax^2 + bx + c$

**Using a Graph**

1. Graph the trinomial and identify the potential zeros. If $a$ is a zero, then one of the factors would be $(x - a)$.
2. If necessary find the other factor by considering the first (F) and the last (L) steps of the FOIL process.
3. Confirm the factorization using FOIL.

**Trial-and-error Method**

1. List all integer factors of $a$, the coefficient of $x^2$.
2. List all integer factors of $c$, the constant term.
3. Try various pairs of the above factors to complete the factorization.

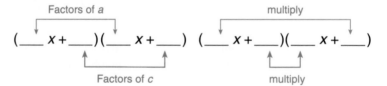

---

You will find that you will use both methods depending on the polynomial, so you should become familiar with each of them.

**EXAMPLE 5**   Factoring a Trinomial

Factor each of the following trinomials. Use a graph for part (a) and the trial-and-error-method for part (b).

a) $2x^2 + x - 3$

b) $15x^2 - 17x - 4$

SOLUTION

a) Figure 5.35 shows a possible root at $x = 1$ and therefore a factor would be $(x - 1)$.

In the factorization of $2x^2 + x - 3$, the product of the first terms must be $2x^2$ and the product of the constant terms must be $-3$.

$$(x \ - \ 1)(\underline{\quad} \ x \ + \ \underline{\quad})$$

**Place factors in the blanks so that the product of the first terms is $2x^2$ and the procuct of the last terms is −3.**

$$(x - 1)(2x + 3)$$

Using the FOIL method, we confirm that

$$(2x + 3)(x - 1) = 2x^2 + x - 3$$

b) In the factorization of $15x^2 - 17x - 4$, the coefficients of $x$ are factors of 15 and the constant terms must be factors of $-4$.

$$(\underline{\quad} \ x \ + \ \underline{\quad})(\underline{\quad} \ x \ - \ \underline{\quad})$$

**Place factors in the blanks so that the middle term coefficient becomes −17.**

$$(5x + 1)(3x - 4)$$

Using the FOIL method, we confirm that

$$(5x + 1)(3x - 4) = 15x^2 - 17x - 4 \qquad \blacksquare$$

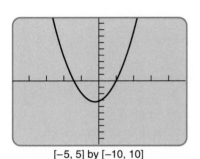

[−5, 5] by [−10, 10]

**FIGURE 5.35** A graph of $y = 2x^2 + x - 3$ shows zeros close to $x = -\frac{3}{2}$ and $x = 1$.

## Some Special Factoring

We end this section by studying several special factoring situations.

Earlier in this section, we observed that a polynomial like $x^2 + 14x + 49$ is a perfect square. In this expression, there is only one variable. An expression with two variables also can be a perfect square, as summarized next.

---

### Perfect Square Trinomials

$$x^2 + 2xy + y^2 = (x + y)^2$$
$$x^2 - 2xy + y^2 = (x - y)^2$$

---

Another familiar pattern is the **difference of two squares**. Using the FOIL method to multiply $(x + y)(x - y)$, we obtain the following equation.

**REMINDER**

It is natural to expect that there might be a way to factor the expression

$$x^2 + y^2$$

using real number coefficients. However, no such factorization exists.

## Difference of Two Squares

$$x^2 - y^2 = (x + y)(x - y)$$

It is also possible to factor both the sum and difference of two cubes. Each of these factorizations can be verified by multiplying the factors.

## Difference and Sum of Two Cubes

$$x^3 - y^3 = (x - y)(x^2 + xy + y^2)$$

$$x^3 + y^3 = (x + y)(x^2 - xy + y^2)$$

**TRY THIS**

Verify the sum of two cubes by multiplying $(x + y)(x^2 - xy + y^2)$.

The difference and sum of two cubes expressions can be verified by the following multiplication. (We illustrate for the difference of two cubes.)

$$
\begin{array}{r}
x^2 + xy + y^2 \\
x - y \\
\hline
-x^2y - xy^2 - y^3 \\
x^3 + x^2y + xy^2 \\
\hline
x^3 \qquad\qquad - y^3
\end{array}
$$

Example 6 illustrates the need to recognize the pattern to use when factoring. Notice if the binomial is a difference; look for the variables to be either perfect squares or perfect cubes. If the binomial is a sum, the variables must be perfect cubes to be factorable.

**REMINDER**

For the variables to be perfect squares, the exponents must be divisible by 2. For the variables to be perfect cubes, the exponent must be divisible by 3.

**EXAMPLE 6**   Factoring Special Polynomials

Factor each of the following polynomials:

a) $4x^2 - 25y^2$

b) $x^3 - 8$

c) $9x^4 - y^6$

d) $y^3 + 8x^3$

SOLUTION

a) Factor using the difference of two squares.

$$4x^2 - 25y^2 = (2x)^2 - (5y)^2$$
$$= (2x + 5y)(2x - 5y)$$

b) Factor using the difference of two cubes.

$$x^3 - 8 = x^3 - (2)^3$$
$$= (x - 2)(x^2 + 2x + 4)$$

c) Factor using the difference of two squares.

$$9x^4 - y^6 = (3x^2)^2 - (y^3)^2$$
$$= (3x^2 - y^3)(3x^2 + y^3)$$

d) Factor using the sum of two cubes.

$$y^3 + 8x^3 = y^3 + (2x)^3$$
$$= (y + 2x)(y^2 - 2xy + 4x^2)$$ ∎

**EXAMPLE 7**   Factoring Special Polynomials

Factor each of these perfect square trinomials.

a) $x^2 - 14x + 49$

b) $4x^2 + 12x + 9$

c) $25x^2 + 40xy + 16y^2$

d) $36x^4 - 12x^2 + 1$

SOLUTION

a) $x^2 - 14x + 49 = (x)^2 - 14x + (7)^2 = (x - 7)^2$

b) $4x^2 + 12x + 9 = (2x)^2 + 12x + (3)^2 = (2x + 3)^2$

c) $25x^2 + 40xy + 16y^2 = (5x)^2 + 40xy + (4y)^2 = (5x + 4y)^2$

d) $36x^4 - 12x^2 + 1 = (6x^2)^2 - 12x^2 + (1)^2 = (6x^2 - 1)^2$ ∎

## Exercises for Section 5.5

In Exercises 1–10, identify whether the trinomial is a perfect square. If so, factor the expression.

1. $x^2 - 4x + 4$

2. $x^2 + 6x + 9$

3. $4y^2 + 12y + 9$

4. $9w^2 - 6w + 1$

5. $x^2 + 8x - 16$

6. $4x^2 + 20x + 25$

7. $x^2 + 4xy + 4y^2$

8. $9x^2 - 6xy + y^2$

9. $x^2 + 2\sqrt{2}x + 2$

10. $x^2 + 2\sqrt{5}x + 5$

In Exercises 11–20, factor the trinomial into a product of linear factors.

11. $x^2 - 10x + 21$

12. $x^2 - 2x - 35$

13. $x^2 + x - 12$

14. $x^2 - 8x + 15$

15. $x^2 - 3x - 28$

16. $x^2 - 14x + 48$

17. $x^2 + 9x + 14$

18. $x^2 - 8x - 33$

19. $x^2 - 14x + 45$

20. $x^2 - x - 42$

In Exercises 21–32, factor each of the following expressions completely.

21. $2x^2 + 5x - 3$

22. $5x^2 - 13x - 6$

23. $6x^2 + 5x - 6$

24. $6x^2 - 5x - 6$

25. $6x^2 - 7x - 5$

26. $3x^2 + 13x - 10$

27. $2x^2 + 3ax - 2a^2$

28. $3y^2 - py - 2p^2$

29. $3r^4 - 5r^2s - 2s^2$

30. $12x^2 - 11xy + 2y^2$

31. $3q^3 - 4qx - 6q^2x + 8x^2$

32. $2x^2(x + y)^2 + xy(x + y)^2 - y^2(x + y)^2$

In Exercises 33–42, solve the equation algebraically.

33. $x^2 + x - 20 = 0$

34. $x^2 - 4x - 21 = 0$

35. $x^2 - 3x - 10 = 0$

36. $x^2 - 5x - 36 = 0$

37. $2x^2 + 7x - 4 = 0$

38. $3x^2 - 14x - 5 = 0$

39. $x^2 - 10x + 25 = 0$

40. $4x^2 + 20x + 25 = 0$

41. $2x^3 + x^2 - 3x = 0$

42. $2x^3 + x^2 - 6x = 0$

In Exercises 43–50, factor each expression completely.

43. $x^3 - b^3$

44. $y^3 - 8x^3$

45. $27x^3 + y^3$

46. $x^4 - y^4$

47. $16x^2 - 49$

48. $y^6 - y^3$

49. $1000k^3 + 125$

50. $8x^3 - 64$

51. **Writing to Learn.** The linear factors of $2x^2 - 5x - 12$ are two of the factors from the list $(2x + 3)$, $(2x - 3)$, $(x + 4)$, and $(x - 4)$. Write a paragraph explaining which two of the possible pairs from this list can be eliminated by considering only the fact that the constant is $-12$.

52. **Writing to Learn.** The linear factors of $3x^2 + 4x - 4$ are two of the factors from the list $(x - 2)$, $(x + 2)$, $(3x - 2)$, and $(3x + 2)$. Write a paragraph explaining which two of the possible pairs from this list can be eliminated by considering only the fact that the coefficient of $x^2$ is 3.

TRANSLATING WORDS TO SYMBOLS

In Exercises 53–56, write an equation that is an algebraic representation of the problem.

53. *Using Algebra in Geometry.* A rectangle is 5 in. longer than it is wide. Its area is 126 sq in.

54. *Using Algebra in Geometry.* A rectangle is 8 in. shorter than it is long. Its area is 768 sq in.

55. *Using Algebra in Geometry.* A triangle has a base that is 3 in. longer than the altitude to this base. Its area is 18 sq in.

56. *Using Algebra in Geometry.* A triangle has a base that is 15 in. longer than the altitude to this base. Its area is 144 sq in.

EXTENDING THE IDEAS

In Exercises 57–60, factor the polynomial into a product of two polynomials, each of degree two.

57. $x^4 - 49$      58. $x^4 - x^2 - 6$

59. $3x^4 - 10x^2 - 8$      60. $4x^4 - 15x^2 - 4$

In Exercises 61–64, factor the polynomial into a product of two binomial factors.

61. $x^3 - 3x^2 + 2x - 6$      62. $2x^3 - 3x^2 + 2x - 3$

63. $2x^3 + 4x^2 - x - 2$      64. $x^3 - 5x^2 + 3x - 15$

In Exercises 65 and 66, factor the expression completely.

65. $x^{2t} - y^{2s}$      66. $x^{3k} - y^{3k}$

67. **Writing to Learn.** Explain how the graph of $y = x^2 + 2x + 5$ gives sufficient information to conclude that $x^2 + 2x + 5$ cannot be factored as $(x - a)(x - b)$, where $a$ and $b$ are real numbers.

68. **Writing to Learn.** Explain how the graph of $y = 3x - x^2 - 4$ gives sufficient information to conclude that $(-x^2 + 3x - 4)$ cannot be factored as $(x - a)(b - x)$, where $a$ and $b$ are real numbers.

LOOKING BACK—MAKING CONNECTIONS

69. Explain how the following graph of $y = 6x^2 + 19x + 15$ provides enough information to decide that $(x + 15)$ and $(6x + 1)$ cannot be the factors of $6x^2 + 19x + 15$.

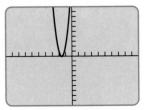

Standard window

For Exercise 69.

70. Explain how the following graph of $y = 2x^2 + 7x + 3$ provides enough information to decide that $(2x - 3)$ and $(x - 1)$ cannot be the factors of $2x^2 + 7x + 3$.

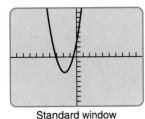

Standard window

For Exercise 70.

---

## 5.6   Solving Polynomial Equations Graphically and Algebraically

Using Algebraic and Graphical Representations in Problem Situations

It is important to note that the factoring methods for solving polynomial equations are practical for only those few polynomials that can be easily factored into linear factors. When an algebraic method fails or is difficult, a graphical solution should be used. For example, the $x$-intercept and the multigraph methods you used to solve linear equations can be applied to polynomial equations.

Example 1 illustrates. Following the accuracy agreement, we shall report the solution with an error of at most 0.01.

### EXAMPLE 1   Solving Equations

Solve the equation $x^3 - 6x^2 + 3x + 3 = 0$.

SOLUTION

A complete graph of $y = x^3 - 6x^2 + 3x + 3$ in Figure 5.36 (a) shows that there are three zeros and hence three solutions to the equation $x^3 - 6x^2 + 3x + 3 = 0$.

To find the solutions, zoom in at each of the $x$-intercepts. Figure 5.36 (b) shows the tick marks as they appear after several zoom-ins. There is a zero between $x = -0.49$ and $x = -0.48$. We report the zero as $x = -0.487$ with an error of at most 0.01.

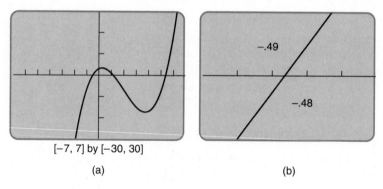

[−7, 7] by [−30, 30]

(a)                                          (b)

**FIGURE 5.36**   a) A graph of $y = x^3 - 6x^2 + 3x + 3$ showing that there are three solutions and (b) after several zoom-ins to find one of the solutions to be $x = -0.487$.

By repeating this process we find the other two solutions. The three solutions to this equation are $x = -0.487$, $x = 1.155$, and $x = 5.332$.   ■

## Using Algebraic and Graphical Representations in Problem Situations

Often a question posed in a problem situation leads to an equation that needs to be solved. The solver needs to decide whether to use an algebraic or a graphical solution method. When the solver is able to factor the polynomial equation into linear factors, it is often reasonable to use an algebraic solution.

Example 4 illustrates a problem that can be reasonably solved algebraically.

**EXAMPLE 2**    THE GARDEN PROBLEM SITUATION: Finding the Length and Width

The area of a rectangular flower garden is 175 sq yd. It is 30 yd longer than it is wide. Find the length and width of the garden.

SOLUTION

Begin by drawing a picture and assigning meaning to variables (see Figure 5.37).

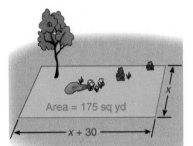

**Width of the garden:** $x$

**Length of the garden:** $x + 30$

Then the equation that represents this problem is

$$x(x + 30) = 175.$$

We solve this equation to find the length and width of the garden.

$$x(x + 30) = 175$$
$$x^2 + 30x - 175 = 0$$
$$(x + 35)(x - 5) = 0$$

$$x + 35 = 0 \qquad x - 5 = 0$$
$$x = -35 \qquad x = 5$$

The solution $x = -35$ has no meaning for this problem since the width is a positive number. Therefore the width is $x = 5$ yd and the length is $5 + 30 = 35$ yd.    ■

**FIGURE 5.37** A rectangle with width $x$ and length $x + 30$.

There are many settings in which it is reasonable to use both algebraic and graphical methods. For example, the **Baseball Problem Situation,** which we now revisit, refers to a dynamic situation—the motion of a ball. The visual understanding provided by the graphical representation of the problem enriches the algebraic representation.

We recommend that a graphical method be used to support the algebraic method of this problem situation, particularly if the solver is not confident of the solution found.

### EXAMPLE 3    THE BASEBALL PROBLEM SITUATION: Finding the Time

A ball is thrown straight up into the air with an initial velocity of 80 ft/sec. The height $h$ after $t$ seconds is $h = -16t^2 + 80t$. At what time after the ball has been thrown is the ball 64 ft above the ground? (See Figure 5.38).

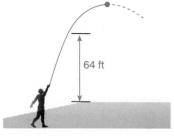

64 ft

**FIGURE 5.38** A ball is thrown straight up into the air.

### SOLUTION

**Solve Algebraically.** The question can be rephrased as follows: For what $t$ is $h = 64$? We need to solve the equation $-16t^2 + 80t = 64$.

$$-16t^2 + 80t = 64$$

$$-16t^2 + 80t - 64 = 0$$

$$-16(t^2 - 5t + 4) = 0$$

$$-16(t - 4)(t - 1) = 0$$

$$t = 4 \quad t = 1$$

The solutions to the equation are $t = 4$ and $t = 1$.

**Support Graphically.** Figure 5.39 provides support for this solution. The ball is 64 ft above the ground after 1 second or 4 seconds. ∎

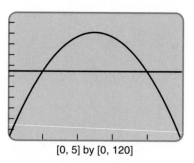

[0, 5] by [0, 120]

**FIGURE 5.39** Graphs of $y = 64$ and $y = -16t^2 + 80t$.

Consider the following problem situation.

---

### THE BOX PROBLEM SITUATION

Squares are cut from the corners of a 20-by-25 in. piece of cardboard, and a box is made by folding up the flaps. (See Figure 5.40.)

---

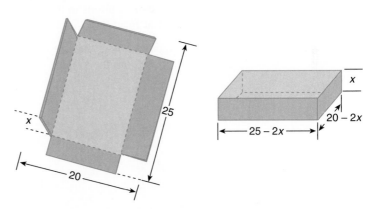

**FIGURE 5.40** Squares are cut from each corner. The flaps are folded up to form a box with no lid.

This Box Problem situation is studied in Example 4.

**EXAMPLE 4**    THE BOX PROBLEM SITUATION: Finding the Width

Find the width $x$ of the square cut from each corner of the cardboard if the volume of the resulting box is 512 sq in.

SOLUTION

**Solve Graphically.**

1. Volume is *length* $\times$ *width* $\times$ *height*. In this case, the width is $20 - 2x$, the length is $25 - 2x$, and the height is $x$. Therefore, an algebraic representation of this problem situation is

$$V = x(20 - 2x)(25 - 2x) = 512.$$

2. Rewrite the equation with one side equal to 0 to solve using the $x$-intercept method.

$$x(20 - 2x)(25 - 2x) = 512$$

$$x(20 - 2x)(25 - 2x) - 512 = 0$$

3. In this problem, $x$ represents the length of the edge of a square. Consequently, it is a positive number. We can see from Figure 5.40 that the largest square that can be cut from a corner of the cardboard has length less than 10. So we restrict our attention to those $x$ satisfying $0 \le x \le 10$. Graph $y_1 = x(20 - 2x)(25 - 2x) - 512$ in a $[0, 10]$ by $[-500, 500]$ window. Figure 5.41 (a) shows a graph of $y_1 = x(20 - 2x)(25 - 2x) - 512$, whose zeros are solutions to this problem. This graph is the graphical representation of the problem situation.

4. Set $X\text{scl} = 0.01$ and use zoom–in to find solutions with an error of at most 0.01. The two solutions to the problem situation are $x = 6.706$ and $x = 1.319$ with error at most 0.01.

If the square corners that are cut from the cardboard have sides that measure 1.319 in. or 6.706 in., the volume of the resulting box is about 512 cu in.

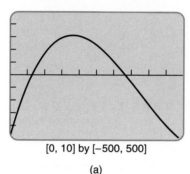

$$1.319 \rightarrow X : X(20 - 2X)$$
$$(25 - 2X)$$
$$512.100489$$
$$6.706 \rightarrow X : X(20 - 2X)$$
$$(25 - 2X)$$
$$511.9477353$$

[0, 10] by [−500, 500]

(a)                    (b)

**FIGURE 5.41**   (a) A graphical representation of the Box Problem with $y = x(20 - 2x)(25 - 2x) - 512$ and (b) the numerical support for the solutions.

**REMINDER**

Notice that a graphical solution is possible for Example 4. The algebraic methods for solution are not possible at the level of Intermediate Algebra in this instance.

**Support Numerically.** Figure 5.41 (b) shows numerical support of the two solutions.   ■

## Exercises for Section  **5.6**

In Exercises 1–14, use an algebraic method to solve the equation.

1. $x^2 - 5x - 24 = 0$

2. $x^2 + 16x + 63 = 0$

3. $x^2 - 4x - 21 = 0$

4. $x^2 - 14x + 45 = 0$

5. $2x^2 + 5x - 12 = 0$

6. $2x^2 - x - 6 = 0$

7. $3x^2 - 10x - 8 = 0$     8. $2x^2 + 5x - 3 = 0$

9. $2x^2 + 3x - 5 = 0$     10. $3x^2 + 11x + 8 = 0$

11. $10x^2 - 11x - 6 = 0$     12. $12x^2 - 23x - 9 = 0$

13. $3x^3 - 7x^2 - 10x = 0$

14. $2x^4 + 13x^3 - 7x^2 = 0$

In Exercises 15–18, use algebraic methods to solve the given equation.

15. $3x^2 + 4x - 3 = x^2 - x$

16. $4x^2 + 3x - 1 = 2x^2 + 4x$

17. $2x^2 - 32x + 90 = 3x + 15$

18. $2x^2 - 42x + 300 = x^2 - 60$

In Exercises 19–24, use a graphical method to find three solutions for the polynomial equations. Find the appropriate graph in the $[-5, 5]$ by $[-10, 10]$ window and zoom in as necessary.

19. $x^3 - 5x^2 + 7x - 2 = 0$

20. $x^3 - 4x^2 + 3x + 2 = 0$

21. $x^3 + 2x^2 - 5x - 4 = 0$

22. $x^3 + x^2 - 3x - 2 = 0$

23. $2x^3 - x^2 - 3x + 1 = 0$

24. $2x^3 - 3x^2 - 2x + 1 = 0$

TRANSLATING WORDS TO SYMBOLS

In Exercises 25–28, write an equation that describes the problem.

25. The product of an integer and five greater than the integer is fourteen.

26. The product of an integer and three greater than the integer is forty.

27. The product of an integer and nine less than the integer is ninety.

28. The product of an integer and seven less than the integer is eighteen.

29. Find the two numbers referred to in Exercise 25.

30. Find the two numbers referred to in Exercise 26.

31. Find the two numbers referred to in Exercise 27.

32. Find the two numbers referred to in Exercise 28.

33. The length of a rectangular room is 10 ft longer than it is wide. If the area of the room is 1200 sq ft, what are the length and width of the room?

34. The length of a rectangular field is 100 yd longer than it is wide. If the area of the field is 300,000 sq ft, what are the length and width of the field?

In Exercises 35–38, solve the given problem.

35. *A Baseball Problem.* A baseball is thrown straight up into the air with an initial velocity of 128 ft/sec.

a) Write an algebraic representation that describes the height of the ball $t$ seconds after the ball is thrown.

b) Write an equation that must be satisfied when the ball is 192 ft off the ground.

c) At what time after the ball has been thrown is it 192 ft off the ground? Solve this problem using either an algebraic or graphical method. Use both if possible.

36. *A Baseball Problem.* A baseball is thrown straight up into the air with an initial velocity of 96 ft/sec.

a) Write an algebraic representation that describes the height of the ball $t$ seconds after the ball is thrown.

b) Write an equation that must be satisfied when the ball is 96 ft off the ground.

c) At what time after the ball has been thrown is the ball 96 ft off the ground? Solve this problem using either an algebraic or graphical method. Use both if possible.

37. *A Box Problem.* A box is made by cutting squares from a 30-by-25 in. piece of cardboard with the flaps being folded up.

a) Write an algebraic representation for the volume of this box when the length of the square cut from the corner is $x$ inches.

b) Write an equation that must be solved to find when the volume of the box is 800 cu in.

c) Find a graphical representation of this problem situation.

d) Find the width $x$ of the square cut from each corner of the cardboard if the volume of the resulting box is 800 cu in.

**38.** *A Picture Frame Problem.* A picture frame is 18 in. by 22 in. How wide should the mat that surrounds the picture be if the area of the picture is 380 sq in.?

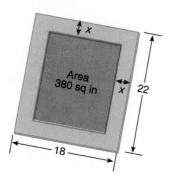

**EXTENDING THE IDEAS**

**39.** *A Box Problem.* Squares are cut from a square piece of cardboard 12 in. on a side.

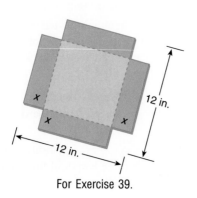

For Exercise 39.

a) Find the algebraic representation $A(x)$ of the area of the bottom of the box and the algebraic representation $V(x)$ of the volume of the box when $x$ is the length of one side of the square cut from the corner.

b) Find a graphical representation of both $A(x)$ and $V(x)$ in the window $[0, 6]$ by $[0, 130]$.

c) What is the length $x$ of the square cut from the corner of the cardboard when the area of the bottom of the box is 70 sq in.?

d) What is the volume of the box when the area of the bottom of the box is 70 sq in.?

e) Are the number of square units of area of the base ever equal to the number of cubic units of volume to the box? Explain how the graphical representations in part (b) help answer this question.

**LOOKING BACK—MAKING CONNECTIONS**

**40.** If the area of the triangle in the following figure is 8 square units, what are the lengths of the sides of the triangle?

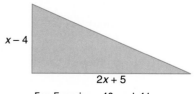

For Exercises 40 and 41.

**41.** Did you use an algebraic or a graphical solution method for Exercise 40? Explain your choice.

**42.** If the area of the shaded region in the following figure is 25 square units, what is the value of $x$? Find the dimensions of the two rectangular regions.

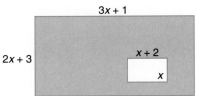

For Exercises 42 and 43.

**43.** Did you use an algebraic or a graphical solution method for Exercise 42? Explain your choice.

## Chapter 5 Summary

| | Exponents | Examples |
|---|---|---|
| $b^n$ | A mathematical expression for any real number $b$ and any positive integer $n$, where $$b^n = \underbrace{b \cdot b \cdot b \cdot \ldots \cdot b}_{n \text{ factors of } b}.$$ $b$ is called the base and $n$ the exponent. | $3^4$ represents the product of $3 \cdot 3 \cdot 3 \cdot 3$. |
| Zero Exponent | $b^0 = 1$ if $b$ is any nonzero real number. | $8^0 = 1$ |
| Negative Exponent | $b^{-n} = \dfrac{1}{b^n}$ if $b$ is any nonzero real number and $n$ is any positive integer. | $4^{-2} = \dfrac{1}{4^2}$ |

| | Properties of Exponents | Examples |
|---|---|---|
| Product Rule for Exponents | $b^n \cdot b^m = b^{n+m}$ if $m$ and $n$ are positive integers, $b$ is any real number, and $b \neq 0$. | $2^5 \cdot 2^4 = 2^9$ |
| Quotient Rule for Exponents | $\dfrac{b^n}{b^m} = b^{n-m}$ or $\dfrac{b^n}{b^m} = \dfrac{1}{b^{m-n}}$ if $m$ and $n$ are integers, $b$ is any real number, and $b \neq 0$. | $\dfrac{6^7}{6^3} = 6^{7-3} = 6^4$ $\dfrac{8^2}{8^5} = \dfrac{1}{8^{5-2}} = \dfrac{1}{8^3}$ |
| Power Rules for Exponents | If $a$ and $b$ are any nonzero real numbers and $n$ is an integer, then the following rules are true: <br> 1. $(ab)^n = a^n \cdot b^n$ | $(2x)^3 = 2^3 x^3$ |

| | **Properties of Exponents** | **Examples** |
|---|---|---|
| | 2. $\left(\dfrac{a}{b}\right)^n = \dfrac{a^n}{b^n}$ | $\left(\dfrac{x}{4}\right)^2 = \dfrac{x^2}{4^2}$ |
| | 3. $(a^n)^m = a^{nm}$ | $(3^5)^2 = 3^{10}$ |

| | **Scientific Notation** | **Examples** |
|---|---|---|
| Using Scientific Notation | A representation of a number in the form $a \times 10^n$, where $1 \le a < 10$ and $n$ is an integer. | 153 is $1.53 \times 10^2$ in scientific notation. |
| Changing a Number to Scientific Notation | 1. Place an arrow "↑" to the right of the first nonzero digit. <br> 2. The number of digits between the arrow and the decimal point is the exponent $n$. <br> 3. If the arrow is to the left of the decimal point, $n$ is positive. If the arrow is to the right of the decimal point, $n$ is negative. | Change 0.000035 to scientific notation: <br><br> 1. Place a "↑" between 3 and 5: 0.00003 ↑5. <br><br> 2. There are five digits between "↑" and ".". <br><br> 3. Because "↑" is to the right of ".", $n$ is negative. Therefore $0.000035 = 3.5 \times 10^{-5}$. |
| Computing with Scientific Notation | 1. Change all numbers to scientific notation. <br> 2. Compute the base-10 calculations with the properties of exponents. Use a grapher to perform operations on the decimal numbers between 1 and 10. <br> 3. Write the answer in either scientific notation or standard notation. | Use scientific notation to compute. <br><br> $\dfrac{(0.25)(40,000)}{0.005} =$ <br><br> $= \dfrac{(2.5 \times 10^{-1}) \cdot (4 \times 10^4)}{5 \times 10^{-3}}$ <br><br> $= \dfrac{(2.5)(4)}{5} \cdot \dfrac{10^{-1} \cdot 10^4}{10^{-3}}$ <br><br> $= 2 \times 10^{-1+4-(-3)}$ <br><br> $= 2 \times 10^6 = 2,000,000$ |

| | **Polynomial Expressions** | **Examples** |
|---|---|---|
| Polynomial Expression in $x$ | An expression that can be written in the form <br><br> $a_n x^n + a_{n-1} x^{n-1} + \cdots + a_0$ | $4x^3 + 3x^2 - 2$ |

| Monomial | A polynomial expression with one term (with perhaps several variables) | $53s^4t$ or $8x^5$ |
|---|---|---|
| Binomial | A polynomial expression with two terms | $2x^5 - 8y^2$ |
| Trinomial | A polynomial expression with three terms | $2x^3 - 8x^2 + 5$ |
| Degree of a Term of a Polynomial in $x$ | The exponent of the variable of the term | The degree of $\sqrt{2}x^5$ is 5. |
| Degree of a Polynomial in $x$ | The degree of the term with the greatest exponent | $5x^2 - 2x^3$ is a polynomial of degree 3. |

| | **Operations on Polynomials** | **Examples** |
|---|---|---|
| Adding Polynomials | To add polynomials, add like terms using the Commutative and Associative Properties of Addition and the Distributive Property. | $(3x^3 - x) + (2x^3 + 4x) =$ <br> $(3x^3 + 2x^3) + (-x + 4x) =$ <br> $(3 + 2)x^3 + (-1 + 4)x =$ <br> $5x^3 + 3x$ |
| Subtracting Polynomials | Change subtraction to addition by using the rule $a - b = a +^- b$. | $(4x + 1) - (2x - 4) =$ <br> $4x + 1 +^- (2x - 4) =$ <br> $4x + 1 - 2x + 4 =$ <br> $(4 - 2)x + (1 + 4) =$ <br> $2x + 5$ |
| Multiplying Binomials | Use the Distributive Property to multiply one binomial by each term in the other binomial and then simplify the result. | $(x + 5)(3x - 1) =$ <br> $(x + 5)(3x) - (x + 5)(1) =$ <br> $3x^2 + 15x - x - 5 =$ <br> $3x^2 + 14x - 5$ |

| | Operations on Polynomials | Examples |
|---|---|---|
| FOIL Method of Multiplying Binomials | "FOIL" represents First, Outer, Inner, and Last. | $(3x - 1)(2x + 2)$ |

F: first
O: outer
I: inner
L: last

$$\overbrace{(3x)(2x)}^{\text{F}} + \overbrace{(3x)(2)}^{\text{O}} +$$

$$\overbrace{(-1)(2x)}^{\text{I}} + \overbrace{(-1)(2)}^{\text{L}} =$$

$$6x^2 + 6x - 2x - 2 =$$

$$6x^2 + 4x - 2$$

| | | |
|---|---|---|
| Square of a Binomial | $(x + y)^2 = x^2 + 2xy + y^2$ <br> $(x - y)^2 = x^2 - 2xy + y^2$ | $(x + 3)^2 = x^2 + 6x + 9$ <br> $(5 - x)^2 = 25 - 10x + x^2$ |
| Product of a Sum and a Difference | $(x + y)(x - y) = x^2 - y^2$ | $(2y + 3)(2y - 3) = 4y^2 - 9$ |

| | Factoring Expressions in Special Forms | Examples |
|---|---|---|
| Factoring Perfect Square Trinomials | $x^2 + 2xy + y^2 = (x + y)^2$ | $4x^2 + 12x + 9 = (2x)^2 + 4x + 3^2 = (2x + 3)^2$ |
| | $x^2 - 2xy + y^2 = (x - y)^2$ | $x^2 - 6xy + 9y^2 = (x - 3y)^2$ |
| Factoring the Difference of Two Squares | $x^2 - y^2 = (x + y)(x - y)$ | $36 - x^2 = (6 + x)(6 - x)$ |
| Factoring the Difference of Two Cubes | $x^3 - y^3 = (x - y)(x^2 + xy + y^2)$ | $8x^3 - 27 = (2x)^3 - 3^3 = (2x - 3)(4x^2 + 6x + 9)$ |
| Factoring the Sum of Two Cubes | $x^3 + y^3 = (x + y)(x^2 - xy + y^2)$ | $x^3 + 125 = x^3 + 5^3 = (x + 5)(x^2 - 5x + 25)$ |

| Factoring by Grouping | **Factoring More General Expressions** | **Examples** |

**Factoring by Grouping**

Given a polynomial with four terms, group in pairs, and factor each group completely. If the two groups then have a common factor, factor it out.

Factor:  $3x^2 - 2x + 15x - 10 =$

$(3x^2 - 2x) + (15x - 10) =$

$x(3x - 2) + 5(3x - 2) =$

$(3x - 2)(x + 5)$

**Factoring Trinomials of the Form $ax^2 + bx + c$**

*Using a Graph*
1. Graph the trinomial and identify the potential zeros. If $a$ is a zero, then one of the factors would be $(x - a)$.
2. If necessary find the other factor by considering the first (F) and the last (L) steps of the FOIL process.
3. Confirm the factorization using FOIL.

*Trial-and-error Method*
1. List all integer factors of $a$, the coefficient of $x^2$.
2. List all integer factors of $c$, the constant term.
3. Try various pairs of the above factors using the FOIL process to complete the factorization.

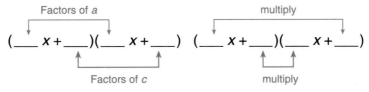

The sum of these products $= b$.

| | **Solving Polynomial Equations** | **Examples** |
|---|---|---|
| Zero-product Property | $a \cdot b = 0$ if and only if $a = 0$ or $b = 0$. | $(x - 3)(x + 5) = 0$ if and only if $x - 3 = 0$ or $x + 5 = 0$. |
| Solving Algebraically | 1. Rewrite the equation, if necessary, so that one side equals zero.<br>2. Combine all like terms and factor the resulting expression, if possible.<br>3. Set each factor equal to zero and use the Zero-product Property to find the solutions. | Solve $3x^2 + 5x = x^2 + 3$:<br>$3x^2 + 5x - x^2 - 3 = 0$<br><br>$2x^2 + 5x - 3 = 0$<br><br>$(2x - 1)(x + 3) = 0$<br><br>$2x - 1 = 0$ or $x + 3 = 0$<br><br>The solutions are $x = \frac{1}{2}$ and $x = -3$. |
| Solving Graphically | The multigraph and $x$-intercept methods can be used to solve an equation graphically.<br><br>A graphical method can be used to provide support for an algebraic method. It also can be used when an algebraic method fails or is difficult. | Solve $x^3 + 5 = 3x$ by finding the graphs of $y = x^3 + 5$ and $y = 3x$. |
| Solving a Polynomial Problem Situation | 1. Identify the quantity that is to be found.<br>2. Write an equation involving $x$ from information in the problem situation.<br>3. Solve the equation algebraically or graphically and use the solution of this equation to solve the problem. | 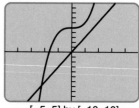<br>[−5, 5] by [−10, 10]<br>Find the $x$ coordinate of the point of intersection. |

## Review Exercises for Chapter 5

In Exercises 1 and 2, evaluate the expression either mentally or using paper and pencil.

**1.** $3^{-2}(5^2 - 7^0)$

**2.** $\dfrac{10^2}{6^{-1}}$

In Exercises 3–6, use properties of exponents to simplify the expression. No negative exponents should be used.

**3.** $(2a^2x)^{-3}(3a^{-3}x^5)^3$

**4.** $(x^{-2}y^{-4})^0(x^3y^{-5})(x^{-2}y^6)^2$

**5.** $\left(\dfrac{a^{-3}b}{x^2y}\right)\left(\dfrac{a^4y^3}{x^{-3}b^3}\right)$   **6.** $\left(\dfrac{2x^4y^{-2}}{x^{-3}y}\right)^2$

**7.** Write 0.0800154 in scientific notation.

**8.** Write $3.2058 \times 10^5$ in standard notation.

**9.** Use scientific notation to compute $\dfrac{(2,950,000)(810)}{0.0003}$.

**10.** Answer the following questions about the polynomial expression $\dfrac{1}{2}x^4 - 35x^8 + 25$:

  a) Is the expression a monomial, binomial, or trinomial?

  b) What is the degree of the expression?

  c) What is the degree of the term $\dfrac{1}{2}x^4$?

**11.** Evaluate $3x^3 - 4x^2 + 5x - 2$ at $x = -1$.

In Exercises 12–14, simplify the expression. Perform the operations horizontally or vertically.

**12.** $(4x^4 - 3x^2 + x) + (2x^4 + 3x^2 - 2x + 5)$

**13.** $(12x^2y^3 - 3xy^2 + 5) - (6x^2y^3 - 5xy^2 + 2)$

**14.** $[2(x^2 - 3x) - 3x(5 - 4x)] + 10x(x - 2)$

In Exercises 15–17, find the product of the expression.

**15.** $(x + 5)(x^2 - 7)$   **16.** $(3 - 2x)(5 - 8x)$

**17.** $(x - 1)(x^2 + 3x + 4)$

**18.** List all factors of $5xy$.

**19.** List all factors of $x^2(x - 3)$.

**20.** Find the greatest common factor of the set of terms $\{20x^2yz^3,\ 28x^3y^3,\ -40x^6y^3z\}$.

**21.** Factor out the greatest common factor for the expression $15x^2(x - 5) + 6x(x - 5)^2$.

In Exercises 22–30, factor the expression completely.

**22.** $x^3 + 9x^2 + 14x$   **23.** $6x^2 - 7x - 20$

**24.** $x^2 - 12x + 36$   **25.** $2s^2 - 3st - 2t^2$

**26.** $8x^2 + 15x - 2$   **27.** $16x^2 - 1$

**28.** $x^3 + 1$   **29.** $9t^2 + 30t + 25$

**30.** $20y^3 + 68y^2 - 7y$

**31.** Factor by grouping: $3x^2 + 5x - 18x - 30$

**32.** Factor by grouping: $x^3y - 3xy + x^2 - 3$

In Exercises 33–36, solve the given equation algebraically and support the solution graphically.

**33.** $2x^2 + 7x + 3 = 0$

**34.** $5x^3 - x^2 = 4x^3 - 2x^2$

**35.** $2x(x - 4) - (x - 4) = 0$

**36.** $(x + 7)(2x + 5)(x - 3) = 0$

In Exercises 37 and 38, solve the equation graphically.

**37.** $x^2 + 5 = 5x$

**38.** $x^3 + x^2 - 3x - 3 = 0$

**39.** *Container Manufacturing.* Which cylinder has the greater volume: one with radius 1.5 in. and height 4 in. or one with radius 2 in. and height 2.5 in.?

**40.** *Container Manufacturing.* Squares are cut from a 24-by-36 in. piece of cardboard with the flaps being folded up to form a box. Write an algebraic representation for the volume of this box if the length of the square cut is $x$ inches.

**41.** The sum of a number and its square is 52.

  a) Write an equation that represents this problem.

  b) Determine the negative number that is a solution to this problem.

## Chapter 5 Test

1. Evaluate using paper and pencil: $(2^3 + 1)^2 - 5^0$.

2. Use the properties of exponents to simplify
$$\frac{x^2 y}{x^{-1} y^2 z}.$$

3. Use the properties of exponents to simplify $(a^2 b^{-3})^2 (2ab)^{-2}$ so that negative exponents appear.

4. Write 21,400 in scientific notation and write $6.27 \times 10^{-4}$ in standard notation.

In Questions 5 and 6, simplify each expression by performing the indicated operations.

5. $2x(x^2 - 4x + 3) - 3(2x^3 + 5x - 4)$

6. $(x + 2)(x^2 - 2x + 5)$

7. Simplify by squaring: $(2x - 3)^2$

8. Find the greatest common factor of the set of terms $\{15r^3 s^2, 25r^4 s^5, 12rs^3\}$.

9. Factor: $2x^2 - x - 10$

10. Factor: $8x^3 - 27$

11. Factor by grouping: $x^2 - bx + ax - ab$

12. Solve algebraically: $x^3 - x = 0$. Support graphically.

13. Solve graphically: $5x^2 - 2x^3 = 5$. Support graphically.

14. The product of an integer and six less than the integer is 27.

  a) Write an equation that describes this statement.

  b) Find all pairs of integers that make this statement true.

15. A rectangular garden is 3 ft longer than twice the width. Find the length and width of the garden if the area measures 464 sq ft.

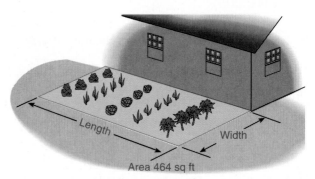

For Question 15.

16. Graph $y = 3x^2 - 19x + 20$ to find a possible linear factor. Then, complete the factorization of $3x^2 - 19x + 20$.

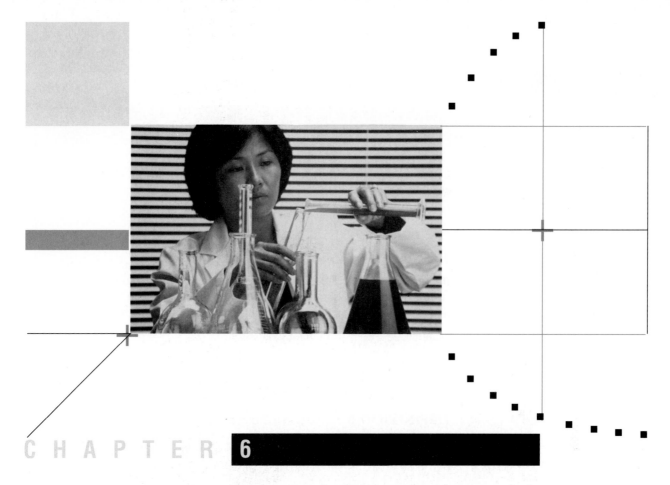

# Rational Expressions and Equations

### AN APPLICATION

Suppose that $x$ liters of distilled water are added to 5 liters of a 30% solution. The resulting solution has a concentration that can be found using the equation

$$P = \frac{1.5}{5 + x}.$$

## 6.1 Simplifying Rational Expressions

Rational Expressions ▪ Equivalent Rational Expressions ▪ Using the Grapher to Simplify a Rational Expression

Until now we have worked with numerical fractions. In this chapter we study algebraic fractions, or rational expressions.

### Rational Expressions

A **rational number** is a quotient of two integers with the denominator not zero. A **rational expression,** like a rational number, is a quotient; however, it is a quotient of two polynomial expressions. Again, the denominator cannot be zero. For example, $\dfrac{2x - 3}{x^2 + 1}$ is a rational expression.

---

**DEFINITION 6.1**   Rational Expression

A **rational expression** is one that can be written in the form

$$\frac{p(x)}{q(x)},$$

where $p(x)$ and $q(x)$ are polynomials and $q(x) \neq 0$.

---

Here are some additional examples of rational expressions:

$$\frac{1}{x}, \quad \frac{2x}{x + 5}, \quad \frac{x^2 + 3x - 1}{x + 7}$$

A rational expression is evaluated when a number replaces the variable of the expression. The **domain** of a rational expression consists of all values for which the expression is defined. That is, all values are in the domain except those that make the denominator zero. When a value is not in the domain of the expression, it is identified by using the $\neq$ symbol as shown.

$$\frac{x^2 + 1}{x - 4}, \quad x \neq 4 \qquad\qquad \frac{x + 1}{(x - 3)(x + 2)}, \quad x \neq 3, x \neq -2$$

## Equivalent Rational Expressions

The following Property of Equivalent Rational Numbers is used to simplify both rational numbers and rational expressions.

**GRAPHER NOTE**

If you attempt to evaluate a rational expression for a number that gives 0 in the denominator, the grapher will give an error message.

---

### Property of Equivalent Rational Expressions

If $a$, $b$, and $c$ represent numbers or algebraic expressions, $b \neq 0$ and $c \neq 0$, then

$$\frac{a \cdot c}{b \cdot c} = \frac{a}{b}.$$

---

A rational number or a rational expression in the form $\frac{a}{b}$ is said to be reduced to **simplest terms,** or **lowest terms,** if the numerator $a$ and denominator $b$ have no factors in common.

Example 1 reviews how this Property of Equivalent Rational Expressions is used to reduce a rational number to lowest terms. The next examples then apply the same property to rational expressions.

**REMINDER**

The greatest common factor (GCF) of two integers is the largest integer that is a factor of each of the numbers. We also can think of it as the largest integer that divides into both numbers evenly.

**EXAMPLE 1**    Reducing a Numerical Fraction

Reduce the following to lowest terms:

a) $\dfrac{4}{6}$

b) $\dfrac{625}{100}$

SOLUTION

Factor the greatest common factor (GCF) from both the numerator and denominator.

a) The GCF$(4, 6) = 2$. Therefore

$$\frac{4}{6} = \frac{2 \cdot 2}{2 \cdot 3} \qquad \textbf{Use the Property of Equivalent Rational Expressions.}$$

$$= \frac{2}{3}.$$

b) The GCF(625, 100) = 25. Therefore

$$\frac{625}{100} = \frac{25 \cdot 25}{25 \cdot 4} \qquad \text{Use the Property of Equivalent Rational Expressions.}$$

$$= \frac{25}{4}.$$

A rational expression is reduced to lowest terms by identifying the GCF of the numerator and the denominator and using the Property of Equivalent Rational Expressions. Example 2 illustrates.

**EXAMPLE 2**    Reducing a Rational Expression

Reduce the following rational expression to lowest terms:

$$\frac{25x^2 y}{5xy}$$

SOLUTION

The numerator and denominator both have a factor of $5xy$.

$$\frac{25x^2 y}{5xy} = \frac{5xy \cdot 5x}{5xy \cdot 1} = 5x$$

In order to apply this property when reducing to lowest terms, we need to identify any factors common to both numerator and denominator. In Example 3, a rational expression is reduced to lowest terms by using the Distributive Property.

**EXAMPLE 3**    Simplifying a Rational Expression

Simplify the following rational expression. Support the solution graphically.

$$\frac{2x^2 - 4x}{2x^2}$$

SOLUTION

**Solve Algebraically.**  Factor the common factor, $2x$, out of the numerator.

$$\frac{2x^2 - 4x}{2x^2} = \frac{2x(x - 2)}{2x^2} \qquad \text{Divide out the common factors.}$$

$$= \frac{x - 2}{x}, \quad x \neq 0$$

**GRAPHER NOTE**

To support our algebra, we will use grapher support as we did in the previous chapters. Most graphs in this chapter will consist of two unconnnected branches, as shown in Figure 6.1. When a second expression appears to have an identical graph, we say that the graphs support a claim that the algebra expressions are equivalent.

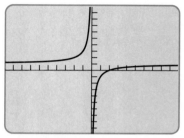

Standard window

**FIGURE 6.1** A graph of $y_1 = (2x^2 - 4x)/2x^2$ and $y_2 = (x - 2)/x$.

**Support Graphically.** Figure 6.1 shows the graph of both $y_1 = (2x^2 - 4x)/2x^2$ and $y_2 = (x - 2)/x$ graphed in the Standard window. They appear to be identical, thus supporting the algebra. ∎

Notice that on the last line of the simplification, the domain of the expression is identified with the statement $x \neq 0$. Throughout the remainder of the text, this type of statement will be assumed and will not be specifically stated.

The fact that the graphs of two different expressions in Figure 6.1 appear to be the same supports our algebraic claim that the two expressions are equivalent. However, it does not ensure that the simplified expression is in lowest terms. Neither does it ensure that there might not be additional common linear factors. It only suggests that the two expressions are equivalent.

**REMINDER**

Parentheses play a vital role in defining rational functions in the grapher.

**EXAMPLE 4**    Simplifying a Rational Expression

Simplify the following rational expression. Support the solution graphically.

$$\frac{2x^2 + x - 15}{3x + 9}$$

SOLUTION

**Solve Algebraically.** Factor the numerator and denominator.

$$\frac{2x^2 + x - 15}{3x + 9} = \frac{(x + 3)(2x - 5)}{3(x + 3)}$$    Use the Property of Equivalent Rational Expressions with a common factor of $x + 3$.

$$= \frac{2x - 5}{3}$$    Simplest form

**Support Graphically.** Figure 6.2 shows the graphs of both $y_1 = (2x^2 + x - 15)/(3x + 9)$ and $y_2 = (2x - 5)/3$ in the 0.1 window. The two graphs appear to be identical, thus providing support for the algebraic simplification. ∎

0.1 window

**FIGURE 6.2** Graphs of $y_1 = (2x^2 + x - 15)/(3x + 9)$ and $y_2 = (2x - 5)/3$ appear to be the same.

## Using the Grapher to Simplify a Rational Expression

A grapher was used in Chapter 5 to help find linear factors of a polynomial. This visual method depends on the equivalence of these statements:

- $x - a$ is a factor of the polynomial;
- $x = a$ is a zero of the polynomial;
- $(a, 0)$ is an $x$-intercept of the graph of the polynomial.

So $x - a$ is a linear factor of both the numerator and denominator if the graphs of both numerator and denominator have $(a, 0)$ as an $x$-intercept. The following exploration asks you to determine visually which rational expressions have common linear factors.

**EXPLORE WITH A GRAPHER**

For each of the following rational functions, graph the numerator and denominator at the same time in the Standard window. Decide whether the given rational expressions have any common linear factors. (Zoom in if necessary.)

1. $\dfrac{x^2 + x - 2}{x^2 + 6x - 7}$

2. $\dfrac{x - 4}{x^2 - 9}$

3. $\dfrac{2x^2 + 3x - 9}{2x^3 - 7x^2 + 9}$

**Generalization:** State in your own words how you can tell visually when the numerator and denominator of a rational expression have any common linear factors.

The graphs of the numerator and the denominator in the exploration's first and third expressions intersect on the $x$-axis. Observe that in these two cases, the numerator and denominator have a common factor. If the graphs of both the numerator and the denominator cross the $x$-axis at the point $x = a$, then $x - a$ is a factor of each of them. This visual technique for finding the common linear factor will be used in Example 5.

You might have concluded that a rational expression is reduced to lowest terms if the graphs of the numerator and the denominator polynomials do not have common $x$-intercepts. This is not necessarily true. It simply means that the numerator and the denominator have no common *linear* factors; they might share a nonlinear factor. You will see an example of sharing a common nonlinear factor in the Extending the Ideas exercises in this section.

**EXAMPLE 5**    Simplifying an Expression: Using a Grapher

Simplify the following rational expression:

$$\frac{x^2 - 25}{x^3 - 125}$$

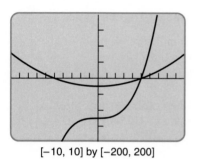

[−10, 10] by [−200, 200]

**FIGURE 6.3** Graphs of $y_1 = x^2 - 25$ and $y_2 = x^3 - 125$ suggest a common $x$-intercept $x = 5$.

Use a grapher to help find any common linear factors. Support the algebraic simplification graphically.

SOLUTION

**Solve Algebraically.** Figure 6.3 shows the graphs of both $y_3 = x^2 - 25$ and $y_4 = x^3 - 125$ in the window $[-10, 10]$ by $[-200, 200]$. Because both curves appear to have an $x$-intercept at $x = 5$, we expect $x - 5$ to be a factor of both the numerator and the denominator. The factorization below confirms that it is true.

$$\frac{x^2 - 25}{x^3 - 125} = \frac{(x - 5)(x + 5)}{(x - 5)(x^2 + 5x + 25)} \qquad \textbf{Divide out the common } (x - 5) \textbf{ factor.}$$

$$= \frac{x + 5}{x^2 + 5x + 25}$$

**Support Graphically.** Figure 6.4 shows graphs of both $y_3 = (x^2 - 25)/(x^3 - 125)$ and $y_4 = (x + 5)/(x^2 + 5x + 25)$. Because both graphs appear to be identical, we have support of the algebraic simplification.    ∎

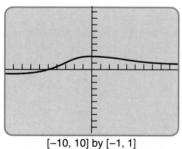

[−10, 10] by [−1, 1]

**FIGURE 6.4** Graphs of $y_3 = (x^2 - 25)/(x^3 - 125)$ and $y_4 = (x + 5)/(x^2 + 5x + 25)$ appear to be identical.

We ask, is the expression

$$\frac{x + 5}{x^2 + 5x + 25}$$

in lowest terms? If not, the numerator $x + 5$ also would be a factor of the denominator. We could graph both $x + 5$ and $x^2 + 5x + 25$ to see if they have the common zero $x = -5$. However, that graph is not necessary. For if $x + 5$ and $x^2 + 5x + 25$ did have a common zero of $x = -5$, it also would have shown up as a common zero of the original graph depicted in Figure 6.3.

We have just illustrated a method of reducing an expression to lowest terms. That method is summarized.

---

### Reducing Rational Expressions to Lowest Terms

1. Factor the numerator and the denominator completely, using a grapher when appropriate to help find common linear factors.
2. Simplify the expression by factoring using the Property of Equivalent Rational Expressions.
3. (Optional) Support graphically by graphing the original expression and the simplified expression in the same window. (If both graphs appear to be identical, we have graphical support for the algebra.)

---

Example 6 shows that a grapher can be used as support for a claim that a rational expression is in lowest terms.

**EXAMPLE 6**    Obtaining Rational Expression in Reduced Form

Use a grapher to obtain graphical support that

$$\frac{x^2 + 5x + 5}{x^2 + 3x + 9}$$

is in lowest terms.

SOLUTION

Figure 6.5 shows a complete graph of both the numerator and denominator. It is visually evident that the numerator and denominator do not have equal zeros. This means that they also do not have any common linear factors. Consequently $\frac{x^2 + 5x + 5}{x^2 + 3x + 9}$ is in lowest terms.  ■

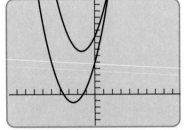

[−10, 10] by [−5, 15]

**FIGURE 6.5** Graphs of $y_1 = x^2 + 5x + 5$ and $y_2 = x^2 + 3x + 9$.

Example 7 also illustrates the technique of factoring a $(-1)$ from an expression to obtain identical factors.

**EXAMPLE 7**    Factors with Opposite Signs

Reduce the following expression to lowest terms, if it is not already reduced. Support the solution graphically.

$$\frac{6x^2 + 7x - 20}{28 - 17x - 3x^2}$$

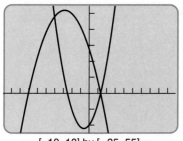

[−10, 10] by [−25, 55]

**FIGURE 6.6** A graph of $y_1 = 6x^2 + 7x - 20$ and $y_2 = 28 - 17x - 3x^2$.

SOLUTION

Figure 6.6 shows the graph of $y_1 = 6x^2 + 7x - 20$ and $y_2 = 28 - 17x - 3x^2$, which appear to have a common $x$-intercept between $x = 1$ and $x = 2$. This suggests that the expression is not reduced to lowest terms. We proceed to factor the numerator and denominator.

$$\frac{6x^2 + 7x - 20}{28 - 17x - 3x^2} = \frac{(3x - 4)(2x + 5)}{(4 - 3x)(7 + x)} \quad \begin{array}{l} \text{Factor } -1 \text{ from } (3x - 4) \\ \text{to obtain the common} \\ \text{factor of } (4 - 3x). \end{array}$$

$$= \frac{-1(4 - 3x)(2x + 5)}{(4 - 3x)(7 + x)}$$

$$= \frac{-(2x + 5)}{(7 + x)}, \quad \text{or}$$

$$= \frac{-2x - 5}{x + 7}$$

It is evident by inspection of this expression that there are no additional common linear factors of the numerator and the denominator. ∎

**REMINDER**

Remember, if two factors have opposite signs throughout, we can factor out $-1$ from one of them in order to have common factors. For example, $(6 - x) = -1(x - 6)$. We can factor the $-1$ out of either the numerator or denominator; however, it is a good practice to factor it out of the factor in the numerator since we tend to leave the negative with the numerator in the final answer.

When a rational expression includes more than one variable, the same algebraic methods are used to reduce it to simplest terms. Grapher methods, however, are not possible. Example 8 illustrates such a case.

**EXAMPLE 8** Reducing Expressions with Two Variables

Simplify the following expression:

$$\frac{x^2 - 2xy + y^2}{3x^2 + 5xy - 8y^2}$$

SOLUTION

Factor the numerator and denominator.

$$\frac{x^2 - 2xy + y^2}{3x^2 + 5xy - 8y^2} = \frac{(x - y)(x - y)}{(x - y)(3x + 8y)} \quad \begin{array}{l} \text{Divide out the common} \\ (x - y) \text{ factor.} \end{array}$$

$$= \frac{x - y}{3x + 8y}$$ ∎

When the expressions contain more than one variable, grapher support cannot be provided. This fact emphasizes that grapher skills will not take the place of algebraic skills.

## Exercises for Section 6.1

In Exercises 1–6, reduce the rational number to lowest terms.

1. $\dfrac{15}{21}$

2. $\dfrac{14}{63}$

3. $\dfrac{42}{48}$

4. $\dfrac{30}{54}$

5. $\dfrac{330}{528}$

6. $\dfrac{385}{210}$

In Exercises 7–10, use a grapher to identify which expression is equivalent to the given rational expression, then confirm algebraically.

7. Is $\dfrac{x^2 - 1}{x - 1}$ equivalent to

   a) $x$, or

   b) $x + 1$?

8. Is $\dfrac{x + 4}{x + 5}$ equivalent to

   a) $\dfrac{4}{5}$, or

   b) it cannot be further simplified?

9. Is $\dfrac{x^2 - 6x + 9}{x - 3}$ equivalent to

   a) $x - 3$, or

   b) $x^2 - 6 - 3$?

10. Is $\dfrac{2x + 6}{2x}$ equivalent to

    a) 6, or

    b) $\dfrac{x + 3}{x}$?

11. The graphs of $y = x^2 - 3x + 2$ and $y = x^2 + x - 6$ are shown in the following figure. Based on this graph, what common linear factor do you expect the numerator and denominator of

$$\frac{x^2 - 3x + 2}{x^2 + x - 6}$$

to have? Confirm algebraically.

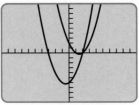

Standard window

For Exercise 11.

12. The graphs of $y = x^2 - 4$ and $y = -x^2 + 2x + 8$ are shown in the following figure. Based on this graph, what common linear factor do you expect the numerator and denominator of

$$\frac{x^2 - 4}{-x^2 + 2x + 8}$$

to have? Confirm algebraically.

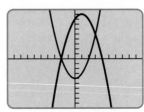

Standard window

For Exercise 12.

In Exercises 13–26, reduce the rational expression to simplest terms. When possible, support your work with a grapher.

13. $\dfrac{12x}{16x^3}$

14. $\dfrac{3x^8}{21x^3}$

15. $\dfrac{x(x - 7)}{x^2 - 49}$

16. $\dfrac{3(x + 2)}{x^2 + 7x + 10}$

17. $\dfrac{(x - 5)(3x + 7)}{x^2 - 10x + 25}$

18. $\dfrac{3x - 5}{6x - 10}$

19. $\dfrac{x - 7}{7 - x}$

20. $\dfrac{2x - 9}{9 - 2x}$

**21.** $\dfrac{x+5}{x+5}$

**22.** $\dfrac{x-9}{x+9}$

**23.** $\dfrac{3x^2y}{12x(x+16)}$

**24.** $\dfrac{5x^2y^3(x-y)}{30xy(x-y)^2}$

**25.** $\dfrac{36x^3y^2}{63x^4y}$

**26.** $\dfrac{21xy^2z^3}{35x^2y^2z}$

In Exercises 27 and 28, find the values of $x$ that make sense in the following problem situations.

**27.** *Average Cost.* A car rental agency charges a flat fee of \$25 plus \$0.28/mi. The average cost per mile for driving a car $x$ miles is

$$\frac{25+0.28x}{x}.$$

**28.** *Restock Charge.* Midwest Supply charges a flat fee of \$15 plus \$1.15/item for restocking any returned items. The average cost of restocking $x$ items is

$$\frac{15+1.15x}{x}.$$

*Using Algebra in Geometry.* In Exercises 29–32, reduce the given ratio to simplest terms.

**29.** $AD/BD$

**30.** $AD/AB$

**31.** $AD/AE$

**32.** $AB/AC$

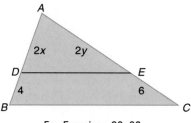

For Exercises 29–32.

In Exercises 33–38, use a grapher to determine if the numerator and the denominator contain any common linear factors. Use the Integer window and Zoom-In if necessary.

**33.** $\dfrac{x^2-2x-15}{x^2+15x+36}$

**34.** $\dfrac{x^2+4x-77}{x^2+8x-105}$

**35.** $\dfrac{x^2-9x-400}{x^2-10x-375}$

**36.** $\dfrac{x^2+15x-450}{x^2+12x-540}$

**37.** $\dfrac{3x^3+7x^2-36x+20}{2x^3-3x^2-17x+30}$

**38.** $\dfrac{3x^3-10x^2-27x+10}{3x^3-14x^2-55x+42}$

In Exercises 39–52, determine if the expression is in simplest terms. If not, reduce it to simplest terms. For Exercises 39–50, support your work with a grapher.

**39.** $\dfrac{(2-x)(3+x)}{(x-2)(x+3)}$

**40.** $\dfrac{(y-6)(y+6)}{(6-y)(7+y)}$

**41.** $\dfrac{2x+3}{4x^2+12x+9}$

**42.** $\dfrac{x^2-64}{x^2-16x+64}$

**43.** $\dfrac{4x-2x^2}{x^2+x-6}$

**44.** $\dfrac{10x-6x^2}{x^2-7x-5}$

**45.** $\dfrac{x^2-x-56}{2x^2-23x+56}$

**46.** $\dfrac{x^3-8}{x^2+2x-8}$

**47.** $\dfrac{125x^3+8}{25x^2-10x+4}$

**48.** $\dfrac{x^3+27}{x^2-9}$

**49.** $\dfrac{x^2+4x+4}{x+2}$

**50.** $\dfrac{4x^2+20x}{x^2+4x-5}$

**51.** $\dfrac{xy-yw+xz-zw}{xy+yw+xz+zw}$

**52.** $\dfrac{xy+xw-zy-zw}{xz+xw-z^2-zw}$

TRANSLATING WORDS TO SYMBOLS

In Exercises 53–56, write the given phrase in symbols.

**53.** A number $x$ divided by a number $y$

**54.** One less than a number $x$, divided by a number $y$

**55.** One more than a number $x$, divided by a number $y$

**56.** Find the reciprocal of $\dfrac{x}{y}$.

**57.** *Industrial Engineering.* The set-up cost for a production line of a new product is $2.3 million. The cost to manufacture each item is $4000.

a) Is the algebraic representation of producing $x$ items $2.3 + 4000x$ or $2.3 + 0.004x$?

b) What is the algebraic representation for the average cost per item of producing $x$ items.

EXTENDING THE IDEAS

**58.** *Research and Development Costs.* Consider the following data on the U.S. population and government expenditures on research and development. (Source: *Statistical Abstract of the United States, 1992*)

| Year | U.S. Population (Millions) | R and D (Billions) |
|------|---------------------------|--------------------|
| 1985 | 237.9 | 113.8 |
| 1986 | 240.2 | 119.5 |
| 1987 | 242.3 | 125.4 |
| 1988 | 244.5 | 133.7 |
| 1989 | 246.8 | 140.5 |
| 1990 | 248.7 | 145.5 |

a) How much was spent per person for research and development in 1990?

b) The U.S. population growth can be modeled between $1985(t = 1)$ and $1990(t = 6)$ by the quadratic polynomial $P(t) = -0.02t^2 + 2.32t + 235.6$. According to this model, what will be the population in 1995?

c) The amount spent on research and development between $1985(t = 1)$ and $1990(t = 6)$ can be modeled by $R(t) = 0.0018t^2 + 6.553t + 106.77$. According to this model, how much will be spent in 1995?

d) Write an algebraic representation (in $t$) for the average cost per person of research and development. (Be careful of the units.)

**59.** Graph both the numerator and denominator of $\dfrac{x^2 + 1}{x^3 + x}$ in the same Standard window.

a) Do they share any common linear factors?

b) Is the fraction reduced to lowest terms? If not, reduce.

c) Do the separate graphs help us to reduce to lowest terms?

**60.** Graph both the numerator and denominator of $\dfrac{x^3 - 1}{x^2 + x + 1}$ in the same Standard window.

a) Do they share any common linear factors?

b) Is the fraction reduced to lowest terms? If not, reduce.

c) Do the separate graphs help us to reduce to lowest terms?

**61.** Find which equation is true by using a grapher to graph each side of the equation separately and then seeing if the two graphs appear to be the same.

a) $\dfrac{x^2 - 9}{x^3 - 27} = \dfrac{x + 3}{x^2 - 3x + 9}$

b) $\dfrac{x^2 - 9}{x^3 - 27} = \dfrac{x + 3}{x^2 + 3x + 9}$

In Exercises 62–67, write each rational expression in simplest form by reducing to lowest terms.

**62.** $\dfrac{x^{2a} - 25}{x^a - 5}$

**63.** $\dfrac{x^{2a} - 4x^a + 4}{x^a - 2}$

**64.** $\dfrac{x^{2b} - 4}{4 - x^{2b}}$

**65.** $\dfrac{x^{3a} - 125}{x^{2a} - 25}$

**66.** $\dfrac{m^{2x} + 5m^x + 6}{m^x + 3}$

**67.** $\dfrac{p^{2x} - p^x - 12}{2p^{2x} + 3p^x - 9}$

**68. Writing to Learn.** In your own words, write how to find common linear factors of two polynomials using a graph.

**69. Writing to Learn.** In your own words, write how to reduce a rational expression to lowest terms and show support graphically.

**70.** The graph of $y = \dfrac{x^2 - x - 2}{x^2 - 3x + 2}$ is shown below in the 0.2 window. (To find the 0.2 window, double the $X$min, $X$max, $Y$min, and $Y$max values of the 0.1 window.)

a) Graph the numerator $y = x^2 - x - 2$ and denominator $y = x^2 - 3x + 2$ simultaneously. What do these graphs tell you about a common factor in the numerator and denominator?

b) Why do you think there is a break at $x = 2$ in the following graph?

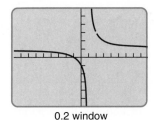

0.2 window

For Exercise 70.

LOOKING BACK—MAKING CONNECTIONS

In Exercises 71–74, find the products or quotients.

71. $\dfrac{25}{36} \cdot \dfrac{42}{15}$

72. $\dfrac{17}{12} \cdot \dfrac{16}{34}$

73. $\dfrac{56}{35} \div \dfrac{64}{15}$

74. $\dfrac{22}{35} \div \dfrac{33}{40}$

In Exercises 75–78, use a grapher. Pick any number at random, say 127, and store $127 \to X$, then verify that the two expressions are indeed not equal.

75. $\dfrac{x+2}{x+7} \neq \dfrac{2}{7}$

76. $\dfrac{2x-3}{2x} \neq -3$

77. $\dfrac{2x^2 + 3x + 5}{2x^2 + 7} \neq \dfrac{3x + 5}{7}$

78. $\dfrac{x(x+y)}{x-y} \neq -x$

In Exercises 79–84, find the products.

79. $[(a+b)+3][(a+b)-3]$

80. $[25+(x-y)][25-(x-y)]$

81. $[(2x+5)+y][(2x+5)-y]$

82. $[16+(3a-b)][16-(3a-b)]$

83. $[12.5-(x-7)][12.5+(x-7)]$

84. $[(a-b)-1][(a-b)+1]$

**6.2**

# Multiplying and Dividing Rational Expressions

Multiplying Rational Expressions ▪ Dividing Rational Expressions

Because multiplication and division are opposite operations, we examine the multiplication process first and then define division in terms of multiplication.

## Multiplying Rational Expressions

When multiplying two numerical fractions, we multiply numerators together and denominators together. To simplify the calculation, we factor each number and use the Principle of Equivalent Rational Expressions, if possible.

$$\frac{6}{35} \cdot \frac{14}{15} = \frac{6 \cdot 14}{35 \cdot 15}$$  **Factor and divide out common factors.**

$$= \frac{2 \cdot 3 \cdot 2 \cdot 7}{5 \cdot 7 \cdot 3 \cdot 5}$$  **Multiply the resulting numerators and denominators together.**

$$= \frac{2 \cdot 2}{5 \cdot 5} = \frac{4}{25}$$

Rational expressions are multiplied in the same way.

---

**DEFINITION 6.2**   Multiplication of Rational Expressions

Let $a$, $b$, $c$, and $d$ be any polynomials such that $b \neq 0$ and $d \neq 0$. Then

$$\frac{a}{b} \cdot \frac{c}{d} = \frac{ac}{bd}.$$

---

In Example 1, the numerators and denominators are monomials. The first step is to look for any common factors in the coefficients, just as we do with numerical factors.

**EXAMPLE 1**   Rational Expressions Containing Monomials

Find the product of the following two rational expressions and reduce to lowest terms:

$$\frac{12x^3z}{5y} \cdot \frac{25y}{4xz^3}$$

SOLUTION

Multiply the numerators and the denominators. Then use the Principle of Equivalent Rational Expressions.

$$\frac{12x^3z}{5y} \cdot \frac{25y}{4xz^3} = \frac{\overset{3}{\cancel{12}}\, x^3z \cdot \overset{5}{\cancel{25}}\, y}{\cancel{5}y \cdot \cancel{4}xz^3}$$  **Use the Principle of Equivalent Rational Expressions to reduce to lowest terms.**

$$= \frac{15x^2}{z^2}$$

**REMINDER**

When multiplying like bases, add the exponents. When dividing like bases, subtract the exponents.

In Example 2, the numerator and denominator polynomials have more than one term. Reducing the product to lowest terms requires that each expression is factored completely.

## EXAMPLE 2    Multiplying Rational Expressions

Find the product of the following two rational expressions and reduce to lowest terms. Support the solution graphically.

$$\frac{x^2 + 3x}{2(x - 7)} \cdot \frac{x^2 - 49}{x^2 - 2x - 15}$$

SOLUTION

**Solve Algebraically.** Factor each numerator and denominator completely.

$$\frac{x^2 + 3x}{2(x - 7)} \cdot \frac{x^2 - 49}{x^2 - 2x - 15} = \frac{x(x + 3)}{2(x - 7)} \cdot \frac{(x - 7)(x + 7)}{(x + 3)(x - 5)} \qquad \text{\textbf{Perform the indicated multiplication.}}$$

$$= \frac{x(x + 3)(x - 7)(x + 7)}{2(x - 7)(x + 3)(x - 5)} \qquad \text{\textbf{Divide out common factors.}}$$

$$= \frac{x(x + 7)}{2(x - 5)}$$

**Support Graphically.** Let $y_1 = (x^2 + 3x)/(2(x - 7))$, $y_2 = (x^2 - 49)/(x^2 - 2x - 15)$, $y_3 = y_1 y_2$, and $y_4 = x(x + 7)/(2(x - 5))$. Figure 6.7 shows the graph of $y_3$ and $y_4$. The two graphs appear to be identical, thus providing graphical support of the solution. Comparing the trace coordinates of the two graphs provides additional support. ■

The procedure followed in Example 2 is summarized next.

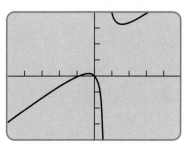

Integer window with vertical dimension [−20, 20]

**FIGURE 6.7** Graph of $y_3$, which is the product of $y_1$ and $y_2$, and $y_4 = x(x + 7)/(2(x - 5))$.

## Multiplying Rational Expressions

1. Completely factor each numerator and each denominator.
2. Perform the indicated multiplication and divide out any factors common to both the numerator and the denominator.
3. Write the product of the numerator factors and the product of the denominator factors (both in factored form). The resulting expression is the factored form in lowest terms.

**EXAMPLE 3**    Multiplying Rational Expressions

Find the product of the following two rational expressions. Support the solution graphically.

$$\frac{x^3 + 8}{x^2 + 8x + 12} \cdot \frac{x + 5}{x^2 - 2x + 4}$$

SOLUTION

**Solve Algebraically.** Factor each numerator and denominator completely. Then divide out common factors.

$$\frac{x^3 + 8}{x^2 + 8x + 12} \cdot \frac{x + 5}{x^2 - 2x + 4} = \frac{\cancel{(x + 2)}\cancel{(x^2 - 2x + 4)}(x + 5)}{\cancel{(x + 2)}(x + 6)\cancel{(x^2 - 2x + 4)}}$$

$$= \frac{x + 5}{x + 6}$$

**Support Graphically.** Let $y_1 = (x^3 + 8)/(x^2 + 8x + 12)$, $y_2 = (x + 5)/(x^2 - 2x + 4)$, $y_3 = y_1 y_2$, and $y_4 = (x + 5)/(x + 6)$. Figure 6.8 shows the graphs of $y_3$ and $y_4$, thus providing support for the algebra. ■

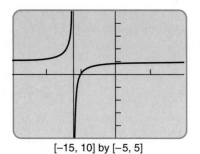

[−15, 10] by [−5, 5]

**FIGURE 6.8** A graph of $y_3$, which is the product of $y_1$ and $y_2$, and $y_4 = (x + 5)/(x + 6)$.

---

### Important Grapher Note

A correct graph of a rational expression like

$$\frac{x + 5}{x + 6}$$

should have two parts like those on the graph in Figure 6.7. However, due to the limited resolution of grapher screens, these two parts on a grapher often appear to be connected with a vertical, or almost vertical, line, as shown in Figure 6.8.

However, grapher support exists whether or not this vertical line appears. Whenever two expressions $y_1$ and $y_2$ appear to have the same graph, we have support that the two expressions are equivalent. Additional support can be obtained by comparing the trace coordinates of the two graphs.

---

When rational expressions include polynomials not in fraction form, avoid unnecessary errors by rewriting the polynomial over 1 to represent it as a fraction. Example 4 illustrates.

**EXAMPLE 4**    Representing a Polynomial as a Fraction

Find the product of the following expressions:

$$\frac{x^2 - 2xy + y^2}{x^4 - y^4} \cdot (x^2 + y^2)$$

SOLUTION

Write both expressions as fractions.

$$\frac{x^2 - 2xy + y^2}{x^4 - y^4} \cdot (x^2 + y^2) = \frac{(x - y)^2}{(x^2 - y^2)(x^2 + y^2)} \cdot \frac{(x^2 + y^2)}{1}$$

$$= \frac{(x - y)(x - y)(x^2 + y^2)}{(x - y)(x + y)(x^2 + y^2)}$$

$$= \frac{x - y}{x + y}$$   ■

## Dividing Rational Expressions

To understand how to divide rational expressions, recall that division is defined in terms of multiplication: $18 \div 3$ is the number such that $3 \times$ (the number) $= 18$; that is,

$$18 \div 3 = x, \quad \text{where } 3x = 18.$$

Similarly,

$$\frac{3}{8} \div \frac{1}{3} = x, \quad \text{where } \frac{1}{3} \cdot x = \frac{3}{8} \quad \text{or} \quad x = \frac{3}{8} \cdot \frac{3}{1} = \frac{9}{8}.$$

Thus $x$ is obtained by "inverting the divisor (the denominator) and multiplying." Division of rational expressions is completed in the same way.

---

**DEFINITION 6.3**    Division of Rational Numbers

Let $a$, $b$, $c$, and $d$ be any polynomials such that $b \neq 0$, $c \neq 0$, and $d \neq 0$. Then

$$\frac{a}{b} \div \frac{c}{d} = \frac{a}{b} \cdot \frac{d}{c} = \frac{ad}{bc}.$$

In other words, to divide one expression by another we multiply by the reciprocal of the divisor. Example 5 illustrates the method.

### EXAMPLE 5  Dividing Rational Expressions

Find the following quotient:

$$\frac{24x^2y^4}{2x^5y^7z} \div \frac{56x^2y^3z^8}{-7x^2yz}$$

SOLUTION

Complete the division by inverting the divisor and multiplying. Then reduce the expression to simplest terms.

$$\frac{24x^2y^4}{2x^5y^7z} \div \frac{56x^2y^3z^8}{-7x^2yz} = \frac{24x^2y^4}{2x^5y^7z} \cdot \frac{-7x^2yz}{56x^2y^3z^8}$$

$$= \frac{(\overset{3}{24}\, x^2y^4)(-7x^2yz)}{(2x^5y^7z)(\underset{\cancel{}}{56}\, x^2y^3z^8)} \qquad \textbf{Use the Property of Equality of Rational Expressions.}$$

$$= \frac{-3x^4y^5z}{2x^7y^{10}z^9} \qquad \textbf{Use the Properties of Exponents.}$$

$$= \frac{-3}{2x^3y^5z^8} \qquad\blacksquare$$

> **REMINDER**
>
> When we are multiplying and dividing fractions, it is often easy to lose a negative sign. In problems such as Example 5, it might be easier to first observe that we are dividing a positive by a negative, so the answer is negative. We then can assign the negative sign to the answer position and ignore the signs as we proceed with the factoring.

Example 6 includes fractions whose numerators or denominators are polynominals with two or more terms. After inverting and multiplying, we factor each numerator and denominator to look for common factors to divide out.

### EXAMPLE 6  Dividing Rational Expressions

Find the following quotient of rational expressions. Support the quotient graphically.

$$\frac{x^2-1}{x^2-6x} \div \frac{x^3-1}{x}$$

SOLUTION

**Solve Algebraically.** Complete the division by inverting the divisor and multiplying. Then factor each expression and reduce the expression to simplest terms.

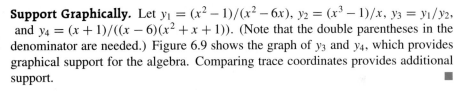

$$\frac{x^2-1}{x^2-6x} \div \frac{x^3-1}{x} = \frac{x^2-1}{x^2-6x} \cdot \frac{x}{x^3-1} \quad \text{Factor and divide out the common factors.}$$

$$= \frac{(x-1)(x+1)x}{x(x-6)(x-1)(x^2+x+1)}$$

$$= \frac{(x+1)}{(x-6)(x^2+x+1)}$$

**Support Graphically.** Let $y_1 = (x^2-1)/(x^2-6x)$, $y_2 = (x^3-1)/x$, $y_3 = y_1/y_2$, and $y_4 = (x+1)/((x-6)(x^2+x+1))$. (Note that the double parentheses in the denominator are needed.) Figure 6.9 shows the graph of $y_3$ and $y_4$, which provides graphical support for the algebra. Comparing trace coordinates provides additional support. ∎

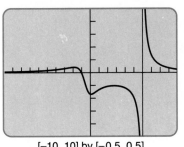

[–10, 10] by [–0.5, 0.5]

**FIGURE 6.9** Graph of $y_3$, which is the quotient of $y_1$ and $y_2$, and $y_4 = (x+1)/((x-6)(x^2+x+1))$.

We can summarize the procedure for finding the quotient of rational expressions as follows.

---

## Dividing Rational Expressions

1. Invert the divisor and change the operation to multiplication.
2. Complete the multiplication and reduce the expression to lowest terms.
3. Support graphically by graphing both the problem and the answer to see if the two graphs appear to be identical. If they are, we can be reasonably sure that the expressions are equivalent.

---

Example 7 illustrates simplifying expressions that contain more than one operation.

**EXAMPLE 7**   Combining Operations

Perform the following operations:

$$\frac{x^2+5x-14}{3x^2-14x-5} \cdot \frac{18x^2-9x-5}{9x^2+65x+14} \div \frac{6x^2-29x+20}{9x^2-43x-10}$$

**SOLUTION**

Complete the division by inverting the divisor and multiplying. Then factor each expression and reduce the expression to simplest terms.

$$\frac{x^2 + 5x - 14}{3x^2 - 14x - 5} \cdot \frac{18x^2 - 9x - 5}{9x^2 + 65x + 14} \div \frac{6x^2 - 29x + 20}{9x^2 - 43x - 10}$$

$$= \frac{x^2 + 5x - 14}{3x^2 - 14x - 5} \cdot \frac{18x^2 - 9x - 5}{9x^2 + 65x + 14} \cdot \frac{9x^2 - 43x - 10}{6x^2 - 29x + 20}$$    Factor each numerator and denominator and divide out the common factors.

$$= \frac{(x+7)(x-2)(3x+1)(6x-5)(x-5)(9x+2)}{(3x+1)(x-5)(x+7)(9x+2)(6x-5)(x-4)}$$

$$= \frac{x-2}{x-4}$$

■

# Exercises for Section 6.2

In Exercises 1–8, find the product of the rational expressions.

**1.** $\dfrac{a}{b^2} \cdot \dfrac{b}{a}$

**2.** $\dfrac{x^2}{y} \cdot \dfrac{yx}{x^2}$

**3.** $\dfrac{x^2 y}{w^2 z} \cdot \dfrac{w^3 z^2}{x^3 y^3}$

**4.** $\dfrac{x^2 y^2}{z^3} \cdot \dfrac{z^2}{x^2 y^2}$

**5.** $\dfrac{2a^2 b^3}{3ab^4} \cdot \dfrac{6ab^3}{10b}$

**6.** $\dfrac{4ab^3 c}{5a^2 b^5 c^2} \cdot \dfrac{25c}{12b^4 c^2}$

**7.** $\dfrac{a^2 - ab}{b} \cdot \dfrac{b^2}{ab - b^2}$

**8.** $\dfrac{1}{b - a} \cdot (a^2 - b^2)$

In Exercises 9–12, find the product of the rational expressions algebraically and support graphically.

**9.** $\dfrac{x^2 - x}{x + 1} \cdot \dfrac{x^2 + x}{x^2 + 2}$

**10.** $\dfrac{1}{x - 1} \cdot \dfrac{x^2 - 1}{x + 1}$

**11.** $\dfrac{4x^2 - 9}{x^2 - x} \cdot \dfrac{x^2 + 4x}{2x^2 + x - 6}$

**12.** $\dfrac{x^4 - 16}{x^2 + x - 6} \cdot \dfrac{x(x + 3)}{x^2 + 4}$

In Exercises 13–15, use the formula for the area of a triangle, $A = \frac{1}{2}bh$.

**13.** *Using Algebra in Geometry.* Find the area of the small triangle in the given figure.

**14.** *Using Algebra in Geometry.* Find the area of the large triangle in the given figure.

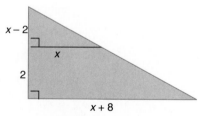

For Exercises 13–15.

**15.** *Using Algebra in Geometry.* Find the ratio of the area of the large triangle to the small triangle, reduced to simplest terms.

In Exercises 16–20, write the division problem as a multiplication problem and then find the product.

**16.** $\dfrac{x^2 - y^2}{(x - y)^2} \div \dfrac{x^2 + y^2}{x - y}$

**17.** $\dfrac{a - b}{ab} \div \dfrac{a + b}{a^2 b^2}$

**18.** $\dfrac{x + y}{x^2 y^2} \div \dfrac{x - y}{x^3 y^3}$

**19.** $\dfrac{x - 5}{x^2 - 49} \div \dfrac{x^2 - 10x + 25}{x - 7}$

**20.** $\dfrac{2a}{a^2 b^2} \div \dfrac{a^3 b^5}{a^3 b^3}$

In Exercises 21 and 22, complete the division with paper and pencil. Then choose the correct graph of the result (shown in the Standard window) from the following figure.

**21.** $\dfrac{x^2 + 1}{x + 3} \div \dfrac{x^2 + 2}{3x + 1}$     **22.** $\dfrac{4}{x - 3} \div \dfrac{2x + 3}{x^2}$

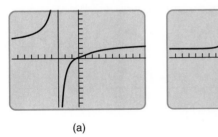

(a)                                        (b)

For Exercises 21 and 22.

Exercises 23 and 24 require that units be converted. It is sometimes helpful to think of a unit like ft/sec as a rational expression. Then different units can be multiplied and divided.

**23.** *Converting Units.* A storage bin is 12 ft long, 5 ft wide, and 4 ft high.

  a) Find the volume in ft$^3$.

  b) Find the volume in yd$^3$.

**24.** *Converting Units.* One cubic foot of water has a mass of 62.4 lb. What is the mass of the water in a container 2 in. × 3 in. × 5 in.?

In Exercises 25–29, find the quotient of the two rational expressions. Support graphically using a grapher.

**25.** $\dfrac{(x - 3)^2}{x + 3} \div \dfrac{x^2 - 9}{x^2 + 5x + 6}$

**26.** $\dfrac{x + 4}{x^2 + 8x + 16} \div \dfrac{x^3 - 64}{x - 4}$

**27.** $\dfrac{x^3 - 1}{x - 2} \div \dfrac{x^2 + x + 1}{x^2 - 4}$

**28.** $\dfrac{x - 3}{2x} \div \dfrac{3 - x}{x^2}$

**29.** $\dfrac{x^2 + 7x + 12}{x - 2} \div \dfrac{x^2 + 5x + 6}{2 - x}$

Exercises 30 and 31 taken together illustrate a way that a grapher can provide numerical support for algebra.

**30.** Multiply and simplify $\dfrac{x + 1}{x} \cdot \dfrac{x^2}{x + 1}$.

**31.** **Writing to Learn.** Define $y_1 = \dfrac{x + 1}{x}$, $y_2 = \dfrac{x^2}{x + 1}$, and $y_3 = y_1 \cdot y_2$. Explain how the following table provides numerical support for the algebra in Exercise 30.

| X | Y1 | Y2 |
|---|------|--------|
| 1 | 2 | .5 |
| 2 | 1.5 | 1.3333 |
| 3 | 1.3333 | 2.25 |
| 4 | 1.25 | 3.2 |
| 5 | 1.2 | 4.1667 |
| 6 | 1.1667 | 5.1429 |
| 7 | 1.1429 | 6.125 |

| X | Y2 | Y3 |
|---|--------|---|
| 1 | .5 | 1 |
| 2 | 1.3333 | 2 |
| 3 | 2.25 | 3 |
| 4 | 3.2 | 4 |
| 5 | 4.1667 | 5 |
| 6 | 5.1429 | 6 |
| 7 | 6.125 | 7 |

$Y_3 = Y_1 {}^* Y_2$

For Exercises 30 and 31.

In Exercises 32–36, find the quotient using algebraic procedures.

**32.** $\dfrac{x^2 - y^2}{x + y} \div \dfrac{(x + y)(x + 2y)}{x^2 + 2xy + y^2}$

**33.** $\dfrac{x^4 - y^4}{x - y} \div \dfrac{x^2 + y^2}{x + y}$

**34.** $\dfrac{1}{4y^2 - 12y + 9} \div \dfrac{1}{4y^2 - 9}$

**35.** $\dfrac{a^2 - b^2}{a^2 - 5a - 6} \div \dfrac{a^2 + 2ab + b^2}{a^2 + a}$

**36.** $\dfrac{9x^2 + 9xy + 2y^2}{3x^2 - 2xy - y^2} \div \dfrac{3x^2 + 5xy + 2y^2}{x^2 - 3xy + 2y^2}$

**37.** *Printing Rate.* A laser printer prints 8 pg/min.

a) How long does it take to print one page?

b) How long does it take to print $x$ pages?

c) At \$0.04/page, how much does it cost to print $x$ pages?

d) If three laser printers are used simultaneously to print $x$ copies, how many minutes will each printer run?

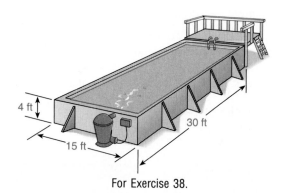

For Exercise 38.

**38.** *Pumping Rate.* A pool is 30 ft long, 15 ft wide, and 4 ft deep. If the filter system cycles all the water in one 24-hr period, the filter system pumps how many gallons per minute? (Water weighs 62.4 lbs/ft$^3$ and 8.33 lbs/gal.)

In Exercises 39–45, perform the indicated operation and simplify if possible.

**39.** $\dfrac{x^2 - 9}{16x^2 y} \cdot \dfrac{4xy}{x^2 + x - 12}$

**40.** $\dfrac{-12x^3 y^5 z}{7x^2 yz^4} \cdot \dfrac{21x^4 y}{6x^5 yz^3}$

**41.** $\dfrac{x^2 + 6x + 5}{x^2 + 2x + 1} \cdot \dfrac{x^2 - 1}{x^2 + 5x}$

**42.** $\dfrac{x - 3}{x - 2} \div \dfrac{x^2 - 6x + 9}{x - 3}$

**43.** $\dfrac{x + 9}{x - 1} \div \dfrac{x^2 - 81}{x^2 - 10x + 9}$

In Exercises 44–50, perform the indicated operation and simplify if possible.

**44.** $\dfrac{x^2 - x - 6}{x^2 - 5x - 14} \cdot \dfrac{2x^2 - 13x - 24}{x^2 - 2x - 3} \div \dfrac{2x^2 - 21x + 40}{x^2 - 6x - 7}$

**45.** $\dfrac{2x^2 - xy - 3y^2}{6x^2 + xy - y^2} \cdot \dfrac{3x^2 - 4xy + y^2}{9x^2 - y^2} \div \dfrac{x^2 - y^2}{6x^2 + 5xy + y^2}$

**46.** $\dfrac{a - b}{a^2 + 2ab + b^2} \cdot \dfrac{a - 2b}{a^2 - b^2} \div \dfrac{a^2 - 4ab + 4b^2}{a^3 + b^3}$

**47.** $\dfrac{4x^2 + 12x + 9}{4x^2 - 4x - 3} \cdot \dfrac{6x^2 + 5x + 1}{2x^2 - 7x - 15} \div \dfrac{3x^2 - 2x - 1}{2x^2 - 13x + 15}$

**48.** $\dfrac{2x^2 + 9x + 10}{x^2 + 5x + 6} \cdot \dfrac{x^2 + 7x + 12}{2x^2 + 3x - 5} \div \dfrac{x^2 - 16}{x^2 - 1}$

**49.** $\dfrac{x^2 + 2x + 1}{x^2 - 2} \cdot \dfrac{x^2 - 5x + 6}{x + 1} \div \dfrac{x - 3}{5x - 20}$

**50.** $\dfrac{x^3 - 125}{x^2 - 10x + 25} \cdot \dfrac{2x^2 + 13x + 15}{3x^2 + 14x - 5} \div \dfrac{x^2 + 5x + 25}{x - 5}$

TRANSLATING WORDS TO SYMBOLS

**51.** Write an algebraic expression for the product of the reciprocals of two consecutive integers. (If $x$ is an integer, what is the next integer?)

**52.** Write an algebraic expression for the quotient of the reciprocals of two consecutive even integers. (If $x$ is an even integer, what is the next even integer?)

EXTENDING THE IDEAS

In Exercises 53–56, perform the indicated operation and simplify if possible.

53. $\dfrac{x^m}{x^{2n}} \cdot \dfrac{x^n}{x^{2m}}$

54. $\dfrac{x^{2m} + 2x^m + 1}{x^n} \cdot \dfrac{x^{2n}}{x^m + 1}$

55. $\dfrac{x^{2m} - 4}{x^{2m} - 9} \div \dfrac{x^m + 2}{x^m - 3}$

56. $\dfrac{x^{2m} - 25}{x^{3m} + 64} \div \dfrac{x^m - 5}{x^{2m} - 4x^m + 16}$

LOOKING BACK—MAKING CONNECTIONS

57. Find the sum of $\dfrac{2}{3} + \dfrac{5}{6}$. Support numerically using a grapher.

58. Find the difference of $\dfrac{3}{10} - \dfrac{2}{15}$. Support numerically using a grapher.

59. Find the sum of $(x^2 + 3x + 5) + (2x^2 - 2x + 9)$.

60. Find the difference of $(3x^2 + 5x + 2) - (2x - 5)$.

---

## 6.3 Adding and Subtracting Rational Expressions

Rational Expressions with Like Denominators • Rational Expressions with Unlike Denominators

We first shall consider adding two rational expressions that have identical denominators. Later in this section, we consider adding rational expressions that do not have identical denominators.

### Rational Expressions with Like Denominators

To add or subtract two rational expressions, we apply the same procedures used for adding or subtracting numerical fractions. That is, to add or subtract two fractions with identical or **common denominators**, we add or subtract the numerators and place the sum or difference over the common denominator. The final answer should be in simplest form, so we reduce to lowest terms. For example,

$$\frac{2}{3} + \frac{4}{3} = \frac{2+4}{3} = \frac{6}{3} = 2 \qquad \text{and} \qquad \frac{2}{3} - \frac{4}{3} = \frac{2-4}{3} = -\frac{2}{3}.$$

> **DEFINITION 6.4** Addition and Subtraction of Rational Expressions
>
> If $a$, $b$, and $c$ represent polynomial expressions, then
>
> $$\frac{a}{c} + \frac{b}{c} = \frac{a+b}{c} \qquad \text{and} \qquad \frac{a}{c} - \frac{b}{c} = \frac{a-b}{c}.$$

Examples 1–4 illustrate adding or subtracting rational expressions with a common denominator. We add or subtract the numerators and the sum or difference is written over the common denominator.

### EXAMPLE 1   Adding Expressions with Common Denominators

Find the sum of the following two rational expressions. Support the solution graphically.

$$\frac{x+3}{x-2} + \frac{2x-1}{x-2}$$

SOLUTION

**Solve Algebraically.** Find the sum of the numerators over the common denominator.

$$\frac{x+3}{x-2} + \frac{2x-1}{x-2} = \frac{(x+3) + (2x-1)}{x-2} \qquad \text{Add like terms to simplify.}$$

$$= \frac{3x+2}{x-2} \qquad \text{Simplest terms}$$

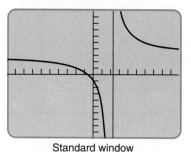

Standard window

**FIGURE 6.10** Graphs of $y_3$ and $y_4$ in Example 1 appear to be identical.

**Support Graphically.** Figure 6.10 shows the graphs of $y_3$ and $y_4$, where $y_1 = (x+3)/(x-2)$, $y_2 = (2x-1)(x-2)$, $y_3 = y_1 + y_2$, and $y_4 = (3x+2)/(x-2)$. Because both graphs appear to be identical, we have graphical support for the algebra.   ■

After addition is completed, reduce the result to lowest terms. Example 2 shows that sometimes additional factoring must be completed before that is possible.

**EXAMPLE 2**    Adding Rational Expressions

Find the following sum of two rational expressions. Support the solution graphically.

$$\frac{3x+7}{(x+3)(x-7)} + \frac{x^2-3x-16}{(x+3)(x-7)}$$

SOLUTION

**Solve Algebraically.** Find the sum of the numerators over the common denominator.

$$\frac{3x+7}{(x+3)(x-7)} + \frac{x^2-3x-16}{(x+3)(x-7)} = \frac{3x+7+x^2-3x-16}{(x+3)(x-7)}$$

$$= \frac{x^2-9}{(x+3)(x-7)}$$

Factor the numerator in search for factors common with the denominator.

$$= \frac{\cancel{(x+3)}(x-3)}{\cancel{(x+3)}(x-7)}$$

Divide out any common factors.

$$= \frac{x-3}{x-7}$$

Simplest form

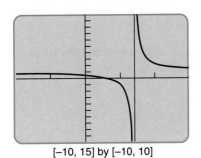

[−10, 15] by [−10, 10]

**FIGURE 6.11** Graphs of $y_3$ and $y_4$ appear to be identical.

**Support Graphically.** Let $y_1 = (3x+7)/((x+3)(x-7))$, $y_2 = (x^2-3x-16)/((x+3)(x-7))$, $y_3 = y_1 + y_2$, and $y_4 = (x-3)/(x-7)$. Figure 6.11 shows the graphs of $y_3$ and $y_4$ providing graphical support for the algebra.  ■

Example 3 illustrates how to find the difference of two rational expressions with common denominators.

**REMINDER**

When subtracting a polynomial of more than one term, it is important to remember to change the signs of all the terms that are being subtracted and then combine like terms. Notice in Example 3 that $-(x-5) = -x+5$.

**EXAMPLE 3**    Subtracting Rational Expressions

Find the following difference of two rational expressions. Support the solution graphically.

$$\frac{2x+7}{3x-5} - \frac{x-5}{3x-5}$$

SOLUTION

**Solve Algebraically.** Subtract the numerators over the common denominator.

$$\frac{2x+7}{3x-5} - \frac{x-5}{3x-5} = \frac{(2x+7)-(x-5)}{3x-5}$$

When subtracting numerators, note the subtraction of the entire polynomial.

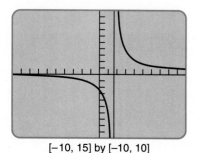

[−10, 15] by [−10, 10]

**FIGURE 6.12** The graph of $y_3 = y_1 - y_2$ and $y_4$ in the same window.

$$= \frac{2x + 7 - x + 5}{3x - 5}$$  Note the change of signs of the expression being subtracted.

$$= \frac{x + 12}{3x - 5}$$

**Support Graphically.** Let $y_1 = (2x + 7)/(3x - 5)$, $y_2 = (x - 5)/(3x - 5)$, $y_3 = y_1 - y_2$, and $y_4 = (x + 12)/(3x - 5)$. Figure 6.12 shows graphs of $y_3$ and $y_4$. Because both graphs appear to be the same, we have grapher support. ■

## Rational Expressions with Unlike Denominators

When two denominators are not equal, we say we have **unlike denominators**. Before two rational expressions with unlike denominators can be added, the expressions must be changed to equivalent expressions that do have like denominators. We call such a denominator a **common denominator.**

When the denominators of the expressions have no factors in common, the common denominator to use is the product of the denominators. Examples 4 and 5 illustrate this method.

**EXAMPLE 4**   Adding Expressions with Unlike Denominators

Find the following sum of two rational expressions:

$$\frac{4}{x} + \frac{5}{y}$$

SOLUTION

Because the denominators have no common factors, a common denominator is the product $x \cdot y$. Multiply the numerator and denominator of each term by the same factor to obtain $xy$ in the denominator.

$$\frac{4}{x} + \frac{5}{y} = \frac{4}{x} \cdot \boxed{\frac{y}{y}} + \frac{5}{y} \cdot \boxed{\frac{x}{x}}$$  The numerator and denominator of each expression is multiplied by whatever factor is required to obtain the denominator $xy$.

$$= \frac{4y}{xy} + \frac{5x}{xy}$$  Add these two terms with like denominators.

$$= \frac{4y + 5x}{xy}$$  ■

The same procedure is followed for subtracting rational expressions with unlike denominators. However, this time we subtract the numerators over the common denominator.

**EXAMPLE 5**  Subtracting Expressions with Unlike Denominators

Find the following difference of two rational expressions. Support the solution graphically.

$$\frac{3}{x-2} - \frac{5}{x+3}$$

SOLUTION

**Solve Algebraically.** Change each term to an equivalent expression with the denominator $(x-2)(x+3)$.

$$\frac{3}{x-2} - \frac{5}{x+3} = \frac{3\,(x+3)}{(x-2)\,(x+3)} - \frac{5\,(x-2)}{(x+3)\,(x-2)}$$

Multiply the numerator and denominator by the appropriate factor to obtain the common denominator $(x-2)(x+3)$.

$$= \frac{3(x+3) - 5(x-2)}{(x-2)(x+3)}$$

Simplify and combine terms in the numerator.

$$= \frac{3x + 9 - 5x + 10}{(x-2)(x+3)}$$

$$= \frac{-2x + 19}{(x-2)(x+3)}$$

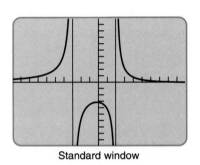

Standard window

**FIGURE 6.13** Graphs of $y_3$ and $y_4$ in the same window appear to be identical.

**Support Graphically.** Let $y_1 = 3/(x-2)$, $y_2 = 5/(x+3)$, $y_3 = y_1 - y_2$, and $y_4 = (-2x + 19)/((x-2)(x+3))$. Figure 6.13 shows the graphs of $y_3$ and $y_4$. Because the two graphs appear to be identical, we have grapher support of the solution. ∎

When the unlike denominators of two rational expressions have common factors, the common denominator that should be chosen is not simply the product of the denominators.

A common denominator of two rational expressions is a multiple of each denominator. Any pair of rational expressions has many common denominators. One of them is called the **least common denominator** (LCD).

We illustrate with a numerical setting. The numbers 12, 24, 36, and 48 are all common multiples of the denominators 2, 3, and 4 of the rational expressions

$$\frac{1}{2}, \frac{2}{3}, \frac{3}{4}.$$

These common multiples are all common denominators. However, 12 is the least common denominator (LCD) since 12 is the smallest of all the common denominators.

The procedure outlined next can be used to find the LCD of either numerical or algebraic expressions.

---

### Finding the Least Common Denominator (LCD)

1. Factor each denominator into prime or irreducible factors.
2. List the highest power of each factor that occurs in the factorization of at least one denominator.
3. Multiply together all the factors that are in the list in step 2.

The product found in step 3 is the **least common denominator** (LCD).

---

Example 6 demonstrates this process of finding the least common denominator.

**EXAMPLE 6**    Finding the LCD

Find the LCD of the following expressions:

a) $\dfrac{2}{15x^2} + \dfrac{1}{7(x-1)}$

b) $\dfrac{x+3}{21x(x-6)} - \dfrac{x-1}{12(x-6)^2}$

SOLUTION

We shall follow the three steps outlined above.

a) The denominators are $15x^2$ and $7(x-1)$.
Step 1: Factor the denominators.
$$15x^2 = 3 \cdot 5 \cdot x^2$$
$$7(x-1) = 7 \cdot (x-1)$$
Step 2: List the highest power for each factor occurring at least once: $3, 5, 7, x^2, (x-1)$.
Step 3: The LCD $= 3 \cdot 5 \cdot 7 \cdot x^2 \cdot (x-1) = 105x^2(x-1)$.

b) The denominators are $21x(x-6)$ and $12(x-6)^2$.
Step 1: Factor the denominators.
$$21x(x-6) = 3 \cdot 7 \cdot x \cdot (x-6)$$
$$12(x-6)^2 = 2^2 \cdot 3 \cdot (x-6)^2$$
Step 2: List the highest power for each factor occurring at least once: $2^2, 3, 7, x, (x-6)^2$.
Step 3: The LCD $= 2^2 \cdot 3 \cdot 7 \cdot x \cdot (x-6)^2 = 84x(x-6)^2$. ∎

The procedure for adding or subtracting rational expressions with unlike denominators can be summarized as follows.

> ## Adding or Subtracting Rational Expressions with Unlike Denominators
>
> 1. Reduce each rational expression to lowest terms.
> 2. Find the LCD of all the denominators.
> 3. Multiply the numerator and the denominator of each expression by whatever factor is needed to obtain the LCD in the denominator.
> 4. The equivalent expressions obtained in Step 2 now have like denominators. Add or subtract the expressions as usual and simplify the result to lowest terms.
> 5. Support with a grapher when necessary by graphing both the original expression and the solution to see if the graphs appear to be identical.

Examples 7, 8, and 9 illustrate how to apply this procedure. Example 7 also illustrates that the terms in the sum or difference should be reduced to lowest terms before the addition or subtraction is completed.

**EXAMPLE 7**   Subtracting with Unlike Denominators

Find the following difference of rational expressions. Support the solution graphically.

$$\frac{3x + 10}{x^2 + 7x + 12} - \frac{x + 2}{x^2 + 5x + 6}$$

SOLUTION

**Solve Algebraically.**  Write the denominators in factored form and find the LCD.

$$\frac{3x + 10}{x^2 + 7x + 12} - \frac{x + 2}{x^2 + 5x + 6}$$

$$= \frac{3x + 10}{(x + 3)(x + 4)} - \frac{x + 2}{(x + 2)\,(x + 3)}$$    Reduce each expression to lowest terms before beginning the addition.

$$= \frac{3x + 10}{(x + 3)(x + 4)} - \frac{1}{(x + 3)}$$    Multiply the numerator and denominator by the appropriate factor to obtain the LCD $.x + 3/.x + 4/:$

$$= \frac{3x + 10}{(x + 3)(x + 4)} - \frac{1}{(x + 3)} \cdot \frac{(x + 4)}{(x + 4)}$$    Combine the numerators over the LCD.

$$= \frac{3x + 10 - (x + 4)}{(x + 3)(x + 4)}$$

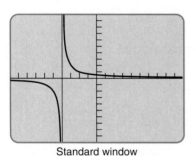

Standard window

**FIGURE 6.14** Graphs of $y_3$ and $y_4$ appear to be identical, which shows grapher support.

**REMINDER**

Opposites are two binomials in the form $(a - b)$ and $(b - a)$. $-1$ times one of the opposites = the other expression: for example, $-1(5 - x) = (x - 5)$.

$$= \frac{2x + 6}{(x + 3)(x + 4)}$$  **Factor the numerator and divide out the common factors.**

$$= \frac{2\,(x + 3)}{(x + 3)\,(x + 4)}$$

$$= \frac{2}{x + 4}$$

**Support Graphically.** Define $y_1 = (3x + 10)/(x^2 + 7x + 12)$, $y_2 = (x + 2)/(x^2 + 5x + 6)$, $y_3 = y_1 - y_2$, and $y_4 = 2/(x + 4)$. Figure 6.14 shows the graphs of both $y_3$ and $y_4$, which provide graphical support of the algebra.  ∎

In Example 8, one denominator is a degree 2 polynomial. The first step in finding the LCD is to factor the denominators. Be on the alert for expressions that are equivalent, such as a factor and its opposite. Using such expressions will keep the algebraic expressions as simple as possible. Example 8 illustrates such a case.

**EXAMPLE 8**  Factors That Are Opposites

Find the following difference of rational expressions. Support the solution graphically.

$$\frac{2}{x^2 - 5x} - \frac{3x}{5 - x}$$

SOLUTION

**Solve Algebraically.** The denominator $x^2 - 5x$ can be factored as $x(x - 5)$. We will take advantage of the fact that $(x - 5)$ and $(5 - x)$ are opposites.

$$\frac{2}{x^2 - 5x} - \frac{3x}{5 - x} = \frac{2}{x(x - 5)} - \frac{3x}{(5 - x)}$$  **Change to an equivalent expression with a denominator $(x - 5)$.**

$$= \frac{2}{x(x - 5)} - \frac{3x}{(5 - x)} \cdot \frac{-1}{-1}$$

$$= \frac{2}{x(x - 5)} - \frac{-3x}{(x - 5)}$$  **Change each denominator to the LCD.**

$$= \frac{2}{x(x - 5)} + \frac{3x}{(x - 5)} \cdot \frac{x}{x}$$

$$= \frac{2}{x(x - 5)} + \frac{3x^2}{x(x - 5)}$$  **Add the numerators over the common denominator.**

$$= \frac{2 + 3x^2}{x(x - 5)}$$

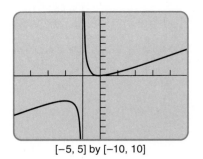

[−5, 15] by [−10, 10]

**FIGURE 6.15** Graphs of the expression defined in $y_3$ and the simplified expression defined in $y_4$ appear to be the same.

**Support Graphically.** Let $y_1 = 2/(x^2 - 5x)$, $y_2 = 3x/(5 - x)$, $y_3 = y_1 - y_2$, and $y_4 = (2 + 3x^2)/(x(x - 5))$. Figure 6.15 shows the graphs of $y_3$ and $y_4$, which provide grapher support, since the two graphs appear to be identical. ∎

**EXAMPLE 9**    Combine and Simplify

Perform the following operations. Support the solution graphically.

$$x - 1 + \frac{1}{x + 1}$$

SOLUTION

**Solve Algebraically.** Begin by representing $x - 1$ as $\dfrac{x - 1}{1}$.

$$x - 1 + \frac{1}{x + 1} = \frac{(x - 1)}{1} + \frac{1}{x + 1}$$

$$= \frac{(x + 1)}{(x + 1)} \cdot \frac{(x - 1)}{1} + \frac{1}{x + 1}$$

$$= \frac{(x + 1)(x - 1) + 1}{x + 1}$$

$$= \frac{x^2 - 1 + 1}{x + 1}$$

$$= \frac{x^2}{x + 1}$$

[−5, 5] by [−10, 10]

**FIGURE 6.16** Graphs of the expression defined in $y_1$ and the simplified expression defined in $y_2$ appear to be the same.

**Support Graphically.** Let $y_1 = x - 1 + 1/(x + 1)$ and $y_2 = x^2/(x + 1)$. Figure 6.16 shows the two graphs, which provide graphical support of the algebra. ∎

## Exercises for Section 6.3

In Exercises 1–10, find the sum or difference. Support the result using a grapher.

1. $\dfrac{x}{5} + \dfrac{3}{5}$

2. $\dfrac{7}{x} + \dfrac{3}{x}$

3. $\dfrac{x}{2} + \dfrac{x + 3}{2}$

4. $\dfrac{x}{y} + \dfrac{x - 2}{y}$

5. $\dfrac{x - 2}{7} + \dfrac{x + 3}{7}$

6. $\dfrac{2x + 6}{15} - \dfrac{x + 3}{15}$

7. $\dfrac{x - 3}{x + 3} + \dfrac{x + 3}{x + 3}$

8. $\dfrac{4x - 3}{x - 5} - \dfrac{2x + 8}{x - 5}$

9. $\dfrac{x^2 + 2x + 4}{(x + 3)(x - 2)} + \dfrac{-x^2 + 5x - 3}{(x + 3)(x - 2)}$

10. $\dfrac{x^2 - 2x - 3}{(x - 4)(x + 1)} - \dfrac{x + 4}{(x - 4)(x + 1)}$

In Exercises 11–16, find the sum or difference. Factor the resulting expression and reduce to lowest terms.

**11.** $\dfrac{3}{x+3} + \dfrac{x}{x+3}$

**12.** $\dfrac{2x}{x-2} + \dfrac{-4}{x-2}$

**13.** $\dfrac{x}{x-2} - \dfrac{2x-2}{x-2}$

**14.** $\dfrac{x^2 - 2x}{x-7} - \dfrac{5x}{x-7}$

**15.** $\dfrac{2x^2 + x + 5}{(x+2)(x-3)} + \dfrac{-x^2 + 3x - 1}{(x+2)(x-3)}$

**16.** $\dfrac{4x^2 + 5}{9x^2 - 24x} - \dfrac{x^2 - x + 29}{9x^2 - 24x}$

In Exercises 17 and 18, complete the given addition with paper and pencil. Then choose the correct graph of the result (shown in the Standard window).

**17.** $\dfrac{1}{x} + \dfrac{x}{1}$

**18.** $\dfrac{8}{x} + \dfrac{8}{x^2}$

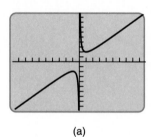
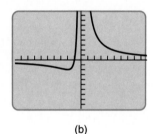

(a)                              (b)

For Exercises 17 and 18.

**19. Writing to Learn.** A student insists that

$$\frac{1}{x} + \frac{1}{2x} = \frac{1}{3x}.$$

Explain three different ways to show that this equation is a false statement.

**20. Writing to Learn.** A student insists that

$$\frac{1}{x} + \frac{1}{x^2} = \frac{2}{x^3}.$$

Explain three different ways to show that this equation is a false statement.

In Exercises 21–26, use the Property of Equivalent Rational Expressions to find the missing numerator.

**21.** $\dfrac{3}{2x} = \dfrac{?}{6x^2}$

**22.** $\dfrac{5y}{3x} = \dfrac{?}{15x^2}$

**23.** $\dfrac{2}{x+4} = \dfrac{?}{x(x+4)}$

**24.** $\dfrac{7}{x-3} = \dfrac{?}{x(x-3)}$

**25.** $\dfrac{x^2 - 2x}{x^3} = \dfrac{?}{x^2}$

**26.** $\dfrac{5x}{(x+3)} = \dfrac{?}{(x+3)(x-3)}$

In Exercises 27–30, write an equivalent fraction with the given denominator.

**27.** $\dfrac{5}{3x^2}$;   denominator   $21x^3$

**28.** $\dfrac{6}{7xyz}$;   denominator   $35x^2y^2z^3$

**29.** $\dfrac{-2}{5x}$;   denominator   $5x(x+3)$

**30.** $\dfrac{x-7}{x+3}$;   denominator   $x^2 + 5x + 6$

In Exercises 31–40, find the sum or the difference.

**31.** $\dfrac{3}{2x} + \dfrac{5}{x}$

**32.** $\dfrac{7}{x} + \dfrac{x+3}{x^2}$

**33.** $\dfrac{3}{14x} - \dfrac{8}{15}$

**34.** $\dfrac{9}{30y} + \dfrac{y}{2}$

**35.** $\dfrac{5}{16x} + \dfrac{3}{5x}$

**36.** $\dfrac{7}{x} - \dfrac{3}{x+4}$

**37.** $\dfrac{6y}{y+3} + \dfrac{9}{y}$

**38.** $\dfrac{5}{x+2} - \dfrac{7x}{x-3}$

**39.** $\dfrac{2x}{x-4} + \dfrac{3x}{x+4}$

**40.** $\dfrac{x-1}{x+1} + \dfrac{x+3}{x+2}$

In Exercises 41–44, find the LCD of the given pair of rational expressions.

**41.** $\dfrac{1}{x(x+1)}$,   $\dfrac{4}{x(x-3)}$

**42.** $\dfrac{x+1}{x(x-2)}$,   $\dfrac{x-3}{x(x+2)}$

**43.** $\dfrac{2}{x^2 - x - 2}$,   $\dfrac{5}{x^2 + 4x + 3}$

**44.** $\dfrac{x+1}{x^2 + 5x + 6}$,   $\dfrac{x-2}{x^2 - 2x - 8}$

In Exercises 45–52, perform the indicated operation. Simplify the result to lowest terms. Support the result using a grapher.

**45.** $\dfrac{3}{x-1} + \dfrac{5}{x+3}$      **46.** $\dfrac{6}{x+4} + \dfrac{1}{x-3}$

**47.** $\dfrac{x}{x-2} - \dfrac{x}{x+1}$      **48.** $\dfrac{5}{x+3} - \dfrac{3}{x+2}$

**49.** $\dfrac{x+3}{(x-2)(x+3)} + \dfrac{3x-4}{(x+3)(x-1)}$

**50.** $\dfrac{2x+3}{(x-1)(x+3)} - \dfrac{x+3}{(x-1)(x+2)}$

**51.** $\dfrac{3x-7}{x^2+4x-5} + \dfrac{2x-6}{x^2-4x+3}$

**52.** $\dfrac{7(x-1)}{2x^2+x-6} - \dfrac{6x+15}{2x^2+9x+10}$

**53.** *Flying Time.* A private jet flys at the rate of $r$ miles per hour. If $t_1$ is the flying time from City A to City B, which are 1,500 miles apart, and $t_2$ the flying time for the return trip, then

$$ t_1 = \frac{1500}{r+18} \quad \text{and} \quad t_2 = \frac{1500}{r-18}. $$

Write an expression that represents the time for the round trip if the rate of the wind is 18 mi/hr. Simplify this expression.

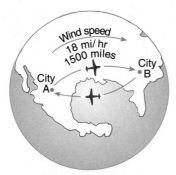

For Exercise 53.

**54.** *Shared Work.* Sarah and John are working together on a large painting project. After working $t$ hours, the fractional part of the work completed by Sarah is $t/3$ and that completed by John is $t/4$. What fractional part of the work has been completed by Sarah and John working together.

TRANSLATING WORDS TO SYMBOLS

**55.** Find the sum of the reciprocals of two consecutive integers.

**56.** Find the sum of the reciprocals of two consecutive odd integers.

**57.** Find the difference of the reciprocals of two consecutive even integers.

**58.** Find the sum of the reciprocals of three consecutive even integers.

In Exercises 59 and 60, use the fact that $d = r \cdot t$.

**59.** *Planning a Trip.* Jack is planning a 300-mi trip. He plans to drive part of the trip on country roads at 55 mph and part of the trip on an interstate highway at 65 mph.

  a) If he drives $x$ miles on country roads, how much time does he drive at 55 mph?

  b) How much time does he drive at 65 mph?

  c) Write an algebraic representation for the total time it takes to drive the 300 mi.

**60.** *Marathon Running.* Jenny is a marathon runner. She runs two paces during the race: one 7 mi/min pace and one 6-mi/min pace.

  a) If she runs $x$ miles at the slower pace, how much time does she run at this pace?

  b) How much time does she run at the faster pace?

  c) Write an algebraic representation for the total time of the race.

**61.** *Multiple Representations on a Grapher.* Use both a numerical table and a graph to show that

$$ \frac{3}{x} + \frac{x}{5} \neq \frac{3+x}{x+5}. $$

**62.** *Multiple Representations on a Grapher.* Use both a numerical table and a graph to show that

$$ \frac{6}{x-1} + \frac{1}{x+1} \neq \frac{7}{2x}. $$

**63.** *Formula from Electronics.* Three resistors with resistance of $R_1$, $R_2$, and $R_3$, respectively, are connected in parallel. The total resistance $R$ satisfies the equation

$$\frac{1}{R} = \frac{1}{R_1} + \frac{1}{R_2} + \frac{1}{R_3}.$$

Write an equivalent equation with a single denominator on the right-hand side.

In Exercises 64–73, perform the indicated operation. Simplify the result.

**64.** $\dfrac{5}{12} + \dfrac{7}{18} - \dfrac{11}{6}$

**65.** $\dfrac{4}{15} - \dfrac{6}{35} + \dfrac{5}{21}$

**66.** $\dfrac{1}{x} + \dfrac{1}{x^2} + 1$

**67.** $\dfrac{1}{x^2} + \dfrac{1}{x} - 1$

**68.** $\dfrac{1}{x - 1} - \dfrac{1}{x^2 - 1}$

**69.** $\dfrac{2}{x - 5} - \dfrac{2}{x + 5}$

**70.** $x - \dfrac{1}{x + 1}$

**71.** $a + \dfrac{1}{a + 1}$

**72.** $\dfrac{3x}{x^2 - x} - \dfrac{2x}{x^2 - 2x + 1}$

**73.** $\dfrac{x}{2x^2 - x - 6} - \dfrac{x - 4}{2x^2 + 5x + 3}$

EXTENDING THE IDEAS

In Exercises 74–77, perform the indicated operation.

**74.** $\dfrac{1}{x + 2} - \dfrac{1}{(x + 2)^2} + \dfrac{1}{(x + 2)^3}$

**75.** $\dfrac{3}{x - 5} - \dfrac{2}{(x - 5)^2} + \dfrac{1}{(x - 5)^3}$

**76.** $\dfrac{1}{x^2 - 4x + 4} + \dfrac{2}{x^2 + 4x + 4} - \dfrac{1}{x^2 - 4}$

**77.** $\dfrac{x}{x - 1} + \dfrac{2}{x^2 + x + 1} + \dfrac{2x^2 + 4}{x^3 - 1}$

**78.** This exercise refers to the function

$$f(x) = \frac{3x + 10}{x^2 + 7x + 12} - \frac{x + 2}{x^2 + 5x + 6}$$

that occurs in Example 7.

**a)** Find the graph of $f(x)$ in the 0.1 window. In this window, you see one part of the complete graph that is shown in Figure 6.14. How many "holes" do you see in the graph?

**b)** Find the trace coordinates for $x = -4$, $x = -3$, and $x = -2$. What do you observe?

**c)** How do the trace coordinates found in part (b) relate to the holes in the graph observed in part (a)?

**d)** Find the graph of $g(x) = \dfrac{2}{x + 4}$ in the 0.1 window. How does it compare to the graph found in part (a)?

**e) Writing to Learn.** Write a paragraph explaining the sense in which the graphs of $f(x)$ and $g(x)$ are identical.

LOOKING BACK—MAKING CONNECTIONS

In Exercises 79–84, write the expression using positive exponents and simplify.

**79.** $x^{-1} + x^{-2}$

**80.** $3x^{-1} + 5x^{-2}$

**81.** $(2x)^{-1} + 3x^{-1}$

**82.** $x^{-1} + y^{-1}$

**83.** $2 + x^{-1} + x^{-2}$

**84.** $(x - 1)^{-1} + x(x + 2)^{-1}$

**85.** *Using Algebra in Geometry.* The following triangles ABC and A'B'C' are similar. Find the length of sides AB and A'C'.

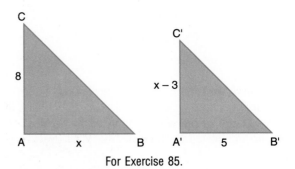

For Exercise 85.

# Complex Fractions

Simplifying Complex Fractions: Method 1 ▪ Simplifying Complex Fractions: Method 2

A **complex fraction** is an expression that has a fraction in either the numerator, the denominator, or both the numerator and denominator. Some examples of complex fractions are the following:

$$\frac{\dfrac{4}{x}}{\dfrac{3}{y}}, \qquad \frac{\dfrac{x-1}{2}}{\dfrac{3x-2}{5}}, \qquad \frac{\dfrac{1}{a}+\dfrac{1}{b}}{\dfrac{1}{a}-\dfrac{1}{b}}, \qquad \frac{\dfrac{1}{x}+2}{x+2}$$

Notice that there are three main components to a complex fraction: the numerator, the denominator, and the fraction line. It is important to be able to distinguish these individual components.

Often it is desirable to simplify complex expressions like those above. There are two methods of simplifying such complex fractions. The method that is easiest varies from one situation to another, so it is important to master both.

## Simplifying Complex Fractions: Method 1

Definition 6.3, the definition for dividing two rational expressions, is the basis for this first method. We interpret the main fraction bar in the expression

$$\frac{\dfrac{a}{b}}{\dfrac{c}{d}}$$

to mean "divide."

Consequently we can interpret the complex fraction as follows:

$$\frac{\dfrac{a}{b}}{\dfrac{c}{d}} = \frac{a}{b} \div \frac{c}{d} = \frac{a}{b} \cdot \frac{d}{c}$$

Example 1 shows how to use Method 1.

**EXAMPLE 1**   Simplifying a Complex Fraction

Simplify the following complex fraction:

$$\frac{\dfrac{x+2}{x}}{\dfrac{x-2}{3x}}$$

SOLUTION

Invert the divisor and multiply.

$$\frac{\dfrac{x+2}{x}}{\dfrac{x-2}{3x}} = \frac{x+2}{\cancel{x}} \cdot \frac{3\cancel{x}}{x-2} \qquad \textbf{Divide out the common factors.}$$

$$= \frac{3(x+2)}{x-2} \qquad\qquad\qquad\qquad\blacksquare$$

When the numerator and/or the denominator of a complex rational expression is itself the sum or difference of rational expressions, we first perform the indicated operations and simplify both the numerator and denominator. Then the division indicated by the main fraction bar is completed. In such cases, method 1 becomes a three-step process.

---

### Simplifying Complex Fractions Using Method 1

In this method, first simplify the numerator and denominator separately and then perform the division. More specifically, follow these steps:

1. Combine terms of the numerator of the complex fraction (if possible) and simplify to one fraction.
2. Combine terms of the denominator of the complex fraction (if possible) and simplify to one fraction.
3. Invert the denominator and multiply times the numerator. Reduce the fraction to lowest terms.

---

Example 2 illustrates that this method is particularly helpful in more complicated problems. The three-step simplification allows you to think of the simplification as three separate problems.

**EXAMPLE 2**     Simplifying Using Method 1

Simplify the following complex fraction:

$$\frac{\dfrac{1}{x+1}-\dfrac{1}{x}}{\dfrac{1}{x}+\dfrac{2}{x+2}}$$

SOLUTION

Simplify the numerator using $x(x+1)$ as LCD. Simplify the denominator using $x(x+2)$ as the LCD.

$$\frac{\dfrac{1}{x+1}-\dfrac{1}{x}}{\dfrac{1}{x}+\dfrac{2}{x+2}}=\frac{\dfrac{x}{x}\cdot\dfrac{1}{(x+1)}-\dfrac{1}{x}\cdot\dfrac{(x+1)}{(x+1)}}{\dfrac{(x+2)}{(x+2)}\cdot\dfrac{1}{x}+\dfrac{2}{(x+2)}\cdot\dfrac{x}{x}}$$

$$=\frac{\dfrac{x-(x+1)}{x(x+1)}}{\dfrac{x+2+2x}{x(x+2)}}$$          **Simplify the numerator and denominator separately.**

$$=\frac{\dfrac{-1}{x(x+1)}}{\dfrac{3x+2}{x(x+2)}}$$          **Rewrite as multiplication by inverting the divisor and multiplying.**

$$=\frac{-1}{x(x+1)}\cdot\frac{x(x+2)}{(3x+2)}$$

$$=\frac{-(x+2)}{(x+1)(3x+2)}$$          ∎

## Simplifying Complex Fractions: Method 2

In method 1, we first simplified the numerator and the denominator of the main fraction bar separately. On the other hand, in method 2 we multiply both the main numerator and denominator by the same expression, which is chosen so that the individual denominators can be eliminated.

Method 2 is sometimes called the LCD method.

---

### Simplifying Complex Fractions Using Method 2: the LCD Method

In this method, multiply the main numerator and denominator by the same expression, which is chosen so that individual denominators can be eliminated.

1. Find the LCD of the denominators of all the individual fractions.
2. Multiply the numerator and the denominator of the complex fraction by this LCD.
3. Simplify the resulting expression.

---

**REMINDER**

The numerator and the denominator must be multiplied by the same number. In this way, the entire fraction is being multiplied by a form of 1.

**GRAPHER NOTE**

Remember that parentheses play a vital role in the entry of fractions into a grapher. When verifying complex fractions, make sure you have parentheses around the entire numerator and around the entire denominator.

Example 3 illustrates the LCD method.

**EXAMPLE 3**   Simplifying Using the LCD Method

Simplify the following complex fraction:

$$\frac{\dfrac{1}{2} + \dfrac{2}{3}}{\dfrac{5}{6} - \dfrac{3}{4}}$$

SOLUTION

**Simplify Algebraically.** The LCD of the denominators $\{2, 3, 4, 6\}$ is 12. Multiply both the numerator and the denominator of the complex fraction by 12.

$$\frac{\dfrac{1}{2} + \dfrac{2}{3}}{\dfrac{5}{6} - \dfrac{3}{4}} = \frac{12\left(\dfrac{1}{2} + \dfrac{2}{3}\right)}{12\left(\dfrac{5}{6} - \dfrac{3}{4}\right)} \qquad \begin{array}{l}\textbf{Multiply the numerator and}\\ \textbf{denominator by the LCD}\\ \textbf{and use the Distributive Property.}\end{array}$$

$$= \frac{6+8}{10-9}$$

$$= \frac{14}{1} \quad \text{or} \quad 14$$

**Support Numerically.** Figure 6.17 shows numerical support of the simplification. ∎

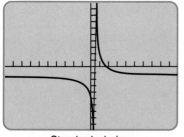

$$(1/2 + 2/3)/(5/6 - 3/4)$$
$$14$$

**FIGURE 6.17** Grapher support for the simplification in Example 3.

Example 4 illustrates using the same method to simplify a complex fraction that contains variables.

**EXAMPLE 4**    Simplifying a Complex Fraction with Variables

Simplify the following complex fraction algebraically and support with a grapher:

$$\frac{\dfrac{3}{a} - \dfrac{5}{a^2}}{\dfrac{1}{a^2} - \dfrac{2}{a}}$$

SOLUTION

**Simplify Algebraically.** Multiply the numerator and denominator by the LCD of $\{a, a^2\}$, which is $a^2$.

$$\frac{\dfrac{3}{a} - \dfrac{5}{a^2}}{\dfrac{1}{a^2} - \dfrac{2}{a}} = \frac{a^2\left(\dfrac{3}{a} - \dfrac{5}{a^2}\right)}{a^2\left(\dfrac{1}{a^2} - \dfrac{2}{a}\right)}$$

Use the Distributive Property in both the numerator and denominator.

$$= \frac{3a - 5}{1 - 2a}$$

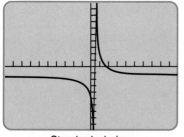

Standard window

**FIGURE 6.18** Grapher support of the algebraic simplification in Example 4. The graphs of $y_3$ and $y_4$ appear to be identical.

**Support Graphically.** Replace $a$ with $x$ and let $y_1 = 3/x - 5/x^2$, $y_2 = 1/x^2 - 2/x$, $y_3 = y_1/y_2$, and $y_4 = (3x - 5)/(1 - 2x)$. Figure 6.18 shows the graphs of both $y_3$ and $y_4$. These two graphs appear to be identical, thus providing graphical support for the algebra. Comparing trace coordinates provides additional support.  ∎

Fractions that contain negative exponents become complex fractions when we eliminate the negative exponent. Example 5 illustrates changing a fraction with negative exponents to a complex fraction and then simplifying.

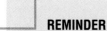

**REMINDER**

Remember that a negative exponent indicates a reciprocal; that is, $a^{-1} = \dfrac{1}{a}$ and $2b^{-1} = \dfrac{2}{b}$.

**EXAMPLE 5**    Negative Exponents within Complex Fractions

Simplify the following rational expression:

$$\frac{x^{-1} + 2y^{-1}}{x^{-1} - y^{-1}}$$

SOLUTION

Begin by eliminating the negative exponents by changing to reciprocal form.

$$\frac{x^{-1} + 2y^{-1}}{x^{-1} - y^{-1}} = \frac{\dfrac{1}{x} + \dfrac{2}{y}}{\dfrac{1}{x} - \dfrac{1}{y}}$$

**Multiply the numerator and denominator by the LCD $xy$.**

$$= \frac{xy\left(\dfrac{1}{x} + \dfrac{2}{y}\right)}{xy\left(\dfrac{1}{x} - \dfrac{1}{y}\right)}$$

**Use the Distributive Property and simplify.**

$$= \frac{xy\left(\dfrac{1}{x}\right) + xy\left(\dfrac{2}{y}\right)}{xy\left(\dfrac{1}{x}\right) - xy\left(\dfrac{1}{y}\right)}$$

$$= \frac{y + 2x}{y - x}$$

∎

Reflect on Example 5. How would you simplify this complex fraction by using the three-step method? Do you have a preference between the two methods?

## Exercises for Section  6.4

In Exercises 1–8, simplify the complex fraction.

1. $\dfrac{\dfrac{2}{x}}{\dfrac{x}{5}}$

2. $\dfrac{\dfrac{x}{7}}{\dfrac{x^2}{14}}$

3. $\dfrac{\dfrac{x}{y}}{\dfrac{x^2}{y^3}}$

4. $\dfrac{x + y}{\dfrac{x}{x - y}}{x}$

5. $\dfrac{\dfrac{x - y}{x + y}}{\dfrac{x + y}{x - y}}$

6. $\dfrac{\dfrac{a + b}{x + y}}{\dfrac{a^2 - b^2}{x^2 - y^2}}$

7. $\dfrac{\dfrac{x + y}{x - y}}{\dfrac{x - y}{x + y}}$

8. $\dfrac{\dfrac{x - 2}{x + 3}}{\dfrac{x^2 - 4}{x^2 - 9}}$

In Exercises 9–14, use method 1 to simplify the complex fraction.

9. $\dfrac{\dfrac{1}{x} + \dfrac{1}{y}}{\dfrac{1}{y} - \dfrac{1}{x}}$

10. $\dfrac{\dfrac{1}{a} - \dfrac{1}{b}}{\dfrac{2}{a} + \dfrac{2}{b}}$

11. $\dfrac{x - \dfrac{15}{x - 2}}{x - \dfrac{20}{x - 1}}$

12. $\dfrac{x + \dfrac{35}{x + 12}}{x - \dfrac{63}{x - 2}}$

13. $\dfrac{\dfrac{2}{3y} + \dfrac{3}{y^2}}{\dfrac{2}{5y} + \dfrac{5}{y}}$

14. $\dfrac{\dfrac{3}{2x} + \dfrac{5}{x^2}}{\dfrac{1}{2x} - \dfrac{3}{x}}$

In Exercises 15 and 16, rewrite the expression to verify that it is a complex fraction.

15. $(x^{-1} + 5^{-1})^{-1}$

16. $2(x^{-1} - 4^{-1})^{-1}$

Exercises 17 and 18 refer to points on a coordinate line. Given points $x_1$ and $x_2$ on a coordinate line, the midpoint $M$ is the point with coordinate $\dfrac{x_1 + x_2}{2}$.

For Exercises 17 and 18.

17. *Using Algebra in Geometry.* If $A$ is a point with coordinate $x_1 = \dfrac{x}{3}$ and $B$ is a point with coordinate $x_2 = \dfrac{x}{5}$, what is the coordinate of the midpoint of segment $AB$?

18. *Using Algebra in Geometry.* If $A$ is a point with coordinate $x_1 = \dfrac{x}{12}$ and $B$ is a point with coordinate $x_2 = \dfrac{x}{4}$, what is the coordinate of the midpoint of segment $AB$?

19. *Multiple Representations on a Grapher.* Use both a numerical table and a graph to show that

$$\dfrac{x^{-2} + 4^{-2}}{x^{-1} + 4^{-1}} \neq x + 4.$$

20. *Multiple Representations on a Grapher.* Use both a numerical table and a graph to show that

$$\dfrac{\dfrac{1}{x} + \dfrac{1}{5}}{\dfrac{2}{x} + \dfrac{3}{5}} \neq \left(\dfrac{1}{x} + \dfrac{1}{5}\right)\left(\dfrac{x}{2} + \dfrac{5}{3}\right).$$

21. *Simplifying Units.* The concepts acceleration, velocity, and time are related by the equation

$$a = \dfrac{v}{t}\dfrac{ft/sec}{sec}.$$

Explain why the units $ft/sec^2$ also are correct for acceleration.

22. *Formula from Optics.* The following formula is familiar in optics:

$$\dfrac{1}{f} = \dfrac{1}{p} + \dfrac{1}{q}.$$

Which of the following complex fraction formulas are equivalent to this formula?

a) $f = (p^{-1} + q^{-1})^{-1}$

b) $f = \dfrac{1}{\dfrac{1}{p} + \dfrac{1}{q}}$

In Exercises 23–26, list the LCD of all the denominators in the complex fraction.

23. $\dfrac{\dfrac{a}{b} + \dfrac{1}{a}}{\dfrac{1}{b^2} - \dfrac{1}{a}}$

24. $\dfrac{\dfrac{1}{a} + \dfrac{1}{2}}{\dfrac{1}{2} - \dfrac{1}{a^2}}$

25. $\dfrac{\dfrac{3}{x} + \dfrac{2}{y}}{\dfrac{2}{x^2} - \dfrac{1}{y^2}}$

26. $\dfrac{\dfrac{2}{x - 1} + \dfrac{1}{x + 2}}{\dfrac{1}{x + 2} - \dfrac{1}{3}}$

In Exercises 27–30, use the LCD method to simplify the complex fraction.

27. Simplify the complex fraction in Exercise 23.

28. Simplify the complex fraction in Exercise 24.

29. Simplify the complex fraction in Exercise 25.

30. Simplify the complex fraction in Exercsie 26.

In Exercises 31–34, simplify the complex fraction. In the even-numbered exercises, support the results using a grapher.

**31.** $\dfrac{\dfrac{1}{x} - 2}{\dfrac{2}{x} + 3}$

**32.** $\dfrac{x - 2}{x - \dfrac{4}{x}}$

**33.** $\dfrac{\dfrac{3}{2x} + \dfrac{5}{x^2}}{\dfrac{1}{3x} - \dfrac{4}{x}}$

**34.** $\dfrac{x + \dfrac{3}{x + 1}}{x - \dfrac{3}{(x - 2)^2}}$

In Exercises 35–40, simplify the complex fraction.

**35.** $\dfrac{\dfrac{a^2}{b} - b}{a - b}$

**36.** $\dfrac{\dfrac{1}{x} - \dfrac{1}{x^2}}{\dfrac{1}{x} + \dfrac{1}{x^2}}$

**37.** $\dfrac{x - \dfrac{5}{y^2}}{x - \dfrac{2}{y}}$

**38.** $\dfrac{\dfrac{1}{2b} + \dfrac{2}{b^2}}{\dfrac{2}{5b} - \dfrac{1}{b}}$

**39.** $\dfrac{2 + \dfrac{x^2 - y^2}{x^2 + y^2}}{2 - \dfrac{x^2 - y^2}{x^2 + y^2}}$

**40.** $\dfrac{x - \dfrac{x + 4}{x + 1}}{\dfrac{x - 2}{x + 1}}$

In Exercises 41–46, state which method you feel is easier to execute to simplify the complex fractions and then simplify.

**41.** $\dfrac{\dfrac{1}{a} - \dfrac{5}{b}}{\dfrac{2}{a} + \dfrac{3}{b}}$

**42.** $\dfrac{\dfrac{1}{x} - \dfrac{1}{x^2}}{\dfrac{2}{x} + \dfrac{3}{x^2}}$

**43.** $\dfrac{\dfrac{1}{x + y} - \dfrac{1}{x - y}}{\dfrac{1}{x - y} + \dfrac{1}{x + y}}$

**44.** $\dfrac{x - \dfrac{1}{a}}{x + \dfrac{1}{a}}$

**45.** $\dfrac{\dfrac{1}{x + 1} - \dfrac{2}{x}}{\dfrac{3}{x + 2} + \dfrac{5}{x + 3}}$

**46.** $\dfrac{\dfrac{1}{a + b} - \dfrac{1}{(a + b)^2}}{\dfrac{1}{2a - b} - \dfrac{1}{a + b}}$

**47.** *Finding Slope.* Line $L$ intersects the graph of $y = \dfrac{1}{x}$ at two points. Show that the slope of line $L$ is $\dfrac{-1}{x_1 x_2}$.

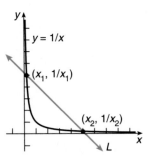

For Exercise 47.

**48.** *Finding Slope.* If line $L$ intersects the graph of $f(x) = \dfrac{5}{1 + x^2}$ at $(x_1, f(x_1))$ and $(-x_1, f(-x_1))$, show that the slope of $L$ is zero.

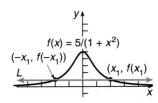

For Exercise 48.

In Exercises 49–56, write the expression without negative exponents and then simplify the complex fraction.

**49.** $\dfrac{x^{-1}}{y^{-1}}$

**50.** $\dfrac{x^{-1}}{x^{-1} + y^{-1}}$

**51.** $\dfrac{x^{-1} + y}{x^{-1} - y}$

**52.** $\dfrac{x^{-1} - y^{-1}}{x^{-1} + y^{-1}}$

**53.** $\dfrac{x^{-1} + 3y^{-1}}{x^{-1} - 2y^{-1}}$

**54.** $\dfrac{a^{-2}}{a^{-2} - b^{-2}}$

55. $\dfrac{2x^{-1} + y^{-1}}{x^{-1} - 3y^{-1}}$

56. $\dfrac{a^{-2} - b^{-2}}{a^{-2}}$

## TRANSLATING WORDS TO SYMBOLS

In Exercises 57–60, write the word description in symbols.

57. The sum of the reciprocals of $a$ and $b$

58. The difference of the reciprocals of $a$ and $b$

59. The reciprocal of the sum of $a$ and $b$

60. The reciprocal of the difference of $a$ and $b$

61. **Writing to Learn.** In your own words, describe a complex fraction.

62. **Writing to Learn.** In your own words, describe which method of simplifying complex fractions you prefer and why.

## EXTENDING THE IDEAS

63. Simplify the following complex fraction:

$$a + \cfrac{1}{a + \cfrac{1}{a + \cfrac{1}{a + \cfrac{1}{a}}}}$$

64. Simplify the following complex fraction:

$$\cfrac{\cfrac{1}{y+1}}{y + 1 + \cfrac{1}{y+1}}$$

## LOOKING BACK—MAKING CONNECTIONS

In Exercises 65–68, simplify the polynomial expression.

65. $(3x^2 + 5x - 3) - (x^2 - 6x + 5)$
66. $(x^2 - 2x + 1) - (3x + 5)$
67. $(2x^2 - 3x + 1) - (2x^2 + x - 8)$
68. $(3x - 5) - (2x + 7)$

In Exercises 69–72, subtract the polynomial expressions vertically.

69. Subtract
$$\begin{array}{r} 3a^2 + 4a + 7 \\ 2a^2 + 2a + 6 \\ \hline \end{array}$$

70. Subtract
$$\begin{array}{r} 4p^2 + 6p + 12 \\ 1.5p^2 + 6p + 3 \\ \hline \end{array}$$

71. Subtract
$$\begin{array}{r} -3x^2 + 2x + 2 \\ 5x^2 - 7x + 17 \\ \hline \end{array}$$

72. Subtract
$$\begin{array}{r} 2x^2 + 3xy + 4y^2 \\ -2x^2 + 2xy + 2y^2 \\ \hline \end{array}$$

## 6.5 Division of Polynomials

Dividing by a Monomial ▪ Dividing by a Binomial ▪ Synthetic Division

In Chapter 5, you learned to add, subtract, and multiply polynomials. In this section, you will learn how to divide one polynomial by another using processes that are similar to division of rational numbers. We begin by considering the case when the divisor is a monomial.

## Dividing by a Monomial

The sum of two or more rational expressions with the same denominator is the sum of the numerators divided by the common denominator. That is,

$$\frac{a}{d} + \frac{b}{d} + \frac{c}{d} = \frac{a+b+c}{d}.$$

Dividing the expression $a + b + c$ by $d$ involves reversing this process and splitting up the expression into three terms. For example, $(4x^4 + 5x^3 - 3x^2) \div x^2$ is completed as follows:

$$\frac{4x^4 + 5x^3 - 3x^2}{x^2} = \frac{4x^4}{x^2} + \frac{5x^3}{x^2} - \frac{3x^2}{x^2}$$

$$= 4x^2 + 5x - 3$$

To divide a polynomial by a monomial, we can divide each term in the numerator by the monomial. Example 1 illustrates this process.

**EXAMPLE 1**    Dividing by a Monomial

Divide $14x^5 - 7x^4 + 3x^3$ by $x^3$ and support the quotient graphically.

SOLUTION

**Simplify Algebraically.** Divide each term in the dividend by the divisor.

$$\frac{14x^5 - 7x^4 + 3x^3}{x^3} = \frac{14x^5}{x^3} - \frac{7x^4}{x^3} + \frac{3x^3}{x^3} \qquad \substack{\textbf{Reduce each term} \\ \textbf{to lowest terms.}}$$

$$= 14x^2 - 7x + 3$$

**Support Graphically.** Let $y_1 = (14x^5 - 7x^4 + 3x^3)/x^3$ and $y_2 = 14x^2 - 7x + 3$. Figure 6.19 shows the graphs of $y_1$ and $y_2$, which provide grapher support for the algebra. ∎

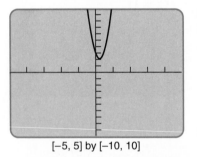

[−5, 5] by [−10, 10]

**FIGURE 6.19** Graphs of both $y_1$ and $y_2$.

When dividing each term of a polynomial by a monomial, a fraction might result. Example 2 illustrates this case.

**EXAMPLE 2**    Dividing by a Monomial

Complete the following division algebraically and support the solution using a grapher:

$$\frac{4x^3 - 6x^2 + 8x - 3}{2x}$$

SOLUTION
**Simplify Algebraically.**

$$\frac{4x^3 - 6x^2 + 8x - 3}{2x} = \frac{4x^3}{2x} - \frac{6x^2}{2x} + \frac{8x}{2x} - \frac{3}{2x}$$

$$= 2x^2 - 3x + 4 - \frac{3}{2x}$$

**Support Graphically.** Let $y_1 = (4x^3 - 6x^2 + 8x - 3)/(2x)$ and $y_2 = 2x^2 - 3x + 4 - 3/(2x)$. Figure 6.20 shows that the graphs of $y_1$ and $y_2$ appear to be the same, thus providing support for the algebra.    ■

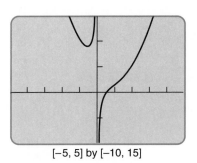

[−5, 5] by [−10, 15]

**FIGURE 6.20**
Graphs of both
$y_1 = (4x^3 - 6x^2 + 8x - 3)/(2x)$
and $y_2 = 2x^2 - 3x + 4 - 3/(2x)$.

## Dividing by a Binomial

An **algorithm** is a step-by-step process. When applying the division algorithm, there is a divisor, dividend, quotient, and remainder. The outcome of the division algorithm can be written as follows:

$$\frac{\text{Dividend}}{\text{Divisor}} = \text{Quotient} + \frac{\text{Remainder}}{\text{Divisor}}$$

Example 3 illustrates this algorithm and uses the "long division" format.

**EXAMPLE 3**    Using a Division Algorithm

Find the quotient and the remainder when $6x^3 + 14x^2 + 10x + 3$ is divided by $3x + 1$. Support the solution graphically.

SOLUTION

**Step 1:** To begin this division process, divide $3x$ into the first term $6x^3$ to obtain $2x^2$. Write this $2x^2$ in the quotient above the horizontal rule.

$$
\begin{array}{r}
2x^2 \qquad\qquad\qquad \leftarrow \text{Quotient} \\
\text{Divisor} \longrightarrow\ 3x + 1 \overline{)\ 6x^3 + 14x^2 + 10x + 3} \quad \leftarrow \text{Dividend}
\end{array}
$$

**Step 2:** Continue by multiplying $2x^2$ by the divisor $3x + 1$ to obtain $6x^3 + 2x^2$, which is subtracted from the dividend and written below another horizontal rule.

$$
\begin{array}{r}
2x^2 \qquad\qquad\qquad\quad \leftarrow \text{Quotient} \\
\text{Divisor} \longrightarrow\ 3x + 1 \overline{)\ 6x^3 + 14x^2 + 10x + 3} \\
\underline{6x^3 +\ 2x^2\ \qquad\qquad} \quad \leftarrow 2x^2 \text{ multiplied by } 3x + 1 \\
12x^2 + 10x + 3 \quad \leftarrow \text{Result of first subtraction}
\end{array}
$$

**Step 3:** Continue this process a second and a third time. This division process ends with a remainder of 1.

$$
\begin{array}{r}
2x^2 + 4x + 2 \quad \longleftarrow \text{Quotient} \\
\text{Divisor} \longrightarrow \quad 3x+1 \overline{\smash{\big)}\ 6x^3 + 14x^2 + 10x + 3} \\
\underline{6x^3 + \phantom{1}2x^2} \\
12x^2 + 10x + 3 \\
\underline{12x^2 + \phantom{1}4x} \quad \longleftarrow 4x \text{ multiplied by } 3x+1 \\
6x + 3 \\
\underline{6x + 2} \\
1 \quad \longleftarrow \text{Remainder}
\end{array}
$$

The quotient is $2x^2 + 4x + 2$ and the remainder is 1. It follows that

$$\frac{6x^3 + 14x^2 + 10x + 3}{3x + 1} = 2x^2 + 4x + 2 + \frac{1}{3x + 1}.$$

**Support Graphically.** Let $y_1 = (6x^3 + 14x^2 + 10x + 3)/(3x + 1)$ and $y_2 = 2x^2 + 4x + 2 + 1/(3x + 1)$. Figure 6.21 shows that the graphs of $y_1$ and $y_2$ appear to be the same, thus supporting the algebra. ■

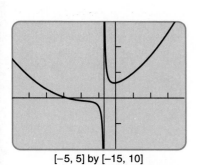

[−5, 5] by [−15, 10]

**FIGURE 6.21** Graphs of both $y_1$ and $y_2$ in Example 3.

When using the long division process, care must be taken that the signs are written correctly in the subtraction step. Recall that to subtract, you use the rule $a - b = a + (-b)$. This rule is sometimes abbreviated as "change the sign of $b$ and add." You will use this rule in the next example.

**EXAMPLE 4**   Dividing by a Binomial

Find the quotient and the remainder when $2x^3 + x^2 - 18x + 10$ is divided by $2x - 5$.

SOLUTION

$$
\begin{array}{r}
x^2 + \phantom{1}3x - \phantom{1}3/2 \\
2x-5 \overline{\smash{\big)}\ 2x^3 + \phantom{1}x^2 - 18x + \phantom{1}10} \\
\underline{{}^{-}2x^3 \pm \phantom{1}5x^2} \quad \longleftarrow \text{Change the sign and add.} \\
6x^2 - 18x + \phantom{1}10 \\
\underline{{}^{-}6x^2 \pm 15x} \quad \longleftarrow \text{Change the sign and add.} \\
-3x + \phantom{1}10 \\
\underline{\pm 3x + 15/2} \\
5/2
\end{array}
$$

The quotient is $x^2 + 3x - 3/2$ and the remainder is 5/2. Using the format

$$\frac{\text{Dividend}}{\text{Divisor}} = \text{Quotient} + \frac{\text{Remainder}}{\text{Divisor}}$$

we can write the fraction as follows:

$$\frac{2x^3 + x^2 - 18x + 10}{2x - 5} = (x^2 + 3x - 3/2) + \frac{5/2}{2x - 5}$$

$$= (x^2 + 3x - 3/2) + \frac{5}{2(2x - 5)} \quad ■$$

When applying the division algorithm to polynomials, it is important to arrange the terms with powers in descending order and to write any "missing term" in the dividend as $0x^k$. Example 5 illustrates why.

**EXAMPLE 5**    Using the Division Algorithm

Find the quotient and the remainder when $3x^3 + x - 2$ is divided by $x^2 + 2x + 1$.

SOLUTION

In the dividend, add the term $0x^2$.

$$
\begin{array}{r}
3x - 6 \quad \longleftarrow \text{Quotient} \\
x^2 + 2x + 1 \overline{\smash{\big)}\ 3x^3 + 0x^2 + x - 2} \\
\underline{-3x^3 \mp 6x^2 \mp 3x} \\
-6x^2 - 2x - 2 \\
\underline{\pm 6x^2 \pm 12x \pm 6} \\
10x + 4 \quad \longleftarrow \text{Remainder}
\end{array}
$$

The quotient is $3x - 6$ and the remainder is $10x + 4$. Therefore

$$\frac{3x^3 + x - 2}{x^2 + 2x + 1} = 3x - 6 + \frac{10x + 4}{x^2 + 2x + 1}. \quad ■$$

It is important that at each step in the division algorithm, each term is written in the column with the same exponent. For example, $x^3$ terms go in the $x^3$ column, $x^2$ terms in the $x^2$ column, and so forth. In this way, like terms are under each other, making the subtraction step easier.

**EXAMPLE 6**    Dividing by a Binomial

Find the quotient and the remainder when $4x^3 + 3x^2 - 4x + 1$ is divided by $x^2 + 1$. Support the conclusion graphically.

SOLUTION
**Simplify Algebraically.**

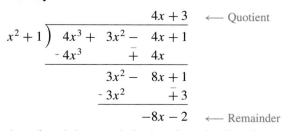

$$
\begin{array}{r}
4x + 3 \quad \longleftarrow \text{Quotient} \\
x^2 + 1 \overline{)\, 4x^3 + 3x^2 - 4x + 1} \\
\underline{-\,4x^3 \qquad\quad +\ 4x} \\
3x^2 - 8x + 1 \\
\underline{-\,3x^2 \qquad +3} \\
-8x - 2 \quad \longleftarrow \text{Remainder}
\end{array}
$$

The quotient is $4x + 3$ and the remainder is $-8x - 2$. Therefore

$$
\frac{4x^3 + 3x^2 - 4x + 1}{x^2 + 1} = 4x + 3 - \frac{8x + 2}{x^2 + 1}.
$$

**Support Graphically.** Figure 6.22 shows that the graphs of $y_1 = (4x^3 + 3x^2 - 4x + 1)/(x^2 + 1)$ and $y_2 = 4x + 3 - (8x + 2)/(x^2 + 1)$ appear to be the same, thus supporting the solution. ■

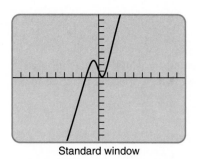

Standard window

**FIGURE 6.22**
Graphs of $y_1 = (4x^3 + 3x^2 - 4x + 1)/(x^2 + 1)$ and $y_2 = 4x + 3 - (8x + 2)/(x^2 + 1)$ appear to be identical.

## Synthetic Division

In this section, we show how to streamline the long division process in the case in which the divisor is a linear factor of the form $x - c$.

Notice that in the following long division process, the variables do not enter into the calculation of the coefficients. They simply serve the need to line up like terms.

$$
\begin{array}{r}
x^2 - x + 2 \\
x - 2 \overline{)\, x^3 - 3x^2 + 4x - 4} \\
\underline{x^3 - 2x^2} \\
-x^2 + 4x \\
\underline{-x^2 + 2x} \\
2x - 4 \\
\underline{2x - 4} \\
0
\end{array}
$$

By not writing the variable and including only the arithmetic on the coefficients that determine the quotient and remainder, the long division process can be streamlined into what we call **synthetic division**.

We explain the synthetic division process in more detail for the above long division problem. List the coefficients in the first row. Because we are dividing by $x - 2$, the 2 is the key number which is indicated by its placement in a bracket on the left of the first row. Each number in the second row is obtained by multiplying the entry in row 3 of the previous column by the number in the bracket, in this case, 2. Each number in the third row is obtained by adding the corresponding entries in rows 1 and 2.

$$
\begin{array}{r}
\text{Divisor} \\
(x-2)
\end{array}
\quad
\underline{2} \begin{array}{|rrrr}
1 & -3 & 4 & -4 \\
 & 2 & -2 & 4 \\
\hline
1 & -1 & 2 & 0
\end{array}
\leftarrow \text{Coefficients of dividend}
$$

Coefficients of quotient        Remainder

**REMINDER**

When applying synthetic division, the coefficient of $x$ in the divisor must be 1.

The quotient is $x^2 - x + 2$ and the remainder is 0.

Example 7 illustrates using the synthetic division procedure with a dividend that has a term missing when it is arranged in descending order.

**EXAMPLE 7**    Using Synthetic Division

Use synthetic division to find the quotient when $3x^4 + 7x^3 + x - 11$ is divided by $x + 3$.

SOLUTION

$$
\begin{array}{r}
\text{Divisor} \\
(x-(-3))
\end{array}
\quad
\underline{-3} \begin{array}{|rrrrr}
3 & 7 & 0 & 1 & -11 \\
 & -9 & 6 & -18 & 51 \\
\hline
3 & -2 & 6 & -17 & 40
\end{array}
$$

$$
\begin{array}{ccccc}
\downarrow & \downarrow & \downarrow & \downarrow & \downarrow \\
3x^3 & -2x^2+6x & -17 & \dfrac{40}{x+3}
\end{array}
$$

The quotient + remainder/divisor is $3x^3 - 2x^2 + 6x - 17 + \dfrac{40}{x+3}$. ■

**EXAMPLE 8**    Using Synthetic Division

Use synthetic division to find the quotient when $x^5 + 2x^4 - 13x^3 + x^2 - 20x + 5$ is divided by $x - 3$.

SOLUTION

$$
\underline{3} \begin{array}{|rrrrrr}
1 & 2 & -13 & 1 & -20 & 5 \\
 & 3 & 15 & 6 & 21 & 3 \\
\hline
1 & 5 & 2 & 7 & 1 & 8
\end{array}
$$

The quotient + remainder/divisor is $x^4 + 5x^3 + 2x^2 + 7x + 1 + \dfrac{8}{x-3}$. ▪

Notice in Example 8 that the remainder is 8. Figure 6.23 shows a grapher evaluation of the polynomial $x^5 + 2x^4 - 13x^3 + x^2 - 20x + 5$ for $x = 3$. The polynomial is equal to 8, which is also the remainder when it is divided by $x - 3$. This is not a coincidence, as the following Remainder theorem illustrates. (The proof is studied in more advanced courses.)

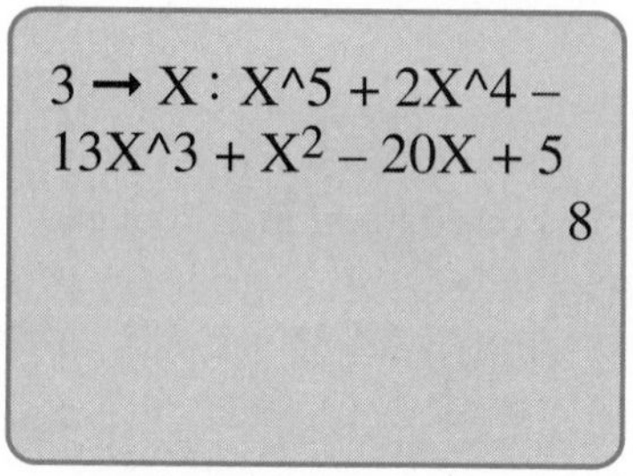

3 → X : X^5 + 2X^4 −
13X^3 + X2 − 20X + 5

           8

**FIGURE 6.23** Grapher support that $f(3) = 8$ in Example 8.

**THEOREM 6.1**  Remainder Theorem

If the polynomial $f(x)$ is divided by $x - c$, the remainder is equal to $f(c)$.

Therefore, if $f(x)$ is divided by $x - c$, $f(c)$ is the last entry in the synthetic division process. Before graphers, this method served as a quick way to evaluate a polynomial. Now we simply use the grapher. Compare the results in Example 8 with those in Figure 6.23.

Example 9 uses the Remainder theorem in conjunction with synthetic division to find a function value.

**EXAMPLE 9**  Finding a Function Value by Synthetic Division

Use synthetic division to find $f(2)$ for $f(x) = x^4 - 5x^3 + 7x^2 - 14x + 5$.

2 → X : X^4 − 5X^3 + 7
X2 − 14X + 5

           -19

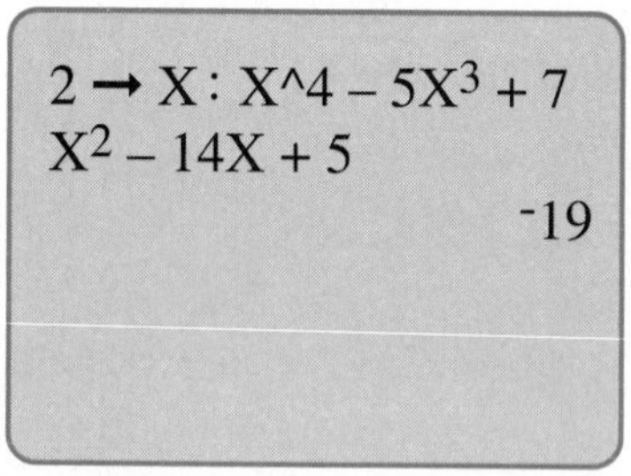

**FIGURE 6.24** Grapher support for Example 9.

SOLUTION

$$
\begin{array}{r|rrrrr}
2 & 1 & -5 & 7 & -14 & 5 \\
  &   & 2 & -6 & 2 & -24 \\
\hline
  & 1 & -3 & 1 & -12 & -19 \\
\end{array}
$$

←  Coefficients of $f(x)$

$\uparrow$
$f(2)$

This synthetic division tells us that $f(2) = -19$.

**Support Numerically.** Figure 6.24 shows numerical support for $f(2) = -19$.

▪

## Exercises for Section   6.5

In Exercises 1–10, complete the division. Support your answers to Exercises 1–8 using a grapher.

1. $\dfrac{3x^4 - 5x^3}{x^2}$

2. $\dfrac{2x^3 - 4x^2}{x^2}$

3. $\dfrac{6x^5 - 12x^3 + 4x}{2x}$

4. $\dfrac{7x^4 - 4x^3 + 2x^2}{x^2}$

5. $\dfrac{8x^2 + 16x - 2}{2x}$

6. $\dfrac{7x^3 + 5x^2 - 3}{x^2}$

7. $\dfrac{3x^3 - 6x^2 + 9x - 12}{3x}$

8. $\dfrac{6x^3 + 27x^2 - 12x + 9}{6x^2}$

9. $\dfrac{x^2y + 3xy^3}{xy}$

10. $\dfrac{6x^3y^2 - 9x^2y}{3x^2y}$

In Exercises 11–16, find the remainder when the polynomial is divided by $2x - 1$.

11. $2x^3 + 5x^2 - 11x + 9$

12. $6x^3 + 5x^2 - 6x + 3$

13. $2x^4 + 7x^3 - 18x^2 + 11x - 2$

14. $4x^4 - 12x^3 + 11x^2 - 7x + 5$

15. $4x^3 - 10x^2 + 14x + 3$

16. $12x^4 + 4x^3 - 13x^2 + 18x - 7$

In Exercises 17–22, find the polynomial that is the quotient when the rational expression is written in the form

$$\text{Quotient} + \frac{\text{Remainder}}{\text{Divisor}}.$$

17. $\dfrac{3x^3 - 13x^2 - 7x + 7}{3x + 2}$

18. $\dfrac{5x^4 + 23x^3 - 25x^2 - 4x + 7}{5x - 2}$

19. $\dfrac{4x^4 + 11x^3 - 14x^2 - 3x + 4}{4x + 3}$

20. $\dfrac{8x^4 + 10x^3 + 13x^2 - 11x + 3}{2x + 1}$

21. $\dfrac{4x^5 - x^4 - 4x^3 + x^2 + 12x - 8}{4x - 1}$

22. $\dfrac{x^6 - 5x^5 + 6x^4 + 7x^3 - 21x^2 + 5x - 12}{x - 3}$

In Exercises 23–28, write the rational expression in the form

$$\text{Quotient} + \frac{\text{Remainder}}{\text{Divisor}}.$$

Support the result with a grapher.

23. $\dfrac{x^4 + x^3 - 2x^2 + 7x + 3}{x^2 - 2x + 3}$

24. $\dfrac{x^5 - x^4 + 4x^2 - 4x + 3}{x^2 - x + 1}$

25. $\dfrac{2x^6 + 3x^5 + 5x^4 - 5x^2 - 2}{2x^2 + 3x + 1}$

26. $\dfrac{3x^5 - 5x^4 + 6x^3 + 2x^2 - 5x + 7}{3x^2 - 2x + 4}$

27. $\dfrac{3x^4 - x^3 + 5x^2 + x + 2}{x^2 + 1}$

28. $\dfrac{3x^4 - 5x^2 + x + 2}{3x^2 - 2}$

In Exercises 29–34, find the quotient and remainder using synthetic division.

29. $(3x^3 - 10x^2 + 2x + 3) \div (x - 3)$

30. $(3x^3 - x^2 - 7x - 2) \div (x + 2)$

31. $(x^4 + 5x^3 + 3x^2 + 3x + 9) \div (x + 1)$

32. $(x^4 - 16) \div (x - 2)$

33. $(2x^4 - 5x^3 + 4x^2 - 4x + 3) \div (x - 1)$

34. $(x^6 - 1) \div (x - 1)$

In Exercises 35–40, use synthetic division to evaluate the given function for the given value.

35. Find $f(-2)$ if $f(x) = x^3 - 3x + 2$.

**36.** Find $f(-3)$ if $f(x) = x^3 + 2x^2 - 2x + 3$.

**37.** Find $f(3)$ if $f(x) = x^6 - 5x^5 + 6x^4 + 7x^3 - 21x^2 + 5x - 12$.

**38.** Find $f(2)$ if $f(x) = x^3 - 8$.

**39.** Find $f(-1)$ if $f(x) = 4x^3 - 7x^2 + x - 17$.

**40.** Find $f(1)$ if $f(x) = x^5 - 3x^4 + 5x^3 - 8x^2 + 2x - 3$.

In Exercises 41 and 42, use a grapher.

**41.** *Using a Grapher to Support Algebra.* Graph both

  a) $y_1 = (x^3 - 5x^2 + 3x - 2)/(x - 4)$ and

  b) $y_2 = x^2 - x - 1 - 6/(x - 4)$.

Match these graphs with the two shown after Exercise 42. What does your observation tell you about the algebra?

**42.** *Using a Grapher to Support Algebra.* Graph both

  a) $y_1 = (x^4 - 5x^2 + 3)/(x^2 + 1)$ and

  b) $y_2 = x^2 - 6 + 9/(1 + x^2)$.

Match these graphs with the following two. What does your observation tell you about the algebra?

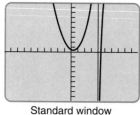 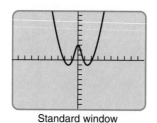

Standard window      Standard window

(a)            (b)

For Exercises 41 and 42.

In Exercises 43–46, use a grapher to determine whether you think the statement is true or false.

**43.** When $x^4 + 5x^2 - 3x + 2$ is divided by $x^2 + 1$, the quotient is $x^2 + 4$ and the remainder is $x + 1$.

**44.** When $x^3 - 5x^2 + 3x - 1$ is divided by $x^2 - 3$, the quotient is $x - 5$ and the remainder is $6x - 16$.

**45.** When $2x^2 - 6x + 17$ is divided by $5x + 1$, the quotient is $0.4x - 1$ and the remainder is 24.

**46.** When $x^3 - 5x^2 + 3x - 1$ is divided by $x^2 - 3$, the quotient is $2.3x - 1$ and the remainder is 2.

**47.** *Using Algebra in Geometry.* The foundation of a house together with a 12-by-15-ft patio has an area of $180 + 48x - 2x^2$. The width of the house is $x$.

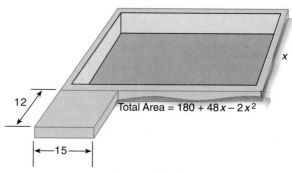

For Exercise 47.

  a) What is the area of the patio?

  b) What is the area of the foundation in terms of $x$, excluding the patio?

  c) What is the length of the house in terms of $x$?

**48.** *Using Algebra in Geometry.* A picture and its frame have an area of $64 + 12x - x^2$. The width of the frame is $x + 4$.

  a) What is the height of the frame?

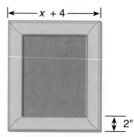

Area of picture & frame = $64 + 12x - x^2$

For Exercise 48.

b) Is the area of the picture (a) $12x - x^2$ or (b) $20x - x^2$? Explain.

## TRANSLATING WORDS TO SYMBOLS

**49.** Let $x$ represent any number and write this expression in symbols: The difference between the number to the fourth power and the number to the second power is divided by one less than the square of the number.

**50.** Let $y$ represent any number and write this expression in symbols: One less than the number to the fourth power is divided by twice the second power of the number.

## EXTENDING THE IDEAS

When the remainder of a division is zero, the divisor is a factor of the dividend. In Exercises 51–56, determine whether $2x + 1$ is a factor of the given polynomial.

**51.** $2x^3 - 5x^2 + x + 2$     **52.** $2x^3 - 9x^2 - 11x - 3$

**53.** $6x^3 - 11x^2 - x + 3$

**54.** $2x^4 + 5x^3 - 8x^2 + x - 2$

**55.** $4x^4 - 6x^3 + 10x^2 + 3x + 3$

**56.** $6x^5 + 3x^4 - 10x^3 - 5x^2 + 6x + 3$

**57. Discover the Pattern.**

$$\frac{x^3 - 1}{x + 1} = x^2 - x + 1 - \frac{2}{x + 1}$$

$$\frac{x^5 - 1}{x + 1} = x^4 - x^3 + x^2 - x + 1 - \frac{2}{x + 1}$$

$$\vdots \qquad \vdots$$

Assuming that the pattern continues, guess what the quotient and remainder are for $\dfrac{x^7 - 1}{x + 1}$. Confirm your guess by completing the division.

**58. Using a Pattern.** Use the pattern discovered in Exercise 57 to write in your own words how to find the quotient and remainder for $\dfrac{x^{23} - 1}{x + 1}$.

## LOOKING BACK—MAKING CONNECTIONS

In Exercises 59–64, factor the polynomial expression.

**59.** $s^2 - 6s + 9$        **60.** $y^2 - 8y + 16$

**61.** $x^2 - 7x + 10$        **62.** $y^2 - y - 12$

**63.** $2x^2 - 5x - 12$        **64.** $6x^2 + x - 2$

In Exercises 65–68, solve the equation.

**65.** $(x - 3) = -3(x + 5)$    **66.** $x^2 - 28 = 3x$

**67.** $2x^2 - 3x + 1 = 0$        **68.** $3x^2 - 2x = 1$

---

**6.6** # Solving Equations Containing Rational Expressions

Solving Rational Expressions Using Graphical Methods ▪ Solving Rational Expressions Using Algebraic Methods ▪ Checking Rational Equations for Extraneous Solutions ▪ Revisiting the Mixture Problem Situation ▪ Additional Problem Situations

In the previous chapters, you learned to solve linear equations by algebraic and graphing methods. We apply those same techniques when we solve equations that contain rational expressions.

## Solving Rational Equations Using Graphical Methods

Although rational equations can be solved graphically by using either the $x$-intercept or the multigraph method, we prefer using the $x$-intercept method because it is usually easier to find all $x$-intercepts of one graph than it is to find all points of intersection between two graphs.

### EXAMPLE 1   Solve Using the $x$-Intercept Method

Use the $x$-intercept method to solve the following equation. Support numerically with a grapher.

$$\frac{2}{x+3} + 15.7 = 25.7$$

SOLUTION

**Solve Graphically.** Rewrite the equation as an equivalent equation with one side equal to zero.

$$\frac{2}{x+3} + 15.7 = 25.7$$

$$\frac{2}{x+3} + 15.7 - 25.7 = 0$$

$$\frac{2}{x+3} - 10 = 0$$

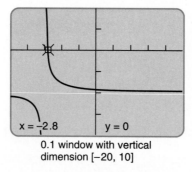

$x = {}^-2.8$    $y = 0$

0.1 window with vertical
dimension [−20, 10]

(a)

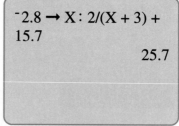

$^-2.8 \rightarrow$ X: 2/(X + 3) + 15.7

25.7

Numerical support
for the solution $x = 2.8$

(b)

**FIGURE 6.25**   (a) A graph of $y = 2/(x+3)-10$ and (b) numerical support for the solution $x = -2.8$ to $2/(x+3) + 15.7 = 25.7$.

Figure 6.25 (a) shows a graph $y = 2/(x + 3) - 10$. Use TRACE to find that the $x$-intercept is $(-2.8, 0)$.

The solution to the equation is $x = -2.8$.

**Support Numerically.** Figure 6.25 (b) provides numerical support for the solution $x = -2.8$. ∎

### EXAMPLE 2    Solve Using the $x$-Intercept Method

Use the $x$-intercept graphing method to solve the following equation. Support the solution numerically.

$$\frac{2}{x + 3} + \frac{5}{x - 4} = \frac{2x + 23}{x^2 - x - 12}.$$

SOLUTION

**Solve Graphically.** To use the $x$-intercept method, we rewrite the equation with one side equal to zero. Thus

$$\frac{2}{x + 3} + \frac{5}{x - 4} = \frac{2x + 23}{x^2 - x - 12}$$

$$\frac{2}{x + 3} + \frac{5}{x - 4} - \frac{2x + 23}{x^2 - x - 12} = 0.$$

Figure 6.26 (a) shows a graph of $y = 2/(x+3)+5/(x-4)-(2x+23)/(x^2-x-12)$.

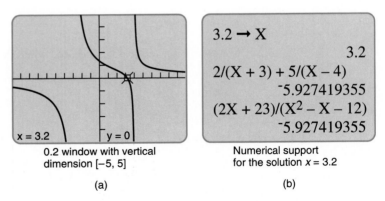

|  |  |
|:--:|:--:|
| 0.2 window with vertical dimension [−5, 5] | Numerical support for the solution x = 3.2 |
| (a) | (b) |

**FIGURE 6.26** (a) A graph of $y = 2/(x + 3) + 5/(x - 4) - (2x + 23)/(x^2 - x - 12)$ showing the zero $x = 3.2$ and (b) numerical support for the solution $x = 3.2$ to the equation $\dfrac{2}{x + 3} + \dfrac{5}{x - 4} = \dfrac{2x + 23}{x^2 - x - 12}$.

Use TRACE to find that the one $x$-intercept is $(3.2, 0)$. The solution to

$$\frac{2}{x+3} + \frac{5}{x-4} = \frac{2x+23}{x^2-x-12}$$

is $x = 3.2$.

**Support Numerically.** Figure 6.26 (b) shows numerical support of $x = 3.2$ as the solution. Numerical support could also be shown by storing $3.2 \rightarrow x$ and evaluating $y_1$ to be 0. ∎

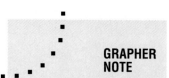

**GRAPHER NOTE**

The 0.2 window referred to in Figure 6.26 can be found by first finding the 0.1 window and then doubling the value for $X$min, $X$max, $Y$min, and $Y$max.

## Solving Rational Equations Using Algebraic Methods

To solve an equation like $\dfrac{x}{4} + \dfrac{1}{2} = \dfrac{7}{8}$, you learned in earlier chapters that you could eliminate the fractions by multiplying both sides of the equation by the least common multiple of the denominators. Note that this is also the LCD. Proceeding algebraically, we find the following:

$$8\left[\frac{x}{4} + \frac{1}{2}\right] = 8 \cdot \frac{7}{8}$$

$$\overset{2}{8}\left(\frac{x}{4}\right) + \overset{4}{8}\left(\frac{1}{2}\right) = 8\left(\frac{7}{8}\right)$$

$$2x + 4 = 7$$

$$2x = 3$$

$$x = 1.5$$

A similar method can be used when the variables are in the denominator. For example, to solve the equation $\dfrac{1}{x} + \dfrac{1}{2} = 3$ begin by multiplying both sides of the equation by $2x$.

$$2x\left[\frac{1}{x} + \frac{1}{2}\right] = 2x \cdot 3$$

$$2x\left(\frac{1}{x}\right) + 2x\left(\frac{1}{2}\right) = 2x(3)$$

$$2 + x = 6x$$

$$2 = 5x$$

$$x = \frac{2}{5}$$

Figure 6.27 shows numerical support on a grapher that $x = 2/5$ is a solution to the equation.

Example 3 uses an algebraic method to find the solution of an equation that includes rational expressions.

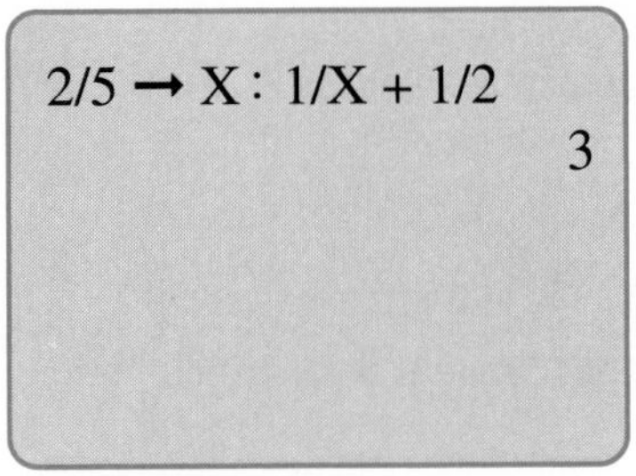

**EXAMPLE 3**   Solve Algebraically and Support Numerically

Solve the following equation using algebraic methods. Support the solution numerically.

$$\frac{1}{2x} + \frac{2}{3x} = \frac{7}{6}$$

SOLUTION

**Solve Algebraically.** Multiply each side of the equation by the LCD $6x$.

$$\frac{1}{2x} + \frac{2}{3x} = \frac{7}{6}$$

$$6x \cdot \left( \frac{1}{2x} + \frac{2}{3x} \right) = 6x \cdot \frac{7}{6} \qquad \text{Use the \textbf{Distributive Property} to eliminate the fractions.}$$

$$\overset{3}{6x} \left( \frac{1}{2x} \right) + \overset{2}{6x} \left( \frac{2}{3x} \right) = 6x \left( \frac{7}{6} \right)$$

$$3 + 4 = 7x$$

$$7 = 7x$$

$$1 = x$$

**Support Numerically.** Figure 6.28 shows numerical support on a grapher that $x = 1$ is the solution to $\frac{1}{2x} + \frac{2}{3x} = \frac{7}{6}$. ∎

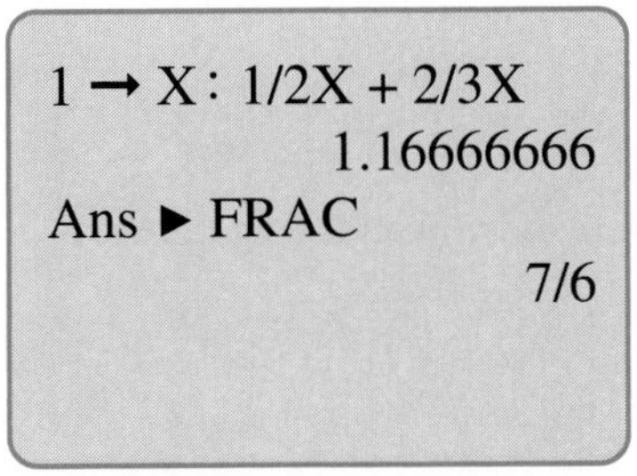

The numerical support with a grapher like the one shown in Figure 6.28 constitutes a check of the solution. When solving an equation with a variable in a denominator, it is *essential* that the solution be checked. The reason is discussed in the next subsection.

## Checking Rational Equations for Extraneous Solutions

Whenever an equation has a variable in a denominator, an algebraic solution method might result in values that in fact fail to be solutions. Such values are called

---

$2/5 \rightarrow X: 1/X + 1/2$

$\qquad\qquad\qquad 3$

**FIGURE 6.27**
Numerical support, using a grapher, of the solution to the equation $\frac{1}{x} + \frac{1}{2} = 3.$

---

$1 \rightarrow X: 1/2X + 2/3X$

$\qquad\qquad 1.16666666$

Ans ▶ FRAC

$\qquad\qquad\qquad 7/6$

**FIGURE 6.28**
Numerical support of Example 3. Some graphers have the FRAC command, as shown here.

**extraneous solutions**. To be sure you have not found an extraneous solution, it is important to check all solutions found by an algebraic method.

Example 4 illustrates an equation that has an extraneous solution.

**EXAMPLE 4**    An Equation with an Extraneous Solution

Use an algebraic method to solve the following equation. Support the solution both numerically and graphically.

$$\frac{1}{x-1} + \frac{1}{x+2} = \frac{3}{x^2+x-2}$$

SOLUTION

**Solve Algebraically.** Multiply both sides of the equation by the LCD $(x-1)(x+2)$.

$$(x-1)(x+2) \cdot \left( \frac{1}{x-1} + \frac{1}{x+2} \right) = (x-1)(x+2) \cdot \frac{3}{x^2+x-2}$$

$$(x-1)(x+2)\left( \frac{1}{x-1} \right) + (x-1)(x+2)\left( \frac{1}{x+2} \right) = (x-1)(x+2) \cdot \frac{3}{(x+2)(x-1)}$$

$$(x+2) + (x-1) = 3$$

$$2x + 1 = 3$$

$$2x = 2$$

$$x = 1$$

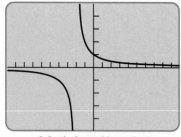

0.2 window with vertical dimension [−5, 5]

**FIGURE 6.29** A graph of $y_1$ provides grapher support that there appears to be no solution to Example 4.

**Support Numerically.** Paper-and-pencil support is provided by replacing the variable in the original equation with $x = 1$.

$$\frac{1}{1-1} + \frac{1}{1+2} = \frac{3}{1^2+1-2}$$

$$\frac{1}{0} + \frac{1}{3} = \frac{3}{0}$$

Because division by 0 is undefined, we see that $x = 1$ is not a solution of the original equation; we call $x = 1$ an extraneous solution. Therefore there is no solution to the original equation.

**Support Graphically.** Figure 6.29 shows a graph of $y_1 = 1/(x-1) + 1/(x+2) - 3/(x^2+x-2)$. It appears that the graph does not intersect the $x$-axis. The trace cursor can be moved farther to the right or left to support that there is no intersection with the $x$-axis.    ∎

We summarize the procedure for solving rational equations by algebraic methods as follows.

---

### Solving Rational Equations by Algebraic Methods

1. Find the LCD of all the denominators in the equation.
2. Multiply each side of the equation by the LCD to eliminate the fractions from the equation.
3. Solve the resulting equation.
4. Support graphically or numerically to show that the solutions are not extraneous solutions.

---

Example 5 illustrates combining both algebraic and graphing methods to solve an equation. The equation is solved using an algebraic method, while the check for extraneous solutions is done numerically.

### EXAMPLE 5 An Equation with Extraneous Solutions

Solve the following rational equation algebraically and support the solution graphically:

$$\frac{x}{x-1} + x = \frac{4x-3}{x-1}$$

SOLUTION

**Solve Algebraically.** Multiply each side of the equation by the LCD $(x-1)$.

$$(x-1) \cdot \left[\frac{x}{(x-1)} + x\right] = (x-1) \cdot \frac{4x-3}{(x-1)}$$

$$(x-1)\left[\frac{x}{(x-1)}\right] + (x-1)x = (x-1) \cdot \frac{4x-3}{(x-1)}$$

$$x + x^2 - x = 4x - 3 \qquad \text{**Combine like terms and arrange in descending order with one side of the equation = 0.**}$$

$$x^2 - 4x + 3 = 0 \qquad \text{**Factor.**}$$

$$(x-3)(x-1) = 0 \qquad \text{**Set each factor = to 0 and solve.**}$$

$$x = 3 \quad \text{or} \quad x = 1$$

**Numerical Check.** Substitute $x = 3$ into the original equation.

$$\frac{3}{3-1} + 3 = \frac{4(3) - 3}{3 - 1}$$

$$\frac{3}{2} + \frac{6}{2} = \frac{9}{2}$$

$$\frac{9}{2} = \frac{9}{2}$$

So $x = 3$ is a solution. Note we see that if $x = 1$ is substituted into the original equation, the denominator is undefined. So $x = 1$ is an extraneous solution.

**Support Graphically.** We use the $x$-intercept method for the graphical check. Rewrite the equation as an equivalent equation with the right-hand side equal to zero, and then define $y_1$ as the left-hand side of the equation.

Figure 6.30 shows a complete graph of $y = x/(x - 1) + x - (4x - 3)/(x - 1)$. Using TRACE, we verify that $x = 3$ is a solution and that $x = 1$ is not a solution. In other words, $x = 1$ is an extraneous solution. ∎

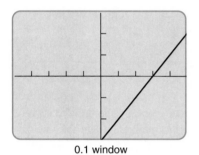

0.1 window

**FIGURE 6.30** Grapher support that $x = 3$ is the solution to the equation in Example 5 and that $x = 1$ is an extraneous solution.

## Revisiting the Mixture Problem Situation

In Section 1.5, you studied the situation in which John adds distilled water to 5 liters of a 30%-acid solution. Recall that $0.30 \times 5 = 1.5$ is the number of liters of acid in the 30% solution. The question asked was how many liters of water should be added to obtain an 18%-acid solution.

The problem was solved by using a numerical method. The following table was completed and an answer to the problem was estimated by studying the table.

| # Liters of water added | Ratio of liters of acid to liters of solution | Percent acid |
|:---:|:---:|:---:|
| 1 | $\dfrac{1.5}{5 + 1} = 0.25$ | 25.0% |
| 2 | $\dfrac{1.5}{5 + 2} = 0.214$ | 21.4% |
| 3 | $\dfrac{1.5}{5 + 3} = 0.188$ | 18.8% |
| 4 | $\dfrac{1.5}{5 + 4} = 0.167$ | 16.7% |
| ⋮ | ⋮ | |
| $x$ | $P(x) = \dfrac{1.5}{5 + x}$ | |

If $x$ represents the number of liters added and $P(x)$ the resulting ratio that is equivalent to the percent solution, we discover from the pattern of the table that

$$P(x) = \frac{1.5}{5 + x}.$$

**EXAMPLE 6**    Revisiting the Mixture Problem

Suppose distilled water is added to 5 liters of a 30%-acid solution. How many liters should be added to obtain an 18.3% solution? Solve the problem graphically and confirm algebraically.

SOLUTION

**Solve Graphically.** From the discussion prior to this example, observe that the algebraic representation of the problem situation is $P(x) = 1.5/(5 + x)$. We need to solve the equation

$$\frac{1.5}{5 + x} = 0.183.$$

To use the multigraph method to solve this equation graphically, find a graph of $y_1 = 1.5/(5 + x)$ and $y_2 = 0.183$ (see Figure 6.31). Zoom in to find that the solution is $x = 3.197$ with an error of at most 0.01.

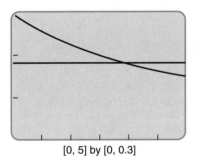

[0, 5] by [0, 0.3]

**FIGURE 6.31** A graph of $y_1 = \dfrac{1.5}{5 + x}$ and $y_2 = 0.183$.

**Confirm Algebraically.**

$$0.183 = \frac{1.5}{5 + x}$$

$$0.183(5 + x) = 1.5$$

$$0.915 + 0.183x = 1.5$$

$$0.183x = 1.5 - 0.915$$

$$x = \frac{0.585}{0.183}$$

$$x = 3.197 \qquad \blacksquare$$

## Additional Problem Situations

In real life, we often encounter problem situations dealing with fractions. Algebraic representations of these problem situations often include rational expressions. Consider the following problem situation.

---

### ROCK CONCERT PROBLEM SITUATION

Tom and his friends spent $200 on a certain number of t-shirts to sell at the Gratefully Living rock concert. Suppose they sell all but three of the number they purchase and that they sell each for $5 more than they paid for them.

---

**EXAMPLE 7**　　THE ROCK CONCERT PROBLEM SITUATION: Finding the Number of Shirts

How many t-shirts must Tom purchase to make a profit of about $100? Decide whether to use a graphical or an algebraic solution method to solve the problem.

SOLUTION

Let $x$ = number of t-shirts purchased.

| | |
|---|---|
| **Cost per t-shirt** | $= \dfrac{200}{x}$ |
| **Sale price per t-shirt** | $= \dfrac{200}{x} + 5$ |
| **Number of t-shirts sold** | $= x - 3$ |
| **Money collected** | $= (x - 3)\left(\dfrac{200}{x} + 5\right)$ |
| **Profit** | $= (x - 3)\left(\dfrac{200}{x} + 5\right) - 200$ |

An algebraic representation for the problem situation is

$$(x - 3)\left(\frac{200}{x} + 5\right) - 200 = 100.$$

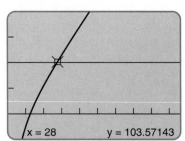

$x = 28 \qquad y = 103.57143$

Integer window with Xmin = 0 and a vertical dimension [−50, 200]

**FIGURE 6.32** Graph of the problem situation in Example 7.

This equation seems like one that would be tedious to solve algebraically, so we choose a graphical method.

Figure 6.32 shows the graph of $y_1 = (x - 3)(200/x + 5) - 200$ and $y_2 = 100$.

Using TRACE, we see that the two cursor positions $(27, 97.78)$ and $(28, 103.57)$ are the closest to being points of intersection with the line $y = 100$.

Therefore Tom must purchase 27 or 28 shirts for a profit of about $100. ∎

Consider the following problem situation.

---

□                                                                    □

### LAWN SERVICE PROBLEM SITUATION

Eric wants to expand his lawn service by hiring more workers. He
decides to test the applicants by having them mow his lawn, which he
knows takes him 2 hr to mow. It takes Greg 3 hr to mow the lawn and
Brian 4 hr to do the same job.

□                                                                    □

---

To analyze this problem situation, consider these points.

1. If several people work together to complete a particular task, then each person
   is doing a fractional portion of the completed job.
2. If finishing a task takes $n$ hours, then after one hour, $1/n$ of the task has been
   completed.

Eric next wants to assess how long it will take them if all three do the job
together.

**EXAMPLE 8**    THE LAWN SERVICE PROBLEM SITUATION: Finding the
Time When Working Together

How long will it take to mow Eric's lawn if Eric, Greg, and Brian work together?

SOLUTION
Consider the fraction of the job that can be done *in 1 hr* by each one working
alone. Adding those fractions represents the fraction of the job that can be done
in 1 hr working together.

**Eric's fraction working alone:** $\dfrac{1}{2}$

**Greg's fraction working alone:** $\dfrac{1}{3}$

**Brian's fraction working alone:** $\dfrac{1}{4}$

**Working together:** $\dfrac{1}{x}$, where $x$ is the time (hours) to complete the job.

$$\frac{1}{x} = \frac{1}{2} + \frac{1}{3} + \frac{1}{4}$$    **This equation represents the
amount of work being done in 1 hr.**

Multiply each side of the equation by the LCD $12x$.

$$12x \frac{1}{x} = 12x \left[ \frac{1}{2} + \frac{1}{3} + \frac{1}{4} \right] \quad \text{Use the Distributive Property and simplify.}$$

$$12 = 6x + 4x + 3x$$

$$12 = 13x$$

$$\frac{12}{13} = x$$

Therefore in $\frac{12}{13}$ of an hour, or 55.385 minutes, they can complete the job. ■

## Exercises for Section 6.6

In Exercises 1–4, find the graph that could be used to solve the given equation using a multigraph method. (The graphs are shown in the Standard window.) How many solutions will the equation have?

1. $\dfrac{8}{1 + x^2} = 3$

2. $\dfrac{2x}{x - 4} = 5$

3. $\dfrac{x^2}{x + 2} = 4$

4. $\dfrac{x^3}{x^2 - 4} = 6$

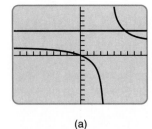

(a)

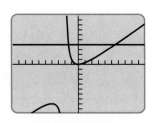

(b)

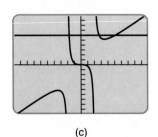

(c)

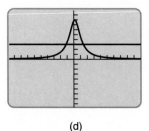

(d)

For Exercises 1–4.

In Exercises 5–14, find the solution to the equation by using the $x$-intercept graphing method with the 0.2 window. (Recall that the 0.2 window is found by doubling the size of the 0.1 window.) Support the solution numerically using a grapher.

5. $\dfrac{3}{x - 1} = \dfrac{6}{x + 5}$

6. $\dfrac{1}{x - 1} = \dfrac{5}{x + 3}$

7. $\dfrac{-3}{x + 1} = 3$

8. $\dfrac{8}{x - 3} = 4$

9. $\dfrac{1}{x} + 3.5 = 5.5$

10. $\dfrac{3}{x + 2} - 5.1 = 2.4$

11. $\dfrac{6}{x - 1} + 7 = 5$

12. $\dfrac{2}{x - 1} + \dfrac{6}{x} = \dfrac{5}{x - 1}$

13. $\dfrac{3}{x - 3} + 4 = \dfrac{x}{x - 3}$

14. $\dfrac{2x - 5}{x - 2} - 2 = \dfrac{3}{x + 2}$

In Exercises 15–20, find the solution to the equation by using an algebraic method. Support the solution using either a graphical or a numerical method.

15. $x + 4 = \dfrac{12}{x}$

16. $3 + x = \dfrac{18}{x}$

17. $\dfrac{1}{x-2} - \dfrac{1}{6} = \dfrac{2}{3x-6}$

18. $\dfrac{x-1}{x^2-4} - \dfrac{2}{x+2} = \dfrac{4}{x-2}$

19. $\dfrac{3}{x+2} - \dfrac{1}{x-2} = \dfrac{2}{x^2-4}$

20. $\dfrac{5}{a+5} + \dfrac{2}{a-5} = \dfrac{8a+10}{a^2-25}$

In Exercises 21–28, use an algebraic method to solve the equation. Determine whether any of your solutions are extraneous.

21. $\dfrac{3}{4x-8} - \dfrac{2}{3x-6} = \dfrac{1}{36}$

22. $\dfrac{3}{x-4} - \dfrac{2}{x-1} = \dfrac{2x+7}{(x-1)(x-4)}$

23. $\dfrac{1}{a+2} - \dfrac{5}{a^2+9a+14} = \dfrac{-3}{a+7}$

24. $\dfrac{2}{x+2} - \dfrac{4}{x+5} = \dfrac{8-x}{x^2+7x+10}$

25. $x - \dfrac{2x}{x+2} = \dfrac{x+6}{x+2}$

26. $\dfrac{x}{(x-1)(x+3)} + \dfrac{x}{x+3} = \dfrac{6-5x}{x^2+2x-3}$

27. $\dfrac{3}{x-2} + \dfrac{1}{x+3} = \dfrac{4x+5}{x^2+x-6}$

28. $\dfrac{2}{x-6} - \dfrac{3}{x-5} = \dfrac{6}{x^2-11x+30}$

In Exercises 29–32, solve the equations using the $x$-intercept graphing method. Begin with the Standard window and zoom in if necessary. Support numerically to check for extraneous solutions.

29. $x - \dfrac{2}{x-3} = \dfrac{x-1}{3-x}$   30. $\dfrac{2x}{x-1} + \dfrac{x-5}{x^2-1} = 1$

31. $\dfrac{x}{x-2} + 5 = \dfrac{2}{x-2}$   32. $\dfrac{x}{x+1} + 9 = \dfrac{-1}{x+1}$

33. **Writing to Learn.** Summarize the several different methods for determining whether solutions found algebraically are extraneous.

34. **Writing to Learn.** When solving an equation involving rational expressions, do you prefer an algebraic solution method or a graphical solution method? Explain when you prefer one and when the other.

In Exercises 35–42, choose either an algebraic method or graphical method to solve the equation. Support an algebraic solution with a grapher.

35. $\dfrac{11}{2x} - \dfrac{2}{3x} = \dfrac{1}{6}$   36. $\dfrac{5x}{x+1} + \dfrac{1}{x} = 5$

37. $\dfrac{2.1}{3x} - \dfrac{5.2}{x} = 2$   38. $\dfrac{x}{x+1} + \dfrac{5}{x-1} = 1$

39. $\dfrac{x-1}{x+1} - \dfrac{2x}{x-1} = -1$

40. $\dfrac{5.3}{5x} - \dfrac{2.6}{x} = 0.77$

41. $\dfrac{4x}{3x-2} + \dfrac{2x}{3x+2} = 2$

42. $\dfrac{a}{a-7} + \dfrac{50}{7-a} = -85$

TRANSLATING WORDS TO SYMBOLS

In Exercises 43–45, write the word description in symbols:

43. The sum of the reciprocals of two consecutive integers

44. The sum of the reciprocals of two consecutive even integers

45. The sum of the reciprocals of two consecutive odd integers

In Exercises 46–50, write an algebraic representation for the problem situation and then solve the problem.

46. Three times a number less twice its reciprocal is $-1$. Find the number.

47. The sum of the reciprocals of two consecutive even integers is $\dfrac{5}{12}$. Find the integers.

48. The difference of the reciprocals of two consecutive odd integers is $\dfrac{2}{35}$. Find the integers.

49. The difference of the reciprocals of two consecutive integers is $\dfrac{1}{30}$. Find the integers.

**50.** The sum of the reciprocals of two consecutive even integers is $\dfrac{7}{24}$. Find the integers.

In Exercises 51–60, solve the problems.

**51.** *Diluting an Acid Solution.* Suppose distilled water is added to 8 liters of a 20%-acid solution. How many liters of distilled water should be added to obtain a 12.5% solution. Solve this problem graphically and confirm algebraically.

**52.** *Diluting an Acid Solution.* Suppose distilled water is added to 6 liters of a 50%-acid solution. How many liters of distilled water should be added to obtain a 7.5% solution. Solve this problem graphically and confirm algebraically.

**53.** *Club Fund Raising.* The Future Teachers Association wants to raise money by selling programs at the school football game. Association members spend $125 for the programs and sell all but 10 of them for $0.50 a piece more than they paid for them. How many must they sell to make a profit?

**54.** *Club Fund Raising.* The Alumni Club wanted to raise money by selling cookbooks. Each would sell for $2.75 more than what the Club paid for it. Members spent $325 on the cookbooks and sold them all.

 a) How many do they have to sell to make a profit of $200?

 b) What is the price per book?

**55.** *Shared Work.* Jack can clean the garage in 2 hr. Matt can do the same job in 5 hr. How long will it take them to clean the garage together?

**56.** *Shared Work.* A swimming pool can be filled using one pipe in 12 hr and by another pipe in 4 hr. How long will it take to fill the pool if both pipes are used?

**57.** *Shared Work.* Kendra and Joenita can paint a room in $3\frac{3}{5}$ hr. Kendra can do the job alone in 6 hr. How long would it take Joenita to paint the room alone?

**58.** *Shared Work.* Kevin built a loft in his dorm room in 6 hr. Trey built the same style loft in 8 hr. They decide to go into business together. How long will it take them, working together, to build each loft?

**59.** *Electrical Engineering.* In an electronic circuit, the total resistance $R$ of two resistors $R_1$ and $R_2$ wired in a parallel circuit is determined by the formula $\dfrac{1}{R} = \dfrac{1}{R_1} + \dfrac{1}{R_2}$.

 a) Find the total resistance of a 200-ohm resistor and a 100-ohm resistor in a parallel circuit.

 b) Find the total resistance of a 500-ohm resistor and a 200-ohm resistor.

 c) If the total resistance is to be 100 ohms, what size resistor (in ohms) should be combined with a 500-ohm resistor?

 d) If the total resistance is to be 300 ohms, what size resistor (in ohms) should be combined with an 800-ohm resistor?

**60.** *Lens Designer.* A convex lens like the one shown in the following diagram refracts light rays to a focal point $F$. If the lens is formed by the arcs of circles of radius $R_1$ and $R_2$ and the distance from the lens to the focal point $F$ is $f$, then $f$, $R_1$, and $R_2$ are related by the equation $\dfrac{1}{f} = \dfrac{1}{R_1} + \dfrac{1}{R_2}$.

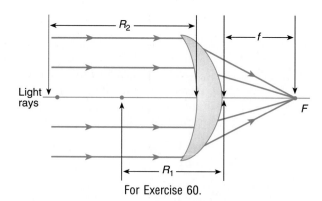

For Exercise 60.

 a) What is the distance from the lens to the focal point of a lens that is constructed with one radius ($R_1$) of 3 cm and the other a radius ($R_2$) of 14 cm?

b) If the focal point is 0.25 cm from the lens that is constructed with one radius of 5/3 cm, what is the second radius used to construct the lens?

EXTENDING THE IDEAS

In Exercises 61–64, solve the equation by a method of your choice.

**61.** $\dfrac{5}{x-3} + \dfrac{2}{x-4} = \dfrac{9}{x^2-7x+12}$

**62.** $\dfrac{5}{x-5} = \dfrac{x}{x-5} + 4$

**63.** $\dfrac{8}{x-3} - \dfrac{4}{x+3} = \dfrac{8}{x+3}$

**64.** $\dfrac{x+1}{x} + \dfrac{x+4}{x} = 6$

**65.** A graph of $y = x/(x-1) + x - (4x-3)/(x-1)$ was found for graphical support of Example 5.

a) Find this graph in the 0.1 window. What is the $y$-coordinate using TRACE for $x = 1$?

b) **Writing to Learn.** Write a paragraph explaining why this graph has a "hole" in it at $x = 1$.

**66.** *Capital Expenditures.* The capital expenditures (in billions of dollars) made by the iron and steel industry between 1985 and 1990 is shown in the following bar graph. (Source: *Statistical Abstract of the United States, 1992.*) These expenditures are modeled by the function

$$y = \dfrac{1}{-0.035t^2 + 0.149t + 0.674} \quad \text{where } t = 1 \text{ for 1985.}$$

a) Use the graph to estimate the expenditures for 1987.

b) Then use the model to confirm the reasonableness of your answer.

LOOKING BACK—MAKING CONNECTIONS

**67.** Solve $A = \dfrac{1}{2}h(b_1 + b_2)$ for $b_2$.

In Exercises 68–71, solve the equation for $x$ and support numerically using the grapher.

**68.** $3(2x-1) + 4(x+1) = 2(3x+4) - 9$

**69.** $(4x-3)(x+5) = 0$

**70.** $5x^2 - 7x = 6$

**71.** $x^2 - 7x = 0$

**72.** *Swimming Pool Construction.* A rectangular swimming pool measures 10 ft by 25 ft. It is surrounded by a deck that has a uniform width all around the pool. If the pool and the deck together have an area of 594 ft$^2$, how wide is the deck?

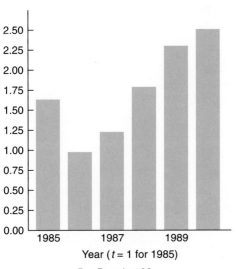

For Exercise 66.

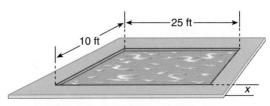

Area of pool and deck = 594 ft$^2$

For Exercise 72.

## Chapter 6 Summary

In this summary $a, b,$ and $c$ represent polynomials.

| | **Rational Expressions** | **Examples** |
|---|---|---|
| Definition of Rational Expression | An expression in the form $\dfrac{a}{b}$, $b \neq 0$. | $\dfrac{2}{x^2 - 4x + 1}$ |
| Property of Equivalent Rational Expressions | $\dfrac{a \cdot c}{b \cdot c} = \dfrac{a}{b}$, $b \neq 0$ and $c \neq 0$. | $\dfrac{3(x + 1)}{20(x + 1)} = \dfrac{3}{20}$ |
| Lowest Terms, Simplest Terms | A form of a rational expression where the numerator and denominator have no common factors | $\dfrac{5y}{6x}$ is in lowest terms; $\dfrac{2x}{7x}$ is not in lowest terms. |
| Reducing a Rational Expression to Lowest Terms | Use the Property of Equivalent Rational Expressions to get the expression in lowest terms | $\dfrac{6x(x - 2)^2}{15x^2(x - 2)} = \dfrac{2(x - 2)}{5x}$ |
| Common Denominator | A polynomial expression that is a multiple of all denominators of several rational expressions | A common denominator of $\dfrac{x}{x + 1}$ and $\dfrac{x - 3}{x^2 - 1}$ is $(x + 1)(x^2 - 1)$. |
| Least Common Denominator (LCD) | A polynomial expression that is the least common multiple of the denominators of several rational expressions | The least common denominator of $\dfrac{x}{x + 1}$ and $\dfrac{x - 3}{x^2 - 1}$ is $(x + 1)(x - 1) = x^2 - 1$. |

| | **Operations on Rational Expressions** | **Examples** |
|---|---|---|
| Multiplication of Rational Expressions | $\dfrac{a}{b} \cdot \dfrac{c}{d} = \dfrac{ac}{bd}$, $b \neq 0$ and $d \neq 0$. | $\dfrac{4}{x} \cdot \dfrac{6y^2}{x} = \dfrac{24y^2}{x^2}$ |

Division of Rational Expressions

$$\frac{a}{b} \div \frac{c}{d} = \frac{a}{b} \cdot \frac{d}{c} = \frac{ad}{bc}, \quad b \neq 0,$$

$$c \neq 0, \text{ and } d \neq 0.$$

$$\frac{x^2 - 3x}{x + 4} \div \frac{x - 3}{x^2 + 5x} =$$

$$\frac{x(x - 3)}{x + 4} \cdot \frac{x(x + 5)}{x - 3} =$$

$$\frac{x^2(x + 5)}{x + 4}$$

Addition and Subtraction of Rational Expressions

$$\frac{a}{c} + \frac{b}{c} = \frac{a + b}{c} \quad \text{and} \quad \frac{a}{c} - \frac{b}{c} =$$

$$\frac{a - b}{c}.$$

$$\frac{1}{2(x - 1)} + \frac{3}{4} =$$

$$\frac{2 + 3(x - 1)}{4(x - 1)} =$$

$$\frac{3x - 1}{4(x - 1)}$$

| Complex Fractions | Examples |
| --- | --- |

Complex Fraction

An expression that has a fraction in either the numerator or the denominator or both

$$\frac{\dfrac{3x}{x^2 - 4x + 3}}{\dfrac{x + 1}{x - 3}}$$

Simplifying Complex Fractions, Method 1

1. Simplify the numerator into one fraction.
2. Simplify the denominator into one fraction.
3. Invert the denominator and multiply times the numerator. Reduce to lowest terms.

Simplifying Complex Fractions, The LCD Method

1. Find the LCD of the denominators of all the individual fractions.
2. Multiply the numerator and the denominator of the complex fraction by this LCD.
3. Simplify the resulting expression.

| | Division of Polynomials | Examples |
|---|---|---|
| Dividing by a Monomial | $\dfrac{a+b+c}{d} = \dfrac{a}{d} + \dfrac{b}{d} + \dfrac{c}{d}$ | $\dfrac{2x^2 - 5x}{x} = \dfrac{2x^2}{x} - \dfrac{5x}{x}$ $= 2x - 5$ |
| Division Algorithm | $\dfrac{\text{Dividend}}{\text{Divisor}} = \text{Quotient} + \dfrac{\text{Remainder}}{\text{Divisor}}$ | $\dfrac{2x^2 - 6}{x + 2} = (2x - 4) + \dfrac{2}{x + 2}$ |
| Synthetic Division | A streamlined process of dividing two polynomials using only coefficients and constants | |
| Remainder Theorem | If a polynomial $f(x)$ is divided by $x - c$, the remainder is equal to $f(c)$. | When $f(x) = 3x^2 + 5x + 6$ is divided by $x + 2$, the remainder is $f(-2) = 8$. |

| | Solving Rational Equations | Examples |
|---|---|---|
| Extraneous Solutions | Possible solutions found algebraically that do not satisfy the original equation | |
| Solving Rational Equations Algebraically | Eliminate the fractions by multiplying all expressions by the LCD and then solve the resulting polynomial expression. Check all solutions in the original equation. Some might be extraneous. | Solve $\dfrac{x+3}{x-1} = x$: $x + 3 = (x - 1)x$ $0 = x^2 - 2x - 3$ $0 = (x - 3)(x + 1)$ Solution: $x = 3$ and $x = -1$ |
| Solving Rational Equations Graphically | Use either the $x$-intercept or the multigraph method to solve a rational equation. | Solve $\dfrac{x+3}{x^2-1} = \dfrac{x}{x+1}$: The graph of $y = \dfrac{x+3}{x^2-1} - \dfrac{x}{x+1}$ shows that $x = 3$ is the only solution. |

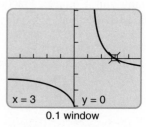

0.1 window

## Review Exercises for Chapter 6

In Exercises 1–5, reduce the rational expression to simplest terms. Support your work with a grapher.

**1.** $\dfrac{660}{1155}$

**2.** $\dfrac{24x^2y^3}{15x^5y}$

**3.** $\dfrac{6x^2 - 11x - 10}{2x^2 - 9x + 10}$

**4.** $\dfrac{4x - x^2}{3x - 12}$

**5.** $\dfrac{x^2 - 9}{x^2 - 6x + 9}$

In Exercises 6–11, perform the indicated operations. Simplify the result to lowest terms.

**6.** $\dfrac{1}{6x^2 + x - 5} \cdot (x + 1)$

**7.** $\dfrac{x^2 - 4}{x^3 y} \cdot \dfrac{x^2 y^2}{x^2 + x - 2}$

**8.** $\dfrac{x - 3}{4x^2} \div \dfrac{x + 3}{2xy}$

**9.** $\dfrac{x^2 + 9x + 14}{x - 3} \div \dfrac{x^2 + 4x + 4}{2x - 6}$

**10.** $\dfrac{x^2 + 6x + 8}{x^2 - x - 2} \cdot \dfrac{x + 3}{x^2 + 3x - 4} \cdot \dfrac{x^2 - 1}{x^2 + 7x + 12}$

**11.** $\dfrac{x + 1}{6x - 12} \cdot \dfrac{x^2 + 5x}{2x^2 - 8x - 10} \div \dfrac{3x + 15}{x - 2}$

**12.** Use the Property of Equivalent Rational Expressions to find the missing denominator for

$$\dfrac{2x - 3}{x + 5} = \dfrac{4x^2 - 9}{?}.$$

**13.** Find the LCD of the following pair of rational expressions:

$$\dfrac{3x - 2}{12x^2 - 4x - 5} \quad \text{and} \quad \dfrac{x + 2}{6x^2 - 5x - 4}$$

In Exercises 14–17, perform the indicated operations. Simplify the result to lowest terms.

**14.** $\dfrac{15x - 1}{6x^2 - 23x - 4} + \dfrac{6x^2 - 2x + 3}{6x^2 - 23x - 4}$

**15.** $\dfrac{2x^2 - 3x - 4}{(x - 5)(x + 1)} - \dfrac{x^2 + 3x - 9}{(x + 1)(x - 5)}$

**16.** $\dfrac{1}{x - 3} - \dfrac{5}{x}$

**17.** $\dfrac{x - 2}{(x - 3)(x + 2)} + \dfrac{2x - 1}{(x + 3)(x - 3)}$

In Exercises 18 and 19, simplify the complex fraction.

**18.** $\dfrac{\dfrac{x^2 - 9}{5x + 5}}{\dfrac{x + 3}{x^2 - 1}}$

**19.** $\dfrac{\dfrac{9x^2 - 1}{x^2 + 2x}}{\dfrac{9x^2 - 6x + 1}{x^2}}$

In Exercises 20 and 21, use method 1 to simplify the complex fraction.

**20.** $\dfrac{1 + \dfrac{5}{x - 4}}{1 - \dfrac{2}{x + 3}}$

**21.** $\dfrac{\dfrac{x}{5} + \dfrac{y}{6}}{\dfrac{4x}{5} + \dfrac{2y}{3}}$

In Exercises 22 and 23, use the LCD method to simplify the complex fraction.

**22.** $\dfrac{\dfrac{1}{x} + \dfrac{1}{4}}{\dfrac{1}{16} - \dfrac{1}{x^2}}$

**23.** $\dfrac{\dfrac{x}{x - 2} - \dfrac{3}{x - 4}}{\dfrac{1}{2} - \dfrac{1}{x - 4}}$

In Exercises 24 and 25, write the expression without negative exponents and then simplify the complex fraction.

**24.** $\dfrac{a^{-1} + 1}{b^{-2}}$

**25.** $\left(x^{-1} - y\right)^{-1}$

In Exercises 26 and 27, complete the division.

**26.** $\dfrac{8x^2 y^2 + 3x^3 y}{xy^2}$

**27.** $\dfrac{15x^4 - 27x^2 + 6}{3x}$

In Exercises 28 and 29, find the quotient and remainder when the division is completed.

Write the division in the form

$$\frac{\text{Dividend}}{\text{Divisor}} = \text{Quotient} + \frac{\text{Remainder}}{\text{Divisor}}.$$

**28.** $(2x^3 - 9x^2 + 19x - 14) \div (2x - 3)$

**29.** $(3x^3 + 4x^2 - 10x - 5) \div (x + 2)$

In Exercises 30 and 31, find the quotient using synthetic division.

**30.** $(4x^3 - 2x^2 + 5x + 30) \div (x + 2)$

**31.** $(x^4 - 3x^2 - 21x - 10) \div (x - 3)$

In Exercises 32 and 33, find the solution to the equation using the $x$-intercept method. Support the solution numerically using a grapher.

**32.** $\dfrac{16 + 3x}{x - 1} = 10$

**33.** $\dfrac{x}{1 - x} + \dfrac{2}{7} = \dfrac{2x}{1 - x}$

In Exercises 34–36, find the solution to each equation using an algebraic method. Determine whether any of the solutions are extraneous. Support the solution graphically.

**34.** $\dfrac{x}{x + 2} = \dfrac{1}{4} - \dfrac{3x}{x + 2}$

**35.** $\dfrac{x}{x - 4} + \dfrac{2}{x + 3} = \dfrac{4x - 2}{x^2 - x - 12}$

**36.** $\dfrac{x}{x - 2} + \dfrac{3}{x - 5} = \dfrac{x + 4}{x^2 - 7x + 10}$

**37.** Choose either an algebraic method or a graphing method to solve the following equation. Verify the solution using a grapher.

$$\frac{x}{x + 3} + \frac{4}{x + 5} = \frac{2x}{x + 3}$$

**38.** Write in symbols: The difference of the reciprocals of $x$ and $x^2$ is equal to the reciprocal of the sum of $x$ and $x^2$.

**39.** Five times a number plus twice the reciprocal of the number is 9.25.

a) Write an algebraic representation for this problem situation.

b) Find the number.

**40.** Linda and Paul take 5 hr to paint a room. If Paul works alone, it takes him 15 hr. How long would it take Linda to paint the room alone?

## Chapter 6 Test

**1.** Reduce $\dfrac{3x^2 + 5x - 2}{2x^2 + 7x + 6}$ to lowest terms.

**2.** Find the LCD of the following pair of rational expressions:

$$\frac{x - 3}{8x^2 + 2x - 15} \quad \text{and} \quad \frac{2x + 5}{12x^2 - 7x - 10}$$

In Questions 3–7, perform the indicated operation. Simplify the result to lowest terms.

**3.** $\dfrac{6x + 3}{x^2 + 2x - 3} \cdot \dfrac{x^2 - x}{4x^2 - 1}$

**4.** $\dfrac{2x - 2y}{x^3 yz} \div \dfrac{y - x}{x^2 z + z}$

**5.** $\dfrac{x(x - 8)}{2x^2 + 5x - 12} + \dfrac{2(x^2 + 10x)}{2x^2 + 5x - 12}$

**6.** $\dfrac{3x}{x + 1} - \dfrac{5}{2x - 3}$

**7.** $\dfrac{2x^3 y^2 - 12xy^2}{4x^2}$

**8.** Simplify the following complex fraction: $\dfrac{\dfrac{x + 2}{x^2 - 16}}{\dfrac{x - 4}{x + 4}}$

**9.** Write the following expression without negative exponents and then simplify the complex fraction:

$$\frac{2x^{-1} - y}{3x^{-1}}$$

**10.** Find the quotient and remainder of the following expression.

Write this division in the specified form

$$(4x^3 - 6x^2 + 3) \div (2x - 1)$$

$$\frac{\text{Dividend}}{\text{Divisor}} = \text{Quotient} + \frac{\text{Remainder}}{\text{Divisor}}.$$

**11.** Find the quotient of the following expression using synthetic division:

$$(2x^3 - 11x^2 + 9x - 6) \div (x - 3)$$

In Questions 12 and 13, find the solution to the equation using an algebraic method. Determine whether or not any of the solutions are extraneous. Support the solution graphically.

**12.** $\dfrac{x + 2}{x - 1} - \dfrac{1}{2} = 3$

**13.** $\dfrac{x + 15}{x^2 - 9} - \dfrac{x}{x + 3} = \dfrac{3}{x - 3}$

**14.** Find the solution to $\dfrac{5}{x - 12} = \dfrac{3}{2x + 1}$ using the $x$-intercept method. Support the solution numerically using a grapher.

**15.** One number is three times another number. Find the numbers if the sum of their reciprocals is 1.

# Rational Exponents, Radicals, and Complex Numbers

## AN APPLICATION

The approximate speed $S$ of a car moments prior to a skidding stop can be estimated using the formula

$$S = 2\sqrt{5L}$$

where $L$ represents the length (in feet) of the skid mark.

# Roots and Radicals

The $n$th Root of a Real Number ▪ Exponents of the Form $1/n$ ▪ Radical Notation ▪ The Pythagorean Theorem ▪ Problem Situations Using Square Roots

Most graphers have the radical symbols $\sqrt{\phantom{x}}$ and $\sqrt[3]{\phantom{x}}$ for square root and cube root. However, to find $n$th roots for $n > 3$ you need to use fractional exponents, as the following exploration illustrates.

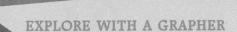

### EXPLORE WITH A GRAPHER

Confirm that each pair is identical:

1. a) $\sqrt{49}$      b) $49 \wedge (1/2)$

2. a) $\sqrt{15}$      b) $15 \wedge (1/2)$

3. a) $\sqrt[3]{214}$      b) $214 \wedge (1/3)$

**Experiment:** How would you find $\sqrt[4]{81}$ or $\sqrt[5]{113}$ on a grapher?

This exploration introduces you to methods of finding $n$th roots on a grapher. We need to introduce the concept of $n$th root.

## The $n$th Root of a Real Number

Because $3^2 = 9$, we call 3 a **square root of 9** and we write

$$\sqrt{9} = 3.$$

It also is true that $(-3)^2 = 9$. So we also call $-3$ a **square root of 9**.

In much the same way, we can define **cube root**, **fourth root**, and, in general, $n$th root. We define a general $n$th root of $a$ as follows.

---

**DEFINITION 7.1** An *n*th Root of a Number

If *n* is an integer greater than 0, then *b* is an **nth root of *a*** if

$$b^n = a.$$

---

Here are some additional examples of roots:

16 is a square root of 256, since $16^2 = 256$.

4 is a cube root of 64, since $4^3 = 64$.

$\dfrac{-2}{3}$ is a square root of $\dfrac{4}{9}$, since $\left(\dfrac{-2}{3}\right)^2 = \dfrac{4}{9}$.

## Exponents of the Form 1/*n*

In Chapters 1 and 5, you learned that in the symbol $b^n$, *n* is a positive whole number called an **exponent**. One of the properties of exponents you learned was

$$(a^n)^m = a^{nm}.$$

We extend the exponent notation to allow exponents to be fractions. We also require that the property of exponents continues to be true. For example,

$$\left(b^{1/2}\right)^2 = b^{(1/2)\cdot 2} = b \qquad \text{and} \qquad \left(b^{1/3}\right)^3 = b^{(1/3)\cdot 3} = b.$$

Because $b^{\frac{1}{2}}$ squared results in *b* and $b^{\frac{1}{3}}$ cubed results in *b*, it is reasonable to declare that

$$b^{1/2} = \sqrt{b} \qquad \text{and} \qquad b^{1/3} = \sqrt[3]{b}.$$

Finally, we need to decide whether $4^{1/2}$ refers to 2 or $-2$, since they are both square roots of 4. This leads to the definition of **principal *n*th root of a number**. We consider the cases *n* even and *n* odd separately.

> **GRAPHER NOTE**
>
> The square root of a negative number is an imaginary number. We will study imaginary numbers later in this chapter. Entering the square root of a negative number in most graphers results in an error message.

---

**DEFINITION 7.2** The Principal *n*th root of *b*, *n* is Even

Let *n* be an even positive integer and *b* a positive real number. Then $b^{1/n}$ is the *positive* real number whose *n*th power is *b*,

$$(b^{1/n})^n = b,$$

and $b^{1/n}$ is the principal *n*th root of *b*.

---

Notice that 2 is the principal square root of 4. The number $-2$ is also a square root but not the principal square root.

In Example 1, we solve for the principal square root.

**EXAMPLE 1**    Finding the Principal Square Root

Find the following roots:

a) $(36)^{1/2}$

b) $-121^{1/2}$

c) $(-121)^{1/2}$

SOLUTION

a) $36^{1/2} = 6$, the principal square root.

b) The principal square root of 121 is 11. Therefore $-121^{1/2} = -(121)^{1/2} = -11$.

c) $(-121)^{1/2}$ is not a real number since no real number squared is $-121$.   ∎

Example 1 part (c) shows that when $n$ is an even integer, the restriction that the base $b$ is positive is necessary in the definition of $b^{1/n}$. With odd values of $n$, however, the base can be either positive or negative. For example,

$$2^3 = 8 \qquad \text{means that} \qquad \sqrt[3]{8} = 2$$

and

$$(-2)^3 = -8 \qquad \text{means that} \qquad \sqrt[3]{-8} = -2.$$

That is, the cube root of 8 is 2 and the cube root of $-8$ is $-2$.

We define the principal $n$th root of a number when $n$ is odd as follows.

**DEFINITION 7.3**    The Principal $n$th root of $b$, $n$ is Odd

If $b$ is a real number and $n$ is an odd positive integer, then $b^{1/n}$ is the real number whose $n$th power is $b$,

$$(b^{1/n})^n = b.$$

Example 2 illustrates finding roots when $n$ is an odd integer. The root and the base have the same sign.

### EXAMPLE 2    Finding Roots When $n$ is Odd

Evaluate the following expressions.   Verify the solutions numerically using a grapher.

a) $125^{1/3}$

b) $(-64)^{1/3}$

c) $\left(-\dfrac{1}{32}\right)^{1/5}$

SOLUTION

a) $125^{1/3} = 5$, since $5^3 = 125$.

b) $(-64)^{1/3} = -4$, since $(-4)^3 = -64$.

c) $\left(-\dfrac{1}{32}\right)^{1/5} = \dfrac{-1}{2}$, since $\left(\dfrac{-1}{2}\right)^5 = -\dfrac{1}{32}$.

We leave the grapher numerical support to the reader.    ■

In the remainder of the text, the principal root is assumed.

**GRAPHER NOTE**

Consult your grapher lab manual for instructions on finding roots with your grapher.

## Radical Notation

The principal $n$th root of a number also can be indicated by using a radical symbol like the one shown here:

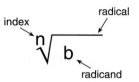

The integer $n > 1$ is called the **index** of the radical and $b$ is called the **radicand**. The symbol $\sqrt{\phantom{x}}$ is the **radical sign**. If no index appears, it is understood to be 2, which is the square root. The radical symbol has the following meaning.

> **DEFINITION 7.4** Radical Notation
>
> If $n$ is a positive integer and $b$ a number for which $b^{1/n}$ is defined, then
> $$\sqrt[n]{b} = b^{1/n}.$$
>
> In addition,
> $$\sqrt[n]{0} = 0^{1/n} = 0.$$

**GRAPHER NOTE**

On some graphers, the number of digits to the right of the decimal that are displayed can be set on the MODE menu. Throughout the rest of this chapter, we will assume a grapher setting of three digits. Accordingly we will write statements like $\sqrt{2} = 1.414$ instead of $\sqrt{2} = 1.414213562\ldots$.

Example 3 illustrates how to use a grapher to evaluate roots. When the root has an index greater than 3, the fractional exponent form is required.

**EXAMPLE 3** Finding Square Roots Using the Grapher

Using a grapher, find the following roots. Note, the $\sqrt{\phantom{x}}$ symbol always refers to the principal root of a number.

a) $\sqrt{4225}$

b) $-\sqrt{625}$

c) $\sqrt[4]{8}$

d) $\sqrt{-64}$.

SOLUTION

a) $\sqrt{4225} = 65$

b) $-\sqrt{625} = -25$

c) $\sqrt[4]{8} = 1.682$

d) $\sqrt{-64}$ is not defined. A grapher shows an error message for $\sqrt{-64}$, indicating it is not a real number.

Figure 7.1 offers grapher support for the solutions to Example 3 (a), (b), and (c). When $\sqrt{-64}$ is entered into the grapher an error message results. ■

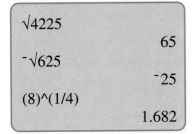

```
√4225
                    65
⁻√625
                   ⁻25
(8)^(1/4)
                 1.682
```

**FIGURE 7.1** The grapher solution for Example 3.

Example 4 illustrates additional cases of using the grapher to obtain decimal approximations for irrational numbers.

**EXAMPLE 4**    Finding Decimal Approximations for Irrational Numbers

Find the three-place decimal approximation for the following radical expressions:

a) $\sqrt[3]{-8.75}$

b) $\sqrt[3]{32.52}$

c) $\sqrt{726}$

SOLUTION

a) $\sqrt[3]{-8.75} = -2.061$

b) $\sqrt[3]{32.52} = 3.192$

c) $\sqrt{726} = 26.944$    ■

## The Pythagorean Theorem

One of the most famous theorems in all of mathematics is the Pythagorean theorem. This theorem states the relationship between the longest side of a right triangle and the lengths of the other two sides. We used this theorem in Chapter 2 to develop the distance formula.

---

### Pythagorean Theorem

If $c$ is the length of the longest side of a right triangle and $a$ and $b$ are the lengths of the shorter sides, then

$$c^2 = a^2 + b^2.$$

The longest side is called the **hypotenuse** and the shorter sides are called **legs**.

---

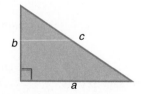

**FIGURE 7.2** The Pythagorean theorem: $c^2 = a^2 + b^2$.

If the lengths of two sides of a right triangle are known, the Pythagorean theorem can be used to find the length of the third side (see Figure 7.2). For example, suppose $a$ and $b$ are known, then $c$ can be found as follows:

$$c^2 = a^2 + b^2$$

$$c = \sqrt{a^2 + b^2}$$

Example 5 illustrates how to use this method to calculate the length of a side of a right triangle.

**EXAMPLE 5**   Finding Missing Sides of a Right Triangle

Use the Pythagorean theorem to find the length of the missing side of the following triangles:

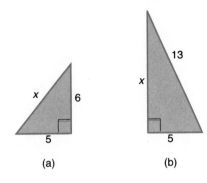

(a)                (b)

SOLUTION
Apply the Pythagorean theorem to each figure.

a) $x^2 = 5^2 + 6^2$

$x^2 = 25 + 36$

$x^2 = 61$            **Find the positive square root of each side.**

$x = \sqrt{61} = 7.810$

Therefore, the missing side has length 7.810.

b) $13^2 = 5^2 + x^2$

$x^2 = 169 - 25$        **Subtract 25 from each side.**

$x^2 = 144$            **Take the square root of each side.**

$x = \sqrt{144} = 12$

Therefore, the missing side has length 12.

## Problem Situations Using Square Roots

Andy, a catcher on the Stars Baseball team, knows that the four bases of an infield are vertices of a square and the distances between home plate and first base and between third base and home are both 90 ft. He also knows that he has farther to throw to get a runner out at second base than to get him out at first or third.

**EXAMPLE 6**   THE BASEBALL PROBLEM SITUATION: Finding Distance

How far is it from home plate to second base?

SOLUTION

Let $x =$ the distance from home base to second base (see Figure 7.3).

$$x^2 = 90^2 + 90^2 \quad \text{Applying the Pythagorean theorem.}$$

$$x^2 = 16200 \quad \text{Take the square root of each side.}$$

$$x = 127.279$$

Therefore Andy needs to throw the ball 127.279 ft to get someone out at second base and only 90 ft to get them out at first or third base.

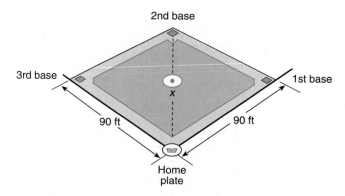

**FIGURE 7.3**   A sketch of the baseball field in Example 6.                ■

Not only are square roots used when the Pythagorean theorem is applied but also when Heron's formula is used. Heron's formula can be used to find the area of any triangle if the lengths of the three sides of the triangle are known:

$$A = \sqrt{s(s-a)(s-b)(s-c)},$$

where $s = \dfrac{a+b+c}{2}$ and $a$, $b$, and $c$ are the lengths of the three sides.

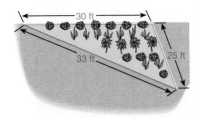

**FIGURE 7.4** Triangular garden problem situation.

### Triangular Garden Problem Situation

A triangular plot of land is used for a vegetable garden. Tracy needs to spread fertilizer on the garden. One bag of fertilizer is needed for each 250 sq ft of garden.

**EXAMPLE 7**    THE TRIANGULAR GARDEN PROBLEM: Finding Area

Find the number of bags of fertilizer Tracy must buy if the lengths of the sides of the garden measure 25 ft, 30 ft, and 33 ft, respectively. (See Figure 7.4.)

$44 \rightarrow X: \sqrt{(X(X-25)(X-30)(X-33))}$

$358.809$

**FIGURE 7.5** Grapher support of Example 7.

SOLUTION
Find $s$ by adding the measure of the sides and dividing the sum by 2.

$$s = \frac{25 + 30 + 33}{2} = 44$$

$$A = \sqrt{44(44-25)(44-30)(44-33)} \quad \textbf{Apply Heron's formula.}$$

$$= 358.809 \text{ sq ft}$$

Because $250 < 358.81 < 500$, we see that Tracy needs to buy more than one bag. Two will be sufficient. Figure 7.5 shows grapher support for this solution. ∎

## Exercises for Section 7.1

In Exercises 1–6, give the exponent form of the expression.

1. $\sqrt{8}$

2. $\sqrt{17}$

3. $\sqrt[3]{12}$

4. $\sqrt[4]{23}$

5. $\sqrt{x}$

6. $\sqrt{ab}$

In Exercises 7–10, give the radical form of the expression.

7. $(3)^{1/2}$

8. $(\pi)^{1/2}$

9. $(6)^{1/4}$

10. $x^{1/3}$

In Exercises 11–22, use a grapher to find a decimal approximation (accurate to three places) for each expression.

11. $\sqrt{192}$

12. $\sqrt{1579}$

13. $\sqrt{0.2476}$

14. $\sqrt{1.2534}$

15. $\sqrt{\dfrac{2}{3}}$

16. $\sqrt{\dfrac{1}{15}}$

17. $\sqrt[3]{147}$

18. $\sqrt[3]{-228}$

19. $\sqrt[3]{\dfrac{-21}{32}}$

20. $\sqrt[3]{\dfrac{5}{12}}$

21. $\sqrt[3]{21.35}$

22. $\sqrt[3]{87.36}$

In Exercises 23–32, find the indicated root without using a grapher.

**23.** $\left(\dfrac{25}{16}\right)^{1/2}$

**24.** $\left(\dfrac{4}{49}\right)^{1/2}$

**25.** $125^{1/3}$

**26.** $(-27)^{1/3}$

**27.** $\sqrt{169}$

**28.** $\sqrt{121}$

**29.** $\sqrt{0.16}$

**30.** $\sqrt{0.04}$

**31.** $\sqrt[3]{\dfrac{-8}{27}}$

**32.** $\sqrt[3]{\dfrac{64}{125}}$

In Exercises 33–38, use a grapher to find the decimal approximation for each irrational number. Round all answers to three decimal places.

**33.** $\sqrt[3]{-327}$

**34.** $\sqrt[3]{832}$

**35.** $\sqrt{721}$

**36.** $\sqrt{51}$

**37.** $32^{1/4}$

**38.** $56^{1/6}$

In Exercises 39–46, evaluate each expression without using a grapher. (Recall that negative exponents have been defined in Chapter 5.)

**39.** $8^{1/3}$

**40.** $27^{-1/3}$

**41.** $49^{1/2}$

**42.** $121^{1/2}$

**43.** $16^{1/4}$

**44.** $32^{-1/5}$

**45.** $64^{-1/3}$

**46.** $1000^{1/3}$

In Exercises 47–52, evaluate the expression using a grapher.

**47.** $81^{1/4}$

**48.** $3125^{1/5}$

**49.** $2401^{1/4}$

**50.** $32768^{1/5}$

**51.** $4913^{1/3}$

**52.** $2744^{1/3}$

In Exercises 53–56, graph the function in the window $[-10, 10]$ by $[-5, 5]$. Choose the correct graph from those shown below.

**53.** $y = \sqrt{x}$

**54.** $y = \sqrt{x + 7}$

**55.** $y = \sqrt[3]{x}$

**56.** $y = \sqrt[3]{x - 4}$

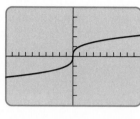

(a)

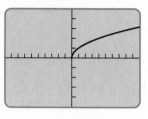

(b)

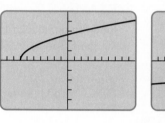

(c)

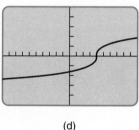

(d)

For Exercises 53–56.

TRANSLATING WORDS TO SYMBOLS

In Exercises 57–60, write the expression symbolically.

**57.** The cube root of eight

**58.** The square root of the fourth power of $x$

**59.** Four to the one-half power

**60.** Negative nine to the one-third power

In Exercises 61–64, determine whether a triangle with the given side lengths $a, b$, and $c$ is a right triangle.

**61.** $a = 5, b = 7, c = 9$     **62.** $a = 5, b = 12, c = 13$

**63.** $a = \sqrt{7}, b = 3, c = 4$

**64.** $a = \sqrt{11}, b = \sqrt{47}, c = 4\sqrt{3}$

**65.** *Using Algebra in Geometry.* Use the Pythagorean theorem to find the length of the missing side of each of the following right triangles:

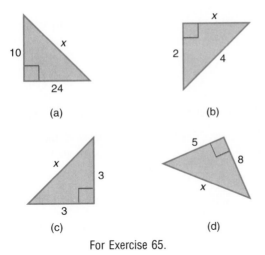

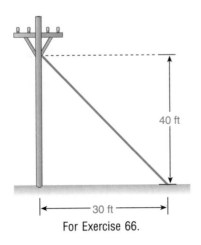

For Exercise 65.

**66.** *Guy Wire Length.* A piece of wire is attached to a telephone pole as a guy wire. Find the length of the guy wire if it is stretched from a position 40 ft above the level ground to a point 30 ft from the base of the pole.

For Exercise 66.

**67.** *Guy Wire Purchase.* Fred needs to calculate how much guy wire to buy from Florida Wire to install 50 telephone poles. Each guy wire stretches from a point on a pole, which is 24 ft above level ground, to a point on the ground 10 ft from the base of the pole. How many feet of wire must he buy?

**68.** *TV Screen Size.* Television sets are sold with the dimension representing the diagonal of the screen. A 20-in. TV means the diagonal of screen is 20 in. Find the measure of one side of a square television screen if the diagonal of the screen is 20 in.

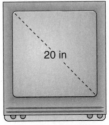

For Exercise 68.

**69.** *TV Screen Size.* The diagonal of a rectangular television screen is 13 in. The length of the screen is 10 in. Find the width.

**70.** Find the area of a triangle if the three sides measure 5 ft, 8 ft, and 10 ft, respectively. Round your answer to three decimal places.

**71.** *Locating a Ladder.* Sun places a 12-ft ladder against the house at a point 10 ft from the ground. How far is the base of the ladder from the house?

### EXTENDING THE IDEAS

Radicals and fractional exponents can be used to define functions. In Exercises 72 and 73, $y_1 = \sqrt{x} = x^{1/2}$ and $y_2 = \sqrt[3]{x} = x^{1/3}$.

**72.** Create the following table either on a grapher or by hand. How are the second and third columns related to the first column?

| X | Y1 | Y2 |
|---|-----|-----|
| 1 | 1.000 | 1.000 |
| 2 | 1.414 | 1.260 |
| 3 | 1.732 | 1.442 |
| 4 | 2.000 | 1.587 |
| 5 | 2.236 | 1.710 |
| 6 | 2.449 | 1.817 |
| 7 | 2.646 | 1.913 |
| X = 1 | | |

For Exercise 72.

**73.** In the following table, why does the word error appear in one column and not the other?

| X | Y1 | Y2 |
|---|---|---|
| −6 | ERROR | −1.817 |
| −5 | ERROR | −1.710 |
| −4 | ERROR | −1.587 |
| −3 | ERROR | −1.442 |
| −2 | ERROR | −1.260 |
| −1 | ERROR | −1.000 |
| 0 | 0.000 | 0.000 |

X = −6

For Exercise 73.

**74. Writing to Learn.** Explain in a few sentences how you multiply $\sqrt{3} \cdot \sqrt[3]{3}$.

**75. Writing to Learn.** Explain how to write $\sqrt{x}$ without using a radical.

LOOKING BACK—MAKING CONNECTIONS

In Exercises 76–81, find the product of the expression. Assume all variables do not equal zero.

**76.** $3x^2 \cdot 5x^3$

**77.** $4x^3 \cdot 2y^5$

**78.** $x^5 \cdot x^{-2} \cdot x^8$

**79.** $(3x^{-1})(5x^2)^3$

**80.** $(4x^2y^{-3})^{-2}$

**81.** $(5xy^{-2})^{-3}(25xy^{-1})^2$

---

## 7.2 Rational Exponents and Radicals

Rational Exponents and Radicals in Algebraic Expressions ▪ Simplifying Expressions with Radicals

**REMINDER**

Properties of Exponents:

1. $b^m \cdot b^n = b^{m+n}$
2. $(b^m)^n = b^{mn}$
3. $(ab)^m = a^m \cdot b^m$
4. $\dfrac{b^m}{b^n} = b^{m-n}, b \neq 0$
5. $\left(\dfrac{b}{a}\right)^m = \left(\dfrac{b^m}{a^m}\right)$
6. $b^{-m} = \dfrac{1}{b^m}$

In Section 7.1, we learned that the symbols $b^{1/n}$ and $\sqrt[n]{b}$ mean the same thing and can be used interchangeably.

We now consider $b^{m/n}$. We assume that the properties of exponents (reviewed in the box to the left) must hold, so the following equation must be true:

$$b^{m/n} = (b^{1/n})^m$$

This equality is the basis for the definition of a rational exponent.

**DEFINITION 7.5** Rational Exponents

If $m$ and $n$ are positive integers with no common factors, then the rational number $m/n$ exponent is defined by

$$b^{m/n} = (b^{1/n})^m,$$

provided $b^{1/n}$ exists. If $b^{1/n}$ does not exist, then $b^{m/n}$ does not exist.

Any expression with a rational exponent can be expressed using radicals. Example 1 illustrates.

**EXAMPLE 1**    Converting from Exponential Form to Radical Form

Show that $7^{2/3}$ is equivalent to both $(\sqrt[3]{7})^2$ and $\sqrt[3]{49}$.

SOLUTION

Begin by using the Definition of Rational Exponents.

$$7^{2/3} = (7^{1/3})^2$$

$$= (\sqrt[3]{7})^2$$

Use the Properties of Exponents to write an equivalent expression.

$$7^{2/3} = (7^2)^{1/3} \quad \textbf{Use the Definition of Rational Exponents.}$$

$$= \sqrt[3]{7^2}$$

$$= \sqrt[3]{49}$$  ■

Notice in the solution to Example 1 that the *denominator* of the rational exponent represents the *root* or the index of the radical and the *numerator* represents the *power* to which the base is being raised. These operations can be performed in either order. Here is another example that illustrates that the orders can be reversed.

$$27^{2/3} = (\sqrt[3]{27})^2 = 3^2 = 9 \quad \text{or} \quad 27^{2/3} = \sqrt[3]{27^2} = \sqrt[3]{729} = 9$$

Usually, it is easiest to take the root first and then raise to the power, as then you are working with smaller numbers.

Conversely, any expression with a radical can be expressed using rational exponents, as Example 2 illustrates.

**REMINDER**

$$\sqrt{9+16} = \sqrt{25} = 5$$

and

$$\sqrt{9} + \sqrt{16} = 3 + 4 = 7.$$

Therefore

$$\sqrt{x+y} \neq \sqrt{x} + \sqrt{y}.$$

**EXAMPLE 2**    Converting from Radical Form to Exponential Form

Change the following expressions to exponential form and simplify if possible:

a) $\sqrt{7}$

b) $\sqrt[3]{2^9}$

c) $\sqrt[5]{x^{10}}$

d) $\sqrt{x^2 + y^2}$

SOLUTION

a) $\sqrt{7} = 7^{1/2}$

b) $\sqrt[3]{2^9} = 2^{9/3} = 2^3 = 8$

c) $\sqrt[5]{x^{10}} = x^{10/5} = x^2$

d) $\sqrt{x^2 + y^2} = (x^2 + y^2)^{1/2}$   ∎

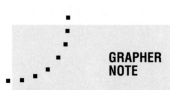

Examples 1 and 2 illustrate that we can now change rational exponents to radicals and vice versa. Here are the general statements:

$$b^{m/n} = (b^{1/n})^m = (\sqrt[n]{b})^m \quad \text{and} \quad (\sqrt[n]{b})^m = \sqrt[n]{b^m} = (b^m)^{1/n}$$

The process of converting back and forth between radical and exponential form occurs frequently in the process of evaluating expressions involving either exponents or radicals. The solution to Example 3 uses both radicals and fractional exponents.

**EXAMPLE 3**   Evaluating Expressions with Rational Exponents

Evaluate the following expressions. Show numerical support for parts (b) and (c).

a) $125^{2/3}$

b) $\left(\dfrac{4}{9}\right)^{3/2}$

c) $8^{-5/3}$

SOLUTION

a) $125^{2/3} = (\sqrt[3]{125})^2 = 5^2 = 25$

b) $\left(\dfrac{4}{9}\right)^{3/2} = \left(\sqrt{\dfrac{4}{9}}\right)^3 = \left(\dfrac{2}{3}\right)^3 = \dfrac{8}{27}$

c) $8^{-5/3} = \dfrac{1}{8^{5/3}} = \dfrac{1}{(\sqrt[3]{8})^5} = \dfrac{1}{2^5} = \dfrac{1}{32}$

**Support Numerically.**   Figure 7.6 shows grapher support for parts (b) and (c).   ∎

```
(4/9)^(3/2)
                    .296
8/27
                    .296
8^(⁻5/3)
                    .031
1/32
                    .031
```

**FIGURE 7.6**
Grapher support for evaluating rational exponents in Example 3.

## Rational Exponents and Radicals in Algebraic Expressions

One benefit of the rational exponent format is that properties of exponents can be used to simplify algebraic expressions. Example 4 illustrates this use of rational exponents.

**EXAMPLE 4**    Simplifying Using Properties of Exponents

Perform the following operations using the properties of exponents:

a) $x^{1/2} \cdot x^{2/3}$

b) $(x^{2/3})^3$

c) $\dfrac{x^{1/4}}{x^{3/2}}$

SOLUTION

a) $x^{1/2} \cdot x^{2/3} = x^{(1/2)+(2/3)} = x^{7/6}$

b) $(x^{2/3})^3 = x^{(2/3)\cdot 3} = x^2$

c) $\dfrac{x^{1/4}}{x^{3/2}} = x^{1/4} \cdot x^{-3/2} = x^{1/4-3/2} = x^{1/4-6/4} = x^{-5/4} = \dfrac{1}{x^{5/4}}$    ■

Another important use of rational exponents is to simplify expressions involving radicals. Often when a radical expression appears to be as simple as possible, you will find that additional simplification occurs when you convert the expression to rational exponent form. Example 5 illustrates.

**EXAMPLE 5**    Changing to Rational Exponents to Simplify

Simplify the following expressions using rational exponents and the properties of exponents. Express your answer in radical form.

a) $\sqrt{x} \cdot \sqrt[3]{x^2}$

b) $\dfrac{\sqrt[3]{x^2}}{\sqrt[4]{x}}$

c) $\sqrt{\sqrt[3]{x^2}}$

SOLUTION

Change each expression to exponential form using rational exponents.

a) $\sqrt{x} \cdot \sqrt[3]{x^2} = x^{1/2} \cdot x^{2/3} = x^{1/2+2/3} = x^{7/6} = \sqrt[6]{x^7}$

b) $\dfrac{\sqrt[3]{x^2}}{\sqrt[4]{x}} = \dfrac{x^{2/3}}{x^{1/4}} = x^{2/3-1/4} = x^{5/12} = \sqrt[12]{x^5}$

c) $\sqrt{\sqrt[3]{x^2}} = \sqrt{x^{2/3}} = (x^{2/3})^{1/2} = x^{1/3} = \sqrt[3]{x}$    ■

## Simplifying Expressions with Radicals

Raising a number to the $n$th power and taking the principal $n$th root of that number appear to be opposite operations. For example,

$$(\sqrt{9^2}) = 9 \quad \text{and} \quad (\sqrt[3]{4^3}) = 4.$$

From these two examples, you might be willing to generalize that

$$\sqrt[n]{x^n} = x \qquad \text{for all } x,$$

which is not true.

We must be careful. If the number $x$ is negative and $n$ is even, this pattern is not true. For example,

$$\sqrt{(-3)^2} \neq -3 \quad \text{and} \quad \sqrt[4]{(-2)^4} \neq -2.$$

However, we can state a valid generalization using absolute value signs.

---

### Powers and Roots Properties

If $b$ is a real number and $n$ is an integer $> 1$, then the following holds:

1. $\sqrt[n]{b^n} = b$, if $n$ is odd.

2. $\sqrt[n]{b^n} = |b|$, if $n$ is even.

---

Example 6 uses these properties in numerical settings and Example 7 uses them in an algebraic setting.

### EXAMPLE 6    Evaluating Radical Expressions

Evaluate the following radical expressions:

a) $\sqrt[3]{16^3}$

b) $\sqrt[5]{(-3.5)^5}$

c) $(\sqrt{5768})^2$

d) $\sqrt{(-14)^2}$

SOLUTION

a) $\sqrt[3]{16^3} = 16$

b) $\sqrt[5]{(-3.5)^5} = -3.5$

c) $(\sqrt{5768})^2 = 5768$

d) $\sqrt{(-14)^2} = |-14| = 14$

■

**EXAMPLE 7**    Simplifying Radical Expressions

Simplify each of the following expressions:

a) $\sqrt[4]{x^8}$

b) $\sqrt{(-x)^2}$

c) $\sqrt[3]{-x^3}$

SOLUTION

a) $\sqrt[4]{x^8} = \sqrt[4]{(x^2)^4} = x^2$

b) $\sqrt{(-x)^2} = |-x| = |x|$

c) $\sqrt[3]{-x^3} = \sqrt[3]{(-x)^3} = -x$

■

## Exercises for Section  7.2

In Exercises 1–8, show that the expression is equivalent to two different radical expressions.

**1.** $16^{5/2}$

**2.** $8^{2/3}$

**3.** $5^{2/3}$

**4.** $3^{3/4}$

**5.** $3^{4/5}$

**6.** $9^{3/2}$

**7.** $27^{2/3}$

**8.** $64^{5/6}$

In Exercises 9–16, change the expression to exponential form.

**9.** $\sqrt{5}$

**10.** $\sqrt{11}$

**11.** $\sqrt[3]{3^2}$

**12.** $\sqrt[3]{2^4}$

**13.** $\sqrt{x^3}$

**14.** $\sqrt[3]{x^2}$

**15.** $\sqrt[4]{(2y)^3}$

**16.** $\sqrt[5]{z^3}$

In Exercises 17–24, change the expression from exponential form to radical form. Evaluate with a paper-and-pencil method and support numerically using a grapher.

**17.** $25^{1/2}$

**18.** $25^{3/2}$

**19.** $16^{1/4}$

**20.** $27^{2/3}$

**21.** $8^{5/3}$

**22.** $125^{5/3}$

**23.** $9^{3/2}$

**24.** $216^{2/3}$

In Exercises 25–34, simplify the expression using the properties of exponents. Write the answer in exponential form.

**25.** $x^{1/2} \cdot x^{3/2}$

**26.** $x^{2/3} \cdot x^{5/3}$

**27.** $x^{1/2} \cdot x^{1/3}$

**28.** $x^{2/3} \cdot x^{3/2}$

**29.** $(x^{2/3})^6$

**30.** $(x^{1/2})^{2/3}$

**31.** $\dfrac{x^{1/2}y^{2/3}}{x^{3/4}y^{5/3}}$

**32.** $\dfrac{x^{2/3}y^{1/5}}{x^3y^2}$

**33.** $\left(\dfrac{x^{1/3}y^{1/4}}{x^{2/3}y^{1/6}}\right)^{12}$   **34.** $\left(\dfrac{x^{3/4}y^{1/4}}{x^{1/2}y^{2/3}}\right)^{8}$

In Exercises 35–44, simplify the expression using the properties of exponents. Write the answer in radical form.

**35.** $x^{1/3} \cdot x^6$

**36.** $x^{1/2} \cdot x^4$

**37.** $(x^{5/3})^{3/2}$

**38.** $(x^{3/4})^{1/2}$

**39.** $\sqrt{x} \cdot \sqrt[3]{x^2}$

**40.** $\sqrt[3]{x} \cdot \sqrt{x}$

**41.** $\dfrac{\sqrt[3]{x^2}}{\sqrt{x}}$

**42.** $\dfrac{\sqrt{x}}{\sqrt[3]{x}}$

**43.** $(\sqrt{x})^{1/2}$

**44.** $(\sqrt[3]{x^2})^{1/3}$

In Exercises 45–48, evaluate the expression without using a grapher.

**45.** $\sqrt[3]{25^3}$

**46.** $\sqrt{144^2}$

**47.** $\sqrt[5]{-123^5}$

**48.** $\sqrt[4]{18^4}$

In Exercises 49–56, change the expression from radical form to exponential form.

**49.** $\sqrt{xy}$

**50.** $\sqrt{5x}$

**51.** $\sqrt{49x^2}$

**52.** $\sqrt[4]{x^3}$

**53.** $\sqrt[3]{x^6 y}$

**54.** $\sqrt[4]{x^2 y^3}$

**55.** $\sqrt{x^2 - y^2}$

**56.** $\sqrt{x^2 + 25}$

In Exercises 57–64, simplify the radical expression.

**57.** $\sqrt{7^2}$

**58.** $\sqrt{23^2}$

**59.** $\sqrt{y^2}$

**60.** $\sqrt{z^6}$

**61.** $\sqrt[3]{x^3}$

**62.** $\sqrt[3]{x^6}$

**63.** $\sqrt[3]{-x^6}$

**64.** $\sqrt[4]{(-x)^4}$

**65.** *Cross-sectional Area of a Cone.* The lateral edge of the following cone has length 12 cm.

a) If the radius $r$ of the base is $r = 2\sqrt{3}$, show that the height $h = \left(12^2 - (2\sqrt{3})^2\right)^{1/2}$.

b) Find the area of the $\triangle ABC$.

**66.** *Volume of a Cone.* Recall that the volume of a cone is found using the formula $V = \frac{\pi}{3}r^2 h$.

a) Find the volume of the following cone when $r = 2\sqrt{3}$.

b) Find the radius of this cone when the height is $h = 9.8$.

c) Find the volume of this cone when $h = 9.8$.

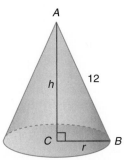

For Exercises 65 and 66.

**67.** *Tool-and-die Design.* A tool-and-die engineer must design a tool with a part shaped as shown in the following figure:

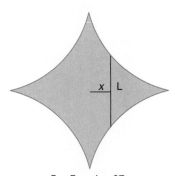

For Exercise 67.

If the cross-sectional length L is $x$ units from the center, then the length of L is $2(5 - x^{2/3})^{3/2}$. Find L when $x = 8$.

**68.** *Multiple Representations on a Grapher.* Use both numerical tables and a graph to show that

$$(\sqrt{x} + \sqrt{5})^2 \neq x + 5.$$

**69.** *Multiple Representations on a Grapher.* Use both numerical tables and a graph to show that

$$\sqrt{x^2 + 3^2} \neq x + 3.$$

**70.** *Multiple Representations on a Grapher.* Use both numerical tables and a graph to show that

$$x^{1/2}(x^{1/3} + x^{1/2}) \neq x^{1/6} + x^{1/4}.$$

Exercises 71–73 will help you become more familiar with your grapher.

**71.** The expressions $8^{2/3}$, $(8^2)^{1/3}$, and $(8^{1/3})^2$ are algebraically equal. Does your grapher recognize them as equal? Check by evaluating each of the following on the Home screen:

a) $8 \wedge (2/3)$

b) $(8 \wedge 2) \wedge (1/3)$

c) $(8 \wedge (1/3)) \wedge 2$

**72.** *Interpreting a Grapher.*

a) Graph $y_1 = x \wedge (2/3)$ in the Standard window. Do you obtain graph (a) or (b)? Repeat for $y_2 = (x \wedge (1/3)) \wedge 2$.

b) What is the domain of the function $y = x^{2/3}$?

c) Which of the following graphs is a complete graph of $y = x^{2/3}$?

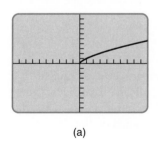 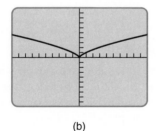

(a)                   (b)

For Exercise 72.

**73.** *Graphing Functions with Rational Exponents.* To find a complete graph of $y = x^{2/5}$, would you enter

a) $y_1 = x \wedge (2/5)$, or

b) $y_2 = (x \wedge (1/5)) \wedge 2$?

Write a paragraph explaining your answer.

TRANSLATING WORDS TO SYMBOLS

In Exercises 74–77, write the expression symbolically.

**74.** The square root of $x$

**75.** The cube root of $x$

**76.** The square of the cube root of $x$

**77.** The square root of the cube of $x$

EXTENDING THE IDEAS

**78.** *Interpreting a Grapher.*

a) What is the domain of the function $y = (x^2)^{1/4}$?

b) What is the domain of the function $y = (x^{1/4})^2$?

c) Would you expect the graphs of $y_2 = x \wedge (1/4) \wedge 2$ and $y_2 = x \wedge 2 \wedge (1/4)$ to be the same? Confirm whether they are.

**79. Writing to Learn.** Does your grapher interpret $y = x^{2/4}$, which you enter as $y_1 = x \wedge (2/4)$, to be identical to the function

a) $y = x^{1/2}$, or

b) $y = (x^2)^{1/4}$?

**80. Writing to Learn.** Would you expect your grapher to produce identical graphs for (a) $y = x^{0.5}$, (b) $y = x^{1/2}$, and (c) $y = x^{2/4}$? Does it?

LOOKING BACK—MAKING CONNECTIONS

In Exercises 81–84, write the expression as a single rational expression.

**81.** $\dfrac{1}{x} + \dfrac{x}{1}$

**82.** $\dfrac{1}{x} - \dfrac{1}{y}$

**83.** $\dfrac{1}{x - y} + \dfrac{1}{y - x}$

**84.** $\dfrac{1}{x - y} + \dfrac{1}{x + y}$

### 7.3

# Multiplication and Division with Radicals

Product Rule for Radicals • Simplifying Square Roots Using Perfect Squares • Simplifying Radicals Using Perfect Powers • Quotient Rule for Radicals • Using the Quotient Rule to Rationalize Denominators • Using Radicals in Problem Situations

Study the patterns suggested in the following exploration. They suggest a product rule for radicals and exponents.

## EXPLORE WITH A GRAPHER

Find a decimal approximation of each of the following radicals:

1. a) $\sqrt{9}\sqrt{6}$,   b) $\sqrt{54}$

2. a) $\sqrt{4}\sqrt{5}$,   b) $\sqrt{20}$

3. a) $\sqrt[3]{3}\sqrt[3]{5}$,   b) $\sqrt[3]{15}$

**State a Generalization:** State a rule for multiplying radicals that is suggested by this exploration.

## Product Rule for Radicals

The above exploration gives us some experience with products of radicals. It appears that the radical of a product is a product of the radicals. That rule is stated more formally next. Because radicals can be expressed as rational exponents, this rule is equivalent to one of the properties of exponents studied earlier.

**REMINDER**

The Product Rule for Radicals holds true only when the radicals have the same indices.

---

### Product Rule for Radicals

If $\sqrt[n]{a}$ and $\sqrt[n]{b}$ are defined, then

$$\sqrt[n]{a} \cdot \sqrt[n]{b} = \sqrt[n]{ab}.$$

---

Here are some examples of how the Product Rule can be used:

$$\sqrt{2} \cdot \sqrt{3} = \sqrt{6}$$

$$\sqrt[3]{4} \cdot \sqrt[3]{5} = \sqrt[3]{20}$$

$$\sqrt[4]{5} \cdot \sqrt[4]{8} = \sqrt[4]{40}$$

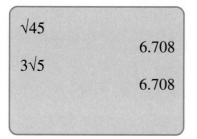

**FIGURE 7.7** Grapher support of the simplification.

## Simplifying Square Roots Using Perfect Squares

A number is a **perfect square** if its square root is a whole number or a fraction. For example, 16, 36, and $\frac{4}{9}$ are all perfect squares since

$$\sqrt{16} = 4, \qquad \sqrt{36} = 6, \qquad \text{and} \qquad \sqrt{\frac{4}{9}} = \frac{2}{3}.$$

We simplify square roots by using the Product Rule and factoring the radicand into two factors such that one is a perfect square. In this way, we are removing the perfect square factor. Here is an example:

$$\sqrt{45} = \sqrt{9} \cdot \sqrt{5} = 3\sqrt{5}$$

Figure 7.7 supports that $\sqrt{45}$ and $3\sqrt{5}$ are equivalent expressions.

**EXAMPLE 1**   Simplifying Radicals

Simplify the following expressions by factoring the radicand and removing any perfect square factors. Support the solutions of parts (a) and (b) using a grapher.

a) $\sqrt{24}$

b) $\sqrt{63}$

c) $\sqrt{98}$

SOLUTION

a) $\sqrt{24} = \sqrt{4} \cdot \sqrt{6} = 2\sqrt{6}$

b) $\sqrt{63} = \sqrt{9} \cdot \sqrt{7} = 3\sqrt{7}$

c) $\sqrt{98} = \sqrt{49} \cdot \sqrt{2} = 7\sqrt{2}$

Figure 7.8 shows decimal approximations for the radicals in parts (a) and (b), indicating numerical support for the solution. ∎

**FIGURE 7.8** Numerical support for parts (a) and (b) of Example 1.

The concept of perfect square also applies to expressions with variables. For example,

$$\sqrt{x^5} \quad \text{can be written as} \quad \sqrt{x^4}\sqrt{x} = x^2\sqrt{x}.$$

Notice that the variable is written as the product of two factors, one of which has a power that is a multiple of the index. Example 2 illustrates this method.

**EXAMPLE 2**    Simplifying Radicals Containing Variables

Simplify the following expressions by factoring the radicand and removing any perfect square factors. In each case, assume $x \geq 0$ and $y \geq 0$.

a) $\sqrt{25x^2}$

b) $\sqrt{18x^4}$

c) $\sqrt{200x^4y^3}$

SOLUTION

a) $\sqrt{25x^2} = \sqrt{25} \cdot \sqrt{x^2} = 5x$

b) $\sqrt{18x^4} = \sqrt{9} \cdot \sqrt{2} \cdot \sqrt{x^4} = 3x^2\sqrt{2}$

c) $\sqrt{200x^4y^3} = \sqrt{100x^4y^2 \cdot 2y}$    **Group together factors that are perfect squares.**

$$= \sqrt{100x^4y^2}\sqrt{2y}$$

$$= 10x^2y\sqrt{2y} \qquad \blacksquare$$

In this chapter, unless otherwise stated, we will assume all variables represent nonnegative real numbers. This assumption avoids the frequent use of absolute value notation.

## Simplifying Radicals Using Perfect Powers

Radicals with indices other that 2 can be simplified in a manner similar to that for square roots. For example, some *perfect power* radicals are the following:

$$\sqrt[3]{x^{12}} = (x^{12})^{1/3} = x^4 \quad \textbf{3 divides into 12 evenly.}$$

$$\sqrt[4]{x^{12}} = (x^{12})^{1/3} = x^3 \quad \textbf{4 divides into 12 evenly.}$$

$$\sqrt{x^{12}} = (x^{12})^{1/2} = x^6 \quad \textbf{2 divides into 12 evenly.}$$

Perfect powers can be used to simplify expressions. For example,

$$\sqrt[3]{x^6} = x^{6/3} = x^2 \quad \text{and} \quad \sqrt{x^2y^4} = x^{2/2}y^{4/2} = xy^2.$$

If the radicand contains variables that are all perfect powers, the radical can be written as rational exponents and then be simplified. Example 3 illustrates this case.

**EXAMPLE 3**    Perfect Power Radicands

Simplify the following expressions by factoring the radicand and removing any powers that are a multiple of the index.

a) $\sqrt{x^4 y^6 z^8}$

b) $\sqrt[3]{27 x^6 y^9 z^{15}}$

SOLUTION

a) $\sqrt{x^4 y^6 z^8} = (x^4 y^6 z^8)^{1/2} = x^2 y^3 z^4$

b) $\sqrt[3]{27 x^6 y^9 z^{15}} = (3^3 x^6 y^9 z^{15})^{1/3} = 3 x^2 y^3 z^5$    ■

What happens if the radicand is not a perfect power? Then we simply write it as a product where one of the factors is a perfect power factor. For example,

$$\sqrt{x^7} = \sqrt{x^6} \cdot \sqrt{x} = (x^{1/2})^6 \cdot \sqrt{x} = x^3 \sqrt{x}.$$

Example 4 uses this algebraic technique for expressions with several variables.

**EXAMPLE 4**    Simplifying Radicals

Simplify the following expressions by factoring the radicand and removing any powers that are a multiple of the index.

a) $\sqrt[3]{8 x^6 y^5}$

b) $\sqrt[4]{32 x^5 y^8}$

SOLUTION

a) $\sqrt[3]{8 x^6 y^5} = \sqrt[3]{(2^3 x^6 y^3) \cdot y^2}$    **Group together factors that are perfect cubes.**

   $= 2 x^2 y \sqrt[3]{y^2}$

b) $\sqrt[4]{32 x^5 y^8} = \sqrt[4]{(2^4 x^4 y^8) \cdot 2x}$    **Group together expressions to the fourth power.**

   $= 2 x y^2 \sqrt[4]{2x}$    ■

To find the product of two radicals of the same index, it is usually easiest to multiply the radicands together and then simplify the resulting radical. Example 5 illustrates finding the product of the radicands as a means to begin simplifying.

**EXAMPLE 5**    Finding the Product of Radicals

Find the following products. Show numerical support of part (a) using the grapher.

a) $\sqrt{3} \cdot \sqrt{8}$

b) $\sqrt{2xy^2} \cdot \sqrt{18x}$

c) $\sqrt[3]{2x^2y} \cdot \sqrt[3]{32xy^2}$

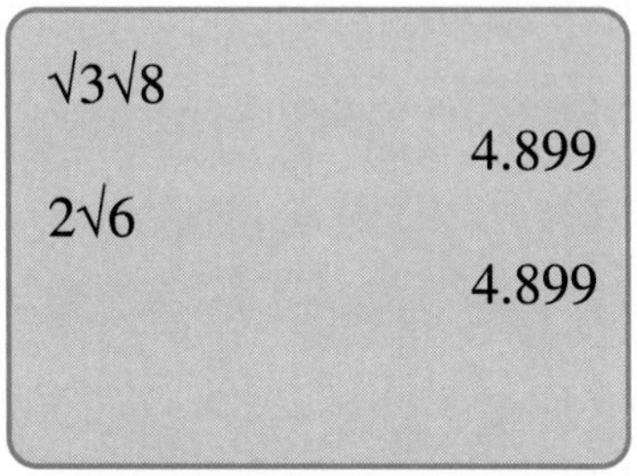

SOLUTION

a) $\sqrt{3} \cdot \sqrt{8} = \sqrt{24} = \sqrt{4} \cdot \sqrt{6} = 2\sqrt{6}$

b) $\sqrt{2xy^2} \cdot \sqrt{18x} = \sqrt{36x^2y^2} = 6xy$

c) $\sqrt[3]{2x^2y} \cdot \sqrt[3]{32xy^2} = \sqrt[3]{64x^3y^3} = 4xy$

**Support Numerically.** Figure 7.9 shows numerical support of part (a).    ■

```
√3√8
              4.899
2√6
              4.899
```

**FIGURE 7.9** Numerical support of the simplification in part (a) of Example 5.

## Quotient Rule for Radicals

There is a quotient rule for radicals that parallels the rule for multiplying radicals. The radical of a quotient is the quotient of the radicals when all expressions involved are defined. Again, this property of radicals is equivalent to one of the properties of exponents stated earlier.

---

### Quotient Rule for Radicals

If $\sqrt[n]{a}$ and $\sqrt[n]{b}$ are defined, $b \neq 0$, then

$$\sqrt[n]{\frac{a}{b}} = \frac{\sqrt[n]{a}}{\sqrt[n]{b}}.$$

---

The Quotient Rule for Radicals, like the corresponding Product Rule, often is used to simplify expressions, both numerical and algebraic. Example 6 illustrates.

**EXAMPLE 6**   Simplifying Quotients

Simplify the following expressions. Support the solution to part (a) using a grapher.

a) $\sqrt{\dfrac{64}{121}}$

b) $\sqrt[3]{\dfrac{8x^4}{y^6}}$

SOLUTION
**Solve Algebraically.**

a) $\sqrt{\dfrac{64}{121}} = \dfrac{\sqrt{64}}{\sqrt{121}} = \dfrac{8}{11}$

b) $\sqrt[3]{\dfrac{8x^4}{y^6}} = \dfrac{\sqrt[3]{8x^3 \cdot x}}{\sqrt[3]{y^6}} = \dfrac{\sqrt[3]{8x^3}\sqrt[3]{x}}{y^2} = \dfrac{2x\sqrt[3]{x}}{y^2}$

Figure 7.10 shows numerical support of the simplification of part (a). ∎

$\sqrt{(64/121)}$

.727

8/11

.727

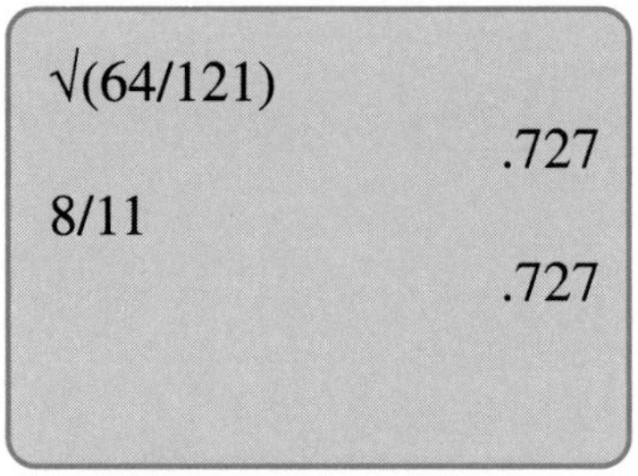

**FIGURE 7.10** The numerical support of the simplification of part (a) of Example 6.

## Using the Quotient Rule to Rationalize Denominators

Example 6 part (b) shows that sometimes after simplifying an expression, a radical remains. In that example, the radical is in the numerator. However, when a radical remains in a denominator, we use a process called **rationalizing the denominator**. We illustrate this process with the example $2/\sqrt{3}$.

$$\frac{2}{\sqrt{3}} = \frac{2}{\sqrt{3}} \cdot \frac{\sqrt{3}}{\sqrt{3}}$$

$$= \frac{2\sqrt{3}}{\sqrt{3} \cdot \sqrt{3}}$$

$$= \frac{2\sqrt{3}}{3}$$

**REMINDER**

Before modern technology, rationalizing the denominator was done because in finding a decimal approximation, it is easier to divide (by hand) by a whole number like 3 than by a long decimal approximation to $\sqrt{3}$. There also are algebraic benefits in doing this.

The crucial property of radicals being used here is $\sqrt{3} \cdot \sqrt{3} = 3$.

A radical expression is simplified when all the following conditions are met:

- All factors that are perfect roots have been removed from the radicand.
- No radical contains a fraction.
- No denominator contains a radical.

Example 7 illustrates the process of rationalizing the denominator.

**EXAMPLE 7** Rationalizing the Denominator

Simplify the following expressions so the denominators do not contain radicals. Support part (a) using a grapher.

a) $\sqrt{\dfrac{5}{7}}$

b) $\sqrt{\dfrac{5}{12}}$

SOLUTION
**Solve Algebraically.**

a) Rationalize the denominator by multiplying by $\sqrt{7}$.

$$\sqrt{\frac{5}{7}} = \frac{\sqrt{5}}{\sqrt{7}} \cdot \frac{\sqrt{7}}{\sqrt{7}} = \frac{\sqrt{35}}{7}$$

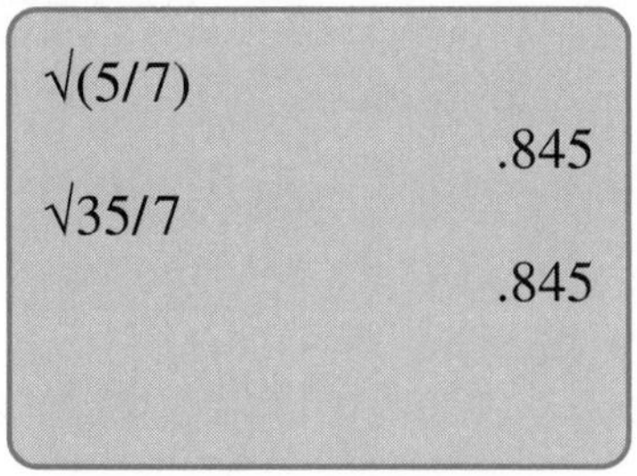

b) Begin by simplifying the denominator and then rationalize.

$$\sqrt{\frac{5}{12}} = \frac{\sqrt{5}}{\sqrt{4} \cdot \sqrt{3}} = \frac{\sqrt{5}}{2\sqrt{3}} \cdot \frac{\sqrt{3}}{\sqrt{3}} = \frac{\sqrt{15}}{6}$$

**Support Numerically.** Figure 7.11 supports part (a) using a grapher. ■

```
√(5/7)
                    .845
√35/7
                    .845
```

**FIGURE 7.11** The grapher support of the solution of Example 7(a).

We can summarize the procedure used to simplify a quotient containing radicals as follows.

---

### Simplifying a Quotient Containing Radicals

1. Reduce the quotient under the radical to lowest terms.
2. To rationalize the denominator, multiply both the numerator and the denominator of the fraction by a radical that will result in the product in the denominator becoming a power that is a multiple of the index.
3. Reduce the resulting expression to lowest terms to simplify the expression.

---

In Example 8, the quotient contains variables as well as constants. The same procedures are used to simplify the expression.

**EXAMPLE 8**   Rationalizing the Denominator

Write the following expressions in equivalent form so the denominators do not contain radicals:

a) $\sqrt{\dfrac{12x}{15x^2y^2}}$

b) $\sqrt[3]{\dfrac{8x^4}{z^2}}$

SOLUTION

a) Begin by reducing to lowest terms.

$$\sqrt{\frac{12x}{15x^2y^2}} = \sqrt{\frac{4}{5xy^2}} = \frac{2}{y\sqrt{5x}} \qquad \text{Rationalize the denominator by making each factor of the radicand a perfect square.}$$

$$= \frac{2}{y\sqrt{5x}} \cdot \frac{\sqrt{5x}}{\sqrt{5x}}$$

$$= \frac{2\sqrt{5x}}{y\sqrt{5^2x^2}} = \frac{2\sqrt{5x}}{5xy}$$

b) Simplify and then rationalize.

$$\sqrt[3]{\frac{8x^4}{z^2}} = \frac{\sqrt[3]{8x^3 \cdot x}}{\sqrt[3]{z^2}} \qquad \text{Rationalize the denominator.}$$

$$= \frac{2x\sqrt[3]{x}}{\sqrt[3]{z^2}} \cdot \frac{\sqrt[3]{z}}{\sqrt[3]{z}}$$

$$= \frac{2x\sqrt[3]{xz}}{\sqrt[3]{z^3}} = \frac{2x\sqrt[3]{xz}}{z} \qquad \blacksquare$$

## Using Radicals in Problem Situations

There are everyday applications that involve radicals. One case is in police investigations. Consider the following problem situation.

**FIGURE 7.12** Car accident problem situation.

□ □

### Car Accident Problem Situation

The approximate speed of a car prior to an accident can be determined from the skid marks left by the car. The investigating officer uses the formula $S = 2\sqrt{5L}$, where $S$ = the speed in miles per hour and $L$ represents the length (in feet) of the skid marks.

□ □

**EXAMPLE 9** CAR ACCIDENT PROBLEM SITUATION: Finding the Speed

An automobile involved in an accident left skid marks 150 ft in length. (See Figure 7.12.) What was the approximate speed the car was traveling?

SOLUTION

$$S = 2\sqrt{5 \cdot 150} = 54.772.$$

The car was traveling approximately 55 mph at the time of the accident. ∎

## Exercises for Section 7.3

In Exercises 1–4, write the product as a single radical. Support on a grapher.

1. $\sqrt{3} \cdot \sqrt{5}$
2. $\sqrt{5} \cdot \sqrt{7}$
3. $\sqrt{7} \cdot \sqrt{3}$
4. $\sqrt{13} \cdot \sqrt{5}$

In Exercises 5–10, simplify the product.

5. $\sqrt{25 \cdot 2}$
6. $\sqrt{7 \cdot 9}$
7. $\sqrt{64 \cdot 3}$
8. $\sqrt{121 \cdot 7}$
9. $\sqrt[3]{8 \cdot 5}$
10. $\sqrt[3]{27 \cdot 3}$

In Exercises 11–16, change the expression to an equivalent expression whose whole number under a radical is as small as possible. Support your answer using a grapher.

11. $\sqrt{75}$
12. $\sqrt{48}$
13. $\sqrt{20}$
14. $\sqrt{27}$
15. $\sqrt[3]{54}$
16. $\sqrt[3]{56}$

In Exercises 17–22, use the Product Rule for Radicals to simplify the number expressed in scientific notation.

17. $\sqrt{8 \times 10^8}$
18. $\sqrt{3 \times 10^{18}}$
19. $\sqrt{5.2 \times 10^{12}}$
20. $\sqrt[3]{8 \times 10^9}$
21. $\sqrt[3]{7.3 \times 10^{15}}$
22. $\sqrt[4]{1 \times 10^{24}}$

In Exercises 23–34, simplify the radical expression. Assume that all variables represent nonnegative real numbers.

23. $\sqrt{y^5}$

24. $\sqrt{x^7}$

25. $\sqrt{x^2 y^4}$

26. $\sqrt{x^4 y^8 z^{10}}$

27. $\sqrt{x^6 y^9}$

28. $\sqrt{x^5 y^8}$

29. $\sqrt[3]{x^6 y^9}$

30. $\sqrt[3]{x^3 y^{12}}$

31. $\sqrt{20x^2 y^4}$

32. $\sqrt{50x^4 y^5}$

33. $\sqrt{32a^5 b^9}$

34. $\sqrt{81a^9 b^6}$

In Exercises 35–46, find the product of the two expressions and simplify. Assume all variables represent nonnegative real numbers.

35. $\sqrt{3} \cdot \sqrt{15}$

36. $\sqrt{6} \cdot \sqrt{8}$

37. $\sqrt{32x} \cdot \sqrt{2x}$

38. $\sqrt{12y} \cdot \sqrt{3y}$

39. $\sqrt{2xy^3} \cdot \sqrt{5x^3 y^5}$

40. $\sqrt{8x^3 y} \cdot \sqrt{5xy^4}$

41. $8\sqrt{6ab} \cdot 3\sqrt{9ab^3}$

42. $9\sqrt{7x^2 y^6} \cdot 2\sqrt{8xy^3}$

43. $\sqrt[3]{3} \cdot \sqrt[3]{9}$

44. $\sqrt[3]{9xy^3} \cdot \sqrt[3]{3x^2 y^6}$

45. $\sqrt[3]{2x^5 y^4} \cdot \sqrt[3]{4x^4 y^5}$

46. $\sqrt[4]{3x^2 y} \cdot \sqrt[4]{27xy^3}$

47. *Multiple Representations on a Grapher.* Use both a numerical table and a graph to show that

$$\sqrt{x^2 - 4} \neq x - 2.$$

48. *Multiple Representations on a Grapher.* Use both a numerical table and a graph to show that

$$\sqrt{(x - 3)^2} \neq x - 3.$$

In Exercises 49–58, simplify the radical expression. Assume all variables represent nonnegative real numbers.

49. $\sqrt{\dfrac{24}{8}}$

50. $\sqrt{\dfrac{9}{16}}$

51. $\sqrt{\dfrac{16}{25}}$

52. $\sqrt{\dfrac{3}{36}}$

53. $\sqrt{\dfrac{16x^6}{4y^4}}$

54. $\sqrt{\dfrac{2x}{18x^5}}$

55. $\sqrt{\dfrac{25x^3 y^4}{100x^5 y^6}}$

56. $\sqrt{\dfrac{3a^2 b^6 z^8}{75a^4 b^{10} z^{12}}}$

57. $\sqrt[3]{\dfrac{2xy^2}{54x^4 y^5}}$

58. $\sqrt[3]{\dfrac{2x^5 y}{250x^2 y^4}}$

In Exercises 59–62, use a grapher to support if the statement is true or false.

59. $\dfrac{1}{\sqrt{2}} = \dfrac{\sqrt{2}}{2}$

60. $\dfrac{3}{\sqrt{5}} = \dfrac{3\sqrt{5}}{5}$

61. $\dfrac{\sqrt{3}}{\sqrt{2}} = \dfrac{\sqrt{6}}{2}$

62. $\dfrac{3\sqrt{5}}{\sqrt{3}} = \sqrt{15}$

In Exercises 63–72, remove all radicals from the denominator. Assume all variables represent nonnegative real numbers.

63. $\dfrac{x}{\sqrt{3}}$

64. $\dfrac{2y}{\sqrt{2}}$

65. $\sqrt{\dfrac{x}{2}}$

66. $\sqrt{\dfrac{3x}{7}}$

67. $\dfrac{2x}{\sqrt{12}}$

68. $\dfrac{2\sqrt{2x}}{\sqrt{5}}$

69. $\sqrt[3]{\dfrac{24x}{7y^2}}$

70. $\sqrt[3]{\dfrac{5a^2}{2c}}$

71. $\sqrt[4]{\dfrac{xy^3}{z}}$

72. $\sqrt[4]{\dfrac{10x}{y^3}}$

In Exercises 73–77, solve the problem.

73. *Police Detective Work.* Using the formula $S = 2\sqrt{5L}$, where $L$ is the length in ft of the skid marks left by a car and $S$ is the speed of the car in mph, find the speed the car was traveling if the length of the skid marks is 200 ft.

74. *Police Detective Work.* Using the formula in Exercise 73, find the approximate length of the skid marks if the car is traveling 75 mph.

75. *Free Fall.* From physics we know that the approximate time $t$ (in seconds) it takes an object to fall a distance $d$ (in feet) is $t = \sqrt{\dfrac{d}{16}}$. Using this

formula, how long will it take a ball dropped from the top of a building 200 ft tall to hit the ground.

**76.** *Free Fall.* Using the formula in Exercise 75, find the time it takes an egg dropped from the roof of a 150 ft building to hit the ground.

**77.** *Pendulum Swing Time.* The time $t$ it takes for a pendulum to complete one period in its swing is approximately $t = 1.11\sqrt{L}$, where $L$ is the length (in feet) of the pendulum. Approximately how long is one period of a pendulum that is 12 ft long?

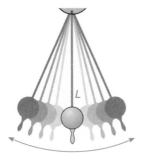

For Exercise 77.

EXTENDING THE IDEAS

**78.** *Using Algebra in Geometry.* A box is three times as long as it is wide and is $\sqrt{3}$ in high.

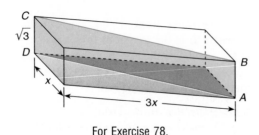

For Exercise 78.

a) Find the area of the rectangular cross section $ABCD$ when $x = 4$.

b) Find the length of the diagonal $AC$ when $x = 5$.

c) Find an algebraic representation $A(x)$ for the area of rectangle $ABCD$ and the length $L(x)$ of the diagonal $AC$ in terms of $x$.

**79.** *Using Algebra in Geometry.* The volume of a prism can be found by using the formula $V = Bh$, where $B$ is the area of the base and $h$ is the perpendicular height. The area of a triangular base can be found using Heron's formula, as follows,

$$A = \sqrt{s(s-a)(s-b)(s-c)},$$

where $a, b,$ and $c$ are the lengths of the sides and $s$ is half the triangle's perimeter. Find the volume of this prism.

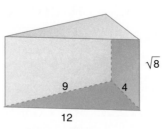

For Exercise 79.

**80. Writing to Learn.** Explain why it is necessary to use the Properties of Exponents to express the product of $\sqrt{3} \cdot \sqrt[3]{4}$ as a single radical.

LOOKING BACK—MAKING CONNECTIONS

In Exercises 81–86, perform the indicated operations to simplify the expression.

**81.** $(x^2 + 5x) + (2x^2 - 3x)$

**82.** $(x^3 - x^2 + 4x + 3) + (2x^2 - 5x - 8)$

**83.** $(4x + 7) - (2x^2 - x + 5)$

**84.** $(x^3 - x^2 + 2) - (x^3 + 2x - 5)$

**85.** $(3x + 7y) - (2x + 3y - 1)$

**86.** $(5x + 3y - 7) - (2x + 6y + 1)$

| 7.4 | **Addition and Subtraction with Radicals** |

Combining Radicals  ▪  Using the Distributive Property with Radicals  ▪  Conjugate Binomial Expressions  ▪  Using Conjugates to Rationalize a Denominator

In Chapter 5, you learned to add and subtract polynomials by adding or subtracting like terms using the Distributive Property. We use a similar procedure when adding or subtracting radicals.

## Combining Radicals

Radical expressions that have the same radicands can be added or subtracted in the same way that we combined like terms when simplifying other algebraic expressions. Consider the following examples:

$$3x + 5x = (3 + 5)x = 8x$$

$$3\sqrt{2} + 5\sqrt{2} = (3 + 5)\sqrt{2} = 8\sqrt{2}$$

$$4x - x = (4 - 1)x = 3x$$

$$4\sqrt{5} - \sqrt{5} = (4 - 1)\sqrt{5} = 3\sqrt{5}$$

In each case, the Distributive Property is used.

Examples 1 and 2 illustrate combining like radicals.

**EXAMPLE 1**    Combining Like Radicals

Simplify the following expressions by combining like radicals. Support the solution to parts (a) and (b) using a grapher.

a) $\sqrt{3} + 5\sqrt{3} - 2\sqrt{3}$

b) $3\sqrt{7} - 2\sqrt{3} + 2\sqrt{7} - \sqrt{3}$

c) $\sqrt[3]{2x} + 5\sqrt[3]{2x} - \sqrt[3]{x} + \sqrt[3]{5}$

SOLUTION
**Solve Algebraically.**

a) Use the Distributive Property to combine like terms.

$$\sqrt{3} + 5\sqrt{3} - 2\sqrt{3} = (1 + 5 - 2)\sqrt{3} = 4\sqrt{3}$$

b) Rearrange terms, using the Commutative Property, in order to combine like terms.

$$3\sqrt{7} - 2\sqrt{3} + 2\sqrt{7} - \sqrt{3} = 3\sqrt{7} + 2\sqrt{7} - 2\sqrt{3} - \sqrt{3}$$
$$= (3+2)\sqrt{7} + (-2-1)\sqrt{3}$$
$$= 5\sqrt{7} - 3\sqrt{3}$$

c) Combine the first two terms, since they are the only like terms.

$$\sqrt[3]{2x} + 5\sqrt[3]{2x} - \sqrt[3]{x} + \sqrt[3]{5} = (1+5)\sqrt[3]{2x} - \sqrt[3]{x} + \sqrt[3]{5}$$
$$= 6\sqrt[3]{2x} - \sqrt[3]{x} + \sqrt[3]{5}$$

**Support Numerically.** Figure 7.13 shows numerical support for parts (a) and (b). ∎

$$\boxed{\begin{array}{l} \sqrt{3} + 5\sqrt{3} - 2\sqrt{3} \\ \qquad\qquad 6.928 \\ 4\sqrt{3} \\ \qquad\qquad 6.928 \\ 3\sqrt{7} - 2\sqrt{3} + 2\sqrt{7} - \sqrt{3} \\ \qquad\qquad 8.033 \\ 5\sqrt{7} - 3\sqrt{3} \\ \qquad\qquad 8.033 \end{array}}$$

**FIGURE 7.13**
Numerical support for parts (a) and (b) of Example 1 using a grapher.

Sometimes it is necessary to simplify a particular radical before the larger expression can be simplified. For example, to simplify $\sqrt{12} + \sqrt{3}$, it is necessary to first simplify $\sqrt{12} = 2\sqrt{3}$. We now have like radicals. Thus

$$\sqrt{12} + \sqrt{3} = 2\sqrt{3} + \sqrt{3} = 3\sqrt{3}.$$

Example 2 illustrates this method.

**EXAMPLE 2**  Simplifying Radical Expressions

Simplify the following radical expression:

$$\sqrt{28x} + \sqrt{63x} - 2\sqrt{7}$$

SOLUTION

Factor each constant in the radicand, with one of the factors being a perfect square.

$$\sqrt{28x} + \sqrt{63x} - 2\sqrt{7} = \sqrt{4\cdot7\cdot x} + \sqrt{9\cdot7\cdot x} - 2\sqrt{7}$$ **Remove perfect squares to simplify each term.**
$$= 2\sqrt{7x} + 3\sqrt{7x} - 2\sqrt{7}$$ **Combine like terms.**
$$= 5\sqrt{7x} - 2\sqrt{7}$$ ∎

When combining fractions, it is easiest to find the least common denominator when the denominators are free of radicals. Example 3 illustrates the need to rationalize the denominator.

**EXAMPLE 3**    Simplifying Radical Expressions

Simplify the following radical expressions.

a) $\sqrt{27} - \dfrac{7}{\sqrt{12}}$

b) $\dfrac{5}{\sqrt{3x}} + \dfrac{\sqrt{12x}}{3x}$

SOLUTION

a) Begin by simplifying the radicals.

$$\sqrt{27} - \dfrac{7}{\sqrt{12}} = \sqrt{9 \cdot 3} - \dfrac{7}{\sqrt{4 \cdot 3}} \qquad \text{Simplify and rationalize the denominator.}$$

$$= 3\sqrt{3} - \dfrac{7}{2\sqrt{3}} \cdot \boxed{\dfrac{\sqrt{3}}{\sqrt{3}}} \qquad \text{Multiply numerator and denominator by } \sqrt{3}.$$

$$= 3\sqrt{3} - \dfrac{7\sqrt{3}}{6} \qquad \text{Rewrite as one fraction over 6, the LCD.}$$

$$= \dfrac{18\sqrt{3} - 7\sqrt{3}}{6} \qquad \text{Simplify the numerator.}$$

$$= \dfrac{11\sqrt{3}}{6}$$

b) Begin by rationalizing the denominator of the first fraction.

$$\dfrac{5}{\sqrt{3x}} + \dfrac{\sqrt{12x}}{3x} = \dfrac{5}{\sqrt{3x}} \cdot \boxed{\dfrac{\sqrt{3x}}{\sqrt{3x}}} + \dfrac{\sqrt{12x}}{3x} \qquad \text{Simplify.}$$

$$= \dfrac{5\sqrt{3x}}{3x} + \dfrac{\sqrt{12x}}{3x} \qquad \text{Add the numerators over the LCD.}$$

$$= \dfrac{5\sqrt{3x} + \sqrt{12x}}{3x} \qquad \text{Simplify the radicals.}$$

$$= \dfrac{5\sqrt{3x} + 2\sqrt{3x}}{3x}$$

$$= \dfrac{7\sqrt{3x}}{3x} \qquad \blacksquare$$

We have added, subtracted, multiplied, and divided radicals. We now apply those basic procedures to more complex problems and look first at extending multiplication.

## Using the Distributive Property with Radicals

Recall that to multiply a polynomial by a monomial, we use the Distributive Property. We use the same procedure when multiplying radicals, as Example 4 demonstrates.

**EXAMPLE 4**     Multiplying by a Monomial

Simplify the following radical expressions. Show numerical support of the solution for part (a) using a grapher.

a) $3\sqrt{2}\left(5\sqrt{3} - 7\right)$

b) $\sqrt[3]{2x}\left(\sqrt[3]{4x^2} - 5\right)$

SOLUTION

**Solve Algebraically.** Use the Distributive Property to remove parentheses and simplify the results.

a) $3\sqrt{2}\left(5\sqrt{3} - 7\right) = 15\sqrt{6} - 21\sqrt{2}$

b) $\sqrt[3]{2x}\left(\sqrt[3]{4x^2} - 5\right) = \sqrt[3]{8x^3} - 5\sqrt[3]{2x} = 2x - 5\sqrt[3]{2x}$

**Support Numerically.** Figure 7.14 shows numerical support for the solution of part (a). ∎

$3\sqrt{2}(5\sqrt{3} - 7)$

7.044

$15\sqrt{6} - 21\sqrt{2}$

7.044

**FIGURE 7.14**
Numerical support of the solution for Example 4(a).

    In the previous chapters, you learned to multiply two binomials together using the FOIL method. We can apply that same procedure using radicals. Example 5 illustrates using the FOIL method to multiply radicals.

**EXAMPLE 5**     Using FOIL with Radicals

Multiply and simplify the following radical expressions. Show numerical support for part (a) using a grapher.

a) $\left(3\sqrt{5} - 2\right)\left(2\sqrt{5} + 6\right)$

b) $\left(2\sqrt{3} + 1\right)\left(\sqrt{2} - 4\right)$

SOLUTION

a) Apply the FOIL method and combine like terms.

$$\left(3\sqrt{5} - 2\right)\left(2\sqrt{5} + 6\right) = 3\sqrt{5}\cdot 2\sqrt{5} + 3\sqrt{5}\cdot 6 - 2\cdot 2\sqrt{5} - 2\cdot 6$$

$$= 6\cdot 5 + 18\sqrt{5} - 4\sqrt{5} - 12$$

$$= 30 + 14\sqrt{5} - 12 = 18 + 14\sqrt{5}$$

b) Begin by applying the FOIL method.

$$\left(2\sqrt{3} + 1\right)\left(\sqrt{2} - 4\right) = 2\sqrt{3}\cdot\sqrt{2} - 2\sqrt{3}\cdot 4 + 1\sqrt{2} - 4\cdot 1$$

$$= 2\sqrt{6} - 8\sqrt{3} + \sqrt{2} - 4$$

**Support Numerically.** Figure 7.15 shows numerical support for part (a). ∎

| |
|---|
| $(3\sqrt{5} - 2)(2\sqrt{5} + 6)$ |
| $\qquad\qquad 49.305$ |
| $18 + 14\sqrt{5}$ |
| $\qquad\qquad 49.305$ |

**FIGURE 7.15**
Numerical support for the solution to Example 5(a).

## Conjugate Binomial Expressions

Real numbers of the form $a + \sqrt{b}$ and $a - \sqrt{b}$ are called **conjugates**. The product of an irrational number like $2 - \sqrt{7}$ and its conjugate is a rational number.

$$\left(2 - \sqrt{7}\right)\left(2 + \sqrt{7}\right) = 4 + 2\sqrt{7} - 2\sqrt{7} - 7 = 4 - 7 = -3$$

Example 6 shows some additional examples of finding the product of two conjugates.

**EXAMPLE 6**   Multiplying Conjugates

Simplify the following radical expressions. Support numerically.

a) $(\sqrt{6} - \sqrt{5})(\sqrt{6} + \sqrt{5})$

b) $(5 + 2\sqrt{13})(5 - 2\sqrt{13})$

c) $(2 - \sqrt{\pi})(2 + \sqrt{\pi})$

SOLUTION

a) $(\sqrt{6} - \sqrt{5})(\sqrt{6} + \sqrt{5}) = (\sqrt{6})^2 - (\sqrt{5})^2 = 6 - 5 = 1$

b) $(5 + 2\sqrt{13})(5 - 2\sqrt{13}) = 5^2 - (2\sqrt{13})^2 = 25 - 4\cdot 13 = 25 - 52 = -27$

c) $(2 - \sqrt{\pi})(2 + \sqrt{\pi}) = 4 - \pi$

**Support Numerically.** Figure 7.16 shows numerical support for the solutions of Example 6. ∎

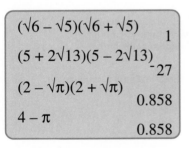

**FIGURE 7.16**
Numerical support of the solution for Example 6.

## Using Conjugates to Rationalize a Denominator

Sometimes we choose to change the form of an expression like $\dfrac{10}{\sqrt{15}-3}$ to obtain an integer in the denominator. This goal can be accomplished by multiplying both the numerator and denominator by the conjugate of the denominator. Example 7 illustrates this procedure.

**EXAMPLE 7**   Rationalizing a Binomial Denominator

Simplify the following radical expressions. Show numerical support for part (a) using a grapher.

a) $\dfrac{10}{\sqrt{15}-3}$

b) $\dfrac{2+\sqrt{3}}{5-2\sqrt{3}}$

**SOLUTION**
**Solve Algebraically.**

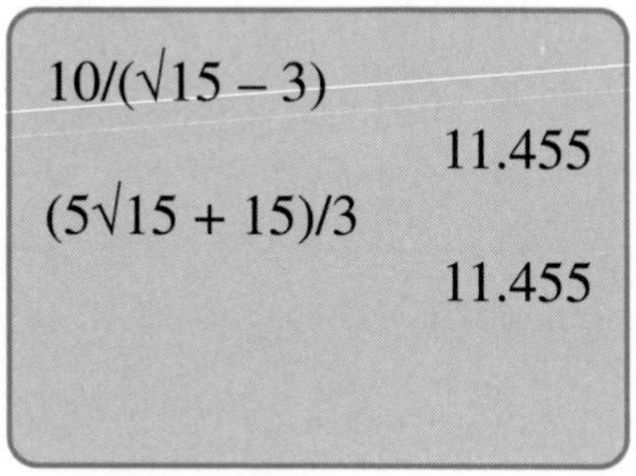

a) Rationalize the denominator by multiplying by the conjugate of the denominator.

$$\frac{10}{\sqrt{15}-3} = \frac{10}{\sqrt{15}-3} \cdot \frac{\sqrt{15}+3}{\sqrt{15}+3}$$

$$= \frac{10(\sqrt{15}+3)}{15-9}$$

$$= \frac{10(\sqrt{15}+3)}{6}$$   Reduce to lowest terms dividing both numerator and denominator by 2.

$$= \frac{5(\sqrt{15}+3)}{3}$$   Use the Distributive Property.

$$= \frac{5\sqrt{15}+15}{3}$$

b) Multiply both the numerator and the denominator by $(5+2\sqrt{3})$.

$$\frac{2+\sqrt{3}}{5-2\sqrt{3}} = \frac{2+\sqrt{3}}{5-2\sqrt{3}} \cdot \frac{5+2\sqrt{3}}{5+2\sqrt{3}}$$   Simplify by using the FOIL method on the numerator and denominator.

10/(√15 − 3)
         11.455
(5√15 + 15)/3
         11.455

**FIGURE 7.17**
Numerical support of the solution for Example 7(a).

$$= \frac{10 + 4\sqrt{3} + 5\sqrt{3} + 2 \cdot 3}{25 - 4 \cdot 3} \quad \text{\small Simplify the numerator} \atop \text{\small and denominator.}$$

$$= \frac{9\sqrt{3} + 16}{13}$$

**Support Numerically.** Figure 7.17 shows numerical support for the solution to part (a).  ■

## Exercises for Section 7.4

In Exercises 1–8, write the expression in terms of a single radical. Use a grapher to support the answers in Exercises 1–4.

**1.** $\sqrt{3} + 5\sqrt{3} - 2\sqrt{3}$    **2.** $6\sqrt{7} - 3\sqrt{7} + 2\sqrt{7}$

**3.** $\frac{3}{5}\sqrt{11} + \frac{2}{5}\sqrt{11}$    **4.** $\frac{3}{4}\sqrt{13} - \frac{1}{2}\sqrt{13} - \sqrt{13}$

**5.** $4\sqrt{x} + 3\sqrt{x}$    **6.** $3\sqrt{2x} + 7\sqrt{2x}$

**7.** $\frac{7}{3}\sqrt{3x} - \frac{1}{6}\sqrt{3x}$    **8.** $2\sqrt{x^2 + 1} - 7\sqrt{x^2 + 1}$

In Exercises 9–18, simplify the radicals as needed and then combine terms. Assume all variables represent nonnegative real numbers.

**9.** $\sqrt{20} + 3\sqrt{5}$

**10.** $\sqrt{48} + \sqrt{12}$

**11.** $3\sqrt{8} + 7\sqrt{72} - 4\sqrt{50}$

**12.** $5\sqrt{27} - 3\sqrt{75} + 2\sqrt{48}$

**13.** $2x\sqrt{20} + 5x\sqrt{125}$

**14.** $3y\sqrt{27} - y\sqrt{108} + 6y\sqrt{48}$

**15.** $4\sqrt[3]{2x} + 8\sqrt[3]{2x}$

**16.** $6\sqrt[3]{54} - 4\sqrt[3]{16}$

**17.** $6\sqrt{28x^2} + 2\sqrt{63x^2}$

**18.** $\sqrt{10x^4} - 3\sqrt{160x^4} + 5\sqrt{90x^4}$

In Exercises 19–21, find the perimeter of the figure.

**19.** *Using Algebra in Geometry.*

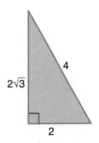

**20.** *Using Algebra in Geometry.*

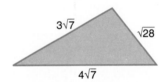

**21.** *Using Algebra in Geometry.*

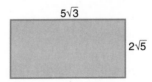

In Exercises 22–35, find the product of the radical expressions. Assume all variables are nonnegative real numbers.

**22.** $\sqrt{3} \cdot \sqrt{12}$    **23.** $\sqrt{2} \cdot \sqrt{32}$

24. $\sqrt{5}(\sqrt{5}+2)$

25. $\sqrt{7}(3-\sqrt{7})$

26. $\sqrt{x}(6+\sqrt{x})$

27. $\sqrt{3x}(2\sqrt{x}+\sqrt{3x})$

28. $(4+\sqrt{2})(3-\sqrt{2})$

29. $(\sqrt{6}+3)(2\sqrt{6}-1)$

30. $(\sqrt{x}+2)(\sqrt{x}-3)$

31. $(\sqrt{y}-4)(\sqrt{y}+5)$

32. $(2\sqrt{x}+1)(3\sqrt{x}-5)$

33. $(15\sqrt{y}+2)(2\sqrt{y}-3)$

34. $(\sqrt{x}+3)^2$

35. $(\sqrt{y}-5)^2$

In Exercises 36–39, select the graph from the following that you think is the graph of the given function. Then use a grapher to check your guess. (All graphs are shown in the [0, 10] by [0, 5] window.)

36. $y=\sqrt{x}+\dfrac{1}{\sqrt{x}}$

37. $y=\dfrac{2\sqrt{x}}{1+x}$

38. $y=\dfrac{1}{\sqrt{x}}+\dfrac{1}{\sqrt{2x}}$

39. $y=\dfrac{\sqrt{x}+1}{\sqrt{x}}$

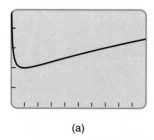

(a)

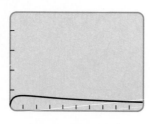

(b)

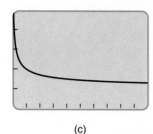

(c)

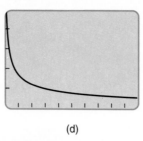

(d)

For Exercises 36–39.

In Exercises 40–49, simplify the expression by rationalizing the denominator and then combining like terms. Assume all variables represent nonnegative real numbers.

40. $5+\sqrt{\dfrac{1}{2}}$

41. $\sqrt{7}-\dfrac{3}{\sqrt{7}}$

42. $\sqrt{18}-\dfrac{\sqrt{48}}{\sqrt{16}}$

43. $\sqrt{12}+\dfrac{\sqrt{27}}{\sqrt{9}}$

44. $\dfrac{\sqrt{3}}{3}+\dfrac{1}{\sqrt{3}}-2$

45. $12-\dfrac{2}{\sqrt{3}}$

46. $\dfrac{1}{\sqrt{5}}+\sqrt{5}$

47. $\dfrac{\sqrt{6}}{2}-\dfrac{2}{\sqrt{6}}$

48. $\dfrac{1}{\sqrt{2x}}+\sqrt{4x}$

49. $\dfrac{\sqrt{3x}}{5}-\dfrac{2}{\sqrt{3x}}$

In Exercises 50–55, write the conjugate of the expression.

50. $\sqrt{3}+5$

51. $\sqrt{2}-3$

52. $6-\sqrt{15}$

53. $7+\sqrt{31}$

54. $\sqrt{x}+5\sqrt{y}$

55. $\sqrt{x}-3y$

In Exercises 56–65, simplify the expression by rationalizing the denominator. Assume all variables are nonnegative real numbers.

56. $\dfrac{3}{\sqrt{11}-5}$

57. $\dfrac{25}{\sqrt{3}+2}$

58. $\dfrac{10}{\sqrt{4}+\sqrt{5}}$

59. $\dfrac{3}{\sqrt{3}+\sqrt{2}}$

60. $\dfrac{2x}{\sqrt{x}+3}$

61. $\dfrac{5y}{\sqrt{y}-7}$

62. $\dfrac{\sqrt{3}-5}{\sqrt{3}+5}$

63. $\dfrac{\sqrt{2}+7}{\sqrt{2}-7}$

64. $\dfrac{\sqrt{x}-2}{\sqrt{x}+5}$

65. $\dfrac{\sqrt{x}-\sqrt{y}}{\sqrt{x}+\sqrt{y}}$

66. **Writing to Learn.** Explain three different methods you can use to show that, in general,

$$\sqrt{a+b}\neq\sqrt{a}+\sqrt{b}.$$

EXTENDING THE IDEAS

67. *Interpreting Graphs.* A trapezoid is inscribed under the graph of $y=\sqrt{x+6}$ down to the $x$-axis.

a) Show that an algebraic representation for the area of this trapezoid is $A=\dfrac{1}{2}x(\sqrt{6}+\sqrt{x+6})$.

b) Find the value of $x$ when the area of the trapezoid is 15.

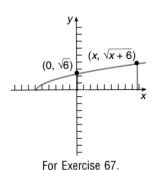

For Exercise 67.

In Exercises 68–71, place the correct symbol ($<$, $>$, or $=$) between the given expressions to make the statement true.

68. $\sqrt{9+16}$ ? $\sqrt{9}+\sqrt{16}$

69. $\sqrt{25-9}$ ? $\sqrt{25}-\sqrt{9}$

70. $\sqrt{36}+\sqrt{64}$ ? $\sqrt{36+64}$

71. $\sqrt{7^2+2^2}$ ? $9$

In Exercises 72–75, find the sum of two fractions by first rationalizing the denominators.

72. $\dfrac{2}{\sqrt{5}}+\dfrac{3}{5}$

73. $\dfrac{1}{2}+\dfrac{3}{\sqrt{3}}$

74. $\dfrac{1}{\sqrt[3]{2x}}+\dfrac{3}{2x}$

75. $\dfrac{1}{2x}+\dfrac{5}{\sqrt[3]{4x^2}}$

TRANSLATING WORDS TO SYMBOLS

In Exercises 76 and 77, write a symbolic expression for the phrase.

76. The sum of the square root of $x$ and the square root of $y$

77. The square root of the sum of $x$ and $y$

LOOKING BACK—MAKING CONNECTIONS

In Exercises 78 and 79, solve the equation.

78. $\dfrac{x-2}{3}+\dfrac{x+3}{5}=6$

79. $\dfrac{2}{x-3}+\dfrac{5}{x+2}=\dfrac{10}{x^2-x-6}$

In Exercises 80–83, find the product of the expression.

80. $2x^2(x^2y+y^3)$

81. $3x^2(2x^2+5x+6)$

82. $(2x+7)(2x-7)$

83. $(3x-5)(3x+5)$

In Exercises 84 and 85, use a graphical method to solve the inequality.

84. $\sqrt{x^2+4}<x+2$

85. $\sqrt[3]{x^3+27}>x+3$

## 7.5 Solving Equations Containing Radicals

Solving Radical Equations Algebraically ▪ Solving Equations with *n*th Roots ▪ Choosing Between Algebraic and Graphical Methods

In Chapter 3, the multigraph and the $x$-intercept methods were used to find solutions to both linear and nonlinear equations. Among the nonlinear examples were some involving radicals. We begin this section with an example that reviews the $x$-intercept method of solving an equation graphically.

### EXAMPLE 1    Solving an Equation with Radicals

Solve the following equation graphically and support the solution numerically:

$$2\sqrt{3x - 2.1} = 15.6$$

SOLUTION

**Solve Graphically.** Prepare the equation for the $x$-intercept method by writing the equation $2\sqrt{3x - 2.1} = 15.6$ in the equivalent form

$$2\sqrt{3x - 2.1} - 15.6 = 0.$$

Figure 7.18(a) shows the graph of $y = 2\sqrt{3x - 2.1} - 15.6$. Zoom in to find that the solution is $x = 20.98$.

**Support Numerically.** Figure 7.18(b) shows numerical support for the solution.

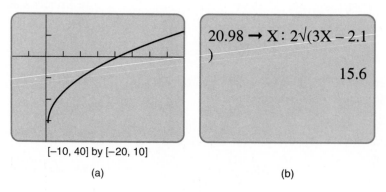

[−10, 40] by [−20, 10]

(a)                                              (b)

**FIGURE 7.18**  (a) A graph of $y = 2\sqrt{3x - 2.1} - 15.6$, and (b) numerical support of the solution of $x = 20.98$.  ■

## Solving Radical Equations Algebraically

When using algebraic methods to solve an equation like

$$\sqrt{x + 3} = 4,$$

it is natural to square both sides of the equation in order to eliminate the radical. However, it is important to be aware that this algebraic step might result in an

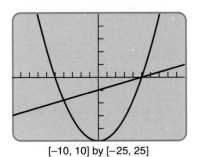

[−10, 10] by [−25, 25]

**FIGURE 7.19** A graph of $y_1 = x - 5$ and $y_2 = x^2 - 25$.

equation with more solutions than the original equation has. Any solutions of the resulting equation that are not solutions of the original equation are called **extraneous solutions.**

To illustrate how this can happen, consider the following simple example:

$$x = 5 \text{ has a single solution.}$$

If we square both sides of the equation, the resulting equation

$$x^2 = 25$$

has the two solutions: $x = 5$ and $x = -5$. The solution $x = -5$, however, is an extraneous solution. Figure 7.19 shows that the graph of $y_1 = x - 5$ has one $x$-intercept at $x = 5$ and $y_2 = x^2 - 25$ has two $x$-intercepts.

When you solve an equation by squaring both sides, it is necessary to complete a numerical and/or a graphical analysis to identify any extraneous solutions.

**EXAMPLE 2**    Solving by an Algebraic Method

Solve the following equation using an algebraic method. Support the solution graphically and numerically.

$$\sqrt{x + 3} = 4$$

SOLUTION

**Solve Algebraically.** Because the radical is already isolated on the left-hand side, we begin by squaring both sides to eliminate the square root.

$$\sqrt{x + 3} = 4 \qquad \textbf{Square both sides of the equation.}$$

$$(\sqrt{x + 3})^2 = 4^2$$

$$x + 3 = 16$$

$$x = 13$$

Therefore the solution to the equation generated by squaring both sides is $x = 13$.

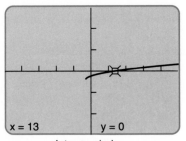

x = 13        y = 0

Integer window

**FIGURE 7.20** A graph of $y = \sqrt{x + 3} - 4$ in Example 2.

**Support Graphically and Numerically.** Figure 7.20 shows the graph of $y = \sqrt{x + 3} - 4$ in the Integer window. The cursor position at $(13, 0)$ provides support that $x = 13$ is a valid solution.

Numerically we see that $\sqrt{13 + 3} = 4$ is indeed a true statement. So $x = 13$ is not an extraneous solution; it is a valid solution.  ∎

Example 3 illustrates a case in which the side of the equation not containing the radical is a binomial.

### EXAMPLE 3    Solving by an Algebraic Method

Solve the following equation using an algebraic method. Support the solution graphically.

$$\sqrt{2x + 13} = x + 5$$

SOLUTION

**Solve Algebraically.** First eliminate the radical by squaring each side of the equation.

| | |
|---|---|
| $\sqrt{2x + 13} = x + 5$ | **Square each side of the equation.** |
| $(\sqrt{2x + 13})^2 = (x + 5)^2$ | |
| $2x + 13 = x^2 + 10x + 25$ | **Make the left-hand side = 0.** |
| $0 = x^2 + 8x + 12$ | **Factor the right-hand side of the equation.** |
| $0 = (x + 6)(x + 2)$ | **Set each factor = 0 and solve.** |
| $x + 6 = 0$   or   $x + 2 = 0$ | |
| $x = -6$   or   $x = -2$ | |

We can confirm that $x = -6$ is an extraneous root by direct substitution.

$$\sqrt{2(-6) + 13} \neq -6 + 5$$

**Support Graphically.** Figure 7.21 shows the graph of $y = \sqrt{2x + 13} - x - 5$. The graph does not intersect the $x$-axis at $x = -6$, which suggests that $x = -6$ is an extraneous solution. The graph also supports that $x = -2$ is a solution, since the $x$-intercept appears to be $-2$.  ∎

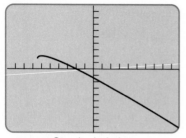

Standard window

**FIGURE 7.21** A graph of $y = \sqrt{2x + 13} - x - 5$ in Example 3.

When more than one radical occurs in an equation, we sometimes need to square both sides more than once in order to free the equations of radicals. Example 4 illustrates this case.

**EXAMPLE 4**    Solving an Equation with Two Radicals

Solve the following equation algebraically and support the solution graphically:

$$\sqrt{4-x} - \sqrt{x+6} = 2$$

SOLUTION

**Solve Algebraically.** Isolate one of the radicals on one side of the equation to begin.

$$\sqrt{4-x} - \sqrt{x+6} = 2$$

$$\sqrt{4-x} = \sqrt{x+6} + 2 \qquad \text{Square both sides of the equation.}$$

$$(\sqrt{4-x})^2 = (\sqrt{x+6} + 2)^2$$

$$4 - x = x + 6 + 4\sqrt{x+6} + 4 \qquad \text{Combine like terms.}$$

$$4 - x = x + 10 + 4\sqrt{x+6} \qquad \text{Isolate the radical on one side of the equation.}$$

$$-6 - 2x = 4\sqrt{x+6} \qquad \text{Simplify by dividing each side by } -2.$$

$$3 + x = -2\sqrt{x+6} \qquad \text{Square both sides to eliminate the radical.}$$

$$(3+x)^2 = (-2\sqrt{x+6})^2$$

$$9 + 6x + x^2 = 4(x+6) \qquad \text{Simplify.}$$

$$9 + 6x + x^2 = 4x + 24 \qquad \text{Make one side of the equation} = 0.$$

$$x^2 + 2x - 15 = 0 \qquad \text{Factor, set each factor} = 0, \text{ and solve.}$$

$$(x+5)(x-3) = 0$$

$$x = -5 \quad \text{or} \quad x = 3$$

We can confirm $x = 3$ is an extraneous root by direct substitution.

$$\sqrt{4-3} - \sqrt{3+6} = \sqrt{1} - \sqrt{9} = 1 - 3 = -2 \quad \text{(not 2)}$$

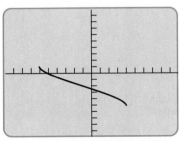

Standard window

**FIGURE 7.22** A graph of $y = \sqrt{4-x} - \sqrt{x+6} - 2$ in Example 4.

**Support Graphically.** Check for extraneous roots using a graph. Figure 7.22 shows a graph of $y = \sqrt{4-x} - \sqrt{x+6} - 2$, which supports that $x = -5$ is a solution and $x = 3$ is an extraneous solution.  ∎

Next, we summarize the algebraic methods that have been used in the last several examples.

---

### Solving Equations That Contain Square Roots Algebraically

1. Isolate a radical on one side of the equation.
2. Square both sides of the equation.
3. If a radical still remains, isolate that radical and repeat step 2.
4. Simplify and solve the resulting equation.
5. Check for extraneous solutions by graphing or direct substitution.

---

## Solving Equations with nth Roots

The five-step procedure outlined above can be modified if the radical is a cube root or any other $n$th root. To eliminate a cube root, raise both sides to the third power. In general to eliminate an $n$th root, raise both sides of the equation to the $n$th power.

Example 5 illustrates solving an equation with a cube root.

**EXAMPLE 5**    Solving an Equation Containing a Cube Root

Solve the following equation using an algebraic method. Support the solution graphically.

$$\sqrt[3]{2x^2 - x - 7} + 5 = 7$$

SOLUTION

**Solve Algebraically.** Begin by isolating the radical.

$$\sqrt[3]{2x^2 - x - 7} + 5 = 7 \qquad \text{Subtract 5 from each side of the equation.}$$

$$\sqrt[3]{2x^2 - x - 7} = 2 \qquad \text{Raise each side of the equation to the third power.}$$

$$(\sqrt[3]{2x^2 - x - 7})^3 = 2^3$$

$$2x^2 - x - 7 = 8 \qquad \text{Make one side of the equation = 0.}$$

$$2x^2 - x - 15 = 0 \qquad \text{Factor the left-hand side.}$$

$$(x - 3)(2x + 5) = 0 \qquad \text{Set each factor} = 0 \text{ and solve.}$$

$$x - 3 = 0 \quad \text{or} \quad 2x + 5 = 0$$

$$x = 3 \quad \text{or} \qquad x = -2.5$$

Algebraic confirmation of these solutions by direct substitution is possible but very messy. Instead we will use graphical support.

**Support Graphically.** Figure 7.23 shows the graph of $y = \sqrt[3]{2x^2 - x - 7} + 5 - 7$. This graph supports that both $x = 3$ and $x = -2.5$ are valid solutions for the equation $\sqrt[3]{2x^2 - x - 7} + 5 = 7$. Neither is an extraneous solution. ■

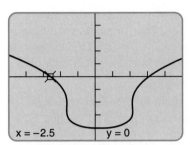

x = −2.5     y = 0

0.1 window with vertical dimension [−5, 5]

**FIGURE 7.23** Grapher support for the solutions $x = -2.5$ and $x = 3$ of Example 5.

Example 5 illustrates that with a graphing method we can visualize that the equation has two solutions. Consequently, we see immediately that neither solution is extraneous.

## Choosing Between Algebraic and Graphical Methods

When several radicals are present in an equation, it is often easier to use a graphing method. In Example 6, we will solve graphically an equation that has two radicals.

**EXAMPLE 6**    Solving an Equation Containing Two Radicals

Solve the following equation using a graphing method. Support the solution numerically.

$$\sqrt{x + 3} - \sqrt{x - 2} = 1$$

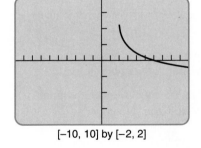

[−10, 10] by [−2, 2]

**FIGURE 7.24** A graph of $y = \sqrt{x + 3} - \sqrt{x - 2} - 1$ in Example 6.

SOLUTION

**Solve Graphically.** The equation $\sqrt{x + 3} - \sqrt{x - 2} = 1$ is equivalent to $\sqrt{x + 3} - \sqrt{x - 2} - 1 = 0$. Use the $x$-intercept method to graph $y = \sqrt{x + 3} - \sqrt{x - 2} - 1$ and zoom in to find that the solution to the equation is $x = 6.00$. See Figure 7.24. We suspect that $x = 6$ is an exact solution.

**Support Numerically.** Substitute $x = 6$ into the original equation.

$$\sqrt{6 + 3} - \sqrt{6 - 2} \stackrel{?}{=} 1$$

$$\sqrt{9} - \sqrt{4} \stackrel{?}{=} 1$$

$$3 - 2 = 1 \qquad \text{True statement}$$

So 6 is the exact solution. ■

## Exercises for Section 7.5

In Exercises 1–10, solve the equation using a graphing method.

1. $x^2 = 4$                      2. $x^2 = 25$

3. $x^2 = -100$                4. $x^2 = -81$

5. $\sqrt{x+1} = 3$           6. $\sqrt{x-2} = 5$

7. $\sqrt{x-7} = 8$           8. $\sqrt{x+5} = 4$

9. $\sqrt{x+1} = -5$        10. $\sqrt{x+15} = -2.5$

In Exercises 11–20, solve the equation by an algebraic method. Check for extraneous roots by examining a graph of the equation.

11. $\sqrt{x-1} = 4$          12. $\sqrt{x+5} = 9$

13. $\sqrt{x-12} = 15$       14. $\sqrt[3]{2x+9} = 3$

15. $\sqrt[4]{2x-1} = 2$      16. $\sqrt{2x+1} = 5$

17. $3\sqrt{2x-1} = 15$      18. $2\sqrt{2x-3} = 12$

19. $(x-7)^2 = 36$           20. $(2x-3)^2 + 12 = 61$

In Exercises 21–30, solve the equation by either an algebraic or a graphical method. Support the solution numerically.

21. $1 + \sqrt{x+5} = \sqrt{2x+5}$

22. $\sqrt{x-3} = \sqrt{x+2} - 1$

23. $\sqrt{x} - \sqrt{x-1} = 1$

24. $5 - x = \sqrt{3x+1} - \sqrt{x+1}$

25. $\sqrt{2x+11} = \sqrt{x+9}$

26. $\sqrt{3x+5} = \sqrt{2x+9}$

27. $\sqrt{x} = 1 - \sqrt{x-1}$

28. $\sqrt{x+4} - 2\sqrt{x-1} = -1$

29. $\sqrt{x-3} + 1 = \sqrt{x+2}$

30. $\sqrt{2x+5} - \sqrt{x+5} = 1$

31. *Comparing Solution Methods.* In this exercise, compare different solution methods for the equation

$$\sqrt{x+5} = x.$$

a) Use the $x$-intercept method on a grapher to find how many solutions this equation has.

b) The algebraic method of squaring both sides of the equation often leads to extraneous solutions. Graph the equation $y = (x+5) - x^2$. Explain how this graph tells if this algebraic method introduces extraneous solutions.

32. *Comparing Solution Methods.* Discuss whether you would use an algebraic or a graphical method to solve each of the following equations and why:

a) $\sqrt{x} = 18$

b) $\sqrt{x+1} = 7$

c) $\sqrt{x+2} + 3 = 2\sqrt{x-4}$

## TRANSLATING WORDS TO SYMBOLS

In Exercises 33–36, write an equation described by the statement.

33. Fifteen is equal to the square root of four less than a number.

34. Eight is equal to the square root of one more than a number.

35. Twenty-five is the square root of two more than the square of a number.

36. The sum of the square root of a number and the square root of one more than the number is five.

In Exercises 37–48, solve the equation using a graphing method. Support the solution numerically.

37. $\sqrt{x-1} = 6$                38. $2\sqrt{x+1} = 6$

39. $\sqrt{10x^2 - 9} = 3x$        40. $\sqrt{5x^2 - 4} = 2x$

41. $\sqrt{x-3} - 1 = \sqrt{x-2}$

42. $\sqrt{2x+1} = \sqrt{x} + 1$

43. $2 - \sqrt{2x+2} = -\sqrt{x-3}$

44. $\sqrt{2x+5} - \sqrt{x+5} = 1$

45. $\sqrt{3x - 5} = \sqrt{2x + 12}$

46. $\sqrt{5x + 7} - \sqrt{3x + 15} = 0$

47. $2\sqrt{4x - 3} = \sqrt{6x + 38}$

48. $\sqrt{2x^2 + 2x} = x + 1$

In Exercises 49–55, solve the problem.

49. *Free Fall.* The approximate time $t$ (in seconds) that it takes an object to fall a distance $d$ (in feet) is $t = \sqrt{d/16}$. A parachutist falls 4 sec before the parachute opens. How far does the parachutist fall during this time?

50. *Free Fall.* A group of eight parachutists plans to do a formation free fall before separating and opening their parachutes. They allow 20 sec to complete their formation and separation before opening their parachutes at 3000 ft. From what altitude must they jump?

51. *Pendulum Swing Time.* The time for one swing of a pendulum is given by $t = 1.11\sqrt{L}$. How long must the pendulum be if one period takes 4 sec.

52. *Pendulum Swing Time.* Is it possible to build a pendulum hanging from the top of a tall building whose swing time is 45 sec? (*Hint:* How tall is the tallest man-made building?)

53. *Car Accident Problem Situation.* Approximately how long will the skid marks be if a car's brakes lock when going 55 mph? (Recall that $S = 2\sqrt{5L}$, where $L$ is measured in feet and $S$ is measured in mph.)

54. *Car Accident Problem Situation.* A police investigator measured the skid marks of a car to be approximately 82 ft. The driver of a car claims that he was not exceeding the 35-mph speed limit. Is this information consistent? Explain.

55. *Right Circular Cone Problem Situation.* In the following figure, $\triangle ABC$ is part of a cross section of the cone.

a) If $x$ is the radius (in inches) of the base of the cone, find an algebraic representation for the altitude $h(x)$.

b) Use the result from part (a) to find an algebraic representation for the area of $\triangle ABC$.

c) Find a complete graph of this problem situation and use it to find the radius of the base if the area of $\triangle ABC$ is 18 sq in.

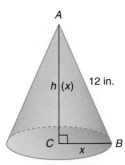

For Exercise 55.

EXTENDING THE IDEAS

56. *Interpreting a Grapher.* Write an equation that could be solved with the following grapher setting. Solve the equation.

$$Y_1 = \sqrt{(X + 3)}$$
$$Y_2 = \sqrt{(X - 1)}$$
$$Y_3 = Y1 - Y2 - 1$$

For Exercise 56.

57. *Interpreting a Grapher.* Write an equation that could be solved with the following grapher setting. Solve the equation.

$$Y_1 = \sqrt{(X^2 + 1)}$$
$$Y_2 = \sqrt{(X + 3)}$$
$$Y_3 = Y1 - Y2 - 1$$

For Exercise 57.

**58. Writing to Learn.** Write a paragraph that explains whether you prefer to use a graphical method or an algebraic method to solve the equation $\sqrt{2x^2 + 6x + 4} + 1 = x + 2$.

**59. Writing to Learn.** Write a paragraph that explains whether you prefer to use a graphical method or an algebraic method to solve the equation $\sqrt{2x + 5} - \sqrt{x + 5} = 1$.

LOOKING BACK—MAKING CONNECTIONS

**60.** A square field with an area of one acre has an area of 43,560 sq ft. What is the length of one side of the field?

**61.** A square is inscribed in a circle that is 16 ft in diameter. What is the length of the square's side?

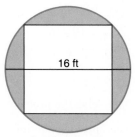

16 ft

For Exercise 61.

**62.** Find the length of the side of a square whose diagonal is 12 ft.

**63.** Find the length of the side of a square with an area of 40 sq in.

**64.** Find the length of the side of a cube whose volume is 216 cu in.

**65.** Write a paragraph describing the domain of the function shown in Figure 7.20.

## 7.6   Complex Numbers

Finding a Solution to $x^2 = -1$ ▪ Numbers of the Form $a + bi$ ▪ Adding, Subtracting, and Multiplying Complex Numbers ▪ Complex Conjugates ▪ Dividing Complex Numbers

An equation like $x^2 = 1$ is one of the simplest second degree equations and its solutions can be easily seen to be $x = 1$ and $x = -1$.

However, an equation like $x^2 = -1$ has *no real solution*, since there is no real number whose square is $-1$. This section explains that we can find a solution to this equation by creating a new system of numbers.

### Finding a Solution to $x^2 = -1$

To find a solution to an equation like $x^2 = -1$, we need to create a new system of numbers. We begin this process by introducing a new symbol $i$ and define that symbol as follows:

$$i = \sqrt{-1} \text{ and } i^2 = -1$$

Observe that $i^2 = -1$ means that $i$ is a solution of the equation $x^2 = -1$.

By combining the definition of $i$ with multiplication properties of radicals, we can find the square root of any negative number. For example,

$$\sqrt{-16} = \sqrt{16} \cdot \sqrt{-1} = 4i.$$

By using the definition of $i$ as $\sqrt{-1}$, we can observe that any square root that has a negative radicand can now be evaluated.

---

### Simplifying the Square Root of a Negative Number

For any positive number $a$,

$$\sqrt{-a} = i\sqrt{a}.$$

---

Example 1 illustrates simplifying square roots that contain negative radicands.

**EXAMPLE 1**    Simplifying Square Roots of Negative Numbers

Write each of the following square roots in simplest form.

a) $\sqrt{-25}$

b) $\sqrt{\dfrac{-49}{36}}$

c) $\sqrt{-72}$

**REMINDER**

$\sqrt{-5}$ is written $i\sqrt{5}$ rather than $\sqrt{5}i$. In this way, the $i$ is not inadvertently included under the radical.

SOLUTION

Begin by writing each radicand as the product of two factors, one of which is $-1$.

a) $\sqrt{-25} = \sqrt{25} \cdot \sqrt{-1} = 5i$

b) $\sqrt{\dfrac{-49}{36}} = \sqrt{\dfrac{49}{36}} \cdot \sqrt{-1} = \dfrac{7}{6}i$

c) $\sqrt{-72} = \sqrt{-1}\sqrt{36}\sqrt{2}$
   $= i \cdot 6 \cdot \sqrt{2}$
   $= 6i\sqrt{2}$

■

## Numbers of the Form $a + bi$

Numbers like $3i$, $-5i$, and $i\sqrt{17}$, which are of the form $bi$ where $b$ is a real number, are sometimes called **imaginary numbers.** The set of all imaginary numbers is a subset of the larger collection of numbers called **complex numbers.** Here is a formal definition of a complex number.

---

**DEFINITION 7.6**   Complex Number

A **complex number** is a number that can be written in the form

$$a + bi,$$

where $a$ and $b$ are real numbers.

   If $a = 0$, then the number is in the form $bi$ and is said to be an **imaginary number.** If $b = 0$, the number is in the form $a + 0i = a$ and is said to be a **real number.**

---

Observe that all real numbers are complex numbers, but a complex number is not necessarily a real number (see Figure 7.25). Some examples of complex numbers are

$$5 + 6i \qquad 2 - i\sqrt{7} \qquad 5i \qquad 9 \qquad -3 - 19i.$$

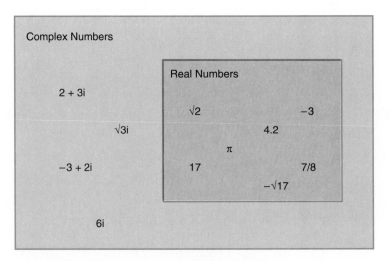

**FIGURE 7.25**   The set of complex numbers.

Two complex numbers in the form $a + bi$ and $c + di$ are equal if and only if $a = c$ and $b = d$. We say their real parts are equal and their imaginary parts are equal.

## Adding, Subtracting, and Multiplying Complex Numbers

We perform the operations of addition, subtraction, and multiplication by treating a complex number as a binomial. To add or subtract two complex numbers, we add or subtract the real parts and then add or subtract the imaginary parts. Thus

$$(a + bi) + (c + di) = (a + c) + (b + d)i$$

and

$$(a + bi) - (c + di) = (a - c) + (b - d)i.$$

This process is similar to that of combining like terms with polynomials.

**EXAMPLE 2**    Adding and Subtracting Complex Numbers

Perform the indicated operations and express your answer in the form $a + bi$.

a) $(5 - 2i) + (3 + 4i)$

b) $(5 - 2i) - (3 + 4i)$

SOLUTION

a) Add the like parts.

$$(5 - 2i) + (3 + 4i) = (5 + 3) + (-2i + 4i)$$
$$= 8 + 2i$$

b) Remove the parentheses and combine the like parts.

$$(5 - 2i) - (3 + 4i) = 5 - 2i - 3 - 4i$$
$$= (5 - 3) + (-2i - 4i)$$
$$= 2 - 6i$$ ■

Here is the procedure to follow when adding or subtracting two complex numbers.

---

**To Add or Subtract Complex Numbers**

1. Combine (add or subtract) the real parts and then combine the imaginary parts.
2. Write the answer in standard $a + bi$ form.

---

The Commutative, Associative, and Distributive Properties hold for complex numbers just as they do for real numbers. Example 3 shows how to use these properties to multiply two complex numbers when one is either real or imaginary.

**EXAMPLE 3**    Multiplying by a Monomial

Simplify the following expressions. Write all answers in standard $a + bi$ form.

a) $5(3 + 4i)$

b) $2i(8 - 5i)$

c) $\sqrt{-16}(2 - i)$

SOLUTION

a) Use the Distributive Property.

$$5(3 + 4i) = 15 + 20i$$

b) Use the Distributive Property.

$$2i(8 - 5i) = 16i - 10i^2 \qquad \textbf{Evaluate } i^2.$$
$$= 16i - 10(-1)$$
$$= 16i + 10 \qquad \textbf{Write in standard form.}$$
$$= 10 + 16i$$

c) Begin by simplifying $\sqrt{-16}(2 - i)$.

$$\sqrt{-16}(2 - i) = 4i(2 - i) \qquad \textbf{Use the Distributive Property.}$$
$$= 8i - 4i^2 \qquad \textbf{Evaluate } i^2.$$
$$= 8i - 4(-1) \qquad \textbf{Simplify and put in standard form.}$$
$$= 4 + 8i$$

**REMINDER**

Always simplify a radical first before performing multiplication, as follows:

$$\sqrt{-3} \cdot \sqrt{-12} = i\sqrt{3} \cdot i\sqrt{12}$$
$$= i^2\sqrt{36} = -6$$

It is incorrect to do the multiplication first, thus $\sqrt{-3} \cdot \sqrt{-12} \neq \sqrt{36} = 6$

In Example 4, we multiply binomials using the FOIL method. We need to simplify and express the answer in standard $a + bi$ form.

**EXAMPLE 4**    Multiplying Complex Numbers

Find the products of the following complex numbers and write the answers in standard $a + bi$ form:

a) $(1 - i)(8 + 7i)$

b) $(-2 + 9i)(3 + i)$

c) $(3 + 4i)^2$

SOLUTION

a) Multiply using the FOIL method.

$$(1 - i)(8 + 7i) = 8 + 7i - 8i - 7i^2 \qquad \text{Evaluate } i^2.$$
$$= 8 + 7i - 8i - 7(-1) \quad \text{Simplify.}$$
$$= 8 - i + 7 \qquad\qquad\quad \text{Write in standard form.}$$
$$= 15 - i$$

b) Multiply using the FOIL method.

$$(-2 + 9i)(3 + i) = -6 - 2i + 27i + 9i^2 \qquad \text{Evaluate } i^2.$$
$$= 6 - 2i + 27i - 9 \qquad \text{Simplify.}$$
$$= -3 + 25i$$

c) Simplify using $(a + b)^2 = a^2 + 2ab + b^2$ to square the binomial.

$$(3 + 4i)^2 = 3^2 + 2(3)(4i) + (4i)^2 \quad \text{Simplify.}$$
$$= 9 + 24i + 16i^2 \qquad\qquad \text{Evaluate } i^2.$$
$$= 9 + 24i - 16 \qquad\qquad \text{Combine like terms.}$$
$$= -7 + 24i \qquad\qquad\qquad\qquad\qquad\quad ■$$

Here is a summary of the process for multiplication of complex numbers.

---

### To Multiply Complex Numbers $a + bi$ and $c + di$

1. Multiply the complex numbers as though they were polynomials.
2. Replace any $i^2$ with $-1$ and simplify.
3. Combine any like terms.
4. Write the answer in standard $a + bi$ form.

---

## Complex Conjugates

Complex numbers $a + bi$ and $a - bi$ are called **complex conjugates**. Notice that the product of a complex number and its complex conjugate is a real number.

$$(a + bi)(a - bi) =$$
$$= a^2 - b^2 i^2$$
$$= a^2 - b^2(-1)$$
$$= a^2 + b^2$$

Summarizing we see that $(a + bi)(a - bi) = a^2 + b^2$. Example 5 illustrates how to use this summary formula to multiply complex conjugates.

**EXAMPLE 5**   Multiplying Complex Conjugates

Find the products of the following complex numbers and write the answers in standard form:

a) $(2 - 3i)(2 + 3i)$

b) $(-2 + 5i)(-2 - 5i)$

c) $(3 + 4i)(3 - 4i)$

SOLUTION

a) $(2 - 3i)(2 + 3i) = 2^2 + 3^2 = 4 + 9 = 13$

b) $(-2 + 5i)(-2 - 5i) = (-2)^2 + 5^2 = 4 + 25 = 29$

c) $(3 + 4i)(3 - 4i) = 3^2 + 4^2 = 9 + 16 = 25$   ■

## Dividing Complex Numbers

To divide a complex number $a + bi$ by another complex number $c + di$, write the quotient in fraction form.

$$(a + bi) \div (c + di) = \frac{a + bi}{c + di}$$

Then multiply the numerator and denominator by the conjugate of the denominator and simplify the result. Example 6 illustrates this process.

**TRY THIS**

Some graphers have complex number functionality. If yours does, use it to perform the division $3 \div (2 + 3i)$. Does your output support the algebraic solution to Example 6?

**EXAMPLE 6**     Dividing Complex Numbers

Find the following quotients and write the answer in standard form:

a) $\dfrac{3}{2 + 3i}$

b) $\dfrac{1 + i}{-2 - 5i}$

SOLUTION

In each part, multiply the numerator and denominator by the complex conjugate of the denominator.

a) $\dfrac{3}{(2 + 3i)} = \dfrac{3}{(2 + 3i)} \cdot \dfrac{(2 - 3i)}{(2 - 3i)}$   Multiply the denominator using the product of complex conjugates is $a^2 + b^2$.

$= \dfrac{3(2 - 3i)}{2^2 + 3^2}$   Simplify the denominator.

$= \dfrac{3(2 - 3i)}{13}$   Because the numerator and denominator have no common factors, simplify the numerator.

$= \dfrac{6 - 9i}{13}$   Write in standard form.

$= \dfrac{6}{13} - \dfrac{9}{13}i$

b) Rationalize the denominator.

$\dfrac{1 + i}{-2 - 5i} = \dfrac{(1 + i)}{(-2 - 5i)} \cdot \dfrac{(-2 + 5i)}{(-2 + 5i)}$   Simplify the numerator using the FOIL method and the denominator using $a^2 + b^2$.

$= \dfrac{-2 + 5i - 2i + 5i^2}{4 + 25}$   Simplify.

$$= \frac{-2 + 5i - 2i - 5}{29} \quad \text{Combine like terms.}$$

$$= \frac{-7 + 3i}{29} \quad \text{Write in standard form.}$$

$$= \frac{-7}{29} + \frac{3}{29}i \quad \blacksquare$$

Here is a summary of the process for dividing complex numbers.

---

### Dividing $a + bi$ by $c + di$

1. Express the quotient as a fraction.
2. Multiply the numerator and denominator of the fraction by the complex conjugate of the denominator.
3. Write the answer in standard $a + bi$ form.

---

## Exercises for Section  7.6

In Exercises 1–10, perform the indicated operation. Write the answers in $a + bi$ form.

1. $(3 + 4i) + (5 - 2i)$     2. $(4 - 6i) + (3 + 8i)$

3. $(6 - 2i) - (8 + 9i)$     4. $(2 + 5i) - (6 + 4i)$

5. $(2 - \sqrt{-4}) + (3 + \sqrt{-9})$

6. $(\sqrt{-4} + 3) + (\sqrt{-16} - 4)$

7. $(\sqrt{12} - \sqrt{-64}) - (3\sqrt{3} + 4i)$

8. $(\sqrt{27} + \sqrt{-81}) + (\sqrt{48} - 5i)$

9. $(\sqrt{16} - \sqrt{-25}) - (3 + 4i)$

10. $(32 + \sqrt{-5}) + (9 - \sqrt{-20})$

In Exercises 11–36, find the product of the two complex numbers. Write the answer in $a + bi$ form.

11. $(3i)(4i)$      12. $(5i)(6i)$

13. $(-2i)(-3i)$      14. $(-5i)(3i)$

15. $4(2 + 3i)$      16. $-1(3 - 4i)$

17. $2(3 - 2i)$      18. $5(2 + i)$

19. $2i(9 - 2i)$      20. $-3i(2 + 3i)$

21. $3i(1 + i)$      22. $-2i(5 - 8i)$

23. $(3 + 4i)(5 - 2i)$      24. $(4 - 6i)(3 + 8i)$

25. $(2 + 7i)(3 - 2i)$      26. $(5 - 3i)(2 + i)$

27. $(2 - 7i)(2 + 7i)$      28. $(5 + 3i)(5 - 3i)$

29. $(3 + 2i)(1 - 2i)$      30. $(1 - 9i)(2 + 5i)$

31. $(1 - 2i)(1 + 2i)$      32. $(3 + i)(3 - i)$

**33.** $(-1 + 2i)^2$          **34.** $(3 + 4i)^2$

**35.** $(5 - 6i)^2$          **36.** $(9 + i)^2$

In Exercises 37–50, find the quotient of the two complex numbers. Write the answer in $a + bi$ form.

**37.** $\dfrac{4}{4 + 5i}$          **38.** $\dfrac{2}{1 + i}$

**39.** $\dfrac{2}{3 - 2i}$          **40.** $\dfrac{5}{2 + i}$

**41.** $\dfrac{i}{2 + 3i}$          **42.** $\dfrac{-i}{5 - 4i}$

**43.** $\dfrac{2 - 7i}{2 + 7i}$          **44.** $\dfrac{5 + 3i}{5 - 3i}$

**45.** $\dfrac{3i}{2 + i}$          **46.** $\dfrac{i}{5 - 3i}$

**47.** $\dfrac{1 + 3i}{2 - 5i}$          **48.** $\dfrac{1 - i}{1 + 3i}$

**49.** $\dfrac{2 + i}{2 - i}$          **50.** $\dfrac{3 + i}{3 - i}$

EXTENDING THE IDEAS

In Exercises 51–54, perform the indicated operation. Write the answer in $a + bi$ form.

**51.** $\dfrac{2}{(1 + i)} + \dfrac{5}{(1 - i)}$          **52.** $\dfrac{3}{(2 - i)} - \dfrac{6}{(3 + i)}$

**53.** $\dfrac{i}{(1 - i)} + \dfrac{2i}{(1 + i)}$          **54.** $\dfrac{2i}{(3 + 2i)} - \dfrac{3i}{(1 + i)}$

In Exercises 55–62, evaluate the expression. Express the answer in $a + bi$ form. The largest power of $i$ in the final answer should be 1.

**55.** $i^{15}$          **56.** $i^{17}$

**57.** $i^{22}$          **58.** $i^{13}$

**59.** $i^{-5}$          **60.** $i^{-2}$

**61.** $i^{-12}$          **62.** $i^{-27}$

**63. Writing to Learn.** In your own words, compare the real numbers and the complex numbers.

LOOKING BACK—MAKING CONNECTIONS

In Exercises 64–69, solve the equation.

**64.** $x^2 - x - 12 = 0$          **65.** $x^2 - 5x - 24 = 0$

**66.** $(2x + 1)(3x - 5) = 0$

**67.** $x(2x - 7) = 0$

**68.** $x^2 = 121$          **69.** $x^2 = 17$

Exercises 70–73 refer to the following geometric representation of the sum of two complex numbers. A complex number $a + bi$ can be represented as a vector in the coordinate plane whose "tail" is the point $(0, 0)$ and whose "head" is the point $(a, b)$. The sum of $a + bi$ and $c + di$ is represented by the vector from $(0, 0)$ to $(a + c, \ b + d)$, the fourth vertex of the parallelogram formed by the two complex numbers.

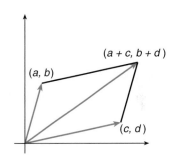

The sum of $a + bi$ and $c + di$.
For Exercises 70–73.

**70.** Draw the figure representing the sum of $2 + 3i$ and $4 - i$.

**71.** Draw the figure representing the sum of $-7 + 3i$ and $2 - 6i$.

**72.** The sum of $5 + 7i$ and $x + yi$ is $2 - 3i$. What is the complex number $x + yi$? Draw a figure of this sum.

**73.** The sum of $1 - 2i$ and $x + yi$ is $-4 + 6i$. What is the complex number $x + yi$? Draw a figure of this sum.

## Chapter 7 Summary

| | **Roots, Radicals and Rational Exponents** | **Examples** |
|---|---|---|
| $n$th root of a number | $b$ is an $n$th root of $a$ if $b^n = a$ and $n$ is a positive integer. | 2 and $-2$ are both fourth roots of 16. |
| Principal $n$th root, $n$ even | If $n$ is even, the principal $n$th root of a positive number $b$ is the real number $b^{1/n}$ that satisfies the equation $\left(b^{1/n}\right)^n = b$. | The principal square root of 36 is 6. |
| Principal $n$th root, $n$ odd | If $n$ is odd, the principal $n$th root of a real number $b$ is the real number $b^{1/n}$ that satisfies the equation $\left(b^{1/n}\right)^n = b$. | The principal cube root of $-125$ is $-5$. |
| Radical | $\sqrt[n]{b}$ is a radical with index $n$ and radicand $b$ and is defined whenever $b^{1/n}$ is defined. Also, $\sqrt[n]{b} = b^{1/n}$ and $\sqrt[n]{0} = 0$. | $\sqrt[3]{7} = 7^{1/3}$ |
| Rational Exponents | If $b^{1/n}$ exists and $m$ and $n$ are positive integers with no common factors, $b^{m/n} = \left(b^{1/n}\right)^m = (b^m)^{1/n}$. | $5^{2/3} = \left(5^{1/3}\right)^2 = \left(5^2\right)^{1/3}$ |

| | **Theorems and Properties** | **Examples** |
|---|---|---|
| Pythagorean Theorem | $a^2 + b^2 = c^2$, given a right triangle with legs of length $a$ and $b$ and hypotenuse of length $c$. | If 7, 24, and 25 are the lengths of the sides of a right triangle, then $7^2 + 24^2 = 25^2$. |
| Powers and Roots Properties | If $b$ is a real number and $n$ is an integer $> 1$, then<br><br>1. $\sqrt[n]{b^n} = b$ if $n$ is odd,<br>2. $\sqrt[n]{b^n} = |b|$ if $n$ is even. | $\sqrt[3]{(-3)^3} = -3$ and $\sqrt[4]{(-2)^4} = 2$ |
| Real-number Conjugates | Real numbers of the form $a + \sqrt{b}$ and $a - \sqrt{b}$ | $3 - 2\sqrt{2}$ and $3 + 2\sqrt{2}$ are conjugates. |

| | Operations on Radicals | Examples |
|---|---|---|
| Product Rule for Radicals | $\sqrt[n]{a} \cdot \sqrt[n]{b} = \sqrt[n]{ab}$, if $\sqrt[n]{a}$ and $\sqrt[n]{b}$ are real numbers. | $\sqrt[3]{6} \cdot \sqrt[3]{2} = \sqrt[3]{12}$ |
| Quotient Rule for Radicals | $\sqrt[n]{\dfrac{a}{b}} = \dfrac{\sqrt[n]{a}}{\sqrt[n]{b}}$, if $\sqrt[n]{a}$ and $\sqrt[n]{b}$ are defined and $b \neq 0$. | $\sqrt{\dfrac{5}{x^2}} = \dfrac{\sqrt{5}}{|x|}$ |
| Addition and Subtraction of Radicals | Use the Distributive Property to combine like radical terms (terms with the same radicand and index). | $\sqrt{3} + 2\sqrt{5} - 3\sqrt{3} =$ $(1-3)\sqrt{3} + 2\sqrt{5} =$ $-2\sqrt{3} + 2\sqrt{5}$ |
| Rationalizing the Denominator | Eliminating a radical from the denominator of a rational expression by multiplying both numerator and denominator by the conjugate of the denominator | Rationalize $\dfrac{1}{\sqrt{x}}$: $\dfrac{1}{\sqrt{x}} = \dfrac{1}{\sqrt{x}} \cdot \dfrac{\sqrt{x}}{\sqrt{x}} = \dfrac{\sqrt{x}}{x}$ |

| | Solving Equations Containing Radicals | Examples |
|---|---|---|
| Solving Algebraically | 1. Isolate a radical on one side of the equation. | Solve $\sqrt{x-3} - 2 = 0$: $\sqrt{x-3} = 2$ |
| | 2. Raise both sides to a power that will eliminate the radical sign from one side of the equation. | $(\sqrt{x-3})^2 = 2^2$ |
| | 3. Repeat Step 1 and 2 if any radicals remain. | $x - 3 = 4$ |
| | 4. Simplify and solve the resulting equation. | $x = 7$ $\sqrt{7-3} - 2 = 0$ |
| | 5. Check for extraneous solutions. | |
| Solving Graphically | Use either the multigraph or $x$-intercept method to determine all solutions with error at most 0.01. | |

| | Complex Numbers | Examples |
|---|---|---|
| $i$ | $i$ is a symbol that represents $\sqrt{-1}$ and $i^2 = -1$. | |
| Complex Number | A number written in the form $a + bi$, where $a$ and $b$ are real numbers and $i = \sqrt{-1}$ | $3 - \sqrt{2}i$ |
| Imaginary Number | A complex number of the form $bi$, where $b$ is a real number and $i = \sqrt{-1}$ | $2i$ |
| Addition and Subtraction of Complex Numbers | Combine all real terms and then combine all imaginary terms, writing the result in $a + bi$ form. | $(3 + i) + (2 - 4i) = (3 + 2) + (1 - 4)i = 5 - 3i$ |
| Multiplication of Complex Numbers | Multiply in the same way as for polynomials. Simplify by replacing $i^2$ with $-1$ and then combine like terms, writing the result in $a + bi$ form. | $(2 + i)(1 - 2i) =$ <br> $2 - 4i + i - 2i^2 =$ <br> $2 - 3i - 2(-1) =$ <br> $4 - 3i$ |
| Division of Complex Numbers | Express in fraction form and multiply both numerator and denominator by the complex conjugate of the denominator, simplifying and writing the result in $a + bi$ form. | $\dfrac{6}{1 - i} = \dfrac{6}{1 - i} \cdot \dfrac{1 + i}{1 + i}$ <br> $= \dfrac{6 + 6i}{1 - i^2}$ <br> $= \dfrac{6 + 6i}{1 - (-1)}$ <br> $= \dfrac{6 + 6i}{2} = 3 + 3i$ |
| Complex Conjugates | Complex numbers of the form $a + bi$ and $a - bi$ | $2 + i$ and $2 - i$ are complex conjugates. |

## Review Exercises for Chapter 7

In this exercise set assume all variables represent nonnegative real numbers.

1. Give the exponent form of $\sqrt[5]{(x - 1)^3}$.

2. Give the radical form of $(2n)^{2/3}$.

In Exercises 3 and 4, evaluate the expression using a grapher. Round the answer to three decimal places.

3. $134^{1/4}$        4. $\sqrt[6]{16}$

In Exercises 5–7, evaluate the expression without using a grapher.

**5.** $49^{-1/2}$

**6.** $\sqrt{\frac{9}{100}}$

**7.** $\sqrt{64 \cdot 10^6}$

In Exercises 8 and 9, simplify the expression.

**8.** $\sqrt[3]{x^5 y^6}$

**9.** $\sqrt{40x^4 y^9}$

In Exercises 10–15, perform the indicated operation and simplify. Express the answer as a single radical.

**10.** $10\sqrt{2} - 3\sqrt{8}$

**11.** $5\sqrt{ab^2} - b\sqrt{9a}$

**12.** $\sqrt[3]{27x^2 y} + \sqrt[3]{125x^2 y}$

**13.** $\sqrt[3]{x^2} \cdot \sqrt[4]{x}$

**14.** $\sqrt{15x} \cdot \sqrt{27x^3}$

**15.** $\sqrt[4]{x^2 yz^2} \cdot \sqrt[4]{x^3 y^2 z^2}$

In Exercises 16–19, perform the indicated operation and simplify.

**16.** $\sqrt{6}(\sqrt{3} + \sqrt{2})$

**17.** $(3 - 2\sqrt{3})(2 - \sqrt{3})$

**18.** $(\sqrt{x} + 4)(\sqrt{x} - 4)$

**19.** $(\sqrt{5} + \sqrt{6})^2$

In Exercises 20–25, evaluate the expression. Write the answer in $a + bi$ form. The largest power of $i$ in the final answer should be 1.

**20.** $i^{-4}$

**21.** $i^3 + i$

**22.** $(7 + 3i) - (4 - 6i)$

**23.** $(12 + \sqrt{-50}) + (15 - \sqrt{-18})$

**24.** $(3 - 5i)^2$

**25.** $(5 + 2i)(4 - 7i)$

In Exercises 26–31, simplify the expression. Rationalize the denominator as needed.

**26.** $\sqrt{\frac{35x^3}{21x^2}}$

**27.** $\frac{18}{\sqrt{2}}$

**28.** $\frac{2}{\sqrt{3} - 1}$

**29.** $\sqrt{27} + \frac{2}{\sqrt{3}}$

**30.** $\frac{2 + i}{3 + i}$

**31.** $\frac{5 - i}{1 - 2i}$

**32.** Find the length of the missing side of the following triangle:

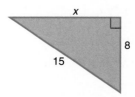

For Exercise 32.

**33.** Determine whether a triangle with side lengths $a = \sqrt{5}$, $b = 2\sqrt{2}$, and $c = \sqrt{13}$ is a right triangle.

In Exercises 34 and 35, solve the equation using a graphical method. Support the solution numerically.

**34.** $\sqrt[3]{2x^2 - 17} = 5$

**35.** $\sqrt{x - 3} + \sqrt{x + 3} = 3$

In Exercises 36 and 37, solve the equation using an algebraic method. Support the solution graphically.

**36.** $\sqrt{3x - 5} = \sqrt{4x + 7}$

**37.** $\sqrt[3]{3x + 2} = -4$

In Exercises 38 and 39, solve the equation using either an algebraic or a graphical method. Support the solution numerically.

**38.** $\sqrt{x + 1.2} = 2$

**39.** $2x - 1 = \sqrt{5x - 4}$

**40.** A car traveling 80 mph is involved in a highway accident and leaves a skid mark. Find the approximate length of the skid mark if the formula is $S = 2\sqrt{5L}$, where $L$ is the length of the skid mark in ft and $S$ is the speed of the car in mph.

## Chapter 7 Test

In this exercise set assume all variables represent nonnegative real numbers.

1. Change $\sqrt[3]{4z}$ to exponential form.

2. Evaluate $16^{3/4}$ without using a grapher.

3. Simplify $\sqrt{80x^8 y^5}$.

4. Write $\sqrt{-63}$ in simplest form.

5. Simplify $3\sqrt{32} - 2\sqrt{18}$.

6. Find the product of $\sqrt{5x^3 y} \cdot \sqrt{15x^4 y^3}$.

7. Find the product of $\sqrt{3x}(\sqrt{6} - \sqrt{2x})$.

8. Simplify $\dfrac{\sqrt{x}}{\sqrt{x} - 2}$ by rationalizing the denominator.

In Questions 9–11, perform the indicated operation. Write the answer in $a + bi$ form.

9. $(7 - 8i) - (3 + 7i)$

10. $(3 - i)^2$

11. $\dfrac{1 - i}{4 + i}$

12. Find the length of the missing side of the following triangle:

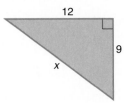

For Question 12.

13. Solve graphically: $\sqrt{2x + 7} + x = 5$.

14. Solve algebraically: $\sqrt{x + 2} = x$.

15. A baseball diamond in a city park has the shape of a square. The distance between first base and second base is 60 ft. What is the distance between first base and third base?

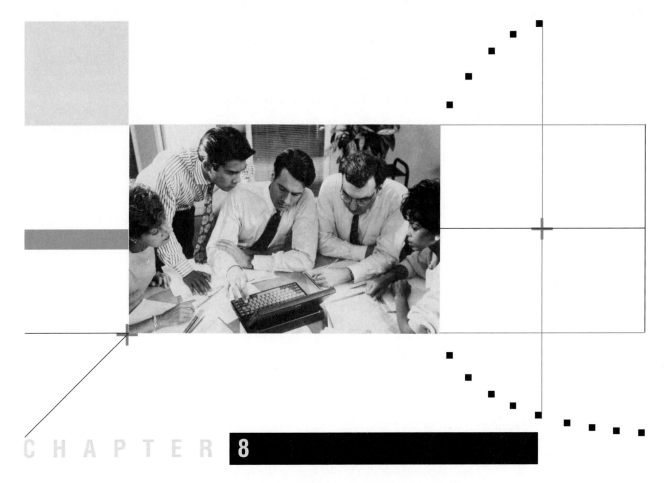

# Quadratic Functions

## AN APPLICATION

Corporate Economists have determined that the profit $P(x)$ of producing $x$ items of one of their production lines is modeled by the equation

$$f(x) = -0.02x^2 + 5.23x + 14.3$$

## 8.1 Solving Quadratic Equations by Completing the Square

Using the Square Root Property ▪ Solving an Equation by Completing the Square

A quadratic equation in $x$ is a special case of a polynomial equation: a polynomial of degree 2. Here is a more formal definition.

---

**DEFINITION 8.1**  Quadratic Equation

An equation is a **quadratic equation in $x$** if it can be written in the form

$$ax^2 + bx + c = 0,$$

where $a$, $b$, and $c$ are real numbers, with $a \neq 0$. This form is called the **standard form of the quadratic equation.**

---

As a special case of a polynomial, a quadratic equation can be solved by factoring, as discussed for polynomials in Section 5.6. However, there are equation-solving methods that apply specifically to quadratic equations. In this section, we will show how the Square Root Property can be used to solve quadratic equations.

In Section 7.1, we learned that positive real numbers have two square roots: one positive and the other negative. For example, since

$$3^2 = 9 \qquad \text{and} \qquad (-3)^2 = 9,$$

we say that 3 is the positive square root of 9 and $-3$ is the negative square root of 9. We write

$$\sqrt{9} = 3 \qquad \text{and} \qquad -\sqrt{9} = -3.$$

In other words, $x = 3$ and $x = -3$ are both solutions to the equation $x^2 = 9$.

The general square root property follows.

---

**Square Root Property**

If $x^2 = a$, where $a$ is any real number, then $x = \sqrt{a}$ or $x = -\sqrt{a}$.

---

## Using the Square Root Property

Example 1 illustrates how the Square Root Property is used in equation solving. In particular, the two solutions to the equation $x^2 = a$ are real numbers if $a$ is a positive real number and nonreal complex numbers if $a$ is a negative real number.

**EXAMPLE 1**   Using the Square Root Property

Solve each of the following equations:

a) $x^2 - 7 = 0$

b) $x^2 + 16 = 0$

SOLUTION

a) $x^2 - 7 = 0$

$$x^2 = 7 \qquad\qquad \textbf{Use the Square Root Property.}$$

$$x = \sqrt{7} \quad \text{or} \quad x = -\sqrt{7}$$

b) $x^2 + 16 = 0$

$$x^2 = -16$$

$$x = \sqrt{-16} \quad \text{or} \quad x = -\sqrt{-16} \quad \textbf{Recall that } \sqrt{-16} = 4i.$$

$$x = 4i \quad \text{or} \quad x = -4i \qquad\qquad\qquad\qquad ■$$

The Square Root Property can be used to solve an equation whenever the variable is part of a perfect square. Example 2 illustrates this concept.

**EXAMPLE 2**   Using the Square Root Property

Solve the equation $(3x - 2)^2 = 15$ algebraically. Support your answer numerically with a grapher.

SOLUTION
**Solve Algebraically.**

$$(3x - 2)^2 = 15$$

$$3x - 2 = \sqrt{15} \qquad\qquad 3x - 2 = -\sqrt{15}$$

$$3x = 2 + \sqrt{15} \qquad \text{or} \qquad 3x = 2 - \sqrt{15}$$

$$x = \frac{2 + \sqrt{15}}{3} \qquad\qquad x = \frac{2 - \sqrt{15}}{3}$$

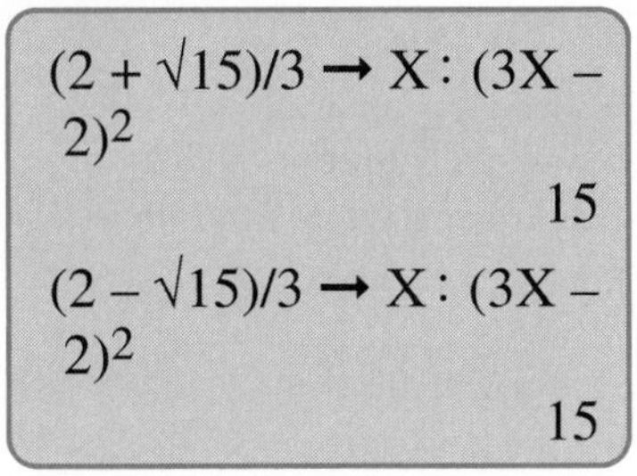

$(2 + \sqrt{15})/3 \rightarrow X : (3X - 2)^2$

$15$

$(2 - \sqrt{15})/3 \rightarrow X : (3X - 2)^2$

$15$

**FIGURE 8.1**  Numerical support for Example 2 on a grapher.

**Support Numerically.** Figure 8.1 shows numerical support on a grapher for the solution.  ∎

## Solving an Equation by Completing the Square

The solution to Example 2 was possible because the variable $x$ was part of a perfect square. So the solution technique illustrated in this example can be used for a quadratic equation in standard form, provided the equation is first written as a perfect square using the method called **completing the square**. To introduce this method, review the following two examples of perfect squares:

$$x^2 + 14x + 49 = (x + \boxed{7})^2$$

$$\frac{1}{2}(14) = \boxed{7} \qquad \text{Note the connection between 14 and 7.}$$

$$x^2 - 24x + 144 = (x - \boxed{12})^2$$

$$\frac{1}{2}(-24) = \boxed{-12} \qquad \text{Note the connection between } -24 \text{ and } -12.$$

When a perfect square trinomial is written as a binomial squared, the constant in the binomial is one-half the coefficient of the $x$-term in the trinomial. This observation becomes the basis for one method of solving a quadratic equation.

---

### Solving a Quadratic Equation by Completing the Square

1. Rewrite the equation (if needed) so that the coefficient of $x^2$ is 1 and only the $x^2$- and $x$-terms are on the left-hand side.
2. Add the square of one–half the coefficient of $x$ to each side of the equation.
3. Write the left-hand side as a square of a binomial and simplify the right-hand side.
4. Use the Square Root Property.
5. Solve for $x$ and simplify if necessary.

---

**EXAMPLE 3**    Completing the Square

Solve $x^2 - 8x + 6 = 0$ by completing the square. Provide numerical support on a grapher.

SOLUTION

**Solve Algebraically.**

$$x^2 - 8x + 6 = 0$$

$$x^2 - 8x = -6 \qquad \text{Step 1}$$

$$x^2 - 8x + \boxed{16} = -6 + \boxed{16} \qquad \text{Step 2:} \ \left(\frac{-8}{2}\right)^2 = 16$$

$$(x - 4)^2 = 10 \qquad \text{Step 3}$$

$$x - 4 = \sqrt{10} \qquad \text{or} \qquad x - 4 = -\sqrt{10} \qquad \text{Step 4}$$

$$x = 4 + \sqrt{10} \qquad \text{or} \qquad x = 4 - \sqrt{10} \qquad \text{Step 5}$$

**Support Numerically.** Figure 8.2 shows numerical support on a grapher for the solution. Sometimes grapher output will be something like $-1E - 13$, which means $-1 \times 10^{-13}$, and is equivalent to 0 relative to the precision of most graphers.  ∎

When the coefficient of $x^2$ is not 1, begin the completing the square process by dividing each side of the equation by the coefficient of $x^2$, as Example 4 illustrates.

**EXAMPLE 4**  Completing the Square

Solve $-2x^2 - 4x + 1 = 0$ by completing the square. Provide both graphical and numerical support for the solution.

SOLUTION

**Solve Algebraically.** Figure 8.3 shows the graph of $y_1 = -2x^2 - 4x + 1$ and suggests there will be two real solutions. Begin by dividing each side of the equation by $-2$ so that the coefficient of $x^2$ is 1.

$$-2x^2 - 4x + 1 = 0$$

$$x^2 + 2x - \frac{1}{2} = 0 \qquad \text{Step 1}$$

$$x^2 + 2x = \frac{1}{2} \qquad \text{Step 1}$$

$$x^2 + 2x + \boxed{1} = \frac{1}{2} + \boxed{1} \qquad \text{Step 2}$$

$$(x + 1)^2 = \frac{3}{2} \qquad \text{Step 3}$$

$$x + 1 = \pm\sqrt{\frac{3}{2}} \qquad \text{Step 4} \qquad \text{The symbol } \pm \text{ is read "plus or minus."}$$

$$x = -1 \pm \sqrt{\frac{3}{2}} \qquad \text{Step 5}$$

---

$4 + \sqrt{10} \to X : X^2 - 8X + 6$

$$0$$

$4 - \sqrt{10} \to X : X^2 - 8X + 6$

$$1E^-13$$

**FIGURE 8.2** Numerical support for Example 3 on a grapher.

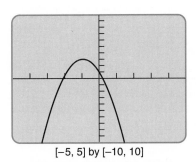

[−5, 5] by [−10, 10]

**FIGURE 8.3** Graphical support for Example 4. The two $x$-intercepts indicate that there are two real solutions to the equation.

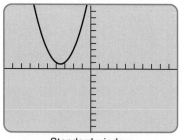

$$^-1 + \sqrt{(3/2)} \rightarrow X : \, ^-2X^2 - 4X + 1$$
$$^-1E\,^-13$$
$$^-1 - \sqrt{(3/2)} \rightarrow X : \, ^-2X^2 - 4X + 1$$
$$0$$

**FIGURE 8.4** Numerical support for the solution to Example 4.

The two solutions are $x = -1 + \sqrt{3/2}$ and $x = -1 - \sqrt{3/2}$, whose decimal representations are $x_1 = 0.225$ and $x_2 = -2.225$.

**Support Graphically and Numerically.** Figure 8.3 provides graphical support for this solution and Figure 8.4 provides numerical support. ∎

Example 5 shows that sometimes the solutions found when completing the square are nonreal complex numbers.

**EXAMPLE 5** Completing the Square

Solve the quadratic equation $x^2 + 7x + 13 = 0$ by completing the square. Provide graphical support for the solutions.

SOLUTION

**Solve Algebraically.**

$$x^2 + 7x + 13 = 0$$
$$x^2 + 7x = -13$$
$$x^2 + 7x + \left(\frac{7}{2}\right)^2 = -13 + \left(\frac{7}{2}\right)^2$$
$$\left(x + \frac{7}{2}\right)^2 = \frac{-52}{4} + \frac{49}{4}$$
$$\left(x + \frac{7}{2}\right)^2 = \frac{-3}{4}$$

Using the Square Root Property, we find two solutions.

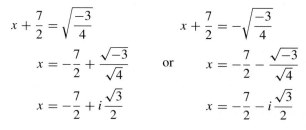

$$x + \frac{7}{2} = \sqrt{\frac{-3}{4}} \qquad\qquad x + \frac{7}{2} = -\sqrt{\frac{-3}{4}}$$
$$x = -\frac{7}{2} + \frac{\sqrt{-3}}{\sqrt{4}} \qquad \text{or} \qquad x = -\frac{7}{2} - \frac{\sqrt{-3}}{\sqrt{4}}$$
$$x = -\frac{7}{2} + i\frac{\sqrt{3}}{2} \qquad\qquad x = -\frac{7}{2} - i\frac{\sqrt{3}}{2}$$

The two solutions to the equation $x^2 + 7x + 13 = 0$ are $x = -\frac{7}{2} + i\frac{\sqrt{3}}{2}$ and $x = -\frac{7}{2} - i\frac{\sqrt{3}}{2}$.

Standard window

**FIGURE 8.5** A graph of $y = x^2 + 7x + 13$ in Example 5. The lack of $x$-intercepts indicates the solution will be two nonreal complex numbers.

**Graphical Support.** Figure 8.5 shows the graph of $y = x^2 + 7x + 13$, which indicates that there are no $x$-intercepts to the graph. This means that there are no real zeros; if there were, the graph would intersect the $x$-axis. The only conclusion possible is that the two solutions must be nonreal complex numbers. ∎

## Exercises for Section   8.1

In Exercises 1–20, use the Square Root Property to solve the equation.

**1.** $x^2 - 25 = 0$

**2.** $x^2 - 144 = 0$

**3.** $3x^2 - 27 = 0$

**4.** $2x^2 - 32 = 0$

**5.** $x^2 + 49 = 0$

**6.** $3x^2 + 48 = 0$

**7.** $x^2 - 2 = 0$

**8.** $x^2 - 11 = 0$

**9.** $x^2 - 32 = 0$

**10.** $x^2 - 12 = 0$

**11.** $(x - 3)^2 = 7$

**12.** $(x - 5)^2 = 19$

**13.** $(x + 5)^2 = 27$

**14.** $(x + 2)^2 = 15$

**15.** $(2x - 5)^2 = 22$

**16.** $(3x - 8)^2 = 24$

**17.** $(3x + 7)^2 = 15$

**18.** $(2x + 5)^2 = 12$

**19.** $(4x + 3)^2 = 32$

**20.** $(2x - 18)^2 = 35$

In Exercises 21–28, solve the equation by completing the square. Include numerical support on a grapher for the solution.

**21.** $x^2 - 4x - 8 = 0$

**22.** $x^2 + 3x - 6 = 0$

**23.** $x^2 - 7x + 2 = 0$

**24.** $x^2 + 2x - 3 = 0$

**25.** $x^2 - 4x - 7 = 0$

**26.** $x^2 - 2x - 1 = 0$

**27.** $x^2 + 6x - 3 = 0$

**28.** $x^2 + 8x - 8 = 0$

In Exercises 29–36, use a graphical method to determine whether the equation will have real or nonreal complex roots. (Refer to Figure 8.5.)

**29.** $x^2 - 4x - 8 = 0$

**30.** $x^2 + 5x - 3 = 0$

**31.** $x^2 - 5x + 7 = 0$

**32.** $3x^2 - 7x + 3 = 0$

**33.** $2x^2 + 5x + 4 = 0$

**34.** $3x^2 - 8x + 6 = 0$

**35.** $2x^2 + 2x + 5 = 1 - x$

**36.** $6x^2 - 6 - x^2 = x + 1$

In Exercises 37–42, solve the equation by completing the square. Include numerical support for the solution. (All solutions are real numbers.)

**37.** $x^2 - 5x - 7 = 0$

**38.** $x^2 + 4x - 6 = 0$

**39.** $2x^2 - 7x + 1 = 0$

**40.** $3x^2 - 6x + 2 = 0$

**41.** $2x^2 - 9x + 4 = 1$

**42.** $7x^2 - 19x + 7 = 2$

In Exercises 43–52, solve the equation by completing the square. Include graphical support for the solution. (Some solutions are nonreal complex numbers.)

**43.** $x^2 + 4x - 3 = 0$

**44.** $x^2 - 5x + 2 = 0$

**45.** $2x^2 + 4x - 1 = 4$

**46.** $3x^2 - 6x + 7 = 3$

**47.** $x^2 - 7x + 13 = 0$

**48.** $2x^2 + 4x + 9 = 0$

**49.** $5x^2 - 15x + 9 = 2$

**50.** $2x^2 - 4x + 3 = x + 2$

**51.** $3x^2 - x + 4 = x^2 + 4x + 1$

**52.** $5x^2 - 2x - 1 = x^2 + 3x + 2$

In Exercises 53–56, assume $a$ is a positive real number and solve for $x$.

**53.** $x^2 = 4a$

**54.** $x^2 + 2x = a$

**55.** $x^2 - 28 = 4a^2$

**56.** $(2x - 3a)^2 = 7$

## TRANSLATING WORDS TO SYMBOLS

In Exercises 57–62, write an algebraic representation in the form of an equation and then solve the equation to find the numbers.

**57.** The sum of the square of a number and its double is 4. Find two numbers that meet this condition.

**58.** The sum of the square of a number and its triple is 2. Find two numbers that meet this condition.

**59.** *Using Algebra in Geometry.* A rectangle whose area is 15 ft² is 3 ft longer than it is wide (see the following figure). Find the dimensions of the rectangle.

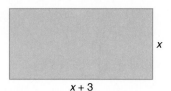

For Exercise 59.

**60.** *Using Algebra in Geometry.* A rectangle whose area is 25 ft$^2$ is 4 ft longer than it is wide. Find the dimensions of the rectangle.

**61.** *Using Algebra in Geometry.* The diagonal of a square is 8 cm longer than the length of its side (see the following figure). Find the length of the side.

For Exercise 61.

**62.** *Using Algebra in Geometry.* The diagonal of a square is 15 cm longer than the length of its side. Find the length of the side.

EXTENDING THE IDEAS

**63.** Graph both $y = x^2 - 3x + 4$ and $y = x + 3$ in the Standard window $[-10, 10]$ by $[-10, 10]$. How are the points of intersection of these graphs related to the $x$-intercepts of the graph of $y = x^2 - 4x + 1$?

**64.** How are the graphs found in Exercise 63 related to the fact that the equations $x^2 - 3x + 4 = x + 3$ and $x^2 - 4x + 1 = 0$ are equivalent?

**65.** Find the graphs of $y = x^2 + 3x + k$ for whole-number values of $k$. Find the smallest whole number $k$ such that $x^2 + 3x + k = 0$ has two nonreal complex roots.

**66.** Find the graphs of $y = x^2 + 3x + k$ for whole-number values of $k$. Find the largest whole number $k$ such that $x^2 + 3x + k = 0$ has two real-number roots.

**67.** Show that if $a + bi$ is a solution to $x^2 + mx + n = 0$, then $a - bi$ also is a solution.

LOOKING BACK—MAKING CONNECTIONS

In Exercises 68–70, explore the relationship between points of intersection of the graphs of two third-degree polynomials and the roots of an associated quadratic equation.

**68.** Find the graphs of $y_1 = x^3 + 2x^2 - 3x + 4$ and $y_2 = x^3 + x^2 + 2x - 4$ in the same window. Are you able to determine graphically whether there are any points of intersection between these two graphs?

**69.** Show that the equations $x^3 + 2x^2 - 3x + 4 = x^3 + x^2 + 2x - 4$ and $x^2 - 5x + 8 = 0$ are equivalent, which means the two equations have identical solutions.

**70.** How does the graph of $y = x^2 - 5x + 8$ tell us that the graphs of the two cubic polynomials in Exercise 68 do not intersect?

## 8.2 Quadratic Formula

Quadratic Formula ▪ Using the Discriminant ▪ Writing an Equation from Its Solution Set ▪ Using Quadratic Equations in Problem Solving

In Section 8.1, we learned that the algebraic method called completing the square can be used to solve any quadratic equation. In some cases, the solutions that result are nonreal complex numbers and in others, they are real numbers.

Suppose the completing-the-square method is used to find the solutions of the general quadratic equation in standard form

$$ax^2 + bx + c = 0,$$

where $a \neq 0$. The result is a formula called the **Quadratic Formula**.

## Quadratic Formula

Apply the completing-the-square method to the general quadratic equation $ax^2 + bx + c = 0$.

$$ax^2 + bx + c = 0 \qquad \text{Step 1: Divide each side by } a.$$

$$x^2 + \frac{b}{a}x + \frac{c}{a} = 0 \qquad \text{Step 1}$$

$$x^2 + \frac{b}{a}x = -\frac{c}{a} \qquad \text{Step 2}$$

$$x^2 + \frac{b}{a}x + \left(\frac{b}{2a}\right)^2 = -\frac{c}{a} + \left(\frac{b}{2a}\right)^2$$

$$\left(x + \frac{b}{2a}\right)^2 = \frac{-4ac + b^2}{4a^2} \qquad \text{Step 3}$$

$$\left(x + \frac{b}{2a}\right)^2 = \frac{b^2 - 4ac}{4a^2} \qquad \text{Step 4}$$

$$x + \frac{b}{2a} = \pm\sqrt{\frac{b^2 - 4ac}{4a^2}} \qquad \text{Step 5}$$

$$x = -\frac{b}{2a} \pm \frac{\sqrt{b^2 - 4ac}}{2a}$$

$$x = \frac{-b \pm \sqrt{b^2 - 4ac}}{2a}$$

---

### Quadratic Formula

The solutions to the equation $ax^2 + bx + c = 0$, where $a \neq 0$, are

$$x = \frac{-b \pm \sqrt{b^2 - 4ac}}{2a}.$$

All quadratic equations can be solved using the Quadratic Formula. When using it, you might need to rewrite the given equation to put it in the form $ax^2 + bx + c = 0$. Example 1 illustrates how to use this formula to solve an equation.

### EXAMPLE 1    Using the Quadratic Formula

Use the Quadratic Formula to solve the equation $2x^2 + 5x - 7 = 0$. Use a grapher to support the solutions numerically.

SOLUTION

**Solve Algebraically.** Because the equation $2x^2 + 5x - 7 = 0$ is in the form $ax^2 + bx + c = 0$, we see that $a = 2$, $b = 5$, and $c = -7$.

$$x = \frac{-b \pm \sqrt{b^2 - 4ac}}{2a} \quad \text{Substitute } a = 2, b = 5, \text{ and } c = -7.$$

$$= \frac{-5 \pm \sqrt{5^2 - 4(2)(-7)}}{2(2)}$$

$$= \frac{-5 \pm \sqrt{25 + 56}}{4}$$

$$= \frac{-5 \pm \sqrt{81}}{4}$$

$$= \frac{-5 \pm 9}{4}$$

The two solutions to this equation are $x = (-5+9)/4 = 1$ and $x = (-5-9)/4 = -7/2$.

**Support Numerically.** Figure 8.6 shows numerical support on a grapher for the two solutions. ∎

```
1 → X: 2X² + 5X − 7
                    0
-7/2 → X: 2X² + 5X −
7
                    0
```

**FIGURE 8.6** Numerical support that $x = 1$ and $x = -7/2$ are solutions to the equation $2x^2 + 5x - 7 = 0$ in Example 1.

In Example 1, the expression $b^2 - 4ac$ is equal to 81, a perfect square. Consequently, the two solutions of the equation $2x^2 + 5x - 7 = 0$ are rational numbers. So we could have used the method of factoring to solve this equation.

Example 2 illustrates the power of the Quadratic Formula when the quadratic equation cannot be factored using integers. It also illustrates how the grapher can be used to provide visual comprehension and estimation of the solutions.

### EXAMPLE 2    Using the Quadratic Formula

Use the Quadratic Formula to solve $3x^2 + 17x - 42 = 0$. Provide graphical support.

SOLUTION

**Solve Algebraically.** In the equation $3x^2 + 17x - 42 = 0$, $a = 3$, $b = 17$, and $c = -42$.

$$x = \frac{-b \pm \sqrt{b^2 - 4ac}}{2a} \qquad \text{Substitute } a = 3, b = 17, \text{ and } c = -42.$$

$$= \frac{-17 \pm \sqrt{17^2 - 4(3)(-42)}}{2(3)}$$

$$= \frac{-17 \pm \sqrt{289 + 504}}{6}$$

$$= \frac{-17 \pm \sqrt{793}}{6}$$

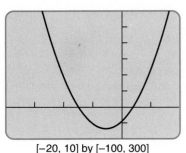

[−20, 10] by [−100, 300]

**FIGURE 8.7** Graphical support for the solution of the equation $3x^2 + 17x - 42 = 0$ in Example 2.

The two solutions are $x = \dfrac{-17 + \sqrt{793}}{6}$ and $x = \dfrac{-17 - \sqrt{793}}{6}$, whose decimal representations are $x = 1.860$ and $x = -7.527$.

**Support Graphically.** The graph of $y = 3x^2 + 17x - 42$ in Figure 8.7 shows graphical support for this algebraic solution. ∎

Example 3 shows that when the expression $b^2 - 4ac$ under the radical is a negative number, the two solutions are nonreal complex numbers.

**EXAMPLE 3**   Using the Quadratic Formula

Solve the equation $2x^2 - 4x + 5 = 0$. Provide graphical support.

SOLUTION

**Solve Algebraically.** In the equation $2x^2 - 4x + 5$, $a = 2$, $b = -4$, and $c = 5$.

$$x = \frac{-b \pm \sqrt{b^2 - 4ac}}{2a} \qquad \text{Substitute } a = 2, b = -4, \text{ and } c = 5.$$

$$= \frac{4 \pm \sqrt{(-4)^2 - 4(2)(5)}}{2(2)}$$

$$= \frac{4 \pm \sqrt{16 - 40}}{4}$$

$$= \frac{4 \pm \sqrt{-24}}{4}$$

The two solutions are these two complex numbers.

$$x = \frac{4 + \sqrt{(-4)(6)}}{4} \qquad x = \frac{4 - \sqrt{(-4)(6)}}{4}$$

$$= \frac{4 + 2i\sqrt{6}}{4} \qquad = \frac{4 - 2i\sqrt{6}}{4}$$

$$= \frac{2 + i\sqrt{6}}{2} \qquad = \frac{2 - i\sqrt{6}}{2}$$

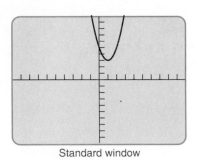

Standard window

**FIGURE 8.8** Graphical support for the solution of $2x^2 - 4x + 5 = 0$ in Example 3.

**Support Graphically.** Figure 8.8 shows that the graph of $y = 2x^2 - 4x + 5$ has no $x$-intercepts. Consequently, the solutions cannot be real numbers, which means that they must be nonreal complex numbers.   ■

## Using the Discriminant

In Examples 1, 2, and 3, we noticed that because the expression $b^2 - 4ac$ is under the radical, it determines whether the solutions are real or nonreal complex numbers.

This expression also can be used to determine the number of real-number solutions. We have already seen that if $b^2 - 4ac$ is positive, there are two real-number solutions. If $b^2 - 4ac$ is zero, there is only the one solution: $-b/2a$. So this expression $b^2 - 4ac$ discriminates among several cases and is consequently called the **discriminant**. Figure 8.9 illustrates these three cases for the situation of $a > 0$. If $a < 0$, the same three cases occur with the parabola opening downward.

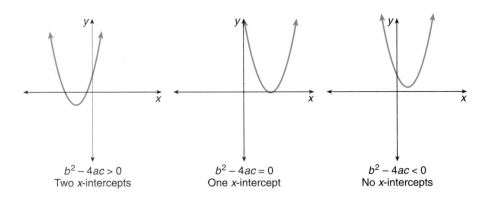

$b^2 - 4ac > 0$
Two $x$-intercepts

$b^2 - 4ac = 0$
One $x$-intercept

$b^2 - 4ac < 0$
No $x$-intercepts

**FIGURE 8.9** Using the discriminant to determine the number of $x$-intercepts.

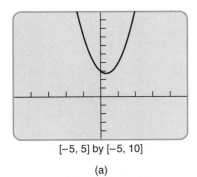

[−5, 5] by [−5, 10]

(a)

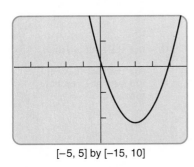

[−5, 5] by [−15, 10]

(b)

**FIGURE 8.10**
Graphs of (a)
$y = 4x^2 - 2x + 3$
indicating two nonreal
complex number solutions
and (b) $y = 3x^2 - 12x + 1$
indicating two real-number
solutions.

> **Discriminant of $ax^2 + bx + c$**
>
> For any quadratic equation $ax^2 + bx + c = 0$, where $a$, $b$, and $c$ are real numbers,
>
> 1. If $b^2 - 4ac > 0$, then there are two real solutions.
> 2. If $b^2 - 4ac = 0$, then there is one real solution.
> 3. If $b^2 - 4ac < 0$, then there are no real solutions. There are, however, two nonreal complex solutions.

**EXAMPLE 4**    Using the Discriminant of a Quadratic Function

Determine how many real-number solutions each of the following equations has. Use both an algebraic and a graphical approach.

a) $4x^2 - 2x + 3 = 0$

b) $3x^2 - 12x + 1 = 0$

SOLUTION
**Solve Algebraically.**

a) In $4x^2 - 2x + 3 = 0$, $a = 4$, $b = -2$, and $c = 3$. Then
$$b^2 - 4ac = (-2)^2 - 4(4)(3) = 4 - 48 = -44.$$

Because the discriminant is negative, there are no real-number solutions and two nonreal complex number solutions to $4x^2 - 2x + 3 = 0$.

b) In $3x^2 - 12x + 1$, $a = 3$, $b = -12$, and $c = 1$. Then
$$b^2 - 4ac = (-12)^2 - 4(3)(1) = 144 - 12 = 132.$$

Because $b^2 - 4ac > 0$, there are two real-number solutions to $3x^2 - 12x + 1 = 0$.

**Support Graphically.** Figure 8.10 shows the graphs of $y = 4x^2 - 2x + 3$ and $y = 3x^2 - 12x + 1$. These graphs support the algebraic conclusions.  ■

**Writing an Equation from Its Solution Set**

If we know that $x = 5/4$ and $x = -2/3$ are the solutions to a quadratic equation, we know that

$$\left( x - \frac{5}{4} \right) \quad \text{and} \quad \left( x + \frac{2}{3} \right)$$

are linear factors of the equation. Therefore, by the Zero Product Property, we reason backwards as follows:

$$\left(x - \frac{5}{4}\right)\left(x + \frac{2}{3}\right) = 0$$

$$x^2 + \frac{2}{3}x - \frac{5}{4}x - \frac{10}{12} = 0$$

$$12x^2 + 8x - 15x - 10 = 0$$

$$12x^2 - 7x - 10 = 0$$

Example 5 illustrates that this same method can be used to find a quadratic equation when the solutions are irrational numbers.

**EXAMPLE 5**   Finding an Equation

Find a quadratic equation whose solutions are $x = 3 + \sqrt{2}$ and $x = 3 - \sqrt{2}$.

SOLUTION

$$\left(x - (3 + \sqrt{2})\right)\left(x - (3 - \sqrt{2})\right) = 0 \quad \text{Use the FOIL method.}$$

$$x^2 - (3 - \sqrt{2})x - (3 + \sqrt{2})x + (3 + \sqrt{2})(3 - \sqrt{2}) = 0 \quad \begin{array}{l}\text{Use the Distributive} \\ \text{Property and combine} \\ \text{terms.}\end{array}$$

$$x^2 - 3x + \sqrt{2}x - 3x - \sqrt{2}x + (9 - 2) = 0$$

$$x^2 - 6x + 7 = 0$$

One equation that has solutions $x = 3 + \sqrt{2}$ and $x = 3 - \sqrt{2}$ is

$$x^2 - 6x + 7 = 0. \qquad \blacksquare$$

**REMINDER**

There are other solutions to Example 5. For example, the equation $2(x - 3 - \sqrt{2})(x - 3 + \sqrt{2}) = 2x^2 - 12x + 14 = 0$ is one of many. How many different equations could there be?

## Using Quadratic Equations in Problem Solving

Solutions to mathematical problems often lead to an equation that needs solving. The following example illustrates a problem that leads to a quadratic equation.

**EXAMPLE 6**   Problem Solving with Quadratic Equations

A fence encloses a park area of 39, 600 ft and the park is 40 ft longer than it is wide. How many feet of fence are needed to surround the park.

SOLUTION

Let $x =$ the width and $x + 40 =$ the length of the park (see Figure 8.11). The equation $x(x + 40) = 39{,}600$ is an algebraic representation that describes the situation.

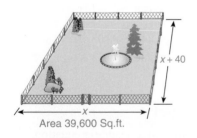

$x + 40$

Area 39,600 Sq.ft.

**FIGURE 8.11** A pictorial representation of the park problem.

$$x(x + 40) = 39600$$

$$x^2 + 40x - 39600 = 0$$

Using the quadratic formula, we obtain the following:

$$x = \frac{-40 \pm \sqrt{40^2 + 4(39600)}}{2}$$

$$= \frac{-40 \pm \sqrt{160000}}{2}$$

$$= \frac{-40 \pm 400}{2}$$

$$= -20 \pm 200$$

So $x = 180$ and $x = -220$ are the two solutions to the algebraic representation $x(x + 40) = 39,600$. However, only $x = 180$ has meaning in this situation, since it doesn't make sense to say one side has a negative length. The park is 180 ft wide and 220 ft long and 800 ft of fence are needed. Figure 8.12 provides graphical support for the solution.

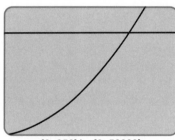

[0, 250] by [0, 50000]

**FIGURE 8.12** Graphical support of the park problem situation in Example 6 using the multigraph method with $y_1 = x(x + 40)$ and $y_2 = 39600$.

## Exercises for Section 8.2

In Exercises 1–12, use the Quadratic Formula to find the solutions to the equation. Use a grapher to provide numerical support for the solutions.

1. $x^2 + 4x - 7 = 0$
2. $x^2 - 3x - 5 = 0$
3. $x^2 - 9x + 12 = 0$
4. $x^2 + 6x - 4 = 0$

5. $3x^2 - 5x - 2 = 0$
6. $3x^2 - 7x - 12 = 0$
7. $2x^2 + 7x - 3 = 0$
8. $4x^2 + 5x - 3 = 0$
9. $3x^2 - 11x + 7 = 0$
10. $5x^2 - 14x + 9 = 0$
11. $5x - 3x^2 + 6 = 0$
12. $9x - 2x^2 + 5 = 0$

In Exercises 13–20, use a graph to decide, before using the Quadratic Formula, whether the solutions of the equation are real numbers or nonreal complex numbers.

**13.** $2x^2 - 5x + 4 = 0$    **14.** $3x^2 + 7x - 2 = 0$

**15.** $4x^2 - 13x + 2 = 0$    **16.** $5x^2 - 11x + 8 = 0$

**17.** $7x - 2x^2 - 9 = 0$    **18.** $-x^2 + 5x - 2 = 0$

**19.** $6x - 4x^2 + 1 = 0$    **20.** $5x - 2x^2 - 4 = 0$

In Exercises 21–32, find the solutions of the equation. Identify the solutions as real numbers or nonreal complex numbers.

**21.** $(r - 1)(r + 2) = 5$    **22.** $3y - 5 = y^2$

**23.** $k^2 + \dfrac{k}{4} = \dfrac{1}{3}$    **24.** $\dfrac{x + 2}{x + 1} = 2x$

**25.** $7v = \dfrac{1}{3} - \dfrac{4v^2}{7}$    **26.** $(k - 3)(k + 5) = -22$

**27.** $\dfrac{2s}{3} - 5 = \dfrac{7s^2}{3}$    **28.** $2t(t - 5) = 7$

**29.** $-3z(7 + z) = 38$    **30.** $\dfrac{t + 7}{t - 1} = 3t$

**31.** $\dfrac{t^2 + 1}{t - 3} = 2t$    **32.** $\dfrac{3t^2}{2} + \dfrac{17}{2} = -\dfrac{2t}{5}$

In Exercises 33–40, use the discriminant to determine whether there is two, one, or zero real-number solutions to the equation. Also provide graphical support.

**33.** $2t^2 = 4t - 5$    **34.** $7x - 3 = 2x^2$

**35.** $14x - 45 = x^2 - 4$    **36.** $2x^2 - 7x + 9 = 0$

**37.** $3x^2 + 5x - 14 = 0$    **38.** $2x + 7 = 4x^2$

**39.** $9 - 4x = 3x^2$

**40.** $2x^2 - 9x + 7 = 3x - 2$

In Exercises 41–48, find a quadratic equation in the form $ax^2 + bx + c = 0$ whose solutions are the given values.

**41.** $x = 5$ and $x = -3$

**42.** $x = -2$ and $x = -6$

**43.** $x = 2/3$ and $x = -5/3$

**44.** $x = -5/2$ and $x = 19/2$

**45.** $x = \sqrt{2}$ and $x = 3\sqrt{2}$

**46.** $x = 2\sqrt{5}$ and $x = 1 - \sqrt{5}$

**47.** $x = 2i$ and $x = -2i$

**48.** $x = 2 + i$ and $x = 2 - i$

In Exercises 49–60, solve the problem.

**49.** The product of two numbers is 1334 and one number is 17 greater than the other. Find the two numbers.

**50.** The product of two numbers is 2214 and one number is 13 greater than the other. Find the two numbers.

**51.** The product of two numbers is 2627. One number is three less than twice the other. Find the two numbers.

**52.** One side of a rectangle is 12 ft longer than another side. The area of the rectangle is 189 sq ft. What are the lengths of the sides of the rectangle?

**53.** *Building a Parking Lot.* A rectangular parking lot is 80 ft longer than it is wide and has an area of 66,000 sq ft. How many feet of fence are needed to build a fence around the parking lot?

**54.** *Building a Livestock Pen.* Larry built a livestock pen with fence on three sides and the barn on the fourth. The pen is 20 ft longer than it is wide and its area is 2925 sq ft. What is the pen's length and width?

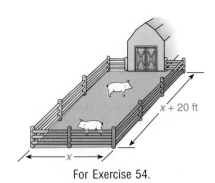

For Exercise 54.

**55.** *Manufacturing.* A computer manufacturer found that when it produced $x$ computers in 1 month, the profit for the month $P(x)$ was approximated by $P(x) = -0.05x^2 + 350x - 36,000$. How many computers were produced if the company profit for 1 month was $225,000?

**56.** *Manufacturing.* The demand for a manufactured item drops as the price goes up. A manufacturer found that the demand $D(x)$ (number of units ordered in 1 month) was approximated by $D(x) = -0.01x^2 + 900$ when the price was $\$x$. What is the price if the demand (in units per month) is 700?

In Exercises 57–59, use the law of physics $h = -16t^2 + v_0t + h_0$, where $h$ is the height above ground level for an object that has been propelled upward with an initial velocity of $v_0$ from an initial height of $h_0$.

**57.** *Projectile Motion.* John throws a ball straight up into the air at 50 ft/sec while standing on the top of a 300-ft tower.

a) Write an algebraic expression that describes the height the ball is above ground level $t$ sec after it is thrown.

b) How far above ground level is the ball 4 sec after it is thrown?

c) How many seconds after the ball is thrown is it 150 ft above ground level?

**58.** *Projectile Motion.* During an experiment in her science course, Marge drops an egg from the top of a 100-ft fire escape and records the time it takes the egg to hit the ground. She measures the time at 2.6 sec. What is the difference between the experimental time and the theoretical time determined by the formula $h = -16t^2 + v_0t + h_0$?

**59.** *Projectile Motion.* A diving board is 10 ft above the water's surface. The diving board propels Jack upward at the rate of 12 ft/sec and he enters the water feet first. How long does it take for Jack to hit the surface of the water?

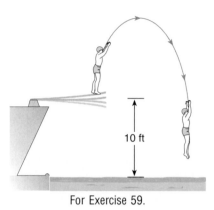

10 ft

For Exercise 59.

**60.** *Picture Frames.* A picture that is 9 in. by 15 in. is surrounded by a frame that is $x$ units wide (see the following figure).

a) If the picture and frame cover an area of 187 sq in., write an algebraic expression that describes the area of the frame only.

b) Find the width of the frame.

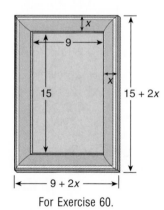

$x$

9

$x$

15

15 + 2x

9 + 2x

For Exercise 60.

EXTENDING THE IDEAS

In Exercises 61–64, $x_1$ and $x_2$ represent the two solutions

$$x_1 = \frac{-b + \sqrt{b^2 - 4ac}}{2a}$$

and

$$x_2 = \frac{-b - \sqrt{b^2 - 4ac}}{2a}$$

of the equation $ax^2 + bx + c = 0$.

**61.** Show that the sum of the two solutions $x_1 + x_2 = -b/a$.

**62.** Show that the product of the two solutions $x_1 \cdot x_2 = c/a$.

**63.** Use the results of Exercises 61 and 62 to select the correct pair of solutions for the equation $6x^2 + 23x - 18 = 0$ from among the following:
(a) $\left\{\frac{2}{3}, \frac{9}{2}\right\}$, (b) $\left\{-\frac{2}{3}, \frac{9}{2}\right\}$, (c) $\left\{\frac{2}{3}, -\frac{9}{2}\right\}$, and
(d) $\left\{-\frac{2}{3}, -\frac{9}{2}\right\}$.

**64.** Use the results of Exercises 61 and 62 to select the correct pair of solutions for the equation $3x^2 - 22x + 35 = 0$ from among the following: (a) $\left\{5, \frac{7}{3}\right\}$, (b) $\left\{\frac{5}{3}, 7\right\}$, (c) $\left\{-5, -\frac{7}{3}\right\}$, and (d) $\left\{-\frac{5}{3}, -7\right\}$.

**65.** Find a value(s) of $b$ so that $3t^2 + bt + 3 = 0$ has only one solution.

**66.** Find a value(s) of $b$ so that $2t^2 + bt + 3 = 0$ has only one solution.

**67.** Find a value(s) of $a$ so that $ax^2 + 5x - 3 = 0$ has only one solution.

**68.** Find a value(s) of $a$ so that $ax^2 + 7x - 2 = 0$ has only one solution.

LOOKING BACK—MAKING CONNECTIONS

**69.** *Shooting Fireworks.* When fireworks are shot into the air from ground level, their height $t$ seconds later is $h$, where $h = rt - 5t^2$ and $r$ is the initial velocity of the object in meters per second.

a) If fireworks are shot into the air with an initial rate of 65 m/sec, what is the algebraic representation that represents the height above the ground $t$ seconds later.

b) How high above the ground are the fireworks 5 sec after being shot?

c) How many seconds after the shooting are the fireworks 150 m above the ground?

d) Assuming that the fireworks explode at approximately the peak of their flight, about how many seconds after they are shot do they explode? (Use your grapher.)

---

**8.3** **Complete Graphs of Quadratic Functions: Solving Equations and Inequalities**

> Sketching the Graph of $y = x^2$ ▪ Finding a Complete Graph of a Quadratic Function ▪ Finding the Number of Solutions by Visual Inspection ▪ Solving Quadratic Inequalities

In Section 8.1, a quadratic equation was defined to be an equation that could be written in the form $ax^2 + bx + c = 0$, where $a$, $b$, and $c$ are real numbers, $a \neq 0$.

When using grapher support in equation solving, as illustrated in Sections 8.1 and 8.2, we were really finding the graph of the quadratic function associated with the given quadratic equation.

> **DEFINITION 8.2**    Quadratic Function
>
> A **quadratic function** $y = f(x)$ is one that can be written in the form
> $$y = ax^2 + bx + c,$$
> where $a$, $b$, and $c$ are real numbers, $a \neq 0$. Constants $a$ and $b$ are called the **coefficients** of the $x^2$- and $x$-terms, respectively, and $c$ is called the **constant term**.

In Sections 8.1 and 8.2, you saw numerous examples of graphs of quadratic functions. In this section, we shall do a more complete study of these graphs. In particular in this and the next several sections, we shall show that the graph of each quadratic function can be obtained by transforming the graph of $y = x^2$ in various ways. So let us first become familiar with the graph of $y = x^2$.

## Sketching the Graph of $y = x^2$

Although a grapher usually will be used for graphing functions, we begin by using the hand-sketch, point-plotting method. A hand-sketch begins by finding a table of values. Example 1 shows such a table.

**EXAMPLE 1**    Finding a Table of Values for the Function $y = x^2$

Use the point-plotting method to find a complete graph of the function $y = x^2$.

SOLUTION

**Finding a Table of Values by Hand.** Often a table of $(x, y)$ pairs can be found by using paper and pencil and a calculator to complete a table like the following:

| $x$ | $-3$ | $-2$ | $-1$ | $0$ | $1$ | $2$ | $3$ |
|---|---|---|---|---|---|---|---|
| $y = x^2$ | 9 | 4 | 1 | 0 | 1 | 4 | 9 |

**Finding a Table of Values with a Grapher.** Some graphers can display a table of function values, as shown in Figure 8.13.

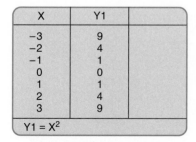

| X | Y1 | |
|---|---|---|
| −3 | 9 | |
| −2 | 4 | |
| −1 | 1 | |
| 0 | 0 | |
| 1 | 1 | |
| 2 | 4 | |
| 3 | 9 | |
| Y1 = X² | | |

**FIGURE 8.13** A grapher-generated table of values.

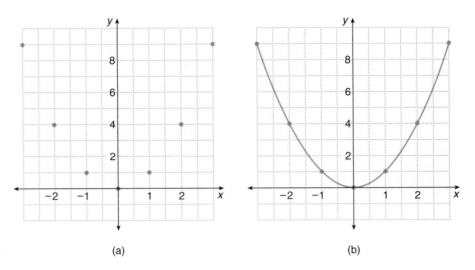

(a)                    (b)

**FIGURE 8.14** A sketch of the graph of $y = x^2$.

Figure 8.14(a) shows the graph of the solution pairs $(x, y)$ from the table of values and Figure 8.14(b) illustrates a completed sketch of the graph of $y = x^2$.  ∎

This function $y = x^2$ is called the **squaring function** because each output value is the square of the corresponding input value. The graph of this equation is called a **parabola**.

When finding the graph of $y = x^2$ on a grapher, the shape of the graph might appear different for different windows, as shown in Figure 8.15.

Choosing an appropriate window is important when finding the graph of a function on a grapher. Fig. 8.15 shows the graph of the same function $y = x^2$ in three different windows.

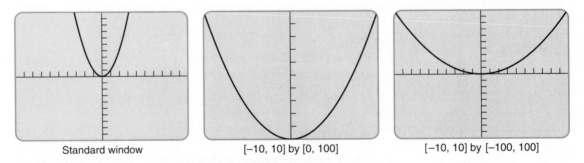

Standard window         [−10, 10] by [0, 100]         [−10, 10] by [−100, 100]

**FIGURE 8.15** Three different windows for the graph of $y = x^2$.

## Finding a Complete Graph of a Quadratic Function

As described in Section 2.5, a **complete graph** is one in which all the important characteristics of a function can be seen. In particular, a complete graph always includes all $x$-intercepts, when they exist.

It can be shown that the graph of each quadratic function, including the simplest function, $y = x^2$, is a parabola, that is, a curve shaped like those in Figure 8.16. If $a > 0$, the parabola opens up and if $a < 0$, the parabola opens down. Consequently, a complete graph of a quadratic function looks like one of the curves in Figure 8.16.

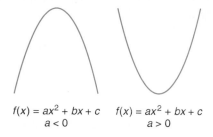

$$f(x) = ax^2 + bx + c \qquad f(x) = ax^2 + bx + c$$
$$a < 0 \qquad\qquad\qquad a > 0$$

**FIGURE 8.16** Shapes of graphs of quadratic functions.

A complete graph of most quadratic functions is not centered in the Standard window like the graph of $y = x^2$. To find a complete graph, it often is helpful to begin with the Standard window and then modify it as necessary until a complete graph has been found. Figure 8.17 shows how a complete graph of $y = x^2 + 15x - 25$ can be found in three steps.

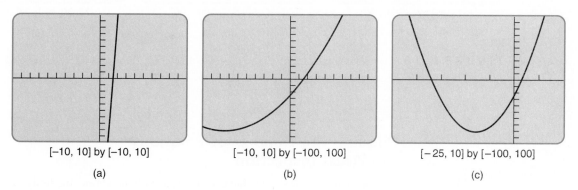

| [−10, 10] by [−10, 10] | [−10, 10] by [−100, 100] | [−25, 10] by [−100, 100] |
| :---: | :---: | :---: |
| (a) | (b) | (c) |

**FIGURE 8.17** Three different windows of the graph of $y = x^2 + 15x - 25$. A complete graph appears in (c).

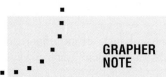
Figure 8.17(a) shows the graph in the Standard window. It is evident that the vertical dimension of the window needs to be larger, so we change the vertical dimension from $[-10, 10]$ to $[-100, 100]$ (see Figure 8.17(b)). The revised graph shows that the vertical dimension $[-100, 100]$ is adequate, but it is now evident that we need to change the horizontal dimension to be larger and to extend to the left. Figure 8.17(c) reflects this change and also shows a complete graph of $y = x^2 + 15x - 25$.

Example 2 shows several more examples of graphs of quadratic functions.

**EXAMPLE 2**    Finding Complete Graphs

Find a complete graph of each of the following quadratic functions:

a)  $f(x) = -x^2 + 18x + 42$

b)  $g(x) = x^2 - 22x - 12$

c)  $h(x) = -1.5x^2 - 62.4x + 143$

SOLUTION
Complete graphs are shown in Figure 8.18.

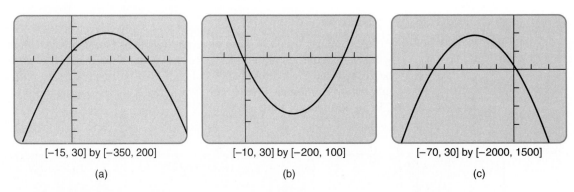

| $[-15, 30]$ by $[-350, 200]$ | $[-10, 30]$ by $[-200, 100]$ | $[-70, 30]$ by $[-2000, 1500]$ |
| :---: | :---: | :---: |
| (a) | (b) | (c) |

**FIGURE 8.18**  Complete graphs of (a) $f(x) = -x^2 + 18x + 42$, (b) $g(x) = x^2 - 22x - 12$, and (c) $h(x) = -1.5x^2 - 62.4x + 143$.  ∎

## Finding the Number of Solutions by Visual Inspection

Throughout this text, we have discussed both algebraic methods and graphical methods of solving equations. When using a graphical method, it is necessary to use a complete graph in order to know that you have found all solutions.

The direction "solve the equation ..." means find all solutions. To solve a quadratic equation graphically, begin by finding a complete graph of the equation situation and follow by finding all solutions. For example, how many solutions are there for the quadratic equation

$$x^2 - 4x + 5 = -x^2 + 2x - 3.$$

A quick glance at Figure 8.19 tells us that there are *no* real solutions because the complete graphs of $y = x^2 - 4x + 5$ and $y = -x^2 + 2x - 3$ do not intersect. Compare this graphical answer to the following algebraic explanation.

### Change to an Equivalent Form.

$$x^2 - 4x + 5 = -x^2 + 2x - 3 \qquad \text{Add } x^2 - 2x + 3 \text{ to each side of the equation.}$$

$$2x^2 - 6x + 8 = 0$$

**Use the Discriminant.** Let $a = 2$, $b = -6$, and $c = 8$ and evaluate the discriminant.

$$b^2 - 4ac = (-6)^2 - 4 \cdot 2 \cdot 8$$

$$= -28$$

We conclude that the equation $x^2 - 4x + 5 = -x^2 + 2x - 3$ has no real solutions because the discriminant is negative.

Comparing the graphical method and the algebraic method, we feel that in the above case the graphical method is quicker and superior to the algebraic method.

### EXAMPLE 3    Determining the Number of Solutions

Use a graphical method to determine the number of real solutions to $x^2 - 5x - 2 = -2x^2 + x + 1$.

SOLUTION

We can use either the $x$-intercept or multigraph method. In this case, we choose the $x$-intercept method. Begin by changing to an equivalent equation with a zero on one side.

$$x^2 - 5x - 2 = -2x^2 + x + 1$$

$$3x^2 - 6x - 3 = 0$$

Figure 8.20 shows a graph of $y = 3x^2 - 6x - 3$. It is evident from the graph that there are two $x$-intercepts. Consequently, the original equation has two real solutions. ■

Example 3 asks us to find the number of real solutions to an equation that is a quadratic on each side of the equality. We transformed the equation to an

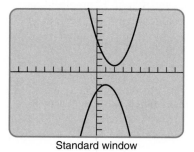

Standard window

**FIGURE 8.19** Complete graphs of $y = x^2 - 4x + 5$ and $y = -x^2 + 2x - 3$.

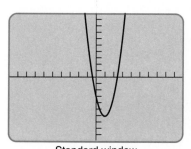

Standard window

**FIGURE 8.20** A graph of $y = 3x^2 - 6x - 3$.

**TRY THIS**

Find graphs of
$y = \frac{1}{2}x^2 + x + 2$ and
$y = x^2 + 4x - 3$ in the
Standard window. Explain
why this graph is not a
complete graph of the
equation
$\frac{1}{2}x^2 + x + 2 = x^2 + 4x - 3$.

equivalent equation

$$3x^2 - 6x - 3 = 0$$

in order to use the $x$-intercept method. It is necessary to find a complete graph to make sure that we have found *all* solutions.

If you use a multigraph approach, it is also necessary to use complete graphs as illustrated by the accompanying Try This activity. The graphs of two distinct quadratic functions intersect in at most two points. The equation in Example 3 can be solved graphically by using zoom-in to find the solutions with an error of at most 0.01.

## Solving Quadratic Inequalities

Quadratic inequalities in one variable can be solved in much the same way as quadratic equations. Examples 4 and 5 illustrate the method.

**EXAMPLE 4**    Solving a Quadratic Inequality Graphically

Solve the inequality $-x^2 + 7x + 4 > 0$.

SOLUTION
Begin by finding a complete graph of $y = -x^2 + 7x + 4$.

Figure 8.21 indicates that the portion of the graph between approximately $x = -0.5$ and $x = 7.5$ lies above the $x$-axis. In other words, when $-0.5 < x < 7.5$ the corresponding $y$-coordinates are positive. Zoom in to find the $x$-intercepts with an error of at most 0.01. Figure 8.22 shows the two $x$-intercepts after zoom-in.

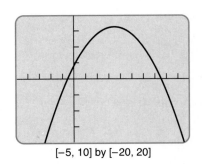

[−5, 10] by [−20, 20]

**FIGURE 8.21** A graph of $y = -x^2 + 7x + 4$ in Example 4.

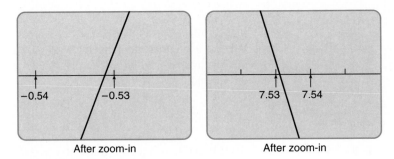

After zoom-in                After zoom-in

**FIGURE 8.22** A graph of $y = -x^2 + 7x + 4$ showing the two $x$-intercepts after zoom-in.

The two $x$-intercepts with an error of at most 0.01 are $x = -0.531$ and $x = 7.531$.

The solution to the inequality $-x^2 + 7x + 4 > 0$ consists of all real numbers in the interval $(-0.531, 7.531)$. The endpoints of the interval are not included because the inequality is $>$ rather than $\geq$.    ■

In Example 5, the inequality compares quadratic and linear expressions.

**EXAMPLE 5**    Solving a Quadratic Inequality Graphically

Solve the inequality $x^2 - 5x - 3 < \dfrac{1}{2}x + 2$.

SOLUTION

We begin by writing an equivalent inequality with a zero on one side.

$$x^2 - 5x - 3 < \frac{1}{2}x + 2$$

$$(x^2 - 5x - 3) - \left(\frac{1}{2}x + 2\right) < 0$$

$$x^2 - 5.5x - 5 < 0$$

Figure 8.23 shows a graph of $y = x^2 - 5.5x - 5$. By zooming in, we find that the $x$-intercepts are $x = -0.794$ and $x = 6.294$. Consequently, $x$ is a solution to the inequality if $-0.794 < x < 6.294$, which is the interval $(-0.794, 6.294)$.    ■

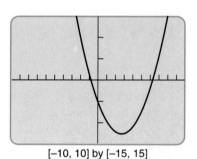

[−10, 10] by [−15, 15]

**FIGURE 8.23**  A graph of $y = x^2 - 5.5x - 5$ in Example 5.

---

## Exercises for Section ▮8.3

In Exercises 1 and 2, complete each part.

1. For the function $y = 3x^2$:

   a) Find a table of $(x, y)$ solution pairs for $x = -4, x = -3, x = -2, x = -1, x = 0,$ $x = 1, x = 2, x = 3,$ and $x = 4$.

   b) Graph each solution pair with paper and pencil.

   c) Sketch a graph of the function by connecting these points with a smooth curve.

2. For the function $y = -2x^2$:

   a) Find a table of $(x, y)$ solution pairs for $x = -4, x = -3, x = -2, x = -1, x = 0,$ $x = 1, x = 2, x = 3,$ and $x = 4$.

b) Graph each solution pair with paper and pencil.

c) Sketch a graph of the function by connecting these points with a smooth curve.

In Exercises 3–6, choose the one grapher window that shows a complete graph of the function.

3. For $f(x) = 2x^2 - 12x + 3$:

   a) $[-5, 5]$ by $[-5, 5]$

   b) $[-10, 10]$ by $[-10, 10]$

   c) $[-10, 10]$ by $[-20, 10]$

4. For $g(x) = -x^2 + 15x - 8$:

   a) $[-10, 10]$ by $[-10, 10]$

b) $[-10, 20]$ by $[-10, 20]$

c) $[-10, 20]$ by $[-10, 50]$

5. For $k(x) = -2x^2 - 28x + 3$:

a) $[-10, 10]$ by $[-10, 10]$

b) $[-20, 10]$ by $[-20, 120]$

c) $[-10, 10]$ by $[-50, 100]$

6. For $f(x) = 3x^2 - 8x - 2$:

a) $[-5, 5]$ by $[-10, 10]$

b) $[-10, 10]$ by $[0, 20]$

c) $[0, 5]$ by $[-10, 10]$

In Exercises 7 and 8, a graph of a quadratic function is given in the accompanying figure that is not a complete graph.

7. **Writing to Learn.** A graph of a quadratic function is shown below. The graph is not a complete graph.

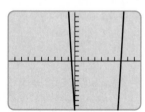

Standard window

For Exercise 7.

Choose the window from (a), (b), and (c) below that is most likely to show a complete graph and then write a paragraph that explains your choice.

a) $[-10, 10]$ by $[-5, 5]$

b) $[-10, 10]$ by $[-40, 10]$

c) $[0, 10]$ by $[-40, 40]$

8. **Writing to Learn.** A graph of a quadratic function is shown below. The graph is not a complete graph.

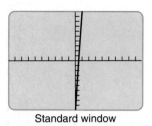

Standard window

For Exercise 8.

Choose the window from (a), (b), and (c) below that is most likely to show a complete graph and then write a paragraph that explains your choice.

a) $[-5, 5]$ by $[-5, 5]$

b) $[-10, 20]$ by $[-40, 40]$

c) $[-10, 10]$ by $[-5, 5]$

In Exercises 9–18, find a complete graph of the quadratic function. To record an answer, sketch the graph and label the window being used.

9. $f(x) = x^2 - 7x + 5$

10. $g(x) = x^2 + 8x - 3$

11. $h(x) = 2x^2 + 12x - 3$

12. $k(x) = 3x^2 - 10x + 2$

13. $f(x) = -4x^2 + 25x + 9$

14. $h(x) = -2x^2 - 33x + 8$

15. $g(x) = -3x^2 + 42x + 5$

16. $k(x) = x^2 + 54x - 5$

17. $f(x) = 2x^2 + 49x + 2$

18. $g(x) = 5x^2 + 124x - 8$

In Exercises 19–25, use a graphical method to determine the number of real solutions of the equation. Use zoom-in if necessary.

19. $x^2 + 5x + 3 = -x^2 - x + 5$

20. $2x^2 + 6x - 3 = -x^2 + x + 4$

21. $2x^2 + 7x + 2 = -x^2 - 2$

22. $x^2 + 3x - 4 = 0.3x^2 + 2x - 5$

23. $2x^2 + 8x - 3 = x^2 - 1$

24. $x^2 + 5x + 3 = -x^2 + 7x - 4$

25. $x^2 + 7x + 4 = -x^2 + 4x - 1$

In Exercises 26–32, use a graphical method to solve the inequality. Find the endpoints of solution intervals with an error of at most 0.01.

26. $x^2 - 7x - 5 < 0$

27. $-x^2 - 4x + 6 > 0$

28. $0.5x^2 + 0.7x - 3 > x + 4$

29. $x^2 + 8x - 6 < 0.5x + 1$

30. $x^2 - 4x - 3 > 0.1x + 3$

31. $2x + 3 < 9x^2 - x - 5$

32. $x^2 - 3x - 2 < -x^2 + 5x + 7$

TRANSLATING WORDS TO SYMBOLS

In Exercises 33 and 34 write the algebraic inequality that has been described in words. Use a grapher to decide if the resulting inequality has a solution.

33. Twice the square of a number is less than five more than the number.

34. Three times a number is less than the square of the number decreased by four.

In Exercises 35–38, solve the problem.

35. *Analysis of a Rectangle.* A rectangle, shown in the following figure, is 5 units longer than it is wide.

a) Write an algebraic expression for the area of the rectangle as a function of $x$.

b) What values of $x$ make sense for this problem?

c) Suppose the area is between 2.3 square units and 7.8 square units. Write an inequality that represents this condition.

d) Find a graph that can be used to solve the inequality in part (c).

e) For what widths $x$ is the area of the rectangle between 2.3 square units and 7.8 square units?

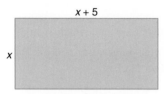

For Exercise 35.

36. *Analysis of a Rectangle.* A rectangle, shown in the following figure, is 8 units longer than it is wide.

a) Write an algebraic expression for the area of the rectangle as a function of $x$.

b) What values of $x$ make sense for this problem?

c) Suppose the area is between 18 square units and 42 square units. Write an inequality that represents this condition.

d) Find a graph that can be used to solve the inequality in part (c).

e) For what widths $x$ is the area of the rectangle between 18 square units and 42 square units?

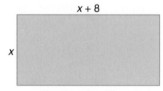

For Exercise 36.

37. *Projectile Motion.* Nick is standing on a cliff 70 ft above a valley and Jane is standing above the valley on a 105-ft cliff. Nick throws a snowball straight up into the air at 60 ft/sec at the same instant Jane throws a snowball up at 35 ft/sec. (Recall the law of physics $h = -16t^2 + v_0 t + h_0$.)

a) Write an algebraic representation for the height of each snowball $t$ seconds after it is thrown.

b) How many feet above the valley will each snowball be 1 sec after it is thrown?

c) Write an inequality whose solution is the interval of time that Jane's snowball is higher than Nick's.

d) During what interval of time is Jane's snowball higher than Nick's?

**38.** *Projectile Motion.* In Exercise 37

a) Whose snowball hits the valley floor first?

b) How many seconds after the first snowball hits the floor does the second snowball hit the valley floor?

c) Do you find it easier to answer Exercise 37 graphically or algebraically? Explain why.

### EXTENDING THE IDEAS

**39.** How many real solutions does the equation $x^2 - 6x + 9 = x^2 + 2x + 1$ have? First use a graphical method, then confirm the solution algebraically. Does either method have any uncertainties? Explain.

**40.** How many real solutions does the equation $x^2 - 4x - 2 = 2x - 11$ have? First use a graphical method, then confirm the solution algebraically. Which method seems most reliable? Why?

### LOOKING BACK—MAKING CONNECTIONS

In Exercises 41 and 42, solve the equation both algebraically and graphically.

**41.** $\dfrac{4}{x+3} = 5$

**42.** $\dfrac{x+1}{(x+3)(x-1)} = 1$

**43.** *Diluting a Mixture.* Pure acid is added to distilled water. The resulting 2-liters of solution is a 15% solution.

a) How many liters of pure acid are needed to make this solution?

b) If an additional 1 liter of distilled water is added to the solution, what percentage solution results?

c) How many liters of distilled water should be added to obtain a 12% acid solution?

**44.** *Distance, Rate, Time.* Antonio is planning an 85-mile trip. (Recall: $d = r \cdot t$.)

a) If the trip takes him 4 hr, what was his average rate of travel?

b) Write an algebraic representation for the rate $r$ as a function of time $t$.

c) How long does the trip take if Antonio averages 50 mph?

---

**8.4**  # Sketching Graphs of Quadratic Functions

Transforming the Graph of $y = x^2$ by a Vertical Slide  ▪  Transforming the Graph of $y = x^2$ by a Horizontal Slide  ▪  Combining Horizontal and Vertical Slides  ▪  Completing the Square and Graph Sketching

In this section, we shall consider families of quadratic functions whose graphs are related to each other. Try to find the link between the visual and the algebraic as we consider these various families.

## Transforming the Graph of $y = x^2$ by a Vertical Slide

Complete the following exploration to discover the relationship between the graph of $y = x^2$ and $y = x^2 + k$, where $k$ is any real number.

### EXPLORE WITH A GRAPHER

Find graphs of $y_1 = x^2$, $y_2 = x^2 + 3$, $y_3 = x^2 + 8$, and $y_4 = x^2 + 12$ in the 0.1 window in which the vertical scale is adjusted to $[-5, 25]$. Graph all four before clearing any one of them.

1. Is the graph of $y_2 = x^2 + 3$ above or below the graph of $y_1 = x^2$? By how many units?
2. Is the graph of $y_3 = x^2 + 8$ above or below the graph of $y_1 = x^2$? By how many units?
3. Predict the location of the graphs of $y = x^2 + 5$ and $y = x^2 - 3$. Confirm your prediction.

**State a Generalization:**  How does the graph of $y = x^2$ compare with the graph of $y = x^2 + k$, where $k$ is any real number?

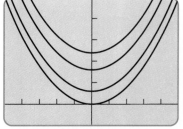

0.1 window with vertical
dimension [−5, 25]

**FIGURE 8.24**  Graphs
of $y_1 = x^2$, $y_2 = x^2 + 3$,
$y_3 = x^2 + 8$, and $y_4 = x^2 + 12$.

This exploration, together with Figure 8.24, suggests that the graph of $y = x^2 + k$ is a parabola and can be obtained by sliding the graph of $y = x^2$ vertically. This conclusion is summarized as follows.

### Vertical Slide by Distance $k$

Let $k$ be a positive real number. Vertical slides of the graph of $y = x^2$ are represented as follows:

1. A vertical slide $k$ units *upward* is $y = x^2 + k$.
2. A vertical slide $k$ units *downward* is $y = x^2 - k$.

A **vertical slide,** sometimes also called a **vertical translation**, changes a graph from one position to another; it transforms the graph. Consequently, it is an example of a **transformation**. We use this transformation in Example 1.

### EXAMPLE 1   Vertical Slide

Describe how the following graphs can be obtained from the graph of $y = x^2$. Support the answer to part (a) graphically.

a) $f(x) = x^2 + 3.8$

b) $g(x) = -12 + x^2$

SOLUTION

a) The function $f(x) = x^2 + 3.8$ can be obtained by sliding the graph of $y = x^2$ up 3.8 units. Figure 8.25 supports this assertion.

b) The quadratic function $g(x) = -12 + x^2$ can be rewritten as $g(x) = x^2 - 12$. So $g(x)$ can be obtained by sliding the graph of $y = x^2$ down 12 units.   ■

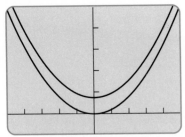

0.1 window with vertical dimension [−5, 25]

**FIGURE 8.25**
Graph of $y_1 = x^2$
and $y_2 = x^2 + 3.8$ in Example 1.

## Transforming the Graph of $y = x^2$ by a Horizontal Slide

Complete the following exploration to discover the relationship between the graph of $y = x^2$ and $y = (x - h)^2$, where $a$ is any real number.

### EXPLORE WITH A GRAPHER

Find graphs of $y_1 = x^2$, $y_2 = (x - 3)^2$, and $y_3 = (x + 4)^2$ in the 0.2 window with a vertical dimension of [−3, 10]. Graph all three before clearing any one of them (see Fig. 8.26).

1. Is the graph of $y_2 = (x - 3)^2$ right or left of $y_1 = x^2$? By how many units?
2. Is the graph of $y_3 = (x + 4)^2$ right or left of $y_1 = x^2$? By how many units?
3. Predict the location of the graphs of $y = (x + 2)^2$ and $y = (x - 5)^2$. Confirm your prediction.

**State a Generalization:**   How does the graph of $y = x^2$ compare with the graph of $y = (x - h)^2$, where $a$ is any real number? What value of $x$ is the zero of $y = (x - h)^2$?

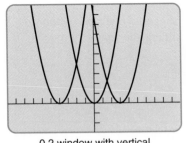

0.2 window with vertical dimension [−3, 10]

**FIGURE 8.26**
Graphs of $y_1 = x^2$,
$y_2 = (x - 3)^2$, and
$y_3 = (x + 4)^2$.

This exploration, along with Figure 8.26, suggests that the graph of any equation of the form $y = (x - h)^2$ is a parabola and can be obtained by sliding the graph of $y = x^2$ horizontally. This conclusion is summarized as follows.

---

### Horizontal Slide by Distance $h$

Let $h$ be a positive real number. Horizontal slides of the graph of $y = x^2$ are represented as follows:

1. A horizontal slide $h$ units *right* is $y = (x - h)^2$.
2. A horizontal slide $h$ units *left* is $y = (x + h)^2$.

---

**EXAMPLE 2**    Horizontal Slide

Describe how the following graphs can be obtained from the graph of $y = x^2$. Support the answer to (b) with a grapher.

a) $f(x) = \left(x - \dfrac{3}{2}\right)^2$

b) $g(x) = (x + 3.4)^2$

SOLUTION

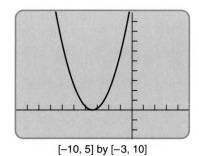

[−10, 5] by [−3, 10]

**FIGURE 8.27**

Graph of $y = (x + 3.4)^2$ in Example 2.

a) The function $f(x) = \left(x - \dfrac{3}{2}\right)^2$ can be obtained by sliding the graph of $y = x^2$ right $\dfrac{3}{2}$ units.

b) The function $g(x) = (x + 3.4)^2$ can be obtained by sliding the graph of $y = x^2$ left 3.4 units. Figure 8.27 supports this claim. ■

## Combining Horizontal and Vertical Slides

Vertical and horizontal slides can be combined. Figure 8.28 on the following page assumes that $h > 0$ and $k > 0$ and shows the result of sliding the graph of $y = x^2$ right $h$ units and up $k$ units. (It is equivalent to first sliding up and then sliding right.)

Notice that after sliding the graph of $y = x^2$ right $h$ units, the equation of the resulting graph is $y = (x - h)^2$. Then after sliding that graph up $k$ units, the equation of the resulting graph is $y = (x - h)^2 + k$.

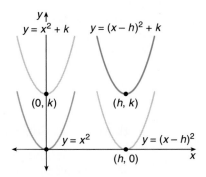

**FIGURE 8.28** Sliding $y = x^2$ right $h$ units and up $k$ units gives $y = (x - h)^2 + k$.

Example 3 illustrates combining horizontal and vertical slides.

**EXAMPLE 3**    Combining Horizontal and Vertical Slides

Sketch a graph of $f(x) = (x + 4)^2 - 8$ without using a grapher.

SOLUTION

The function $f(x) = (x + 4)^2 - 8$ can be obtained from the graph of $y = x^2$ as follows:

1. Slide left 4 units.
2. Slide down 8 units.

The sketch of the graph of $y = (x + 4)^2 - 8$ is shown in Figure 8.29.  ∎

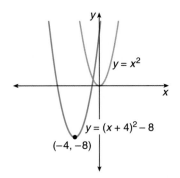

**FIGURE 8.29** Graphs of $y = x^2$ and $y = (x + 4)^2 - 8$ in Example 3.

In Example 3, the equation of the function was given and you were asked to sketch the graph. Example 4 asks for an equation of a graph that has been obtained from a sequence of slides.

**EXAMPLE 4**    Transforming the Parabola $y = x^2$

The graph of $y = x^2$ is transformed by a horizontal slide 5 units right and a vertical slide 3 units down. Write an algebraic representation of the resulting graph.

SOLUTION

1. Suppose $G_1$ is the graph of $y = x^2$. Slide $G_1$ right 5 units to obtain the graph $G_2$, which is defined by $y = (x - 5)^2$.

2. Next slide the graph $G_2$ down 3 units to obtain the graph $G_3$.
3. Graph $G_3$ is the graph defined by $y = (x - 5)^2 - 3$.

   $y = (x - 5)^2 - 3$ is an algebraic representation of the resulting graph.   ∎

## Completing the Square and Graph Sketching

Suppose a quadratic function is written in the form

$$y = x^2 + bx + c.$$

How can we sketch its graph? Examples 3 and 4 illustrate that when a quadratic equation is written in the form $y = (x - h)^2 + k$, its graph can be sketched by thinking of sliding the graph $y = x^2$ horizontally and vertically. So rewriting an equation in the form

$$y = (x - h)^2 + k$$

helps us sketch its graph.

For example, the function $f(x) = x^2 - 14x + 49$ is a perfect square. So if we write $f$ as

$$f(x) = x^2 - 14x + 49 = (x - 7)^2,$$

we see that the graph of $f$ can be obtained by sliding the graph of $y = x^2$ to the right 7 units (see Figure 8.30).

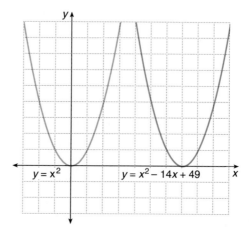

**FIGURE 8.30**   Graph of $y = x^2 - 14x + 49 = (x - 7)^2$.

But suppose the quadratic function is not a perfect square? To illustrate this situation, consider the example

$$g(x) = x^2 - 14x + 50.$$

We know that if the constant were 49 rather than 50, the quadratic would be a perfect square. So we write 50 as $49 + 1$ and obtain a perfect square as follows:

$$g(x) = x^2 - 14x + 50 = (x^2 - 14x + 49) + 1$$
$$= (x - 7)^2 + 1$$

Thus the graph of $y = g(x)$ can be obtained from $y = x^2$ by a horizontal slide right 7 units followed by a slide up 1 unit (see Figure 8.31).

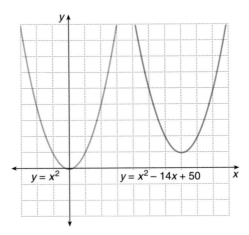

**FIGURE 8.31** Graph of $y = x^2 - 14x + 50$.

**EXAMPLE 5**   Completing a Square

Write each of the following quadratic functions in the form $y = (x - h)^2 + k$ by completing a square:

a)  $f(x) = x^2 + 18x + 83$

b)  $g(x) = x^2 - 6x + 7$

SOLUTION

a) Reason as follows: $x^2 + 18x + c$ is a perfect square if the constant $c$ is equal to $\left(\frac{1}{2} \cdot 18\right)^2 = 81$. So write 83 as $81 + 2$ and regroup to form a perfect square, as shown here:

$$f(x) = x^2 + 18x + 83 = x^2 + 18x + (81 + 2)$$
$$= (x^2 + 18x + 81) + 2$$
$$= (x + 9)^2 + 2$$

b) Reason as follows: $x^2 - 6x + c$ is a perfect square if the constant $c$ is equal to $\left(\dfrac{1}{2} \cdot -6\right)^2 = 9$. So write 7 as $9 - 2$ and regroup to form a perfect square, as shown here:

$$g(x) = x^2 - 6x + 7 = x^2 - 6x + (9 - 2)$$
$$= (x^2 - 6x + 9) - 2$$
$$= (x - 3)^2 - 2$$   ■

Example 6 shows how to use this completing-the-square method in graphing.

**EXAMPLE 6**    Finding the Graph of a Quadratic Function

Complete a square for each of the following functions and determine a horizontal and/or vertical slide that can be used to find the graph of the equation. Provide grapher support for part (b).

a)  $f(x) = x^2 + 10x + 33$

b)  $g(x) = x^2 - 8x + 13$

SOLUTION

a) Because $x^2 + 10x + 25$ is a perfect square, write 33 as $25 + 8$.

$$f(x) = x^2 + 10x + 33 = x^2 + 10x + (25 + 8)$$
$$= (x^2 + 10x + 25) + 8$$
$$= (x + 5)^2 + 8$$

   The graph of $f(x)$ can be obtained by sliding the graph of $y = x^2$ left 5 units and up 8 units.

b) Because $x^2 - 8x + 16$ is a perfect square, write 13 as $16 - 3$.

$$g(x) = x^2 - 8x + 13 = x^2 - 8x + (16 - 3)$$
$$= (x^2 - 8x + 16) - 3$$
$$= (x - 4)^2 - 3$$

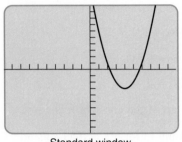

Standard window

**FIGURE 8.32**
A graph of part (b)
$y = x^2 - 8x + 13$ in
Example 6.

The graph of $g(x)$ can be obtained by sliding $y = x^2$ right 4 units and down 3 units to obtain the graph of $y = x^2 - 8x + 13$. Figure 8.32 supports this algebra. ■

Observe that because the graph of $y = x^2 + 10x + 33$ in Example 6(a) does not have any $x$-intercepts, the solutions to $x^2 + 10x + 33 = 0$ are complex numbers.

## Exercises for Section 8.4

In Exercises 1–20, describe how the graph of the function can be obtained from the graph of $y = x^2$. Support your answer with a grapher.

**1.** $y = x^2 - 8$  **2.** $y = x^2 + 3$

**3.** $y = x^2 + 17$  **4.** $y = x^2 - 4$

**5.** $y = (x - 3)^2$  **6.** $y = (x + 2)^2$

**7.** $y = (x + 5)^2$  **8.** $y = (x - 14)^2$

**9.** $y = x^2 - 48$  **10.** $y = (x + 3.5)^2$

**11.** $y = (x - \sqrt{2})^2$  **12.** $y = x^2 - \sqrt{5}$

**13.** $y = (x - 2)^2 + 4$  **14.** $y = (x + 1)^2 - 3$

**15.** $y = (x + 3)^2 + 6$  **16.** $y = (x - 8)^2 - 5$

**17.** $y = (x - 0.5)^2 + \frac{3}{2}$  **18.** $y = (x + \sqrt{2})^2 - 3$

**19.** $y = (x - \sqrt{5})^2 - \sqrt{2}$

**20.** $y = (3 - x)^2 + 4$

In Exercises 21–28, find the value of $c$ that makes the trinomial a perfect square and then complete a square.

**21.** $x^2 + 8x + c$  **22.** $x^2 + 24x + c$

**23.** $x^2 - 4x + c$  **24.** $x^2 - 12x + c$

**25.** $x^2 + 5x + c$  **26.** $x^2 - 7x + c$

**27.** $x^2 - 19x + c$  **28.** $x^2 + 41x + c$

In Exercises 29–36, complete a square for the quadratic function and determine how its graph can be obtained from the graph of $y = x^2$.

**29.** $y = x^2 - 6x + 5$  **30.** $y = x^2 + 4x + 16$

**31.** $y = x^2 + 10x + 18$  **32.** $y = x^2 - 12x - 4$

**33.** $y = x^2 + 5x - 12$  **34.** $y = x^2 + 3x - 8$

**35.** $y = 8x + x^2 - 3$  **36.** $y = 5x + x^2 + 2$

### TRANSLATING WORDS TO SYMBOLS

In Exercises 37–43, write an algebraic representation of the graph obtained from the change described.

**37.** Slide the graph of $y = x^2$ right 3 units.

**38.** Slide the graph of $y = x^2$ right 8 units.

**39.** Slide the graph of $y = (x - 2)^2$ up 3 units.

**40.** Slide the graph of $y = x^2$ down 7 units.

**41.** The graph of $y = x^2$ slides 4 units right and 6 units down.

**42.** The graph of $y = x^2$ slides 3 units left and 8 units up.

**43.** Slide the graph of $y = x^2$ left 1 unit and up 3 units.

In Exercises 44 and 45, solve the given problems.

**44.** *Projectile Motion.* The functions $h_1 = -16(t - 3.5)^2 + 196$ and $h_2 = -16(t - 5)^2 + 196$ can be used as algebraic representations of two balls thrown up into the air from ground level, one 1.5 sec after the other.

a) Find the graph of each function in the window [0, 10] by [0, 200].

b) How many seconds after the first ball is thrown do the two balls reach the same height?

**45.** *Projectile Motion.* Suppose the function $h_1 = -16(t - 3.5)^2 + 196$ is the algebraic representation of a ball thrown up into the air from ground level. A second ball is then thrown up with the same initial velocity 3.5 sec later.

a) Would the algebraic representation of the second ball thrown be $h_2 = -16(t + 7)^2 + 196$ or $h_2 = -16(t - 7)^2 + 196$?

b) How many seconds after the first ball is thrown will the two balls have the same height above ground level?

EXTENDING THE IDEAS

**46.** Show algebraically that functions $y = (x - 3)^2$ and $y = (3 - x)^2$ are identical.

In Exercises 47–54, look at the equation and develop a mental image of its graph. Decide whether the graph crosses the $x$-axis. Support your answer with a grapher.

**47.** $y = (x + 5)^2 - 17$     **48.** $y = (x - 3)^2 + 1$

**49.** $y = x^2 + 18$     **50.** $y = (x - 3)^2 - 11$

**51.** $y = (x - 3)^2 + 8$     **52.** $y = -x^2 + 3$

**53.** $y = 17 - (x - 1)^2$     **54.** $y = (x + 5)^2 + 7$

**55.** What are the conditions on the constant $k$ if the graph of $y = x^2 - k$ does not intersect the $x$-axis.

**56.** What is the $y$-intercept of the graph of $y = (x - h)^2$?

**57.** What are the conditions on the constant $h$ if the graph of $y = (x - h)^2$ can be obtained from $y = x^2$ by a slide left?

**58.** Suppose we slide the graph of $y = (x + 3)(x - 3)$ right one unit. Will the resulting graph be the graph of $y = (x + 2)(x - 4)$ or $y = (x + 4)(x - 2)$? Support your answer with a grapher.

**59.** Suppose we slide the graph of $y = (x + 5)(x - 5)$ right one unit. What is the equation of the resulting graph? Support your answer with a grapher.

**60.** Suppose we slide the graph of $y = (x + 2)(x - 3)$ one unit right. What is the equation of the resulting graph? Support your answer with a grapher.

LOOKING BACK—MAKING CONNECTIONS

**61.** *Analysis of a Right Triangle.* The long leg of the following right triangle is 1 unit longer than the short leg and the hypotenuse is 3 units longer than the short leg.

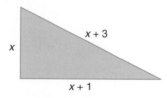

For Exercise 61.

a) Write an algebraic expression that results from applying the Pythagorean theorem. (Recall the Pythagorean theorem: $c^2 = a^2 + b^2$, where $a$ and $b$ are the lengths of the legs and $c$ is the length of the hypotenuse.)

b) Find the length of the triangle's sides.

c) Did you use an algebraic method or a graphical method? Explain why you made the choice you did.

**62.** *Analysis of a Right Triangle.* The long leg of the following right triangle is 3 units longer than the short leg and the hypotenuse is 9 units longer than the short leg.

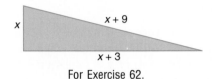

For Exercise 62.

a) Write an algebraic expression that results from applying the Pythagorean theorem.

b) Do you think there exists a triangle with the above conditions? Find the lengths of the sides if such a triangle exists.

**63. Writing to Learn.** Explain how we know that the graph of $y = (x + 3)(x - 5)$ can be obtained by applying slides to the graph of $y = x^2$.

In Exercises 64–66, solve the system of equations. Do you prefer a graphical method or an algebraic method? Explain why.

**64.** $\begin{cases} y = x^2 - 3 \\ y = 2x + 3 \end{cases}$

**65.** $\begin{cases} y = x^2 - 2 \\ y = -x^2 + 3 \end{cases}$

**66.** $\begin{cases} y = x^2 + 5x - 2 \\ y = -x^2 + 3 \end{cases}$

## 8.5 More on Graph Sketching of Quadratic Functions

Transforming the Graph of $y = x^2$ by Vertical Stretch and Shrink Factors ▪ Combining Vertical Stretch, Shrink, and Slide and Reflection in the $x$-axis ▪ Completing the Square and Graph Sketching ▪ Summarizing Graph Sketching for Quadratic Functions

Section 8.4 discussed the graphs of quadratic functions of the form $y = x^2 + bx + c$, which can be rewritten as one of the following:

$$y = x^2 + k, \qquad y = x^2 - k, \qquad y = (x - h)^2, \qquad \text{and} \qquad y = (x + h)^2$$

These graphs can be obtained from $y = x^2$ by vertical and horizontal slides.

Notice that throughout Section 8.4, the coefficient of $x^2$ was always 1. In this section, we investigate how the coefficient of $x^2$ impacts the graph.

### Transforming the Graph of $y = x^2$ by Vertical Stretch and Shrink Factors

Complete the following exploration to discover the relationship between the graph of $f(x) = x^2$ and $g(x) = ax^2$, where $a$ is any real number.

Find graphs of $y_1 = x^2$, $y_2 = 2x^2$, $y_3 = 3x^2$, and $y_4 = 0.5x^2$ in the 0.1 window in which the vertical scale is adjusted to $[-5, 40]$ (see Figure 8.33). Graph all four before clearing any one of them.

1. Move the trace cursor from graph to graph to compare the $y$-coordinates of the points $(2, y_1)$, $(2, y_2)$, $(2, y_3)$, and $(2, y_4)$. Repeat this activity for the points $(3, y_1)$, $(3, y_2)$, $(3, y_3)$, and $(3, y_4)$.
2. Which graph would you describe as being obtained from the graph of $y = x^2$ by "stretching up."
3. Which graph would you describe as being obtained from the graph of $y = x^2$ by "shrinking down."
4. Predict the location of the graphs of $y = 4x^2$ and $y = 0.3x^2$. Confirm your prediction.

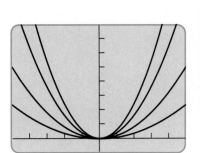

0.1 window with vertical dimension $[-5, 40]$

**FIGURE 8.33** Graphs of $y_1 = x^2$, $y_2 = 2x^2$, $y_3 = 3x^2$, and $y_4 = 0.5x^2$.

Notice in Figure 8.33 that the graph of $y = x^2$ seems to be "stretched up" to obtain the graph of $y = 2x^2$ and "shrunk down" to obtain the graph of $y = 0.5x^2$. In each case, we say that the graph of $y = x^2$ has been transformed by a vertical stretch factor. Definition 8.3 introduces this terminology.

---

**DEFINITION 8.3**   Vertical Stretch by Factor $a$

The graph of $y = ax^2$ can be obtained from the graph of $y = x^2$ by

1. a **vertical stretch**  if $a > 1$, and
2. a **vertical shrink**  if $0 < a < 1$.

   The positive number $a$ is called the **stretch factor** (if $a > 1$) or the **shrink factor** (if $0 < a < 1$).

---

**EXAMPLE 1**   Finding the Graph of $y = ax^2$

Find the following graphs. Are they obtained from $y = x^2$ by a stretch or shrink? What is the factor in each case? Support graphically.

a) $y_1 = 3.5x^2$

b) $y_2 = 0.7x^2$

SOLUTION

a) Because $3.5 > 1$, the graph of $y = 3.5x^2$ is obtained from $y = x^2$ by a vertical stretch. The stretch factor is 3.5 (see Figure 8.34(a)).

b) Because $0.7 < 1$, the graph of $y = 0.7x^2$ is obtained from $y = x^2$ by a vertical shrink. The shrink factor is $a = 0.7$ (see Figure 8.34(b)).

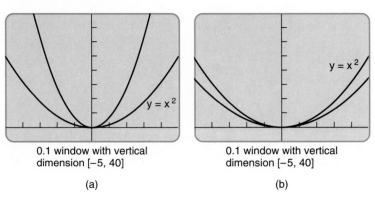

0.1 window with vertical dimension [−5, 40]

(a)

0.1 window with vertical dimension [−5, 40]

(b)

**FIGURE 8.34** Graphs of (a) $y_1 = 3.5x^2$ and $y_2 = x^2$ and (b) $y_1 = 0.7x^2$ and $y_2 = x^2$. ■

The next example can be done without using a grapher.

**EXAMPLE 2**  Comparing Graphs

Figure 8.35 shows the following graphs. Match the function with the graph.

a) $y = \sqrt{2}x^2$

b) $y = \pi x^2$

c) $y = (1/\sqrt{3})x^2$

SOLUTION

We need only compare the coefficients of $x^2$ in the three equations.

a) $\sqrt{2}$ is approximately 1.414.

b) $\pi$ is approximately 3.142.

c) $1/\sqrt{3}$ is approximately 0.577.

Because $\pi > \sqrt{2} > 1/\sqrt{3}$, we conclude that $y = \pi x^2$ is matched with graph (1), $y = \sqrt{2}x^2$ is matched with graph (2), and $y = (1/\sqrt{3})x^2$ with graph (3). ■

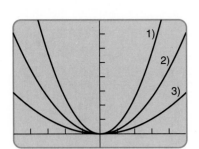

0.1 window with vertical dimension of [−5, 40]

**FIGURE 8.35** Graphs of (1) $y = \pi x^2$, (2) $y = \sqrt{2}x^2$, and (3) $y = (1/\sqrt{3})x^2$.

In the equations considered so far in this section, the coefficient of $x^2$ has been a positive number. Example 3 probes what happens when the coefficient of $x^2$ is negative by comparing $y_1 = x^2 + 1$ and $y_2 = -x^2 - 1$.

**EXAMPLE 3**   Comparing $y = x^2 + 1$ and $y = -x^2 - 1$

Find the graphs of $y_1 = x^2 + 1$ and $y_2 = -x^2 - 1$ in the 0.1 window with the vertical dimensions changed to $[-15, 15]$. How do $y_1$ and $y_2$ compare for the same value of $x$?

SOLUTION
**Solve Algebraically.**

$$y_2 = -x^2 - 1 = -(x^2 + 1) = -y_1$$

For a given value of $x$, the values $y_1$ and $y_2$ are opposites of each other.

**Support Graphically.** Figure 8.36(a) shows the trace cursor at a point on the graph of $y_1$, and Figure 8.36(b) shows the trace cursor on the graph of $y_2$ at a point with the same $x$-coordinate. Notice that the $y$-coordinates are the opposite of each other.

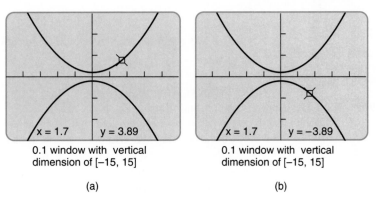

| x = 1.7 | y = 3.89 |
| 0.1 window with vertical dimension of [−15, 15] | |
| (a) | |

| x = 1.7 | y = −3.89 |
| 0.1 window with vertical dimension of [−15, 15] | |
| (b) | |

**FIGURE 8.36**   Graphs of $y = x^2 + 1$ and $y = -x^2 - 1$.   ∎

We call the graphs of $y_1 = x^2 + 1$ and $y_2 = -x^2 - 1$ in Figure 8.36(a) and (b) reflections of each other in the $x$-axis. Notice that if point $(a, b)$ is on one graph, $(a, -b)$ is on the other graph.

---

### Reflection in the $x$-axis

The graph of $y = -f(x)$ is called the **reflection in the $x$-axis** of the graph of $y = f(x)$.

A point $(a, b)$ is on the graph of a function $y = f(x)$ if and only if the point $(a, -b)$ is on the graph of the function $y = -f(x)$.

---

Example 4 illustrates the concept of reflection in the $x$-axis.

**EXAMPLE 4**   Reflection in the $x$-axis

Find complete graphs of $f(x) = x^2$, $g(x) = 3x^2$, and $h(x) = -3x^2$ in the 0.1 window with the vertical dimensions changed to $[-20, 20]$. Identify both algebraically and graphically which graphs are reflections of each other in the $x$-axis.

SOLUTION
**Solve Algebraically.**

$$h(x) = -3x^2 = -(3x^2) = -g(x)$$

We conclude that the graphs of $g$ and $h$ are reflections of each other in the $x$-axis.

**Support Graphically.** Figure 8.37 shows the graphs of the three functions $f(x)$, $g(x)$, and $h(x)$.

It is visually evident that the graphs of $g$ and $h$ are reflections of each other in the $x$-axis. We also can observe that the graph of $g$ can be obtained from the graph of $f$ by a vertical stretch with a stretch factor of 3. The graph of $h$ can be obtained from the graph of $f$ by a vertical stretch with a stretch factor of 3, followed by a reflection in the $x$-axis. ■

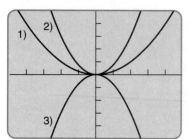

0.1 window with vertical dimension $[-20, 20]$

**FIGURE 8.37** Graphs of (1) $f(x) = x^2$, (2) $g(x) = 3x^2$, and (3) $h(x) = -3x^2$ in Example 4.

## Combining Vertical Stretch, Shrink, and Slide and Reflection in the $x$-axis

The four transformation types—vertical slide, horizontal slide, vertical stretch or shrink, and reflection in the $x$-axis—can be combined. The next several examples illustrate various combinations.

**EXAMPLE 5**    Combining Transformations

Sketch the graph of the following without using a grapher:

a) $f(x) = -(x + 2)^2$

b) $g(x) = -(x - 3)^2 + 2$

SOLUTION

a) The graph of $f$ can be obtained from the graph of $y = x^2$ in two stages. First, slide the graph of $y = x^2$ left 2 units to obtain the graph of $y = (x+2)^2$ (graph 2). Then reflect this graph in the $x$-axis to obtain the graph of $f(x) = -(x+2)^2$ (graph 3). Figure 8.38(a) shows the graphs in this two-step process.

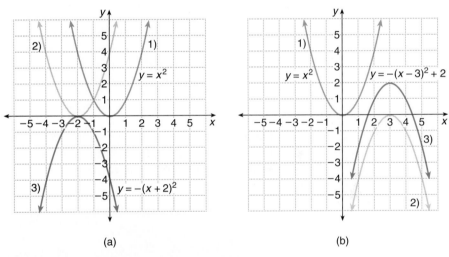

(a)                                    (b)

**FIGURE 8.38** Graphs of (a) $y = x^2$ and $f(x) = -(x + 2)^2$ and (b) $y = x^2$ and $g(x) = -(x - 3)^2 + 2$.

b) The graph of $g$ can be obtained from the graph of $y = x^2$ in three stages. First, slide the graph of $y = x^2$ right 3 units. Second, reflect this graph in the $x$-axis to obtain the graph of $y = -(x - 3)^2$ (graph 2). Finally, slide the second stage graph up 2 units to obtain the graph of $y = -(x - 3)^2 + 2$ (graph 3). Figure 8.38(b) shows the graphs of $y = x^2$ and $y = -(x - 3)^2$ and the graph of $g(x) = -(x - 3)^2 + 2$.    ■

In Example 5, functions were defined in terms of their equations and their graphs were obtained by performing slide and reflection transformations. In Example 6, this process is reversed—a sequence of transformations is described and you are asked to find the algebraic representation.

**EXAMPLE 6**   Combining Transformations

The graph of a function $y = f(x)$ is obtained from the graph of $y = x^2$ by the following sequence.

a) A horizontal slide left 5 units, followed by
b) a vertical stretch with a stretch factor of 3, followed by
c) a vertical slide down 6 units

Find an algebraic representation for $y = f(x)$. Also provide graphical support for the result.

SOLUTION
**Solve Algebraically.**

a) Suppose $G_1$ is the graph of $y = x^2$. Slide $G_1$ left 5 units to obtain the graph $G_2$, which is defined by $y = (x + 5)^2$.
b) Let $G_3$ be the graph obtained from $G_2$ by a vertical stretch with stretch factor 3. The graph of $G_3$ is defined by $y = 3(x + 5)^2$.
c) Let $G_4$ be the graph obtained from $G_3$ by a slide down 6 units. The graph of $G_4$ is defined by $y = 3(x + 5)^2 - 6$.

$y = 3(x + 5)^2 - 6$ is the equation of the parabola $G_4$.

**Support Graphically.** Figure 8.39(a) shows the equations that define graphs $G_1$, $G_2$, $G_3$, and $G_4$. Figure 8.39(b) shows these graphs in the 0.2 window with a vertical dimension of $[-15, 15]$.

$$Y1 = X^2$$
$$Y2 = (X + 5)^2$$
$$Y3 = 3(X + 5)^2$$
$$Y4 = 3(X + 5)^2 - 6$$

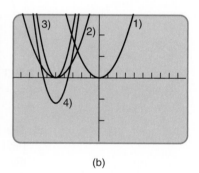

(a)                                          (b)

**FIGURE 8.39**   The graph of $y = x^2$ is transformed into the graph of $y = 3(x + 5)^2 - 6$ in Example 6.   ∎

The examples in Sections 8.4 and so far in Section 8.5 demonstrate that a quadratic function that is written in the form

$$f(x) = a(x - h)^2 \pm k \quad \text{or} \quad f(x) = a(x + h)^2 \pm k$$

is a particularly convenient form for sketching the graph of the function. We shall call this form the **perfect-square form** of a quadratic function. So there is benefit in changing from the standard form $f(x) = ax^2 + bx + c$ to the perfect-square form $f(x) = a(x - h)^2 \pm k$.

## Completing the Square and Graph Sketching

Completing the square is the process that allows us to change from the standard form $y = ax^2 + bx + c$ to the perfect square form $y = a(x - h)^2 + k$.

In Section 8.4, the completing-the-square method was applied only to cases in which the coefficient of $x^2$ was 1. Here we illustrate a three-step process for completing the square for a function like $f(x) = 3x^2 + 18x + 8$.

1. Factor the 3 from the $x^2$- and $x$-terms and separate the constant 8 from the first two terms.

$$f(x) = 3x^2 + 18x + 8$$
$$= 3(x^2 + 6x \quad\ ) + 8$$

2. Use methods of Section 8.4 to find the value that completes the square of $x^2 + 6x$. We use $\triangle$ to indicate this value.

$$f(x) = 3(x^2 + 6x \quad\ ) + 8$$
$$= \underbrace{3(x^2 + 6x + \triangle)} + 8 - \underbrace{3 \times \triangle} \qquad \text{In this case, } \triangle \text{ is 9.}$$
$$= 3(x^2 + 6x + 9) + 8 - 27$$

Adding the value $\triangle$ inside the parentheses is really adding $(3 \times \triangle)$, which also must be subtracted to maintain the equality.

3. The expression $x^2 + 6x + 9$ is a perfect square. Write it in this form.

$$f(x) = 3(x^2 + 6x + 9) + 8 - 27$$
$$= 3(x + 3)^2 - 19$$

We see that $f(x) = 3x^2 + 18x + 8$ is equivalent to the perfect-square form $f(x) = 3(x + 3)^2 - 19$. This form can be used to sketch the graph of $f$.

**EXAMPLE 7**  Changing to Perfect-square Form

Write each of the following quadratic functions in perfect-square form:

a) $f(x) = 2x^2 + 16x + 9$

b) $g(x) = -x^2 + 6x - 5$

SOLUTION

a) $f(x) = 2x^2 + 16x + 9 = 2(x^2 + 8x \qquad) + 9$

$$= 2(x^2 + 8x + \triangle) + 9 - 2 \times \triangle$$  **Choose $\triangle$ to complete the square.**

$$= 2(x^2 + 8x + 16) + 9 - 32$$

$$= 2(x + 4)^2 - 23$$

b) $g(x) = -x^2 + 6x - 5 = -(x^2 - 6x \qquad) - 5$

$$= -(x^2 - 6x + \triangle) - 5 + \triangle$$

$$= -(x^2 - 6x + 9) - 5 + 9$$

$$= -(x - 3)^2 + 4 \qquad\qquad \blacksquare$$

## Summarizing Graph Sketching for Quadratic Functions

**REMINDER**

The order of the transformations described in this summary box can be done in a variety of orders except that the vertical slide must be done last.

---

### Sketching Quadratic Functions

A graph of $f(x) = ax^2 + bx + c$ can be sketched using the following steps:

1. Write $f(x)$ in the perfect square form $f(x) = a(x \pm h)^2 \pm k$, where $h$ and $k$ are real positive numbers.
2. If $a > 0$, obtain the graph of $f$ from the graph of $y = x^2$ by
   a) a horizontal slide, followed by
   b) a vertical stretch or shrink, followed by
   c) a vertical slide.
3. If $a < 0$, obtain the graph of $f$ from the graph of $y = x^2$ by
   a) a horizontal slide, followed by
   b) a vertical stretch or shrink, followed by
   c) a reflection in the $x$-axis, followed by
   d) a vertical slide.

---

Example 8 illustrates how the completing-the-square method is used in sketching the graph of a quadratic function.

**EXAMPLE 8** Sketching the Graph of a Quadratic Function

Write $f(x) = 2x^2 - 18x + 38$ in perfect-square form and describe the transformations that can be used to obtain the graph of $f$ from $y = x^2$. Provide graphical support.

SOLUTION
**Solve Algebraically.**

$$f(x) = 2x^2 - 18x + 38 = 2(x^2 - 9x + \triangle) + 38 - 2 \times \triangle$$

$$= 2\left(x^2 - 9x + \left(\frac{9}{2}\right)^2\right) + 38 - 2 \times \frac{81}{4}$$

$$= 2\left(x - \frac{9}{2}\right)^2 + \frac{76}{2} - \frac{81}{2}$$

$$= 2\left(x - \frac{9}{2}\right)^2 - \frac{5}{2}$$

From this perfect-square form, we conclude that the graph of $f$ can be obtained from the graph of $y = x^2$ by

a) a slide right by 9/2 units, followed by
b) a vertical stretch with a stretch factor of 2, followed by
c) a slide down by 5/2 units.

**Support Graphically.** Figure 8.40 shows the graph of both $y = x^2$ and $f(x) = 2x^2 - 18x + 38$. It is visually evident that $f(x) = 2x^2 - 18x + 38$ can be obtained from $y = x^2$ by a slide right, a vertical stretch, and a slide down. ∎

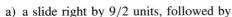

Standard window

**FIGURE 8.40**
Graphs of $y = x^2$ and $f(x) = 2x^2 - 18x + 38$ in Example 8.

---

## Exercises for Section 8.5

In Exercises 1–8, find a complete graph of both $y = x^2$ and the given function in the same viewing window. Determine whether the function is obtained from $y = x^2$ by a stretch or a shrink.

1. $y = 3x^2$    2. $y = \frac{1}{2}x^2$

3. $y = \sqrt{2}x^2$    4. $y = 2x^2$

5. $y = 0.6x^2$    6. $y = 1.3x^2$

7. $y = 15x^2$    8. $y = 0.15x^2$

In Exercises 9–12, three quadratic functions are given. In each case, an appropriate grapher window can be found so that the graphs look like the following figure. Match the functions with the graphs in the figure.

9. (a) $y = 4x^2$, (b) $y = 0.6x^2$, and (c) $y = 2x^2$

10. (a) $y = 0.2x^2$, (b) $y = 1.3x^2$, and (c) $y = 0.6x^2$

11. (a) $y = 12x^2$, (b) $y = 6x^2$, and (c) $y = 2x^2$

**12.** (a) $y = 30x^2$, (b) $y = 4x^2$, and (c) $y = 14x^2$

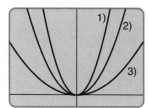

For Exercises 9–12.

**13.** Find a window for which the functions in Exercise 9 look like the given figure.

**14.** Find a window for which the functions in Exercise 10 look like the given figure.

**15.** Find a window for which the functions in Exercise 11 look like the given figure.

In Exercises 16–23, predict how the two graphs will compare to each other. Use a grapher to provide graphical support.

**16.** $y = x^2 + 3$ and $y = -(x^2 + 3)$

**17.** $y = -x^2 + 3$ and $y = x^2 - 3$

**18.** $y = x^2 + 4$ and $y = -x^2 - 4$

**19.** $y = 2(x - 3)^2$ and $y = (x - 3)^2$

**20.** $y = 5 - x^2$ and $y = x^2 - 5$

**21.** $y = 2x^2 - 1$ and $y = 4x^2 - 2$

**22.** $y = 2x^2 + 1$ and $y = x^2 + 0.5$

**23.** $y = -12x^2 + 3$ and $y = 12x^2 + 3$

In Exercises 24–31, describe a sequence of transformations that change the graph of $y = x^2$ to the graph of the function given in the exercise.

**24.** $y = 2x^2 - 3$　　　**25.** $y = 3x^2 + 4$

**26.** $y = 3(x - 4)^2$　　　**27.** $y = 0.3(x + 2)^2$

**28.** $y = -2(x + 1)^2$　　　**29.** $y = -3(x - 2)^2$

**30.** $y = -3(x - 1)^2 + 2$　**31.** $y = -(x + 3)^2 - 2$

In Exercises 32–39, sketch the graph of the function without using a grapher. Then support your sketch with a grapher.

**32.** $y = 2x^2 - 3$　　　**33.** $y = 3x^2 + 1$

**34.** $y = -(x + 3)^2$　　　**35.** $y = -(x - 4)^2$

**36.** $y = 3(x + 1)^2$　　　**37.** $y = 2(x - 3)^2$

**38.** $y = -2(x - 3)^2 - 1$　**39.** $y = -(x - 1)^2 + 3$

TRANSLATING WORDS TO SYMBOLS

In Exercises 40–43, write an algebraic representation $y = f(x)$ for the function that is obtained from $y = x^2$ by the sequence of transformations described in each exercise.

**40.** The graph of a function $y = f(x)$ is obtained from the graph of $y = x^2$ by the sequence (a) a vertical stretch by factor 2, followed by (b) a slide down 3 units.

**41.** The graph of a function $y = f(x)$ is obtained from the graph of $y = x^2$ by the sequence (a) a reflection in the $x$-axis, followed by (b) a slide up 4 units.

**42.** The graph of a function $y = f(x)$ is obtained from the graph of $y = x^2$ by the sequence (a) a vertical shrink by a factor of $1/2$, followed by (b) a reflection in the $x$-axis.

**43.** The graph of a function $y = f(x)$ is obtained from the graph of $y = x^2$ by the sequence (a) a slide right 3 units, followed by (b) a reflection in the $x$-axis, followed by (c) a vertical stretch by 5 units.

In Exercises 44–57, write the quadratic function in perfect-square equivalent form.

**44.** $y = x^2 + 2x - 6$　　**45.** $y = x^2 + 4x - 7$

**46.** $y = x^2 - 6x - 3$　　**47.** $y = x^2 - 8x + 2$

**48.** $y = x^2 + 3x + 5$　　**49.** $y = x^2 - 5x - 3$

**50.** $y = 3x^2 + 27x + 12$　**51.** $y = 2x^2 - 6x - 7$

**52.** $y = 4x^2 + 28x - 21$

**53.** $y = 17x^2 - 51x - 19$

**54.** $y = 2x^2 - 3x + 8$    **55.** $y = 3x^2 + 5x - 3$

**56.** $y = 4x^2 + 2x + 9$    **57.** $y = 3x^2 - x - 4$

In Exercises 58–67, write the function in perfect-square form and describe how the graph can be obtained from $y = x^2$.

**58.** $y = 2x^2 + 4x - 7$    **59.** $y = 2x^2 - 8x + 7$

**60.** $y = 3x^2 - 9x + 12$    **61.** $y = 5x^2 + 25x - 14$

**62.** $y = -3x^2 + 9x - 17$

**63.** $y = -2x^2 + 6x - 8$

**64.** $y = 2x^2 + 5x - 3$    **65.** $y = 2x^2 - 7x + 11$

**66.** $y = \frac{1}{2}x^2 + 4x - 20$    **67.** $y = \frac{1}{3}x^2 + 3x + 24$

630 ft

For Exercise 68.

**68.** *Free Fall.* When an object is dropped at $t = 0$ from a height of $k$ feet, its distance above the ground (ignoring air resistance) after $t$ seconds of fall is $d = -16t^2 + k$.

a) Suppose a ball is dropped from the top of the St. Louis arch (630 ft high). Write an algebraic representation for the distance above the ground after $t$ seconds.

b) Find a complete graph of the problem situation from the time the ball is dropped until it hits the ground.

c) Use a graphical method to find the time it takes for the ball to hit the ground.

d) Describe how the graph of this problem situation can be obtained from the graph of $y = x^2$.

e) If a first ball is dropped at $t = 0$ and a second ball is dropped 1 sec later at $t = 1$, is the algebraic representation of the second ball $y = -16(t + 1)^2 + 630$ or $y = -16(t - 1)^2 + 630$? How far apart are the two balls 1 sec after the second ball is dropped?

## EXTENDING THE IDEAS

**69.** Describe the conditions on constant $a$ to guarantee that the graph of $y = ax^2$ can be obtained from the graph of $y = x^2$ by a vertical stretch.

**70.** Describe the conditions on constant $a$ to guarantee that the graph of $y = ax^2$ can be obtained from the graph of $y = x^2$ by a reflection in the $x$-axis.

**71.** Describe the conditions on constant $k$ to guarantee that the graph of $y = 3(x-1)^2 + k$ can be obtained from the graph of $y = x^2$ by a sequence of transformations that include a slide up.

**72.** Describe the conditions on constant $k$ to guarantee that the graph of $y = 2x^2 + 4x + k$ can be obtained from the graph of $y = x^2$ by a sequence of transformations that include a slide up 3 units.

**73.** For what value of $k$ does the graph of $y = 2x^2 + 3x + k$ pass through the point $(1, 8)$.

**74.** For what value of $a$ does the graph of $y = ax^2 - 3$ pass through the point $(2, 5)$.

**75. Writing to Learn.** Write a paragraph that explains how you know that the graph of $y = (3x - 2)(x + 5)$ can be obtained by applying slides to the graph of $y = 3x^2$.

## LOOKING BACK—MAKING CONNECTIONS

**76.** *Volume of a Container.* A box with volume of $230\ cm^3$ is 5 cm high and 6 cm longer than it is wide (see the following figure).

a) Write an algebraic representation for the volume of this box.

b) Find the width of the box.

For Exercise 76.

**77.** If the graph of $y = 4(x - 3)(x + 2)$ is obtained by applying transformations to the graph of $y = x^2$, is a vertical stretch factor of 2, 3, or 4 needed?

**78.** If the graph of $y = 2(3x - 2)(x + 5)$ is obtained by applying transformations to the graph of $y = x^2$, is a vertical stretch factor of 2, 3, or 6 needed?

**79.** If the graphs of the following equations are each obtained from the graph of $y = x^2$, which one(s) require a reflection in the $x$-axis: (a) $y = 3(x - 7)$ $(x + 2)$, (b) $y = 2(4 - x)(7 - x)$, and (c) $y = 6(x - 4)(3 - x)$?

**80.** Nolan can throw a ball with a velocity of 120 ft/sec. The Statue of Liberty is 305 ft high. Use a graphical method to predict if Nolan can throw a ball over the Statue of Liberty. If not, what velocity is needed?

---

## 8.6 Vertex and Symmetry of a Parabola

Locating the Vertex of a Parabola ▪ Analyzing the General Quadratic Function ▪ Revisiting the Baseball and Garden Problems

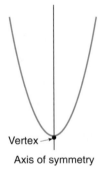

Vertex →

Axis of symmetry

**FIGURE 8.41** Each parabola has a vertex and an axis of symmetry.

In section 8.5 we saw that a quadratic function of the form $f(x) = ax^2 + bx + c$ could be rewritten in perfect-square form $f(x) = a(x \pm h)^2 \pm k$, where $h$ and $k$ are real numbers.

Furthermore, the graph of a quadratic function in perfect-square form can be obtained from the graph of $y = x^2$ by some combination of horizontal and vertical slides, a vertical stretch or shrink, and possibly a reflection in the $x$-axis. The graph of $y = x^2$ is a parabola and each transformation preserves that property of being a parabola.

Figure 8.41 shows the vertex and the axis of symmetry for a parabola. Locating the vertex of a parabola identifies the position of its graph.

### Locating the Vertex of a Parabola

We begin by introducing a form of a quadratic equation known as the **vertex form**. At first glance, vertex form might appear to be identical to perfect-square form. They are similar but not necessarily identical.

---

**DEFINITION 8.4**    Vertex Form of a Quadratic Equation

A quadratic function $f(x) = ax^2 + bx + c$ is written in **vertex form** if real numbers $h$ and $k$ are found such that

$$f(x) = a(x - h)^2 + k.$$

---

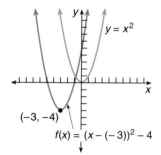

$y = x^2$

$(-3, -4)$

$f(x) = (x - (-3))^2 - 4$

**FIGURE 8.42**
Graphs of $y = x^2$ and $f(x) = (x - (-3))^2 - 4$.

Notice that the quadratic function $f(x) = (x+3)^2 - 4$ is in perfect-square form but it is not in vertex form. For vertex form, a $-$ sign is needed in the parentheses and a $+$ sign is needed before the constant. Here is the comparison.

$$\underbrace{f(x) = (x + 3)^2 - 4}_{\uparrow} \qquad \underbrace{f(x) = \left(x - (-3)\right)^2 + (-4)}_{\uparrow}$$

Perfect-square form          Vertex form

The graph of $f$ in Figure 8.42 is obtained from the graph of $y = x^2$ by a slide left 3 units and a slide down 4 units. Doing this places the vertex of the graph at the point $(-3, -4)$. Notice the correlation between the coordinates of the vertex, the line of symmetry, and the constants in vertex form. We summarize the relationship next.

---

## Vertex and Line of Symmetry of a Parabola

The point $(h, k)$ is the **vertex** of the parabola, which is the graph of

$$f(x) = a(x - h)^2 + k.$$

The vertical line $x = h$ is the **line of symmetry** of the parabola.

---

**EXAMPLE 1**    Finding the Vertex of a Parabola

Find the vertex and the equation of the line of symmetry of each of the following parabolas. Support part (c) graphically.

a) $f(x) = (x + 8.3)^2 - 12$

b) $g(x) = 3(x - \pi)^2 + \sqrt{2}$

c) $h(x) = 2x^2 - 32x + 7$

SOLUTION
**Solve Algebraically.**

a) Rewrite $f(x)$ in vertex form $f(x) = (x - (-8.3))^2 + (-12)$. The vertex of $y = f(x)$ is $(-8.3, -12)$ and the line of symmetry equation is $x = -8.3$.

b) This is already given in vertex form. The vertex of $y = g(x)$ is $(\pi, \sqrt{2})$ and the line of symmetry equation is $x = \pi$.

c) First complete the square.

$$h(x) = 2x^2 - 32x + 7 = 2(x^2 - 16x + ?) + (7 - 2 \times ?)$$

$$= 2(x^2 - 16x + 64) + (7 - 2(64))$$

$$= 2(x - 8)^2 + (-121)$$

The vertex of $y = h(x)$ is $(8, -121)$ and the line of symmetry equation is $x = 8$.

**Support Graphically.** Figure 8.43 shows the graph of $y = 2x^2 - 32x + 7$ in the window $[-3, 18]$ by $[-130, 50]$. ∎

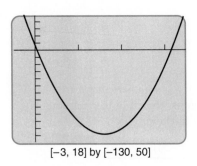

[−3, 18] by [−130, 50]

**FIGURE 8.43** Graph of $y = 2x^2 - 32x + 7$ in Example 1.

## Analyzing the General Quadratic Function
## $f(x) = ax^2 + bx + c$

To find the vertex and the line of symmetry of a specific quadratic function, you can always complete the square and write the function in vertex form. However, rather than repeat this algebra for each specific quadratic function, we shall perform this algebra for the general quadratic function $f(x) = ax^2 + bx + c$. The results of this general solution can then be applied to each specific case.

$$f(x) = ax^2 + bx + c = a\left(x^2 + \frac{b}{a}x + ?\right) + (c - a \times ?)$$

$$= a\left(x^2 + \frac{b}{a}x + \left(\frac{b}{2a}\right)^2\right) + \left(c - a \times \left(\frac{b}{2a}\right)^2\right)$$

$$= a\left(x + \frac{b}{2a}\right)^2 + \left(c - \frac{b^2}{4a}\right)$$

$$= a\left(x + \frac{b}{2a}\right)^2 + \left(\frac{4ac - b^2}{4a}\right)$$

So if $f(x) = ax^2 + bx + c$ is written in vertex form $f(x) = a(x - h)^2 + k$, we see that

$$h = -\frac{b}{2a} \quad \text{and} \quad k = \frac{4ac - b^2}{4a}.$$

There is a second way to characterize the constant $k$. To see this, we use the form

$$f(x) = a\left(x + \frac{b}{2a}\right)^2 + \left(\frac{4ac - b^2}{4a}\right)$$

and evaluate $f\left(-\frac{b}{2a}\right)$.

$$f\left(-\frac{b}{2a}\right) = a\left(-\frac{b}{2a} + \frac{b}{2a}\right)^2 + \left(\frac{4ac - b^2}{4a}\right)$$

$$= 0 + \left(\frac{4ac - b^2}{4a}\right)$$

Rather than trying to remember the expression $\frac{4ac - b^2}{4a}$ for $k$, it is probably easier to remember that

$$k = f\left(-\frac{b}{2a}\right).$$

---

**THEOREM 8.1**   Vertex and Line of Symmetry for the General Quadratic Function $f(x) = ax^2 + bx + c$

The quadratic function $f(x) = ax^2 + bx + c$ is called the **general quadratic function**. Its vertex and line of symmetry equation are as follows:

**Vertex:** $\left(-\frac{b}{2a}, \frac{4ac - b^2}{4a}\right)$   or   $\left(-\frac{b}{2a}, f\left(-\frac{b}{2a}\right)\right)$

**Line of Symmetry:** $x = -\frac{b}{2a}$

---

In Example 1, we found the vertex of the parabola by completing the square of the given quadratic function. In Example 2, we shall use Theorem 8.1 to find the vertex of the parabola.

**EXAMPLE 2**   Finding the Vertex and Line of Symmetry

Find the line of symmetry and the vertex of the quadratic function

$$f(x) = 7x^2 - 12x + 17.$$

Provide graphical support.

**SOLUTION**

**Solve Algebraically.** For $f(x) = 7x^2 - 12x + 17$, $a = 7$, $b = -12$, and $c = 17$. Therefore the line of symmetry equation is

$$x = -\frac{b}{2a} = -\frac{-12}{14} = \frac{6}{7}.$$

The vertex is found using three methods.

a) Evaluate $f\left(\dfrac{6}{7}\right)$.

$$f\left(\frac{6}{7}\right) = 7\left(\frac{6}{7}\right)^2 - 12\left(\frac{6}{7}\right) + 17$$

$$= \frac{36}{7} - \frac{72}{7} + 17$$

$$= -\frac{36}{7} + 17 = -5\frac{1}{7} + 17$$

$$= 11\frac{6}{7}$$

The vertex of the parabola is the point $\left(\dfrac{6}{7}, 11\dfrac{6}{7}\right)$.

b) Evaluate $\dfrac{4ac - b^2}{4a}$.

$$\frac{4ac - b^2}{4a} = \frac{4(7)(17) - (-12)^2}{4(7)}$$

$$= \frac{476 - 144}{28} = \frac{83}{7} = 11\frac{6}{7}$$

c) The calculation in part (b) was completed using a paper-and-pencil method, but it also could have been done on a grapher. The grapher display shown in Figure 8.44 shows the grapher calculation. The vertex of the parabola is (0.857, 11.857).

**Support Graphically.** Figure 8.45 shows the graph of $f(x) = 7x^2 - 12x + 17$ in the window $[-5, 5]$ by $[-5, 50]$. Use TRACE to support that the coordinates of the vertex are (0.857, 11.857).    ■

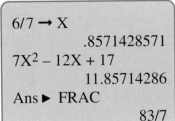

6/7 → X
                .8571428571
7X² – 12X + 17
                11.85714286
Ans ► FRAC
                        83/7

**FIGURE 8.44** A grapher computation of the vertex for Example 2. Some graphers have the FRAC command that allows the output to be represented in fraction form.

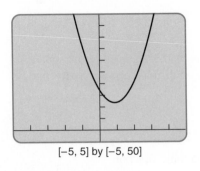

$[-5, 5]$ by $[-5, 50]$

**FIGURE 8.45**
A graph of $f(x) = 7x^2 - 12x + 17$ for Example 2.

## Revisiting the Baseball and Garden Problems

In the **Baseball Problem Situation,** a ball is thrown straight up into the air with an initial velocity of 88 ft/sec (see Figure 8.46). The algebraic representation of

the problem situation is

$$h = -16t^2 + 88t.$$

Because this is a quadratic function, we know its graph is a parabola. The maximum height of the ball can be found with a grapher using the techniques discussed in Section 2.3. Algebraic solution methods are used in Example 3.

**FIGURE 8.46** A baseball is thrown straight up into the air in Example 3.

**EXAMPLE 3**    Revisiting the Baseball Problem

Find the maximum height that a ball reaches if it is thrown straight up into the air with an initial velocity of 88 ft/sec.

SOLUTION

The height of the ball above the ground depends upon time and is given by $h = -16t^2 + 88t$. Figure 8.47 (a) shows this (time, height) curve as a parabola $y = ax^2 + bx + c$, with $a = -16$, $b = 88$, and $c = 0$. We use Theorem 8.1 to calculate the coordinates of the vertex of this parabola. Figure 8.47 (b) shows that the grapher can be used to evaluate $-\dfrac{b}{2a}$ to be 2.75 and then find the corresponding $y$-coordinate, $f(2.75)$.

$$\left(-\frac{b}{2a}, f\left(-\frac{b}{2a}\right)\right) = \left(-\frac{88}{2(-16)}, f(2.75)\right)$$

$$= (2.75, 121)$$

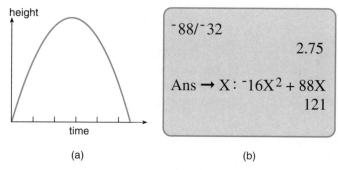

(a)　　　　　　　　　　(b)

**FIGURE 8.47**  (a) A graph of $y = -16t^2 + 88t$ and (b) evaluating the coordinates of the vertex using a grapher.

The vertex of this parabola is (2.75, 121), so the maximum height of the ball is 121 ft. ∎

### EXAMPLE 4   THE GARDEN PROBLEM SITUATION: Finding Maximum Area and Dimensions

The algebraic representation of the **Garden Problem** in Section 2.3 is

$$A = w(24 - 2w) = 24w - 2w^2.$$

Find the maximum area of the garden and the dimensions of that maximum area.

SOLUTION

The coordinates of the vertex of the parabola $y = 24x - 2x^2$ provide the solution. In this equation $a = -2, b = 24$, and $c = 0$. Figure 8.48 shows a graph of this parabola. We use Theorem 8.1 to calculate the coordinates of the vertex of this parabola by first finding $-\dfrac{b}{2a}$ to be 6.

$$\left(-\frac{b}{2a}, f\left(-\frac{b}{2a}\right)\right) = \left(-\frac{24}{2(-2)}, f(6)\right)$$

$$= (6, 72)$$

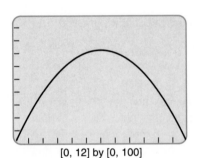

[0, 12] by [0, 100]

**FIGURE 8.48** A graph of $y = 24x - 2x^2$ in Example 4.

The vertex of this parabola is (6, 72). The maximum area of the garden is 72 sq ft and occurs when the width is 6 ft and the length is 12 ft. ∎

### EXAMPLE 5   Application: Determining the Profit Curve

The economists of a large corporation collected production cost data and determined that the profit $P(x)$ for producing $x$ items (in thousands) could be represented by the quadratic function $P(x) = -0.02x^2 + 5.23x + 14.3$. How many items should be produced to obtain a maximum profit?

SOLUTION

The quadratic function $P(x) = -0.02x^2 + 5.23x + 14.3$ has coefficients $a = -0.02, b = 5.23$, and $c = 14.3$. The maximum value of this function is the $y$-coordinate of the vertex of the parabolic graph. Because the vertex falls on the line of symmetry, it is sufficient to find the line's $x$-coordinate. The line of symmetry equation is

$$x = -\frac{b}{2a} = -\frac{5.23}{-0.04} = 130.75.$$

Figure 8.49(a) shows a graph of this quadratic function, while Figure 8.49(b) shows grapher calculations that illustrate that the vertex of this parabola is (130.75, 356.211). To maximize profit, the company should produce 130,750 items. ∎

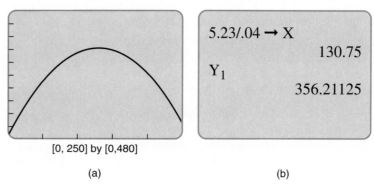

**FIGURE 8.49** (a) A graph of $y_1 = -0.02x^2 + 5.23x + 14.3$ and (b) the grapher computation.

## Exercises for Section 8.6

In Exercises 1–6, determine whether the equation is written in vertex form.

**1.** $y = 3(x - 2)^2 + 5$     **2.** $y = 2(x + 4)^2 + 3$

**3.** $y = 4x^2 + 2$     **4.** $y = 2x^2 + 5x - 1$

**5.** $y = 3x^2 - 4x + 2$     **6.** $y = -4(x + 2)^2 + 7$

In Exercises 7–16, identify the vertex and line of symmetry for the parabola.

**7.** $y = 3(x - 2)^2 + 4$     **8.** $y = 2(x + 1)^2 - 2$

**9.** $y = -4(x + 2)^2 - 5$     **10.** $y = -2(x - 3) + 2$

**11.** $y = 17 - 3(x + 2)^2$     **12.** $y = 5 - (x + 3)^2$

**13.** $y = 2x^2 + 8x - 2$     **14.** $y = 3x^2 - 9x + 18$

**15.** $y = 2(x + 2\pi)^2 - \sqrt{3}$

**16.** $y = \sqrt{7} - 4(x - 7)^2$

In Exercises 17 and 18, write equations for two distinct parabolas with the given vertex.

**17.** $(1, 2)$

**18.** $(-3, 1)$

In Exercises 19–24, use Theorem 8.1 to find the vertex and line of symmetry for the quadratic function without completing the square.

**19.** $f(x) = 3x^2 - 4x + 5$

**20.** $g(x) = 2x^2 + 7x - 3$

**21.** $h(x) = -4x^2 - 11x + 2$

**22.** $f(x) = 12x^2 - 8x + 24$

**23.** $g(x) = 7x^2 - 5x + 15$

**24.** $h(x) = -3x^2 + 6x - 2$

**25.** *Projectile Motion.* A baseball is thrown straight up into the air with an initial velocity of 88 ft/sec. Recall that the height above ground level of the ball is given by the algebraic representation $h = -16t^2 + 88t$.

a) Find a complete graph of this problem situation. What viewing window do you use?

b) What is the height of the ball 2 seconds after it was thrown into the air?

c) What is the maximum height of the ball? How

long does it take for the ball to reach that height?

d) What is the relationship between the time it takes the ball to reach its maximum height and the time it takes the ball to reach the ground. How long does it take for it to hit the ground?

**26.** *Projectile Motion.* When a ball is thrown straight up into the air with an initial velocity of $v_0$ ft/sec, $h = -16t^2 + v_0t$ is an algebraic representation that describes the height of the ball $t$ seconds after the ball is thrown.

a) Find an algebraic representation for a ball thrown with an initial velocity of 110 ft/sec.

b) What is the vertex of the parabolic graph of this algebraic representation.

c) What is the maximum height of the ball?

**27.** *Projectile Motion.* Suppose a baseball is thrown straight up into the air from the top of a 120-ft tower with an initial velocity of 88 ft/sec.

a) The following graph represents the problem situation of throwing a ball from the tower straight up with an initial velocity of 88 ft/sec.

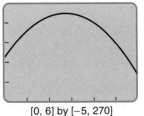

[0, 6] by [−5, 270]

For Exercise 27.

How can a graph that represents the ball thrown from the ground be obtained from the graph in this figure?

b) How does the maximum height for the ball thrown from the 120-ft tower compare to the maximum height of the ball thrown straight up into the air from ground level?

c) Write an algebraic representation for the problem situation of the ball thrown from the tower.

**28.** *Maximizing Profit.* Suppose the profit $(P)$ for producing $x$ items (in thousands) is modeled by the profit function $P(x) = -0.05x^2 + 7.25x + 15.5$. How many items should be produced to maximize profits?

**29. Writing to Learn.** Write a paragraph that explains how the transformation of "slide up" relates to throwing a ball straight up into the air with an initial velocity of 88 ft/sec (a) from ground level and (b) from on top of a 120-ft tower.

**30. Writing to Learn.** Write a paragraph that explains how the transformation of "slide right" relates to throwing two balls straight up into the air from ground level with an initial velocity of 88 ft/sec from ground level one when $t = 0$ and the other one second later.

**31. Writing to Learn.** Write a paragraph that explains how the phrases "opens up" and "opens down" relate to the graphs of quadratic functions and to the form of their equations.

TRANSLATING WORDS TO SYMBOLS

**32.** Write an equation of the quadratic function in standard form whose graph has vertex of $(3, 2)$ and whose vertical stretch factor is 5.

**33.** The graph $G$ of a quadratic function whose vertical stretch factor is 2 has vertex $(-2, -4)$. Suppose $G$ slides three units right and 4 units down. Write an equation of this quadratic function.

EXTENDING THE IDEAS

**34.** Find the value of $a$ if $y = a(x - h)^2 + k$ has vertex $(1, 3)$ and passes through the point $(2, 5)$.

**35.** Find the value of $a$ if $y = a(x - h)^2 + k$ has vertex $(-2, 1)$ and passes through the point $(-1, -2)$.

**36.** Find an equation of a parabola whose line of symmetry is $x = -4$ and whose graph can be obtained from the graph of $y = x^2$ by a horizontal slide and a vertical stretch with a factor of 2.

**37.** Suppose $a$ and $b$ are any real numbers. Find the equation for the line of symmetry of the quadratic equation $y = (x - a)(x - b)$.

LOOKING BACK—MAKING CONNECTIONS

**38. Writing to Learn.** If $a$, $h$, and $k$ are any real numbers, explain how the graphs of $y = a(x - h)^2 - k$ and $y = -a(h - x)^2 + k$ are related.

**39. Writing to Learn.** Explain how Theorem 8.1 can be used to help determine a viewing window that will show a complete graph of $y = 4x^2 - 123x + 150$.

## Chapter 8 Summary

| | Quadratic Equations | Examples |
|---|---|---|
| Quadratic Equation in $x$ in Standard Form | $ax^2 + bx + c = 0$, where $a$, $b$, and $c$ are real numbers with $a \neq 0$ | $-x^2 - 3x + \sqrt{2} = 0$ |
| Discriminant of $ax^2 + bx + c = 0$ | The expression $b^2 - 4ac$ | The discriminant of $6x^2 - 5x + 2 = 0$ is $(-5)^2 - 4(6)(2) = -23$. |
| Using the Discriminant to Determine Number of Solutions | 1. If $b^2 - 4ac > 0$, then there are two real solutions. <br> 2. If $b^2 - 4ac = 0$, then there is one real solution. <br> 3. If $b^2 - 4ac < 0$, then there are no real solutions, but there are two non-real complex solutions. | The equation $6x^2 - 5x + 2 = 0$ has no real solutions, since its discriminant is $-23$. |

| | Algebraic Methods of Solving Quadratic Equations | Examples |
|---|---|---|
| Square Root Property | If $a > 0$, then $x^2 = a$ has the two solutions $x = \sqrt{a}$ or $x = -\sqrt{a}$. | $x^2 = 3$ has solutions $x = \sqrt{3}$ and $x = -\sqrt{3}$. |

| | Algebraic Methods of Solving Quadratic Equations | Examples |
|---|---|---|
| Completing the Square for $ax^2 + bx + c = 0$ | 1. Rewrite so that only $ax^2$ and $bx$ are on the left-hand side. If $a \neq 1$, divide all terms by $a$ so that $a = 1$.<br>2. Add the square of $\dfrac{b}{2}$ to each side.<br>3. Write the left-hand side as a square of a binomial and simplify the right-hand side.<br>4. Use the Square Root Property and then solve for $x$. | Solve $x^2 - 6x = 112$:<br>$$x^2 - 6x + 9 = 112 + 9$$<br>$$(x - 3)^2 = 121$$<br>$$x - 3 = \pm 11$$<br>$$x = 3 \pm 11$$<br>$$x = 14 \text{ or } x = -8$$ |
| Quadratic Formula | The solutions to the equation $ax^2 + bx + c = 0$, where $a \neq 0$ are<br>$$x = \frac{-b \pm \sqrt{b^2 - 4ac}}{2a}.$$ | Solve $2x^2 + 3x - 5 = 0$:<br>$$x = \frac{-3 \pm \sqrt{3^2 - 4(2)(-5)}}{2(2)}$$<br>$$= \frac{-3 \pm \sqrt{49}}{4}$$<br>$$= \frac{-3 \pm 7}{4}$$<br>$$x = -\tfrac{5}{2} \text{ or } x = 1$$ |

| | Quadratic Functions and Parabolas | Examples |
|---|---|---|
| Quadratic Function | $f(x) = ax^2 + bx + c$, where $a$, $b$, and $c$ are real numbers $a \neq 0$ | $f(x) = -5x^2 + 4x - 1$ |
| Vertex | The point $(h, k)$ of the parabola, which is the graph of $f(x) = a(x - h)^2 + k$<br><br>The point $\left( -\dfrac{b}{2a}, f\left(\dfrac{-b}{2a}\right) \right)$ of the parabola, which is the graph of $f(x) = ax^2 + bx + c$ | The vertex of $y = 2(x - 4)^2 + (-1)$ is $(4, -1)$. |
| Line of Symmetry | The line $x = h$ of the parabola, which is the graph of $f(x) = a(x - h)^2 + k$<br><br>The line $x = -b/2a$ of the parabola, which is the graph of $f(x) = ax^2 + bx + c$ | The line of symmetry of $y = -2x^2 + 5x - 8$ is $x = \dfrac{5}{4}$. |

Vertex Form of a Quadratic Function

A quadratic function $f(x) = ax^2 + bx + c$ written in the form $f(x) = a(x-h)^2 + k$, where $h$ and $k$ are real numbers

$f(x) = (x-(-3))^2 + 2$ is the vertex form of $f(x) = x^2 + 6x + 11$.

---

**A Graphical Solution to a Quadratic Inequality**

**Examples**

Solving a Quadratic Inequality Graphically

To solve an inequality like $x^2 - x - 2 > 0$, find a complete graph of $y = x^2 - x - 2$ and find the $x$-intercepts of the graph.

   The solution to the inequality consists of all $x$-coordinates of points that lie above the $x$-axis (or below, if the inequality sign is $<$). If $\leq$ or $\geq$ is used, the $x$-intercepts are included in the solution.

The $x$-intercepts of a graph of $y = x^2 - x - 2$ are $x = -1$ and $x = 2$.

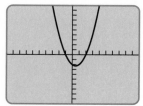

The graph of $y = x^2 - x - 2$ in the Standard window.

The solution to the inequality is $(-\infty, -1) \cup (2, \infty)$.

---

**Transformations of Graphs of Quadratic Functions**

**Examples**

Horizontal Slide by Distance $h$

A horizontal slide of the graph of $y = x^2$ results in

1. $y = (x - h)^2$ if the slide is $h$ units right, and
2. $y = (x + h)^2$ if the slide is $h$ units left.

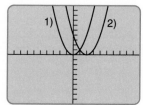

Graph (2) is a graph of $y = (x-2)^2$ which can be obtained by sliding a graph of $y = x^2$ two units right.

| | Transformations of Graphs of Quadratic Functions | Examples |
|---|---|---|

**Vertical Slide by Distance $k$**

A vertical slide of the graph of $y = x^2$ results in

1. $y = x^2 + k$ if the slide is $k$ units upward, and
2. $y = x^2 - k$ if the slide is $k$ units downward.

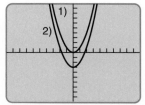

Graph (2) is a graph of $y = x^2 - 3$, which can be obtained by sliding a graph of $y = x^2$ three units down.

**Vertical Stretch/Shrink by a Factor of $a$**

A vertical stretch or shrink of the graph of $y = x^2$ results in $y = ax^2$.

1. If $a > 1$, it is a vertical stretch with a stretch factor of $a$.
2. If $0 < a < 1$, it is a vertical shrink with a shrink factor of $a$.

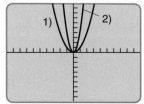

Graph (2) is the graph of $y = 4x^2$, which can be obtained by stretching a graph of $y = x^2$ by a stretch factor of 4.

**Reflection in the $x$-axis**

A reflection in the $x$-axis of the graph $y = x^2$ results in $y = -x^2$.

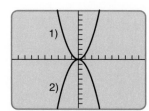

Graph (2) is a graph of $y = -x^2$, which can be obtained by reflecting a graph of $y = x^2$ in the $x$-axis.

## Review Exercises for Chapter 8

In Exercises 1 and 2, solve the equation by using the Square Root Property.

**1.** $3x^2 + 48 = 0$    **2.** $(x + 8)^2 = 35$

In Exercises 3–6, solve the equation by completing the square.

**3.** $x^2 - 6x + 7 = 0$    **4.** $x^2 + x - 3 = 0$

**5.** $-x^2 - 5x + 1 = 0$      **6.** $4x^2 - 9x + 5 = 0$

In Exercises 7–10, use the Quadratic Formula to find the solutions to the equation.

**7.** $4x^2 - 7x - 8 = 0$      **8.** $-x^2 + 7x + 5 = 0$

**9.** $2x^2 - x = 10x^2 - 2x - 4$

**10.** $3x = \dfrac{x - 2}{x + 3}$

In Exercises 11 and 12, use a graphical method to solve the inequality. Find the endpoints of solution intervals with an error of at most 0.01.

**11.** $-x^2 + 3x - 1 < 0$

**12.** $x^2 - 3x > 2x^2 - 5x - 4$

In Exercises 13 and 14, use the discriminant to determine the number of real-number solutions to the equation. Also provide graphical support.

**13.** $0.25x^2 + 4x + 16 = 0$

**14.** $-x^2 + 6 = 10x$

In Exercises 15 and 16, find a quadratic equation whose solutions are the given values of $x_1$ and $x_2$.

**15.** $x_1 = -\dfrac{3}{2}$ and $x_2 = \dfrac{1}{3}$

**16.** $x_1 = 2\sqrt{6}$ and $x_2 = -\sqrt{6}$

**17.** Choose from the following the one grapher window that shows a complete graph of $f(x) = 3x^2 + 32x - 14$:

a) $[-10, 10]$ by $[-10, 10]$

b) $[-15, 5]$ by $[-50, 10]$

c) $[-15, 5]$ by $[-100, 20]$

**18.** Find a complete graph of $y = -x^2 - 18x + 9$.

In Exercises 19 and 20, write an algebraic representation $y = f(x)$ for the function that is obtained from $y = x^2$ by the sequence of transformations described in each exercise.

**19.** a) A vertical stretch with a stretch factor of 3, followed by

b) A horizontal slide left 2 units, followed by

c) A vertical slide up 10 units.

**20.** a) A reflection in the $x$-axis, followed by

b) A horizontal slide right 6 units, followed by

c) A vertical slide down 3 units.

In Exercises 21–23, describe how the graph of the function can be obtained from the graph of $y = x^2$. Support your answer with a grapher.

**21.** $y = \frac{1}{5}(x - 2)^2 + 8$      **22.** $y = -3x^2 + 7$

**23.** $y = -2(x + 4)^2$

In Exercises 24–26, complete the square for the quadratic function and determine how its graph can be obtained from the graph of $y = x^2$.

**24.** $y = x^2 + 14x + 32$      **25.** $y = -x^2 - 2x + 4$

**26.** $y = 3x^2 - 12x + 4$

In Exercises 27–30, sketch the graph of the function without using a grapher. Then support your sketch with a grapher.

**27.** $y = 4x^2 - 3$      **28.** $y = 2(x - 5)^2 + 1$

**29.** $y = -0.5(x + 4)^2 - 2$

**30.** $y = -(x + 2)^2$

In Exercises 31 and 32, identify the vertex and line of symmetry for the parabola.

**31.** $y = 2(x - 1)^2 - 6$      **32.** $y = -(x + 3)^2 + 4$

In Exercises 33 and 34, find the vertex and line of symmetry for the parabola without completing the square. Use Theorem 8.1.

**33.** $y = 2x^2 + 9x - 5$      **34.** $y = x^2 + 6x + 3$

**35.** Assume $a$ is a positive real number. Solve the equation $x^2 - 4x = 9a$ for $x$.

**36.** What are the conditions on constant $a$ if the graph of $y = ax^2$ can be obtained from the graph of $y = x^2$ by a vertical shrink?

**37.** Find the value(s) of $a$ so that $ax^2 - 4x - 6 = 0$ has only one real solution.

**38.** The product of two numbers is 405 and one number is 4.5 greater than the other. Find the two numbers.

**39.** The length of a rectangular garden is 3 ft longer than the garden's width. The area of the garden is 208 sq ft. Determine the garden's dimensions.

**40.** A ball is thrown straight up into the air at an initial velocity of 70 ft/sec. The algebraic representation of the height $h$ of this ball at $t$ seconds is $h(t) = -16t^2 + 70t$.

a) Determine algebraically the maximum height of the ball. Support your answer graphically.

b) Determine when the maximum height of the ball is reached.

## Chapter 8 Test

**1.** Use the Square Root Property to solve $(x - 4)^2 = 10$.

**2.** Solve $2x^2 + x - 15 = 0$ by completing the square.

**3.** Use the Quadratic Formula to find the solutions to $-2x^2 + 2x - 5 = 0$.

**4.** Find the value of the discriminant of the following equation and determine the number of real solutions it has:

$$8x^2 - 3x + 2 = 0$$

**5.** Find a quadratic equation whose solutions are $x_1 = \dfrac{5}{4}$ and $x_2 = -1$.

**6.** Choose from the following the one grapher window that shows a complete graph of $f(x) = -0.2x^2 - 3x + 15$:

a) $[-10, 10]$ by $[-50, 50]$

b) $[-25, 15]$ by $[-30, 30]$

c) $[-20, 20]$ by $[-20, 20]$

**7.** Use a graphical method to solve $-2x^2 - 3x + 15 > 0$. Find the endpoints of the intervals with an error of at most 0.01.

**8.** Determine how the graph of $y = 4(x + 3)^2 - 5$ can be obtained from the graph of $y = x^2$.

**9.** Complete the square for $y = -x^2 + 2x + 2$ and determine how its graph can be obtained from the graph of $y = x^2$.

**10.** Write an algebraic representation of the graph obtained from the graph of $y = x^2$ as a result of the following changes:

a) A vertical shrink by a factor of $\dfrac{1}{3}$, followed by

b) A horizontal slide right 4 units, followed by

c) A vertical shift down 2 units.

**11.** Find the value(s) of $c$ so that $-2x^2 + 7x + c = 0$ has only one solution.

**12.** Sketch the graph of $y = -(x + 3)^2 - 6$ without using a grapher. Describe a sequence of transformations that change the graph of $y = x^2$ to the graph of this function.

**13.** Find the vertex and line of symmetry for $y = 3(x - 2)^2 - 5$.

**14.** Find the vertex and line of symmetry for $y = 5x^2 + 8x - 4$.

**15.** The sum of the square of a number and its double is 19.25. Find two numbers that satisfy this condition.

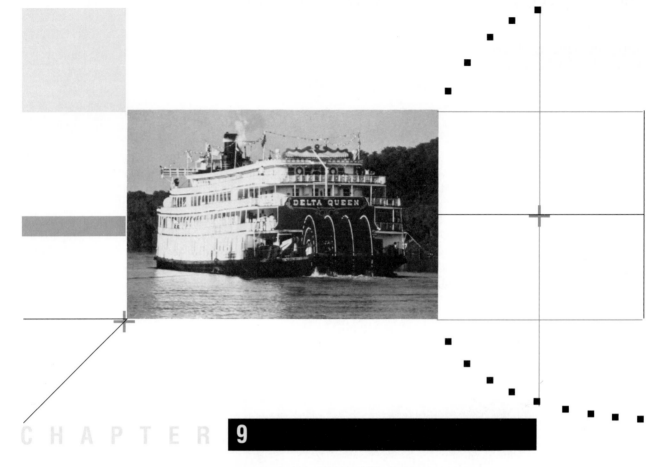

# Higher-Order Systems of Equations and Matrices

## AN APPLICATION

A riverboat takes less time to go 5 miles downstream than it does to go 5 miles upstream. Knowing the time of each part of the trip, you can find both the speed of the boat in still water and the speed of the river current by finding the product

$$A^{-1}B$$

where both $A$ and $B$ represent matrices.

# Solving Systems of Equations in Three Variables

Visualizing a Solution to a System  ▪  Solving a System by the
Elimination Method  ▪  Problem Situations Using Systems of Equations

In Sections 3.5 and 3.6, we saw that a particular solution to a system of two linear equations in two variables is an ordered pair $(x, y)$ of real numbers and that the solution can be graphed as a point in a rectangular coordinate plane.

This chapter focuses on systems of equations with *three* variables. A particular solution to such a system is an ordered triple $(x, y, z)$, which can be graphed as a point in a three-dimensional, rectangular coordinate system. Figure 9.1 shows point $(6, 8, 4)$ in such a system. It is 6 units from the $yz$-plane, 8 units from the $xz$-plane, and 4 units from the $xy$-plane.

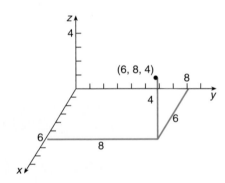

**FIGURE 9.1**  A three-dimensional, rectangular coordinate system.

We define linear equations in three variables next.

---

**DEFINITION 9.1**    A Linear Equation in Three Variables

If $A, B, C$, and $D$ are any real numbers with at least one of $A, B$, or $C$ not zero, then

$$Ax + By + Cz = D$$

is called a linear equation in the three variables $x, y$, and $z$.

---

Example 1 illustrates how to use paper and pencil to confirm numerically that a particular ordered triple is a solution to a system of equations.

**EXAMPLE 1**   Confirm a Solution Algebraically

Confirm that the ordered triple $(5, -2, 1)$ is a solution to the following system of equations:

$$x + \phantom{3}y + 3z = 6$$
$$-3x - \phantom{5}y + 5z = -8$$
$$x + 6y + 4z = -3$$

Support numerically using a grapher.

SOLUTION

**Confirm Algebraically.**  We substitute $x = 5$, $y = -2$, and $z = 1$ into each of the three equations.

$$x + y + 3z = 6 \quad \textbf{Substitute } x = 5, y = -2, \text{ and } z = 1.$$

$$5 + (-2) + 3(1) = 3 + 3$$

$$= 6$$

$$-3x - y + 5z = -8$$

$$-3(5) - (-2) + 5(1) = -15 + 2 + 5$$

$$= -8$$

$$x + 6y + 4z = -3$$

$$5 + 6(-2) + 4(1) = 5 - 12 + 4$$

$$= -3$$

We see that $(5, -2, 1)$ is a solution to the system.

**Support Numerically.**  Figure 9.2 shows grapher support that $(5, -2, 1)$ is a solution to the system. ■

$5 \rightarrow X: \,^-2 \rightarrow Y: 1 \rightarrow Z:$

$X + Y + 3Z$

$\phantom{XXXXXXXXXX} 6$

$^-3X - Y + 5Z$

$\phantom{XXXXXXXXX} ^-8$

$X + 6Y + 4Z$

$\phantom{XXXXXXXXX} ^-3$

**FIGURE 9.2**  Numerical support of the solution to Example 1.

## Visualizing a Solution to a System

The set of all solution points to a linear equation like $3x + 4y + 2z = 12$ is a plane in a three-dimensional, rectangular coordinate system. Figure 9.3 on the following page shows a triangular portion of the plane of all solution points for the equation $3x + 4y + 2z = 12$.

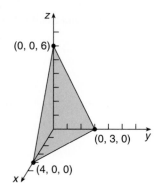

**FIGURE 9.3** A portion of the solution plane for the equation $3x + 4y + 2z = 12$.

Because the set of all solution points for each equation in a system with 3 variables is a plane, there are three planes associated with a system of three linear equations. How are these planes related to each other? There are four possibilities, as follows:

1. They might intersect in a common point.
2. They might intersect in a common line.
3. They might have no points in common.
4. They might coincide.

Figure 9.4 illustrates these four possibilities.

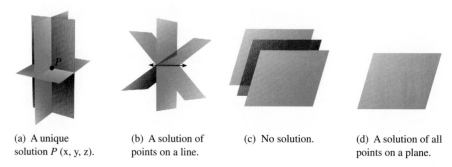

(a) A unique solution $P$ (x, y, z).

(b) A solution of points on a line.

(c) No solution.

(d) A solution of all points on a plane.

**FIGURE 9.4** Types of solutions for a system of three linear equations in $x$, $y$, and $z$.

Examples 2 and 3 illustrate the first of these cases, the case of a unique solution.

## Solving a System by the Elimination Method

To solve systems in the past, we relied on paper-and-pencil analytic methods like those illustrated in the next two examples. Example 2 illustrates how to add two different pairs of equations to eliminate a variable and reduce the three-equation system to a two-equation system.

**EXAMPLE 2**  Solving a System of Three Linear Equations

Solve the following system of equations algebraically and support numerically.

$$x + y + z = 13 \qquad (1a)$$

$$3x + 2y + 2z = 28 \qquad (1b)$$

$$x - y + z = 6 \qquad (1c)$$

SOLUTION
**Solve Algebraically.**

1. Eliminate the $y$-variable by adding Eqs. (1$a$) and (1$c$).

$$x + y + z = 13$$
$$\underline{x - y + z = 6}$$
$$2x \phantom{+ y} + 2z = 19 \qquad (2a)$$

2. Next use Eqs. (1$a$) and (1$b$) to eliminate the *same* variable $y$. Multiply (1$a$) by $-2$ so that the coefficients of $y$ differ only by a sign. Then add.

$$-2x - 2y - 2z = -26$$
$$\underline{3x + 2y + 2z = \phantom{-2}28}$$
$$x \phantom{+ 2y + 2z} = \phantom{-2}2$$

3. Substitute $x = 2$ into Eq. (2$a$) and solve for $z$.

$$2(2) + 2z = 19$$
$$4 + 2z = 19$$
$$2z = 15$$
$$z = 7.5$$

4. Substitute $x = 2$ and $z = 7.5$ into any of the original equations to solve for y. We will substitute into Eq. (1$a$).

$$2 + y + 7.5 = 13$$
$$y + 9.5 = 13$$
$$y = 3.5$$

**Support Numerically.** The solution to the system is the ordered triple $(2, 3.5, 7.5)$. Figure 9.5 shows that $x = 2$, $y = 3.5$, and $z = 7.5$ satisfy all three equations. ∎

```
2 → X: 3.5 → Y: 7.5 → Z:
X + Y + Z
                  13
3X + 2Y + 2Z
                  28
X – Y + Z
                   6
```

**FIGURE 9.5**
Numerical support of the solution to Example 2.

In Example 2, when $-2$ times Eq. (1$a$) was added to Eq. (1$c$), both variables $y$ and $z$ were eliminated. This was a convenient coincidence, which usually does not occur. Example 3 shows a more typical elimination method.

**EXAMPLE 3**    Solving a System of Three Linear Equations

Solve the following system algebraically and support the solution numerically.

$$x - 2y - z = 4 \qquad (1a)$$
$$2x + 3y + z = 13 \qquad (1b)$$
$$3x - y + 2z = 7 \qquad (1c)$$

**TRY THIS**

Solve Example 3 by choosing to eliminate the $x$-terms first. The solution should be identical to that in Example 3. Which plan is easier to execute? Often one is easier than another. Take time to devise the most efficient plan.

SOLUTION
**Solve Algebraically.**

1. Add Eqs. (1a) and (1b) to eliminate variable $z$. Name the resulting Eq. (2a).

$$\begin{cases} x - 2y - z = 4 \\ 2x + 3y + z = 13 \end{cases} \quad \text{reduces to} \quad 3x + y = 17 \qquad (2a)$$

Multiply Eq. (1b) by $-2$ and add the result to Eq. (1c).

$$\begin{cases} -4x - 6y - 2z = -26 \\ 3x - y + 2z = 7 \end{cases} \quad \text{reduces to} \quad -x - 7y = -19 \qquad (2b)$$

2. Multiply Eq. (2a) by 7 so that the coefficients of $y$ differ only in sign in the system (2a) and (2b). That is

$$\begin{cases} 7(3x + y) = 7(17) \\ -x - 7y = -19 \end{cases} \quad \text{becomes the system} \quad \begin{aligned} 21x + 7y &= 119 \qquad (3a) \\ -x - 7y &= -19 \qquad (3b) \end{aligned}$$

3. Add Eqs. (3a) and (3b) to obtain

$$20x = 100$$
$$x = 5.$$

4. Substitute $x = 5$ into Eq. (2a) or (2b). We will substitute into Eq. (2a).

$$3(5) + y = 17$$
$$15 + y = 17$$
$$y = 2$$

5. To find $z$, substitute $x = 5$ and $y = 2$ into any equation containing $z$ and solve. We will substitute into Eq. (1a).

$$5 - 2(2) - z = 4$$
$$1 - z = 4$$
$$-z = 3$$
$$z = -3$$

The solution to the system is the ordered triple $(5, 2, -3)$.

**Support Numerically.** Figure 9.6 shows numerical support of the solution using a grapher. ∎

```
5 → X : 2 → Y : ⁻3 → Z :
X - 2Y - Z
                    4
2X + 3Y + Z
                   13
3X - Y + 2Z
                    7
```

**FIGURE 9.6** Numerical support of the solution to Example 3.

Here is a summary of one method used for solving a system of three equations in three variables that has a unique solution.

> ## Solving Systems in Three Variables by the Elimination Method
>
> 1. Eliminate one variable from a pair of equations. The result is an equation in two variables.
> 2. Eliminate the *same* variable from a different pair of equations. The result is an equation in two variables.
> 3. Steps 1 and 2 result in a system of two equations in two variables. Use either the elimination or substitution method to solve this new system for the two variables.
> 4. Substitute the two variables into one of the original equations to find the value of the third variable.
> 5. The solution is an ordered triple. Support this solution numerically using the grapher.

Sometimes there is no common point of intersection, as in Figure 9.4(c). Example 4 illustrates this case.

**EXAMPLE 4**    Solving a System with No Solution

Solve the following system:

$$x + 3y - 2z = 6 \tag{1a}$$

$$3x + 9y - 6z = 4 \tag{1b}$$

$$5x - 2y - 3z = 7 \tag{1c}$$

SOLUTION

**Solve Algebraically.**

1. Multiply Eq. (1*a*) by $-3$ and add to Eq. (1*b*) to eliminate the variable $x$.

$$\begin{cases} -3\,(x + 3y - 2z) = -3\,(6) \\ 3x + 9y - 6z = 4 \end{cases} \quad \text{becomes} \quad \begin{array}{l} -3x - 9y + 6z = -18 \\ 3x + 9y - 6z = 4 \end{array}$$

2. Add the two equations. The resulting equation $0 = -14$ is a false statement. There is no solution to the system since Eqs. (1a) and (1b) have no common solution.  ∎

Example 5 is an example of the case illustrated in Figure 9.4(d), in which the three planes coincide.

### REMINDER

The solution of a system of equations consists of all points common to all three planes. If two planes are parallel and therefore have no points in common, it is impossible for all three planes to have any points in common. Thus it is not necessary to continue once two planes are found parallel.

**EXAMPLE 5**    Solving a System of Equivalent Equations

Solve the following system:

$$x - 2y + 3z = 5 \tag{1a}$$

$$-2x + 4y - 6z = -10 \tag{1b}$$

$$3x - 6y + 9z = 15 \tag{1c}$$

SOLUTION

**Solve Algebraically.**

1. Multiply Eq. (1a) by 2 and add to Eq. (1b) to eliminate the variable $x$.

$$\begin{cases} 2\,(x - 2y + 3z) = 2\,(5) \\ -2x + 4y - 6z = -10 \end{cases} \quad \text{becomes} \quad \begin{aligned} 2x - 4y + 6z &= 10 \\ -2x + 4y - 6z &= -10 \end{aligned}$$

2. Add the two equations. The resulting equation $0 = 0$ is a true statement. This shows that Eqs. (1a) and (1b) are equivalent equations and have the same solution.

3. Equation (1a) multiplied by 3 is equivalent to Eq. (1c). Therefore all three equations are equivalent and the solution to any one is the same for all three. Thus the solution is all ordered triples $(x, y, z)$ that are solutions of $x - 2y + 3z = 5$. Note that we selected Eq. (1a), but we could have selected any of the three equations in the system, since they are all equivalent.    ■

## Problem Situations Using Systems of Equations

The following problem situation deals with three unknown quantities.

---

☐                                                                ☐

**NUT MIXTURE PROBLEM SITUATION**

A nut mixture contains peanuts worth $1.50/lb, cashews worth $3.25/lb, and walnuts worth $4.00/lb.

☐                                                                ☐

---

Problem situations of this type can be represented algebraically by letting a different variable represent each of the unknown quantities. Then write an algebraic representation in the form of a system of equations. Example 6 illustrates this method.

**EXAMPLE 6**   Nut Mixture Problem: Finding the Amounts of Each

Suppose there are 2 more pounds of peanuts than cashews in a 25-lb mixture. If the mixture is worth $2.76/lb, how many pounds of each type of nut are in the mixture? Support numerically.

SOLUTION

**Algebraic Representation of the Problem:**
Let $x$ = number of pounds of peanuts at $1.50/lb.
Let $y$ = number of pounds of cashews at $3.25/lb.
Let $z$ = number of pounds of walnuts at $4.00/lb.

**The total cost of peanuts in cents** $= 150x$

**The total cost of cashews in cents** $= 325y$

**The total cost of walnuts in cents** $= 400z$

$$x + y + z = 25$$

**This equation represents the total number of pounds.**

$$150x + 325y + 400z = 276(25)$$

**This equation represents the total cost in cents.**

$$x = y + 2$$

**This equation compares the number of pounds of peanuts and cashews.**

**Solve Algebraically.**

1. The third equation already contains only two variables, so we will substitute this expression for $x$ into the other two equations. Replace $x$ with $y + 2$ in the first equation.

$$(y + 2) + y + z = 25$$

$$2y + z + 2 = 25 \quad \textbf{Simplify.}$$

$$2y + z = 23 \tag{1a}$$

Replace $x$ with $y + 2$ in the second equation.

$$150(y + 2) + 325y + 400z = 276(25)$$

$$150y + 300 + 325y + 400z = 6900 \quad \textbf{Simplify.}$$

$$475y + 400z = 6600 \tag{1b}$$

2. Solve this resulting system of two equations with two variables.

$$2y + z = 23 \tag{1a}$$

$$475y + 400z = 6600 \tag{1b}$$

We can solve Eq. (1a) for $z$.

$$z = 23 - 2y$$

Replace $z$ with this expression in ($1b$) and solve for $y$.

$$\begin{aligned} 475y + 400(23 - 2y) &= 6600 \\ 475y + 9200 - 800y &= 6600 \\ -325y + 9200 &= 6600 \\ -325y &= -2600 \\ y &= 8 \end{aligned}$$

Substitute $y = 8$ into Eq. ($1a$).

$$\begin{aligned} 2(8) + z &= 23 \\ 16 + z &= 23 \\ z &= 7 \end{aligned}$$

3. Substitute the values for $y$ and $z$ into any of the three original equations to solve for $x$.

$x + 8 + 7 = 25$ **Replace $y$- and $z$-values into the first equation, $x + y + z = 25$**

$x + 15 = 25$ **Simplify.**

$x = 10$

The solution (10, 8, 7) indicates that there are 10 lb of peanuts, 8 lb of cashews, and 7 lb of walnuts. Figure 9.7 shows numerical support of the solution using a grapher. ■

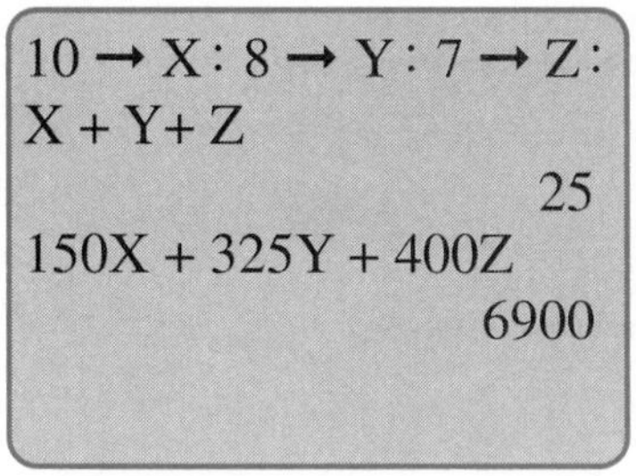
10 → X: 8 → Y: 7 → Z:
X + Y+ Z
25
150X + 325Y + 400Z
6900

**FIGURE 9.7**
Numerical support of the solution to the **Nut Mixture Problem Situation.**

The paper-and-pencil method illustrated in this section is at best very tedious. It is not practical for larger systems. In Section 9.4, you will learn a matrix method in which the computation can be done with a grapher.

## Exercises for Section 9.1

In Exercises 1–4, verify that the given ordered triple is a solution to the system of equations.

**1.** (3, 2, −4); $\begin{cases} x + y + z = 1 \\ 2x - 3y + 6z = -24 \\ 5x + y - z = 21 \end{cases}$

**2.** (2, 1, 0); $\begin{cases} x - 3y + 2z = -11 \\ 2x - 4y + 3z = -15 \\ 3x - 5y - 4z = 5 \end{cases}$

**3.** (0, −3, 3); $\begin{cases} x - 3y + 2z = -11 \\ 2x - 4y + 3z = -15 \\ 3x - 5y - 4z = 5 \end{cases}$

4. $(7, -2, 6)$;
$$\begin{cases} x - y + 2z = 21 \\ 2x + 4y + 3z = -15 \\ x - 5y - z = 11 \end{cases}$$

In Exercises 5–8, find the missing number in the ordered triple that is the solution of the system of equations.

5. $(-2, ?, 9)$;
$$\begin{cases} x - 3y - 5z = -56 \\ 4x + y - z = -14 \\ x - y - z = -14 \end{cases}$$

6. $(?, -7, 1)$;
$$\begin{cases} x - y + z = 8 \\ 2x + 3y - 5z = -26 \\ 3x + 4y - z = -29 \end{cases}$$

7. $(1, 2, ?)$;
$$\begin{cases} x - 3y + 2z = -11 \\ 2x - 4y + 3z = -15 \\ 3x - 5y - 4z = 5 \end{cases}$$

8. $(?, -2, 2)$
$$\begin{cases} 3x - 4y + z = 19 \\ 2x + 4y + z = 0 \\ x - 2y + 5z = 17 \end{cases}$$

In Exercises 9–14, solve the system of equations using either the elimination or substitution method.

9.
$$\begin{cases} x + y + z = 0 \\ x - y + z = -2 \\ 2x - 3y - 2z = 3 \end{cases}$$

10.
$$\begin{cases} 2x + y - z = 8 \\ x - y + 2z = 9 \\ 3x + 2y - 2z = 11 \end{cases}$$

11.
$$\begin{cases} x + z = 7 \\ x + y - z = 3 \\ 2x - 3y + z = -16 \end{cases}$$

12.
$$\begin{cases} x - 2y + z = 4 \\ 2x + 3y - 2z = 4 \\ y + 3z = 3 \end{cases}$$

13.
$$\begin{cases} x - y + z = -3 \\ 2x + 2y - z = 35 \\ x - 3y - 2z = -2 \end{cases}$$

14.
$$\begin{cases} x - y + z = 5 \\ 3x - 2y + 5z = 19 \\ 2x + 5y - z = -13 \end{cases}$$

15. *Candy Mixture.* A grocer wants to mix three kinds of hard candy to sell for $2.40/lb. He needs 50 pounds of candy altogether. He mixes sour balls worth $3.50/lb, butterballs worth $2.50/lb, and starlight mints worth $1.75/lb. He mixes twice as many butterballs as sour balls.

a) Let $x$ equal the number of pounds of sour balls, $y$ the number of pounds of butterballs, and $z$ the number of pounds of starlight mints. Write an algebraic representation for the problem situation in the form of a system of equations.

b) Find the number of pounds of each kind of candy he mixes together.

16. *Interest on an Investment.* Amy split her $25,000 inheritance among three investments by placing twice as much in a savings account paying 5% annual interest as in a mutual fund paying 6% annually. The remainder of the money was invested in a savings bond paying 7% annually. At the end of the year, she had earned $1580.

a) Let $x$ equal the amount of money invested in the account paying 5%, $y$ the amount invested at 6%, and $z$ the amount invested at 7%. Write an algebraic representation for the problem situation in the form of a system of equations.

b) Solve the system of equations to find how much money Amy put into each account.

c) Find the total amount of interest earned if Amy had invested the same amount in each account but the rate of interest was 5.5%, 6.5%, and 7.5%, respectively.

In Exercises 17–22, solve the problem situation by first writing a system of equations and then solving the system.

**17.** *Individual Cost.* Donny, Dan, and Matt stopped at Cheesy Cheeseburgers after class. Donny bought two cheeseburgers, one large drink, and two orders of fries for $3.95. Dan bought three cheeseburgers and two large drinks for $4.05. Matt bought one cheeseburger, one large drink, and one order of fries for $2.47. What is the price of each cheeseburger, large drink, and order of fries?

**18.** *Investment Problem.* Beth invests $22,000 in stocks, bonds, and a mutual fund. She receives a 10%, 9%, and 11% return on each investment, respectively. The sum of the money invested in stocks and bonds is $1000 more than twice the amount in mutual funds. The total return annually is $2170. How much did Beth invest at each rate?

**19.** The sum of three numbers is 23. The second number is two less than the first number. The sum of the first two numbers is three more than three times the third number. Find the numbers.

**20.** The sum of three integers is 45. The sum of the first two integers is nine more than the third integer. The sum of the first and third integer is twice the second. Find the integers.

**21.** *Coin Problem.* Tanesha has $4.20 in her bank in nickels, dimes, and quarters. She has 44 coins altogether. The number of nickels is one more than twice the number of dimes. Find the number of each kind of coin she has in the bank.

**22.** *Coin Problem.* Patrick has stashed away $16.60 in nickels, dimes, and quarters in his sock drawer. The sum of the nickels and dimes is four less than three times the number of quarters. The total number of coins is 136. Find the number of each kind of coin.

**23.** **Writing to Learn.** Write in your own words the procedure you would use to reduce the following

system of three equations in three variables to a system of two equations with two variables:

$$\begin{cases} x - y + z = 51 \\ 3x - 4y + 5z = 20 \\ 6x + 5y - z = -18 \end{cases}$$

### EXTENDING THE IDEAS

**24.** Find the values of $a$, $b$, and $c$ so that the following points lie on the graph of the equation $y = ax^2 + bx + c$: $(1, 2)$, $(-3, -2)$, and $(5, 38)$.

**25.** Find the values of $a$, $b$, and $c$ so that the following points lie on the graph of the equation $y = ax^2 + bx + c$: $(2, 8)$, $(-5, 8)$, and $(1, -4)$.

### LOOKING BACK—MAKING CONNECTIONS

**26.** The sum of the measures of three angles of a triangle is $180°$. The measure of the second angle is three times the smallest angle. The measure of the third angle is $5°$ more than the second angle. Find the three angles.

In Exercises 27 and 28, look at the following pattern and see if you can complete the chart:

$$15^2 = 225$$
$$25^2 = 625$$
$$35^2 = 1225$$
$$45^2 = 2025$$
$$55^2 = 3025$$
$$65^2 = ?$$
$$75^2 = ?$$

For Exercises 27 and 28.

**27.** Can you generalize how to mentally square a number that ends in 5?

**28.** Can you explain algebraically why your method

works? (*Hint:* Each number can be written in general as $(10n + 5)$. What happens when you square this binomial?)

---

## 9.2 Matrix Algebra

**REMINDER**

Matrices provide an efficient way to record data and thus can be used in a variety of business problems. Computers are designed to handle large matrices. Thus most computer software libraries include a spreadsheet program, which is basically an array of numbers. Spreadsheets, like matrices, can be used to manipulate large sets of numbers.

The **Nut Mixture Problem Situation** studied in Section 9.1 involved a mixture of peanuts, cashews, and walnuts. The Tasty Nut Company, which manufactures this nut mixture, has plants and warehouses at two locations. On one particular day, their inventory (in tons) was as follows:

$$
\begin{array}{cccc}
 & \text{Peanuts} & \text{Cashews} & \text{Mixed nuts} \\
\text{Site A} & \left[\begin{array}{ccc} 2.5 & 3.8 & 1.2 \\ \text{Site B} \quad 4.2 & 6.1 & 3.8 \end{array}\right]
\end{array}
$$

There are many settings, like the inventory control office of the Tasty Nut Company, in which it is natural and helpful to record data in a rectangular array. Such a rectangular array is called a **matrix**.

The entries in a matrix are arranged in rows and columns, and a matrix is usually named by a capital letter; for example, the matrix in the above inventory example is named $T$ for Tasty and is as follows:

$$
T = \begin{bmatrix} 2.5 & 3.8 & 1.2 \\ 4.2 & 6.1 & 3.8 \end{bmatrix}
$$

The **dimension** of a matrix is described by the number of rows and columns. Sometimes the dimension is incorporated in the matrix name, such as in $T_{2\times3}$. In this case, the first number in the subscript stands for the number of rows in the matrix and the second number stands for the number of columns. We indicate the dimension in a problem only when necessary.

A **square matrix** has the same number of rows as columns. The following matrix A is an example of a square matrix:

$$
A = \begin{bmatrix} -5 & 10 \\ 15 & 3 \end{bmatrix}
$$

**EXAMPLE 1**　Determining the Dimension of a Matrix

Determine the dimensions of the following matrices:

$$a)\ A = \begin{bmatrix} 1 & -3 & 4 \\ 2 & -6 & 7 \end{bmatrix} \qquad b)\ B = \begin{bmatrix} 0 & 3 \\ 1 & 0 \\ 5 & 2 \\ 7 & 1 \end{bmatrix} \qquad c)\ C = \begin{bmatrix} 0 & 3 & 8 \\ 1 & 0 & -2 \\ 5 & 2 & 6 \end{bmatrix}$$

SOLUTION

a) $A$ is a $2 \times 3$ matrix, since it has 2 rows and 3 columns.
b) $B$ is a $4 \times 2$ matrix, since it has 4 rows and 2 columns.
c) $C$ is a $3 \times 3$ matrix, since it has 3 rows and 3 columns. ■

## Matrix Addition and Subtraction

The Tasty Nut Company uses a matrix $J$ to record the total number of nuts in tons sold in January and matrix $F$ for that sold in February.

$$J = \begin{array}{c} \\ \text{Site A} \\ \text{Site B} \end{array} \begin{array}{ccc} \text{Peanuts} & \text{Cashews} & \text{Walnuts} \\ \begin{bmatrix} 2.1 & 1.8 & 2.3 \\ 3.1 & 4.1 & 3.4 \end{bmatrix} \end{array}$$

$$F = \begin{array}{c} \\ \text{Site A} \\ \text{Site B} \end{array} \begin{array}{ccc} \text{Peanuts} & \text{Cashews} & \text{Walnuts} \\ \begin{bmatrix} 2.5 & 2.7 & 3.2 \\ 4.2 & 2.1 & 1.8 \end{bmatrix} \end{array}$$

A new matrix named $J + F$ records the total number of tons sold in the two months combined.

$$J + F = \begin{array}{c} \\ \text{Site A} \\ \text{Site B} \end{array} \begin{array}{ccc} \text{Peanuts} & \text{Cashews} & \text{Walnuts} \\ \begin{bmatrix} 4.6 & 4.5 & 5.5 \\ 7.3 & 6.2 & 5.2 \end{bmatrix} \end{array}$$

The context of the problem makes it clear that corresponding entries of the two matrices should be added to find the sum. That is, the **sum of two matrices** of the same size is the matrix found by adding the corresponding entries of the two matrices. The **difference of two matrices** of the same size is defined in a similar way, that is, subtract the corresponding entries.

**REMINDER**

Usually, labels of rows and columns are not included when writing matrices, so they are not included in the remainder of this section.

**EXAMPLE 2**　Adding and Subtracting Matrices

Find

a) The sum $A + B$
b) The difference $A - B$

$$\text{where,} \qquad A = \begin{bmatrix} 4 & -3 \\ 2 & -6 \end{bmatrix} \qquad \text{and} \quad B = \begin{bmatrix} 2 & 5 \\ 1 & 0 \end{bmatrix}.$$

**GRAPHER NOTE**

Consult your grapher lab manual to learn how to enter matrices and to add and subtract them. The grapher we use names a matrix by using square brackets around an $A$, $B$, or $C$. Whenever we refer to matrices on a grapher, we will denote them $[A]$, $[B]$, or $[C]$.

SOLUTION

**Paper-and-Pencil Solution.**

a) Add the corresponding entries.

$$A + B = \begin{bmatrix} 4+2 & -3+5 \\ 2+1 & -6+0 \end{bmatrix} = \begin{bmatrix} 6 & 2 \\ 3 & -6 \end{bmatrix}$$

b) Subtract the corresponding entries.

$$A - B = \begin{bmatrix} 4-2 & -3-5 \\ 2-1 & -6-0 \end{bmatrix} = \begin{bmatrix} 2 & -8 \\ 1 & -6 \end{bmatrix}$$

Most graphers can add and subtract matrices. On the grapher we use, matrices can be entered using the Matrix edit menu. Figure 9.8 shows the Matrix edit menu after matrices $A$ and $B$ of Example 2 have been entered.

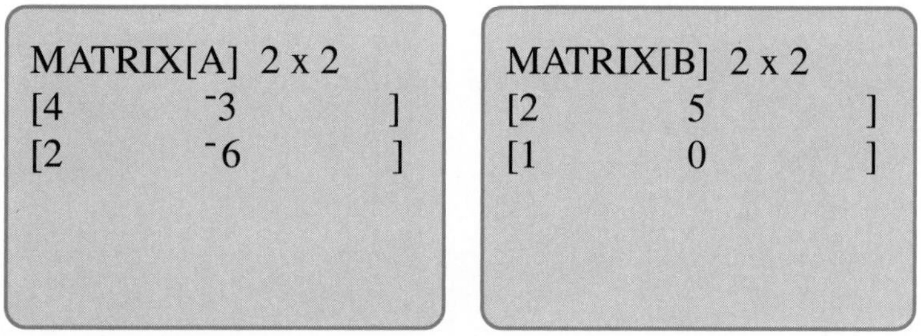

```
MATRIX[A] 2 x 2
[4        ⁻3        ]
[2        ⁻6        ]
```
(a)

```
MATRIX[B] 2 x 2
[2        5        ]
[1        0        ]
```
(b)

**FIGURE 9.8** (a) Matrix $A$ and (b) Matrix $B$ screens after entries are entered in the grapher.

Example 3 shows how to find the sum of two matrices using a grapher.

**EXAMPLE 3**   Adding and Subtracting Matrices Using a Grapher

Use a grapher to find the sum $A + B$ and the difference $A - B$, where

$$A = \begin{bmatrix} 4 & -3 \\ 2 & -6 \end{bmatrix} \quad \text{and} \quad B = \begin{bmatrix} 2 & 5 \\ 1 & 0 \end{bmatrix}.$$

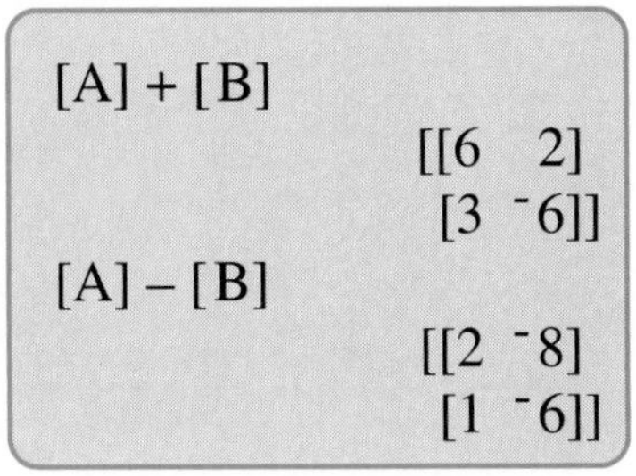

[A] + [B]
$$[[6 \quad 2]$$
$$[3 \ {}^-6]]$$

[A] − [B]
$$[[2 \ {}^-8]$$
$$[1 \ {}^-6]]$$

**FIGURE 9.9** Grapher computation of $A + B$ and $A − B$.

SOLUTION

**Grapher Solution.** Figure 9.9 shows the grapher screen on which $A + B$ and $A − B$ were found. ■

Readers should be very familiar with adding and subtracting numbers. Addition and subtraction, and later, multiplication, of matrices is less familiar, however; it is not obvious whether all the familiar properties of the basic operations are satisfied when working with matrices. Example 4 illustrates the Property of Commutativity of Addition for matrices.

**EXAMPLE 4**   Adding Matrices

Given the following matrices $A$, $B$, and $C$, use a grapher to find the sums in (a)–(d).

$$A = \begin{bmatrix} 5 & -2 \\ 8 & 12 \end{bmatrix} \qquad B = \begin{bmatrix} 1 & 15 \\ 7 & -3 \end{bmatrix} \qquad C = \begin{bmatrix} 3 & 8 \\ 0 & -2 \end{bmatrix}$$

a) $A + B$
b) $B + A$
c) $B + C$
d) $C + B$

**Grapher Solution.** Figure 9.10 shows the grapher screens that indicate the matrix sums.

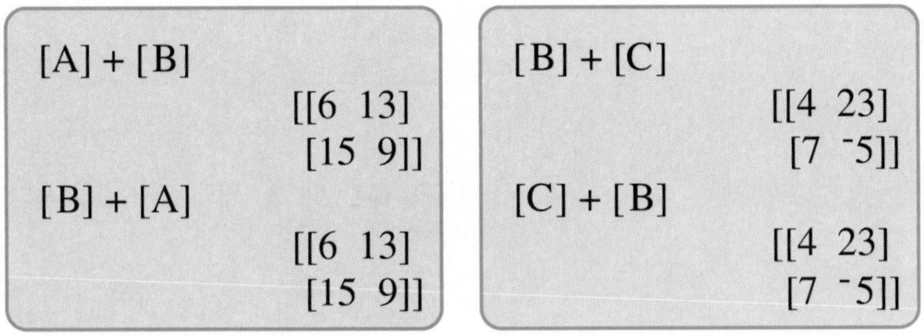

[A] + [B]
$$[[6 \ 13]$$
$$[15 \ 9]]$$

[B] + [A]
$$[[6 \ 13]$$
$$[15 \ 9]]$$

[B] + [C]
$$[[4 \ 23]$$
$$[7 \ {}^-5]]$$

[C] + [B]
$$[[4 \ 23]$$
$$[7 \ {}^-5]]$$

**FIGURE 9.10** Grapher computation of the sums in Example 4. ■

Example 4 supports that the Commutative Property holds for addition of matrices. We now turn to multiplication of matrices. In this section, we concentrate only on multiplying matrices by a constant, which in matrix vocabulary is called multiplying by a **scalar**.

## Multiplication of a Matrix and a Scalar

The product of a matrix and a scalar can be found by multiplying each entry of the matrix by the scalar. The process is called **scalar multiplication**. Most graphers can multiply a matrix by a scalar.

**GRAPHER NOTE**

The grapher we use has the capability of multiplying a scalar and a matrix. Read your grapher lab manual to learn how to complete such a product.

**EXAMPLE 5**    Scalar Multiplication

Find the product of 3 and matrix $A$ if

$$A = \begin{bmatrix} 2 & 1 & 5 \\ 6 & -1 & 3 \end{bmatrix}.$$

SOLUTION

**Paper-and-Pencil Solution.** Multiply each entry of the matrix by 3.

$$3 \begin{bmatrix} 2 & 1 & 5 \\ 6 & -1 & 3 \end{bmatrix} = \begin{bmatrix} 6 & 3 & 15 \\ 18 & -3 & 9 \end{bmatrix}$$

**Support Numerically.** Figure 9.11 shows support of the product using a grapher. ∎

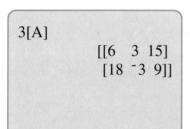

```
3[A]
        [[6   3  15]
        [18  ⁻3  9]]
```

**FIGURE 9.11** Grapher support of the product in Example 5.

**EXAMPLE 6**    Scalar Multiplication

Find $3A$ and $-2A$ if

$$A = \begin{bmatrix} 1 & -4 \\ 6 & 10 \end{bmatrix}.$$

SOLUTION

**Paper-and-Pencil Solution.** Multiply each entry of the matrix by the appropriate scalar.

$$3A = 3 \begin{bmatrix} 1 & -4 \\ 6 & 10 \end{bmatrix} = \begin{bmatrix} 3 & -12 \\ 18 & 30 \end{bmatrix}$$

$$-2A = -2 \begin{bmatrix} 1 & -4 \\ 6 & 10 \end{bmatrix} = \begin{bmatrix} -2 & 8 \\ -12 & -20 \end{bmatrix}$$

**Support Numerically.** Figure 9.12 shows numerical support for the solution using a grapher. ∎

```
3[A]
        [[3   ⁻12]
        [18   30]]
⁻2[A]
        [[⁻2    8]
        [⁻12  ⁻20]]
```

**FIGURE 9.12** Grapher support of the product in Example 6.

Next, we revisit the Tasty Nut Company and use the matrices that the company provided for January and February to predict its March sales. Recall that the matrix $J + F$ represents the January and February sales combined.

|  |  | Peanuts | Cashews | Walnuts |
|---|---|---|---|---|
| $J + F =$ | Site A | 4.6 | 4.5 | 5.5 |
|  | Site B | 7.3 | 6.2 | 5.2 |

---

### NUT MIXTURE PROBLEM SITUATION REVISITED

The Tasty Nut Company is located in Miami, Florida, and March is the height of the tourist season for that area. The company has found that in past years, their March sales are, on an average, 25% higher than their January and February sales combined.

---

**REMINDER**

Recall that 100% of a number is the number itself. A number increased by 25% is then 125% of the number. Similarly a number decreased by 20% is 80% of the original number.

**EXAMPLE 7**   Finding a Percentage Using Matrices

Predict the number of tons of each kind of nut the company should sell in March.

SOLUTION

Represent an increase of 25% as 125% of the January and February sales.

$$1.25(J + F) = 1.25\begin{bmatrix} 4.6 & 4.5 & 5.5 \\ 7.3 & 6.2 & 5.2 \end{bmatrix} = \begin{bmatrix} 5.75 & 5.625 & 6.875 \\ 9.125 & 7.75 & 6.5 \end{bmatrix}$$

Therefore at Site A, they should sell 5.75 tons of peanuts, 5.625 tons of cashews, and 6.875 tons of walnuts. At Site B, they should sell 9.125 tons of peanuts, 7.75 tons of cashews, and 6.5 tons of walnuts. ∎

As you can see, matrices are useful to record data for any period of time. The numbers in the array can be manipulated using any operation for real numbers. The scalar can be in the form of an integer, a fraction, or a percentage, and our grapher can quickly multiply each entry in the matrix by the scalar.

## Exercises for Section   9.2

In Exercises 1–12, consider the following matrices $A$, $B$, and $C$, perform the indicated operations using a paper-and-pencil method, and support the solution using a grapher.

$$A = \begin{bmatrix} 2 & 3 \\ 5 & -7 \end{bmatrix} \quad B = \begin{bmatrix} -1 & 0 \\ -2 & 5 \end{bmatrix} \quad C = \begin{bmatrix} -5 & 0 & 2 \\ -3 & 6 & 1 \\ 1 & 0 & 4 \end{bmatrix}$$

1. Find the dimension of $A$.

2. Find the dimension of $B$.

3. Find the dimension of $C$.

4. What is the dimension of $A + B$?

5. Find $A + B$ and $B + A$.

6. Find $A - B$ and $B - A$.

7. Can you find $A + C$? Why?

**8.** Can you find $B + C$? Why?

**9.** $2A + 3B$        **10.** $3B + 2A$

**11.** $-3A - B$        **12.** $-1B + B$

**13.** Find 80% of $A$ if $A = \begin{bmatrix} 225 & 300 \\ 500 & 175 \end{bmatrix}$.

**14.** Find 110% of $B$ if $B = \begin{bmatrix} 30 & 200 \\ 360 & 540 \end{bmatrix}$.

**15.** *Inventory Records.* The Grove Park Library has two branches (north and south). It recorded in the following matrix the number of books checked out at each branch in the month of April according to fiction, nonfiction, and historical categories:

$$A = \begin{array}{c} \\ \text{Site N} \\ \text{Site S} \end{array} \begin{array}{ccc} \text{Fiction} & \text{Nonfiction} & \text{Historical} \\ \begin{bmatrix} 325 & 278 & 110 \\ 230 & 186 & 134 \end{bmatrix} \end{array}$$

Library employees have projected an increase of 15% in the number of books checked out in each category in the month of May. Find the projected number of books in matrix [M].

**16.** *Sales Records.* The Dixie Winery has recorded in the following matrices the number of bottles of white, red, and blush wines sold for the month of July and August at two of their sites:

$$J = \begin{array}{c} \\ \text{Site A} \\ \text{Site B} \end{array} \begin{array}{ccc} \text{White} & \text{Red} & \text{Blush} \\ \begin{bmatrix} 525 & 234 & 75 \\ 798 & 562 & 125 \end{bmatrix} \end{array}$$

$$A = \begin{array}{c} \\ \text{Site A} \\ \text{Site B} \end{array} \begin{array}{ccc} \text{White} & \text{Red} & \text{Blush} \\ \begin{bmatrix} 605 & 351 & 124 \\ 725 & 478 & 156 \end{bmatrix} \end{array}$$

a) Using matrix addition, find the total number of bottles sold in each category at each location.

b) Using matrices, find a matrix comparing the number of bottles sold at Site A with the number sold at Site B.

**17.** *Inventory Records.* The Tasty Nut Company records the number of tons of peanuts, cashews, and walnuts consumed for March and April using the following matrices $M$ and $A$ for sites A and B:

$$M = \begin{array}{c} \\ \text{Site A} \\ \text{Site B} \end{array} \begin{array}{ccc} \text{Peanuts} & \text{Cashews} & \text{Walnuts} \\ \begin{bmatrix} 2.2 & 1.5 & 2.8 \\ 2.5 & 3.1 & 3.2 \end{bmatrix} \end{array}$$

$$A = \begin{array}{c} \\ \text{Site A} \\ \text{Site B} \end{array} \begin{array}{ccc} \text{Peanuts} & \text{Cashews} & \text{Walnuts} \\ \begin{bmatrix} 2.5 & 3 & 2.6 \\ 2 & 2.4 & 3 \end{bmatrix} \end{array}$$

a) Find the matrix representation $M + A$ of the total number of tons of each kind of nut sold in March and April.

b) Find the difference $M - A$. At which site and for which kind of nut was there the largest increase in March over April?

**18.** *Sales Records.* The Red and White Supermarket needs to record the number of crates of celery, lettuce, and tomatoes sold over a three-day period at three different stores. The information is recorded as an array of numbers as follows. $M$ is Monday's record, $T$ is Tuesday's, and $W$ is Wednesday's. The columns represent the number of crates of celery, lettuce, and tomatoes, respectively.

$$M = \begin{array}{c} \\ \text{Site A} \\ \text{Site B} \\ \text{Site C} \end{array} \begin{array}{ccc} C & L & T \\ \begin{bmatrix} 9 & 5 & 6 \\ 8 & 3 & 5 \\ 6 & 2 & 4 \end{bmatrix} \end{array} \quad T = \begin{array}{c} \\ \text{Site A} \\ \text{Site B} \\ \text{Site C} \end{array} \begin{array}{ccc} C & L & T \\ \begin{bmatrix} 9 & 4 & 5 \\ 7 & 6 & 7 \\ 9 & 5 & 6 \end{bmatrix} \end{array}$$

$$W = \begin{array}{c} \\ \text{Site A} \\ \text{Site B} \\ \text{Site C} \end{array} \begin{array}{ccc} C & L & T \\ \begin{bmatrix} 8 & 3 & 5 \\ 8 & 7 & 5 \\ 8 & 4 & 3 \end{bmatrix} \end{array}$$

a) Find the total number of crates sold each day by finding the sum $M + T + W$.

b) Which store sold the most total number of crates combined on Mondays?

**19.** *Matrix Record Keeping.* Mary, Matthew, and Lisa are taking Biology II this term. The instructor informed them that if they chose to attend an optional week of intense labs at the end of the term, all grades earned on their three exams could be raised 20%. The instructor keeps the grades on a spreadsheet as follows:

| | Test 1 | Test 2 | Test 3 |
|---|---|---|---|
| Mary | 85 | 60 | 85 |
| Matthew | 75 | 85 | 80 |
| Lisa | 70 | 85 | 75 |

a) Write an algebraic representation of the present grades in the form of a matrix.

b) Find the proposed grades by finding 120% of the matrix written in Exercise 19(a). Do you think it is advantageous for them to attend the extra lab sessions?

### EXTENDING THE IDEAS

In Exercises 20–23, consider the following matrices $A$, $B$, and $C$, perform the indicated operations using a paper-and-pencil method, and support the solution using a grapher.

$$A = \begin{bmatrix} 25 & 32 \\ 45 & -15 \end{bmatrix} B = \begin{bmatrix} -12 & 18 \\ -17 & 35 \end{bmatrix} C = \begin{bmatrix} -51 & 22 \\ -15 & 20 \end{bmatrix}$$

**20.** Find $(A + B) + C$ and $A + (B + C)$.

**21.** Find $(A - B) - C$ and $A - (B - C)$.

**22.** Find $2(A + B) + C$ and $2A + 3(B + C)$.

**23.** Find $2(A - B) + 5C$ and $3A - 2(B + C)$.

### LOOKING BACK—MAKING CONNECTIONS

In Exercises 24–33, simplify the given expression.

**24.** $(x + 2)^2 + (x^2 - 25)$

**25.** $(x - 5)^2 - (x^2 + x - 6)$

**26.** $[(a + b) - 4][(a + b) + 4]$

**27.** $[(x + 3) - 5y][(x + 3) - 5y]$

**28.** $[3 - (x + y)][3 + (x + y)]$

**29.** $[x - (y + 5)][x + (y + 5)]$

**30.** $(x^2 + 12) - [(x^2 + 5) - (x - 3)(x + 5)]$

**31.** $(2x^2 + x) - [(3x^2 + 5) - (2x + 1)(x - 4)]$

**32.** $(x + 3)(4x - 1) - (x^2 - 2x - 5)$

**33.** $(x + 2)(x - 6) - (x^2 + 3x + 2)$

---

## 9.3 Multiplication of Matrices

> Multiplication of Two 2 × 2 Matrices ▪ Multiplication of Larger Matrices
> ▪ Identity Matrices ▪ Inverse Matrices

### Multiplication of Two 2 × 2 Matrices

It seems natural to define the product of two matrices of the same size similar to addition and subtraction of matrices—that is, by multiplying corresponding entries together. However, this is not the standard definition of multiplication.

In Section 9.4 matrix multiplication will be applied to solving systems of linear equations. At that time, it will become clearer why matrix multiplication is defined as it is.

We next define matrix multiplication $A \cdot B$ for $2 \times 2$ matrices.

**DEFINITION 9.2**  Matrix Multiplication for $2 \times 2$ Matrices

The product of the two $2 \times 2$ matrices

$$A = \begin{bmatrix} a_{11} & a_{12} \\ a_{21} & a_{22} \end{bmatrix} \qquad B = \begin{bmatrix} b_{11} & b_{12} \\ b_{21} & b_{22} \end{bmatrix}$$

is denoted by $AB$ and defined as follows:

$$AB = \begin{bmatrix} a_{11} & a_{12} \\ a_{21} & a_{22} \end{bmatrix} \cdot \begin{bmatrix} b_{11} & b_{12} \\ b_{21} & b_{22} \end{bmatrix}$$

$$= \begin{bmatrix} a_{11}b_{11} + a_{12}b_{21} & a_{11}b_{12} + a_{12}b_{22} \\ a_{21}b_{11} + a_{22}b_{21} & a_{21}b_{12} + a_{22}b_{22} \end{bmatrix}$$

Example 1 shows the product of two $2 \times 2$ matrices $A$ and $B$ written out as it is done with paper and pencil. However, it is also possible to calculate the product of matrices with a grapher.

**EXAMPLE 1**  Finding a Product of Two Matrices

Find the product $A \cdot B$, where

$$A = \begin{bmatrix} 3 & -4 \\ 2 & -8 \end{bmatrix} \qquad \text{and} \qquad B = \begin{bmatrix} 4 & -2 \\ -1 & 9 \end{bmatrix}.$$

Use a grapher to show numerical support of the solution.

SOLUTION

**Paper-and-Pencil Solution.**

$$\begin{bmatrix} 3 & -4 \\ 2 & -8 \end{bmatrix} \cdot \begin{bmatrix} 4 & -2 \\ -1 & 9 \end{bmatrix} = \begin{bmatrix} 3 \cdot 4 + (-4)(-1) & 3(-2) + (-4)9 \\ 2 \cdot 4 + (-8)(-1) & 2(-2) + (-8)9 \end{bmatrix}$$

$$= \begin{bmatrix} 16 & -42 \\ 16 & -76 \end{bmatrix}$$

**Support Numerically.** Figure 9.13 shows numerical support of the product using a grapher.

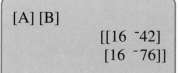

```
[A] [B]
          [[16  ⁻42]
           [16  ⁻76]]
```

**FIGURE 9.13** The product of $A \cdot B$ as completed on a grapher.

Multiplication of real numbers is commutative, that is, $ab = ba$ for all real numbers. Complete the following exploration to investigate whether matrix multiplication is commutative.

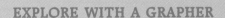

## EXPLORE WITH A GRAPHER

Enter the matrices $[A]$, $[B]$, and $[C]$ using the matrix menu.

$$A = \begin{bmatrix} 1 & -3 \\ 2 & -6 \end{bmatrix} \qquad B = \begin{bmatrix} 0 & 3 \\ 1 & 0 \end{bmatrix} \qquad C = \begin{bmatrix} 3 & 8 \\ 0 & -2 \end{bmatrix}$$

Find the following products:

1. $AB$ and $BA$.
2. $BC$ and $CB$.
3. $AC$ and $CA$.

**Generalize.**  Do you think matrix multiplication is commutative? Experiment: Find one pair of matrices $D$ and $E$ such that $D \cdot E = E \cdot D$.

## Multiplication of Larger Matrices

Notice that when completing the product $A \cdot B$ in Example 1 with paper and pencil, rows of matrix $A$ are matched with columns of matrix $B$ and the number of columns of $A$ must equal the number of rows of $B$. In fact, this is the only condition that must be satisfied in order for the product of two matrices to be defined. The product $A \cdot B$ also can be defined for matrices of larger sizes if the number of *columns* of $A$ equals the number of *rows* of $B$.

---

**DEFINITION 9.3**    Matrix Multiplication

Let $A$ be an $m \times r$ matrix and $B$ be an $r \times n$ matrix. Then the **matrix product**, $AB$, is the $m \times n$ matrix $C$ whose entries are found as follows: To find the entry $c_{ij}$ in row $i$ and column $j$, pair row $i$ of $A$ with column $j$ of $B$ and add products of corresponding entries. For example, if $r = 3$, we have

$$c_{ij} = (a_{i1} \quad a_{i2} \quad a_{i3}) \cdot \begin{pmatrix} b_{1j} \\ b_{2j} \\ b_{3j} \end{pmatrix} = a_{i1}b_{1j} + a_{i2}b_{2j} + a_{i3}b_{3j}.$$

For other values of $r$, the same pattern prevails.

---

Example 2 illustrates how to multiply a $2 \times 3$ matrix by a $3 \times 2$ matrix, with the product a $2 \times 2$ matrix.

### EXAMPLE 2    Multiplying Two Matrices

Find the product $AB$ by a paper-and-pencil method and support the solution using your grapher, where

$$A = \begin{bmatrix} 2 & 1 & -3 \\ 0 & 1 & 2 \end{bmatrix} \quad \text{and} \quad B = \begin{bmatrix} 1 & -4 \\ 0 & 2 \\ 1 & 0 \end{bmatrix}.$$

### SOLUTION

**Paper-and-Pencil Solution.** Because the number of columns of the first matrix equals the number of rows of the second matrix, it is possible to multiply $A$ and $B$. The product is a $2 \times 2$ matrix.

$$AB = \begin{bmatrix} 2 & 1 & -3 \\ 0 & 1 & 2 \end{bmatrix} \cdot \begin{bmatrix} 1 & -4 \\ 0 & 2 \\ 1 & 0 \end{bmatrix}$$

$$= \begin{bmatrix} (2)(1) + (1)(0) + (-3)(1) & 2(-4) + (1)(2) + (-3)(0) \\ (0)(1) + (1)(0) + (2)(1) & (0)(-4) + (1)(2) + (2)(0) \end{bmatrix}$$

$$= \begin{bmatrix} -1 & -6 \\ 2 & 2 \end{bmatrix}$$

```
[A] [B]
          [[¯1  ¯6]
           [2    2]]
```

**FIGURE 9.14** The product of $A \cdot B$ in Example 2 as completed on a grapher.

**Numerical Support.** Enter matrices $A$ and $B$ and find $A \cdot B$. Figure 9.14 shows the product as it appears on a grapher. ∎

## Identity Matrices

In arithmetic the number 1 is called the multiplicative identity because $a \cdot 1 = 1 \cdot a$ for all real numbers $a$. Similarly, there are identity matrices.

An $n \times n$ matrix, $I_n$, is an **identity matrix** for all $n \times n$ matrices, $A$, if $AI_n = I_n A = A$. Each of the following are identity matrices:

$$I_2 = \begin{bmatrix} 1 & 0 \\ 0 & 1 \end{bmatrix} \qquad I_3 = \begin{bmatrix} 1 & 0 & 0 \\ 0 & 1 & 0 \\ 0 & 0 & 1 \end{bmatrix} \qquad I_4 = \begin{bmatrix} 1 & 0 & 0 & 0 \\ 0 & 1 & 0 & 0 \\ 0 & 0 & 1 & 0 \\ 0 & 0 & 0 & 1 \end{bmatrix}$$

Notice that the entries on the main diagonal of an identity matrix are ones and all other entries in the matrix are zeros.

### EXAMPLE 3    Confirming an Identity

Show that $I_2$ is an identity $2 \times 2$ matrix.

SOLUTION

**Paper-and-Pencil Solution.** We must show that $I_2 \cdot A = A \cdot I_2$ for any $2 \times 2$ matrix $A$. Let

$$A = \begin{bmatrix} a & b \\ c & d \end{bmatrix}$$

be any $2 \times 2$ matrix. Then

$$\begin{bmatrix} 1 & 0 \\ 0 & 1 \end{bmatrix} \cdot \begin{bmatrix} a & b \\ c & d \end{bmatrix} = \begin{bmatrix} 1(a) + 0(c) & 1(b) + 0(d) \\ 0(a) + 1(c) & 0(b) + 1(d) \end{bmatrix} = \begin{bmatrix} a & b \\ c & d \end{bmatrix}.$$

In a similar fashion, it also can be shown that $A \cdot I_2 = A$.

$$\begin{bmatrix} a & b \\ c & d \end{bmatrix} \cdot \begin{bmatrix} 1 & 0 \\ 0 & 1 \end{bmatrix} = \begin{bmatrix} (a)1 + (b)0 & (a)0 + (b)1 \\ (c)1 + (d)0 & (c)0 + (d)1 \end{bmatrix} = \begin{bmatrix} a & b \\ c & d \end{bmatrix}$$ ■

## Inverse Matrices

The definition for identity matrix was modeled after a similar concept in arithmetic. In arithmetic we call $\frac{1}{5}$ and 5 multiplicative inverses because $\frac{1}{5} \cdot 5 = 5 \cdot \frac{1}{5} = 1$. Inverse matrices are defined in a similar manner.

---

**DEFINITION 9.4**   Inverse Matrices

A matrix, denoted $A^{-1}$, is the **inverse of an $n \times n$ matrix $A$** if

$$A \cdot A^{-1} = A^{-1} \cdot A = I_n.$$

---

**REMINDER**

The $-1$ in the symbol $A^{-1}$ does not mean "reciprocal" as it does in arithmetic. Read it as "inverse." So $A^{-1}$ is read "$A$ inverse."

Notice that if $AB = BA = I_n$, $A$ is the inverse of $B$ and $B$ is the inverse of $A$.

**EXAMPLE 4**   Confirming Inverse Matrices

Show that

$$A = \begin{bmatrix} 5 & 1 \\ 4 & 1 \end{bmatrix} \quad \text{and} \quad B = \begin{bmatrix} 1 & -1 \\ -4 & 5 \end{bmatrix}$$

are matrices which are inverses of each other.

SOLUTION

**Paper-and-Pencil Solution.**

$$AB = \begin{bmatrix} 5 & 1 \\ 4 & 1 \end{bmatrix} \cdot \begin{bmatrix} 1 & -1 \\ -4 & 5 \end{bmatrix} = \begin{bmatrix} 1 & 0 \\ 0 & 1 \end{bmatrix}$$

and

$$BA = \begin{bmatrix} 1 & -1 \\ -4 & 5 \end{bmatrix} \cdot \begin{bmatrix} 5 & 1 \\ 4 & 1 \end{bmatrix} = \begin{bmatrix} 1 & 0 \\ 0 & 1 \end{bmatrix}$$

**Grapher Solution.**

1. Enter the elements of $A$ and $B$.
2. Find $AB$ and $BA$. Both are equal to the identity matrix, which verifies that $A$ and $B$ are matrices (see Figure 9.15) that are inverses of each other. ∎

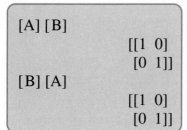

**FIGURE 9.15** Grapher support that A and B are inverse matrices in Example 4.

After verifying that $AB = BA = I_n$, as done in Example 4, it is correct to write

$$A^{-1} = \begin{bmatrix} 1 & -1 \\ -4 & 5 \end{bmatrix} \quad \text{and} \quad B^{-1} = \begin{bmatrix} 5 & 1 \\ 4 & 1 \end{bmatrix}.$$

Not all matrices have inverses. A matrix that has an inverse is called an **invertible matrix**. Theorem 9.1 shows when a $2 \times 2$ matrix has an inverse and how to find it.

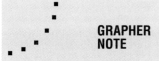

**GRAPHER NOTE**

On the grapher we use

1. *det* appears on the matrix menu and can be used to calculate the determinant of any square matrix within the capacity of the grapher. *det A* on a grapher looks like *det [A]*.
2. The inverse of a square matrix $A$ can be found by using the $x^{-1}$ key, $[A]^{-1}$.
3. An error message usually occurs when the inverse of a matrix does not exist.

**THEOREM 9.1** Inverses of $2 \times 2$ Matrices

A $2 \times 2$ matrix $A = \begin{bmatrix} a & b \\ c & d \end{bmatrix}$ has an inverse if and only if $ad - bc \neq 0$. The inverse of matrix $A$ is

$$A^{-1} = \frac{1}{ad - bc} \begin{bmatrix} d & -b \\ -c & a \end{bmatrix}.$$

The expression $ad - bc$ associated with matrix $\begin{bmatrix} a & b \\ c & d \end{bmatrix}$ is important because it determines whether $A$ has an inverse. Consequently, this expression is called the **determinant** of $A$ and is denoted by either

$$\det A \quad \text{or} \quad \begin{vmatrix} a & b \\ c & d \end{vmatrix}.$$

The two equations are

$$\det A = ad - bc \quad \text{or} \quad \begin{vmatrix} a & b \\ c & d \end{vmatrix} = ad - bc.$$

The following theorem is included in Theorem 9.1 for $2 \times 2$ matrices. However, it also applies to all square matrices.

---

■                                                                                           ■

**THEOREM 9.2**   Inverse of an $n \times n$ Matrix

A matrix has an inverse matrix if and only if det $A \neq 0$.

■                                                                                           ■

---

Theorem 9.1 can be used to find the determinant of a $2 \times 2$ matrix with paper and pencil; it is of course faster to use a grapher. Note that the details explaining how to calculate the determinant of matrices larger than $2 \times 2$ have not been included in this text, so you will need to use a grapher to calculate the determinant of larger matrices.

**EXAMPLE 5**   Finding Inverse Matrices

Use a grapher to determine whether each matrix is invertible. If so, find its inverse matrix.

a) $A = \begin{bmatrix} 4 & 1 \\ 7 & 2 \end{bmatrix}$     b) $B = \begin{bmatrix} 2 & -1 \\ -6 & 3 \end{bmatrix}$

SOLUTION

**Grapher Solution.**

a) Figure 9.16(a) shows that det $A = 1$. So $A^{-1}$ exists and

$$A^{-1} = \begin{bmatrix} 2 & -1 \\ -7 & 4 \end{bmatrix}.$$

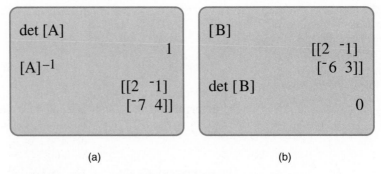

(a)                                                      (b)

**FIGURE 9.16**   Grapher screen for Example 5 showing matrix computations for determining both determinants and the inverse of a matrix if the determinant is not equal to zero.

b) Figure 9.16(b) shows that det $B = 0$. Consequently, $B$ does not have an inverse. ∎

The final example of this section uses a grapher to verify that $AA^{-1} = I$ for a $3 \times 3$ matrix.

**EXAMPLE 6**    Finding Inverse Matrices

Find the inverse of the matrix $A$ and verify that $AA^{-1} = I$ for

$$A = \begin{bmatrix} 4 & -1 & -3 \\ 1 & 2 & -1 \\ 0 & -2 & 2 \end{bmatrix}.$$

SOLUTION

**Grapher Solution.** Figure 9.17(a) shows the grapher output for finding $A^{-1}$ and Figure 9.17(b) shows the output for showing that $A * A^{-1} = I$.

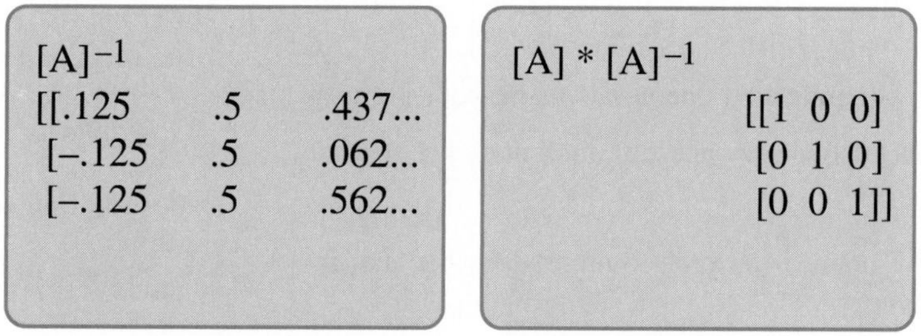

$[A]^{-1}$
$[[.125 \qquad .5 \qquad .437...$
$[-.125 \qquad .5 \qquad .062...$
$[-.125 \qquad .5 \qquad .562...$

$[A] * [A]^{-1}$
$[[1 \ 0 \ 0]$
$[0 \ 1 \ 0]$
$[0 \ 0 \ 1]]$

(a)

(b)

**FIGURE 9.17**    Grapher output for Example 6. ∎

## Exercises for Section 9.3

In Exercises 1–20, use the following matrices $A$ and $B$:

$$A = \begin{bmatrix} 1 & 0 \\ 5 & -3 \end{bmatrix} \quad B = \begin{bmatrix} -5 & 2 \\ -1 & 6 \end{bmatrix}$$

In Exercises 1–6, use a paper-and-pencil method to find the following:

1. $A \cdot B$

2. $B \cdot A$

**3.** $A^2$

**4.** $B^2$

**5.** $A \cdot (A + B)$

**6.** $B(A - B)$

In Exercises 7–20, use a grapher to find the following:

**7.** $A^{-1}$

**8.** $5A + 3B$

**9.** $2A - 3B$

**10.** $2AB$

**11.** $3A^2$

**12.** $(A + B)^2$

**13.** $6A - 2B$

**14.** $3A + B$

**15.** det $A$

**16.** det $B$

**17.** $B^{-1}$

**18.** $B^{-1}A^{-1}$

**19.** $(A \cdot B)^{-1}$

**20.** $A \cdot B^{-1}$

In Exercises 21–35, use the following matrices $A$, $B$, and $C$:

$$A = \begin{bmatrix} 1 & 5 \\ -8 & -4 \end{bmatrix} B = \begin{bmatrix} -5 & 6 & 9 \\ -1 & 2 & 0 \end{bmatrix} C = \begin{bmatrix} -1 & 4 & 8 \\ -2 & 0 & 7 \\ 0 & 3 & 1 \end{bmatrix}$$

Perform the indicated operations using your grapher.

**21.** How many rows and columns does the matrix $B \cdot C$ have?

**22.** How many rows and columns does the matrix $C \cdot B$ have?

**23.** $C^2$

**24.** $B^2$

**25.** $2A \cdot 3B$

**26.** $5A \cdot 2B$

**27.** $A^2B$

**28.** $BC$

**29.** det $A$

**30.** det $B$

**31.** $A^{-1}$

**32.** $B^{-1}$

**33.** $A \cdot A^{-1}$

**34.** $B \cdot B^{-1}$

In Exercises 35–38, write out your response to the situation.

**35. Writing to Learn.** Explain why the product $A \cdot B$ is defined but the product $B \cdot A$ is not. Matrix $A$ and matrix $B$ are defined above Exercise 21.

**36. Writing to Learn.** Write a paragraph explaining why the following statement is false: "A product of two matrices is defined only if the two matrices are the same size."

**37. Writing to Learn.** Enter the following matrices into a grapher:

$$A = \begin{bmatrix} -2 & 1 & 9 \\ -5 & 3 & 1 \end{bmatrix} B = \begin{bmatrix} -3 & 1 & 7 \\ -5 & 1 & 0 \end{bmatrix}$$

Try to find the product $AB$ on the grapher. Explain why an error message occurs.

**38. Writing to Learn.** Tell how you could alter matrix $A$ and matrix $B$ in Exercise 37 so that multiplication of the matrices could take place.

EXTENDING THE IDEAS

In Exercises 39–48, state whether the statement is true or false.

**39.** A matrix is a rectangular array of numbers.

**40.** Addition of matrices is commutative.

**41.** Subtraction of matrices is commutative.

**42.** Multiplication of matrices is commutative.

**43.** All matrices have an inverse matrix.

**44.** All matrices have an identity matrix associated with them.

**45.** A matrix times its inverse equals the identity matrix.

**46.** Any matrix can be multiplied by a scalar.

**47.** A square matrix has only perfect squares as entries.

**48.** The sum of two $3 \times 2$ matrices is also a $3 \times 2$ matrix.

LOOKING BACK—MAKING CONNECTIONS

In Exercises 49–54, solve the system of equations algebraically and support numerically using a grapher.

**49.** $\begin{cases} 2x + 3y = -1 \\ 3x - 2y = -8 \end{cases}$

**50.** $\begin{cases} x + 3y = 5 \\ 3x - y = -5 \end{cases}$

**51.** $\begin{cases} 2x - 3y = -1 \\ -4x + 6y = 2 \end{cases}$

**52.** $\begin{cases} x + y = 24 \\ x - y = -6 \end{cases}$

**53.** $\begin{cases} y = x \\ y = x^2 + 2x - 12 \end{cases}$

**54.** $\begin{cases} y = x + 3 \\ y = x^2 + 6x + 9 \end{cases}$

**55.** *Study Time.* In a particular week, Mike and Wendy studied 96 hr for an algebra course. The difference of the total number of hours they studied individually is 6 hr. If Wendy studied longer than Mike, find a system of equations representing the problem situation and solve the system to find how long each one studied.

---

**9.4** # Solving a System of Equations Using Matrices

Writing a System of Linear Equations as a Matrix Product • Solving Systems of Linear Equations Using Matrices • Solving Problems with Systems of Equations

In this section, the matrix theory studied in Sections 9.2 and 9.3 will be used to solve systems of linear equations. We begin by considering the following system as an example:

$$3x - 4y = -11$$
$$2x + 7y = 41$$

The matrix $A = \begin{bmatrix} 3 & -4 \\ 2 & 7 \end{bmatrix}$ is called the **matrix of coefficients** of the system. Consider the product $A \begin{bmatrix} x \\ y \end{bmatrix}$.

$$A \begin{bmatrix} x \\ y \end{bmatrix} = \begin{bmatrix} 3 & -4 \\ 2 & 7 \end{bmatrix} \cdot \begin{bmatrix} x \\ y \end{bmatrix} = \begin{bmatrix} 3x - 4y \\ 2x + 7y \end{bmatrix}$$

This product is a $2 \times 1$ matrix that appears like the left-hand side of the given system of equations.

The matrix $B = \begin{bmatrix} -11 \\ 41 \end{bmatrix}$ is called the **matrix of constants** of the system.

## Writing a System of Linear Equations as a Matrix Product

Two matrices are equal if each pair of corresponding elements are equal. Using this definition of equality, we rewrite the system

$$3x - 4y = -11$$

$$2x + 7y = 41$$

as a matrix equation:

$$\begin{bmatrix} 3x - 4y \\ 2x + 7y \end{bmatrix} = \begin{bmatrix} -11 \\ 41 \end{bmatrix}$$

$$\begin{bmatrix} 3 & -4 \\ 2 & 7 \end{bmatrix} \cdot \begin{bmatrix} x \\ y \end{bmatrix} = \begin{bmatrix} -11 \\ 41 \end{bmatrix}$$

$$A \begin{bmatrix} x \\ y \end{bmatrix} = B$$

where $A$ is the matrix of coefficients of the system and $B$ is the matrix of constants.

**EXAMPLE 1**   Writing a System as a Matrix Product

Write the following system of equations as a matrix equation that includes the expression $A \begin{bmatrix} x \\ y \end{bmatrix}$, where $A$ is the matrix of coefficients of the following system:

$$7x + 3y = 23$$

$$x - 12y = 19$$

SOLUTION

$$\begin{bmatrix} 7 & 3 \\ 1 & -12 \end{bmatrix} \cdot \begin{bmatrix} x \\ y \end{bmatrix} = \begin{bmatrix} 23 \\ 19 \end{bmatrix}$$

∎

This method of writing a system of linear equations as a matrix equation is not restricted to systems of two equations and two variables. In a system of three equations and three variables, the matrix of coefficients is simply a $3 \times 3$ matrix.

**EXAMPLE 2**     Writing a System as a Matrix Product

Write the following system of equations as a matrix equation that includes a matrix product:

$$4x - 2y + 3z = 3$$

$$2x + \ y - 4z = 8$$

$$-x + 3y - 2z = 5$$

SOLUTION

$$\begin{bmatrix} 4 & -2 & 3 \\ 2 & 1 & -4 \\ -1 & 3 & -2 \end{bmatrix} \cdot \begin{bmatrix} x \\ y \\ z \end{bmatrix} = \begin{bmatrix} 3 \\ 8 \\ 5 \end{bmatrix}$$

■

## Solving Systems of Linear Equations Using Matrices

Examples 1 and 2 showed that a system of linear equations with an equal number of equations and variables can be written as a matrix equation. For example, a $2 \times 2$ system can be written

$$A \cdot \begin{bmatrix} x \\ y \end{bmatrix} = B, \tag{1}$$

where $A$ is a square matrix of coefficients and $B$ is a matrix of constants. To compute this product using a grapher, we enter the coefficient matrix into matrix A and the constant matrix into matrix B in the Matrix edit menu. Thus

$$B = \begin{bmatrix} b_1 \\ b_2 \end{bmatrix}.$$

**REMINDER**

Matrix multiplication is not commutative. Because we want to multiply $A \cdot \begin{bmatrix} x \\ y \end{bmatrix}$ on the left by $A^{-1}$, we must multiply $B$ on the left also.

We know that $A^{-1}$ does not always exist, but when it does, we can multiply each side of Eq. (1) on the left by $A^{-1}$ to obtain

$$A^{-1} \cdot A \cdot \begin{bmatrix} x \\ y \end{bmatrix} = A^{-1} \cdot B$$

$$\begin{bmatrix} x \\ y \end{bmatrix} = A^{-1} \cdot B.$$

We see that the solution to the system is equal to the product of matrices $A^{-1}$ and $B$.

Example 3 demonstrates this method of solving a $2 \times 2$ system of equations.

**EXAMPLE 3**     Solving a System of Linear Equations Using Matrices

Solve the following system of equations using matrix multiplication:

$$4x - \ 3y = -26$$

$$15x + 11y = \ \ 36$$

SOLUTION
**Solve Algebraically.**

1. Write the system as a matrix equation.

$$\begin{bmatrix} 4 & -3 \\ 15 & 11 \end{bmatrix} \begin{bmatrix} x \\ y \end{bmatrix} = \begin{bmatrix} -26 \\ 36 \end{bmatrix}$$

2. Solve the matrix equation for $\begin{bmatrix} x \\ y \end{bmatrix}$.

$$\begin{bmatrix} 4 & -3 \\ 15 & 11 \end{bmatrix}^{-1} \begin{bmatrix} 4 & -3 \\ 15 & 11 \end{bmatrix} \begin{bmatrix} x \\ y \end{bmatrix} = \begin{bmatrix} 4 & -3 \\ 15 & 11 \end{bmatrix}^{-1} \cdot \begin{bmatrix} -26 \\ 36 \end{bmatrix}$$

$$\begin{bmatrix} x \\ y \end{bmatrix} = \begin{bmatrix} 4 & -3 \\ 15 & 11 \end{bmatrix}^{-1} \cdot \begin{bmatrix} -26 \\ 36 \end{bmatrix}$$

```
[A]⁻¹ * [B]
                    [[⁻2]
                     [ 6 ]]
```

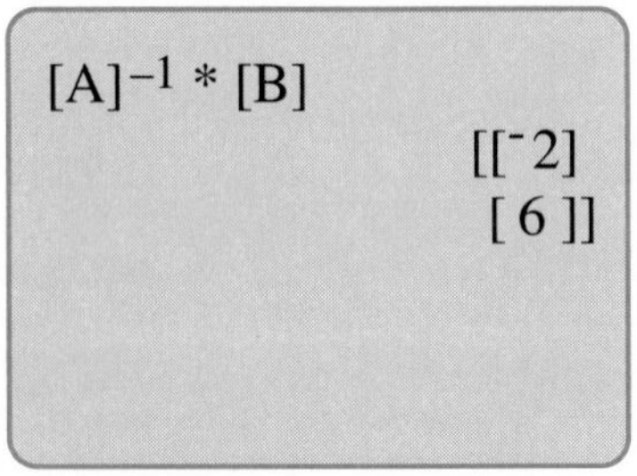

**FIGURE 9.18** The matrix calculation for Example 3 using a grapher.

3. Enter into a grapher the matrix of coefficients $A = \begin{bmatrix} 4 & -3 \\ 15 & 11 \end{bmatrix}$ and the matrix $B = \begin{bmatrix} -26 \\ 36 \end{bmatrix}$. Find the matrix product as shown in Figure 9.18. The solution to the system is $x = -2$ and $y = 6$.

**Support Numerically.** Figure 9.19 shows numerical support on a grapher of the solution. ∎

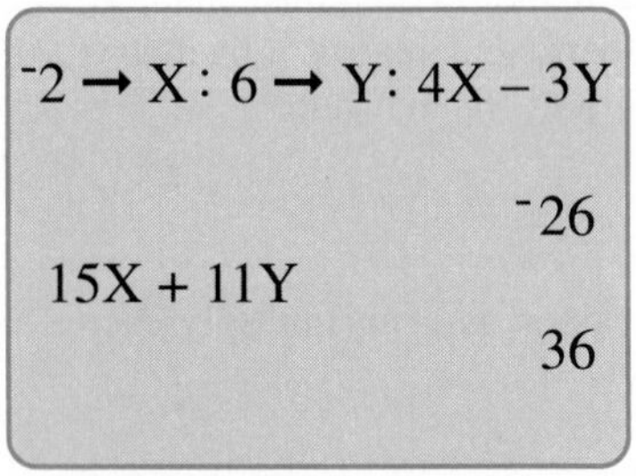

```
⁻2 → X : 6 → Y : 4X − 3Y

                       ⁻26
15X + 11Y
                        36
```

**FIGURE 9.19** Grapher support of the solution to the system of equations in Example 3.

The method illustrated by Example 3 can be used to solve any system with $n$ equations and $n$ variables. The method is summarized in Theorem 9.3.

---

**THEOREM 9.3**   Solving a System of Equations

Suppose a system of $n$ equations and $n$ variables is written as the matrix equation $AX = B$, where $A$ is the matrix of coefficients of the system and $X$ and $B$ are $n \times 1$ matrices of variables and constants, respectively.

If $\det A \neq 0$, then the system

$$AX = B$$

has the unique solution

$$X = A^{-1}B.$$

Theorem 9.3 is used in Example 4 to solve a system of three equations with three variables.

**EXAMPLE 4**   Solving a System of Three Equations

Use matrix multiplication to solve the following system. Support the solution numerically using a grapher.

$$2x - 3y + z = 1.1$$
$$5x + 2y - 3z = 3.2$$
$$3x - 5y + z = -2.4$$

SOLUTION

**Solve Algebraically.**

1. Write the system as a matrix equation.

$$\begin{bmatrix} 2 & -3 & 1 \\ 5 & 2 & -3 \\ 3 & -5 & 1 \end{bmatrix} \cdot \begin{bmatrix} x \\ y \\ z \end{bmatrix} = \begin{bmatrix} 1.1 \\ 3.2 \\ -2.4 \end{bmatrix}$$

2. Let

$$A = \begin{bmatrix} 2 & -3 & 1 \\ 5 & 2 & -3 \\ 3 & -5 & 1 \end{bmatrix} \quad \text{and} \quad B = \begin{bmatrix} 1.1 \\ 3.2 \\ -2.4 \end{bmatrix}.$$

3. Find the matrix product $A^{-1} \cdot B$ on a grapher as shown in Figure 9.20.

$$\begin{bmatrix} x \\ y \\ z \end{bmatrix} = \begin{bmatrix} 2.5 \\ 3 \\ 5.1 \end{bmatrix}$$

[A]$^{-1}$ * [B]
[[2.5]
[ 3 ]
[5.1]]

**FIGURE 9.20** The matrix calculation on a grapher.

**Support Numerically.** Figure 9.21 shows numerical support, using a grapher, that the unique solution to this system is the ordered triple $(2.5, 3, 5.1)$.   ■

A system of $n$ equations in $n$ variables can contain equations with fewer than $n$ variables. In that event, fill the position of the missing variable with a 0 in the matrix array. Example 5 solves a system of three equations in which individual equations have only two variables.

2.5 → X: 3 → Y: 5.1 → Z

2X – 3Y + Z          1.1

5X + 2Y – 3Z          3.2

3X – 5Y + Z          ‾2.4

**FIGURE 9.21** The numerical support of the solution to the system of equations in Example 4.

**EXAMPLE 5**   Solving a System That Has Missing Terms

Use matrix multiplication to solve the following system. Support numerically using a grapher.

$$5.4x \qquad + 3.2z = 47.06$$
$$3y + 6.5z = -165.85$$
$$6.8x - 4.5y \qquad = 126.7$$

$[A]^{-1} * [B]$

$$[[ 27.5 ]$$
$$[ 13.4 ]$$
$$[{}^{-}31.7]]$$

**FIGURE 9.22** The matrix calculation of Example 5 using a grapher.

$27.5 \rightarrow X : 13.4 \rightarrow Y :$
${}^{-}31.7 \rightarrow Z : 5.4X + 3.2Z$
$$47.06$$
$3Y + 6.5Z$
$${}^{-}165.85$$
$6.8X - 4.5Y$
$$126.7$$

**FIGURE 9.23**
Numerical support of the solution to the system of equations in Example 5.

**SOLUTION**

**Solve Algebraically.**

1. Write the system as a matrix equation.

$$\begin{bmatrix} 5.4 & 0 & 3.2 \\ 0 & 3 & 6.5 \\ 6.8 & -4.5 & 0 \end{bmatrix} \cdot \begin{bmatrix} x \\ y \\ z \end{bmatrix} = \begin{bmatrix} 47.06 \\ -165.85 \\ 126.7 \end{bmatrix}$$

2. Let

$$A = \begin{bmatrix} 5.4 & 0 & 3.2 \\ 0 & 3 & 6.5 \\ 6.8 & -4.5 & 0 \end{bmatrix} \quad \text{and} \quad B = \begin{bmatrix} 47.06 \\ -165.85 \\ 126.7 \end{bmatrix}.$$

3. Find the matrix product $A^{-1} \cdot B$ on a grapher as shown in Figure 9.22.

Therefore

$$\begin{bmatrix} x \\ y \\ z \end{bmatrix} = \begin{bmatrix} 27.5 \\ 13.4 \\ -31.7 \end{bmatrix}.$$

**Support Numerically.** Figure 9.23 shows support on a grapher that the unique solution to this system is the ordered triple $(27.5, 13.4, -31.7)$. ■

## Solving Problems with Systems of Equations

Consider the following new problem situation.

□ □

### BOAT/CURRENT PROBLEM SITUATION

Haylan has a motorboat that he wants to take 15 mi down the Black Creek River and back. The speed of the boat as recorded on its speedometer is constant going down and up the river. He finds that it takes him only 1 hr to go downstream but 1 hr 40 min going upstream.

□ □

An algebraic representation of this problem situation is found in Example 6. The key formula for this problem is

$$\text{distance} = \text{rate} \times \text{time}.$$

or,

$$d = r \times t.$$

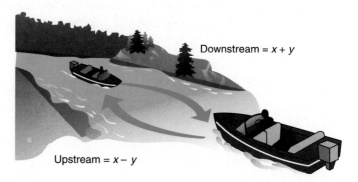

Downstream = $x + y$

Upstream = $x - y$

**FIGURE 9.24** The **Boat/Current Problem Situation.**

**EXAMPLE 6**   Solving the Boat/Current Problem:  An
Algebraic Representation

Find the algebraic represention of the boat's rate going both downstream and upstream.

SOLUTION

$d = 15$ mi both up- and downstream.  Going downstream, the river adds speed.
Going upstream, the river subtracts speed.
Let $x =$ rate of boat in still water.
Let $y =$ rate of current.

**Boat's rate downstream** $= x + y$

**Boat's rate upstream** $= x - y$

Apply the $d = r \cdot t$ formula twice—once for upstream and once for downstream—
to obtain the following pair of equations:

$$1(x + y) = 15 \quad \text{This equation represents } d = r \times t \text{ going downstream.}$$

$$\frac{5}{3}(x - y) = 15 \quad \text{This equation represents } d = r \times t \text{ going upstream.}$$

This pair of linear equations is an algebraic representation of the problem
situation. ■

Example 7 continues the discussion of the **Boat/Current Problem Situation.**

**EXAMPLE 7**    Solving the Boat/Current Problem (cont'd)

Find the rate of the boat in still water and the rate of the river's current.

SOLUTION

1. The algebraic representation of the problem situation is the following system of linear equations developed in Example 6.

$$1(x + y) = 15$$

$$\frac{5}{3}(x - y) = 15$$

2. Rather than introducing a rounding error by approximating 5/3 as a decimal, multiply both sides of the second equation by 3/5 to eliminate the fraction. The system becomes as follows:

$$1(x + y) = 15 \qquad\qquad x + y = 15$$

$$\text{becomes}$$

$$\frac{3}{5} \cdot \frac{5}{3}(x - y) = \frac{3}{5} \cdot \frac{15}{1} \qquad x - y = 9$$

3. This system is equivalent to the matrix equation

$$\begin{bmatrix} 1 & 1 \\ 1 & -1 \end{bmatrix} \cdot \begin{bmatrix} x \\ y \end{bmatrix} = \begin{bmatrix} 15 \\ 9 \end{bmatrix}.$$

4. Because the determinant of the matrix of coefficients is not zero, by Theorem 9.3 the solution is found as the matrix product

$$\begin{bmatrix} x \\ y \end{bmatrix} = \begin{bmatrix} 1 & 1 \\ 1 & -1 \end{bmatrix}^{-1} \cdot \begin{bmatrix} 15 \\ 9 \end{bmatrix}.$$

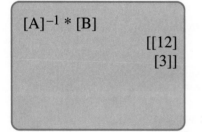

[A]⁻¹ * [B]

[[12]
[3]]

**FIGURE 9.25** The matrix calculation of Example 7 using a grapher.

Figure 9.25 shows that $A^{-1} \cdot B = \begin{bmatrix} 12 \\ 3 \end{bmatrix}$ and the unique solution to this system is the ordered pair (12, 3).

Therefore the boat's rate in still water is 12 mph and the rate of the river's current is 3 mph.

**Support Numerically.** $12 + 3 = 15$ represents the rate of the boat going downstream, where the time was 1 hr; thus the distance is $1 \cdot 15 = 15$. With the boat going upstream, the rate is $12 - 3 = 9$. Since it took 5/3 hrs upstream, the distance is $9 \cdot (5/3) = 15$.

In the previous chapters, you learned to solve investment problems using various methods. Example 8 illustrates an investment problem situation using a system of three equations in three variables. We choose to solve the system using matrices.

☐                                                                                                    ☐

## Investment Problem Situation

Chris decided to invest $25,000 in three different types of bonds paying 6%, 7%, and 8.5%, respectively. The total amount of interest earned at the end of 1 year was $1870. He invested $5000 more in the 8.5% investment than in the 7% investment.

☐                                                                                                    ☐

An algebraic representation for the problem situation is found in Example 8 in the form of a system of equations.

**EXAMPLE 8**    Solving the Investment Problem

How much did Chris invest at each rate?

SOLUTION

1. Find the algebraic representation.
   Let $x =$ the amount invested at 6%.

   Let $y =$ the amount invested at 7%.

   Let $z =$ the amount invested at 8.5%.

   $$x + y + z = 25000$$     **This equation represents the total amount of money invested.**

   $$0.06x + 0.07y + 0.085z = 1870$$     **This equation represents the total amount of interest.**

   $$z = y + 5000$$     **The investment at 8.5% is $5000 more than the 7% investment.**

2. Write as a system of equations.

   $$x + y + z = 25000$$

   $$0.06x + 0.07y + 0.085z = 1870$$

   $$- \quad y + z = 5000$$

$[A]^{-1} * [B]$

$$[[ 6000 ]$$
$$[ 7000 ]$$
$$[12000]]$$

**FIGURE 9.26** The matrix calculation of Example 8 using a grapher.

3. Write this as the matrix equation.

$$\begin{bmatrix} 1 & 1 & 1 \\ 0.06 & 0.07 & 0.085 \\ 0 & -1 & 1 \end{bmatrix} \cdot \begin{bmatrix} x \\ y \\ z \end{bmatrix} = \begin{bmatrix} 25000 \\ 1870 \\ 5000 \end{bmatrix}$$

and enter $A = \begin{bmatrix} 1 & 1 & 1 \\ 0.06 & 0.07 & 0.085 \\ 0 & -1 & 1 \end{bmatrix}$ and $B = \begin{bmatrix} 25000 \\ 1870 \\ 5000 \end{bmatrix}$ into a grapher.

The solution is the matrix product $A^{-1} \cdot B$, assuming $A^{-1}$ exists.
Figure 9.26 shows that the unique solution to the system is the ordered triple (6000, 7000, 12,000). Therefore Chris invested $6000 in the 6% bonds, $7000 in the 7% bonds, and $12,000 in the 8.5% bonds. ∎

## Exercises for Section 9.4

In Exercises 1–4, write the system of equations as a matrix equation.

1. $\begin{cases} 2x - 3y = 9 \\ 5x + 4y = 13 \end{cases}$

2. $\begin{cases} 2.1x + 3.7y = 9.5 \\ 0.6x - 2.5y = 13.9 \end{cases}$

3. $\begin{cases} 5.1x - 8.6y + 3.1z = 19.7 \\ 3.2x + 2.7y + 6.8z = 18.3 \\ 4.5x - 5.6y + 0.3z = 20.9 \end{cases}$

4. $\begin{cases} 2x \quad\ - z = 9 \\ 3x + 4y + 5z = 16 \\ \quad\ 4y - 6z = 12 \end{cases}$

In Exercises 5–8, write the matrix equation as a system of linear equations.

5.

$$\begin{bmatrix} 2 & -3 \\ 5 & -6 \end{bmatrix} \cdot \begin{bmatrix} x \\ y \end{bmatrix} = \begin{bmatrix} 12 \\ 15 \end{bmatrix}$$

6.

$$\begin{bmatrix} 4.1 & 6.5 \\ -3.5 & -9.2 \end{bmatrix} \cdot \begin{bmatrix} x \\ y \end{bmatrix} = \begin{bmatrix} 2.7 \\ 8.6 \end{bmatrix}$$

7.

$$\begin{bmatrix} 5.6 & -3.1 & 0 \\ -7.2 & 0 & 8.8 \\ 2.3 & 5.1 & 0 \end{bmatrix} \cdot \begin{bmatrix} x \\ y \\ z \end{bmatrix} = \begin{bmatrix} 15.4 \\ 19.7 \\ 18.3 \end{bmatrix}$$

8.

$$\begin{bmatrix} 5 & 8 & 0 \\ 0 & 10 & 7 \\ 3 & -2 & -6 \end{bmatrix} \cdot \begin{bmatrix} x \\ y \\ z \end{bmatrix} = \begin{bmatrix} 12 \\ 26 \\ 18 \end{bmatrix}$$

In Exercises 9–26, solve the system of linear equations on your grapher using matrices.

9. $\begin{cases} 5x - 6y = -15 \\ 2x + 3y = 21 \end{cases}$

10. $\begin{cases} 3x + 7y = 74 \\ 2x - 8y = 76 \end{cases}$

11. $\begin{cases} -4x + 5y = 25 \\ 3x - y = 6 \end{cases}$

12. $\begin{cases} 2.1x - 3.2y = 12.9 \\ 5.2x + 2.4y = 30 \end{cases}$

13. $\begin{cases} 3x - 2y = -8.7 \\ x + 6y = 51.1 \end{cases}$

**14.** $\begin{cases} x - 4y = -35.2 \\ 5x + 6y = 8.6 \end{cases}$

**15.** $\begin{cases} 2x - 3y = 9 \\ 5x + 4y = 13 \end{cases}$

**16.** $\begin{cases} 5x - 4y + z = -26 \\ 3x + y - 5z = 34 \\ 6x - 5y + 2z = -36 \end{cases}$

**17.** $\begin{cases} 7x - 3y + 4z = 25 \\ -3x + 5y - z = 1 \\ 2x + y + z = 10 \end{cases}$

**18.** $\begin{cases} x + y + z = 2 \\ x + 3z = -5 \\ 2y - z = 11 \end{cases}$

**19.** $\begin{cases} 2x - 3y + 5z = 22 \\ 3x - 6z = 15 \\ 2y + 4z = -8 \end{cases}$

**20.** $\begin{cases} 2x - 5y + z = -4.3 \\ 3x - 4y + 2z = 10.3 \\ 5x + 2y + 7z = 80.5 \end{cases}$

**21.** $\begin{cases} 5x - 4y + z = -26 \\ 3x + y - 5z = 34 \\ 6x - 5y + 2z = -36 \end{cases}$

**22.** $\begin{cases} 2.1x - 3.7y + 6.5z = 25.56 \\ 3.4x + 5.2y - z = 11.74 \\ 6.8x + 2.3y - 5.6z = -14.77 \end{cases}$

**23.** $\begin{cases} x + y + z = 5.1 \\ 2.3x - 5.8z = -39.26 \\ x + 5y = -23.9 \end{cases}$

**24.** $\begin{cases} 5x - 3y + z = 0 \\ 2x + 6y + 5z = 28.5 \\ -3x - 2y + 7z = -29.5 \end{cases}$

**25.** $\begin{cases} 4x - 8y + 3z = 25 \\ 5x - 2y - z = 23 \\ -2x - 7y + 3z = 9 \end{cases}$

**26.** $\begin{cases} 8.1x + 3.2y + z = 15.2 \\ -3.5x - y + 2z = -4 \\ 6.1x + 2.23y - 4z = 2.6 \end{cases}$

## EXTENDING THE IDEAS

**27.** Write a system of equations with $(-4, 5, 7)$ as its only solution. (There is more than one correct answer.)

**28.** Write a system of equations with $(17, -3.5, 9.2)$ as its only solution. (There is more than one correct answer.)

**29.** Determine $a$, $b$, and $c$ so that the graph of $y = ax^2 + bx + c$ passes through the points $(-1, 4)$, $(1, 2)$, and $(2, 7)$.

**30.** Determine $a$, $b$, and $c$ so that the graph of $y = ax^2 + bx + c$ passes through the points $(-1, -4)$, $(0, -2)$, and $(3, -8)$.

**31.** Find the solution to the following system of four equations with four variables by using matrices:

$$\begin{cases} 3w + 2x - 4y + z = -26 \\ 8w - x + 2y = 60 \\ 3w + 2x - 7y - z = -48 \\ 5w + x - y + z = 15 \end{cases}$$

**32.** *Combined Rates.* The *Annabelle Lee* paddleboat can travel from St. Augustine to Jacksonville, Fl, (a distance of 20 mi) up the St. John's River with the current in 2 hr. It takes 2.5 hr to return the same distance going against the current.

a) Write an algebraic representation for the problem situation in the form of a system of linear equations.

b) Find the rate of the current and the boat's rate in still water.

For Exercise 32.

In Exercises 33–38, write a system of equations as an algebraic representation for the problem situation and then solve the system using matrices.

**33.** *Finding Rates.* A boat traveling downstream, with the current, can cover 21 mi in the same time (3 hr) that it takes to travel 9 mi upstream against the current. Find the rate of the boat in still water and the rate of the current.

**34.** *Finding Rates.* A boat travels 45 mi downstream on a river in 3 hr and makes the return trip in 5 hr. Find the rate of the current and the boat's rate in still water.

**35.** *Finding Rates.* A plane flies 2800 mi with the wind in 4 hr and makes the return trip against the wind in 5 hr. Find the rate of the wind and the plane's rate in still air.

**36.** *Investment Analysis.* Kyronda invests $20,000 in three different stocks that pay 9.5%, 7%, and 6.5% simple interest, respectively. If she invests twice as much at 9.5% as 6.5% and the total interest at the end of a year is $1634, how much does she invest at each rate?

**37.** *Pipe Cutting.* A 40-ft pipe is cut into three pieces. The largest piece is three times longer than the smallest piece and the other piece is 2 ft shorter than the largest piece. Find the length of each piece.

**38.** *Angle Measurement.* The measure of one angle of a triangle is 10° more than twice the measure of a second angle. The measure of the third angle is the sum of the measures of the other two angles. Find the measure of each angle.

LOOKING BACK—MAKING CONNECTIONS

**39.** Show that the triangle determined by the points $(-3, -3)$, $(3, -12)$, and $(6, 3)$ is a right triangle.

**40.** Show that the four-sided figure determined by the points $(-2, -2)$, $(2, 8)$, $(8, 17)$, and $(4, 7)$ is a parallelogram.

**41.** Determine $m$ and $b$ so that the graph of $y = mx + b$ passes through the points $(2, -1)$ and $(3, 5)$.

## Chapter 9 Summary

| | Systems of Equations; Higher-order Equations | Examples |
| --- | --- | --- |
| Linear Equation in Three Variables | $Ax + By + Cz = D$, where $x$, $y$, and $z$ are variables and $A, B, C,$ and $D$ are real numbers and not all of $A, B,$ and $C$ are zero. | $3x + 5y - 2z = 10$ |

| System of Higher-order Equations | A system of equations with at least three variables | A higher-order system: |
|---|---|---|

A higher-order system:
$$3x - 4y + z = 1$$
$$x + y - z = 2$$
$$x - 2y + 3z = 0$$

| Ordered Triple | A solution to a system of equations in three variables, written in the form $(x, y, z)$ | The ordered triple $(2, 0, 1)$ is a solution to the following: |
|---|---|---|

$$x + 2y - z = 1$$
$$2x - 5y + 3z = 7$$
$$x - 3y + 2z = 4$$

---

| | **Methods for Finding Solutions to a System** | **Examples** |
|---|---|---|
| Solving a System of Three Equations in Three Variables by the Elimination Method | 1. Eliminate one of the variables from a pair of equations.<br>2. Eliminate the same variable from a different pair of equations.<br>3. Solve the resulting system of equations for the two remaining variables.<br>4. Substitute the two values into one of the original equations to find the value of the third variable.<br>5. Support the result numerically. | |
| Solving an $n \times n$ System Using Matrices | 1. Find the matrix of coefficients $A$, matrix of constants $B$, and matrix of variables $X$, such that $AX = B$.<br>2. If $\det A \neq 0$, a unique solution exists that is $X = A^{-1}B$, where $A^{-1}$ is the inverse matrix of $A$. | Consider, for example, the following system:<br>$$2x - y = 3$$<br>$$-x + y = 2$$<br>For this system,<br>$$A = \begin{bmatrix} 2 & -1 \\ -1 & 1 \end{bmatrix}, B = \begin{bmatrix} 3 \\ 2 \end{bmatrix},$$<br>$$A^{-1}B = \begin{bmatrix} 1 & 1 \\ 1 & 2 \end{bmatrix} \cdot \begin{bmatrix} 3 \\ 2 \end{bmatrix} = \begin{bmatrix} 5 \\ 7 \end{bmatrix}.$$<br>The solution is the ordered pair $(5, 7)$. |

| | Matrix Algebra Terminology | Examples |
|---|---|---|
| Matrix | A rectangular array of data whose entries are arranged in rows and columns | The matrix $\begin{bmatrix} 3 & -5 & 2 \\ -1 & 0 & 3 \end{bmatrix}$ has 2 rows and 3 columns. |
| Dimension, $m \times n$ | The number of rows and columns of a matrix, where $m$ is the number of rows and $n$ is the number of columns of the matrix | $\begin{bmatrix} 3 & -5 & 2 \\ -1 & 0 & 3 \end{bmatrix}$ is a $2 \times 3$ matrix. |
| Square Matrix | A matrix that has the same number of rows as columns | $\begin{bmatrix} 1 & -2 \\ 4 & -6 \end{bmatrix}$ is a $2 \times 2$ square matrix. |
| Determinant of a $2 \times 2$ Matrix | Given $A = \begin{bmatrix} a & b \\ c & d \end{bmatrix}$, the determinant is $ad - bc$, denoted by $\begin{vmatrix} a & b \\ c & d \end{vmatrix}$ or det $A$. | Given $A = \begin{bmatrix} 5 & 3 \\ 1 & 2 \end{bmatrix}$, det $A = 7$. |

| | Operations with Matrices | Examples |
|---|---|---|
| Sum of Two $m \times n$ Matrices | The sum is found by adding the corresponding entries of two matrices that have the same dimension. Corresponding entries are entries in different matrices that are located in the same row and same column. | $\begin{bmatrix} 3 \\ 1 \\ -1 \end{bmatrix} + \begin{bmatrix} 2 \\ -1 \\ 5 \end{bmatrix} = \begin{bmatrix} 5 \\ 0 \\ 4 \end{bmatrix}$ |
| Difference of Two $m \times n$ Matrices | The difference is found by subtracting the corresponding entries of two matrices that have the same dimension. | $\begin{bmatrix} 6 \\ 0 \end{bmatrix} - \begin{bmatrix} 2 \\ -1 \end{bmatrix} = \begin{bmatrix} 4 \\ 1 \end{bmatrix}$ |
| Scalar Multiplication of a Matrix | The scalar product is found by multiplying all entries in a matrix by a scalar, a real-number constant. | $2\begin{bmatrix} 3 & -2 \\ 4 & 0 \end{bmatrix} = \begin{bmatrix} 6 & -4 \\ 8 & 0 \end{bmatrix}$ |

| Multiplication of Two Matrices, $A_{m \times n}$ and $B_{n \times r}$ | For each $ij$ entry in the product matrix, add the products of each corresponding entry of row $i$ in matrix $A$ and column $j$ in matrix $B$. | $\begin{bmatrix} 0 & 1 \\ 2 & 3 \end{bmatrix} \cdot \begin{bmatrix} 4 & 5 \\ 6 & 7 \end{bmatrix} =$ $\begin{bmatrix} 0+6 & 0+7 \\ 8+18 & 10+21 \end{bmatrix} =$ $\begin{bmatrix} 6 & 7 \\ 26 & 31 \end{bmatrix}$ |
|---|---|---|

|  | **Important Matrix Concepts** | **Examples** |
|---|---|---|
| Identity Matrix | An $n \times n$ matrix, $I_n$, such that $A \cdot I = I \cdot A = A$, where $A$ is any $n \times n$ matrix. Entries of $I$ are 1's on the main diagonal and 0's elsewhere. | $I_2 = \begin{bmatrix} 1 & 0 \\ 0 & 1 \end{bmatrix}$ |
| Inverse Matrix | A matrix $A^{-1}$, in which $A \cdot A^{-1} = A^{-1} \cdot A = I_n$ for an $n \times n$ matrix $A$. Not every $n \times n$ matrix has an inverse. | If $A = \begin{bmatrix} 3 & 2 \\ 1 & 1 \end{bmatrix}$, $A^{-1} = \begin{bmatrix} 1 & -2 \\ -1 & 3 \end{bmatrix}$. |
| Invertible Matrix | A matrix that has an inverse |  |

## Review Exercises for Chapter 9

In Exercises 1 and 2, find the missing number in the ordered triple that is the solution of the system of equations.

**1.** $(2, 0, ?)$; $\begin{cases} 5x - 3y + 2z = 4 \\ 2x + 5y - 4z = 16 \\ 8x + 2y + 3z = 7 \end{cases}$

**2.** $(?, -4, 2)$; $\begin{cases} 2x - 3y - 3z = 4 \\ 2x + 3y + 8z = 2 \\ 3x - y + 2z = 5 \end{cases}$

**3.** Write a system of equations with $(2, -2, 5)$ as the solution.

In Exercises 4–15, consider the following matrices $A$, $B$, and $C$. Perform the indicated operations using a paper-and-pencil method and support the solution using a grapher.

$$A = \begin{bmatrix} 6 & 4 \\ 2 & 0 \end{bmatrix} \quad B = \begin{bmatrix} 3 & -1 \\ 2 & 1 \end{bmatrix} \quad C = \begin{bmatrix} 1 & 2 & -1 \\ 0 & 3 & 1 \end{bmatrix}$$

**4.** Find the dimension of $A + B$.

**5.** Find the dimension of $A \cdot C$.

**6.** det $A$

**7.** det $B$

**8.** $A + B$

**9.** $B - A$

**10.** $5C$

**11.** 150% of $A$

**12.** $3A + 2B$

**13.** $2A - B$

**14.** $A \cdot B$

**15.** $B \cdot C$

In Exercises 16–21, consider the following matrices $A$, $B$, and $C$. Perform the indicated operations using a grapher.

$$A = [3 \quad 5] \quad B = \begin{bmatrix} 1 & 7 \\ 3 & 8 \end{bmatrix} \quad C = \begin{bmatrix} 2 & -2 \\ 1 & -1 \end{bmatrix}$$

**16.** $B + C$

**17.** $2B - 3C$

**18.** $A \cdot B$

**19.** $B \cdot C$

**20.** $B^2$

**21.** $\det [C]$

**22.** Find the inverse matrix of $A = \begin{bmatrix} 3 & 5 \\ 4 & 7 \end{bmatrix}$.

**23.** Find the inverse matrix of $A = \begin{bmatrix} 3 & -1 \\ -4 & 2 \end{bmatrix}$.

**24.** Determine the $4 \times 4$ identity matrix $I_4$.

In Exercises 25 and 26, determine the matrix of coefficients and the matrix of constants for the system of equations.

**25.** $\begin{cases} 4x + 6y = 7 \\ -3x + \ y = 0 \end{cases}$

**26.** $\begin{cases} 6x = 15 \\ x - 5y = 12 \end{cases}$

In Exercises 27 and 28, write the system of equations as a matrix equation.

**27.** $\begin{cases} 2x - 5y = 0 \\ x + 8y = 10 \end{cases}$

**28.** $\begin{cases} 12x + 15y = 7 \\ 9x - \ 3y = 8 \end{cases}$

In Exercises 29 and 30, write the matrix equation as a system of linear equations.

**29.** $\begin{bmatrix} -1 & 3 \\ 0 & 2 \end{bmatrix} \cdot \begin{bmatrix} x \\ y \end{bmatrix} = \begin{bmatrix} 2 \\ -1 \end{bmatrix}$

**30.** $\begin{bmatrix} 3 & -3 & 1 \\ 5 & 0 & 2 \\ 1 & 2 & -1 \end{bmatrix} \cdot \begin{bmatrix} x \\ y \\ z \end{bmatrix} = \begin{bmatrix} 1 \\ 3 \\ -1 \end{bmatrix}$

In Exercises 31–34, solve the system of equations using the elimination or substitution method.

**31.** $\begin{cases} 2x - \ y - 3z = 5 \\ 3x - 2y - 6z = 5 \\ x + 2y + 8z = 13 \end{cases}$

**32.** $\begin{cases} 4x - 3y - 5z = 6 \\ 3x + 2y + 2z = 0 \\ 2x + 2y - \ z = 9 \end{cases}$

**33.** $\begin{cases} 5x - 3y - 3z = 5 \\ 9x + 2y - 4z = -4 \\ -2x + 6y - 5z = 18 \end{cases}$

**34.** $\begin{cases} x - 3y - 3z = 2 \\ 7x + 5y + \ z = 6 \\ 6x - \ y - 4z = 6 \end{cases}$

In Exercises 35–38, solve the system of linear equations using matrices.

**35.** $\begin{cases} 3x - \ 4y - 2z = 3 \\ 6x + \ 8y + 7z = 3 \\ 2x + 10y - 9z = 16 \end{cases}$

**36.** $\begin{cases} 2x - 3y + 3z = 3 \\ -x + 2y - \ z = 8 \\ 5x + \ y \quad\ \ = -7 \end{cases}$

**37.** $\begin{cases} 5x + \ y - 4z = 18 \\ 4x - 2y + \ z = 20 \\ 2x - \ y + \ z = 15 \end{cases}$

**38.** $\begin{cases} 2x - \ y - 3z = -6 \\ 8x - 7y - 5z = 0 \\ 6x + 2y - 8z = 94 \end{cases}$

**39.** *Integers Problem.* The sum of three integers is zero. The sum of the first two integers is six more than the third integer. If the third integer is doubled, it is triple the second integer. Find the integers.

**40.** *Travel Problem.* A tour boat travels 21.75 mi upstream on a river in 2.9 hr and makes the return trip in 1.5 hr. Find the rate of the current and the boat's rate in still water.

## Chapter 9 Test

In Questions 1–6, consider the following matrices $A$, $B$, and $C$. Perform the indicated operations using paper and pencil.

$$A = \begin{bmatrix} 2 & 1 \\ 0 & -1 \\ -3 & 2 \end{bmatrix} \quad B = \begin{bmatrix} 5 & -2 \\ 3 & 0 \end{bmatrix} \quad C = \begin{bmatrix} 3 & 4 \\ 1 & 7 \end{bmatrix}$$

1. Find the dimension of $A$.

2. Find the determinant of $C$.

3. Evaluate $B + C$.

4. Evaluate $B - C$.

5. Evaluate $3B + 2C$.

6. Evaluate $A \cdot B$.

7. Use a grapher to find 75% of $[D]$ if

$$D = \begin{bmatrix} 1.5 & -2 \\ 5.6 & 0.44 \end{bmatrix}.$$

8. Use a grapher to determine the inverse matrix of

$$\begin{bmatrix} 3 & 4 \\ -1 & 2 \end{bmatrix}.$$

9. Write $\begin{cases} 5x - 10y = 6 \\ x + 6y = 3 \end{cases}$ as a matrix equation.

10. Write $\begin{bmatrix} 1 & 0 & 4 \\ 3 & -2 & 2 \\ 0 & 5 & -1 \end{bmatrix} \cdot \begin{bmatrix} x \\ y \\ z \end{bmatrix} = \begin{bmatrix} 1 \\ 8 \\ 3 \end{bmatrix}$ as a system

of linear equations.

In Questions 11 and 12, solve the system of equations using the elimination or substitution method.

11. $\begin{cases} 2x - 5y + z = 1 \\ 3x + y + 5z = 0 \\ x - 4y + 2z = 5 \end{cases}$ 12. $\begin{cases} 3x - 2y + z = -1 \\ 2x - y - z = 5 \\ 2x \quad\quad + z = 8 \end{cases}$

In Questions 13 and 14, solve the system of linear equations using matrices.

13. $\begin{cases} 5x - 9y \quad\quad = 2 \\ 2x - 5y + z = 3 \\ 6x + y - 4z = 6 \end{cases}$ 14. $\begin{cases} 3x - 6y + 2z = 2 \\ 5x + 2y + z = 4 \\ 2x + 4y - 5z = -3 \end{cases}$

15. Chun Lee invests \$1200 in two bank accounts. His savings account yields a 5.5% annual rate of interest and his checking account a 2.5% annual rate of interest. The total amount earned in the first year is \$56.10. Let $x$ represent the amount of money put into the savings account and $y$ the amount of money in the checking account.

a) Write an algebraic representation for the problem situation in the form of a system of linear equations.

b) Use matrices to solve the system found in part (a) to determine the amount invested at each rate.

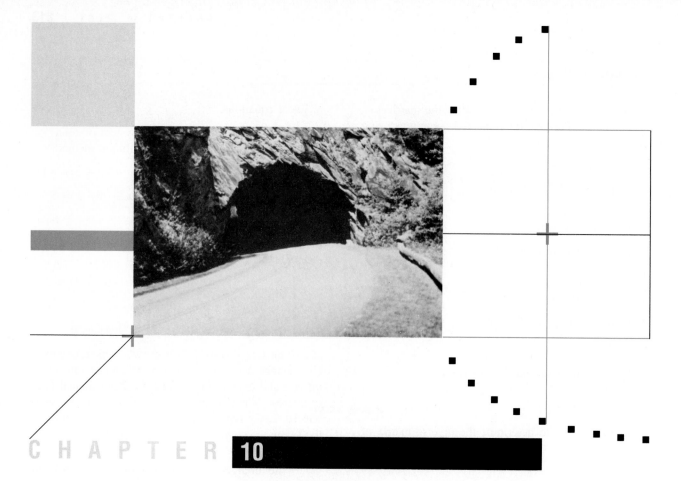

# CHAPTER 10

## Conic Sections

AN APPLICATION

Suppose that a tunnel entrance is shaped like an arc of an ellipse, 22 ft high at its highest point and 40 ft across. The height $h$ at a point $x$ feet from the edge of the tunnel can be found using the formula

$$h = 22\sqrt{\left(1 - \frac{x^2}{20^2}\right)}.$$

Finding the Equation of a Circle Centered ▪ at the Origin (0, 0)
▪ Finding the Graph of the Equation $x^2 + y^2 = r^2$ ▪ Circles Not
Centered at the Origin ▪ Graphing a Circle Not Centered at the Origin

A **conic section** is the intersection of a plane and a cone. Observe the shape of the slice that results. The angle at which the cone is sliced produces three different types of conic sections, as illustrated in Figure 10.1.

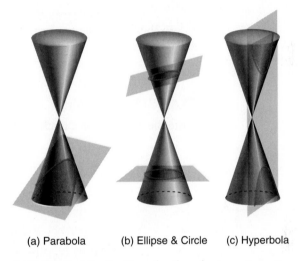

(a) Parabola  (b) Ellipse & Circle  (c) Hyperbola

**FIGURE 10.1** How the three basic conic sections are derived from the intersection of a right circular cone and a plane.

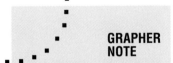

**GRAPHER NOTE**

Consult your grapher lab manual for information on using your grapher to graph conic sections.

There are three basic conic sections: **parabolas, ellipses,** and **hyperbolas**. Note that **circles** are a special case of the ellipse. In Chapter 8, you studied one type of conic section, the parabola. In this chapter, we will study the parabola further, as well as study the ellipse and hyperbola.

## 10.1 Circles

A **circle** consists of all points equidistant from a selected point, the **center**. The distance from the center to any point on the circle is constant and is called the **radius** of the circle.

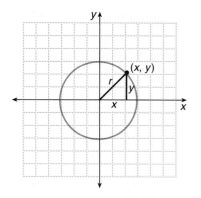

**FIGURE 10.2** An equation of a circle with center $(0,0)$ and radius $r$ is $x^2 + y^2 = r^2$.

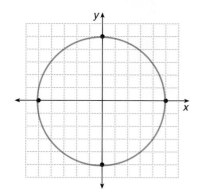

**FIGURE 10.3** A sketch of the graph of $x^2 + y^2 = 25$.

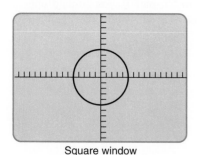

Square window

**FIGURE 10.4** A circle, with center at $(0,0)$ and radius 5, obtained using a grapher.

## Finding the Equation of a Circle Centered at the Origin (0, 0)

The Pythagorean theorem can be used to find the equation of a circle with center at $(0,0)$ and radius $r$. If a perpendicular line is drawn from any point $(x, y)$ on the circle to the $x$-axis, a right triangle is formed. The length of the hypotenuse is $r$, the radius, while the legs have lengths $x$ and $y$. Applying the Pythagorean theorem $a^2 + b^2 = c^2$ to the circle graphed in Figure 10.2, we have the general equation of a circle centered at $(0,0)$

$$x^2 + y^2 = r^2.$$

Example 1 illustrates writing an equation of a circle with center at origin and the given radius.

### EXAMPLE 1    Writing an Equation of a Circle

Write the equation of a circle with radius 3 and center at $(0,0)$.

SOLUTION
Using the equation $x^2 + y^2 = r^2$ and $r = 3$,

$$x^2 + y^2 = 3^2$$
$$x^2 + y^2 = 9.$$                              ■

## Finding the Graph of the Equation $x^2 + y^2 = r^2$

To find a graph of the equation $x^2 + y^2 = 25$ using a grapher, we first must solve the equation for $y$.

$$x^2 + y^2 = 25$$
$$y^2 = 25 - x^2 \qquad \textbf{Take the square root}$$
$$\qquad\qquad\qquad \textbf{of each side of the equation.}$$
$$y = \pm\sqrt{25 - x^2}$$

Figure 10.3 shows the sketch of a graph of the equation $x^2 + y^2 = 25$ with center at $(0,0)$ and radius 5. Can this sketch be supported using a grapher? Yes, by graphing $y = \sqrt{25 - x^2}$ and $y = -\sqrt{25 - x^2}$. To enter these equations into the grapher, define $y_1 = \sqrt{(25 - x^2)}$ and $y_2 = -y_1$. Graph both equations in a Square window (see Figure 10.4).

Sometimes on a grapher the circle does not appear to be complete. In certain windows, this gap will occur; therefore it is important to examine the equation

**GRAPHER NOTE**

The Square window is useful when graphing a circle. Due to the resolution of the pixels on the screen, gaps between the graphs of $y_1$ and $y_2$ might appear using some windows. Can you explain why?

itself in order to recognize which type of conic section is being graphed. Also, if the window is not a Square window, the circle might appear to be distorted.

## Circles Not Centered at the Origin

The distance formula can be used to find the equation of any circle in the Cartesian coordinate system. Figure 10.5 illustrates a circle with center at $(h, k)$ and radius $r$.

Applying the distance formula to the circle in Figure 10.5, we find that the distance $r$ between any point $(x, y)$ on the circle and the center $(h, k)$ is

$$r = \sqrt{(x - h)^2 + (y - k)^2}.$$

Square each side of this equation to remove the radical.

$$r^2 = (x - h)^2 + (y - k)^2$$

This equation is the standard form of a circle.

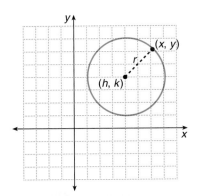

**FIGURE 10.5** A circle with center at $(h, k)$ and radius $r$.

---

### Standard Form of a Circle

The **standard form for an equation of a circle** with center $(h, k)$ and radius $r$ is

$$(x - h)^2 + (y - k)^2 = r^2.$$

Conversely, the graph of an equation in this form is a circle with center $(h, k)$ and radius $r$.

---

Example 2 illustrates how to find an equation of a circle when we know its center and radius.

**EXAMPLE 2**    Finding an Equation of a Circle

Write an equation of a circle centered at $(3, -5)$ with radius 6.

SOLUTION
Substitute $h = 3$, $k = -5$, and $r = 6$ in the standard equation of a circle.

$$(x - 3)^2 + (y - (-5))^2 = 6^2 \quad \textbf{Simplify.}$$

$$(x - 3)^2 + (y + 5)^2 = 36 \qquad ■$$

## Graphing a Circle Not Centered at the Origin

Example 3 illustrates how to use a paper-and-pencil method to sketch a graph of a circle, given the equation.

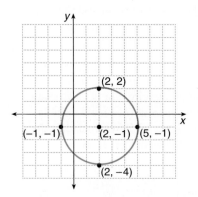

**FIGURE 10.6** The circle with center at $(2, -1)$ and radius 3 in Example 3.

**EXAMPLE 3**  Sketching a Graph of a Circle

Sketch the graph of $(x - 2)^2 + (y + 1)^2 = 9$ using paper and pencil.

SOLUTION
Begin by changing the equation to standard form.

$$(x - 2)^2 + (y + 1)^2 = 9 \qquad \text{Rewrite the equation in standard form.}$$

$$(x - 2)^2 + (y - (-1))^2 = 3^2$$

The center is at $(2, -1)$, not $(0, 0)$, and the radius is 3. Figure 10.6 shows a sketch of this circle. ∎

Example 4 illustrates that an equation in the standard form of a circle can be easily sketched. However, to graph a circle using a grapher it is necessary to solve for $y$ and enter one half of the circle at a time.

**EXAMPLE 4**  Graphing a Circle

Sketch the graph of the equation $(x - 3)^2 + (y + 1)^2 = 36$. Support using a grapher.

SOLUTION
Figure 10.7 shows a sketch of the circle centered at $(3, -1)$ and radius 6.

**FIGURE 10.7** A circle with center at $(3, -1)$ and radius 6.

**Support Graphically.** To prepare the equation for a grapher, solve for $y$.

$$(x - 3)^2 + (y + 1)^2 = 36$$

$$(y + 1)^2 = 36 - (x - 3)^2 \qquad \text{Take the square root of each side of the equation.}$$

$$y + 1 = \pm\sqrt{36 - (x - 3)^2}$$

$$y = -1 \pm \sqrt{36 - (x - 3)^2}, \text{ or}$$

$$y = -1 + \sqrt{36 - (x - 3)^2} \quad \text{and} \quad y = -1 - \sqrt{36 - (x - 3)^2}$$

To simplify entering each of these equations into the grapher, we begin by entering the term with the radical as $y_1 = \sqrt{(36 - (x - 3)^2)}$, $y_2 = -1 + y_1$, and $y_3 =$

$-1 - y_1$. Figure 10.8(a) shows all three equations on the Function-defining menu. Deselect $y_1$ and graph $y_2$ and $y_3$ in a Square window.

Figure 10.8(b) shows a graph of the circle.

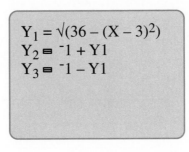

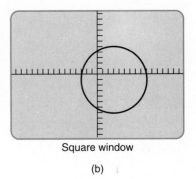

Square window

(a)                                              (b)

**FIGURE 10.8** (a) The Function-defining menu for this circle, and (b) a graph of $(x - 3)^2 + (y + 1)^2 = 36$, shown on a grapher. ∎

Sometimes an equation is not in standard form. To write an equivalent equation in standard form, it is often necessary to use a completing-the-square process, which we studied in Chapter 8. Example 5 uses this process.

**EXAMPLE 5** Using Standard Form to Sketch a Graph

Sketch a graph of the equation $x^2 + 10x + y^2 - 4y - 8 = 12$ using a paper-and-pencil method.

SOLUTION

We begin by completing the square for the $x$-terms grouped together and the $y$-terms grouped together.

$$x^2 + 10x + y^2 - 4y - 8 = 12$$ Group the x-terms, y-terms, and constants.

$$(x^2 + 10x + \quad) + (y^2 - 4y + \quad) = 8 + 12$$ Complete the squares, adding the additional terms to each side of the equation.

$$(x^2 + 10x + \boxed{25}) + (y^2 - 4y + \boxed{4}) = 8 + 12 + \boxed{25} + 4$$ Rewrite in factored form.

$$(x + 5)^2 + (y - 2)^2 = 49$$

From this standard form, we see that the center is $(-5, 2)$ and the radius is 7. So locate the center and then plot the four points 7 units left, right, up, and down from the center. Sketch the circle through these points (see Figure 10.9 on the following page).

**REMINDER**

When you are completing the square, once the coefficient of the $x^2$ term is one, take one half of the coefficient of the $x$-term and square it. This is the constant term to add on to have a perfect square trinomial. That is, $x^2 + 10x$ becomes $x^2 + 10x + 25$, which now can be factored into a binomial squared, $(x + 5)^2$.

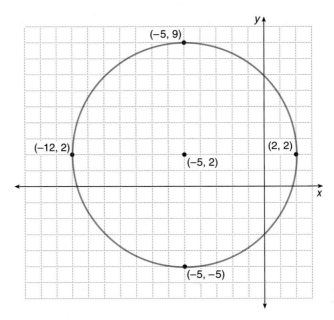

**FIGURE 10.9**  A circle with the center at $(-5, 2)$ and radius 7, sketched using paper and pencil.   ∎

## Exercises for Section  10.1

In Exercises 1–10, write an equation of the circle with the given center and radius.

**1.** Center $(0, 0), r = 4$    **2.** Center $(0, 0), r = 7$

**3.** Center $(0, 0), r = 2$    **4.** Center $(0, 0), r = 9$

**5.** Center $(1, 5), r = 12$

**6.** Center $(25, 34), r = \sqrt{15}$

**7.** Center $(4, -9), r = \sqrt{27}$

**8.** Center $(-15, 20), r = \sqrt{8}$

**9.** Center $(0, -2.1), r = 25$

**10.** Center $(-10, -8), r = 8$

In Exercises 11–16, find the center and the radius of the circle described by the equation.

**11.** $x^2 + y^2 = 81$    **12.** $x^2 + (y - 2)^2 = 49$

**13.** $x^2 + y^2 = 19$    **14.** $x^2 + (y + 3)^2 = 35$

**15.** $(x + 7)^2 + (y - 5)^2 = 56$

**16.** $(x + 2)^2 + (y + 9)^2 = 70$

In Exercises 17–22, use a grapher to find a graph of the circle determined by the equation.

**17.** $y = \pm\sqrt{25 - x^2}$    **18.** $y = \pm\sqrt{121 - x^2}$

**19.** $x^2 + y^2 = 36$    **20.** $x^2 + y^2 = 81$

**21.** $(x - 5)^2 + (y - 15)^2 = 49$

**22.** $(x + 12)^2 + (y + 10)^2 = 64$

In Exercises 23–30, sketch a graph of the equation using a paper-and-pencil method. Support using a grapher.

**23.** $x^2 + y^2 = 100$    **24.** $x^2 + y^2 = 121$

**25.** $x^2 + y^2 = 15$    **26.** $x^2 + y^2 = 85$

**27.** $(x - 6)^2 + (y + 5)^2 = 64$

**28.** $(x - 10)^2 + (y + 4)^2 = 25$

**29.** $x^2 + (y + 8)^2 = 36$

**30.** $(x - 12)^2 + y^2 = 144$

In Exercises 31–34, write an equation of the circle shown in the sketch.

**31.**

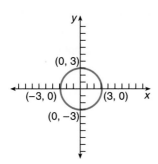

**32.**

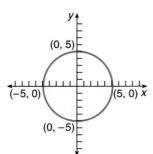

**33.**

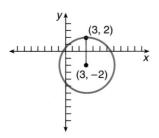

**34.**

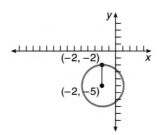

In Exercises 35–44, write an equation in the standard form of a circle by completing the square and then sketch its graph.

**35.** $x^2 + 6x + y^2 - 16 = -9$

**36.** $x^2 + 10x + y^2 - 16y = 55$

**37.** $x^2 + 12x + 9 + y^2 - 8y + 15 = 8$

**38.** $x^2 - 4x + 5 + y^2 + 18y = -31$

**39.** $x^2 + y^2 - 6x + 12y - 22 = -3$

**40.** $x^2 + y^2 - 18y = -32$

**41.** $x^2 - 14x + y^2 + 22y = -70$

**42.** $x^2 - 8x + 5 + y^2 + 10y = 0$

**43.** $x^2 + y^2 + 8y = 9$

**44.** $x^2 - 4x + y^2 + 10y = -25$

In Exercises 45–47, write a response to the stated situation.

**45. Writing to Learn.** Write a few sentences explaining why the graph of a circle might not look perfectly round if graphed with a grapher.

**46. Writing to Learn.** Explain why the equation $x^2 + y^2 = -36$ is not the equation of a circle.

**47. Writing to Learn.** Write a paragraph explaining the differences in the graphs of the following equations: $x^2 + y^2 = 25$, $(x - 3)^2 + y^2 = 25$, $x^2 + (y + 1)^2 = 25$, and $(x - 3)^2 + (y + 1)^2 = 25$.

**48.** *Computer-assisted Design.* The automobile industry uses computers to design parts and new models. A coordinate system is positioned on a computer screen so that the center of an 8-in. radius rear wheel is at the origin.

a) Write an equation whose graph is the rear wheel.

b) Write an equation whose graph is the front wheel.

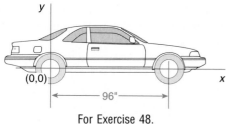

For Exercise 48.

**49.** *Industrial Engineering.* An airplane hangar with a semicircular cross section with a 25-ft radius has a rectangular door on one end. The bottom right-hand corner of the door is 5 ft from the corner of the hanger. What is the height ($h$) of the door?

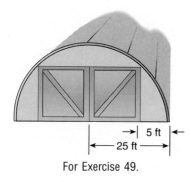

For Exercise 49.

EXTENDING THE IDEAS

**50.** Without graphing, determine whether the points $(3, 4)$, $(-3, 4)$, $(-3, -4)$, and $(3, -4)$ are on the circle $x^2 + y^2 = 25$.

**51.** Determine $r$ so that the circle $(x - 3)^2 + (y + 3)^2 = r^2$ passes through the point $(0, 0)$.

**52.** Determine $k$ so that the circle $(x - 3)^2 + (y - k)^2 = 25$ passes through the point $(-1, 6)$.

**53.** Write an equation of the circle with the center at $(3, -2)$ that passes through the point $(8, -2)$.

**54.** Write an equation of the circle with the center at origin that passes through the point $(4, 3)$.

**55.** Write a program for graphing a circle with center $(h, k)$ and radius $= r$.

LOOKING BACK—MAKING CONNECTIONS

In Exercises 56 and 57, determine whether the triangle formed by joining the points $A$, $B$, and $C$ is a right triangle.

**56.** $A(-2, -2)$, $B(5, 7)$, and $C(7, 1)$

**57.** $A(-4, 1)$, $B(2, 5)$, and $C(4, 2)$

In Exercises 58 and 59, determine whether the quadrilateral formed by joining the points $A$, $B$, $C$, and $D$ is a square.

**58.** $A(-6, 1)$, $B(-1, -4)$, $C(4, 1)$, and $D(-1, 6)$

**59.** $A(0, 0)$, $B(0, 5)$, $C(5, 5)$, and $D(5, 0)$

---

## 10.2 The Ellipse

Finding an Equation of an Ellipse Centered at the Origin • Sketching a Graph of an Ellipse • Using a Grapher to Graph an Ellipse • Graphing an Ellipse Not Centered at the Origin • Using a Graph of an Ellipse in a Problem Situation

An **ellipse** is determined by a constant and two fixed points, each called a **focus point** (plural is **foci**). The constant must be greater than the distance between the

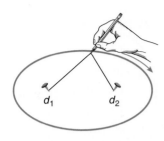

**FIGURE 10.10** An ellipse drawn using a pencil holding a string taut and anchored at the two foci. The sum of the distances $d_1 + d_2$ is constant.

two foci. The ellipse determined by the two foci and the given constant is the set of points $(x, y)$ in a plane such that the sum of the distances from $(x, y)$ to the two foci is equal to the constant.

An ellipse can be constructed using a pencil, two thumbtacks, and a piece of string. Place the two thumbtacks at a distance apart that is smaller than the length of the string. Attach each end of the string to a thumbtack. Hold the pencil against the string, pull it taut as shown in Figure 10.10, and then move the pencil. The path of the point describes an ellipse because the sum of the distances $d_1$ and $d_2$ is a constant.

## Finding an Equation of an Ellipse Centered at the Origin

Figure 10.11 shows the ellipse determined by foci $(-3, 0)$ and $(3, 0)$ and constant 10. That is, the sum of the distances from $(x, y)$ to the foci is 10, which means that

$$\sqrt{(x - 3)^2 + (y - 0)^2} + \sqrt{(x + 3)^2 + (y - 0)^2} = 10.$$

This equation can be simplified to

$$\frac{x^2}{25} + \frac{y^2}{16} = 1.$$

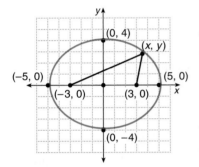

**FIGURE 10.11** An ellipse with the center at $(0, 0)$ and foci $(-3, 0)$ and $(3, 0)$.

The $x$-intercepts of this curve are $(-5, 0)$ and $(5, 0)$ and the $y$-intercepts are $(0, 4)$ and $(0, -4)$. These four points are called the **vertices** of the ellipse, and the line segments through opposite vertices are called **axes**. The **major axis** is the axis that contains the foci; the other axis is the **minor axis**.

This example fits the following standard form for an equation of an ellipse.

---

### Standard Form of an Ellipse Centered at the Origin

The **standard form for an equation of an ellipse centered at the origin** is

$$\frac{x^2}{a^2} + \frac{y^2}{b^2} = 1.$$

The $x$-intercepts of the ellipse are $(\pm a, 0)$ and the $y$-intercepts of the ellipse are $(0, \pm b)$. The intercepts are the **vertices** of the ellipse.

---

The major axis of an ellipse can be either horizontal or vertical, as seen in Figure 10.12 on the following page. For an ellipse centered at the origin, the $x$-intercepts are $(-a, 0)$ and $(a, 0)$ and the $y$-intercepts are $(0, -b)$ and $(0, b)$.

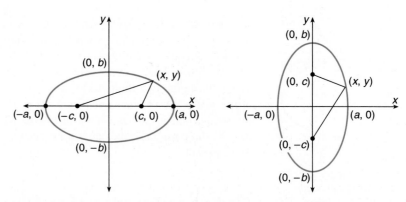

**FIGURE 10.12** An ellipse with center at the origin, foci at $(\pm c, 0)$ or $(0 \pm c)$, horizontal vertices at $(\pm a, 0)$, and vertical vertices $(0, \pm b)$. The foci are located on the major axis and are $c$ units from the center. It can be shown that $c^2 = a^2 - b^2$, if $a^2 > b^2$, or $c^2 = b^2 - a^2$, if $b^2 > a^2$.

## Sketching a Graph of an Ellipse

To sketch a graph of an ellipse centered at the origin, plot the four vertices and then draw a smooth curve through these points. Example 1 illustrates sketching the graph using a paper-and-pencil method.

**EXAMPLE 1**    Sketching a Graph of an Ellipse

Use paper and pencil to sketch the graph of the equation

$$\frac{x^2}{16} + \frac{y^2}{9} = 1.$$

SOLUTION
We determine from the form of this equation that the ellipse is centered at the origin. We want to find the vertices of the ellipse.

$$\frac{x^2}{16} + \frac{y^2}{9} = 1 \qquad \substack{\textbf{Rewrite in the general} \\ \textbf{form of the ellipse.}}$$

$$\frac{x^2}{4^2} + \frac{y^2}{3^2} = 1$$

The ellipse has vertices $(-4, 0)$, $(4, 0)$, $(0, -3)$, and $(0, 3)$. Figure 10.13 shows a paper-and-pencil sketch of the ellipse. ∎

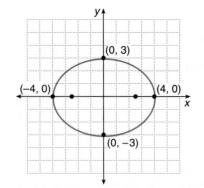

**FIGURE 10.13** An ellipse with center at origin, $x$-intercepts $(\pm 4, 0)$, and $y$-intercepts $(0, \pm 3)$.

We can summarize the procedure for sketching an ellipse that is centered at the origin as follows.

> ### Sketching a Graph of an Ellipse Centered at $(0, 0)$
>
> 1. Write the equation in standard form.
>
> $$\frac{x^2}{a^2} + \frac{y^2}{b^2} = 1$$
>
> 2. Plot the vertices at $(-a, 0)$, $(a, 0)$, $(0, -b)$, and $(0, b)$.
> 3. Sketch the graph of the ellipse through the four vertices.
> 4. (Optional) The foci are $c$ units from the origin on the major axis, where $a$, $b$, and $c$ satisfy the equation $a^2 - b^2 = c^2$, if $a^2 > b^2$, or $b^2 - a^2 = c^2$, if $b^2 > a^2$.

## Using a Grapher to Graph an Ellipse

Can an ellipse be drawn using a grapher? Yes; however, because an ellipse is not the graph of a function we need to enter one half of the ellipse at a time. Example 2 illustrates rewriting the equation to enter it into the grapher.

**EXAMPLE 2**    Graphing an Ellipse on a Grapher

Use a grapher to graph the equation

$$\frac{x^2}{15} + \frac{y^2}{8} = 1.$$

SOLUTION

To prepare the equation to enter into the grapher, solve for $y$.

$$\frac{x^2}{15} + \frac{y^2}{8} = 1$$

$$\frac{y^2}{8} = 1 - \frac{x^2}{15} \qquad \text{Simplify the fraction and multiply each side by 8.}$$

$$y^2 = \frac{8(15 - x^2)}{15} \qquad \text{Take the square root of each side.}$$

$$y = \pm\sqrt{\frac{8(15 - x^2)}{15}}$$

Use a Square window to graph $y_1 = \sqrt{(8(15 - x^2)/15)}$ and $y_2 = -y_1$.

Figure 10.14 shows the ellipse on a grapher. ∎

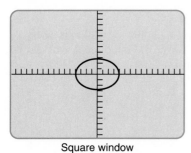

Square window

**FIGURE 10.14** The ellipse in Example 2 with center at the origin and vertices at $(\pm\sqrt{15}, 0)$ and $(0, \pm\sqrt{8})$.

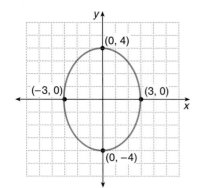

**FIGURE 10.15** The ellipse in Example 3 with center at the origin and vertices at $(\pm 3, 0)$ and $(0, \pm 4)$.

Example 3 illustrates the need first to determine that an equation is of an ellipse and then to rewrite the equation in the standard form.

**EXAMPLE 3** Sketching a Graph of an Ellipse

Sketch the graph of the equation $16x^2 + 9y^2 = 144$. Support using a grapher.

SOLUTION
Rewrite the equation into the standard form for an equation of an ellipse.

$$16x^2 + 9y^2 = 144 \quad \textbf{Divide each side by 144.}$$

$$\frac{x^2}{9} + \frac{y^2}{16} = 1$$

$$\frac{x^2}{3^2} + \frac{y^2}{4^2} = 1$$

Because $b = 4$, the horizontal vertices are $(\pm 3, 0)$ and the vertical vertices are $(0, \pm 4)$. Figure 10.15 shows the sketch of the curve through the four vertices.

**Support Graphically.** Prepare the equation to enter into a grapher by solving for $y$.

$$16x^2 + 9y^2 = 144$$

$$9y^2 = 144 - 16x^2$$

$$y^2 = \frac{144 - 16x^2}{9}$$

$$y = \pm\sqrt{\frac{144 - 16x^2}{9}}$$

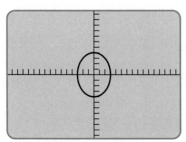

Square window

**FIGURE 10.16** The ellipse in Example 3 with center at origin and vertices at $(-3, 0)$, $(3, 0)$, $(0, 4)$, and $(0, -4)$.

Graph $y_1 = \sqrt{(144 - 16x^2)/9}$ and $y_2 = -y_1$ in a Square window. Figure 10.16 shows the graph of the ellipse on a grapher. ■

Notice that the major axis in the ellipse in Example 3 is the vertical axis. This is because when the equation is written in standard form, the denominator of the $y^2$-term is greater than the denominator of the $x^2$-term.

## Graphing an Ellipse Not Centered at the Origin

Example 4 focuses on the equation

$$4x^2 + y^2 + 24x - 10y + 45 = 0,$$

which is an equation for an ellipse that is not centered at the origin. The center of such an ellipse can be determined by using the method of completing the square that was employed in Section 10.1 for circles. Using this method, we can rewrite this equation into the standard form described next.

---

### Standard Form of an Ellipse

The **standard form for an equation of an ellipse with center at** $(h, k)$ is

$$\frac{(x-h)^2}{a^2} + \frac{(y-k)^2}{b^2} = 1.$$

The distances from the center to each vertex on the horizontal axis is $a$ units and to each vertex on the vertical axis is $b$ units.

---

Example 4 illustrates the following three steps for finding a sketch of an ellipse:

1. Complete the squares to obtain an equation in standard form.
2. Identify the center of the ellipse.
3. Plot the center and the four vertices, and then sketch the ellipse.

**EXAMPLE 4**    Graphing an Ellipse Not Centered at the Origin

Sketch a graph of the equation $4x^2 + y^2 + 24x - 10y + 45 = 0$.

SOLUTION

Begin by completing the squares to obtain an equation in the standard form of an ellipse.

$$4x^2 + y^2 + 24x - 10y + 45 = 0 \qquad \text{Group the } x\text{- and } y\text{-terms.}$$

$$(4x^2 + 24x) + (y^2 - 10y) = -45 \qquad \text{Factor, making the coefficient of each squared term} = 1.$$

$$4(x^2 + 6x \quad) + (y^2 - 10y \quad) = -45 \qquad \text{Complete the square.}$$

$$4(x^2 + 6x + 9) + (y^2 - 10y + 25) = -45 + 36 + 25 \qquad \text{Factor.}$$

$$4(x + 3)^2 + (y - 5)^2 = 16 \qquad \text{Divide each side by 16.}$$

$$\frac{(x+3)^2}{4} + \frac{(y-5)^2}{16} = 1 \qquad \text{Put in standard form.}$$

$$\frac{(x-(-3))^2}{2^2} + \frac{(y-5)^2}{4^2} = 1$$

From this standard form, we see that the center of the ellipse is $(-3, 5)$.

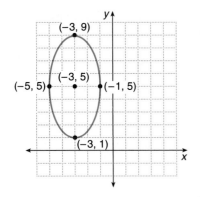

**FIGURE 10.17** The ellipse in Example 4 with center at $(-3, 5)$ and vertices at $(-5, 5)$, $(-1, 5)$, $(-3, 9)$, and $(-3, 1)$.

The horizontal distance from the center to a vertex is 2 units; the vertical distance from the center to a vertex is 4 units. Consequently, the vertices are $(-5, 5)$, $(-1, 5)$, $(-3, 9)$, and $(-3, 1)$.

Figure 10.17 shows a sketch of the ellipse. ■

When an equation of an ellipse is given in standard form, its graph can be sketched using a paper-and-pencil method. Example 5 illustrates how to graph the ellipse using a grapher to support the sketch.

**EXAMPLE 5**    Supporting a Sketch Using a Grapher

Use a grapher to support the sketch of the equation

$$\frac{(x-1)^2}{4} + \frac{(y+2)^2}{9} = 1.$$

SOLUTION

Figure 10.18 shows a sketch of $\dfrac{(x-1)^2}{4} + \dfrac{(y+2)^2}{9} = 1$, with the center $(1, -2)$. The horizontal distance from the center to a vertex is 2 units, and the vertical distance from the center to a vertex is 3 units. The vertices are $(-1, -2)$, $(3, -2)$, $(1, 1)$, and $(1, -5)$.

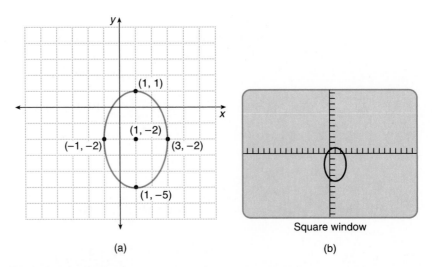

Square window

(a)                    (b)

**FIGURE 10.18** (a) A sketch of the ellipse in Example 5 and (b) grapher support of the sketch.

**Support Graphically.** Begin with the standard form and solve for $y$ in preparation for entering the equation into the $Y =$ menu.

$$\frac{(x-1)^2}{4} + \frac{(y+2)^2}{9} = 1$$

$$\frac{(y+2)^2}{9} = 1 - \frac{(x-1)^2}{4} \qquad \text{Simplify by writing as one fraction.}$$

$$\frac{(y+2)^2}{9} = \frac{4-(x-1)^2}{4} \qquad \text{Multiply each side by 9.}$$

$$(y+2)^2 = 9\left(\frac{4-(x-1)^2}{4}\right) \qquad \text{Take the square root of each side.}$$

$$(y+2) = \pm\frac{3\sqrt{4-(x-1)^2}}{2}$$

$$y = -2 \pm \frac{3\sqrt{4-(x-1)^2}}{2}$$

Define $y_1 = 3\sqrt{(4-(x-1)^2)}/2$, $y_2 = -2 + y_1$, and $y_3 = -2 - y_1$. Deselect $y_1$, and graph $y_2$ and $y_3$ in a Square window. Figure 10.18(b) shows the resulting graph. ∎

## Using a Graph of an Ellipse in a Problem Situation

Consider the following problem situation.

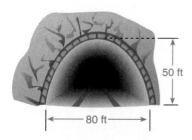

**FIGURE 10.19** The **Tunnel Problem Situation**.

---

### TUNNEL PROBLEM SITUATION

A tunnel that goes through the Blue Ridge Mountains is shaped in a semi-elliptical arch. See Figure 10.19. It is 80 ft wide and 50 ft high.

---

There are many problems related to this problem situation. Example 6 considers several of them.

**EXAMPLE 6**    Solving the Tunnel Problem

Find a graph that models the tunnel opening in the **Tunnel Problem Situation**. Then find the height of the arch both 8 ft and 16 ft from the edge of the tunnel.

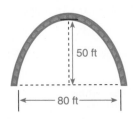

**FIGURE 10.20** A sketch of the **Tunnel Problem Situation.**

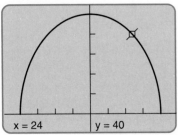

x = 24     y = 40

Integer window with
vertical dimension [−10, 55]

**FIGURE 10.21** Graph of $y_1 = 50\sqrt{(1 - x^2/40^2)}$.

SOLUTION
Figure 10.20 shows the dimensions of the tunnel. If we superimpose a coordinate system on the tunnel entrance with the origin at the center of the road, the elliptical tunnel can be represented by the equation

$$\frac{x^2}{40^2} + \frac{y^2}{50^2} = 1.$$

To prepare this equation for the grapher, solve the equation for $y$.

$$\frac{x^2}{40^2} + \frac{y^2}{50^2} = 1$$

$$\frac{y^2}{50^2} = 1 - \frac{x^2}{40^2} \qquad \textbf{Multiply each side by } 50^2.$$

$$y^2 = 50^2(1 - \frac{x^2}{40^2}) \qquad \begin{array}{l}\textbf{Take the square root}\\ \textbf{of each side of the equation.}\end{array}$$

$$y = \pm 50\sqrt{(1 - \frac{x^2}{40^2})}$$

To graph the top half of this ellipse, define $y_1 = 50\sqrt{(1 - x^2/40^2)}$ on a grapher.

Figure 10.21 shows this graph in the Integer window with the vertical dimension adjusted to $[-10, 55]$. Because the outer edge of the tunnel corresponds to $x = 40$, the points 8 ft and 16 ft from this edge will correspond to $x = 32$ and $x = 24$, respectively.

Using TRACE, you can determine that points (24, 40) and (32, 30) are on this ellipse. Therefore, the arch is 30 ft high 8 ft from the tunnel edge, and the arch is 40 ft high 16 ft from the edge. ■

## Exercises for Section   10.2

In Exercises 1–6, give an equation of the ellipse, and list the coordinates of the center and the vertices.

1. $\dfrac{x^2}{36} + \dfrac{y^2}{4} = 1$     2. $\dfrac{x^2}{9} + \dfrac{y^2}{16} = 1$

3. $\dfrac{x^2}{25} + \dfrac{y^2}{81} = 1$     4. $\dfrac{x^2}{4} + \dfrac{y^2}{64} = 1$

5. $\dfrac{x^2}{5} + \dfrac{y^2}{16} = 1$     6. $\dfrac{x^2}{7} + \dfrac{y^2}{13} = 1$

In Exercises 7–12, sketch a graph of the ellipse using a paper-and-pencil method. Then support the graph using a grapher.

7. $\dfrac{x^2}{64} + \dfrac{y^2}{9} = 1$     8. $\dfrac{x^2}{25} + \dfrac{y^2}{4} = 1$

9. $\dfrac{x^2}{25} + \dfrac{y^2}{49} = 1$     10. $\dfrac{x^2}{81} + \dfrac{y^2}{16} = 1$

11. $\dfrac{x^2}{9} + \dfrac{y^2}{15} = 1$     12. $\dfrac{x^2}{4} + \dfrac{y^2}{10} = 1$

In Exercises 13–20, list the coordinates of the ellipse's center and the vertices.

13. $\dfrac{(x - 1)^2}{4} + \dfrac{(y - 5)^2}{9} = 1$

14. $\dfrac{(x+3)^2}{25} + \dfrac{(y-6)^2}{36} = 1$

15. $\dfrac{(x+1)^2}{49} + \dfrac{(y+2)^2}{25} = 1$

16. $\dfrac{(x-2)^2}{1} + \dfrac{(y+4)^2}{16} = 1$

17. $\dfrac{x^2}{49} + \dfrac{(y-3)^2}{4} = 1$

18. $\dfrac{(x-1)^2}{64} + \dfrac{y^2}{16} = 1$

19. $\dfrac{(x-5)^2}{7} + \dfrac{(y+3)^2}{12} = 1$

20. $\dfrac{(x+7)^2}{6} + \dfrac{(y-1)^2}{5} = 1$

In Exercises 21–24, sketch a graph of the ellipse using a paper-and-pencil method. Support the graph using a grapher.

21. $\dfrac{(x-2)^2}{9} + \dfrac{(y-1)^2}{49} = 1$

22. $\dfrac{(x+5)^2}{16} + \dfrac{(y-3)^2}{25} = 1$

23. $\dfrac{x^2}{81} + \dfrac{(y-1)^2}{36} = 1$     24. $\dfrac{(x-7)^2}{4} + \dfrac{y^2}{49} = 1$

In Exercises 25–32, write the equation in the standard form of the ellipse. Then list the coordinates of the center and vertices and determine the length of the major and minor axes.

25. $9x^2 + 4y^2 = 36$      26. $x^2 + 9y^2 = 225$

27. $2x^2 + 8y^2 = 72$      28. $4x^2 + y^2 = 16$

29. $x^2 + 4y^2 - 10x + 32y = -73$

30. $x^2 + 4y^2 - 4x - 8y = -4$

31. $x^2 - 6x + y^2 + 20y - 12 = 0$

32. $x^2 - 12x + y^2 - 14y = 15$

In Exercises 33–36, write the equation in standard form of the ellipse that is centered at $(0, 0)$ and satisfies the given criteria.

33. Vertices: $(\pm 2, 0)$, $(0, \pm 5)$

34. Vertices: $(\pm\sqrt{3}, 0)$, $(0, \pm\sqrt{7})$

35. The major axis is horizontal with length 10. The minor axis has length 8.

36. The major axis is vertical with length 25. The minor axis has length 12.

In Exercises 37–40, determine an equation of the ellipse from the graph.

37.

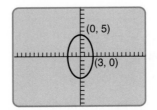

38.

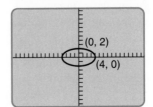

39.

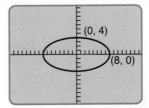

40.

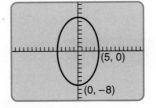

In Exercises 41–48, identify the coordinates of the center and vertices of the ellipse and then graph using a paper-and-pencil method.

41. $x^2 + 4y^2 = 36$      42. $9x^2 + 15y^2 = 225$

43. $4x^2 + 9y^2 = 36$     44. $16x^2 + 4y^2 = 16$

**45.** $x^2 + 4y^2 + 2x - 24y = 27$

**46.** $9x^2 + 16y^2 + 54x - 32y = 47$

**47.** $x^2 - 6x + y^2 + 8y - 11 = 0$

**48.** $x^2 - 14x + y^2 + 10y - 26 = 0$

In Exercises 49–56, graph the equation using a grapher.

**49.** $x^2 + 4y^2 = 16$      **50.** $9x^2 + 4y^2 = 36$

**51.** $x^2 + y^2 = 20$      **52.** $x^2 + y^2 = 15$

**53.** $x^2 + 10x + y^2 = 75$    **54.** $x^2 - 8x + y^2 = 36$

**55.** $(x - 5)^2 + 4(y + 2)^2 = 16$

**56.** $9(x + 3)^2 + (y - 7)^2 = 81$

**57.** *Analysis of a Tunnel.* For the tunnel described in Example 6, find the height of the arch 20 ft from the outside edge.

**58.** *Analysis of a Tunnel.* For the tunnel described in Example 6, find how many feet from the outer edge is the tunnel arch 25 feet high.

**59.** *Finding the Height of a Monument.* The Arc de Triomphe located in Paris, France, is a semi-elliptical arch that has an opening 154 ft tall and 48 ft wide.

a) Find a graph of the problem situation.

b) Find the height of the opening at a distance 10 ft from the edge of the opening.

EXTENDING THE IDEAS

In Exercises 60–65, write an equation of the ellipse that satisfies the given criteria.

**60.** Vertices: $(-2, 0)$, $(2, 0)$, $(0, -3)$, and $(0, 3)$

**61.** Vertices: $(-5, 0)$, $(5, 0)$, $(0, -6)$, and $(0, 6)$

**62.** Vertices: $(-4, -5)$, $(8, -5)$, $(2, -1)$, and $(2, -9)$

**63.** Vertices: $(-8, 7)$, $(2, 7)$, $(-3, 5)$, and $(-3, 9)$

**64.** Center at $(8, -10)$ and goes through the point $(8, -4)$ and the length of the minor axis is 8

**65.** Center at $(5, 7)$ and goes through the point $(7, 7)$ and the length of the major axis is 12

LOOKING BACK—MAKING CONNECTIONS

In Exercises 66–71, identify the graph from its equation. List the important characteristics of the graph, then graph the equation using a paper-and-pencil method. Support the graph using a grapher.

**66.** $x^2 + y^2 = 36$      **67.** $x^2 + y^2 = 49$

**68.** $y = x^2 + 3$      **69.** $y = x^2 - 5$

**70.** $y = (x - 5)^2 + 1$    **71.** $y = (x + 3)^2 - 4$

---

## 10.3   Parabolas with a Horizontal Line of Symmetry

> Sketching the Graph of a Parabola ▪ Using a Grapher to Graph a Parabola in Vertex Form ▪ Changing an Equation to Standard Vertex Form

In Chapter 8, you studied the quadratic functions of the form $y = ax^2 + bx + c$ and found that by completing the square, you could rewrite them in the vertex form

$y = a(x - h)^2 + k$. The point $(h, k)$ is the vertex and the vertical line $x = h$ is the line of symmetry of the parabola. Parabolas of this form open either up or down, as shown in Figure 10.22.

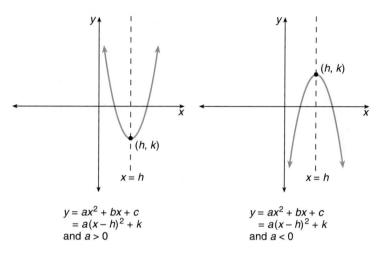

$$y = ax^2 + bx + c$$
$$= a(x - h)^2 + k$$
$$\text{and } a > 0$$

$$y = ax^2 + bx + c$$
$$= a(x - h)^2 + k$$
$$\text{and } a < 0$$

**FIGURE 10.22**   The two basic types of parabolas studied in Chapter 8 open either up or down.

In this chapter, we will study equations of the form $x = ay^2 + by + c$. We begin with the simplest of all such equations,

$$x = y^2.$$

Example 1 illustrates how to find a graph of this equation on a grapher. Notice that $x = y^2$ does not determine a function of $x$, so it is again necessary to graph one half of the graph using $y_1$ and the other using $y_2$.

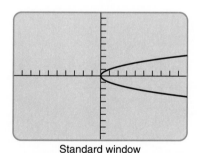

Standard window

**FIGURE 10.23**  A graph of $x = y^2$ in Example 1 using a grapher.

**EXAMPLE 1**     Using a Grapher

Graph $x = y^2$ using a grapher.

SOLUTION

To prepare the equation for entering into a grapher, solve the equation for $y$.

$$x = y^2$$     **Take the square root of each side of the equation.**

$$\sqrt{x} = y \quad \text{or} \quad -\sqrt{x} = y$$

Graph $y_1 = \sqrt{x}$ and $y_2 = -y_1$ in the Standard viewing window.

This graph is a parabola that opens to the right, as shown in Figure 10.23.

Other parabolas that open right or left can be obtained from the graph of $x = y^2$ by using the transformations studied in Chapter 8. These types of parabolas have characteristics similar to those that open up or down. If their vertex, line of symmetry, and one other point are known, a hand-sketch can be completed.

We summarize the relationship between the vertex, the line of symmetry, and the graph as follows.

---

### Vertex and the Line of Symmetry of a Parabola

A graph of the equation

$$x = a(y - k)^2 + h$$

is a parabola. Its **vertex** is $(h, k)$ and its **line of symmetry** is the horizontal line $y = k$.

---

It can be shown that the graph of $x = a(y - k)^2 + h$ can be obtained from the graph of $x = y^2$ by a horizontal stretch or shrink (with factor $|a|$) and a reflection in the $y$-axis (only if $a$ is negative), followed by a horizontal and vertical slide.

Figure 10.24 illustrates the two categories of parabolas that are graphs of equations of the form $x = a(y - k)^2 + h$.

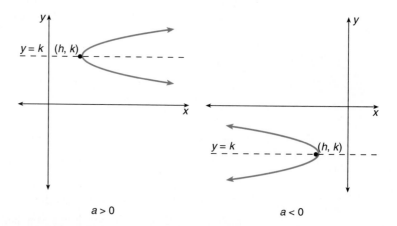

$$a > 0 \qquad\qquad a < 0$$

**FIGURE 10.24** Graphs of the two types of parabolas with horizontal lines of symmetry.

Example 2 shows how an equation in vertex form can be used to find the vertex and the line of symmetry of a parabola.

**EXAMPLE 2**   Finding the Vertex and Line of Symmetry

Find the vertex and the line of symmetry of the graphs of the following equations:

a) $x = (y - 3.5)^2 + 8$

b) $x = -2(y + 6)^2 - \sqrt{7}$

SOLUTION

a) The vertex of the graph of $x = (y - 3.5)^2 + 8$ is the point $(8, 3.5)$ and the line of symmetry is $y = 3.5$.

b) Rewrite $x = -2(y + 6)^2 - \sqrt{7}$ as $x = -2(y - (-6))^2 - \sqrt{7}$. Because this is an equation in vertex form, it is evident that the vertex is $(-\sqrt{7}, -6)$ and the line of symmetry is $y = -6$.   ■

## Sketching the Graph of a Parabola

We can hand-sketch the graph of a parabola whose equation is written in the form $x = a(y - k)^2 + h$ by identifying a vertex, an axis of symmetry, and an $x$-intercept. Example 3 illustrates this method.

**EXAMPLE 3**   Sketching a Graph of a Parabola

Sketch a graph of the equation $x = (y - 3)^2 + 5$.

SOLUTION

1. Plot the vertex $(5, 3)$ and the line of symmetry, $y = 3$, which is the horizontal line through this vertex.
2. Find the $x$-intercept by letting $y = 0$ and solve the resulting equation.

$$x = (0 - 3)^2 + 5$$

$$= 9 + 5 = 14$$

3. Plot the point $(14, 0)$, which is a point on the graph. The point $(14, 6)$ is the same distance from the axis on the opposite side of the line of symmetry and is also a point on the graph.
4. Sketch the graph through these points with a smooth curve (see Figure 10.25 on the following page).

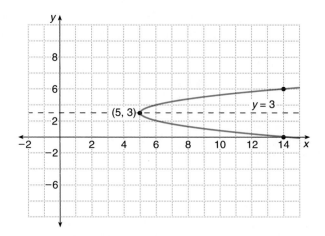

**FIGURE 10.25** A sketch of the graph of $x = (y - 3)^2 + 5$ in Example 3.

The following summary focuses on the relationship between an equation of a parabola and the distinguishing characteristics of its graph.

---

### Parabolas with Equations in Vertex Form

The graph of the equation $y = a(x - h)^2 + k$ is a parabola with

- vertex $(h, k)$ and
- line of symmetry the vertical line $x = h$.

The parabola opens upward if $a > 0$ and downward if $a < 0$.

The graph of the equation $x = a(y - k)^2 + h$ is a parabola with

- vertex $(h, k)$ and
- line of symmetry the horizontal line $y = k$.

The parabola opens to the right if $a > 0$ and to the left if $a < 0$.

---

## Using a Grapher to Graph a Parabola in Vertex Form

Example 4 illustrates that any parabola that opens right or left can be graphed using a grapher. However, when the parabola opens right or left its equation does not define a function of $x$. Consequently, the upper and lower halves of the curve must be graphed as separate functions.

**EXAMPLE 4**    Graphing a Parabola Opening to the Left

Find a graph of the equation $x = -2(y + 5)^2 - 3$ using a grapher.

| REMINDER |
|---|

Because the equation in Example 4 is in vertex form, we know the parabola opens left, has vertex $(-3, -5)$, and has a horizontal line of symmetry $y = -5$.

SOLUTION

To prepare the equation for entering into a grapher, solve for $y$.

$$x = -2(y + 5)^2 - 3$$

$$x + 3 = -2(y + 5)^2$$

$$\frac{x + 3}{-2} = (y + 5)^2 \qquad \textbf{Take the square root of each side.}$$

$$\pm\sqrt{\frac{x + 3}{-2}} = y + 5$$

$$y = -5 \pm \sqrt{\frac{x + 3}{-2}}$$

$$y = -5 + \sqrt{\frac{x + 3}{-2}} \quad \text{or} \quad y = -5 - \sqrt{\frac{x + 3}{-2}}$$

Define $y_1 = \sqrt{(x + 3)/-2}$, $y_2 = -5 + y_1$, and $y_3 = -5 - y_1$. Deselect $y_1$ and graph $y_2$ and $y_3$ in the $[-15, 5]$ by $[-10, 10]$ window. Figure 10.26 shows a graph of the equation.

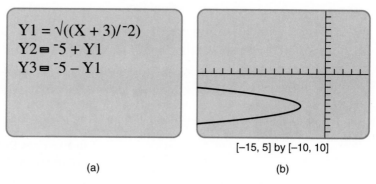

$$Y1 = \sqrt{((X + 3)/^-2)}$$
$$Y2 = {}^-5 + Y1$$
$$Y3 = {}^-5 - Y1$$

$[-15, 5]$ by $[-10, 10]$

(a)                                    (b)

**FIGURE 10.26**    (a) The Function-defining menu and (b) a graph of $x = -2(y + 5)^2 - 3$, both shown on a grapher.

## Changing an Equation to Standard Vertex Form

Example 4 demonstrates how to find the graph of a parabola using a grapher once its equation is in vertex form. If the equation is in the form $x = ay^2 + by + c$, we can use the method of completing the square to transform the equation to an equivalent vertex form.

**EXAMPLE 5**    Graphing an Equation in the Form $x = ay^2 + by + c$

Find the graph of $x = y^2 - 6y + 3$ using a grapher.

SOLUTION

1. In order to enter this equation into a grapher, transform the equation to the vertex form by completing the square.

$$x = y^2 - 6y + 3 \qquad \text{Group the y-terms and complete the square.}$$

$$x = (y^2 - 6y + 9) - 9 + 3$$

$$x = (y - 3)^2 - 6$$

2. To prepare the equation for entering into a grapher, solve for $y$.

$$x = (y - 3)^2 - 6$$

$$x + 6 = (y - 3)^2 \qquad \text{Take the square root of each side.}$$

$$\pm\sqrt{x + 6} = y - 3 \qquad \text{Add 3 to each side.}$$

$$3 + \sqrt{x + 6} = y \quad \text{and} \quad 3 - \sqrt{x + 6} = y$$

3. Define $y_1 = \sqrt{(x + 6)}$, $y_2 = 3 + y_1$, and $y_3 = 3 - y_1$. Figure 10.27 shows the graph of $y_2$ and $y_3$ in the Standard window.    ∎

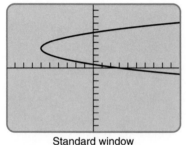

Standard window

**FIGURE 10.27** A graph of $x = (y - 3)^2 - 6$ in Example 5.

---

## Exercises for Section   **10.3**

In Exercises 1–8, state the direction in which the parabola opens.

**1.** $y = (x - 3)^2 - 5$    **2.** $y = -2(x + 2)^2 + 3$

**3.** $x = -(y - 5)^2 - 7$    **4.** $x = (y - 4)^2 + 1$

**5.** $y = -x^2 + 2x - 5$    **6.** $y = 2x^2 + 3x + 1$

**7.** $x = y^2 - 3y - 10$    **8.** $x = -y^2 + 3x + 2$

In Exercises 9–16, state the vertex and the line of symmetry of the parabola.

**9.** $x = y^2 + 5$    **10.** $x = -y^2 - 3$

**11.** $x = (y - 7)^2$    **12.** $x = (y + 5)^2$

**13.** $x = (y + 1)^2 + 4$     **14.** $x = (y + 3)^2 - 5$

**15.** $x = -(y - 5)^2 + 8$     **16.** $x = -(y - 4)^2 + 1$

In Exercises 17–22, hand sketch a graph.

**17.** $y = (x - 5)^2 - 9$     **18.** $y = -2(x + 1)^2 + 2$

**19.** $x = (y - 7)^2 - 3$     **20.** $x = -(y - 1)^2 + 12$

**21.** $x = (y + 4)^2 - 7$     **22.** $x = -(y + 5)^2 - 6$

In Exercises 23–28, find a graph of the equation on a grapher.

**23.** $x = 2(y - 4)^2 + 3$     **24.** $x = 3(y - 1)^2 + 7$

**25.** $x = -3(y + 2)^2 - 4$     **26.** $x = -(y + 3)^2 - 6$

**27.** $x = 2(y - 8)^2 + 12$     **28.** $x = -(y + 9)^2 - 18$

In Exercises 29–36, transform the equation of the parabola to the standard vertex form by completing the square.

**29.** $y = x^2 + 2x - 7$     **30.** $y = x^2 - 6x + 9$

**31.** $x = y^2 - 8y - 12$     **32.** $x = y^2 + 4y - 10$

**33.** $y = 3x^2 + 6x - 14$     **34.** $y = 2x^2 + 16x - 5$

**35.** $x = 2y^2 - 12y - 7$     **36.** $x = -2y^2 + 20y - 15$

In Exercises 37–44, graph the parabola using a grapher.

**37.** $y = x^2 + 2x - 5$     **38.** $y = x^2 - 6x - 8$

**39.** $x = y^2 - 8y + 3$     **40.** $x = y^2 + 4y - 5$

**41.** $x = y^2 - 5y - 14$     **42.** $x = y^2 + 3y - 9$

**43.** $x = y^2 - 7y + 10$     **44.** $x = y^2 + 3y - 5$

In Exercises 45 and 46, write an explanation for the question.

**45. Writing to Learn.** How does the equation of a parabola determine whether a parabola has a vertical or horizontal axis of symmetry?

**46. Writing to Learn.** Why is the graph of $x = ay^2 + by + c$ not the graph of a function of $x$?

EXTENDING THE IDEAS

**47.** Find the value of $k$ if $(0, 3)$ is a point on the graph of $x = 3y^2 + ky$.

**48.** Find the value of $k$ if $(1, 4)$ is a point on the graph of $x = 2y^2 + ky - 3$.

In Exercises 49–52, consider the following: In Section 10.3, you learned to prepare an equation like $x = y^2 - 4y + 7$ for graphing on a grapher by (a) completing the square to find vertex form and (b) solving the resulting equation for $y$. An alternate method is to write the equation as $y^2 - 4y + (7 - x) = 0$ and use the quadratic formula to solve for $y$. Note that the term $(7 - x)$ must be treated as the constant term. Use the quadratic formula to solve the equation for $y$.

**49.** $x = y^2 - 4y + 2$     **50.** $x = y^2 - 2y + 7$

**51.** $x = 2y^2 + 4y + 3$     **52.** $x = 3y^2 + 6y + 2$

LOOKING BACK—MAKING CONNECTIONS

In Exercises 53–60, describe transformations that can be used to obtain the given graph from the graph of $x = y^2$. Support your answer using a grapher.

**53.** $x = y^2 - 3$     **54.** $x = -y^2 + 2$

**55.** $x = (y - 4)^2$     **56.** $x = (y + 3)^2$

**57.** $x = (y + 4)^2 + 2$     **58.** $x = (y + 1)^2 - 3$

**59.** $x = -(y - 6)^2 + 12$     **60.** $x = -(y - 5)^2 + 11$

**61. Rocket Engineering.** Find the maximum height a rocket reaches if it is shot straight up with an initial velocity of 78 ft/sec using the formula $h = -16t^2 + v_0 t$.

**62. Rocket Engineering.** At Space Camp, Michael and Susan built model rockets to launch. Both rockets were launched simultaneously, with Michael's rocket traveling at an initial velocity $v_0$ of 80 ft/sec and Susan's at 60 ft/sec. Using the law of physics $h = -16t^2 + v_0 t$, answer the following questions:

a) Write an algebraic representation for the distance each rocket travels in $t$ seconds after they are launched.

b) Find a complete graph of this problem situation.

c) What is the maximum height each rocket reaches?

d) Whose rocket reaches its maximum height first?

e) Whose rocket hits the ground first and how many seconds later does the other rocket hit the ground?

## 10.4 The Hyperbola

Sketching the Graph of a Hyperbola ▪ Graphing a Hyperbola on a Grapher ▪ Graphing a Hyperbola Not Centered at the Origin

We begin this section with an exploration.

### EXPLORE WITH A GRAPHER

Find the graph of $x^2 - y^2 = 100$ by defining the following:

$$y_1 = \sqrt{(x^2 - 100)}$$

$$y_2 = -\sqrt{(x^2 - 100)}$$

1. Use the Integer window and begin with the trace cursor at $(10, 0)$. Move the cursor to the right and compare the difference between the $x$- and $y$-coordinates. Continue moving the trace cursor to the right as the graph scrolls off the screen. What happens to the difference between the $x$- and $y$-coordinates?

2. Next, define $y_3 = x$ and $y_4 = -x$ and graph $y_1$, $y_2$, $y_3$, and $y_4$ in the Integer window.

**Generalization** Explain how the visual image supports your observation about the numerical pattern.

A **hyperbola**, like an ellipse, is determined by a constant and two fixed points, each called a **focus** point. However, instead of the *sum* of two distances being a constant, as with an ellipse, in the case of the hyperbola, the *difference* between two distances is constant.

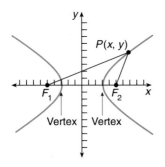

**FIGURE 10.28** A hyperbola has two foci. The difference $F_1P - F_2P$ is a constant.

More specifically, the hyperbola determined by the foci and the given constant is the set of points $(x, y)$ in the plane such that the absolute value of the difference between the distances from $(x, y)$ to the two foci is equal to the constant (see Figure 10.28).

A hyperbola has two branches. The two branches can open either right and left or up and down, depending on whether the foci are on the $x$- or $y$-axis. Each branch of the hyperbola has a vertex and the **center of the hyperbola** is the midpoint of the line segment joining the vertices.

Using the distance formula and the fact that the difference of the distances from each point to the foci is constant, it can be shown that an equation for a hyperbola centered at the origin fits one of the following standard forms.

---

### Standard Form of a Hyperbola Centered at the Origin

The **standard form for an equation of a hyperbola centered at the origin** is one of the following:

$$\frac{x^2}{a^2} - \frac{y^2}{b^2} = 1 \qquad \text{or} \qquad \frac{y^2}{b^2} - \frac{x^2}{a^2} = 1$$

For the form on the left, the vertices are $(-a, 0)$ and $(a, 0)$ and the branches open to the left and right. For the form on the right, the vertices are $(0, -b)$ and $(0, b)$ and the branches open up and down.

---

Figure 10.29 shows two hyperbolas—one opening right and left and the other opening up and down.

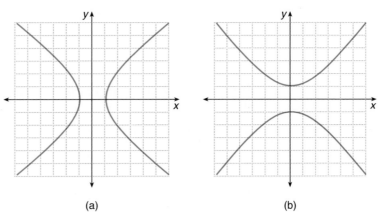

(a)                              (b)

**FIGURE 10.29** (a) A sketch of a hyperbola with equation $\frac{x^2}{a^2} - \frac{y^2}{b^2} = 1$ and (b) a sketch of a hyperbola with equation $\frac{y^2}{b^2} - \frac{x^2}{a^2} = 1$.

Observe from these forms that the location of a hyperbola's vertices are determined by the positive coefficients of either the $x^2$- or $y^2$-term.

**EXAMPLE 1**    Finding the Vertices of a Hyperbola

Find the vertices for the hyperbolas determined by the following equations:

a) $\dfrac{x^2}{5^2} - \dfrac{y^2}{7^2} = 1$    b) $\dfrac{y^2}{3^2} - \dfrac{x^2}{8^2} = 1$

SOLUTION

a) The graph of $\dfrac{x^2}{5^2} - \dfrac{y^2}{7^2} = 1$ opens left and right. Its vertices are $(-5, 0)$ and $(5, 0)$.

b) The graph of $\dfrac{y^2}{3^2} - \dfrac{x^2}{8^2} = 1$ opens up and down and its vertices are $(0, -3)$ and $(0, 3)$. ∎

## Sketching the Graph of a Hyperbola

There are two lines through the center of a hyperbola, called **asymptotes,** that are key to sketching a hyperbola by hand. As you move away from the origin, the asymptotes of a hyperbola are the two lines to which the hyperbola gets closer and closer. They are the diagonals of a rectangle that we call **the fundamental rectangle** of the hyperbola, described as follows.

---

### Fundamental Rectangle of a Hyperbola Centered at the Origin

The **fundamental rectangle** of the hyperbola that is the graph of

$$\frac{x^2}{a^2} - \frac{y^2}{b^2} = 1$$

is the rectangle whose vertices are $(a, b)$, $(-a, b)$, $(-a, -b)$, and $(a, -b)$. The **asymptotes** of this hyperbola are the lines with equations

$$y = \frac{b}{a}x \quad \text{and} \quad y = -\frac{b}{a}x.$$

---

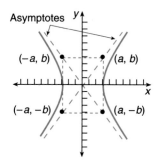

Asymptotes

$(-a, b)$    $(a, b)$

$(-a, -b)$    $(a, -b)$

**FIGURE 10.30** A hyperbola and its fundamental rectangle and asymptotes.

Figure 10.30 shows both the fundamental rectangle and the two asymptotes for a hyperbola.

**EXAMPLE 2**    Finding the Asymptotes of a Hyperbola

Find the asymptotes of the hyperbola determined by the graph of

$$\frac{x^2}{4} - \frac{y^2}{16} = 1.$$

SOLUTION

The given equation is in standard form, with $b^2 = 16$ and $a^2 = 4$. The asymptotes are given by the following equations:

$$y = \frac{b}{a}x \qquad \text{and} \qquad y = -\frac{b}{a}x \qquad \textbf{Substitute for } \textbf{\textit{a}} \textbf{ and } \textbf{\textit{b}}.$$

$$y = \frac{4}{2}x = 2x \qquad\qquad y = -\frac{4}{2}x = -2x$$

■

**TRY THIS**

Graph the hyperbola in Example 2 and then zoom out. Compare the graph that results and the equations of the asymptotes, $y = 2x$ and $y = -2x$.

In Example 1, we found the vertex of a hyperbola; in Example 2, we found asymptotes. Both are important for sketching a hyperbola by hand. The complete process is as follows.

---

**Graphing a Hyperbola Centered at (0,0)**

1. Plot the vertices of the hyperbola.
2. Sketch the fundamental rectangle.
3. Sketch the asymptotes as diagonals of the fundamental rectangle.
4. Sketch the graph. Each branch goes through the vertex and approaches each asymptote.

---

Example 3 illustrates this process of sketching a hyperbola.

**EXAMPLE 3**    Sketching a Hyperbola

Sketch the graph of the equation

$$\frac{x^2}{4} - \frac{y^2}{16} = 1.$$

SOLUTION

This equation is one of the standard forms for a hyperbola centered at the origin. Because the coefficient of $x^2$ is positive, the hyperbola opens right and left. In this equation in standard form, $4 = a^2$ and $16 = b^2$. Therefore

- (2, 0) and (−2, 0) are the vertices and
- $y = 2x$ and $y = -2x$ are the equations of the asymptotes.

Figure 10.31 shows a sketch of the hyperbola.

**REMINDER**

When sketching a hyperbola, make sure the branches approach, but never touch, the asymptotes.

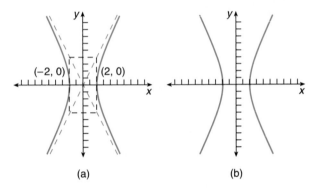

(a)        (b)

**FIGURE 10.31**   (a) A sketch of the hyperbola $\dfrac{x^2}{4} - \dfrac{y^2}{16} = 1$ in Example 4 with the fundamental rectangle and asymptotes and (b) the hyperbola itself.

Example 4 illustrates sketching a hyperbola with vertical branches. Notice that the $y^2$-term is positive in this case.

**EXAMPLE 4**   Graphing a Hyperbola

Sketch a graph of the equation

$$\frac{y^2}{9} - \frac{x^2}{25} = 1.$$

SOLUTION

This equation is a standard form for a hyperbola centered at the origin. Because the coefficient of $y^2$ is positive, the hyperbola opens up and down. In this equation

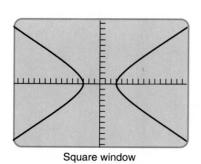

**FIGURE 10.32** The sketch of the hyperbola $\dfrac{y^2}{9} - \dfrac{x^2}{25} = 1$ in Example 4.

in standard form, $a^2 = 25$ and $b^2 = 9$. Therefore

- $(0, 3)$ and $(0, -3)$ are the vertices and
- $y = \dfrac{3}{5}x$ and $y = -\dfrac{3}{5}x$ are the equations of the asymptotes.

Plot the following corners of the fundamental rectangle: $(5, 3)$, $(-5, 3)$, $(-5, -3)$, and $(5, -3)$ (see Figure 10.32). ∎

## Graphing a Hyperbola on a Grapher

A hyperbola can be graphed using a grapher by first solving the equation for $y$ and then applying the same techniques that were used to graph an ellipse.

**EXAMPLE 5**  Graphing a Hyperbola Using a Grapher

Find a graph of $4x^2 - 9y^2 = 36$ using a grapher.

SOLUTION

To prepare for entering into a grapher, solve the equation for $y$.

$$4x^2 - 9y^2 = 36$$

$$4x^2 - 36 = 9y^2$$

$$\frac{4x^2 - 36}{9} = y^2$$

$$\pm\sqrt{\frac{4x^2 - 36}{9}} = y$$

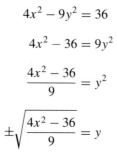

Square window

**FIGURE 10.33** A graph of the hyperbola $4x^2 - 9y^2 = 36$.

Graph both $y_1 = \sqrt{((4x^2 - 36)/9)}$ and $y_2 = -y_1$ in a Square window (see Figure 10.33). ∎

## Graphing a Hyperbola Not Centered at the Origin

When a hyperbola is shifted from the origin horizontally or vertically, an equation for the resulting hyperbola fits a format similar to that of an ellipse and terms are separated by a minus sign instead of a plus sign. This format is summarized next.

## The Standard Form of a Hyperbola

The **standard form for an equation of a hyperbola** with the center at $(h, k)$ is one of the following:

$$\frac{(x-h)^2}{a^2} - \frac{(y-k)^2}{b^2} = 1 \quad \text{or} \quad \frac{(y-k)^2}{b^2} - \frac{(x-h)^2}{a^2} = 1$$

The asymptotes are the lines through the center with slopes $m = b/a$ and $m = -b/a$.

To sketch a graph of a hyperbola not centered at the origin, we first must write the equation in the standard form of a hyperbola, as Example 6 illustrates.

**EXAMPLE 6**    Writing the Equation in Standard Form

Rewrite the following equation in the standard form of a hyperbola and locate the center and the coordinates of the vertices.

$$9(x-3)^2 - 16(y+1)^2 = 144$$

SOLUTION
Change the equation to standard form.

$$9(x-3)^2 - 16(y+1)^2 = 144 \qquad \textbf{Divide each side by 144.}$$

$$\frac{(x-3)^2}{16} - \frac{(y+1)^2}{9} = 1$$

From this standard form, we see that the center of the hyperbola is $(3, -1)$. Because the $x^2$-term is positive, the vertices are located at $(-1, -1)$ and $(7, -1)$.    ∎

In Example 7, we sketch a hyperbola that is not centered at the origin.

**EXAMPLE 7**    Sketching a Hyperbola

Sketch a graph of the equation

$$\frac{(x+5)^2}{4} - \frac{(y+2)^2}{9} = 1.$$

SOLUTION

The standard form of the equation is

$$\frac{(x - (-5))^2}{4} - \frac{(y - (-2))^2}{9} = 1.$$

From this form, we see that the center of the hyperbola is $(-5, -2)$. The upper right vertex of the fundamental rectangle is 2 units right and 3 units up from this center. The four corners of the fundamental rectangle are $(-7, 1)$, $(-7, -5)$, $(-3, 1)$, and $(-3, -5)$. Plot them and draw the two asymptotes (see Figure 10.34(a)).

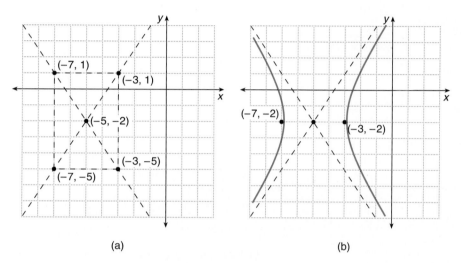

(a)                                        (b)

**FIGURE 10.34** (a) A sketch of the fundamental rectangle and the asymptotes of the hyperbola $\dfrac{(x + 5)^2}{4} - \dfrac{(y + 2)^2}{9} = 1$ and (b) a sketch of the hyperbola in Example 7.

Complete the sketch by drawing the branches from the two vertices $(-3, -2)$ and $(-7, -2)$ to the two asymptotes. Figure 10.34 (b) shows a completed sketch. ∎

To sketch the graph of a hyperbola on a grapher, the equation must first be prepared for the grapher. This preparation can be done either by completing the square or by using the quadratic formula. Examples 8 and 9 illustrate these two methods.

**EXAMPLE 8**     Rewriting in Standard Form and Graphing

Graph the equation of $9y^2 + 36y - x^2 + 6x = 54$ using a grapher.

SOLUTION

Rewrite the equation by completing the square and solving for $y$.

$$9y^2 + 36y - x^2 + 6x = 54 \qquad \text{Group like terms.}$$

$$9(y^2 + 4y \quad) - (x^2 - 6x \quad) = 54 \qquad \text{Complete the square.}$$

$$9(y^2 + 4y + 4) - (x^2 - 6x + 9) = 54 + 36 \quad - 9 \qquad \text{Factor.}$$

$$9(y + 2)^2 - (x - 3)^2 = 81 \qquad \begin{array}{l}\text{Eliminate the } x\text{-term} \\ \text{from the left side.}\end{array}$$

$$9(y + 2)^2 = 81 + (x - 3)^2 \qquad \text{Divide each side by 9.}$$

$$(y + 2)^2 = \frac{81 + (x - 3)^2}{9} \qquad \begin{array}{l}\text{Take the square root} \\ \text{of each side of the} \\ \text{equation.}\end{array}$$

$$(y + 2) = \pm\sqrt{\frac{81 + (x - 3)^2}{9}}$$

$$y = -2 \pm \sqrt{\frac{81 + (x - 3)^2}{9}}$$

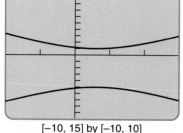

[−10, 15] by [−10, 10]

**FIGURE 10.35** A graph of the hyperbola $9y^2 + 36y - x^2 + 6x = 54$ in Example 8.

Define $y_1 = \sqrt{((81 + (x - 3)^2)/9)}$, $y_2 = -2 + y_1$, and $y_3 = -2 - y_1$. Deselect $y_1$ and graph $y_2$ and $y_3$ in the $[-10, 15]$ by $[-10, 10]$ window. Figure 10.35 shows a graph of the hyperbola. ∎

In Example 9 we graph the equation in Example 8 but this time using the quadratic formula instead of completing the square to prepare the equation for the grapher.

**EXAMPLE 9**    Using the Quadratic Formula to Graph a Hyperbola

Graph the equation $9y^2 + 36y - x^2 + 6x = 54$ using a grapher.

SOLUTION

Rewrite the equation in the form $ay^2 + by + c = 0$.

$$9y^2 + 36y - x^2 + 6x = 54$$

$$9y^2 + 36y + (-x^2 + 6x - 54) = 0$$

In this form, $a = 9$, $b = 36$, and $c = -x^2 + 6x - 54$.

Substitute these values in the quadratic formula,

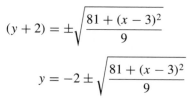

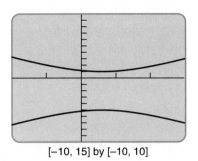

[−10, 15] by [−10, 10]

**FIGURE 10.36** A graph of the hyperbola $9y^2 + 36y - x^2 + 6y = 54$ in Example 9.

$$= \frac{-36 \pm \sqrt{36^2 - 4(9)(-x^2 + 6x - 54)}}{2(9)}$$

$$= \frac{-36 \pm \sqrt{36^2 - 36(-x^2 + 6x - 54)}}{18}$$

It is not necessary to simplify. Define $y_1 = \sqrt{(36^2 - 36(-x^2 + 6x - 54))}$, $y_2 = (-36 + y_1)/18$, $y_3 = (-36 - y_1)/18$. Deselect $y_1$ and graph $y_2$ and $y_3$ in the $[-10, 15]$ by $[-10, 10]$ window. Figure 10.36 shows a graph of $y_1$ and $y_2$.  ∎

## Exercises for Section 10.4

In Exercises 1–8, find the center and the vertices of the graph of the equation.

1. $\dfrac{x^2}{9} - \dfrac{y^2}{25} = 1$

2. $\dfrac{x^2}{4} - y^2 = 1$

3. $\dfrac{y^2}{16} - \dfrac{x^2}{36} = 1$

4. $\dfrac{y^2}{25} - x^2 = 1$

5. $\dfrac{y^2}{81} - \dfrac{x^2}{64} = 1$

6. $\dfrac{x^2}{36} - \dfrac{y^2}{49} = 1$

7. $\dfrac{x^2}{36} - \dfrac{y^2}{121} = 1$

8. $\dfrac{x^2}{64} - \dfrac{y^2}{9} = 1$

In Exercises 9–16, state whether the graph of the equation opens right and left or up and down. Also find the vertices of the fundamental rectangle and the equations of the asymptotes.

9. $\dfrac{x^2}{16} - \dfrac{y^2}{36} = 1$

10. $\dfrac{x^2}{4} - \dfrac{y^2}{49} = 1$

11. $\dfrac{y^2}{25} - x^2 = 1$

12. $\dfrac{y^2}{64} - \dfrac{x^2}{4} = 1$

13. $\dfrac{x^2}{4} - \dfrac{y^2}{9} = 1$

14. $y^2 - \dfrac{x^2}{4} = 1$

15. $\dfrac{y^2}{18} - \dfrac{x^2}{32} = 1$

16. $\dfrac{x^2}{21} - \dfrac{y^2}{34} = 1$

In Exercises 17–24, change the equation to standard form. Then find the vertices of the fundamental rectangle for a graph of the equation.

17. $9x^2 - y^2 = 81$

18. $25x^2 - y^2 = 25$

19. $4y^2 - x^2 = 144$

20. $9y^2 - 16x^2 = 144$

21. $4x^2 - y^2 = 1$

22. $9x^2 - 36y^2 = 36$

23. $y^2 - x^2 = 1$

24. $9y^2 - 4x^2 = 1$

In Exercises 25–28, give the coordinates of the center and vertices of the hyperbola.

25. $\dfrac{(x-3)^2}{4} - \dfrac{(y-1)^2}{16} = 1$

26. $\dfrac{(x+4)^2}{16} - \dfrac{y^2}{25} = 1$

27. $\dfrac{(y+1)^2}{81} - \dfrac{(x-2)^2}{4} = 1$

28. $\dfrac{(y+3)^2}{49} - \dfrac{x^2}{16} = 1$

In Exercises 29–40, sketch a graph of the hyperbola.

29. $\dfrac{x^2}{16} - \dfrac{y^2}{4} = 1$

30. $\dfrac{x^2}{64} - \dfrac{y^2}{36} = 1$

**31.** $\dfrac{y^2}{4} - \dfrac{x^2}{81} = 1$      **32.** $\dfrac{y^2}{9} - x^2 = 1$

**33.** $\dfrac{(x-3)^2}{4} - \dfrac{(y-1)^2}{16} = 1$

**34.** $\dfrac{(x+4)^2}{16} - \dfrac{x^2}{9} = 1$

**35.** $\dfrac{(y+1)^2}{81} - \dfrac{(x-2)^2}{4} = 1$

**36.** $\dfrac{(y+3)^2}{49} - \dfrac{x^2}{36} = 1$

**37.** $x^2 - 9y^2 = 81$      **38.** $9x^2 - 9y^2 = 81$

**39.** $y^2 - x^2 = 1$      **40.** $25y^2 - 9x^2 = 225$

In Exercises 41–46, find a graph of the equation using a grapher.

**41.** $9x^2 - 16y^2 = 144$      **42.** $4x^2 - 9y^2 = 36$

**43.** $2x^2 - 3y^2 = 12$      **44.** $9y^2 - 4x^2 = 12$

**45.** $5y^2 - 3x^2 = 8$      **46.** $9y^2 - 8x^2 = 36$

**47.** *Architectural Design.* Twin tower apartment buildings each have a garden in front of them that is shaped like a hyperbola, as shown in the following figure. The equation of the hyperbola is $25x^2 - 81y^2 = 50625$. Find the distance in meters between the two gardens at their closest point.

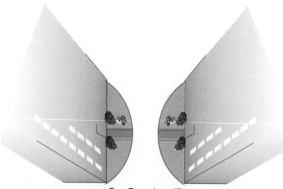

For Exercise 47.

In Exercises 48 and 49, compare the graphs of the two equations and list their similarities and differences.

**48. Writing to Learn.** Compare the graphs.

$$\frac{x^2}{9} - \frac{y^2}{4} = 1$$

$$\frac{y^2}{9} - \frac{x^2}{4} = 1$$

**49. Writing to Learn.** Compare the graphs.

$$\frac{x^2}{9} - \frac{y^2}{4} = 1$$

$$\frac{(x-3)^2}{9} - \frac{(y+5)^2}{4} = 1$$

EXTENDING THE IDEAS

In Exercises 50–53, find a graph of the equation on a grapher. Prepare the equation for the grapher by using the quadratic formula to solve for $y$ (in terms of $x$). For example, the equation $x^2 - 2x - y^2 + 5y - 8 = 0$ can be thought of as the following quadratic equation (in variable $y$):

$$y^2 - 5y - (x^2 - 2x - 8) = 0$$

The expression $x^2 - 2x - 8$ should be thought of as a constant term.

**50.** $x^2 + 4x - 4y^2 + 4y + 1 = 0$

**51.** $x^2 + 6x - y^2 - 4y + 18 = 0$

**52.** $3x^2 + 4xy + 3y^2 - 12x + 2y + 7 = 0$

**53.** $2x^2 - xy + 3y^2 - 3x + 4y - 6 = 0$

LOOKING BACK—MAKING CONNECTIONS

In Exercises 54–63, state the type of graph for the equation.

**54.** $x^2 + 9y^2 = 81$      **55.** $9x^2 + 9y^2 = 36$

**56.** $\dfrac{x^2}{36} - \dfrac{y^2}{9} = 1$    **57.** $\dfrac{x^2}{16} - \dfrac{y^2}{25} = 1$    **60.** $y = x^2 + 6x - 15$    **61.** $x = y^2 + 2x - 5$

**58.** $\dfrac{y^2}{9} + \dfrac{x^2}{81} = 1$    **59.** $\dfrac{y^2}{9} + x^2 = 1$    **62.** $y = (x - 7)^2 + 5$    **63.** $x = (y + 1)^2 + 7$

## 10.5  Nonlinear Systems of Equations

### Solving a Nonlinear System

Recall that an equation in two variables that is not a linear equation is called a **nonlinear equation** in two variables. For example, each of the following are nonlinear equations:

$$y = x^2 + 3 \qquad y^2 + x^2 = 25 \qquad x^2 - 3y^2 + 24x - 5y = 18$$

A system of equations is called a **nonlinear system** if it contains at least one nonlinear equation.

Here is an example of a nonlinear system of equations.

$$y = x^2 - 9$$
$$y = x + 3$$

Substitution and eliminating the variable were two algebraic methods used in Chapter 3 for solving a system of equations. We also solved systems by graphical methods. All three methods can be used for solving nonlinear systems.

### Solving a Nonlinear System

To find the solution of a nonlinear system of equations, we recommend you simultaneously graph each equation in the system. The solution to the system then consists of the points of intersection of these equations.

Example 1 illustrates how to solve such a system graphically. The solution is supported numerically.

**EXAMPLE 1**    Solving a Nonlinear System Graphically

Solve the following system of equations graphically. Confirm the solution numerically.

$$y = x^2 - 9$$
$$y = x + 3$$

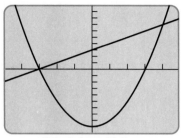

0.1 window with the
vertical dimension
changed to [−10, 10]

**FIGURE 10.37** The
solution to Example 1 consists
of the points of intersection of
the parabola and the line.

SOLUTION

**Solve Graphically.** Figure 10.37 shows the graphs of $y_1 = x^2 - 9$ and $y_2 = x + 3$. These curves intersect at two points. Use TRACE to determine that the points of intersection are $(-3, 0)$ and $(4, 7)$.

**Confirm Numerically.**

1. Substitute $x = -3$ and $y = 0$ into each equation to see if it is a true statement.

$$0 = (-3)^2 - 9 \qquad \text{True statement that results after substituting the values into } y_1.$$

$$0 = -3 + 3 \qquad \text{True statement that results after substituting the values into } y_2.$$

2. Substitute $x = 4$ and $y = 7$ into each equation.

$$7 = 4^2 - 9 \qquad \text{True statement that results using } y_1.$$

$$7 = 4 + 3 \qquad \text{True statement that results using } y_2.$$

The solution to the system are the points $(-3, 0)$ and $(4, 7)$. ■

In Example 1, it was convenient to use a 0.1 window because the coordinates of the solutions were integer values. However, most often when working with nonlinear equations, we recommend beginning with the Standard window and then changing it as necessary. Usually zoom-in methods will be necessary, as we illustrate in Example 2.

**EXAMPLE 2**  Solving a Nonlinear System Graphically

Solve the following system of equations graphically and support the solution numerically.

$$x^2 + y^2 = 16$$

$$y = 2x^2 - 2$$

SOLUTION

**Solve Graphically.** To prepare the first equation for a grapher, solve for $y$. Thus $y = \pm\sqrt{16 - x^2}$. Figure 10.38 shows the following graphs in a Square window:

$$y_1 = \sqrt{16 - x^2}$$

$$y_2 = -y_1$$

$$y_3 = 2x^2 - 2$$

Figure 10.38 (a) shows that there are two points of intersection. Use zoom-in to find that the points of intersection are $(1.678, 3.631)$ and $(-1.678, 3.631)$, with an error of at most 0.01.

**Support Numerically.** Store 1.678 as variable $x$ and 3.631 as variable $y$, then evaluate $x^2 + y^2$. The result is 15.999845. The value of $2x^2 - 2$ is 3.631368. These outcomes support the claim that $(1.678, 3.631)$ is a solution. In a similar manner, we can obtain support that $(-1.678, 3.631)$ is also a solution.

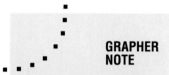

**GRAPHER NOTE**

When zooming in to find the points of intersection in Example 2, it is easiest to deselect $Y_2$ since the graphs $Y_1$ and $Y_3$ are intersecting. This speeds up using TRACE to find the common point of intersection.

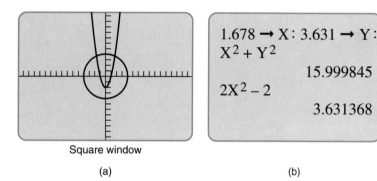

Square window

(a)

$1.678 \rightarrow X: 3.631 \rightarrow Y:$
$X^2 + Y^2$

$\qquad 15.999845$

$2X^2 - 2$

$\qquad 3.631368$

(b)

**FIGURE 10.38** The solution to Example 2 are the points of intersection of the parabola and the circle, and (b) shows numerical support of the solution.

Example 3 illustrates how to use a graph of a system to determine the number of intersections and then solve the system algebraically.

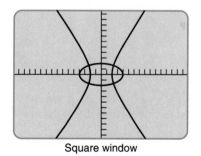

Square window

**FIGURE 10.39** The solution to Example 3 is the four ordered pairs representing the points of intersection of the ellipse and the hyperbola.

**EXAMPLE 3**  Solving a Nonlinear System Algebraically

Solve the following system of equations algebraically:

$$9x^2 - 4y^2 = 36$$
$$x^2 + 4y^2 = 16$$

SOLUTION

**Solve Algebraically.** Figure 10.39 shows a graph of this system of equations, and makes it evident that there are four points of intersection. Solve the system using the elimination method. Add the two equations to eliminate the $y$-terms.

$$9x^2 - 4y^2 = 36$$

$$x^2 + 4y^2 = 16 \qquad \text{Add the equations to eliminate the } y^2\text{-terms.}$$

$$10x^2 = 52$$

$$x^2 = 5.2$$

$$x = \pm\sqrt{5.2}$$

$$x = \pm 2.280$$

Substitute $x^2 = 5.2$ into either equation to solve for $y^2$. We choose the second equation.

$$5.2 + 4y^2 = 16$$

$$4y^2 = 10.8$$

$$y^2 = 2.7$$

$$y = \pm\sqrt{2.7}$$

$$y = \pm 1.643$$

Therefore the point of intersection in the first quadrant is $(2.280, 1.643)$. Because the hyperbola and ellipse are symmetric with respect to both axes, we see that the intersection in the second quadrant is $(-2.280, 1.643)$, in the third quadrant $(-2.280, -1.643)$, and in the fourth quadrant $(2.280, -1.643)$. ∎

Example 4 illustrates solving the system algebraically using substitution.

### EXAMPLE 4    Solve a Nonlinear System Algebraically

Solve the following system of equations algebraically and support numerically:

$$x^2 + 9y^2 = 9$$

$$y = x^2 - 1$$

SOLUTION

We first find a graph of the system to see how many points of intersection are in the solution. Figure 10.40 shows that there appear to be three points of intersection and therefore three solutions.

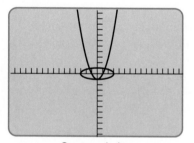

Square window

**FIGURE 10.40** The solution to Example 4 is the three ordered pairs representing the points of intersection of the ellipse and the parabola.

**Solve Algebraically.** Solve using the substitution method. One equation is solved for $y$, so

$$x^2 + 9y^2 = 9$$

$$y = x^2 - 1 \qquad \text{Substitute into the first equation to solve for } y.$$

$$x^2 + 9(x^2 - 1)^2 = 9 \qquad \text{Square the binomial.}$$

$$x^2 + 9(x^4 - 2x^2 + 1) = 9 \qquad \text{Use the Distributive Property.}$$

$$x^2 + 9x^4 - 18x^2 + 9 = 9 \qquad \text{Simplify.}$$

$$9x^4 - 17x^2 = 0 \qquad \text{Factor } x^2 \text{ out of each term.}$$

$$x^2(9x^2 - 17) = 0 \qquad \text{Set each factor} = 0 \text{ and solve for } x.$$

$$x^2 = 0 \quad \text{or} \quad 9x^2 = 17$$

$$x = 0 \quad \text{or} \quad x^2 = \frac{17}{9}$$

$$x = \pm \frac{\sqrt{17}}{3} \qquad \begin{array}{l}\text{Substitute into the second} \\ \text{equation to solve for } y.\end{array}$$

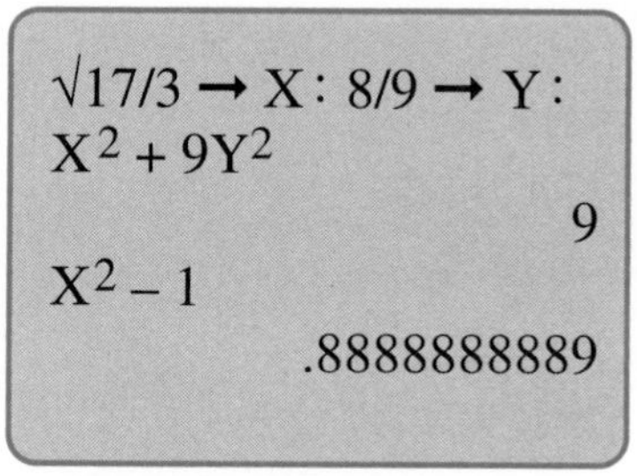

$$y = 0^2 - 1 \quad \text{or} \quad y = \left(\frac{\pm\sqrt{17}}{3}\right)^2 - 1$$

$$y = -1 \quad \text{or} \quad y = \frac{17}{9} - 1 = \frac{8}{9}$$

The solution consists of the ordered pairs $(0, -1)$, $\left(\dfrac{\sqrt{17}}{3}, \dfrac{8}{9}\right)$, and $\left(-\dfrac{\sqrt{17}}{3}, \dfrac{8}{9}\right)$.

**Support Numerically.** Figure 10.41 shows grapher numerical support for the intersection point $\left(\dfrac{\sqrt{17}}{3}, \dfrac{8}{9}\right)$ as a solution. The intersection $(0, -1)$ can be evaluated mentally to confirm $0^2 + 9(-1)^2 = 9$ and $0^2 - 1 = -1$ are true statements. ■

---

$\sqrt{17}/3 \rightarrow \text{X}: 8/9 \rightarrow \text{Y}:$

$\text{X}^2 + 9\text{Y}^2$

$\qquad\qquad\qquad 9$

$\text{X}^2 - 1$

$\qquad\qquad .8888888889$

**FIGURE 10.41**
Numerical support that the pair $(\frac{\sqrt{17}}{3}, 8/9)$ is in a solution of the system in Example 4. Notice that 0.8888888889 is a decimal approximation of 8/9.

---

## Exercises for Section  **10.5**

In Exercises 1–6, graph the system of equations and identify the number of points of intersection.

1. $\begin{cases} y = -x + 2 \\ y = 2x^2 - 9 \end{cases}$

2. $\begin{cases} y = -2x^2 + 8 \\ y = x - 5 \end{cases}$

3. $\begin{cases} x^2 + y^2 = 36 \\ y = x^2 + 7 \end{cases}$

4. $\begin{cases} 4x^2 + y^2 = 49 \\ y = 2x^2 - 8 \end{cases}$

5. $\begin{cases} x^2 - y^2 = 1 \\ x^2 + y^2 = 100 \end{cases}$

6. $\begin{cases} y^2 - 4x^2 = 1 \\ x^2 + 4y^2 = 100 \end{cases}$

In Exercises 7–14, use a graphical method to find a solution to the system of equations.

7. $\begin{cases} y = x + 2 \\ y = x^2 - 3 \end{cases}$

8. $\begin{cases} y = -x + 5 \\ y = x^2 - 8 \end{cases}$

9. $\begin{cases} y = x^2 - 5 \\ y = x + 1 \end{cases}$

10. $\begin{cases} y = -2x + 3 \\ y = \quad x^2 - 1 \end{cases}$

11. $\begin{cases} y = -x^2 + x + 5 \\ y = \quad x^2 - 7 \end{cases}$

12. $\begin{cases} y = -2x^2 + 4 \\ y = \quad x^2 - x + 2 \end{cases}$

13. $\begin{cases} x^2 - y^2 = 1 \\ x^2 + y^2 = 25 \end{cases}$

14. $\begin{cases} 9y^2 - 4x^2 = 1 \\ \quad y^2 + x^2 = 64 \end{cases}$

In Exercises 15–24, find the solution to the system algebraically and support graphically.

15. $\begin{cases} y = x + 5 \\ y = x^2 - 7 \end{cases}$

16. $\begin{cases} y = -2x + 4 \\ y = \quad x^2 - 11 \end{cases}$

17. $\begin{cases} y = x \\ x^2 + y^2 = 36 \end{cases}$

18. $\begin{cases} y = -x + 3 \\ y^2 - x^2 = 36 \end{cases}$

19. $\begin{cases} y = x + 1 \\ x^2 + y^2 = 25 \end{cases}$

20. $\begin{cases} y = -x + 3 \\ x^2 + y^2 = 36 \end{cases}$

21. $\begin{cases} x^2 + y^2 = 244 \\ x^2 - y^2 = 44 \end{cases}$

22. $\begin{cases} 2x^2 + y^2 = 353 \\ x^2 - 2y^2 = -386 \end{cases}$

23. $\begin{cases} y = x^2 - 15 \\ x^2 - 2y^2 = -25 \end{cases}$

24. $\begin{cases} y = x^2 - 10 \\ x^2 + y^2 = 52 \end{cases}$

25. **Writing to Learn.** Consider the set of all systems of equations of the form

$$\begin{cases} ax^2 + by^2 = ab \\ \quad y = cx^2 + d \end{cases}$$

where $a, b, c,$ and $d$ are constants. Write a paragraph explaining the possible number of points of intersections for the curves in this system.

In Exercises 26–28, write a system of equations that describes the problem. Solve the system either algebraically or graphically.

26. The sum of two integers is 24 and the sum of their squares is 306. Find the integers.

27. The difference of two integers is 8. The sum of their squares is 544. Find the integers.

28. The difference of two integers is 1. The difference of their squares is 15. Find the integers.

29. *Using Algebra in Geometry.* The hypotenuse of a right triangle is 13 (see the following figure). The difference of the lengths of the legs is 7. Find the length of each leg.

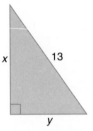

For Exercise 29.

30. *Using Algebra in Geometry.* One leg of a right triangle is 12 (see the following figure). The hypotenuse is 6 more than the other leg. Find the hypotenuse and the other leg.

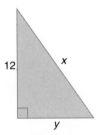

For Exercise 30.

31. *Using Algebra in Geometry.* The total surface area of a rectangular box with a lid is 52 in$^2$ (see the following figure). The area of the base is 12 in$^2$; the height of the box is 2 in. Find the length and the width.

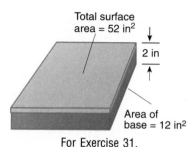

Total surface
area = 52 in²

2 in

Area of
base = 12 in²

For Exercise 31.

## EXTENDING THE IDEAS

**32.** a) How many points of intersection are in a graph of the following system of equations?

$$\begin{cases} x^2 + y^2 = 65 \\ y = x^3 + 2x^2 - 5x - 6 \end{cases}$$

b) Find the solution to this system.

c) Discuss whether you used an algebraic or a graphical method to solve this system and why.

**33.** How many points are in the solution to the following system of equations?

$$\begin{cases} 3x^2 + y^2 = 72 \\ y = x^3 + 4x^2 - 11x - 30 \end{cases}$$

**34.** How many points are in the solution to the following system of equations?

$$\begin{cases} 2x^2 + y^2 = 72 \\ y = x^3 + 4x^2 - 11x - 30 \end{cases}$$

**35. Writing to Learn.** Discuss how the solution to the system

$$\begin{cases} \dfrac{x^2}{4} + \dfrac{y^2}{9} = 1 \\ y = x^2 - 8 \end{cases}$$

compares to the solution to the system

$$\begin{cases} \dfrac{(x-5)^2}{4} + \dfrac{(y+3)^2}{9} = 1 \\ y + 3 = (x-5)^2 - 8 \end{cases}.$$

(*Hint*: You should not need to find the graphs of these systems or actually find their solutions.)

## LOOKING BACK—MAKING CONNECTIONS

In Exercises 36 and 37, describe the transformations used to obtain the indicated graph.

**36.** $y = x^2 - 5x + 7$ from the graph of $y = x^2$

**37.** $y = 3x^2 + 9x - 14$ from the graph of $y = x^2$

## Chapter 10 Summary

| | Conic Sections | Examples |
|---|---|---|
| Definition of a Conic Section | The intersection of a plane and a cone. This intersection takes the shape of a parabola, an ellipse, or a hyperbola. | |

| | Circles | Examples |
|---|---|---|
| Definition of a Circle | A conic section determined by all points equidistant from a selected point (the center) | |

| | Circles | Examples |
|---|---|---|
| Radius | The distance from the center to any point on the circle | |
| Standard Form for an Equation of a Circle | center: $(h, k)$<br>radius: $r$<br>standard form equation:<br>$(x - h)^2 + (y - k)^2 = r^2$ | If the center $= (-3, 5)$ and the radius $= 4$, the standard form of the equation is $(x - (-3))^2 + (y - 5)^2 = 16$. |
| Sketching the Graph of a Circle with Paper and Pencil | Write the equation in standard form. Determine the center $(h, k)$ and radius $r$ and draw the circle. | |
| Graphing a Circle on a Grapher | Prepare the equation for a grapher by solving the equation for $y$. The result is an equation of the form<br>$y = k \pm \sqrt{\text{an expression in } x}$.<br>Graph these two equations. | Graph the circle<br>$(x - 2)^2 + (y - 3)^2 = 20$.<br>Then<br>$y = 3 \pm \sqrt{20 - (x - 2)^2}$.<br>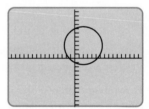<br>The graphs of $y_2 = 3 + y_1$ and $y_3 = 3 - y_1$ in a Square window, where $y_1 = \sqrt{(20 - (x - 2)^2)}$. |

| | Ellipses | Examples |
|---|---|---|
| Definition of an Ellipse | A conic section determined by the set of all points $(x, y)$ whose distances to two fixed points add up to a constant value | |
| Foci | The two fixed points that, with a given constant, determine an ellipse | |
| Major Axis | The line segment through an ellipse that contains the foci | |
| Minor Axis | A line segment, perpendicular to the major axis, that bisects the ellipse | |

| Center | The point inside an ellipse at which the major and minor axes intersect | |
|---|---|---|
| Vertices | The four points at which the major and minor axes intersect the ellipse | |
| Standard Form for an Equation of an Ellipse | center: $(h, k)$<br>distance from center to vertex on horizontal axis: $a$<br>distance from center to vertex on vertical axis: $b$<br>standard form equation:<br>$$\frac{(x-h)^2}{a^2} + \frac{(y-k)^2}{b^2} = 1$$ | If the center $= (-3, 5)$, $a = 3$, and $b = 4$, the standard form of the equation is<br>$$\frac{(x-(-3))^2}{3^2} + \frac{(y-5)^2}{4^2} = 1.$$ |
| Sketching a Graph of an Ellipse with Paper and Pencil | Write the equation in standard form. Plot the point $(h, k)$ and plot the four vertices $(h \pm a, k)$, $(h, k \pm b)$. Connect the vertices in the shape of an ellipse. | |
| Graphing an Ellipse on a Grapher | Prepare the equation for a grapher by solving the equation for $y$. The result is an equation of the form<br>$y = k \pm \sqrt{\text{an expression in } x}$.<br>Graph the two equations. | Graph the ellipse<br>$$\frac{(x+3)^2}{9} + \frac{(y-5)^2}{16} = 1.$$<br>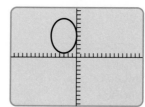<br>The graphs of $y_2 = 5 + y_1$ and $y_3 = 5 - y_1$ in a Square window, where $y_1 = \frac{4}{3}\sqrt{9 - (x+3)^2}$. |

| | **Parabolas with a Horizontal Line of Symmetry** | **Examples** |
|---|---|---|
| Vertex Form of a Parabola | vertex: $(h, k)$<br>line of symmetry: $y = k$<br>vertex form of the equation:<br>$x = a(y - k)^2 + h$ | If the vertex $= (-1, 3)$ and the line of symmetry is $y = 3$, the vertex form of the equation is<br>$x = 2(y - 3)^2 + (-1)$. |

| | **Parabolas with a Horizontal Line of Symmetry** | **Examples** |
|---|---|---|
| Sketching a Graph of a Parabola with Paper and Pencil | Write the equation in vertex form. Plot the point $(h, k)$ and sketch the line $y = k$. Next plot the $x$-intercept $(x_1, 0)$ and the point $(x_1, 2k)$ symmetric about the line of symmetry. Connect the three points in the shape of a parabola. | |
| Graphing a Parabola on a Grapher | Prepare the equation for a grapher by solving the equation for $y$. The result is an equation in the form $y = k \pm \sqrt{\text{an expression in } x}$. Graph the two equations. | Graph the parabola $x = (y + 4)^2 + 2$. 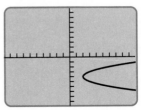 The graphs of $y_2 = -4 + y_1$ and $y_3 = -4 - y_1$ in the Standard window, where $y_1 = \sqrt{(x - 2)}$. |

| | **Hyperbolas** | **Examples** |
|---|---|---|
| Definition of a Hyperbola | A conic section determined by two fixed points and a constant. It consists of the set of all points $(x, y)$ determined as follows. Find the distances from $(x, y)$ to the two fixed points and consider the absolute value of the difference of these distances. This value must be the constant. | |
| Foci | The two fixed points that, with a given constant, determine a hyperbola | |
| Center | The midpoint of the line segment that joins the vertices | |

| | | |
|---|---|---|
| Vertices | The two points at which the line through the foci intersect the hyperbola | |

| | | |
|---|---|---|
| Standard Form for an Equation of a Hyperbola | There are two different forms: center: $(h, k)$ vertices: $(h \pm a, k)$ $$\frac{(x - h)^2}{a^2} - \frac{(y - k)^2}{b^2} = 1$$ center: $(h, k)$ vertices: $(h, k \pm b)$ $$\frac{(y - k)^2}{b^2} - \frac{(x - h)^2}{a^2} = 1$$ | If the center is $(5, 1)$ and the vertices are $(5, 5)$ and $(5, -3)$, the standard form of the equation is $$\frac{(y - 1)^2}{4^2} - \frac{(x - 5)^2}{5^2} = 1.$$ |

| | | |
|---|---|---|
| Fundamental Rectangle of a Hyperbola | A rectangle used when sketching a hyperbola. Its vertices, determined by the center $(h, k)$ and the values of $a$ and $b$ in the standard form of the equation, are $(h + a, k + b)$, $(h - a, k + b)$, $(h - a, k - b)$, and $(h + a, k - b)$. | The fundamental rectangle of $$\frac{(x - 5)^2}{5^2} - \frac{(y - 1)^2}{4^2} = 1 \text{ has}$$ vertices $(10, 5)$, $(0, 5)$, $(0, -3)$, and $(10, -3)$. |

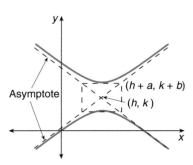

| | |
|---|---|
| Asymptotes | The diagonals of the fundamental rectangle. Asymptotes intersect at the center of the hyperbola and have slopes $m = \dfrac{b}{a}$ and $m = -\dfrac{b}{a}$. |

| | Hyperbolas | Examples |
|---|---|---|
| Sketching the Graph of a Hyperbola with Paper and Pencil | Write the equation in standard form. Plot the center, the vertices of the hyperbola, and the vertices of the fundamental rectangle. Draw the two asymptotes. Draw the hyperbola through each vertex, approaching the asymptotes. | |
| Graphing a Hyperbola on a Grapher | Prepare the equation for a grapher by solving the equation for $y$. The result is an equation of the form $$y = k \pm \sqrt{\text{an expression in } x}.$$ Graph these two equations. | Graph the hyperbola $$\frac{(y-2)^2}{4} - \frac{(x+2)^2}{9} = 1.$$ 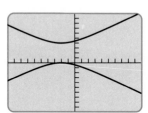 The graphs of $y_2 = 2 + y_1$ and $y_3 = 2 - y_1$ in the Standard window, where $y_1 = \frac{2}{3}\sqrt{9 + (x+2)^2}$. |

| | Nonlinear Systems | Examples |
|---|---|---|
| Nonlinear Equation | An equation with more than one variable and one or more terms of degree two or higher | $y = 3x^2 - 4$ |
| Nonlinear System of Equations | A system of two or more equations in which at least one equation is nonlinear | $x + y = 5$ <br> $x^2 + y = 7$ |
| Solving Nonlinear Systems | Use the multigraph method to visualize the points of intersection. Then confirm these values by solving algebraically, using either the elimination or substitution method. | |

## Review Exercises for Chapter 10

In Exercises 1 and 2, write an equation of the circle that has the given center and radius.

**1.** Center $(2, -3)$, $r = 10$

**2.** Center $(-4, 5)$, $r = 6$

In Exercises 3 and 4, find the center and radius for the circle.

**3.** $x^2 + (y - 4)^2 = 4$

**4.** $(x - 5)^2 + (y + 2)^2 = 15$

In Exercises 5 and 6, find the center and vertices for the ellipse.

**5.** $\dfrac{(x + 10)^2}{49} + \dfrac{(y + 7)^2}{25} = 1$

**6.** $\dfrac{(x + 2)^2}{81} + \dfrac{y^2}{100} = 1$

In Exercises 7 and 8, find the vertex and line of symmetry for the parabola.

**7.** $x = -(y + 2)^2 - 6$     **8.** $x = 3(y - 1)^2 + 2$

In Exercises 9 and 10, find the center and vertices for the hyperbola.

**9.** $\dfrac{y^2}{9} - \dfrac{(x - 11)^2}{64} = 1$

**10.** $\dfrac{(x + 7)^2}{25} - \dfrac{(y - 9)^2}{4} = 1$

In Exercises 11–14, graph the equation using a grapher. Identify which conic section is graphed.

**11.** $y^2 + 10y = -x - 21$

**12.** $y^2 = -x^2 - 6x + 9$

**13.** $y^2 = 3x^2 + 12$

**14.** $y^2 + 4y = -4x^2 + 32x - 52$

In Exercises 15–18, identify the conic section from its equation. Then sketch the equation using a paper-and-pencil method. Support using a grapher.

**15.** $\dfrac{(x + 8)^2}{9} + \dfrac{(y - 2)^2}{25} = 1$

**16.** $\dfrac{(y - 5)^2}{16} - \dfrac{x^2}{4} = 1$

**17.** $(x - 1)^2 + (y + 6)^2 = 1$

**18.** $x = (y - 2)^2 - 8$

In Exercises 19–26, write the equation in the standard form for the conic section indicated and then sketch the graph.

**19.** $x^2 + y^2 + 6x + 12y + 9 = 0$, circle

**20.** $x^2 + 4y^2 - 2x + 48y + 129 = 0$, ellipse

**21.** $4x^2 - 25(y - 2)^2 - 100 = 0$, hyperbola

**22.** $2x^2 - 4x - y = 7$, parabola

**23.** $25x^2 + 4y^2 - 24y = 64$, ellipse

**24.** $x^2 - 8x + y^2 = 25$, circle

**25.** $y^2 + 16y - x + 60 = 0$, parabola

**26.** $10x^2 - 7y^2 - 70 = 0$, hyperbola

**27.** Determine the value of $h$ so that the circle $(x - h)^2 + (y - 1)^2 = 169$ passes through the point $(3, 13)$.

**28.** Write in standard form the equation of the ellipse that is centered at $(0, 0)$ and has a vertical major axis with length 18 and a minor axis with length 12.

**29.** Find the value of $a$ if $(0, 3)$ is a point on the graph of the parabola $x = a(y - 4)^2 + 2$.

**30.** Determine the vertices of the fundamental rectangle and the equations of the asymptotes of the hyperbola $\dfrac{(x - 2)^2}{16} - \dfrac{(y + 1)^2}{36} = 1$.

In Exercises 31 and 32, graph the system of equations and identify the number of points of intersection.

**31.** $\begin{cases} y = 2(x-5)^2 + 2 \\ y = 6x - 10 \end{cases}$  **32.** $\begin{cases} x^2 + y^2 = 25 \\ 36x^2 + y^2 = 36 \end{cases}$

In Exercises 33–35, use a graphical method to find a solution to the system of equations.

**33.** $\begin{cases} y^2 - 4x^2 = 1 \\ y = -3x - 4 \end{cases}$  **34.** $\begin{cases} y = 2x^2 - 3 \\ y = 2 - x \end{cases}$

**35.** $\begin{cases} x^2 + y^2 = 25 \\ y = -2(x-3)^2 - 4 \end{cases}$

In Exercises 36–38, find the solution to the system algebraically and support graphically.

**36.** $\begin{cases} y^2 - x^2 = 16 \\ x^2 + y^2 = 24 \end{cases}$  **37.** $\begin{cases} 4x^2 + 25y^2 = 100 \\ y = x + 2 \end{cases}$

**38.** $\begin{cases} x - y^2 = 3 \\ y = x^2 - 6x + 9 \end{cases}$

**39.** A memorial to veterans is shaped like a semi-elliptical arch. It is 20 ft wide and 25 ft high. Determine the height of the arch 2 ft from the edge of the arch.

**40.** The difference of two integers is 23. The difference of their squares is 2277.

a) Write a system of equations that describes the problem.

b) Find the two integers by solving the system found in part (a) algebraically and support graphically.

## Chapter 10 Test

**1.** Find the center and radius for the circle described by the equation $(x+4)^2 + (y-5)^2 = 16$.

**2.** Find the center and vertices for the ellipse described by the equation $\dfrac{(x-8)^2}{36} + \dfrac{(y+6)^2}{64} = 1$.

**3.** Find the vertex and line of symmetry for the parabola described by the equation $x = -2(y-3)^2 - 4$.

**4.** Find the center and vertices for the hyperbola described by the equation $\dfrac{(x-2)^2}{49} - \dfrac{(y+1)^2}{16} = 1$.

In Questions 5–8, identify the conic section from its equation. Then sketch a graph of the equation using a paper-and-pencil method.

**5.** $\dfrac{x^2}{25} + \dfrac{(y-10)^2}{4} = 1$

**6.** $(x-6)^2 + (y+3)^2 = 36$

**7.** $x = (y+5)^2 + 2$

**8.** $\dfrac{(x-4)^2}{81} - \dfrac{(y-1)^2}{100} = 1$

In Questions 9–12, write the equation in the standard form for the conic section indicated. Graph using a grapher.

**9.** $x^2 + y^2 - 4x + 6y - 3 = 0$, circle

**10.** $4x^2 + y^2 - 100 = 0$, ellipse

**11.** $y^2 + 12y - x + 25 = 0$, parabola

**12.** $(y+1)^2 - x^2 - 4 = 0$, hyperbola

**13.** Use a graphical method to find the solution to the following system of equations:

$$\begin{cases} x^2 - y^2 = 1 \\ y = (x - 2)^2 - 1 \end{cases}$$

**14.** Use an algebraic method to find the solution to the following system of equations:

$$\begin{cases} y = -x + 1 \\ \dfrac{x^2}{4} + y^2 = 1 \end{cases}$$

**15.** Determine the vertices of the fundamental rectangle and the equations of the asymptotes of the hyperbola

$$\frac{(y + 5)^2}{36} - \frac{x^2}{25} = 1.$$

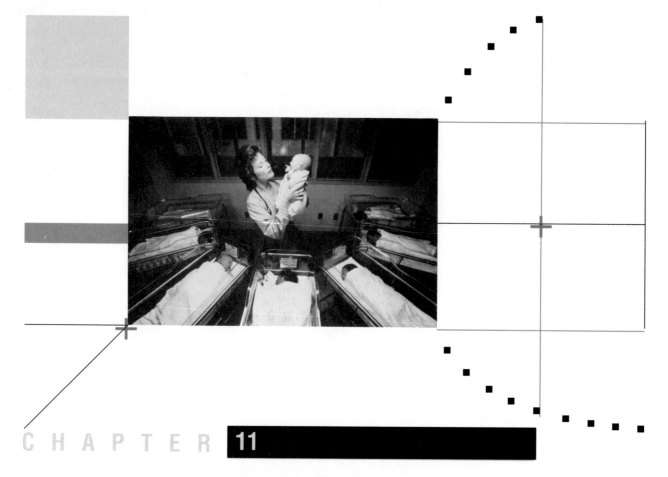

CHAPTER **11**

# Exponential and Logarithmic Functions

## AN APPLICATION

A city with a population of 150,000 that is growing at a rate of 2% each year will have a population $P(t)$ in $t$ years that can be approximated by the formula

$$P(t) = 150000(1 + 0.02)^t.$$

## 11.1 Exponential Functions

Two new families of functions that we study in this chapter, exponential and logarithmic functions, are important because they model growth and decay problems. It is often instructive to compare a new concept with one that is already familiar. For example, compare the expressions

$$x^2 \quad \text{and} \quad 2^x.$$

Both have a *base* and an *exponent*, but the positions of the variable and constant have been reversed. For $x^2$, the base is the variable $x$; for $2^x$, the base is the constant 2. For $x^2$, the exponent is the constant 2; for $2^x$, the exponent is the variable $x$. Expressions like $2^x$ are used in defining the concept of exponential function.

### Defining and Graphing Exponential Functions

When the variable is in the exponent position and the base is a positive real number not equal to 1, such as $y = 2^x$, the function is called an **exponential function**. Examples of exponential functions are as follows:

$$y = 3^x \qquad y = 4^x \qquad y = \left(\frac{1}{2}\right)^x$$

---

**DEFINITION 11.1**  Exponential Function

Let $a$ be a positive real number $a \neq 1$. Then,

$$y = a^x$$

is called an **exponential function with base $a$**.

---

Do you see why each of the following are not exponential functions?

$$y = x^3 \qquad y = \frac{2}{x} \qquad y = 2^5$$

In Example 1, we use a point-plotting method to sketch the graph of the function $y = 2^x$.

## EXAMPLE 1   Graphing an Exponential Function

Complete a hand-sketch of the function $y = 2^x$.

SOLUTION

We begin by completing a table of solution pairs, as follows, and then sketch the graph, shown in Figure 11.1.

| $x$ | $-3$ | $-2$ | $-1$ | 0 | 1 | 2 | 3 | 4 |
|---|---|---|---|---|---|---|---|---|
| $y = 2^x$ | $\frac{1}{8}$ | $\frac{1}{4}$ | $\frac{1}{2}$ | 1 | 2 | 4 | 8 | 16 |

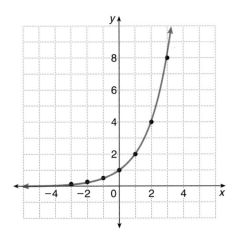

**FIGURE 11.1**  A sketch of $y = 2^x$ in Example 1.

In the table in Example 1, $x$ can be *any* real number. However, $y$ is always a *positive* number. So, for the function $y = 2^x$,

■ the domain is $(-\infty, \infty)$, and
■ the range is $(0, \infty)$.

The function $y = 2^x$ is sometimes referred to as the doubling function. Study the table of values in Example 1 until you understand why the name doubling function is an appropriate name for this function.

In Example 2, we sketch the graph of the function that might be called the halving function.

**EXAMPLE 2**   Graphing an Exponential Function

Complete a hand-sketch of the function $y = \left(\frac{1}{2}\right)^x$.

SOLUTION
We begin by completing a table of solution pairs, as follows, and then sketch the graph, as shown in Figure 11.2.

| $x$ | $-3$ | $-2$ | $-1$ | $0$ | $1$ | $2$ | $3$ | $4$ |
|---|---|---|---|---|---|---|---|---|
| $y = \left(\frac{1}{2}\right)^x$ | $8$ | $4$ | $2$ | $1$ | $\frac{1}{2}$ | $\frac{1}{4}$ | $\frac{1}{8}$ | $\frac{1}{16}$ |

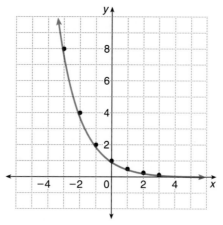

**FIGURE 11.2**   A sketch of $y = \left(\frac{1}{2}\right)^x$ in Example 2.   ■

Observe that the domain and range of the function $y = \left(\frac{1}{2}\right)^x$ are identical to that of $y = 2^x$; that is,

- the domain is $(-\infty, \infty)$, and
- the range is $(0, \infty)$.

## Graphing Exponential Functions on a Grapher

You will discover that the graph of *any* exponential function looks, more or less, like the graph in either Example 1 or Example 2. Complete the following exploration to discover a condition that can be used to predict whether the graph of the equation $y = a^x$ will look like the graph of either $y = 2^x$ or $y = \left(\dfrac{1}{2}\right)^x$.

### EXPLORE WITH A GRAPHER

1. Graph each of the following pairs of equations in the $[-10, 10]$ by $[-2, 10]$ window:
   a)   $y = 3^x$ and $y = 0.6^x$
   b)   $y = 0.5^x$ and $y = 1.4^x$
   c)   $y = 1.1^x$ and $y = 0.85^x$
2. Predict what graphs for each of the following equations will look like. Confirm your prediction with the grapher.
   a)   $y = 1.6^x$
   b)   $y = 0.8^x$
   c)   $y = 0.65^x$
   d)   $y = 1.25^x$

**Generalize**   If $a > 1$, will the graph of $y = a^x$ look more like the graph of Example 1 or of Example 2? What if $0 < a < 1$?

The following generalizations summarize the basic characteristics of the graphs of exponential functions.

---

### Graphs of Exponential Functions

- If $a > 1$, the graph of $y = a^x$ has a shape like the graph of $y = 2^x$.
- If $0 < a < 1$, the graph of $y = a^x$ has a shape like the graph of $y = \left(\frac{1}{2}\right)^x$.
- For all base values $a > 0$, the $y$-intercept for $y = a^x$ is the point $(0, 1)$.
- There are *no* $x$-intercepts of the graph $y = a^x$ when $a > 0$.

---

Is $y = 3^{-x}$ an exponential function? Does the expression $3^{-x}$ fit the form described in the definition of exponential function? At first one might say no because there is a negative sign before the variable. However, recall that

$$a^{-1} = \frac{1}{a}.$$

Therefore

$$3^{-x} = (3^{-1})^x = \left(\frac{1}{3}\right)^x.$$

Consequently, we see that $y = 3^{-x}$ and $y = \left(\frac{1}{3}\right)^x$ are identical exponential functions. This fact can be supported visually by graphing $y = 3^{-x}$ and $y = \left(\frac{1}{3}\right)^x$.

**EXAMPLE 3**    Graphing an Exponential Function

Graph $y = 3^x$ and $y = 3^{-x}$ in the 0.1 window. Find the domain and range of each function and the $y$-intercept of its graph.

SOLUTION

Figure 11.3 gives the graphs of $y_1 = 3^x$ and $y_2 = 3^{-x}$ in the 0.1 window and shows the following to be true:

**domain for both:**  $(-\infty, \infty)$
**range for both:**  $(0, \infty)$
**$y$-intercept for both:**  $(0, 1)$    ■

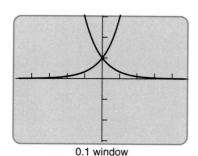

0.1 window

**FIGURE 11.3** Graphs of $y = 3^x$ and $y = 3^{-x}$ in Example 3. Notice each graph is the reflection of the other about the $y$-axis.

## Solving Exponential Equations

In equations studied earlier, the variable has never been in the exponent position. Some equations having a variable as an exponent are simple enough that they can be solved by inspection, as Example 4 demonstrates.

**EXAMPLE 4**    Solving Equations

Use mental calculations to find a solution for each of the following equations:

a) $2^x = 8$

b) $3^x = 9$

c) $2^x = \frac{1}{4}$

SOLUTION

a) $2^x = 8$ if and only if $x = 3$.

b) $3^x = 9$ if and only if $x = 2$.

c) $2^x = \frac{1}{4}$ if and only if $x = -2$.    ■

Suppose we want to solve an equation like

$$2^x = 0.$$

Example 1 showed that the graph of $y = 2^x$ gets close to the $x$-axis as $x$ approaches negative infinity but never intersects the $x$-axis. This means that $2^x$ is never zero and the equation $2^x = 0$ has no solution. In general, if $a$ is any real number,

$$a^x = 0 \text{ has no solution.}$$

However, if $b > 0$ equations like $a^x = b$ do have a solution. The following property can be used to solve some exponential equations algebraically.

---

### Property of Exponential Equations

If $a > 0$ and $a \neq 1$, then $a^x = a^y$ if and only if $x = y$.

---

Sometimes by writing each side of an equation in exponential form with the same base, we can use the above property to solve the equation. Example 5 illustrates.

**EXAMPLE 5**    Solving Equations

Solve each of the following equations and confirm the answer numerically:

a) $4^x = 2^3$

b) $\left(\dfrac{1}{3}\right)^x = 9$

SOLUTION

a)                    $4^x = 2^3$    **Write each side of the equation with base 2.**

$(2^2)^x = 2^3$

$2^{2x} = 2^3$    **The bases are equal, so the exponents must be equal.**

$2x = 3$

$x = \dfrac{3}{2}$

b) $\qquad \left(\dfrac{1}{3}\right)^x = 9$    **Write each side of the equation with base 3.**

$$(3^{-1})^x = 3^2$$

$$3^{-x} = 3^2$$

$$-x = 2$$

$$x = -2$$

**Confirm Numerically.**

a) $4^{3/2} = (\sqrt{4})^3 = 2^3$

b) $\left(\dfrac{1}{3}\right)^{-2} = 3^2 = 9$    ■

## Using Exponential Functions to Solve Problems

Numerous situations occur when something either increases or decreases at a constant rate over a period of years. For example, suppose a city's population $P$ increases at a constant rate of 3% (or 0.03) each year. The population at the end of the next several years would be as follows:

$$P + P(0.03) = P(1 + 0.03) \qquad \text{**Population 1 yr later**}$$

$$P(1 + 0.03) + P(1 + 0.03)0.03 = P(1 + 0.03)^2 \qquad \text{**Population 2 yr later**}$$

$$\vdots$$

$$P(1 + 0.03)^{t-1} + P(1 + 0.03)^{t-1}0.03 = P(1 + 0.03)^t \qquad \text{**In general, the population** } t \text{ **years later**}$$

These are examples of problems concerning exponential growth and exponential decay found in biology, chemistry, business, and other social and physical sciences.

☐ ☐

**POPULATION GROWTH PROBLEM SITUATION**

A city has a population of 50,000 that is increasing at a rate of 3% per year.

☐ ☐

The exponential function that represents the population $t$ years later is

$$f(t) = 50{,}000(1 + 0.03)^t.$$

**EXAMPLE 6**    Solving a Population Growth Problem

In how many years will the population be 200,000? Solve this problem graphically and provide numerical support.

SOLUTION

**Solve Graphically.** We shall use the multigraph method. Graph $y_2 = 200{,}000$ and $y_1 = 50000(1 + .03)^x$ in the $[-40, 75]$ by $[-50000, 300000]$ window. Zoom in to find the solution $x = 46.900$ with an error of at most 0.01.

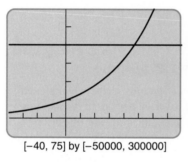

| X | Y1 | Y2 |
|---|----|----|
| 46.897 | 199985 | 200000 |
| 46.898 | 199991 | 200000 |
| 46.899 | 199997 | 200000 |
| 46.9 | 200003 | 200000 |
| 46.901 | 200009 | 200000 |
| 46.902 | 200015 | 200000 |
| 46.903 | 200020 | 200000 |

X = 46.899

$[-40, 75]$ by $[-50000, 300000]$        Grapher-generated table

(a)        (b)

**FIGURE 11.4**    (a) Graphs of $y_1 = (50{,}000)(1 + .03)^x$ and $y_2 = 200{,}000$ and (b) a table showing numerical support for Example 6.

**REMINDER**

Growth could mean "increasing" in the usual sense, or negative growth implies decreasing and is called decay.

**Support Numerically.** The table in Figure 11.4 shows that the population reaches 200,000 between 46.899 and 46.9 yr later, therefore the solution is the population will be 200,000 in about 46.9 years.    ■

In Example 6, the population was people. But the mathematics is the same whether what is growing are people, bacteria, dollars, or the decay rate of radioactive material. In all cases, a population $P$ growing at a constant rate $r$ (represented as a decimal per unit of time) will have a total population ($S$) at $t$ units of time later, that is,

$$S = P(1 + r)^t.$$

In Example 7, we apply this algebraic representation to a bacterial growth situation.

### EXAMPLE 7   Predicting Bacterial Growth

Suppose the number of bacteria in a petri dish doubles every hour. In how many hours will 100 bacteria increase to 250,000.

SOLUTION

When the population doubles, there has been a 100% increase and $r = 1$. So $t$ hours later, the total population is

$$S = 100(1 + 1)^t = 100(2^t).$$

**Solve Graphically.** Figure 11.5 shows the graphs of the algebraic representations $y_1 = 100(2^x)$ and $y_2 = 250,000$. Using zoom-in, we find that the $x$-value at the point of intersection is 11.288, with an error of at most 0.01.

**Support Numerically.** Figure 11.5 also shows numerical support for the solution. When $x = 11.288$, $y_1$ has the value 250,049.84. This observation supports the following conclusion: The bacterial growth will be approximately 250,000 after 11.288 hours.

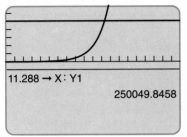

11.288 → X: Y1

250049.8458

[0, 20] by [−1000, 300000]

**FIGURE 11.5** Graphs of $y_1 = 100(2^x)$ and $y_2 = 250,000$ in Example 7.

## Exercises for Section 11.1

In Exercises 1–4, complete a hand-sketch of the exponential function.

1. $y = 3^x$

2. $y = 10^x$

3. $y = 2^{-x}$

4. $y = 3^{-x}$

In Exercises 5–10, solve the exponential equation by mental calculation.

5. $2^x = 4$

6. $3^x = 27$

7. $10^x = 1000$

8. $\left(\dfrac{1}{2}\right)^x = \dfrac{1}{4}$

9. $3^x = \dfrac{1}{3}$

10. $2^x = \dfrac{1}{8}$

In Exercises 11–14, solve the exponential equation.

11. $4^x = 2^6$

12. $9^x = 3^4$

13. $8^x = 2^9$

14. $\left(\dfrac{1}{2}\right)^x = 8$

In Exercises 15–22, functional notation is used for the exponential functions $f(x) = 2^x$, $g(x) = \left(\dfrac{1}{3}\right)^{x+1}$, and $h(x) = 10^x$. Evaluate the expression.

15. $f(3)$

16. $g(2)$

17. $g(0)$

18. $h(3)$

19. $g(-1) + h(2)$

20. $f(3) + g(-3)$

21. $h(3) + h(-3)$

22. $f(2) + f(-2)$

In Exercises 23 and 24, graph the functions in the 0.1 window and find the point common to all three graphs.

23. $y_1 = 2^x$    $y_2 = 3^x$    $y_3 = 5^x$

24. $y_1 = 2^{-x}$    $y_2 = (3.5)^{-x}$    $y_3 = 6^{-x}$

In Exercises 25 and 26, determine which two of the three functions are equivalent. Support your answer with a grapher.

**25.** a) $y_1 = 2^{-x}$   b) $y_2 = -2^x$   c) $y_3 = \left(\dfrac{1}{2}\right)^x$

**26.** a) $y_1 = \left(\dfrac{1}{4}\right)^x$   b) $y_2 = 4^{-x}$   c) $y_3 = -4^x$

In Exercises 27–30, match the equation with one of the following graphs using a Standard window. Do not use your grapher.

**27.** $y = -2^x$   **28.** $y = -2^{-x}$

**29.** $y = -0.5^x - 3$   **30.** $y = 2^x$

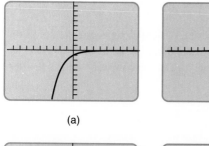

(a)

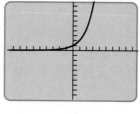

(b)

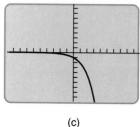

(c)

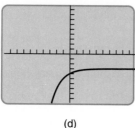

(d)

For Exercises 27–30.

In Exercises 31–34, use properties of exponents to select the pair of functions that are equivalent.

**31.** a) $y_1 = 5^{2x+6}$   b) $y_2 = 5^{2(x)} + 6$
   c) $y_3 = 25^{x+3}$

**32.** a) $y_1 = 3^{-(x-4)}$   b) $y_2 = \left(\dfrac{1}{3}\right)^{x-4}$
   c) $y_3 = -3^{x-4}$

**33.** a) $y_1 = 4^{x-2}$   b) $y_2 = 2^{2x-4}$
   c) $y_3 = -2^{2-x}$

**34.** a) $y_1 = 2^{-(x-4)}$
   b) $y_2 = (0.5)^{x-4}$
   c) $y_3 = -2^{4-x}$

**35.** *Population Growth.* The population of Orangedale is increasing at the rate of 4% per year. The population in 1990 was 50,000 people. Find the population in 1991, 1992, and 1995, respectively.

**36.** *Bacterial Growth.* Assume a bacterial culture is growing at the rate of 20%/hr. If there are 2000 bacteria present at 9:00 a.m., find the number present at 10:00 a.m., 12:00 noon and 6:00 p.m., respectively.

**37.** *Population Growth.* The population $P$ of a town is 325,000 and is increasing at the rate of 2.5% per year.

a) Find an algebraic representation for $P$ as a function of time.

b) When will the population of the town be 1 million?

**38.** *Population Growth.* The population of a small town in the year 1890 was 6000. Assume that the population increased at the rate of 2.5% per year.

a) What was the population in 1915?

b) What was it in 1940?

**39.** *Compound Interest.* If $5000 are invested in an account that pays 5% interest compounded annually, the total dollars ($T$) after $t$ years is $T = 5000(1 + 0.05)^t$.

a) Assuming there are no other deposits or withdrawals, how much will be in the account after 6 yr?

b) How long will it take until the account has $10,000?

**40.** *Compound Interest.* If $8000 are invested in an account that pays 5.25% interest compounded annually, the total dollars ($T$) after $t$ years is $T = 8000(1 + 0.0525)^t$.

a) How much will be in the account after 4 yr assuming no other deposits or withdrawals are made?

b) How long will it take until the account has $25,000?

**41.** *Population Decline.* The population $P$ of a town is 250,000 and is decreasing at the rate of 2% per year.

a) Find a graph of the algebraic representation $P(t) = 250,000(1 - 0.02)^t$ as a function of time (in years).

b) Determine when the population of the town will be 100,000.

**42.** *Car Depreciation.* A compact car that costs $10,000 new is known to depreciate at the rate of 15% per year. Find the value of the car when it reaches the following ages:

a) 1 yr old

b) 5 yr old

c) $t$ years old

**43.** *Car Depreciation.* A luxury car that costs $22,000 new is known to depreciate at the rate of 25% per year. Find the value of the car when it reaches the following ages:

a) 1 yr old

b) 5 yr old

c) $t$ years old

**44.** *Car Depreciation.* Assume the cars in Exercises 42 and 43 were purchased the same year. In how many years will the cost of the compact car be worth more than the luxury car?

## TRANSLATING WORDS TO SYMBOLS

In Exercises 45–48, write the equation that corresponds to the phrase.

**45.** Five to some power $x$ is fifteen.

**46.** One third to a power is one eighth.

**47.** A function of $x$ is the sum of twice the number and five.

**48.** A function of $x$ is the difference of three times the number and six.

## EXTENDING THE IDEAS

In Exercises 49 and 50, find the points of intersection of the graphs of the equations. Use both of the following methods and then compare the two. Which do you prefer?

**Method 1.** Graph each equation in the same window and then use zoom-in to find the points at which the two graphs intersect.

**Method 2.** Find the zeros of the difference of the two functions. For example, in Exercise 49 the difference is $y = 2^x - x^2$. Use the zeros to find the point at which the graphs cross.

**49.** $y = x^2$ and $y = 2^x$

**50.** $y = 3^x$ and $y = x^3$

**51.** *Circulation Growth.* The Northville Free Press had a circulation of 14,560 in 1990 and 15,785 in 1991. Assume that the circulation grows at the same annual growth rate experienced in 1990 and 1991. What will be the circulation at the beginning of 1996? (*Hint:* The percentage of growth is determined by finding the amount of growth in 1 yr divided by the original circulation. Use three decimal places.)

## WRITING TO LEARN

In Exercises 52 and 53, write a paragraph explaining how to use the graph of $y = 2^x$ to graph the equation using transformations.

**52.** $y = 2^{x+1} - 3$

**53.** $y = -2^x + 5$

## LOOKING BACK—MAKING CONNECTIONS

**54.** Find 8-3/4% of $20,000.

**55.** The wholesale price of an automobile is marked up 15% by the dealer to get the selling price. Find the selling price if the wholesale price is $10,500.

**56.** Joshua receives a 25% employee discount on purchases at Hart's Bicycle Shop. Find his cost, before taxes, to purchase a bike retail priced at $125.

**57.** Lynette needs to increase the size of a picture using a photocopying machine to fit a frame with an area of 15 in². The area of the picture originally was 6 in². What is the percent of increase at which to set the copy machine in order to increase the picture to the desired size?

## 11.2 Inverse Functions

### Finding the Inverse Function

To understand the concept of logarithmic functions, we first develop the concept of inverse function.

The *inverse* of a process is an "undoing" of the process. For example, cubing a number is the inverse of taking the cube root, illustrated as follows:

$$2^3 = 8 \qquad 3^3 = 27 \qquad (-2)^3 = -8$$

$$\sqrt[3]{8} = 2 \qquad \sqrt[3]{27} = 3 \qquad \sqrt[3]{-8} = -2$$

It follows that the function $y = x^3$ is the inverse of the function $y = \sqrt[3]{x}$.

A special condition must be met by a function if it is to have an inverse that is also a function. Study the two functions illustrated in Figure 11.6. One of them has the necessary condition; the other does not.

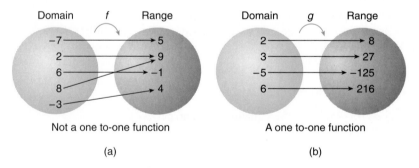

**FIGURE 11.6** Function $f$ is not one-to-one, and function $g$ is one-to-one.

Notice in Figure 11.6 (a) that $f(8) = f(2) = 9$. The two domain numbers 2 and 8 are associated with the same range number 9. On the other hand, the function $g$ has no pair of domain numbers associated with the same range number. We say that $g$ is **one-to-one** (see Figure 11.6 (b) and $f$ is **not one-to-one** (see Figure 11.6 (a)).

In order for a function to have an inverse that is also a function, it must be a **one-to-one function**.

---

■                                                                                                    ■

### DEFINITION 11.2   Inverse Function

If a function $f$ is one-to-one, then the **inverse function of f**, denoted $f^{-1}$, consists of all ordered pairs $(y, x)$, where the pair $(x, y)$ belongs to $f$.

■                                                                                                    ■

---

**REMINDER**

The symbol $f^{-1}(x)$ refers to the inverse function. It does not refer to the multiplicative inverse $1/f(x)$.
Notice that $f^{-1}(x)$ and $\left(f(x)\right)^{-1}$ do not represent the same thing: $\left(f(x)\right)^{-1} = 1/f(x)$.

It often is difficult to see that a function is one-to-one by looking at its equation. However, we use a method called the **horizontal line test** to tell, by looking at a function's graph, if the function is one-to-one. Consequently, the horizontal line test can be used to verify that an inverse function exists.

---

### Horizontal Line Test

A function is a one-to-one function if *every* horizontal line intersects the graph of the function at most once.

---

If there is one horizontal line $y = k$ that intersects the graph of a function $f$ in two points, $(x_1, k)$ and $(x_2, k)$, there are two pairs in the function $f$ with the same second coordinate; consequently, $f$ is *not* one-to-one.

Example 1 illustrates using the horizontal line test to see if a function is one-to-one; therefore, its inverse is a function.

**EXAMPLE 1**   Using the Horizontal Line Test

Sketch a graph of the following functions.

a) $f(x) = x^2$
b) $g(x) = x^3$

Use the horizontal line test to determine if $f$ and $g$ are one-to-one functions.

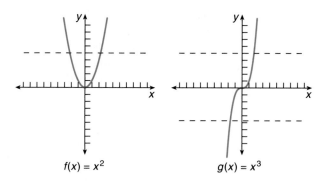

$f(x) = x^2$          $g(x) = x^3$

**FIGURE 11.7**  A horizontal line intersects the graph of $f(x) = x^2$ more than once. Each horizontal line intersects the graph of $g(x) = x^3$ only one time.

SOLUTION

Figure 11.7 shows the graphs of $f$ and $g$.

We see from the horizontal line test that $f$ is *not* a one-to-one function. We also see from the test that $g$ *is* a one-to-one function. ■

## Finding the Inverse Function

The inverse of a function is found by interchanging the $x$- and $y$-values of the ordered pairs of the function. So interchanging $x$ and $y$ is a crucial step in finding the equation of the inverse. To do this, follow the following four-step process.

---

### Finding the Inverse Function $f^{-1}(x)$

If $f(x)$ is a one-to-one function, you can find its inverse $f^{-1}(x)$.

1. Replace $f(x)$ in the equation with $y$ (if necessary).
2. Interchange $x$ and $y$.
3. Solve for $y$.
4. Replace $y$ with $f^{-1}(x)$.

---

Example 2 illustrates how to use this process to find the inverse.

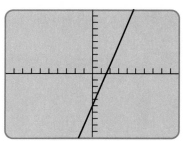

Standard window

**FIGURE 11.8** A graph of $f(x) = 3x - 5$. Any horizontal line will intersect the graph one time.

**EXAMPLE 2**     Finding the Equation of the Inverse

Show that $f(x) = 3x - 5$ is a one-to-one function and find $f^{-1}(x)$.

SOLUTION

We see in Figure 11.8 that the graph of $f(x) = 3x - 5$ satisfies the horizontal line test. Consequently, $f^{-1}(x)$ is also a function.

Use the four-step process to find $f^{-1}(x)$.

$$f(x) = 3x - 5 \qquad \text{Replace } f(x) \text{ in the equation with } y.$$

$$y = 3x - 5 \qquad \text{Interchange } x \text{ and } y.$$

$$x = 3y - 5 \qquad \text{Solve for } y.$$

$$\frac{x + 5}{3} = y \qquad \text{Replace } y \text{ with } f^{-1}(x).$$

$$\frac{x + 5}{3} = f^{-1}(x)$$

Therefore $f(x) = 3x - 5$ and $f^{-1}(x) = \dfrac{x + 5}{3}$. ∎

In Example 2, we found the inverse of a linear function. In Example 3, we investigate a third-degree polynomial.

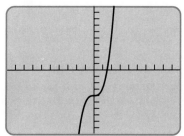

Standard window

**FIGURE 11.9** A graph of $f(x) = x^3 - 4$ in Example 3. Any horizontal line will intersect the graph one time.

**EXAMPLE 3**     Finding the Equation of the Inverse

Show that $f(x) = x^3 - 4$ is a one-to-one function and find $f^{-1}(x)$.

SOLUTION

We see that the complete graph of $f(x) = x^3 - 4$ in Figure 11.9 satisfies the horizontal line test. Consequently, $f^{-1}(x)$ is also a function. Use the four-step process to find $f^{-1}(x)$.

$$f(x) = x^3 - 4 \qquad \text{Replace } f(x) \text{ in the equation with } y.$$

$$y = x^3 - 4 \qquad \text{Interchange } x \text{ and } y.$$

$$x = y^3 - 4 \qquad \text{Solve for } y.$$

$$x + 4 = y^3$$

$$\sqrt[3]{x + 4} = y \qquad \text{Replace } y \text{ with } f^{-1}(x).$$

$$\sqrt[3]{x + 4} = f^{-1}(x)$$

Therefore $f(x) = x^3 - 4$ and $f^{-1}(x) = \sqrt[3]{x + 4}$. ∎

**EXAMPLE 4**    Finding Function Values

If $f(x) = x^3 - 4$ and $f^{-1}(x) = \sqrt[3]{x + 4}$, find the following:

a) $f(2)$

b) $f^{-1}(4)$

SOLUTION

a) $f(x) = x^3 - 4$

$\quad f(2) = 2^3 - 4$

$\qquad = 8 - 4$

$\qquad = 4$

b) $f^{-1}(x) = \sqrt[3]{x + 4}$

$\quad f^{-1}(4) = \sqrt[3]{4 + 4}$

$\qquad = \sqrt[3]{8} = 2.$    ■

**TRY THIS**

Using the equations for $f(x)$ and $f^{-1}(x)$ in Example 4, verify that

a) $f(f^{-1}(x)) = x$ and

b) $f^{-1}(f(x)) = x$.

The two relationships are always true for a function and its inverse.

We see in Example 4 that $(2, 4)$ is a pair in $f$ and $(4, 2)$ is a pair in $f^{-1}$. This observation is consistent with our method of finding an inverse $f^{-1}$. Because we interchange $x$ and $y$ in finding the inverse, we should not be surprised to find that interchanging a pair in $f$ should result in a pair in $f^{-1}$. Figure 11.10

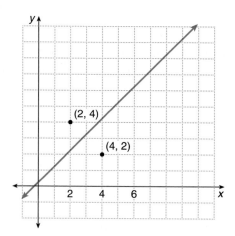

**FIGURE 11.10** The point $(4, 2)$ can be obtained by reflecting the point $(2, 4)$ over the graph of $y = x$.

also shows that each point of the pair $(2, 4)$ and $(4, 2)$ can be obtained from the other by a reflection in the line $y = x$.

Example 5 reinforces this relationship between the graphs of a function $f$ and its inverse $f^{-1}$.

**EXAMPLE 5** Finding a Graph of the Inverse

Find the graph of each of the following in the 0.1 window: $f(x) = x^3$, $y = f^{-1}(x)$, and $y = x$.

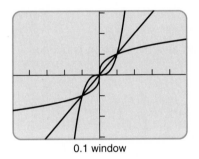

0.1 window

**FIGURE 11.11** A graph of $f(x) = x^3$ and $f^{-1}(x) = \sqrt[3]{x}$ and $y = x$ in Example 5.

SOLUTION

We begin by finding $f^{-1}(x)$ using the four-step process.

$$f(x) = x^3 \qquad \textbf{Replace } f(x) \textbf{ with } y.$$

$$y = x^3 \qquad \textbf{Interchange } x \textbf{ and } y.$$

$$x = y^3 \qquad \textbf{Solve for } y.$$

$$\sqrt[3]{x} = y \qquad \textbf{Replace } y \textbf{ with } f^{-1}(x).$$

$$\sqrt[3]{x} = f^{-1}(x)$$

Figure 11.11 shows the graph of $y_1 = x^3$, $y_2 = \sqrt[3]{x}$, and $y_3 = x$. Notice that the graphs of $y_1$ and $y_2$ are reflections of one another in the line $y = x$. ■

## Exercises for Section 11.2

In Exercises 1 and 2, determine whether the functions $f$ and $g$ are one-to-one.

1.

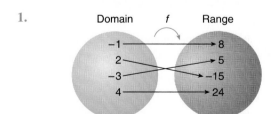

2.

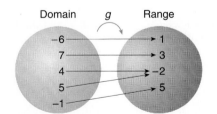

In Exercises 3–6, state what the horizontal line test tells us about the function.

3.                             4.

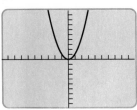    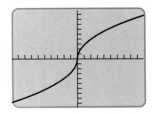

5.                             6.

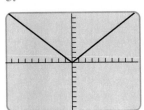    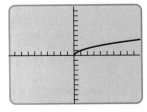

In Exercises 7–12, use a grapher to decide whether the given function is one-to-one.

7. $f(x) = 0.5x + 3$    8. $f(x) = \left(\dfrac{3}{2}\right)x - 4$

9. $f(x) = x^2 + 2x - 4$    10. $f(x) = 3x - x^2 + 1$

11. $f(x) = 2x^3 - 5x$    12. $f(x) = 0.5x^2 - 7$

In Exercises 13–20, the function is one-to-one. Find $f^{-1}(x)$.

13. $f(x) = 2x + 3$    14. $f(x) = 5x - 2$

15. $f(x) = -3x + 8$    16. $f(x) = -2x + 1$

17. $f(x) = 1/x$    18. $f(x) = -3/x$

19. $f(x) = x^3 + 6$    20. $f(x) = \sqrt[3]{x + 3}$

In Exercises 21–24, graph $f$, $f^{-1}$, and the line $y = x$ in the 0.1 window.

21. $f(x) = 2x^3 - 1$    22. $f(x) = \sqrt[3]{x + 2}$

23. $f(x) = 1/x^3$    24. $f(x) = x^3 + 2$

25. Graph $f(x) = 2x + 4$ in the Integer window.

a) Is the point $(0, 4)$ on the graph?

b) Is the point $(4, 0)$ on the graph of the inverse?

c) Find a point $(a, b)$ that is on both the graph of the function and the graph of the inverse.

26. Graph $f(x) = x/3 + 2$ in the Integer window.

a) Is the point $(0, 2)$ on the graph?

b) Is the point $(2, 0)$ on the graph of the inverse?

c) Find a point $(a, b)$ that is on both the graph of the function and the graph of the inverse.

EXTENDING THE IDEAS

In Exercises 27–30, show that the function is one-to-one and find $f^{-1}(x)$.

27. $f(x) = \dfrac{2x - 7}{x + 3}$    28. $f(x) = \dfrac{3x + 1}{x + 2}$

29. $f(x) = \dfrac{x + 3}{x + 1}$    30. $f(x) = \dfrac{x - 5}{x + 2}$

In Exercises 31 and 32, use the parametric equation feature of a grapher. Set the grapher to parametric mode and enter the given equations. Use the following WINDOW settings for Tmin, Tmax, and Tstep, and use whatever $x$ and $y$ settings are needed to obtain a square window on your grapher.

```
WINDOW
Tmin  = -3
Tmax  = 3
Tstep = .05
Xmin  =
Xmax  =
Xscl  = 1
Ymin  =
Ymax  =
Yscl  = 1
```

31. Enter the following equations and find the resulting graph in a square window. Describe what you observe.

$$X_{1t} = t, \quad Y_{1t} = t^3$$

$$X_{2t} = t^3, \quad Y_{2t} = t$$

$$X_{3t} = t, \quad Y_{3t} = t$$

**32.** Enter the following equations and find the resulting graph. Explain why the resulting graphs are not all graphs of functions of $x$.

$$X_{1t} = t, \qquad Y_{1t} = t^3 - 3t$$

$$X_{2t} = t^3 - 3t, \qquad Y_{2t} = t$$

$$X_{3t} = t, \qquad Y_{3t} = t$$

LOOKING BACK—MAKING CONNECTIONS

In Exercises 33–40, let $f(x) = x^2 + 5$ and $g(x) = 2x - 3$ and evaluate the expression.

**33.** $f(4)$

**34.** $f(-1)$

**35.** $f(5) - g(3)$

**36.** $f(2) - g(-3)$

**37.** $f(x + h)$

**38.** $g(x + h)$

**39.** $f(x + h) - f(x)$

**40.** $g(x + h) - g(x)$

## 11.3    Logarithmic Functions

> Evaluating a Logarithm ▪ Evaluating Logarithms with a Grapher ▪ Solving Logarithmic Equations ▪ Using Common and Natural Logarithms in Problem Situations

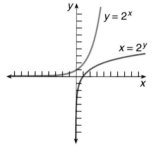

**FIGURE 11.12** A graph of $y = 2^x$ and its inverse function.

The graph of $y = 2^x$ is the curve shown in color in Figure 11.12. It satisfies the horizontal line test, which tells us that $y = 2^x$ is a one-to-one function. Consequently, it has an inverse. The graph of the inverse is obtained by reflecting in the line $y = x$ the graph of $y = 2^x$, as shown in Figure 11.12.

Suppose we want to find the algebraic equation for the inverse of $y = 2^x$ using the method developed in Section 11.2.

$$y = 2^x \qquad \textbf{Reverse the } x \textbf{ and } y.$$

$$x = 2^y$$

$$? = ? \qquad \textbf{Solve for } y. \textbf{ But with methods studied so far, there is no way to solve for } y.$$

After interchanging the $x$ and $y$, we should solve for $y$. However, using the methods we have studied so far, we have no way to solve the equation, $x = 2^y$, for $y$. We need to introduce a new definition, that of a **logarithm**.

---

■                                                                                    ■

### DEFINITION 11.3    Logarithmic Functions

For all positive numbers $a$, with $a \neq 1$, and for all positive numbers $x$,

$$y = \log_a x \qquad \text{means} \qquad x = a^y.$$

The value of $\log_a x$ is called the **logarithm of $x$ with base $a$**.

■                                                                                    ■

A logarithm is an exponent.

Logarithm      Exponent

$$y = \log_a x \qquad x = a^y$$

Base      Base

Being able to change back and forth from exponential form to logarithmic form is an important skill. The following are examples of an expression written in both logarithmic and exponential forms.

| **Logarithmic Form** | **Exponential Form** |
|---|---|
| $\log_3 1 = 0$ | $3^0 = 1$ |
| $\log_5 25 = 2$ | $5^2 = 25$ |
| $\log_{\frac{1}{2}} \frac{1}{8} = 3$ | $\left(\frac{1}{2}\right)^3 = \frac{1}{8}$ |
| $\log_2 \frac{1}{16} = -4$ | $2^{-4} = \frac{1}{16}$ |

Example 1 illustrates changing from logarithmic form to exponential form. Remember that a logarithm is an exponent.

**EXAMPLE 1**   Changing to Exponential Form

Write each of the following logarithmic equations in exponential form:

a) $\log_8 64 = 2$

b) $\log_4 \frac{1}{64} = -3$

c) $\log_{\frac{1}{2}} \left(\frac{1}{32}\right) = 5$

SOLUTION

a) $\log_8 64 = 2$    **A logarithm is an exponent. So 2 is the exponent used with base 8.**

$8^2 = 64$

b) $\log_4 \frac{1}{64} = -3$

$4^{-3} = \frac{1}{64}$

c) $\log_{\frac{1}{2}} \left(\frac{1}{32}\right) = 5$

$\left(\frac{1}{2}\right)^5 = \frac{1}{32}$

In Example 1, the logarithmic form was given and we changed it to the exponential form. In Example 2, we reverse the process: We change an exponential form to the logarithmic form. Again, remember that a logarithm is an exponent.

**EXAMPLE 2**    Changing to Logarithmic Form

Write each of the following equations in logarithmic form:

a) $6^4 = 1296$

b) $5^{-3} = \dfrac{1}{125}$

c) $\left(\dfrac{1}{4}\right)^2 = \dfrac{1}{16}$

SOLUTION

**TRY THIS**

Graph $y_1 = 10^x$, $y_2 = x$, and $y_3 = \log x$ in the 0.1 window. This is visual support that $y = 10^x$ and $y = \log x$ are inverse functions.

a)                    $6^4 = 1296$          **The exponent 4 is the value of the logarithm with base 6.**

$$\log_6 1296 = 4$$

b)                    $5^{-3} = \dfrac{1}{125}$

$$\log_5 (1/125) = -3$$

c)                    $\left(\dfrac{1}{4}\right)^2 = \dfrac{1}{16}$

$$\log_{\frac{1}{4}} (1/16) = 2$$

## Evaluating a Logarithm

The reasoning used in Examples 1 and 2 can be used for logarithmic functions of all bases.

Again we emphasize that $\log_a x$ is equal to an exponent. It is the exponent $y$ such that $a^y = x$. We use this generalization in the following two examples.

**GRAPHER NOTE**

To enter $\log_a x$ into a grapher, use the following fact:

$$\log_a x = \frac{\log x}{\log a}$$

where $\log x$ means $\log_{10} x$. Use this fact to support the solutions to Examples 3 and 4.

**EXAMPLE 3**    Evaluating a Logarithm

Find the value of $\log_3 9$ without using a grapher.

SOLUTION

$\log_3 9 = y$          **$y$ is the exponent to be used with base 3.**

$3^y = 9$          **Think mentally. What is the value of $y$?**

$y = 2$

Therefore $\log_3 9 = 2$.

Example 3 was simple enough that the second step could be done mentally. However, Example 4 illustrates that sometimes it is helpful to rewrite a number like 1/64 as a power of the base.

**EXAMPLE 4**    Evaluating a Logarithm

Find the value of $\log_2 \left( \dfrac{1}{64} \right)$ without using a grapher.

SOLUTION

$$\log_2(1/64) = x \qquad \text{\textbf{x is the exponent to be used with the base 2.}}$$

$$2^x = \frac{1}{64} \qquad \text{\textbf{Rewrite 64 as a power of 2.}}$$

$$2^x = \frac{1}{2^6} \qquad \text{\textbf{Rewrite as in exponential form with base 2.}}$$

$$2^x = 2^{-6}$$

$$x = -6$$

Therefore $\log_2(1/64) = -6$.  ■

## Evaluating Logarithms with a Grapher

In Examples 3 and 4, logarithms were evaluated with paper-and-pencil methods. These evaluations were possible without using a grapher because the number being evaluated was a power of the base. In other words, the evaluation method used in the last two examples cannot be used in most cases. Most often we need to rely on a grapher or other scientific calculator to evaluate logarithms.

The logarithm with base 10 is commonly used, since our decimal numeration system is a base 10 system. Consequently, $y = \log_{10} x$ is called the **common logarithm**. It is indicated on most graphers as the **LOG** key.

Another important base is the number $e$, an irrational number that is approximately 2.718. In calculus, this number $e$ is more thoroughly defined. The function $y = \log_e x$ is called the **natural logarithm** and is indicated on a grapher as the **LN** key.

Often when using common and natural logarithms, we omit their bases (10 and $e$, respectively) from the notation and write instead

$$y = \log x \quad \text{for} \quad y = \log_{10} x$$

and

$$y = \ln x \quad \text{for} \quad y = \log_e x.$$

Example 5 illustrates using a grapher to evaluate the common logarithm.

**GRAPHER NOTE**

Most graphers have a key to evaluate the functions $y = \log x$ and $y = \ln x$.

**EXAMPLE 5**    Finding Logarithms

Use a grapher to find the following logarithms:

a) $\log 125$

b) $\log 35.17$

c) $\ln 12.3$

SOLUTION

See Figure 11.13.

log 125
                2.096910013
log 35.17
                1.546172368
ln 12.3
                2.509599262

**FIGURE 11.13**   Grapher
evaluation of logarithms.

## Solving Logarithmic Equations

It is often necessary in applications to solve a logarithmic equation like

$$\log x = 100.$$

To solve this equation, change the equation from logarithmic form to exponential form. Example 6 illustrates.

**EXAMPLE 6**    Solving Logarithmic Equations

Solve the following equations for $x$:

a) $\log x = 2.5$

b) $\log_3 (x - 3) = 2$

c) $\log_x 81 = 2$

d) $\log_x \left( \dfrac{1}{32} \right) = -5$

SOLUTION

a) $\log_{10} x = 2.5$

$\qquad 10^{2.5} = x$    **Use a grapher to evaluate $x$
accurate to hundredths.**

$\qquad 316.23 = x$

b)  $\log_3(x - 3) = 2$      **Change to exponential form.**

$$3^2 = x - 3$$

$$9 = x - 3$$

$$12 = x$$

c)  $\log_x 81 = 2$      **Change to exponential form.**

$$x^2 = 81$$

$$x = 9$$      **The base of a logarithm must be positive.**

d)  $\log_x \left(\dfrac{1}{32}\right) = -5$      **Change to exponential form.**

$$x^{-5} = \frac{1}{32}$$

$$x^{-5} = 2^{-5}$$

$$x = 2$$

## Using Common and Natural Logarithms in Problem Situations

After any major earthquake, its intensity is reported as a certain number on the Richter scale. But what is this Richter scale? It is determined as a common logarithm, as explained next.

---

### EARTHQUAKE PROBLEM SITUATION

The magnitude $M$ of an earthquake (on the Richter Scale) is found using the formula

$$M = \log_{10} I,$$

where $I$ represents the intensity of the earthquake.

---

An earthquake's intensity is measured using a seismograph. An important question regarding the Richter scale is, What is the comparison between a quake with magnitude 3 compared to one with magnitude 2? Example 7 asks a related question.

**EXAMPLE 7**    Measuring the Intensity of an Earthquake

How many times more intense is an earthquake that measures 6 on the Richter scale than one that measures 4 on the scale?

SOLUTION

Let $I_1$ be the intensity of an earthquake that measures 6 and $I_2$ the intensity of one that measures 4 on the Richter scale.

An earthquake that measures 6 on the scale is measured as follows:

$$\log_{10} I_1 = 6$$

$$10^6 = I_1$$

The earthquake that measures 4 on the scale is measured as follows:

$$\log_{10} I_2 = 4$$

$$10^4 = I_2$$

Therefore the earthquake measuring 6 is $10^2$, or 100, times greater than the one measuring 4 on the Richter scale.    ∎

One reason the logarithm with base $e$ is called the natural logarithm is that it models phenomenon in the natural world. One example is bacterial growth.

**BACTERIAL GROWTH PROBLEM SITUATION**

In a certain bacterial culture, the number $B$ of bacteria present is represented by

$$B = 15{,}000e^{0.3t},$$

where $t$ represents the time in hours.

A variety of questions can be asked about this situation, for example:

1. If the bacterial count is 15,000 at 8 a.m., how many hours later will the count be 25,000?
2. If the bacterial count is 15,000 at 8 a.m., what will be the count 3 hr later?

The problems posed and solved in Example 8 are related to this second question.

**EXAMPLE 8**    Measuring Bacterial Growth

Let $t = 0$ represent the amount of bacteria present at 8 a.m. Find the number of bacteria present at the following times and express the answer correct to two places:

a) 10 a.m.
b) 6 p.m.

SOLUTION

a) Let $t = 0$ represent the time at 8 a.m. and $t = 2$ the time at 10 a.m.

$$B = 15{,}000e^{0.3(2)} \quad \text{An algebraic representation of the number of bacteria at 10 a.m.}$$

$$= 27331.78201$$

Therefore there are about 27,332 bacteria present at 10 a.m.

b)

$$B = 15{,}000e^{0.3(10)} \quad \text{An algebraic representation of the number of bacteria at 6 p.m.}$$

$$= 301{,}283.0538$$

Therefore there are about 301,283 bacteria present at 6 p.m. ∎

## Exercises for Section    11.3

In Exercises 1–10, write the logarithmic equation in exponential form.

**1.** $\log_2 8 = 3$

**2.** $\log_2 32 = 5$

**3.** $\log_3 9 = 2$

**4.** $\log_4 64 = 3$

**5.** $\log_{10} 1000 = 3$

**6.** $\log_{10} 100000 = 5$

**7.** $\log_{\frac{1}{2}}\left(\dfrac{1}{16}\right) = 4$

**8.** $\log_2\left(\dfrac{1}{8}\right) = -3$

**9.** $\log_5\left(\dfrac{1}{125}\right) = -3$

**10.** $\log_{\frac{1}{4}}\left(\dfrac{1}{16}\right) = 2$

In Exercises 11–18, write the exponential equation in logarithmic form.

**11.** $10^2 = 100$

**12.** $10^3 = 1000$

**13.** $3^4 = 81$

**14.** $2^6 = 64$

**15.** $5^3 = 125$

**16.** $4^5 = 1024$

**17.** $9^3 = 729$

**18.** $7^4 = 2401$

In Exercises 19–28, evaluate the logarithm without using a grapher.

**19.** $\log_2 16$

**20.** $\log 100$

**21.** $\log_4 64$

**22.** $\log_9 81$

**23.** $\log_{1/2} 8$

**24.** $\log_3 27$

**25.** $\log_5\left(\dfrac{1}{125}\right)$

**26.** $\log_3(9^2)$

**27.** $\log 0.001$

**28.** $\log 0.01$

In Exercises 29–32, find the common log using a grapher.

**29.** $\log 35$

**30.** $\log 50$

**31.** $\log 3.25$

**32.** $\log 2.7$

In Exercises 33–36, find the natural log using a grapher.

**33.** $\ln 25$

**34.** $\ln 56$

**35.** $\ln 4.25$

**36.** $\ln 8.5$

In Exercises 37–44, solve the equation without using a grapher. Support the solution using a grapher.

**37.** $\log_8 x = 2$

**38.** $\log_4 x = 5$

**39.** $\log_x (1/27) = -3$

**40.** $\log_5 (x + 1) = 3$

**41.** $\log_x 625 = 2$

**42.** $\log_7 x^2 = -2$

**43.** $\log_8 x = 1$

**44.** $\log_6 (x + 5) = 0$

**45.** *Predicting Bacteria Growth.* The number $B$ of bacteria present in a certain culture is given by the formula $B = 50{,}000e^{0.5t}$, where $t$ is measured in hours and $t = 0$ occurs at noon. Find the number of bacteria present at 3 p.m.

**46.** *Ant Population Growth.* The number $N$ of ants in a colony are increasing according to the relationship $N = 200e^{0.04t}$, where $t$ is the number of days and $t = 0$ the first of the month. What will the population be on the fifteenth of the month.

**47.** *Money Growth.* Your father gives you $1 on your first birthday with a pledge to double the gift each succeeding year. That is, you receive $2 on the second birthday, $4 on the third, and so on. On what birthday would your gift exceed a million dollars?

TRANSLATING WORDS TO SYMBOLS

In Exercises 48–51, write the equation represented by the sentence.

**48.** The log of nine base three is two.

**49.** The log of $x$ base five is three.

**50.** The log of 625 base $x$ is two.

**51.** The common log of 100 is two.

EXTENDING THE IDEAS

In chemistry, the $pH$ of a solution is defined to be $pH = -\log(H^+)$, where $H^+$ is the hydrogen ion concentration of the solution in moles per liter. In Exercises 52–54, find the $pH$ of a solution with hydrogen ion concentration of the given moles per liter.

**52.** $10^{-6}$

**53.** $10^{-3}$

**54.** $6.32 \cdot 10^{-12}$

**55. Writing to Learn.** Explain in a paragraph how many times greater an earthquake that measures 8 on the Richter scale is than one that measures 3.

**56. Writing to Learn.** Given the exponential function $y = a^x$ with points $\left(-3, \dfrac{1}{8}\right)$, $(-1, 0.5)$, $(2, 4)$, and $(3, 8)$, name the coordinates of four points on the graph of $y = \log_a x$. Write a paragraph explaining your answer.

**57. Writing to Learn.** Write a paragraph comparing the graphs of $y = a^x$ and $y = \log_a x$.

**58. Writing to Learn.** Write a paragraph explaining why $y = \log_a x$, with $a > 0$ and $a \neq 1$, is undefined for $x \leq 0$.

LOOKING BACK—MAKING CONNECTIONS

In Exercises 59–64, use the properties of exponents to simplify the expression.

**59.** $2^{3x+5} \cdot 2^{-2x}$

**60.** $5^{n-2} \cdot 5^{n+6}$

**61.** $(3^2)^{x-5}$

**62.** $(-3^3)^{x+2}$

**63.** $\dfrac{5^{3x+5}}{5^{x-1}}$

**64.** $\dfrac{3^{2x+1}}{3^{x+1}}$

## 11.4 | Properties of Logarithms

We can find visual information about properties of logarithms by graphing logarithmic functions. The next exploration provides a visual way to discover whether two expressions are equal.

### EXPLORE WITH A GRAPHER

The following equations *might* or *might not* be true. Find the graph of each side of each equation in the viewing rectangle $[-3, 10]$ by $[-2, 3]$. Decide whether you think the equation is true for all values of $x$.

1. $\log(x + 5) = \log x + \log 5$

2. $\log x^2 = (\log x)^2$

3. $\log \dfrac{x}{3} = \dfrac{\log x}{\log 3}$

Which of these equations are true?

In this exploration, you might have discovered that properties you guessed to be true are in fact not true. It is also important to learn those properties that *are* true. The following statements describe such properties. A graphing utility can be used to support and demonstrate them.

---

**Basic Properties of Logarithms**

Let $a > 0$ and $a \neq 1$. Then the following statements are true:

a) $a^{\log_a x} = x$ for every positive real number $x$

b) $\log_a a = 1$

c) $\log_a 1 = 0$

---

Each of these three properties can be verified by considering both the logarithm and its equivalent exponential form.

a) $y = \log_a x$ is equivalent to $x = a^y$. Therefore

$$x = a^y \qquad \textbf{Substitute } y = \log_a x.$$

$$= a^{\log_a x}.$$

b) $a^1 = a$ is equivalent to $\log_a a = 1$.
c) $a^0 = 1$ is equivalent to $\log_a 1 = 0$.

Example 1 illustrates how these formulas are used.

**EXAMPLE 1**    Using the Properties

Evaluate each of the following expressions:

a) $3^{\log_3 9}$

b) $\log_5 5$

c) $\log_4 1$

SOLUTION

a) $3^{\log_3 9} = 9$

b) $\log_5 5 = 1$

c) $\log_4 1 = 0$                                                     ■

Remembering that a logarithm is an exponent, it is not surprising that for each rule regarding exponents there is a similar rule for logarithms. We begin with the Product Rule.

---

**Product Rule for Logarithms**

For positive real numbers $x$, $y$, and $b$, $b \neq 1$,

$$\log_b xy = \log_b x + \log_b y.$$

---

We summarize this rule in words as follows:

*The log of a product is the sum of the logs.*

To see why this rule is true, let $\log_b x = u$ and $\log_b y = v$. Then, using the exponential form, we have $x = b^u$ and $y = b^v$.

$$xy = b^u b^v \qquad \text{Use the Product Rule for exponents.}$$

$$xy = b^{u+v} \qquad \text{Write this in an equivalent logarithmic form.}$$

$$\log_b xy = u + v \qquad \text{Substitute for } u \text{ and } v.$$

$$= \log_b x + \log_b y$$

Here are some examples of this rule.

$$\log(2 \cdot 3) = \log 2 + \log 3$$

$$\log 3x = \log 3 + \log x$$

$$\log_3 x^2 = \log_3 x + \log_3 x$$

A second rule is the Quotient Rule.

---

## Quotient Rule for Logarithms

For positive real numbers $x$, $y$, and $b$, $b \neq 1$,

$$\log_b \frac{x}{y} = \log_b x - \log_b y.$$

---

We summarize this rule in words as follows:

*The log of a quotient is the difference of the log of the numerator and the log of the denominator.*

To see why this rule is true, let $\log_b x = u$ and $\log_b y = v$. Then, using the exponential form, we have $x = b^u$ and $y = b^v$.

$$\frac{x}{y} = \frac{b^u}{b^v} \qquad \text{Use the Quotient Rule for exponents.}$$

$$\frac{x}{y} = b^{u-v} \qquad \text{Write this in an equivalent logarithmic form.}$$

$$\log_b \frac{x}{y} = u - v \qquad \text{Substitute for } u \text{ and } v.$$

$$= \log_b x - \log_b y$$

Here are some examples of this rule.

$$\log_2 \frac{x}{8} = \log_2 x - \log_2 8$$

$$= \log_2 x - 3$$

$$\ln\left(\frac{x-2}{x}\right) = \ln(x-2) - \ln x$$

Example 2 shows how both the Product and Quotient Rules can be used to expand a logarithm expression.

**REMINDER**

Throughout the remainder of this chapter we assume all expressions are defined.

**EXAMPLE 2**    Expanding a Logarithmic Expression

Expand each of the following expressions assuming that $x > 0$:

a) $\log \dfrac{3(x+2)}{x}$

b) $\log_2 \dfrac{4x^2 + 12x}{8(x+1)}$

SOLUTION

a)  Begin with the Quotient Rule.

$$\log \frac{3(x+2)}{x} = \log 3(x+2) - \log x \qquad \textbf{Use the Product Rule.}$$

$$= \log 3 + \log(x+2) - \log x$$

b)  Begin by simplifying the fraction.

$$\log_2 \frac{4x^2 + 12x}{8(x+1)} = \log_2 \frac{4x(x+3)}{8(x+1)}$$

$$= \log_2 \frac{x(x+4)}{2(x+1)} \qquad \textbf{Use the Quotient Rule.}$$

$$= \log_2 x(x+4) - \log_2 2(x+1) \qquad \textbf{Use the Product Rule.}$$

$$= \log_2 x + \log_2(x+4) - (\log_2 2 + \log_2(x+1))$$

$$= \log_2 x + \log_2(x+4) - 1 - \log_2(x+1) \qquad \blacksquare$$

The last rule to consider is the Power Rule.

---

### Power Rule for Logarithms

For positive real numbers $x$, $y$, and $b$, $b \neq 1$, then for every real number $c$,

$$\log_b x^c = c \log_b x.$$

---

We summarize this rule in words as follows:

*The log of a number to a power is the exponent*
*times the log of the number.*

To see why this rule is true, let $\log_b x = u$. Then, using the exponential form, we have $x = b^u$.

$$x^c = \left(b^u\right)^c \qquad \text{Use the Power Rule for exponents.}$$

$$x^c = b^{uc} \qquad \text{Write this in an equivalent logarithmic form.}$$

$$\log_b x^c = uc = cu \qquad \text{Substitute for } u.$$

$$= c \log_b x$$

Here are some examples of this Power Rule.

$$\log_2 2^3 = 3 \qquad \text{The log of the base raised to a power equals the power.}$$

$$\log x^3 = 3 \log x$$

Example 3 illustrates how to use the Power Rule to expand a logarithmic expression.

### EXAMPLE 3    Expanding a Logarithmic Expression

Expand each of the following expressions assuming that $x > 2$:

a) $\log_3 3(x - 2)^2$

b) $\log_5 8x^2$

SOLUTION

a) Rewrite using the Product Rule.

$$\log_3 3(x - 2)^2 = \log_3 3 + \log_3 (x - 2)^2 \qquad \text{Use the Power Rule.}$$

$$= 1 + 2\log_3 (x - 2)$$

b) Rewrite using the Product Rule.

$$\log_5 8x^2 = \log_5 8 + \log_5 x^2 \quad \textbf{Use the Power Rule.}$$

$$= \log_5 8 + 2\log_5 x \qquad \blacksquare$$

Sometimes all three rules—the product, the quotient, and the power—are used to expand an expression. Example 4 illustrates this case.

**EXAMPLE 4**   Expanding a Logarithmic Expression

Expand each of the following expressions assuming that $x > 0$ in (a) and $x > 5$ in (b):

a) $\log_2 \dfrac{2^7(x+3)}{x^3}$

b) $\log \dfrac{4(x-5)}{x^3(x+6)}$

SOLUTION

a) Rewrite using all three rules as needed.

$$\log_2 \frac{2^7(x+3)}{x^3} = \log_2 2^7(x+3) - \log_2 x^3$$

$$= \log_2 2^7 + \log_2(x+3) - 3\log_2 x$$

$$= 7\log_2 2 + \log_2(x+3) - 3\log_2 x$$

b) Rewrite using all three rules as needed.

$$\log \frac{4(x-5)}{x^3(x+6)} = \log 4(x-5) - \log x^3(x+6)$$

$$= (\log 4 + \log(x-5)) - (3\log x + \log(x+6))$$

$$= \log 4 + \log(x-5) - 3\log x - \log(x+6) \qquad \blacksquare$$

In Examples 2 to 4, a single logarithmic expression was expanded to a sum and/or a difference of several simpler logarithmic expressions. Sometimes we need to reverse that process and change an expanded form to a single logarithm expression. Example 5 illustrates this process.

**EXAMPLE 5**   Changing from Expanded Form

Rewrite each of the following expressions so there is a single logarithmic term assuming that $x > 0$:

a) $3\log x - \log(x+1)$

b) $2\log_2 x - \log_5 x + \log_2(x+1) - \log_5 \dfrac{5}{x}$

SOLUTION

a) Begin by using the Power Rule.

$$3\log x - \log(x+1) = \log x^3 - \log(x+1) \quad \text{Use the Quotient Rule.}$$

$$= \log \frac{x^3}{(x+1)}$$

b) Because all of the terms do not have the same base, group together terms with like bases.

$$2\log_2 x - \log_5 x + \log_2(x+1) - \log_5 \frac{5}{x}$$

$$= \log_2 x^2 + \log_2(x+1) - \left(\log_5 x + \log_5 \frac{5}{x}\right)$$

$$= \log_2 x^2(x+1) - \log_5 \left(x \cdot \frac{5}{x}\right) \quad \begin{array}{l}\textbf{Note that we}\\\textbf{cannot combine}\\\textbf{terms with}\\\textbf{different bases.}\end{array}$$

$$= \log_2 x^2(x+1) - \log_5 5$$

$$= \log_2 x^2(x+1) - 1 \qquad \qquad ■$$

## Exercises for Section  11.4

In Exercises 1–6, compute without using a grapher.

**1.** $2^{\log_2 16}$

**2.** $10^{\log_{10} 8}$

**3.** $\log_3 1$

**4.** $\log_8 1$

**5.** $\log_4 4$

**6.** $\log_{1/2} 0.5$

In Exercises 7–22, use properties of logarithms to write the expression as a sum, difference, or product of logarithms. Assume that values of $x$ are restricted so that all terms are defined.

**7.** $\log_2 3x$

**8.** $\log_3 5y$

**9.** $\log_3 5(x+3)$

**10.** $\log_2 8(x-1)$

**11.** $\log_2(xy^2)$

**12.** $\log_3(x^3y)$

**13.** $\log_5 \dfrac{x^2}{y^3}$

**14.** $\log_{10} \dfrac{(x-3)^2}{5}$

**15.** $\log_4 \dfrac{x}{9}$

**16.** $\log_{10} \dfrac{100}{x}$

**17.** $\log_8 x^2$

**18.** $\log_6 7(x+2)^3$

**19.** $\log_4 5000x^5$

**20.** $\log_6 x^3\sqrt{2}$

**21.** $\log_3 \dfrac{5x^6}{2y^3}$

**22.** $\log_9 \dfrac{2x^3}{7y^2}$

In Exercises 23–28, use properties of logarithms to write the sum, difference, or product of logarithms as the log of a single expression. Assume that all variables are restricted so that all terms are defined.

**23.** $\log_2 x + 5\log_2 y$

**24.** $\log_4 8 - \log_4 y$

**25.** $\log_3(x-4) - \log_3(x+5)$

**26.** $\log_{10}(x^2+2x-3) - 6\log_{10} x$

**27.** $2\log_4 125 + 5\log_4(x+2)$

**28.** $3\log_2 x + 4\log_2 y$

In Exercises 29–38, tell whether the equation is true or false.

**29.** $\log 35 = \log 5 \cdot \log 7$

**30.** $\log 48 = \log 6 + \log 8$

**31.** $\log\left(\dfrac{3}{4}\right) = \log 3 - \log 4$

**32.** $\log\left(\dfrac{5}{6}\right) = \dfrac{\log 5}{\log 6}$

**33.** $\log 25^2 = 2 \log 25$

**34.** $\log 5^3 = \log 125$

**35.** $\ln(3 + 9)^2 = 2 \ln 12$

**36.** $\ln(8 \cdot 3) = 2 + \log 12$

**37.** $\log(10 + 15) = 2 \log 5$

**38.** $\log(10 + 15) = \log 10 + \log 15$

In Exercises 39–46, use properties of logarithms to replace the ? with either $=$ or $\neq$. Assume that all terms are defined.

**39.** $\log_{10} 5 + \log_{10} 8$ ? $\log_{10} 13$

**40.** $\log_5 3 - \log_5 2$ ? $\log_5 1$

**41.** $\log_2 x - \log_2 9$ ? $\log_2 \dfrac{x}{9}$

**42.** $3 \log_4 y + \log_4 z$ ? $\log_4 y^3 z$

**43.** $2 \log_a(x - 3) + 5 \log_a x$ ? $\log_a x^5(x - 3)^2$

**44.** $\log_5(x + 1) + \log_5 x$ ? $\log_5(x + 1)x$

**45.** $\log(x + 8) - \log(x + 2)$ ? $\log(x + 8)(x + 2)$

**46.** $3 \log(x - 7) - 2 \log(x + 1)$ ? $\log \dfrac{3(x - 7)}{2(x + 1)}$

In Exercises 47–54, solve the equation without using a calculator.

**47.** $\log_3 x = 2$

**48.** $\log_2 x = 5$

**49.** $\log_x\left(\dfrac{1}{125}\right) = -3$

**50.** $\log_3(x + 1) = 2$

**51.** $\log_x 64 = 3$

**52.** $\log_2 x = -2$

**53.** $\log_3 x^2 = -2$

**54.** $\log_6(2x + 5) = 0$

In Exercises 55–62, change the equation from exponential to logarithmic form.

**55.** $y = 5^x$

**56.** $y = 2^x$

**57.** $xy = 3^x$

**58.** $xy = 4^5$

**59.** $\dfrac{x}{y} = 2^x$

**60.** $\dfrac{x}{y} = 2^5$

**61.** $x + y = 3^4$

**62.** $x - y = 4^{-2}$

In Exercises 63–66, change the equation from logarithmic form to exponential form.

**63.** $\log_3 9 = 2$

**64.** $\log_2 x = y$

**65.** $\log_5(x/y) = -3$

**66.** $\log_2(x^2 + 5x) = -3$

## TRANSLATING WORDS TO SYMBOLS

In Exercises 67–70, write the numerical representation of the sentence.

**67.** The log of the product of 2 and $x$ is equal to the sum of the log of 2 and the log of $x$.

**68.** The log of the quotient of $(x + 2)$ and $y$ is equal to the difference of the log of $(x + 2)$ and the log of $y$.

**69.** The log of $x$ to the power three is 3 times the log of $x$.

**70.** The log of the quotient $3x/y$ is equal to the log of $3x$ minus the log of $y$.

## EXTENDING THE IDEAS

In Exercises 71 and 72, show without using a grapher that the statement is true.

**71.** $\log_4 64 = \log_3 27$

**72.** $\log_9 81 = \log_3 9$

**73. Writing to Learn.** Explain why, when you evaluate $\log(-25.5)$, you get an error message on the grapher.

**74. Writing to Learn.** Research the historical background and importance of logarithms. Logs were invented in the seventeenth century by a Scottish mathematician in order to simplify calculations. Write a few paragraphs about the history.

**75. Writing to Learn.** Write a paragraph explaining how the emphasis on logarithms changed with the invention of the computer and scientific calculator.

LOOKING BACK—MAKING CONNECTIONS

Solve the following equations for $x$.

**76.** $\sqrt{x+3} = 8$       **77.** $\sqrt{2x-5} - 9 = 16$

**78.** $\sqrt{x+2} = 3 + \sqrt{x-1}$

**79.** $\sqrt{2x+5} - \sqrt{x-6} = 3$

---

## 11.5 Solving Logarithmic Equations

Solving Logarithmic Equations Graphically ▪ Using Logarithms to Solve Exponential Equations ▪ Solving Problem Situations Using Logarithms

We begin this section studying algebraic techniques for solving logarithmic equations.

Some logarithmic equations can be solved by simply changing from logarithmic form to exponential form. This method was illustrated in Section 11.3. For example, to solve $\log_2 x = 12$, we reason as follows:

$\log_2 x = 12$       **Remember that a logarithm is an exponent.**

$2^{12} = x$

$x = 4096$

For more complicated equations, it is convenient to use the following property of logarithms. This property is true because the logarithmic functions are one-to-one functions.

---

### Property of Logarithmic Equations

If $x$ and $y$ are both positive, then

$$\log_b x = \log_b y \qquad \text{if and only if} \qquad x = y.$$

Example 1 illustrates how properties of logarithms studied in Section 11.4 together with the Property of Logarithmic Equations can be combined to solve these equations.

### EXAMPLE 1   Solve Algebraically

Solve $\log_2(x + 3) - \log_2 5 = \log_2 7$.

SOLUTION

$$\log_2(x + 3) - \log_2 5 = \log_2 7 \qquad \text{Isolate the term with variable } x \text{ on the left-hand side.}$$

$$\log_2(x + 3) = \log_2 7 + \log_2 5$$

$$\log_2(x + 3) = \log_2(7 \cdot 5) \qquad \text{Use the Property of Logarithmic Equations.}$$

$$x + 3 = 35$$

$$x = 32$$

Because $\log_2(x + 3)$ is defined when $x = 32$, we conclude that $x = 32$ is the unique solution to the equation. ■

The equation in Example 1 has a unique solution. Example 2 shows an example of an equation which at first appears to have two solutions; however, one proves to be an extraneous solution.

### EXAMPLE 2   Solve Algebraically

Solve $\log_5(x + 3) + \log_5(x - 4) = \log_5 8$.

SOLUTION

$$\log_5(x + 3) + \log_5(x - 4) = \log_5 8 \qquad \text{Use the Product Rule.}$$

$$\log_5(x + 3)(x - 4) = \log_5 8 \qquad \text{Use the Property of Logarithmic Equations.}$$

$$x^2 - x - 12 = 8$$

$$x^2 - x - 20 = 0$$

$$(x - 5)(x + 4) = 0$$

$$x = 5 \quad \text{and} \quad x = -4$$

Recall that the logarithm of a negative number is undefined. When $x = -4$ is substituted back into the equation, both terms $\log_5(x + 3)$ and $\log_5(x - 4)$

are undefined. Therefore $x = -4$ is an extraneous solution; the only solution is $x = 5$. ∎

## Solving Logarithmic Equations Graphically

A logarithmic equation can be solved graphically using the same techniques studied earlier in this text. Also, algebraic methods can be supported graphically.

Graphical methods require that we find the graphs of logarithmic functions. We will restrict our discussion in this section to logarithmic equations with base 10 or base $e$ because these functions can be graphed using the built-in function keys for LOG and ln found on most graphers.

**EXAMPLE 3**   Solving a Logarithmic Equation

Solve $\ln x = -1.5$ graphically. Confirm the solution algebraically.

SOLUTION

**Solve Graphically.** We use the $x$-intercept method. Figure 11.14(a) shows the graph of $y_1 = \ln x + 1.5$ in the Standard window. After one zoom-in (Figure 11.14(b)), it is clear that there is a solution. Continuing the zoom-in, we can show that the solution is $x = 0.223$ with an error of at most 0.01.

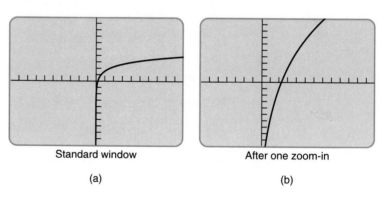

|                 |                  |
|-----------------|------------------|
| Standard window | After one zoom-in |
| (a)             | (b)              |

**FIGURE 11.14**   $f(x) = \ln x + 1.5$ in Example 3.

**Confirm Algebraically.** Change from logarithmic form to exponential form.

$$\ln x = -1.5$$
$$x = e^{-1.5}$$

Use the exponentiation key on a grapher to show that $x = 0.2231301601$, therefore supporting the $x = 0.223$ solution. ■

The solution in Example 3 was confirmed algebraically. Sometimes, however, it is convenient to support the solution numerically, which we demonstrate in Example 4.

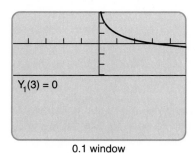

Y₁(3) = 0

**FIGURE 11.15** A graphical solution and numeric support for Example 4.

**EXAMPLE 4**   Solve Graphically and Support Numerically

Solve $\frac{1}{2}\log(x + 6) = \log x$.

**SOLUTION**

**Solve Graphically.** We use the $x$-intercept method. So graph $y_1 = 0.5\log(x + 6) - \log x$ in the 0.1 window. Figure 11.15 suggests that $x = 3$ is the one zero of this function.

**Support Numerically.** To complete numerical support, evaluate $Y_1$ at $x = 3$. That is, show that $Y_1(3) = 0$. (See the lower portion of the split screen shown in Figure 11.15.) ■

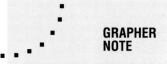

**GRAPHER NOTE**

Some graphers cannot evaluate $Y_1$ at 3 by using the notation $Y_1(3)$. Check in your grapher lab manual.

**EXAMPLE 5**   Solve Algebraically and Support Numerically

Solve $\log x^2 - \log 5 = \log 7 + \log 2x$. Support both graphically and numerically.

**SOLUTION**
**Solve Algebraically.**

$$\log x^2 - \log 5 = \log 7 + \log 2x$$

$$\log \frac{x^2}{5} = \log(7 \cdot 2x) \qquad \text{If logs are equal,}$$
$$\text{then numbers are equal.}$$

$$\frac{x^2}{5} = 14x$$

$$x^2 = 70x$$

$$x^2 - 70x = 0$$

$$x(x - 70) = 0$$

$$x = 0 \quad \text{and} \quad x = 70$$

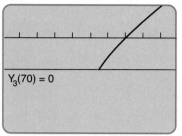

$Y_3(70) = 0$

[0, 100] by [−0.25, 0.25]

**FIGURE 11.16**
Numerical and graphical support for Example 5.

Because logarithmic functions are not defined for $x = 0$, we see that $x = 0$ is an extraneous root. The solution to the equation is $x = 70$.

**Support Graphically and Numerically.** Let $y_1 = \log x^2 - \log 5$ and $y_2 = \log 7 + \log 2x$ and find the graph of $y_3 = y_1 - y_2$. Figure 11.16 provides graphical and numerical support for the solution $x = 70$. ■

## Using Logarithms to Solve Exponential Equations

When an exponential equation has different bases, it often is impossible to express them both in terms of the same bases. Therefore there is a need to take the log of each side of the equation. Example 6 illustrates this case.

**EXAMPLE 6**    Solve an Exponential Equation That Has Different Bases

Solve $4^{3x-3} = 25$. Support numerically.

SOLUTION
**Solve Algebraically.**

$$4^{3x-3} = 25 \qquad \text{Take the log of each side of the equation.}$$

$$\log 4^{3x-3} = \log 25 \qquad \text{Use the Power Rule to simplify.}$$

$$(3x - 3)\log 4 = \log 25$$

$$3x - 3 = \frac{\log 25}{\log 4}$$

$$3x = \frac{\log 25}{\log 4} + 3$$

$$x = \frac{1}{3}\left(\frac{\log 25}{\log 4} + 3\right)$$

$$x = 1.773976032$$

Rounding to three decimal places, we obtain the solution $x = 1.774$.

$1.774 \to X : 4^{\wedge}(3x - 3)$
    $25.00249217$

**FIGURE 11.17**
Numerical support for Example 6.

**Support Numerically.** Figure 11.17 shows numerical support for the solution $x = 1.774$. ■

## Solving Problem Situations Using Logarithms

We can use logarithms to solve the following interest problem situation.

---

☐                                                                              ☐

### COMPOUND INTEREST PROBLEM SITUATION

John invests some money at 7% interest compounded annually. The algebraic representation

$$S = P(1.07)^x$$

describes the total value $S$ of an investment of $P$ dollars after $x$ years.

☐                                                                              ☐

---

**EXAMPLE 7**    Finding When an Investment Triples

In how many years will John's investment triple? Support numerically.

SOLUTION

**Solve Algebraically.** If John invests $P$ dollars, he will have $3P$ dollars after his investment triples. Therefore the algebraic representation

$$S = P(1.07)^x$$

becomes $3P = P(1.07)^x$.

| | |
|---|---|
| $3P = P(1.07)^x$ | **Divide each side by P.** |
| $3 = (1.07)^x$ | **Take the log base 10 of each side.** |
| $\log 3 = \log(1.07)^x$ | **Use the Power Rule for logs.** |
| $\log 3 = x \log(1.07)$ | |
| $\dfrac{\log 3}{\log(1.07)} = x$ | |
| $16.23757367 = x$ | |
| $16.24 = x$ | |

Therefore the investment will triple in value in about 16.24 yr.

| X | Y1 | |
|------|--------|---|
| 16.2 | 2992.4 | |
| 16.21 | 2994.4 | |
| 16.22 | 2996.4 | |
| 16.23 | 2998.5 | |
| 16.24 | 3000.5 | |
| 16.25 | 3002.5 | |
| 16.26 | 3004.6 | |

Y1 = 1000(1 + .07)^X

**FIGURE 11.18** A grapher-generated table for Example 7, with $x$ = number of years and $y_1$ = amount of money accrued in $x$ yr.

**Support Numerically.** Pick an arbitrary amount of money to invest, say $1000. Figure 11.18 shows that when John invests $1000 at 7%, the amount of money accrued after 16.24 yr is more than $3000. The grapher-generated table shows support with $x$ = the number of years the money is invested and $y_1$ = the amount of money accrued. ■

## Exercises for Section 11.5

In Exercises 1–10, solve the equation algebraically.

**1.** $\log_9 27 = x$

**2.** $\log_{27} 9 = x$

**3.** $\log_4(x + 2) = 2$

**4.** $\log_8(x + 3) = 2$

**5.** $3 \log_2 x = 6$

**6.** $2 \log_3 x = 4$

**7.** $\log_5 x = 4$

**8.** $\log_4(x - 3) = 3$

**9.** $\log_3(x + 3) + \log_3(x - 2) = \log_3 14$

**10.** $\log_4(x + 1) = \log_4 5 - \log_4(x - 3)$

In Exercises 11–16, solve the equation graphically and support the solution numerically.

**11.** $\log x + \log 2 = 3$

**12.** $\log x + \log(x + 3) = 2$

**13.** $\log(x + 1) - \log x = 1$

**14.** $\log x + \log(x - 3) = 1$

**15.** $x \log 1.08 = \log 3$     **16.** $x \log 5 = \log 8$

In Exercises 17–20, choose between an algebraic or a graphical method and solve the equation. Explain why you chose the method you did and discuss whether the other method would work also.

**17.** $3 \log(x - 1) - 2 \log 4 = 0$

**18.** $\log(x - 1) + \log(x + 2) = 2$

**19.** $\ln(x + 1) + \ln x = \ln(x + 5) - \ln(x + 3)$

**20.** $\ln x + \ln(x + 5) = 2 \ln(x + 2)$

In Exercises 21–32, solve algebraically and support the solution numerically.

**21.** $5^{x-1} = 3^x$

**22.** $9^{x-4} = 7.13$

**23.** $2^x = 46$

**24.** $3.6^x = 52.5$

**25.** $3^{2x} = 7^{x-1}$

**26.** $12^{x-4} = 3^{x-2}$

**27.** $4^x = 6^{x-2}$

**28.** $2.2^{x-5} = 9.32$

**29.** $x \log 5 = \log 7 - 2x$     **30.** $x \log 3 = \log 5$

**31.** $(x + 1) \log 2 = \log 8$     **32.** $(x - 5) \log 4 = \log 2$

**33.** *Money Growth.* Stella invests some money at 5% interest compounded annually. How long does she need to invest the money to double her investment?

**34.** *Interest on Investment.* Amelia deposits $500 in a bank that pays 7% interest annually. Assume she makes no other deposits or withdrawals. How much will she accumulate after 5 yr?

**35.** *Cell Growth.* A single-cell organism divides into two organisms every 3 hr. How many hours does it take for one organism to increase to 1000?

**36.** *Population Growth.* The population of a certain country is 28 million and growing at the rate of 3% annually. Assuming the population is continuously growing, the population P (in millions) $t$ years from now is determined by the formula $P = 28e^{0.03t}$. In how many years will the population be 40 million?

**37.** *Population Growth.* The population in a certain country is 20 million and growing at the rate of 5% annually. Assuming the population is continuously growing, the population P (in millions) $t$ years from now is determined by the formula $P = 20e^{0.05t}$. In how many years will the population be 40 million?

EXTENDING THE IDEAS

In Exercises 38–45, use the equation

$$\log_b x = \frac{\log_{10} x}{\log_{10} b},$$

which can be shown to be true. Evaluate on a grapher.

**38.** $\log_3 2$

**39.** $\log_2 5$

**40.** $\log_5\left(\frac{1}{2}\right)$

**41.** $\log_2\left(\frac{1}{8}\right)$

**42.** $\log_{\frac{1}{2}} 6$

**43.** $\log_{\frac{1}{3}} 7$

**44.** $\log_{0.1}(0.02)$

**45.** $\log_{0.2}(1.5)$

In Exercises 46–48, use the equation

$$\log_b x = \frac{\log_{10} x}{\log_{10} b}.$$

**46.** Find the graph of $y = \log_2 x$ and use it to solve the equation $\log_2 x = 2.5$.

**47.** Find a graphical solution to the equation $\log_2 x + 1 = \log_{\frac{1}{2}} 2$.

**48.** Find a graphical solution to the equation $\log_9(x + 4) = \log_2 x$.

**49. Writing to Learn.** Find the graphs of $y = \log_2 x$, $y = \log_{1.3} x$, and $y = \log_{1.6} x$ both sequen-

tially and simultaneously in the window $[-2, 10]$ by $[-2, 10]$. What one point do they have in common? Conjecture a generalization about the relationship between the graphs of $y = \log_a x$ and $y = \log_b x$.

LOOKING BACK—MAKING CONNECTIONS

In Exercises 50–53, solve the equation graphically and support numerically.

**50.** $(1.05)^x = 4.1$          **51.** $(1.5)^x = 3000$

**52.** $(1.3)^x = 25$          **53.** $3^{-x} = 5$

## Chapter 11 Summary

| | **Exponential Functions** | **Examples** |
|---|---|---|
| Definition of Exponential Function | If $a$ is a positive real number and $a \neq 1$, then $y = a^x$ is an exponential function with base $a$. | $y = 3^x$ is an exponential function. |
| Doubling Function | The exponential function $y = 2^x$ | |
| Graphs of Exponential Functions | If $0 < a < 1$, the graph of $y = a^x$ has a shape like that in the graph of $y = \left(\frac{1}{2}\right)^x$.<br>If $a > 1$, the graph of $y = a^x$ has a shape like that in the graph of $y = 2^x$.<br>All exponential functions of the form $y = a^x$ have $y$-intercept $(0, 1)$. | 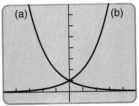<br>The graphs of (a) $y = \left(\frac{1}{2}\right)^x$ and (b) $y = 2^x$ in the 0.1 window with a vertical dimension of $[-1; 8]$. |
| Property of Exponential Equations | If $a^x = a^y$, then $x = y$, where $a > 0$ and $a \neq 1$. | If $3^4 = 3^{2y}$, then $4 = 2y$. |

### Exponential Functions

| | |
|---|---|
| Exponential Growth | $S = P(1 + r)^t$, where $S$ is total population, $P$ is the initial population, $r$ is the growth rate per period, and $t$ is the number of growth periods. |

### Inverse Functions

| | | Examples |
|---|---|---|
| One-to-One Function | A function that has exactly one distinct number in its domain associated with exactly one distinct number in its range | The function $y = x$ is one-to-one; the function $y = x^2$ is not. |
| Horizontal Line Test for One-to-One Functions | A function is one-to-one if every horizontal line intersects the graph of the function in at most one point. | |
| Inverse Function | The inverse function of $f$, denoted $f^{-1}$, consists of all ordered pairs $(y, x)$, where $(x, y)$ belongs to $f$. $f$ must be one-to-one in order for $f^{-1}$ to exist. | If $f(x) = x - 4$, then $f^{-1}(x) = x + 4$. Notice that $(1, -3)$ belongs to $f$ and $(-3, 1)$ belongs to $f^{-1}$. |
| Finding an Inverse Function | To find the inverse function $f^{-1}(x)$ of a one-to-one function $f(x)$, follow these steps: <br> 1. Replace $f(x)$ in the function equation with $y$. <br> 2. Interchange $x$ and $y$. <br> 3. Solve for $y$. <br> 4. Replace $y$ with $f^{-1}(x)$. | To find the inverse of $f(x) = 3x - 9$, $$y = 3x - 9$$ $$x = 3y - 9$$ $$x + 9 = 3y$$ $$1/3x + 3 = y.$$ So $f^{-1}(x) = \frac{1}{3}x + 3$. |

### Logarithms

| | | Examples |
|---|---|---|
| Definition of Logarithm | A logarithm of $x$ with base $a$ is $\log_a x$, where $x > 0$, $a > 0$, and $a \neq 1$ and $y = \log_a x$ means $x = a^y$ | |
| Common Logarithm | A logarithm that has base 10, most often written without the base notation | $\log 4$ is a common log, where $\log 4 = \log_{10} 4$. |

| | | |
|---|---|---|
| Natural Logarithm | A logarithm that has base $e$, an irrational number. The notation for $\log_e$ is ln. | $\ln 5$ is a natural log, where $\ln 5 = \log_e 5$. |
| Property of Logarithmic Equations | If $x > 0$ and $y > 0$, $\log_b x = \log_b y$ if and only if $x = y$. | $\log_4(x + 1) = \log_4 5x$ if and only if $x + 1 = 5x$. |
| Solving Logarithmic Equations Algebraically | Simplify the equation as much as possible and solve using the Property of Logarithmic Equations or the definition of logarithm. | Solve $$2\log_3 x = \log_3 8 - \log_3 2,$$ as follows: $$\log_3 x^2 = \log_3 \frac{8}{2}$$ $$\log_3 x^2 = \log_3 4$$ $$x^2 = 4$$ $$x = \pm 2$$ The only solution is $x = 2$. |
| Solving Logarithmic Equations Graphically | Use the multigraph or $x$-intercept method to solve. | |

| | **Properties of Logarithms** | **Examples** |
|---|---|---|
| Basic Properties | For $a > 0$ and $a \neq 1$, the follows holds: <br> 1. $a^{\log_a x} = x$ for every positive real number $x$ <br> 2. $\log_a a = 1$ <br> 3. $\log_a 1 = 0$ | $9^{\log_9 x} = x$ <br> $\log_7 7 = 1$ <br> $\log_{12} 1 = 0$ |
| Product Rule | For positive real numbers $x$, $y$, and $b$, $b \neq 1$, $$\log_b xy = \log_b x + \log_b y.$$ | $\ln 3x = \ln 3 + \ln x$ |
| Quotient Rule | For positive real numbers $x$, $y$, and $b$, $b \neq 1$, $$\log_b \frac{x}{y} = \log_b x - \log_b y.$$ | $\log \dfrac{a}{2} = \log a - \log 2$ |

| | **Properties of Logarithms** | **Examples** |
|---|---|---|
| Power Rule | For positive real numbers $x$, $y$, and $b$, $b \neq 1$ and every real number $c$, | $\log_5 3^2 = 2 \log_5 3$ |
| | $\log_b x^c = c \log_b x.$ | |

## Review Exercises for Chapter 11

In Exercises 1 and 2, solve the exponential equation by mental calculations.

**1.** $4^x = 16$

**2.** $3^x = \dfrac{1}{9}$

In Exercises 3 and 4, solve the exponential equation.

**3.** $\left(\dfrac{1}{5}\right)^x = 25$

**4.** $2^x = 8^2$

**5.** Use properties of exponents to select from the following the pair of functions that are equivalent.

a) $y_1 = \left(\dfrac{1}{2}\right)^{x-3}$   b) $y_2 = -2^{x-3}$   c) $y_3 = 2^{-(x-3)}$

**6.** Use properties of exponents to select from the following the pair of functions that are equivalent.

a) $y_1 = 9 \cdot 3^{4x}$   b) $y_2 = 3^{4x-2}$   c) $y_3 = 9^{2x-1}$

In Exercises 7 and 8, use a grapher to decide whether the function is one-to-one.

**7.** $y = x^3 + 2x$

**8.** $y = -3\sqrt{x}$

In Exercises 9–12, the equation is one-to-one. Find the equation of the inverse of $f$.

**9.** $f(x) = 5 - 7x$

**10.** $f(x) = -4x - 8$

**11.** $f(x) = \sqrt[3]{x-1}$

**12.** $f(x) = (x-3)^3$

In Exercises 13–15, write the logarithmic equation in exponential form.

**13.** $\log_5 x = 3$

**14.** $\log_x 4 = 2$

**15.** $\log_2(3x + 4) = 6$

In Exercises 16–18, write the exponential equation in logarithmic form.

**16.** $7^4 = 2401$

**17.** $20 = 3^x$

**18.** $3x - y = 2^4$

In Exercises 19 and 20, evaluate the logarithm without using a grapher.

**19.** $\log_2 \dfrac{1}{4}$

**20.** $\log_5 25$

In Exercises 21 and 22, find the common or natural log using a grapher.

**21.** $\ln 7.389$

**22.** $\log 0.5$

In Exercises 23 and 24, solve the equation without using a grapher.

**23.** $\log_x 324 = 2$

**24.** $\log_2(x - 1) = 3$

In Exercises 25 and 26, compute without using a grapher.

**25.** $4^{\log_4 4}$

**26.** $\log_6 1$

In Exercises 27 and 28, use the properties of logarithms to write the expression as a sum, difference, or product of simple logarithms. Assume that all terms are defined.

**27.** $\log_8 \left(\dfrac{7x^2}{3}\right)$

**28.** $\log_3 \left(\dfrac{a-1}{a+1}\right)$

In Exercises 29 and 30, use the properties of logarithms to write the sum, difference, or product of logarithms as the log of a single expression. Assume that all terms are defined.

**29.** $2\log_7 a + 3\log_7 b$    **30.** $3\ln x - \ln 3$

In Exercises 31 and 32, tell whether the equation is true or false.

**31.** $\log_2 \dfrac{3}{5} = \dfrac{\log_2 3}{\log_2 5}$

**32.** $\log_5 70 = \log_5 7 + \log_5 10$

In Exercises 33–36, solve the equation algebraically.

**33.** $\log_3(2x - 1) = 5$    **34.** $3\log_8 x = 2$

**35.** $3^{2x+1} = 10$    **36.** $5^x = 4^{x-1}$

In Exercises 37 and 38, solve the equation graphically and support the solution numerically.

**37.** $2\ln 2x = \ln(x + 2)$

**38.** $\log 15 - \log(8x - 7) = \log x$

**39.** If \$2000 is invested in an account that pays 8% annually, the total dollars $S$ after $t$ years is $S = 2000(1.08)^t$. Assume there are no other deposits and no withdrawals.

    a) How much is in the account after 10 yr?

    b) How long will it take for the money in the account to double?

**40.** In a certain bacteria culture, the number $B$ of bacteria present is represented by $B = 30{,}000e^{0.25t}$, where $t$ represents the time in hours. If $t = 0$ represents the amount of bacteria present at 1 p.m., find the number of bacteria present at 10 p.m.

---

## Chapter 11 Test

**1.** Solve the exponential equation $2^5 = 4^x$.

**2.** Use the properties of exponents to select from the following the pair of functions that are equal.

    a) $y_1 = 3^{-(x+2)}$         b) $y_2 = -3^{x+2}$

    c) $y_3 = -\left(\dfrac{1}{3}\right)^{-(x+2)}$

**3.** Explain why $f(x) = \dfrac{1}{2}x - 6$ is one-to-one. Then find the equation of the inverse of $f$.

**4.** Write $\log_8(x - 1)^2 = 2$ in exponential form.

**5.** Write $3^6 = 729$ in logarithmic form.

**6.** Evaluate $\log_4 16$ without using a grapher.

**7.** Solve $\log_4(x - 3) = 0$ without using a grapher.

**8.** Use the properties of logarithms to write the expression $\log_5 2x^6$ as a sum, difference, or product of simple logarithms. Assume that $x > 0$.

**9.** Use the properties of logarithms to write the expression $2\log 5 + \log 2x$ as the log of a single expression.

**10.** Solve $\log_2(3x + 4) = 6$ algebraically.

**11.** Solve $2^{2x} = 3^{x+1}$ algebraically.

**12.** Solve $\log_x 30 = 2$ algebraically.

**13.** Solve $\log x + \log(2x - 1) = \log 3$ graphically and support the solution numerically.

**14.** The population of Cherry Creek is increasing at a rate of 3% per year. The population in 1990 was 20,200. Find the population in 1999 if this rate of growth continues.

**15.** If \$800 is invested in an account that pays 6.5% compounded annually, the total dollars $T$ after $t$ years is $T = 800(1.065)^t$. Assuming there are no withdrawals and no other deposits, how long will it take until the account has \$1200?

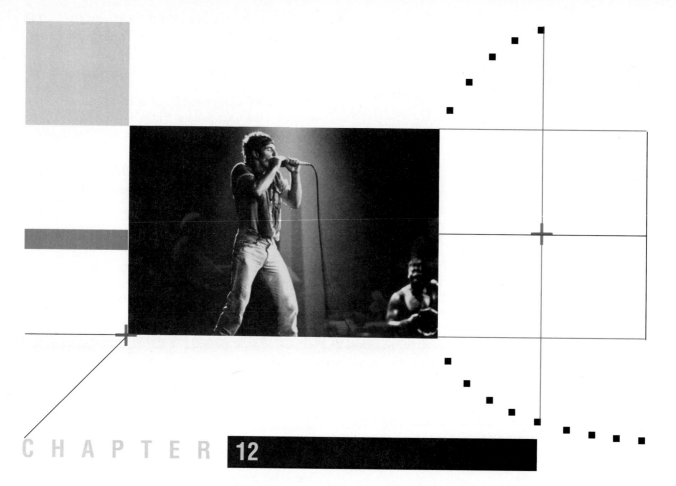

# Sequences and Series, and the Binomial Theorem

### AN APPLICATION

Suppose that a recording star signs a contract in which they earn $100,000 for the first release of that recording and half as much on each succeeding release of the same recording. After $n$ releases their total earnings ($T$) can be found by the formula

$$T = \frac{100,000(1 - 0.5^n)}{1 - 0.5}.$$

| 12.1 | # Arithmetic Sequences and Series |
|---|---|

Arithmetic Sequences  ▪  Arithmetic Series  ▪  Applications of
Arithmetic Series

Ordered lists of numbers, called **sequences**, occur frequently in life and are often used to describe some physical or social phenomenon. Each number of the sequence is called a **term**.

Here are several examples of sequences.

$$1, 2, 3, 4, \cdots \quad \text{\bf The natural numbers}$$
$$1, 3, 5, 7, \cdots \quad \text{\bf Odd numbers}$$
$$2, 4, 6, 8, 10 \quad \text{\bf Even numbers less than 12}$$

Subscripts often are used to identify the specific terms of a sequence. For example, $a_1 = 1, a_2 = 4, a_3 = 7$ identifies the first three terms of the sequence

$$1, 4, 7, 10, 13, \cdots, a_n, \cdots$$

The three dots in an infinite sequence indicate that the pattern continues in the same manner indefinitely.

Notice that each positive integer $n$ is associated with the number $a_n$, precisely the characteristic of a function. That is, $n$ represents an element of the domain, and $a_n$ represents an element of the range.

---

**DEFINITION 12.1**   Infinite Sequence

An **infinite sequence** is a function whose domain is the set of positive integers.

---

**TRY THIS**

Some graphers can generate a sequence in the following manner: $1 \to$ X ENTER, $X+2 \to$ X ENTER. Repeated pressing of the ENTER key produces the terms of the sequence. Find the tenth term of this sequence.

## Arithmetic Sequences

There are different types of sequences. In an **arithmetic sequence,** a specific number called the **common difference** $d$ is added to each term to obtain the next term. An arithmetic sequence is determined once the first term $a_1$ and the common difference $d$ are selected. Some examples of arithmetic sequences are the following:

$$1, 3, 5, 7, \cdots \qquad a_1 = 1 \text{ and } d = 2$$
$$5, 10, 15, 20, 25, \cdots \qquad a_1 = 5 \text{ and } d = 5$$
$$12, 10, 8, 6, 4, \cdots \qquad a_1 = 12 \text{ and } d = -2$$

**DEFINITION 12.2** Arithmetic Sequence

A sequence $a_n$ is an **arithmetic sequence** if there is a real number $d$ such that

$$a_n = a_{n-1} + d \qquad n \geq 2.$$

The number $d$ is called the common difference of the arithmetic sequence.

**REMINDER**

The common difference can be obtained by subtracting a term from the succeeding term, that is,

$$a_n - a_{n-1}.$$

Because $a_n$ is defined in terms of the preceding terms, $a_{n-1}$, we say that the sequence $\{a_n\}$ is defined **recursively.** The equation $a_n = a_{n-1} + d$ is called a **recursive equation.**

Consider the following arithmetic sequence defined by $a_1 = 3$ and $d = 2$. The remaining terms of the sequence can be found by using the definition of arithmetic sequence.

$$a_1 = 3$$
$$a_2 = a_1 + 2 = 3 + 2 = 5$$
$$a_3 = a_2 + 2 = 5 + 2 = 7$$
$$a_4 = a_3 + 2 = 7 + 2 = 9$$
$$\vdots$$

**REMINDER**

The sequence $a_1, a_2, a_3, \cdots,$ $a_n$ can be denoted as $\{a_n\}$.

By rewriting this sequence, we can find another pattern, as follows:

$$a_1 = 3$$

$$a_2 = 3 + 2$$ The common difference 2 has been added to the first term one time.

$$a_3 = 3 + \underbrace{2 + 2}_{2}$$ 2 has been added on two times.

$$a_4 = 3 + \underbrace{2 + 2 + 2}_{3}$$ 2 has been added on three times.

$$\vdots$$

$$a_n = a_1 + \underbrace{2 + \cdots + 2}_{n-1}$$ 2 has been added on $n - 1$ times.

**GRAPHER NOTE**

Some graphers can generate sequences by using the command **seq(**. Check your grapher lab manual for more information.

This pattern suggests the following theorem.

> **THEOREM 12.1**  $a_n$ in an Arithmetic Sequence
>
> If $a_n$ is an arithmetic sequence, then
>
> $$a_n = a_1 + (n-1)d,$$
>
> where $n$ is the number of the term, $a_1$ is the first term, and $d$ is the common difference.

Example 1 illustrates finding the $n$th term.

**EXAMPLE 1**    Finding a Particular Term of a Sequence

Find the 6th term of the following arithmetic sequence:

$$23, 26, 29, \cdots$$

| X | Y1 | |
|---|----|---|
| 1 | 23 | |
| 2 | 26 | |
| 3 | 29 | |
| 4 | 32 | |
| 5 | 35 | |
| 6 | 38 | |
| 7 | 41 | |
| Y1 = 23 + (X − 1)3 | | |

**FIGURE 12.1** The grapher-generated table for Example 1 with $y_1 = 23 + (x-1)3$.

SOLUTION

**Solve Algebraically.**  Find $d = 29 - 26 = 26 - 23 = 3$ and $a_1 = 23$. Substitute these values in the formula $a_n = a_1 + (n-1)d$.

$$a_6 = 23 + (5)3$$
$$= 23 + 15$$
$$= 38$$

**Support Numerically.**  Figure 12.1 shows numerical support with a grapher-generated table. The table is generated using $y_1 = 23 + (x-1)(3)$, with $x_1 = 23$. We see that the value of $y_1$ when $x = 6$ is 38.  ∎

Sometimes it is convenient to use a graph to find a particular term in a sequence. Example 2 uses a graph to find the same term in the same sequence; with this method, however, you can find any term in the sequence.

**EXAMPLE 2**    Finding Terms of a Sequence Graphically

Use a graphical method to find the 6th, 30th, and 56th terms of the following arithmetic sequence:

$$23, 26, 29, \cdots$$

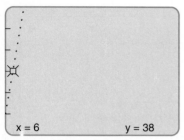

Integer window in
the first quadrant

**FIGURE 12.2** A graph
of $y = 23 + (x - 1)(3)$ in
Example 2.

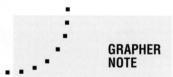

**GRAPHER
NOTE**

When displaying sequences
on a grapher set the mode
to Dot mode instead of
Connected mode. Consult
your grapher lab manual to
find on your grapher the
dimensions for the Integer
window in Example 2.

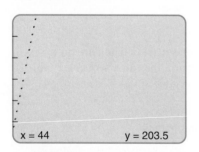

**FIGURE 12.3** A graph
of $y = 10 + (x - 1)(4.5)$ in
Example 3.

SOLUTION

We found in Example 1 that $a_1 = 23$ and $d = 3$. We find a graph of $y_1 = 23 + (x - 1)(3)$ in the Integer window with the origin $(0, 0)$ at the bottom left-hand corner of the grapher screen.

The trace cursor in Figure 12.2 supports that the 6th term of the sequence is 38. Similarly, the trace cursor can be used to find that $y = 110$ when $x = 30$ and $y = 188$ when $x = 56$.

Therefore the 30th and the 56th terms are 110 and 188, respectively. ■

**EXAMPLE 3**    Finding the Number of the Term

In the following arithmetic sequence, which term has the value of 203.5?

$$10, 14.5, 19, 23.5, \cdots$$

SOLUTION

Observe that $a_1 = 10$ and $d = 14.5 - 10 = 4.5$. Graph $y_1 = 10 + (x - 1)(4.5)$ in an Integer window for the first quadrant. Figure 12.3 shows a graph of the sequence.

The TRACE feature (with the trace cursor going off the screen) shows that $x = 44$ when $y = 203.5$.

Therefore 203.5 is the 44th term of the sequence. ■

## Arithmetic Series

When the domain of a sequence consists of only the first $n$ positive integers, we call the sequence a **finite sequence**. With a finite sequence, it is possible to find the sum $a_1 + a_2 + a_3 + \cdots + a_n$. Such a sum is called a **finite series**. However, when the sequence is *infinite*, it is not possible to find the sum by direct addition. Other methods are necessary as illustrated by the discussion of infinite geometric series in Section 12.2.

Consider the finite arithmetic series $1 + 3 + 5 + 7$ and reverse the order of the terms and then add those terms to the original series.

$$
\begin{array}{ccccccc}
1 & + & 3 & + & 5 & + & 7 \\
7 & + & 5 & + & 3 & + & 1 \\
\hline
8 & + & 8 & + & 8 & + & 8
\end{array}
$$

The pattern is clear. The sum of each column is equal to the sum of the first and last term of the sequence, that is, the sum $a_1 + a_n$. Therefore the sum of twice the series is the number of terms in the series, $n$, times the sum $a_1 + a_n$. Consequently, if $S_n = a_1 + a_2 + \cdots + a_n$, then

$$S_n = \frac{n}{2}(a_1 + a_n).$$

This formula can be rewritten as follows:

$$S_n = \frac{n}{2}(a_1 + a_n)$$

$$S_n = \frac{n}{2}[a_1 + a_1 + (n-1)d]$$

$$= \frac{n}{2}[2a_1 + (n-1)d]$$

So there are two formulas that can be used to find the sum of the first $n$ terms of an arithmetic sequence.

---

### The Sum of the First $n$ Terms of an Arithmetic Sequence

The sum of the first $n$ terms of the arithmetic sequence $a_n = a_1 + (n-1)d$ is

$$S_n = \frac{n}{2}[a_1 + a_n]$$

or

$$S_n = \frac{n}{2}[2a_1 + (n-1)d].$$

---

The common difference in an arithmetic sequence can be a positive or a negative value. Example 4 illustrates how a negative value of $d$ affects the values of the individual terms and the sum of the terms.

**EXAMPLE 4**  Finding a Particular Term of a Sequence

Find the 8th term of the following arithmetic sequence.

$$25, 20, 15, 10, \cdots, a_8 \cdots, a_n.$$

Also find a value of $n$ that makes the sum $a_1 + a_2 + a_3 + \cdots + a_n$ the largest value.

SOLUTION
We see that $a_1 = 25$ and $d = -5$. Consequently, the $n$th term of the sequence is defined by the formula $a_n = 25 + (n-1)(-5)$. We evaluate this expression for $n = 8$ to see that

$$a_8 = 25 + (8-1)(-5) = -10.$$

| X | Y1 | |
|---|---|---|
| 1 | 25 | |
| 2 | 45 | |
| 3 | 60 | |
| 4 | 70 | |
| 5 | 75 | |
| 6 | 75 | |
| 7 | 70 | |

Y1 = (X/2)(2(25) + (X − 1)(⁻5))

**FIGURE 12.4** A table of sum values where $X =$ the number of terms and $Y_1$ is the sum.

The sum, $S_n$, is determined by the formula

$$S_n = (n/2)(2(25) + (n-1)(-5)).$$

Find a table of sums $S_n$.

Figure 12.4 shows a table for the values for $x = 1$ to $x = 7$. We see that largest sum is 75 and occurs both when $n = 5$ and when $n = 6$. ■

## Applications of Arithmetic Series

There are often times when you want to know the sum of a sequence. For example, consider the following problem situation.

**TRY THIS**

In the Total Salary Problem Situation, find the percent of the raise for each of the first ten years. What observations can you make regarding this percent?

☐                                                                           ☐

## TOTAL SALARY PROBLEM SITUATION

John is paid a salary of $28,500 for his first year of employment. His employer agrees to increase his salary $3500 at the end of each full year of employment.

☐                                                                           ☐

**EXAMPLE 5**     THE TOTAL SALARY PROBLEM SITUATION: Finding Total Salary

What will John's total earnings be at the end of the following periods:

a) 5 yr
b) 8 yr
c) 20 yr

SOLUTION

The sequence of successive years' salaries is

$$28{,}500, \ 32{,}000, \ 35{,}500, \ 39{,}000, \cdots.$$

The sum $S_n$ of the first $n$ terms of this sequence is

$$S_n = \frac{n}{2}[2(28{,}500) + (n-1)3500].$$

To find these values for $n = 5$, $n = 8$, and $n = 20$, define

$$y_1 = (x/2)(2(28{,}500) + (x-1)3500).$$

Figure 12.5 shows the total salary, $y_1$, evaluated for these three values. John's total earnings will total $177,500 after 5 yr, $326,000 after 8 yr, and $1,235,000 after 20 yr. ■

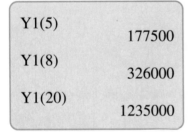

Y1(5)
                177500
Y1(8)
                326000
Y1(20)
                1235000

**FIGURE 12.5**
Values of $y_1 =$ $(x/2)(2(28{,}500) + (x-1)3500)$ for $x = 5$, $x = 8$, and $x = 20$ in Example 5.

## Exercises for Section 12.1

In Exercises 1–6, write the first five terms of the arithmetic sequence with the given first term and common difference.

**1.** $a_1 = 5, d = 2$
**2.** $a_1 = 24, d = 3$

**3.** $a_1 = 35, d = -5$
**4.** $a_1 = -3, d = -2$

**5.** $a_1 = 0.5, d = 1.5$
**6.** $a_1 = -2.5, d = -0.5$

In Exercises 7–12, write the next five terms of the arithmetic sequence.

**7.** $5, 9, 13, \cdots$
**8.** $-9, -2, 5, \cdots$

**9.** $1.5, 3, 4.5, \cdots$
**10.** $-3.2, -3, -2.8, \cdots$

**11.** $x, x + 3, x + 6, \cdots$

**12.** $x, -x, -3x, \cdots$

In Exercises 13–18, find the indicated term of the arithmetic sequence algebraically.

**13.** $a_1 = 10, d = 6$; find $a_7$.

**14.** $a_1 = 12, d = 2$; find $a_{10}$.

**15.** $a_1 = -5, d = -7$; find $a_{12}$.

**16.** $a_1 = -20, d = -4$; find $a_8$.

**17.** $a_1 = 0.5, d = 1.5$; find $a_{20}$.

**18.** $a_1 = -10, d = -2.5$; find $a_{10}$.

In Exercises 19–24, find the indicated term of the arithmetic sequence using a graphing method.

**19.** $12, 20, 28, \cdots$; find $a_{25}$

**20.** $15, 17, 19, \cdots$; find $a_{35}$

**21.** $-5, -13, -21, \cdots$; find $a_{43}$

**22.** $-2, -4, -6, \cdots$; find $a_{65}$

**23.** $2.5, 7, 11.5, \cdots$; find $a_9$

**24.** $-3.8, -6.2, -8.6, \cdots$; find $a_{20}$

In Exercises 25–30, find the indicated sum of the arithmetic sequence algebraically or using a graph and a formula.

**25.** $12, 15, 18, \cdots$; find $S_{25}$.

**26.** $11, 4, -3, \cdots$; find $S_{35}$.

**27.** $-35, -27, -19, \cdots$; find $S_{12}$.

**28.** $-18, -12, -6, \cdots$; find $S_{24}$.

**29.** Find $n$ if $a_1 = 10, a_n = 55, S_n = 325$.

**30.** Find $n$ if $a_1 = 11, d = -7$, and $S_n = -205$.

**31.** Find the 6th term of the arithmetic sequence $-2 + \sqrt{3}, -1, -\sqrt{3}, \cdots$.

**32.** Find $n$ for the sequence for which $a_n = 633$, $a_1 = 9$, and $d = 24$.

**33.** Form an arithmetic sequence in which $a_1 = 36$, $a_3 = 48$, and $a_2$ is the average of $a_1$ and $a_3$.

**34.** Find the 8th term of the arithmetic sequence $1 + i, 2 - i, 3 - 3i, \cdots$.

**35.** For the arithmetic series $30 + 28 + 26 + \cdots$, find the value or values of $n$ when the sum of the first $n$ terms, $S_n$, is the largest.

**36.** For the arithmetic series $16, 13, 10, \cdots$, find the value or values of $n$ when the sum of the first $n$ terms is the largest.

**37.** For the arithmetic series $120 + 100 + 80 + \cdots$, find the value or values of $n$ when the sum of the first $n$ terms is the largest.

**38.** For the arithmetic series $36 + 24 + 12 + \cdots$, find the value or values of $n$ when the sum of the first $n$ terms is the largest.

In Exercises 39–46, find the sum of the sequence of numbers.

**39.** The integers from 1 to 100.

**40.** The odd integers from 1 to 99.

**41.** The even integers from 2 to 100.

**42.** The first 1000 odd integers.

**43.** The first 1000 even integers.

**44.** The multiples of 5 from 50 to 100.

**45.** The integers between 25 and 75.

**46.** The integers between $-100$ and 0.

In Exercises 47 and 48, use the given table to find $a_n$ for the arithmetic sequence represented in the $Y_1$ column.

**47.**

| X | Y1 | |
|---|-----|---|
| 1 | -15 | |
| 2 | -7 | |
| 3 | 1 | |
| 4 | 9 | |
| 5 | 17 | |
| 6 | 25 | |
| 7 | 33 | |
| X = 1 | | |

**48.**

| X | Y1 | |
|---|-----|---|
| 3 | 17 | |
| 4 | 13 | |
| 5 | 9 | |
| 6 | 5 | |
| 7 | 1 | |
| 8 | -3 | |
| 9 | -7 | |
| X = 3 | | |

**49.** *Triangular Display.* Stacey is stacking oranges in a triangle for a window display. There are 21 oranges in the bottom row. How many oranges are in the triangular display?

For Exercise 49.

**50.** *Assessing Gift Money.* Jacqlyn's parents gave her a special birthday bank with $10 in it on her first birthday. On her second birthday, they put in $20, on her third birthday $30, and so forth. No other money was put in the bank. How much money did she have in the bank on her twentieth birthday? (Assuming she counted it after they deposited her twentieth birthday present.)

**51.** *Predicting Total Sales.* Lasertronics, Inc. had sales of $50,000 the first year in business. If its sales increase steadily by $15,000/yr for the first 6 yr, find the amount of sales in the sixth year.

**52.** *Figuring Total Salary.* Jenny earns a salary of $26,000 her first year and receives a guaranteed raise of $1700/yr thereafter. How long will she have to work for the company to double her starting pay?

**53.** *Car Depreciation.* A car cost $15,000 new and depreciates at the rate of $1800/yr for the first 3 yr. How much is the car worth at the end of 3 yr?

**54.** *Computer Depreciation.* A certain computer system costs $28,000 new and depreciates at the rate of $2500/yr for the first 5 yr. How much is the computer worth at the end of 5 yr?

**55. Writing to Learn.** Explain the difference between an infinite and a finite sequence. What is the difference in listing the terms?

**56. Writing to Learn.** What is meant by the common difference in an arithmetic sequence?

**57. Writing to Learn.** Explain how to find a particular term of a sequence using a graphing method.

**58. Writing to Learn.** When finding a particular term by the graphing method, explain how the graph changes when the common difference is negative.

EXTENDING THE IDEAS

In Exercises 59 and 60, answer the questions about the given arithmetic sequence.

**59.** 19, 16, 13, 10, $\cdots$

a) Find the 10th term using a graphing method.

b) Use a graph to find the sum of the first 10 terms.

c) Find how many terms are needed to give the largest sum.

d) Explain what the intersection of the two graphs in parts (a) and (b) tells you about the sequence.

60. $-11, -3, 5, \cdots$

a) Find the 8th term using a graphing method.

b) Use a graph to find the sum of the first six terms.

c) Find how many terms are needed to give the largest sum.

d) Explain what the intersection of the two graphs in parts (a) and (b) tells you about the sequence.

LOOKING BACK—MAKING CONNECTIONS

In Exercises 61–70, simplify the expression. Express all answers with positive exponents only.

61. $x^5 \cdot x^{-9}$

62. $x^8 \cdot x^{-5}$

63. $x^m \cdot x^{-2m}$

64. $x^{-3m} \cdot x^{-2m}$

65. $\dfrac{x^{-4}}{x^{-7}}$

66. $\dfrac{x^{-6}}{x^{-8}}$

67. $\dfrac{x^{-3m}}{x^{-5m}}$

68. $\dfrac{x^{-m}}{x^{-3m}}$

69. $\dfrac{x^{m+1}}{x^{m-1}}$

70. $\dfrac{x^{-m-1}}{x^{-m+2}}$

## 12.2 Geometric Sequences and Series

Geometric Sequences ▪ Alternating Geometric Sequences ▪ Geometric Series ▪ Infinite Geometric Series ▪ Problem Situations Involving Geometric Sequences and Series

### Geometric Sequences

A sequence is called a **geometric sequence** if each each term is obtained by multiplying the previous term by a specific number, called the **common ratio**.

---

**DEFINITION 12.3**   Geometric Sequence

A sequence $a_n$ is called a **geometric sequence** if there is a nonzero real number $r$ such that

$$a_n = r \cdot a_{n-1}, \qquad n \geq 2.$$

The number $r$ is called the **common ratio** of the geometric sequence.

---

Notice $a_n$ is defined in the terms of the preceding term, $a_{n-1}$. Therefore, the sequence $\{a_n\}$ is defined recursively. The equation $a_n = r \cdot a_{n-1}$ is a recursive equation.

Here are some examples of geometric sequences.

$$1, 2, 4, 8, \cdots \qquad a_1 = 1 \text{ and } r = 2$$

$$9, -27, 81, -343, \cdots \qquad a_1 = 9 \text{ and } r = -3$$

$$\frac{1}{2}, \frac{1}{4}, \frac{1}{8}, \frac{1}{16}, \frac{1}{32} \qquad a_1 = \tfrac{1}{2} \text{ and } r = \tfrac{1}{2}$$

$$0.1, 0.05, 0.025, 0.0125 \qquad a_1 = 0.1 \text{ and } r = 0.5$$

Consider the following geometric sequence defined by $a_1 = 2$ and $r = 3$:

$$a_1 = 2$$

$$a_2 = r \cdot a_1 = 3 \cdot 2 = 6$$

$$a_3 = r \cdot a_2 = 3 \cdot 6 = 18$$

$$\vdots$$

By rewriting this sequence, we can find another pattern, as follows:

$$a_1 = 2$$

$$a_2 = 2 \cdot 3 \qquad \text{The first term is multiplied by the common ratio 3 one time.}$$

$$a_3 = 2 \cdot \underbrace{3 \cdot 3}_{2 \text{ factors}} \qquad \text{The first term is multiplied by the common ratio 3 two times.}$$

$$a_4 = 2 \cdot \underbrace{3 \cdot 3 \cdot 3}_{2 \text{ factors}} \qquad \text{The first term is multiplied by the common ratio 3 three times.}$$

$$\vdots$$

$$a_n = a_1 \cdot \underbrace{3 \cdot 3 \cdots 3}_{n-1 \text{ factors}} \qquad \text{The first term is multiplied by the common ratio } n-1 \text{ times.}$$

This pattern suggests the following theorem.

---

**THEOREM 12.2**  A Term $a_n$ in a Geometric Sequence

If $a_n$ is a geometric sequence with common ratio $r$, then

$$a_n = a_1 \cdot r^{n-1},$$

where $n$ is the number of the term, $a_1$ is the first term, and $r$ is the common ratio.

Example 1 shows a sequence when the first term and common ratio is known.

### EXAMPLE 1    Finding a Geometric Sequence

Write the first five terms of a geometric sequence if the first term is 2 and the common ratio is $-3$.

SOLUTION

If $a_1 = 2$ and $r = -3$, then the following results:

$$a_1 = 2$$
$$a_2 = r \cdot a_1$$
$$= -3 \cdot 2 = -6$$
$$a_3 = r \cdot a_2$$
$$= -3 \cdot -6 = 18$$
$$a_4 = r \cdot a_3$$
$$= -3 \cdot 18 = -54$$
$$a_5 = r \cdot a_4$$
$$= -3 \cdot -54 = 162$$

The first five terms of the sequence are $2, -6, 18, -54,$ and $162$. ∎

Example 2 illustrates finding a geometric sequence given two terms in the sequence.

### EXAMPLE 2    Finding the First and $n$th Term

The second and third terms of a geometric sequence are 20 and 80, respectively. Determine the first term and a formula for the $n$th term.

SOLUTION

The numbers 20 and 80 are the second and third numbers in the sequence. The common ratio can be found by dividing a term by the preceding term.

$$r = \frac{a_3}{a_2} = \frac{80}{20} = 4$$

Because $a_2 = r \cdot a_1$, we can solve for $a_1$ to obtain

$$a_1 = \frac{a_2}{r} = \frac{20}{4} = 5.$$

Therefore, using Theorem 12.2, we can conclude that the nth term is

$$a_n = 5(4)^{n-1}.$$

The sequence is $5, 20, 80, \cdots, 5(4)^{n-1}, \cdots$. ∎

---

**TRY THIS**

Some graphers can generate this geometric sequence by doing the following:

2 → X ENTER,

−3X → X ENTER.

By repeatedly pressing ENTER, successive terms of the sequence can be found. Find the sixth term of this sequence.

## Alternating Geometric Sequence

When a geometric sequence has terms that alternate between positive and negative values, the sequence is called an **alternating sequence**. For an alternating sequence, the common ratio is a negative number.

Example 3 illustrates finding a specific term in an alternating sequence.

**EXAMPLE 3**   Finding a Term in an Alternating Sequence

Use the TRACE feature of a grapher to find the 7th and 10th term of the following geometric sequence. Show numerical support.

$$3, -6, 12, -24, \cdots$$

SOLUTION

**Solve Graphically.** The common ratio is found by considering the ratio of successive terms of the sequence. For this sequence, the ratio $r$ is

$$r = \frac{-24}{12} = \frac{12}{-6} = \frac{-6}{3} = -2.$$

Because $a_1 = 3$ and $r = -2$, the sequence is defined by

$$a_n = 3(-2)^{n-1}.$$

Figure 12.6 shows the graph of $y_1 = 3(-2)^{x-1}$ in the Integer window for the first and fourth quadrant with vertical dimensions $[-1000, 1000]$. (Note, the grapher must be in Dot mode or strange-looking phenomena can occur.)

Use the TRACE feature to find that $y = 192$ when $x = 7$. Therefore the 7th term is 192. We also find $y = -1536$ when $x = 10$. Thus the 10th term is $-1536$.

**Support Numerically.** Figure 12.6(b) shows a table showing the 5th to 11th terms of the sequence. This table supports that $a_7 = 192$ and $a_{10} = -1536$.   ■

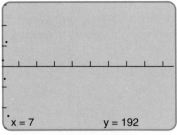

| X | Y1 | |
|---|---|---|
| 5 | 48 | |
| 6 | -96 | |
| 7 | 192 | |
| 8 | -384 | |
| 9 | 768 | |
| 10 | -1536 | |
| 11 | 3072 | |

x = 7          y = 192          Y1 = 3(-2)^(x − 1)

An Integer window

(a)                                        (b)

**FIGURE 12.6**  (a) A graph of $y = 3(-2)^{x-1}$ and (b) a table showing terms of the sequence in Example 3.

In Example 3, we used a graph to find the individual terms of a geometric sequence. Example 4 illustrates finding a term using an algebraic method.

**EXAMPLE 4**    Finding a Term in a Geometric Sequence

Use an algebraic method to find the 8th term of the following geometric sequence. Support the solution numerically.

$$5, 10, 20, 40, \cdots$$

| X | Y1 | |
|---|----|---|
| 2 | 10 | |
| 3 | 20 | |
| 4 | 40 | |
| 5 | 80 | |
| 6 | 160 | |
| 7 | 320 | |
| 8 | 640 | |
| Y1 = 5(2)^(X − 1) | | |

**FIGURE 12.7** A table showing support for Example 4.

SOLUTION

**Solve Algebraically.** We see that $a_1 = 5$ and $r = 2$. Therefore the $n$th term of the sequence is defined by $a_n = a_1 \cdot r^{n-1}$.

Then

$$a_8 = 5 \cdot 2^7$$
$$= 5 \cdot 128$$
$$= 640.$$

**Support Numerically.** Figure 12.7 gives a table of values showing that $a_8 = 640$. ■

It often is tedious to find the terms of a sequence using paper-and-pencil algebraic methods. Sometimes it is more convenient to evaluate a defining expression using a grapher. Example 5 illustrates.

**EXAMPLE 5**    Finding the $n$th Term in a Geometric Sequence

Find a formula for the $n$th term, $a_n$, for the following geometric sequence. Use this expression to find $a_{20}$.

$$1, -2, 4, -8, \cdots$$

Y1(20)

⁻524288

FIGURE 12.8
Evaluating the expression
$y_1 = 1(-2)^{x-1}$ for $x = 20$ in Example 5.

SOLUTION

We see that $a_1 = 1$, and comparing successive terms of the sequence, we find $r$,

$$r = \frac{-8}{4} = -2$$

so $a_n = a_1 \cdot r^{n-1} = 1(-2)^{n-1}$. Figure 12.8 shows the value of $y_1 = 1(-2)^{x-1}$ for $x = 20$.

Therefore $a_{20} = -524{,}288$. ■

Example 6 shows an algebriac method that can be used to find the terms of a geometric sequence whenever two terms of the sequence are known.

**EXAMPLE 6**    Finding a Term in a Geometric Sequence

Find the first eight terms of the geometric sequence for which $a_3 = 4$ and $a_8 = \dfrac{1}{256}$.

SOLUTION

The first eight terms of this geometric sequence fit the pattern

$$\_\_, \_\_, 4, \_\_, \_\_, \_\_, \_\_, \frac{1}{256}.$$

Deleting the first two terms we have another geometric sequence that begins $4, \_\_, \_\_, \_\_, \_\_, \dfrac{1}{256}$ with the same common ratio. For this second sequence, $a_1 = 4$ and $a_6 = \dfrac{1}{256}$ in the formula $a_n = a_1 \cdot r^{n-1}$.

$$\frac{1}{256} = 4r^5 \qquad \text{Multiply each side of the equation by } \tfrac{1}{4}\text{:}$$

$$\frac{1}{4} \cdot \frac{1}{256} = \frac{1}{4} \cdot 4r^5$$

$$\frac{1}{1024} = r^5 \qquad \text{Take the fifth root of each side of the equation.}$$

$$\frac{1}{4} = r$$

The original sequence now can be completed by finding $a_1$ from $a_3$. We know that $r = \dfrac{1}{4}$ and

$$a_3 = a_1 r^2 = 4 \qquad \text{Replace } r \text{ with } \tfrac{1}{4}.$$

$$a_1 \left( \frac{1}{16} \right) = 4$$

$$a_1 = 64 \qquad \text{Find } a_2.$$

$$a_2 = 64 \left( \frac{1}{4} \right) = 16$$

Therefore, the original sequence is

$$64, 16, 4, 1, \frac{1}{4}, \frac{1}{16}, \frac{1}{64}, \frac{1}{256} \qquad \blacksquare$$

## Geometric Series

Sometimes the solution of a problem requires that we find the sum of the first $n$ terms of the geometric sequence

$$a_1, a_1 r, a_1 r^2, a_1 r^3, \cdots, a_1 r^{n-1}.$$

Our task is to find a formula for the sum, $S_n$, of this finite series

$$S_n = a_1 + a_1 r + a_1 r^2 + a_1 r^3 + \cdots + a_1 r^{n-1}.$$

The following algebraic argument derives such a formula:

$$S_n = a_1 + a_1 r + a_1 r^2 + a_1 r^3 + \cdots + a_1 r^{n-1}$$

    **Multiply each side of the equation by $r$.**

$$r S_n = \phantom{a_1 +} a_1 r + a_1 r^2 + a_1 r^3 + \cdots + a_1 r^{n-1} + a_1 r^n$$

    **Subtract the two equations.**

    **Factor out $S_n$ from the left-hand side and $a_1$ from the right-hand side of the equation.**

$$S_n - r S_n = a_1 - a_1 r^n$$

$$S_n(1 - r) = a_1 - a_1 r^n$$

$$S_n = \frac{a_1(1 - r^n)}{1 - r}, r \neq 1$$

We summarize this formula as follows.

---

### The Sum of the First $n$ Terms of a Geometric Sequence

If $S_n$ is the sum of the first $n$ terms of the geometric sequence

$$S_n = a_1 + a_1 r + a_1 r^2 + \cdots + a_1 r^{n-1},$$

then

$$S_n = \frac{a_1(1 - r^n)}{1 - r}, r \neq 1.$$

---

Example 7 uses this formula to find the sum $S_8$ given a term and the common ratio.

**EXAMPLE 7**   Finding a Sum of a Finite Geometric Sequence

Find $S_8$ for the geometric sequence $10, 5, 2.5, \cdots$. Show numerical support for the answer.

SOLUTION

**Solve Algebraically.** Use the formula $S_n = a_1 \dfrac{1 - r^n}{1 - r}$ with $a_1 = 10$ and $r = \dfrac{1}{2}$.

$$S_8 = 10\left(\frac{1 - 0.5^8}{1 - 0.5}\right)$$

$$= 19.921875$$

| X | Y1 | |
|---|---|---|
| 2 | 15 | |
| 3 | 17.5 | |
| 4 | 18.75 | |
| 5 | 19.375 | |
| 6 | 19.688 | |
| 7 | 19.844 | |
| 8 | 19.922 | |

Y1 = 10((1 − .5^X)/(1 − .5))

**FIGURE 12.9** A table showing support for the solution in Example 7.

**Support Numerically.** Figure 12.9 shows a table with $y_1 = 10(1 − 0.5^x)/(1 − 0.5)$ supporting the sum of the first eight terms as being 19.921875. ∎

## Infinite Geometric Series

An infinite sum like

$$a_1 + a_2 r + a_3 r^2 + \cdots + a_n r^{n-1} + \cdots$$

is called an **infinite geometric series**. Its sum cannot be evaluated by direct addition because it is not possible to add an infinite number of terms. However, as shown above

$$S_n = a_1 + a_2 r + \cdots + a_1 r^{n-1}$$

$$= \frac{a_1(1 − r^n)}{1 − r}$$

The following exploration uses this formula to examine what happens as $n$ gets larger.

### EXPLORE WITH A GRAPHER

Investigate the sum $S_n = a_1 + a_2 r + a_3 r^2 + \cdots + a_n r^{n-1}$ for increasing values of $n$, where $a_1 = 12$ and $r = 0.5$, by graphing the expression

$$y_1 = \frac{a_1(1 − r^n)}{1 − r}$$

in Dot mode, replacing $n$ with $x$. (Use an Integer window in the first quadrant; on the grapher we use, this window is [0, 94] by [0, 62].)

Use the TRACE feature to do the following:

1. Find $S_4$, $S_5$, and $S_6$.
2. Find the value of $n$ for which $S_n = 23.99707$.
3. Find the smallest three consecutive values of $n$ that appear to have equal values for $S_n$.
4. As $n$ gets larger, $S_n$ gets closer to a specific value. What is that value?

Repeat steps 3 and 4 for $r = 0.7$ and $r = 0.8$. What general patterns do you observe?

The above exploration helps us to visualize that as $n$ gets larger, the sum

$$S_n = 12 + 12\left(\frac{1}{2}\right) + 12\left(\frac{1}{4}\right) + \cdots + 12\left(\frac{1}{2}\right)^{n-1}$$

approaches the value 24.

In general if $0 < |r| < 1$, the sum $S_n$ approaches a specific value as $n$ gets larger. We summarize as follows.

---

## Infinite Geometric Series

If $|r| < 1$, then the sum $S$ of an infinite geometric series with the first term $a_1$ and the common ratio $r$ is

$$S = \frac{a_1}{1 - r}.$$

If $|r| \geq 1$, then the sum does not exist.

---

In Example 8, the common ratio is $\frac{1}{5}$, so the sum exists.

**EXAMPLE 8**    Finding a Sum of an Infinite Geometric Sequence

Find the sum, if it exists, of the following infinite geometric series:

$$75 + 15 + 3 + 0.6 + \cdots$$

SOLUTION
The ratio $r$ of consecutive terms is

$$\frac{0.6}{3} = \frac{3}{15} = \frac{15}{75} = \frac{1}{5}.$$

Therefore, the infinite sum exists. Here $a_1 = 75$. To find the sum, we evaluate the following:

$$S = \frac{75}{1 - \frac{1}{5}}$$

$$= \frac{75}{\frac{4}{5}}$$

$$= 93.75$$

Therefore the sum of the infinite series $75 + 15 + 3 + 0.6 + \cdots$ is 93.75.

$$75 \rightarrow A: 1/5 \rightarrow R: Y1$$
$$93.75$$

**FIGURE 12.10**
Numerical support for the
solution in Example 8.

**Support Numerically.** Figure 12.10 shows this computation done on a grapher
with $y_1 = a/(1 - r)$. ■

## Problem Situations Involving Geometric Sequences and Series

---

### CONTRACT NEGOTIATIONS PROBLEM SITUATION

The country music group, Nashville Alive, agrees to a contract by which
they make \$60,000 for their first recording and half as much on each
succeeding release of that recording.

---

**EXAMPLE 9**    THE CONTRACT NEGOTIATIONS PROBLEM SITUATION:
Finding Individual and Total Royalties

How much money does the group Nashville Alive make on the fifth release of
their first recording in the **Contract Negotiations Problem Situation?** What is
the total amount they have made in the five releases?

SOLUTION

**Solve Algebraically.** Because $a_1 = 60{,}000$ and $r = \dfrac{1}{2}$, we find

$$a_5 = 60{,}000 \left(\frac{1}{2}\right)^4$$

$$= 3750$$

Therefore, they will make \$3750 on their fifth recording.

The total amount of money made for the 5 releases is the sum

$$S_5 = 60{,}000 + 60{,}000 \left(\frac{1}{2}\right) + \cdots + 60{,}000 \left(\frac{1}{2}\right)^4.$$

We find this sum by evaluating

$$\frac{60{,}000(1 - 0.5^5)}{1 - 0.5} = 116{,}250.$$

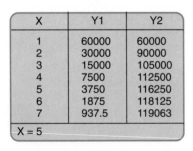

| X | Y1 | Y2 |
|---|------|--------|
| 1 | 60000 | 60000 |
| 2 | 30000 | 90000 |
| 3 | 15000 | 105000 |
| 4 | 7500 | 112500 |
| 5 | 3750 | 116250 |
| 6 | 1875 | 118125 |
| 7 | 937.5 | 119063 |

X = 5

**FIGURE 12.11**
Numerical support for the
solution in Example 9.

**Support Numerically.** Figure 12.11 shows numerical support using a grapher-
generated table where $y_1 = 60000(1/2)^{x-1}$ and $y_2 = 60000(1 - 0.5^x)/(1 - 0.5)$.

The fifth release produces an income of \$3750 and the total amount earned for
all five releases of their first recording is \$116,250. ■

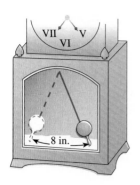

**FIGURE 12.12** The pendulum in Example 10. The swing of the pendulum traces an arc of 8 in. in the first swing.

---

☐            ☐

### THE PENDULUM PROBLEM SITUATION

Each swing of a pendulum is 75% of its previous swing. For example, if the tip traces an arc of 8 in. in the first swing to the right, it will swing through a 6-in. arc on the swing back to the left.

☐            ☐

---

Example 10 looks at the swings of a pendulum.

**EXAMPLE 10**    THE PENDULUM PROBLEM SITUATION: Finding the Distance a Pendulum Travels

Determine the total distance the pendulum has traveled after 10 swings if the tip traces an arc of 8 in. in the first swing (see Figure 12.12). After how many swings does the pendulum appear to stop?

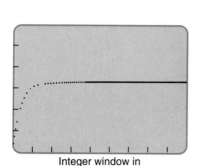

Integer window in
the first quadrant

**FIGURE 12.13** A graph of the sum of the terms of the series in Example 10.

SOLUTION

**Solve Graphically.** The desired sum is $8 + 8(0.75) + 8(0.75)^2 + \cdots + 8(0.75)^9$. Thus, we need to find $S_{10}$ for the sequence with $a_1 = 8$ and $r = 0.75$.

1. Graph $y_1 = 8(1 - 0.75^x)/(1 - 0.75)$ in an Integer window in the first quadrant.
2. Use the TRACE feature to find $y = 30.197968$ when $x = 10$. So, after 10 swings the total distance traveled is 30.198 in.
3. Move the trace cursor to the right and observe that after $x = 63$, the sum remains 32, indicating that the total distance traveled is unchanged. Therefore the pendulum appears to have stopped, or is moving only an insignificant amount, after 63 swings.

Figure 12.13 shows a graph of the sum of the series with $y_1 = 8(1 - 0.75^x)/(1 - 0.75)$. The sum is 32 in. after 63 swings of the pendulum. There is actually insignificant change after the 29th swing, when the sum is 31.99238. ■

---

## Exercises for Section   **12.2**

In Exercises 1–6, write the first five terms of the geometric sequence with the given first term and common ratio.

1. $a_1 = -5, r = 2$      2. $a_1 = 3, r = 4$

3. $a_1 = 3, r = -0.5$      4. $a_1 = -3, r = -2$

5. $a_1 = 0.5, r = 0.5$      6. $a_1 = -2.5, r = -5$

In Exercises 7–12, write the next five terms of the geometric sequence.

**7.** $16, 8, 4, \cdots$

**8.** $-9, 3, -1, \cdots$

**9.** $\dfrac{1}{2}, \dfrac{1}{4}, \dfrac{1}{8}, \cdots$

**10.** $9, 6, 4, \cdots$

**11.** $x, x^2 + x, x(x+1)^2 \cdots$

**12.** $x, -1, \dfrac{1}{x}, \cdots$

In Exercises 13–18, find the indicated term of the geometric sequence algebraically and support numerically.

**13.** $a_1 = 5$, $r = 3$; find $a_8$.

**14.** $a_1 = 8$, $r = 2$; find $a_{20}$.

**15.** $a_1 = -5$, $r = -2$; find $a_{12}$.

**16.** $a_1 = -2$, $r = -4$; find $a_{15}$.

**17.** $a_1 = 0.5$, $r = -0.2$; find $a_{20}$.

**18.** $a_1 = -10$, $r = -0.5$; find $a_{10}$.

In Exercises 19–24, find the indicated term of the geometric sequence using a graphing method and support numerically.

**19.** $4, 2, 1, \cdots$ ; find $a_{25}$

**20.** $6, 30, 150, \cdots$ ; find $a_{35}$

**21.** $-5, 2.5, -1.25, \cdots$ ; find $a_{43}$

**22.** $-2, 4, -8, \cdots$ ; find $a_{15}$

**23.** $2, 3, 4.5, \cdots$ ; find $a_{12}$

**24.** $-3.5, 7, -14, \cdots$ ; find $a_{20}$

In Exercises 25 and 26, use the information in the given table to find $Y_1$, where $Y_1$ is the $n$th term of a geometric sequence.

**25.**

| X | Y1 | |
|---|---|---|
| 3 | 48 | |
| 4 | 192 | |
| 5 | 768 | |
| 6 | 3072 | |
| 7 | 12288 | |
| 8 | 49152 | |
| 9 | 196608 | |
| X = 3 | | |

**26.**

| X | Y1 | |
|---|---|---|
| 1 | .25 | |
| 2 | .125 | |
| 3 | .0625 | |
| 4 | .03125 | |
| 5 | .01563 | |
| 6 | .00781 | |
| 7 | .00391 | |
| X = 1 | | |

In Exercises 27–30, find the indicated sum of the geometric sequence either algebraically or using a graph.

**27.** $12 + 6 + 3 + \cdots$ ; find $S_{25}$.

**28.** $-2 + 2 + (-2) + 2 + \cdots$ ; find $S_{35}$.

**29.** $5 + (-10) + 20 + (-40) + \cdots$ ; find $S_{12}$.

**30.** $-4 + (-24) + (-144) + (-864) + \cdots$ ; find $S_9$.

In Exercises 31–34, find the sum, if it exists, of each of the following infinite geometric series. If the sum does not exist, so state.

**31.** $12 + (-6) + 3 + (-1.5) + \cdots$

**32.** $2 + 6 + 18 + \cdots$

**33.** $0.5 + 0.05 + 0.005 + 0.0005 + \cdots$

**34.** $14 + 7 + 3.5 + \cdots$

**35.** If the first term of a geometric sequence is 12 and the common ratio is $\dfrac{3}{2}$, find the next five terms.

**36.** Find the 10th term of the geometric sequence $\sqrt{3}, 3, 3\sqrt{3}, \cdots$.

**37.** Find the sum of the first 10 terms of the geometric series $\dfrac{1}{2} + \dfrac{1}{4} + \dfrac{1}{8} + \cdots$.

**38.** Find the sum of the first six terms of the geometric series $\dfrac{5}{3} + \dfrac{5}{9} + \dfrac{5}{27} + \cdots$.

**39.** Find the sum of the first 10 terms of the geometric series $\sqrt{5} + 5 + 5\sqrt{5} + \cdots$.

**40.** *Height of a Bouncing Ball.* A ball is dropped from the top of a building that is 75 ft above the ground.

On each bounce, it rebounds $\frac{2}{3}$ of the height it fell. How high is the ball at the top of the sixth bounce?

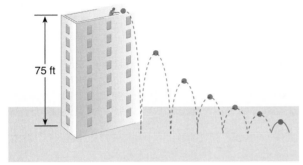

For Exercise 40.

**41.** *Finding the Total Amount of Gift Money.* Nicole's grandmother promises to pay her $1 on the first day of December and double the money every day until Christmas.

a) How much does Nicole receive on Christmas Day?

b) What is the total amount of money her grand-mother has given her, including the Christmas Day gift?

**42.** *Monetary Incentive.* Nathaniel has a rich uncle who is trying to motivate him to lose weight. The uncle tells Nathaniel that he will give him $1 on the first day of June and double the amount each day if he stays on the diet for a month, or he will give him $100 the first day and increase the amount by $25/day for the month of June. Which plan should Nathaniel agree to in order to acquire the most money?

**43.** *Predicting the Total Number of Customers.* A utility company has 30,000 customers and expects to have a 5% increase in customers each year for the next 6 yr. How many customers will the company have at the end of the sixth year?

**44.** *Car Depreciation.* Jean-Paul bought a new car costing $16,000. It depreciates 20% per year. How much is the car worth after 5 yr? (*Hint:* If it depreciates 20% per year, it is worth 80% of its value for the previous year.)

**45.** *Predicting Half-Life.* A radioactive substance has a half-life of 800 yr. If 320 g are present today, how much substance will be present in 4000 yr?

**46.** *Bacterial Growth.* A certain bacterium doubles every 3 hrs. If 200 bacteria were present to begin with, how many bacteria are present after 48 hrs?

**47.** *Arc Length of a Pendulum.* The length of the arc through which a pendulum swings is 60% of its previous swing. If the length of the arc of the original swing is 60 cm, find

a) the length of the arc of the fourth swing and

b) the total distance the tip of the pendulum travels after the fourth swing.

**48. Writing to Learn.** Explain how to find the common ratio of a geometric sequence or series.

**49. Writing to Learn.** Explain how it is possible to find the sum of an infinite geometric series.

**50. Writing to Learn.** Explain why it is impossible to find the sum of an infinite geometric series if the common ratio is greater than 1.

EXTENDING THE IDEAS

**51.** Find the sum of the geometric series $1 + 3 + 9 + 27 + \cdots + 2187$ and the number of terms in the series.

**52.** Find the sum of the geometric series $5 + 10 + 20 + 40 + \cdots + 327,680$ and the number of terms in the series.

**53.** Graph the sum of the geometric series $6 + 3 + 1.5 + 0.75 + \cdots$ and explain where there appears to be a horizontal asymptote. Why does this asymptote exist? (*Hint:* This is best seen using connected mode.)

LOOKING BACK—MAKING CONNECTIONS

In Exercises 54–61, expand or simplify the expression, whichever is appropriate.

**54.** $(5x - 3)(5x + 3)$       **55.** $(3x - y)(3x + y)$       **58.** $(4x - 5)^2$       **59.** $(7x + 1)^2$

**56.** $(x - 7)^2$       **57.** $(x + 12)^2$       **60.** $[(x + y) - 3]^2$       **61.** $[(x + 2y) + 4]^2$

---

## 12.3 Binomial Expansion

Using a Pattern • Binomial Coefficients

### Using a Pattern

In Chapter 5, you learned to use a pattern to find the square of a binomial. In this section, we examine a pattern that can be used to find a binomial raised to any power.

To begin we consider a binomial raised to the following powers:

$$n = 0 : \quad (x + y)^0 = 1$$

$$n = 1 : \quad (x + y)^1 = x + y$$

$$n = 2 : \quad (x + y)^2 = x^2 + 2xy + y^2$$

$$n = 3 : \quad (x + y)^3 = x^3 + 3x^2y + 3xy^2 + y^3$$

$$n = 4 : \quad (x + y)^4 = x^4 + 4x^3y + 6x^2y^2 + 4xy^3 + y^4$$

$$n = 5 : \quad (x + y)^5 = x^5 + 5x^4y + 10x^3y^2 + 10x^2y^3 + 5xy^4 + y^5$$

Observe that for each value of $n$, each of the following statements is true:

1. There are $n + 1$ terms.
2. The sum of the exponents of each term is $n$.
3. The first term is $x^n$ and the powers of $x$ decrease by 1 in each term.
4. The last term is $y^n$ and the powers of $y$ decrease by 1 from the last to the first term, $x^n y^0$.
5. There is symmetry among the coefficients.
6. The pattern is as follows:

$$(x + y)^n = x^n + \underline{?}\ x^{n-1}y^1 + \underline{?}\ x^{n-2}y^2 + \cdots + \underline{?}\ xy^{n-1} + y^n$$

The coefficients of each term that should replace the question marks can be found in the rows of the famous triangle of numbers known as **Pascal's Triangle**, shown next.

| | | | | | | | | | | | |
|---|---|---|---|---|---|---|---|---|---|---|---|
| $n = 0$ | | | | | | 1 | | | | | |
| $n = 1$ | | | | | 1 | | 1 | | | | Coefficients of $(x + y)^1$ |
| $n = 2$ | | | | 1 | | 2 | | 1 | | | Coefficients of $(x + y)^2$ |
| $n = 3$ | | | 1 | | 3 | | 3 | | 1 | | Coefficients of $(x + y)^3$ |
| $n = 4$ | | 1 | | 4 | | 6 | | 4 | | 1 | Coefficients of $(x + y)^4$ |
| $n = 5$ | 1 | | 5 | | 10 | | 10 | | 5 | 1 | Coefficients of $(x + y)^5$ |

Notice that each line starts with a 1 and that each entry of a line is the sum of the two entries above it in the previous line. Following this pattern, the coefficients for $(x + y)^6$ would be

$$1 \quad 6 \quad 15 \quad 20 \quad 15 \quad 6 \quad 1.$$

Example 1 illustrates how Pascal's Triangle can be used.

**EXAMPLE 1**    Finding a Binomial Expansion Using Pascal's Triangle
Expand $(x + y)^6$.

SOLUTION

1. Begin by finding the terms without the coefficients.

$$(x + y)^6 = x^6 + \underline{?} \ x^5 y^1 + \underline{?} \ x^4 y^2 + \underline{?} \ x^3 y^3 + \underline{?} \ x^2 y^4 + \underline{?} \ xy^5 + y^6$$

2. Fill in the blanks using coefficients from the seventh line of Pascal's Triangle.

$$(x + y)^6 = 1x^6 + 6x^5 y^1 + 15x^4 y^2 + 20x^3 y^3 + 15x^2 y^4 + 6xy^5 + 1y^6$$

Therefore the final expression is

$$(x + y)^6 = x^6 + 6x^5 y + 15x^4 y^2 + 20x^3 y^3 + 15x^2 y^4 + 6xy^5 + y^6. \qquad \blacksquare$$

Finding a power of a binomial that is a difference like $(x - 2)$ can be done by changing it to a sum: $x - 2 = x + (-2)$. For example, $(x - 2)^3$ can be found as follows:

$$(x - 2)^3 = (x + (-2))^3$$

$$= x^3 + 3x^2(-2) + 3x(-2)^2 + (-2)^3$$

$$= x^3 - 6x^2 + 12x - 8$$

The outcome is the same as $(x+2)^3$, except that the signs of the terms alternate. We use this pattern of alternating signs in Example 2.

**EXAMPLE 2** Finding a Binomial Expansion

Expand $(x - 3)^4$. Support the answer graphically.

SOLUTION

1. Begin by finding the terms without the coefficients.

$$(x - 3)^4 = x^4 - \underline{?}\ x^3(3)^1 + \underline{?}\ x^2(3)^2 - \underline{?}\ x(3)^3 + (3)^4$$

2. The coefficients are obtained from the fourth line of Pascal's Triangle.

$$1 \quad 4 \quad 6 \quad 4 \quad 1$$

Thus

$$(x - 3)^4 = 1x^4 - 4x^3(3)^1 + 6x^2(3)^2 - 4x(3)^3 + 1(3)^4.$$

3. Simplifying each term, we have

$$(x - 3)^4 = x^4 - 4x^3(3) + 6x^2(9) - 4x(27) + 81$$

$$= x^4 - 12x^3 + 54x^2 - 108x + 81.$$

**Support Graphically.** Graph $y_1 = (x - 3)^4$ and $y_2 = x^4 - 12x^3 + 54x^2 - 108x + 81$. Figure 12.14 shows that the graphs appear to be identical and therefore support the solution. ■

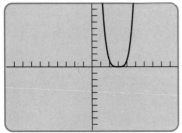

Standard window

**FIGURE 12.14** A graph of $y_1 = (x - 3)^4$ and $y_2 = x^4 - 12x^3 + 54x^2 - 108x + 81$.

**GRAPHER NOTE**

Consult your grapher lab manual for the location of the factorial on your grapher.

FACTORIAL NOTATION

Pascal's Triangle is easy to remember and therefore provides a quick and efficient way to find the coefficients when the powers are small. For larger powers of $(x + y)^n$, the coefficients can be obtained with a formula involving **factorial notation** denoted $n!$.

---

■                                                  ■

**DEFINITION 12.4** Factorial Notation $n!$

If $n$ is any positive integer, then

$$n! = n(n - 1)(n - 2) \cdots 3 \cdot 2 \cdot 1.$$

We define $0! = 1$.

■                                                  ■

---

Some examples of factorials are

$$3! = 3 \cdot 2 \cdot 1 \quad \text{and} \quad 5! = 5 \cdot 4 \cdot 3 \cdot 2 \cdot 1.$$

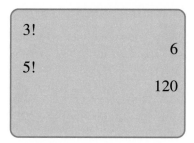

**FIGURE 12.15**
Evaluating factorials using a grapher.

Figure 12.15 shows evaluation of a factorial using a grapher. It is important to consult your grapher lab manual to learn how to evaluate factorials.

**EXAMPLE 3**   Evaluating a Factorial Expression

Evaluate each of the following expressions using a paper-and-pencil method and support using a grapher.

a) $\dfrac{6!}{2!}$

b) $\dfrac{8! \cdot 5!}{4! \cdot 3!}$

SOLUTION
**Paper-and-Pencil Method.**

a) $\dfrac{6!}{2!} = \dfrac{6 \cdot 5 \cdot 4 \cdot 3 \cdot 2 \cdot 1}{2 \cdot 1} = 6 \cdot 5 \cdot 4 \cdot 3 = 360$

b) $\dfrac{8! \cdot 5!}{4! \cdot 3!} = \dfrac{8 \cdot 7 \cdot 6 \cdot 5 \cdot 4! \cdot 5 \cdot 4 \cdot 3!}{4! \cdot 3!} = 8 \cdot 7 \cdot 6 \cdot 5 \cdot 5 \cdot 4 = 33,600$

**Support Numerically.** Figure 12.16 shows numerical support for the solutions using a grapher. ∎

```
6!/2!
                360
8! 5!/(4! 3!)
              33600
```

**FIGURE 12.16**
Numerical support for the solutions to Example 3.

## Binomial Coefficients

We use the notation $\binom{n}{r}$ to represent a specific number that depends on the values of $n$ and $r$. This number turns out to be related to both binomial expansions and counting methods that will be discussed in the next section.

We begin by defining the symbol $\binom{n}{r}$, which can also be written $_nC_r$.

**GRAPHER NOTE**

Most graphers can calculate the binomial coefficient directly. Refer to your grapher lab manual to confirm the calculations in Example 4.

---

**DEFINITION 12.5**   Binomial Coefficient $\binom{n}{r}$

If $n$ and $r$ are two nonnegative integers, $r \le n$, the number called the **binomial coefficient $n$ choose $r$**, denoted $\binom{n}{r}$, is defined by

$$_nC_r = \binom{n}{r} = \frac{n!}{r!(n-r)!}.$$

---

Example 4 illustrates how to compute binomial coefficients using a paper-and-pencil method and how to support numerically on a grapher using the $_nC_r$ notation.

**EXAMPLE 4**   Finding the Binomial Coefficients

Find each of the following binomial coefficients by paper-and-pencil method and support using a grapher:

a) $\dbinom{5}{3}$

b) $\dbinom{9}{5}$

SOLUTION

a) $\dbinom{5}{3} = \dfrac{5!}{3!(5-3)!} = \dfrac{5 \cdot 4 \cdot 3!}{3! \cdot 2!} = 10$

b) $\dbinom{9}{5} = \dfrac{9!}{5!(9-5)!} = \dfrac{9 \cdot 8 \cdot 7 \cdot 6 \cdot 5!}{5! \cdot 4 \cdot 3 \cdot 2 \cdot 1} = 126$

**Support Numerically.** Figure 12.17 shows evaluating the $\binom{n}{r}$ constant using a grapher. ∎

```
5 nCr 3
                  10
9 nCr 5
                 126
```

**FIGURE 12.17**
Numerical support for
Example 4.

Notice that $\binom{n}{r}$ is the $(r+1)$ entry in the $n$th row of Pascal's Triangle. Use a grapher to evaluate each of the following coefficients to see that this triangular array and Pascal's Triangle are identical:

$$\binom{0}{0}$$

$$\binom{1}{0} \qquad \binom{1}{1}$$

$$\binom{2}{0} \qquad \binom{2}{1} \qquad \binom{2}{2}$$

$$\binom{3}{0} \qquad \binom{3}{1} \qquad \binom{3}{2} \qquad \binom{3}{3}$$

$$\binom{4}{0} \qquad \binom{4}{1} \qquad \binom{4}{2} \qquad \binom{4}{3} \qquad \binom{4}{4}$$

$$\binom{5}{0} \qquad \binom{5}{1} \qquad \binom{5}{2} \qquad \binom{5}{3} \qquad \binom{5}{4} \qquad \binom{5}{5}$$

$$\cdot \qquad \cdot \qquad \cdot \qquad \cdot \qquad \cdot \qquad \cdot \qquad \cdot$$

The relationship between the binomial coefficients and Pascal's Triangle is summarized in the theorem known as the binomial theorem.

---

**THEOREM 12.3**   Binomial Theorem

For any positive integer $n$,

$$(x+y)^n = \binom{n}{0}x^n + \binom{n}{1}x^{n-1}y + \binom{n}{2}x^{n-2}y^2 + \cdots + \binom{n}{n-1}xy^{n-1} + \binom{n}{n}y^n.$$

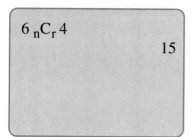

**FIGURE 12.18**
Grapher evaluation of $\binom{6}{4}$ in Example 5.

Example 5 illustrates how to use the binomial theorem to calculate a particular term in an expansion; Example 6 uses the theorem to expand a binomial expression.

**EXAMPLE 5**    Using the Binomial Theorem

Find the 5th term of the expansion of $(x + 3y)^6$.

SOLUTION
Use grapher to evaluate $\binom{6}{4} = 15$ (see Figure 12.18). Then the 5th term is

$$\binom{6}{4}x^{6-4}(3y)^4 = 15x^2(81y^4)$$

$$= 1215x^2y^4.$$

**EXAMPLE 6**    Using the Binomial Theorem

Expand $(2x + 3y)^5$ using the binomial theorem.

SOLUTION

$$(2x + 3y)^5 = \binom{5}{0}(2x)^5 + \binom{5}{1}(2x)^4(3y) + \binom{5}{2}(2x)^3(3y)^2 + \binom{5}{3}(2x)^2(3y)^3 +$$

$$\binom{5}{4}(2x)(3y)^4 + \binom{5}{5}(3y)^5$$

$$= 32x^5 + (5)(16x^4)(3y) + 10(8x^3)(9y^2) + 10(4x^2)(27y^3) +$$

$$5(2x)(81y^4) + 243y^5$$

$$= 32x^5 + 240x^4y + 720x^3y^2 + 1080x^2y^3 + 810xy^4 + 243y^5$$

## Exercises for Section **12.3**

In Exercises 1 and 2, find the indicated row of Pascal's Triangle.

**1.** The seventh row.

**2.** The tenth row.

In Exercises 3–8, expand the expression using Pascal's Triangle to find the coefficients of the term.

**3.** $(x + y)^4$

**4.** $(x - y)^4$

**5.** $(x - 2y)^3$

**6.** $(x + 3y)^3$

**7.** $(2x + 5y)^5$

**8.** $(3x - 2y)^5$

In Exercises 9–14, evaluate the factorial expression.

**9.** $5!$

**10.** $8!$

**11.** $\dfrac{5!}{3!2!}$

**12.** $\dfrac{6!}{4!2!}$

**13.** $\dfrac{8!}{5!3!}$

**14.** $\dfrac{9!}{7!2!}$

In Exercises 15–24, tell whether the statement is true or false.

15. $5! - 3! = 2!$

16. $7! - 4! = 3!$

17. $5 \cdot 4! = 5!$

18. $6 \cdot 5! = 6!$

19. $(6 - 4)! = 6! - 4!$

20. $(8 - 5)! = 8! - 5!$

21. $\dfrac{20!}{4!} = 5!$

22. $\dfrac{12!}{4!} = 3!$

23. $\dfrac{4!}{4} = \dfrac{5!}{5}$

24. $\dfrac{7!}{7!} = \dfrac{3!}{3}$

In Exercises 25–30, evaluate the binomial coefficients.

25. $\dbinom{5}{3}$

26. $\dbinom{9}{4}$

27. $\dbinom{12}{9}$

28. $\dbinom{13}{8}$

29. $\dbinom{10}{7}$

30. $\dbinom{18}{15}$

In Exercises 31–38, find the indicated term of an expansion using the binomial theorem.

31. $(x + y)^{13}$; 4th term

32. $(x - y)^{10}$; 6th term

33. $(2x + 5)^3$; 2nd term

34. $(3x + 7)^4$; 3rd term

35. $(2x - 8y)^5$; 2nd term

36. $(3x + y)^7$; 5th term

37. $(x + 3y)^7$; 5th term

38. $(x - 2y)^8$; 6th term

In Exercises 39–46, expand the binomial using the binomial theorem.

39. $(a + b)^7$

40. $(a - b)^5$

41. $(x + 2y)^5$

42. $(x + 5)^7$

43. $(2x - 5)^4$

44. $(3x + 4)^3$

45. $(3x + 2y)^6$

46. $(2x - 3y)^4$

## EXTENDING THE IDEAS

In Exercises 47–50, use the binomial theorem to expand the expression.

47. $(\sqrt{x} + 5)^4$

48. $(\sqrt{x} - 2)^3$

49. $(\sqrt{x} - \sqrt{y})^6$

50. $(\sqrt{x} + 2\sqrt{y})^6$

## LOOKING BACK—MAKING CONNECTIONS

In Exercises 51–58, solve the equation or inequality using either an algebraic or a graphing method.

51. $|2x + 7| = 9$

52. $|5x - 4| = 11$

53. $|2x + 7| \geq 9$

54. $|5x - 4| \geq 11$

55. $6x^2 - 19x - 7 = 0$

56. $5x^2 + 2x - 3 = 0$

57. $6x^2 - 19x - 7 < 0$

58. $5x^2 + 2x - 3 < 0$

---

**12.4** # Counting Principle, Permutations, and Combinations

Counting Principle ▪ Permutations ▪ Combinations

## Counting Principle

With the availability of computers, an area of mathematics that is rapidly growing is statistics. Counting is a topic that is vital to any study involving data record keeping. A counting problem asks how many ways an event can occur. Sometimes it is easiest to simply list all possibilities, as is done in Example 1.

**EXAMPLE 1** Counting by Listing

How many different three-digit license plates can be made consisting of the digits 1, 2, 3, and 4 if no digit is repeated?

SOLUTION

| | | | | | |
|---|---|---|---|---|---|
| 123 | 124 | 132 | 134 | 142 | 143 |
| 213 | 214 | 231 | 234 | 241 | 243 |
| 312 | 314 | 321 | 324 | 341 | 342 |
| 412 | 413 | 421 | 423 | 431 | 432 |

There are $6 \cdot 4 = 24$ possibilities. ∎

Sometimes it is impossible to write the entire list of possible ways an event can occur. In such cases, we use counting principles. The most important of these is the **Fundamental Counting Principle,** which follows.

---

### Fundamental Counting Principle

Let $S_1$, $S_2$ be two events, neither influencing the outcome of the other. If $S_1$ can occur in $m$ different ways and $S_2$ can occur in $n$ different ways, then the number of ways the two events can occur is $m \cdot n$.

---

This principle can be extended to three or more events as well.

**EXAMPLE 2** Using the Fundamental Counting Principle

A certain voice mail system asks for the user's identification number, which must be a five-digit number. Find the number of possible user ID numbers.

SOLUTION

Because the digits can be repeated, there are 10 possible choices for each of the five digits.

| 0, 1, 2, 3, 4, 5, 6, 7, 8, 9 | 0, 1, 2, 3, 4, 5, 6, 7, 8, 9 | 0, 1, 2, 3, 4, 5, 6, 7, 8, 9 | 0, 1, 2, 3, 4, 5, 6, 7, 8, 9 | 0, 1, 2, 3, 4, 5, 6, 7, 8, 9 |
|---|---|---|---|---|
| **10** | **10** | **10** | **10** | **10** |

Therefore the number of possible user ID's is

$$10 \cdot 10 \cdot 10 \cdot 10 \cdot 10 = 10^5 = 100,000.$$ ■

The Fundamental Counting Principle also can be used in cases in which letters, people, things, etc., can be rearranged. Example 3 illustrates how to make letter and number arrangements.

**EXAMPLE 3**   Using the Fundamental Counting Principle

A certain state uses both letters and numbers on its license plates in the pattern of two letters followed by three digits. How many different license plates are possible?

SOLUTION
The license plate consists of five separate symbols. We can think of these symbols as being chosen one at a time. There are 26 letters of the alphabet, so 26 possibilities exist to fill each of the first two slots in the license plates. And because there are 10 digits, (0–9), 10 possibilities exist to fill the last three slots.

Thus the number of license plates meeting these conditions is

$$26 \cdot 26 \cdot 10 \cdot 10 \cdot 10 = 26^2 \cdot 10^3 = 676,000.$$ ■

## Permutations

When $n$ distinct objects are arranged in some order, we often are interested in learning how many different orderings are possible. In this type of problem, we are studying **permutations**.

---

■                                                                               ■

### DEFINITION 12.6    Permutation

A **permutation** of $n$ elements is an ordering of the $n$ elements.

■                                                                               ■

---

Example 4 illustrates finding the possible ways members of a team can be arranged. When one member is chosen for the first slot, there are four left for slot 2. Once that slot is filled, there are three left for the next slot, and so on.

### EXAMPLE 4    Solving a Permutation

Five hurdle jumpers are lined up in order. How many different ways can they be arranged?

SOLUTION

The possible ways to fill each of the five slots is

$$5 \cdot 4 \cdot 3 \cdot 2 \cdot 1 = 5! = 120.$$

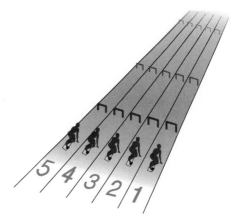

The setting of Example 4 can be generalized. Theorem 12.4 describes the general situation.

**THEOREM 12.4**   Number of Permutations of $n$ Elements

There are $n!$ different permutations of $n$ elements.

Example 5 illustrates a situation in which not all members of the set are to be included in the arrangement. In that example, only four members are selected from a group of 50 possibilities. The Fundamental Counting Principle is used to solve the problem.

**EXAMPLE 5**   Another Permutation

A fraternity that has 50 members is electing a president, vice president, secretary, and treasurer. In how many different ways can these offices be filled if everyone in the fraternity is eligible for each office but an individual can hold only one office?

SOLUTION
The number of choices to fill each position is as follows:

| President | 50 choices |
| Vice-President | 49 choices |
| Secretary | 48 choices |
| Treasurer | 47 choices |

Therefore there are $50 \cdot 49 \cdot 48 \cdot 47 = 5,527,200$ possible ways to fill the officer positions.

A special notation can be used to describe the situation illustrated by Example 5 where there were 50 members from which 4 officers were chosen. In general, there are $n$ objects from which $r$ are chosen. The following definition introduces the notation.

---

**DEFINITION 12.7**    Permutation Notation $_nP_r$

A permutation of a set of $n$ elements taken $r$ at a time can be denoted $_nP_r$.

---

In Example 5, we were interested in finding the number $_{50}P_4$. This number is $50 \cdot 49 \cdot 48 \cdot 47$, which can be written using the factorial notation as follows:

$$50 \cdot 49 \cdot 48 \cdot 47 = \frac{50!}{46!}$$

The general situation is the following.

---

**THEOREM 12.5**    Evaluation of $_nP_r$

If $n$ and $r$ are integers and $0 \le r \le n$, then

$$_nP_r = \frac{n!}{(n-r)!}.$$

---

Most graphers can evaluate $_nP_r$, as Example 6 illustrates.

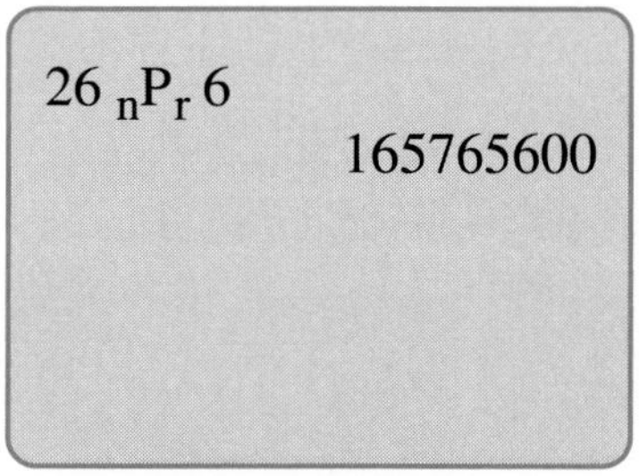

**EXAMPLE 6**    Counting the Number of Permutations

How many different six-letter passwords can be given for possible access to a particular computer program if the letters cannot be repeated? (The letters do not necessarily have to make a word.)

SOLUTION

Evaluating the number of permutations of 26 letters taken six at a time, we have

$$_{26}P_6 = \frac{26!}{(26-6)!} = \frac{26 \cdot 25 \cdot 24 \cdot 23 \cdot 22 \cdot 21 \cdot 20!}{20!} = 165{,}765{,}600.$$

**Support Numerically.** Figure 12.19 shows the evaluation using a grapher.    ∎

26 nPr 6

165765600

**FIGURE 12.19**
Numerical support for the solution to Example 6.

## Combinations

Example 1 found the number of possible three-digit license plates consisting of the digits 1, 2, 3, and 4. In the case of license plates, the plate with 123 and the plate 132 are considered two distinct plates.

On the other hand, suppose a fraternity formed committees of three chosen from the four officers. How many committees could be formed? Here we see that the committee consisting of the president, vice president, and secretary is the same committee as the one consisting of vice president, secretary, and president. The order of the listing does not make a different committee.

Finding the number of subsets of $r$ elements that can be selected from $n$ elements is called **counting the combinations of $n$ things taken $r$ at a time.** Whenever two different orderings of the same elements represent the same entity, we are counting combinations, not permutations.

Example 7 illustrates finding the number of 3 person committees taken from 4 officers.

### EXAMPLE 7   Counting the Number of Committees

How many committees of three can be formed from the four officers of the fraternity?

#### SOLUTION

Let the four officers be represented by $P$, $V$, $S$, and $T$. There are the following 24 possible ways of ordering three letters selected from the four letters:

| | | | | | |
|---|---|---|---|---|---|
| $PVS$ | $PVT$ | $PSV$ | $PST$ | $PTV$ | $PTS$ |
| $VPS$ | $VPT$ | $VSP$ | $VST$ | $VTP$ | $VTS$ |
| $SPV$ | $SPT$ | $SVP$ | $SVT$ | $STP$ | $STV$ |
| $TPV$ | $TPS$ | $TVP$ | $TVS$ | $TSP$ | $TSV$ |

However, each of the six listings

| | | | | | |
|---|---|---|---|---|---|
| $PVS$ | $VPS$ | $PSV$ | $VSP$ | $SPV$ | $SVP$ |

represents the committee of president, vice president, and secretary. The same is true for any other committee. There are six listings for each.

Consequently, the total number of committees is equal to the number of different listings $_4P_3$ divided by the 3! different orderings for each committee. That is,

$$\frac{_4P_3}{3!} = \frac{4!}{(4-3)!3!}$$

$$= \frac{4!}{1! \cdot 3!}$$

$$= 4.$$

Therefore, there are 4 committees of 3 persons that can be formed by the four officers. ■

Example 7 is a special case of a more general setting. When the counting problem requires counting the number of different subsets, we are dealing with a combination problem. Different orderings of the same elements are counted as the same entity.

Here is the general situation. To find the number of subsets of $r$ people selected from a larger group of $n$ people, do the following:

1. Find $_nP_r = \dfrac{n!}{(n-r)!}$.

2. Divide this number by $r!$, the number of different orderings of each subset of $r$.

The resulting number,

$$\frac{n!}{r!(n-r)!},$$

describes the number of subsets that are possible. The following theorem states the general case.

---

**THEOREM 12.6**   The Number of Combinations of $n$ Elements Taken $r$ at a Time

If $n$ and $r$ are nonnegative integers, where $r \le n$, then the number of combinations of $n$ elements taken $r$ at a time is

$$_nC_r = \frac{n!}{r!(n-r)!},$$

the binomial coefficient defined in Section 12.3.

---

Example 8 illustrates finding the number of combinations.

### EXAMPLE 8   Finding the Number of Starters

The Stars baseball team has 22 players on the roster; however, only nine players can start a game. How many different nine-person starting teams are possible (assuming all players can play all positions)?

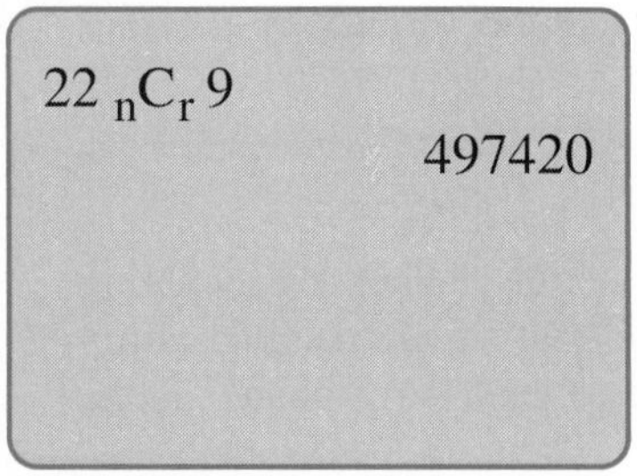

$22 \, _nC_r \, 9$

$497420$

**FIGURE 12.20**
Numerical support for the solution to Example 8.

SOLUTION

Here we need to know the number of combinations of 22 players taken nine at a time. Therefore $n = 22$ and $r = 9$.

$$_{22}C_9 = \frac{22!}{9!(22 - 9)!}$$

$$= 497420$$

Therefore, there are 497,420 different starting teams. See Figure 12.20.    ■

## Exercises for Section  12.4

In Exercises 1–8, evaluate the number of permutations or combinations using a grapher.

**1.** $_7P_5$            **2.** $_6P_3$

**3.** $_{15}P_{11}$         **4.** $_{12}P_{10}$

**5.** $_7C_4$            **6.** $_8C_6$

**7.** $_{16}C_{12}$         **8.** $_{24}C_{16}$

**9.** *Number of Access Codes.* A certain computer program requires a three-digit access code using the digits 7, 8, and 9. How many different codes are possible?

**10.** *Number of Access Codes.* Access to student records on a computer requires a five-letter access code. How many different codes are possible to gain access?

**11.** *Number Selection.* How many different ways can four digits be selected for the lottery's Play-Four game if the digits 0 through 9 are possible and each can be repeated?

**12.** *Number Selection.* How many different ways can five digits be selected for the lottery's Fantasy Five game if the digits 0 through 9 are possible and each can be repeated?

**13.** *Arranging a Display.* How many different ways can eight cans of different vegetables be arranged on a display shelf?

**14.** *Arranging a Display.* How many different ways can a librarian display 10 books in a row?

**15.** *Number of Area Codes.* How many different three-digit area codes are possible? (Remember that the first digit cannot be 0 or 1 and the middle digit must be a 0 or 1.)

**16.** *Number of Social Security Numbers.* How many different social security numbers are possible with the format XXX-XX-XXXX?

**17.** *Number of License Plates.* Find the number of possible license plates that have three letters followed by three digits if digits and letters can be repeated.

**18.** *Number of License Plates.* Find the number of possible license plates that have three letters followed by three digits if neither digits nor letters can be repeated.

**19.** *Number of Relay Teams.* Find the number of possible four-person relay teams that can be made from a 22-person swim team.

**20.** *Number of Bridge Hands.* There are 52 cards in a deck of cards used for playing bridge. How many possible 13-card hands can be dealt?

**21.** *Number of Pinochle Hands.* There are 48 cards in a deck of cards used for playing Pinochle. How many possible 12-card hands can be dealt?

**22.** *Committee Selection.* There are 25 instructors in the mathematics department. How many different five-person committees can be formed consisting of mathematics instructors?

### EXTENDING THE IDEAS

**23.** *Committee Selection.* There are 250 seniors and 175 juniors. How many different five-person committees can be formed consisting of three seniors and two juniors?

**24.** *Outfit Selection.* Amelia has six dresses, five pairs of shoes, and three jackets. How many choices of outfits consisting of a dress, a pair of shoes, and a jacket does she have?

**25.** *Seating Options.* There are 20 students that meet in a classroom with 32 chairs. How many ways is it possible for the students to be seated?

**26.** Telephone numbers are listed with an area code followed by seven digits, such as (617)-XXX-XXXX. Remember the first digit cannot be 0 or 1. How many telephone numbers are possible in a particular area code?

**27.** How many five-digit zip codes are possible that begin with the number 3?

### LOOKING BACK—MAKING CONNECTIONS

**28.** An ant population is growing at a rate of 2.5 times the total population of the previous generation. If there are 300 ants in the first generation, how many ants will there be in the fifth?

In Exercises 29 and 30, simplify the expression.

**29.** $\dfrac{x^{-1}+2}{4-x^{-1}}$  **30.** $\dfrac{3x^{-2}+1}{3x^{-2}-1}$

In Exercises 31 and 32, use the completing-the-square method to solve the equation.

**31.** $x^2 - 6x + 3 = 0$  **32.** $x^2 + 8x - 5 = 0$

In Exercises 33 and 34, use the quadratic formula to solve the equation.

**33.** $8x^2 + 6x = 9$  **34.** $4x^2 - x = 5$

## Chapter 12 Summary

| | Sequences and Series | Examples |
|---|---|---|
| Infinite Sequence | A function whose domain is the set of positive integers; an ordered list of numbers. | $5, 10, 15, 20, \ldots$ |
| Finite Sequence | A sequence with a finite number of terms. | $10, 100, 1000, 10000$ |
| Term | A number in a sequence. | 8 is one term in the sequence $2, 5, 8, 11$ |
| Finite Series | The sequence $S_n$ where $S_n = a_1 + \ldots + a_n$. | |

|  | **Arithmetic Sequences** | **Examples** |
|---|---|---|
| Arithmetic Sequence | A sequence in which the first term is given and each successive term is obtained by adding a specific number $d$ to the previous term. | The sequence $6, 16, 26, \ldots$ is arithmetic with $d = 10$. |
| Common Difference | The number $d$ that is the difference of two successive terms of an arithmetic sequence. | In the arithmetic sequence $4, 7, 10, d = 3$. |
| Term $a_n$ of an Arithmetic Sequence | $a_n = a_1 + (n - 1)d$, where $n$ is the number of the term, $a_1$ is the first term, and $d$ is the common difference. | Given $a_1 = 2$ and $d = 5$, $a_4 = 2 + 3(5) = 17$. |
| Sum of an Arithmetic Sequence (Finite Arithmetic Series) | The sum $S_n$ of the first $n$ terms of an arithmetic sequence is $S_n = \frac{n}{2}[a_1 + a_n]$ or $S_n = \frac{n}{2}[2a_1 + (n - 1)d]$. | The sum of the first 6 odd numbers $S_6 = \frac{6}{2}[1 + 11] = 36$ or $S_6 = \frac{6}{2}[2 + 5(2)] = 36$. |

|  | **Geometric Sequences** | **Examples** |
|---|---|---|
| Geometric Sequence | A sequence in which the first term is given and each successive term is obtained by multiplying the previous term by a nonzero real number $r$. | The sequence $1, 2, 4, 8, 16, 32, \ldots$ is geometric with $r = 2$. |
| Alternating Sequence | A geometric sequence with terms that alternate between positive and negative values. | The sequence $5, -10, 20, -40, \ldots$ is alternating. |
| Common Ratio | The number $r$ that multiplies a term of a geometric sequence to obtain the next term in the sequence. | In the sequence $1, 0.1, 0.01, 0.001, \ldots r = 0.1$. |
| Term $a_n$ of a Geometric Sequence | $a_n = a_1 \ldots r^{n-1}$, where $n$ is the number of the term, $a_1$ is the first term, and $r$ is the common ratio. | Given $a_1 = 1$ and $r = 3$, $a_5 = 1 \cdot 3^4 = 81$. |
| Partial Sum of a Geometric Sequence (Finite Geometric Series) | The sum $S_n$ of the first $n$ terms of a geometric sequence is $S_n = \frac{a_1(1 - r^n)}{1 - r}$. | Given the sequence $1, 2, 4, 8, \ldots$ the sum $S_4 = \frac{1(1 - 2^4)}{1 - 2} = 15$. |

| | | |
|---|---|---|
| Infinite Geometric Series | The sum $S$ of an infinite geometric series with the first term $a_1$ and the common ratio $r$ with $0 < |r| < 1$ is $S = \dfrac{a_1}{1-r}$. (If $|r| \geq 1$, the sum does not exist.) | The sum $S$ of the series $1 + 0.1 + 0.01 + \ldots$ is $S = \dfrac{1}{1 - \dfrac{1}{10}}$ $= \dfrac{1}{\dfrac{9}{10}} = \dfrac{10}{9} = 1.111.$ |

| | **Binomial Expansion** | **Examples** |
|---|---|---|
| Binomial Expansion | The result of raising a binomial to an integer power $n$, where $n > 1$. | |
| Pascal's Triangle | A famous triangle of numbers with each row showing the coefficients of the terms in each binomial expansion and each entry is the sum of the two numbers above it. | |
| Factorial Notation (!) | For any positive integer $n$, $n! = n(n-1)(n-2)\cdots 3 \cdot 2 \cdot 1$. We define $0! = 1$. | $4! = 4 \cdot 3 \cdot 2 \cdot 1 = 24$ |
| Binomial Coefficient | For nonnegative integers $n$ and $r$, $r \leq n$, $$_nC_r = \binom{n}{r} = \frac{n!}{r!(n-r)!}$$ | $\dbinom{6}{2} = \dfrac{6!}{2!4!} = 15$ |
| Binomial Theorem | For any positive integer $n$: $$(x+y)^n = \binom{n}{0}x^n + \binom{n}{1}x^{n-1}y + \binom{n}{2}x^{n-2}y^2 + \cdots \binom{n}{n-1}xy^{n-1} + \binom{n}{n}y^n$$ | |

| | **Counting Principle, Permutations and Combinations** | **Examples** |
|---|---|---|
| Fundamental Counting Principle | Given two events, $S_1$ and $S_2$, with $S_1$ occurring in $m$ different ways and $S_2$ occurring in $n$ | |

| | Counting Principle, Permutations and Combinations | Examples |
|---|---|---|
| | different ways, the number of ways the two events can occur is $m \cdot n$. | |
| Permutation of $n$ Elements | An ordering of $n$ elements. | |
| Permutation Notation | The number of permutations of a set of $n$ elements taken $r$ at a time is written as $_nP_r$. | |
| Evaluation of $_nP_r$ | $_nP_r = \dfrac{n!}{(n-r)!}$ for integers $r$ and $n$ where $0 \leq r \leq n$. | $_7P_2 = \dfrac{7!}{5!} = 7 \cdot 6 = 42$ |
| Number of Permutations of $n$ Elements | There are $n!$ different permutations of $n$ elements. | Five elements can be arranged in $5! = 120$ ways. |
| Number of Permutations of $n$ Elements Taken $r$ at a Time | There are $_nP_r$ different permutations of $n$ elements taken $r$ at a time. | Five elements taken three at a time can be arranged in $_5P_3 = \dfrac{5!}{2!} = 60$ ways. |
| Combination | A subset of a collection of elements. | |
| Combination Notation | The number of combinations of $n$ elements taken $r$ at a time is written as $_nC_r$. | |
| Evaluation of $_nC_r$ | $_nC_r = \dfrac{n!}{r!(n-r)!}$ for integers $r$ and $n$ where $0 \leq r \leq n$. | $_6C_2 = \dfrac{6!}{2! \cdot 4!} = 15$ |
| Number of Combinations of $n$ Elements Taken $r$ at a Time | There are $_nC_r$ different combinations of $n$ elements taken $r$ at a time. | Five elements taken three at a time without regard to order can be grouped in $_5C_3 = 10$ ways. |

## Review Exercises for Chapter 12

In Exercises 1 and 2, write the next three terms of the given arithmetic sequence.

**1.** $4.5, 6, 7.5, \ldots$     **2.** $-3, -7, -11, \ldots$

In Exercises 3 and 4, write the next three terms of the given geometric sequence.

**3.** $3, -6, 12, \ldots$     **4.** $9801, 3267, 1089, \ldots$

In Exercises 5 and 6, write the first five terms of the arithmetic sequence with the given first term and common difference.

**5.** $a_1 = 8, d = -3$     **6.** $a_1 = 1, d = 0.6$

In Exercises 7 and 8, write the first five terms of the geometric sequence with the given first term and common ratio.

**7.** $a_1 = 5, r = 10$        **8.** $a_1 = -2, r = 4$

In Exercises 9 and 10, use an algebraic method to find the indicated term of the arithmetic sequence with the given first term and common difference. Support graphically.

**9.** $a_1 = 5, d = 1.4$, find $a_{12}$

**10.** $a_1 = 14, d = 48$, find $a_{25}$

In Exercises 11 and 12, use an algebraic method to find the indicated term of the geometric sequence with the given first term and common ratio. Support graphically.

**11.** $a_1 = 6, r = 4$, find $a_{16}$

**12.** $a_1 = 1000, r = \dfrac{1}{2}$, find $a_{10}$

In Exercises 13 and 14, find the indicated sum of the arithmetic sequence algebraically. Support graphically.

**13.** $a_1 = 4, d = 0.1, S_{25}$    **14.** $a_1 = 3, d = 22, S_{10}$

In Exercises 15 and 16, use an algebraic method to find the indicated sum, if it exists, of the geometric series algebraically. Support graphically.

**15.** $1 + 8 + 64 + \ldots$ , find $S_7$

**16.** $1 + 0.6 + 0.36 + \ldots$ , find $S_{20}$

**17.** Find the sum of the series $\dfrac{1}{3} + \dfrac{1}{6} + \dfrac{1}{12} + \ldots$ .

**18.** For the series $31 + 27 + 23 + \ldots$ , find the value or values of $n$ for which the sum of the series is the largest.

**19.** Find $n$ for the arithmetic sequence for which $a_n = 535, a_1 = 3$ and $d = 7$.

**20.** Find the sixth row of Pascal's Triangle.

In Exercises 21 and 22, expand each expression using Pascal's Triangle to find the coefficients of each term.

**21.** $(2x + y)^5$            **22.** $(a - 3)^6$

In Exercises 23 and 24, expand each expression using the binomial theorem.

**23.** $(x - 2a)^4$            **24.** $(3x + 1)^7$

In Exercises 25 and 26, find the indicated term of an expansion using the binomial theorem.

**25.** $(4x - 5)^6$; third term  **26.** $(2a - 3b)^8$; fourth term

In Exercises 27 and 28, evaluate the following factorial expressions.

**27.** $\dfrac{10!}{6! \cdot 4!}$                **28.** $\dfrac{8! \cdot 5!}{9! \cdot 2!}$

In Exercises 29 and 30, evaluate the following binomial coefficients.

**29.** $\dbinom{11}{6}$            **30.** $\dbinom{8}{5}$

In Exercises 31 and 32, evaluate the number of permutations or combinations using a grapher.

**31.** $_{10}C_8$

**32.** $_{10}P_8$

In Exercises 33 and 34, tell whether the following statements are true or false.

**33.** $_{10}C_9 =_{10} C_1$

**34.** $_8P_1 =_8 C_1$

**35.** Two children have designed an activity using a checkerboard where the first person places one penny on a square, the second person places two pennies on the square next to the first, the first person places four pennies on the next square, and so on, until all 64 squares are covered. When the 64 squares are covered, how much money is on the checkerboard?

**36.** The length of arc through which a pendulum swings is 80% of its previous swing. The length of the arc of the original swing is 20 cm.

a) Determine the length of the arc of the fifth swing.

b) Determine the total distance the pendulum has traveled after 15 swings.

**37.** An access code is designed to accept two digits followed by one letter. How many possible codes are there?

**38.** At a scholarship pageant, three finalists are asked to sit on stage where five chairs are provided. How many different seating arrangements are there?

**39.** There are eight members of a local chapter of a national organization. Two members will be chosen to represent the chapter at a national meeting. If all members are interested in attending, how many different choices are possible for the two representatives?

## Chapter 12 Test

**1.** Write the first five terms of the geometric sequence
$$\frac{4}{3}, \frac{2}{3}, \frac{1}{3}, \ldots.$$

**2.** Write the first five terms of the arithmetic sequence if the first term is 100 and the common difference is $-30$.

**3.** Write the first five terms of the geometric sequence if the first term is 9 and the common ratio is 1.5.

**4.** Use algebraic means to find the value of $a_{30}$ of the arithmetic sequence defined by $a_1 = 1$ and $d = 19$.

**5.** Use algebraic means to find the value of $a_{12}$ of the geometric sequence defined by $a_1 = 2$ and $r = 4$.

**6.** Use algebraic means to find the sum $S_{40}$ of the arithmetic sequence defined by $a_1 = 20$ and $d = 3.5$.

**7.** Use algebraic means to find the sum $S_{10}$ of the geometric series $3 + 12 + 48 + \ldots$ .

**8.** Find the sum of the geometric series $5 + 4 + 3.2 + \ldots$ .

**9.** Find the sum of all odd integers from 25 to 75.

**10.** What is the value of $d$ in the arithmetic sequence which has terms $a_1 = 8$ and $a_{10} = 53$?

**11.** Use the binomial theorem to expand $(x + 3)^6$.

**12.** Find the third term of the expansion of $(3x - 2)^5$.

**13.** Evaluate $\dfrac{10!}{5! \cdot 3!}$.

**14.** Alicia is paid a salary of $21,000 for her first year of employment and is guaranteed an increase in salary of $1680 per year at the end of each full year of employment.

a) What will Alicia's salary be at the end of 10 years?

b) What will Alicia's total earnings be at the end of 10 years?

**15.** How many different ways can four participants in a panel discussion be seated in a row?

# A N S W E R S

## Chapter 1

### SECTION 1.1

**1.** $0.\overline{714285}$; repeating

**3.** 0.85; terminating

**5.** $1.8\overline{3}$; repeating

**7.** 2.904; terminating

**9.** Integer, rational number

**11.** Whole number, integer, rational number

**13.** Rational number

**15.** Irrational number

**17.** Rational number

**19.** False

**21.** False

**23.** True

**25.**

**27.**

**29.**

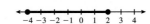

**31.**

**33.**

**35.** $<$ or $\leq$

**37.** $>$ or $\geq$

**39.** $<$ or $\leq$

**41.** 200

**43.** 2352

**45.** $\frac{4}{9}$

**47.** $\frac{29}{2}$

**49.** 6

**51.** 14

**53.** 54

**55.** 503

**57.** 17

**59.** $\frac{19}{2}$

**61.** 29

**63.** 15,625

**65.** Ambiguous

**67.** $5^2 + 3^2$ and $(5 + 3)^2$. These do not have the same value.

**69.**

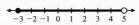

$$\begin{array}{ccccccccc} \bullet & | & | & | & | & | & | & | & \circ \\ -3 & -2 & -1 & 0 & 1 & 2 & 3 & 4 & 5 \end{array}$$

**71.** Oral Exercise

**73.** $3813

**75.** Written Exercise

**77.** 92.476%

**79.** (d)

**81.** 125 ft$^2$; $356.25

**83.** 54 ft in Exercise 81; 86.2 ft in Exercise 82

**SECTION 1.2**

**1.** 15

**3.** $^-8$

**5.** $^-13$

**7.** 16

**9.** $^-3$

**11.** 23

**13.** 7

**15.** 0

**17.** 7

**19.** 0

**21.** 144.9

**23.** $^-48.2$

**25.** 63.9

**27.** 125.76

**29.** Commutative Property of Addition

**31.** Associative Property of Multiplication

**33.** Commutative and Associative Properties of Addition

**35.** Distributive Property

**37.** Associative Property of Multiplication

**39.** $6 \cdot 21 = 6 \cdot (20 + 1) = 6 \cdot 20 + 6 \cdot 1 = 120 + 6 = 126$

**41.** $7 \cdot 12 = 7 \cdot (10 + 2) = 7 \cdot 10 + 7 \cdot 2 = 70 + 14 = 84$

**43.** $5 \cdot 62 = 5 \cdot (60 + 2) = 5 \cdot 60 + 5 \cdot 2 = 300 + 10 = 310$

**45.** $3 \cdot 91 = 3 \cdot (90 + 1) = 3 \cdot 90 + 3 \cdot 1 = 270 + 3 = 273$

**47.** True

**49.** False

**51.** opposite, subtraction: $-2$

**53.** opposite, subtraction, opposite: $-7$

**55.** opposite, opposite, subtraction: 11

**57.** opposite, opposite, subtraction, opposite: 16

**59.** 8 yd

**61.** $295.46°\,K$

**63.** a)

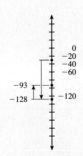

b) 96 ft

c) 93 ft below the surface

d) 22,128 ft

**65.** $|12.3 + (^-5.8)|$

**67.** $|^-13.2| - |^-12.6|$

**69.** $^-1 \cdot x = 2$

$^-2 \cdot x = 4$

**71.** Written Exercise

**73.** $0.60, \frac{3}{5}$

**75.** $33\frac{1}{3}\%, 0.\overline{3}$

**77.** $0.375, \frac{3}{8}$

**79.** $80\%, 0.80$

**SECTION 1.3**

**1.** $x$

**3.** $3x$

**5.** $\triangle$

**7.** $3x, 2y^2$

**9.** 5

**11.** 0

**13.** 7

**15.** 31

**17.** 4

**19.** 30

**21.** 31

**23.** 11

**25.** 13

**27.** 85,821.8611

**29.** $-195,157.2035$

**31.** 149.6766

**33.** $-\dfrac{460}{1521}$

**35.** 740.5707

**37.** 16

**39.** 46

**41.** 4822

**43.**

| °F | °C |
|-----|--------|
| −12 | −24.44 |
| 25 | −3.89 |
| 48 | 8.89 |
| 68 | 20 |
| 105 | 40.56 |

**45.** a) $19,690

b) Profit goes up by $65

c) April; $275 more

**47.** $3x + 4$

**49.** $3.7y - 5x$

**51.** $2x$

**53.** $146.135, in millions

**55.** Area: 16.619 sq in.
Circumference: 14.451 in.

**57.** 268 sq in.

**59.** Written Exercise

SECTION 1.4

**1.** Yes

```
29 → X
                    29
5X − 13
                   132
```

**3.** No

```
9.5 → X
                    9.5
X²+17X
                  251.75
```

**5.** Yes

```
3 → X
                     3
5/X+X/3
               2.666666667
8/3
               2.666666667
```

**7.** Yes

```
5.2 → X
                    5.2
X²+1.5
                  28.54
```

**9.** $x = 25$

**11.** $x = -15$

**13.** $x = 5$

**15.** $x = 8$

**17.** $x = 110$

**19.** $x = 1.5$

**21.** $x = 14, \ x = -14$

**23.** $x = 9$

**25.** $x = \frac{19}{2}$

**27.** $x = -\frac{68}{3}$

**29.** $x = -\frac{3}{2}$

**31.** $x = 4$

**33.** $x = -\frac{40}{71}$

**35.** $x = -\frac{4}{11}$

**37.** $x = -12$

**39.** $x = \frac{25}{6}$

**41.** $x = 2$

**43.** $q = 1$

**45.** $t = 14$

**47.** $k = -\frac{29}{3}$

**49.** $(-\infty, 16)$

**51.** $(-\infty, 13]$

**53.** $[19, \infty)$

**55.** $(6, \infty)$

**57.** $\left(-\infty, \frac{8}{3}\right)$

**59.** $\left(-\infty, \frac{29}{3}\right]$

**61.** $(-\infty, 5 + \sqrt{2})$

**63.** $x = \frac{7}{4}$

**65.** $x = -6, \ x = 6$

**67.** $3x - 2 = 75$

**69.** $70 = 3x - 2$

**71.** $3x = 2x + 2$

**73.** $x - 6 = 2x$

**75.** Contradiction

**77.** Not a contradiction

**79.** Answers will vary.

**81.** 7, 21, 21

**83.** 2412.743 cu ft

**85.** 523.599 cu in.

## SECTION 1.5

**1.** $0.17x$

**3.** $A = W(W + 3)$

**5.** a) Answers will vary.

b)

| Width | Length | Area |
|-------|--------|------|
| 1 | 16 | 16 |
| 2 | 14 | 28 |
| 3 | 12 | 36 |
| 4 | 10 | 40 |
| 5 | 8 | 40 |
| 6 | 6 | 36 |
| 7 | 4 | 28 |
| 8 | 2 | 16 |

c) Widths 4 and 5. Width 4.5 gives area 40.5 sq ft

**7.** a)

| Width | Length | Area |
|-------|--------|------|
| 1 | 14 | 14 |
| 2 | 13 | 26 |
| 3 | 12 | 36 |
| 4 | 11 | 44 |
| 5 | 10 | 50 |
| 6 | 9 | 54 |
| 7 | 8 | 56 |
| 8 | 7 | 56 |
| 9 | 6 | 54 |
| 10 | 5 | 50 |
| 11 | 4 | 44 |
| 12 | 3 | 36 |
| 13 | 2 | 26 |
| 14 | 1 | 14 |

b) Widths 7 and 8. Width 7.5 gives area 56.25 sq ft

c) The largest area is found when length = width i.e. when the rectangle is a square.

**9.** a) 1 liter

b)

| Liters Added | Ratio | Percent Acid |
|--------------|-------|--------------|
| 1 | 1/6 | 16.67% |
| 2 | 1/7 | 14.29% |
| 3 | 1/8 | 12.50% |
| 4 | 1/9 | 11.11% |
| 5 | 1/10 | 10% |
| 6 | 1/11 | 9.10% |
| 7 | 1/12 | 8.33% |
| 8 | 1/13 | 7.69% |

c) 15 liters

d) 11.11%

e) 5.3 liters

**11.** a)

| Seconds | Distance Traveled |
|---------|-------------------|
| 1 | 66 |
| 2 | 132 |
| 3 | 198 |
| 4 | 264 |
| 5 | 330 |

b) $66t$ ft

c) 2.5 sec

**13.** a) 7 sec

b) Maximum height is 196 ft after 3.5 sec.

c) No, the Astrodome is 208 ft high.

**15.** a) $L - 2$

b) $30 = 2W + 2L - 4$

c) $A = LW - 4$

**17.** Oral Exercise

**19.** $0.\overline{142857}$

**21.** $-0.0\overline{6}$

## SECTION 1.6

**1.** $x + 12$

**3.** $2(x - 2)$

**5.** $x - y$

**7.** $5 + 2x$

**9.** $0.15(x + 5)$

**11.** $\frac{x}{5}$

**13.** $2x = 7$

**15.** $2x = x - 4$

**17.** $2(x + 3) = 125$

**19.** $\frac{1}{2}(x + 30) = 25$

**21.** a) $0.18x + x = 35.50$

b) \$30.08

c) \$3008

**23.** \$83.23

**25.** Apple: \$218,500,000; IBM: \$341,500,000

**27.** 1.818 gal

**29.** 1.333 liters of 25% solution

**31.** a) $0.08x$; $0.065(12,000 - x)$

b) $0.08x + 0.065(12,000 - x)$

c) \$7000 at 8%; \$5000 at 6.5%

**33.** \$12,500

**35.** \$2988.24

**37.** 233.333 sec (or 3 min 53.333 sec)

**39.** a) $4x + 2\left(\frac{2x}{3}\right) = 738$

b) $W = 138.375$;

$L = 276.75$

c) 38,295.281 sq ft;

$\frac{38,295.281}{43,560} = 0.879$

**41.** $12 + 38 + x = 180$

## REVIEW EXERCISES

**1.** $0.\overline{125}$

**2.** 3.515625

**3.**

**4.**

**5.**

**6.**

**7.** Rational number

**8.** Integer, rational number

**9.** Whole number, integer, rational number

**10.** Rational number

**11.** 31

**12.** $-20$

**13.** $-4.1$

**14.** 32

**15.** 14

**16.** 29.8

**17.** Distributive property

**18.** Commutative property of multiplication, associative property of multiplication

**19.** 8

**20.** $-\frac{3}{4}$

**21.** 10

**22.** 81.681 in.

**23.** $y = -\frac{4}{3}$

**24.** $k = -\frac{18}{7}$

**25.** $x = 8.2$

**26.** $x = 9$

**27.** $x = 4$

**28.** $x = 6.5$

**29.** $(-14, \infty)$

**30.** $(-\infty, -3]$

**31.** $[16, \infty)$

**32.** $(-\infty, 6)$

**33.** $|-3| + |6| = x$

**34.** $2x = x^2 - 3$

**35.** $R = x + 0.35x$ or $R = 1.35x$

**36.** $x = -12, x = 12$

**37.** Maximum height: 49 ft at 1.75 sec

**38.** \$9320

**39.** 187.5 sec

**40.** 60 ml

CHAPTER 1 TEST

**1.** Rational number

**2.** Irrational number

**3.**

**4.**

**5.** $-3$

**6.** 28

**7.** $\frac{8}{5}$

**8.** $-5.62$

**9.** Associative property of addition

**10.** Distributive property

**11.** $x = -1$

**12.** $x = \frac{2}{5}$

**13.** $[11, \infty)$

**14.** $|3x| + |-7.5| = 55$

**15.** a) $27.95C + 42.95L$

   b) \$2565.35

   c) \$2525.10

   d) Monday; \$40.25

## Chapter 2

SECTION 2.1

**1.** Straight line

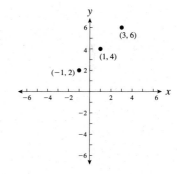

**3.** Not a straight line

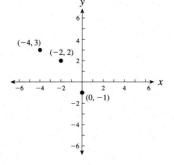

**5.** Not a straight line

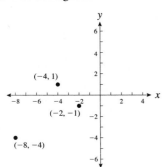

**7.** $(-4, 5)$

**9.** $(-4, -3)$

**11.** c

**13.** $X\min = -10$

$X\max = 15$

$X\text{scl} = 5$

$Y\min = -8$

$Y\max = 10$

$Y\text{scl} = 1$

**15.** $X\min = -300$

$X\max = 450$

$X\text{scl} = 100$

$Y\min = 30$

$Y\max = 50$

$Y\text{scl} = 5$

**17.** Answers will vary, depending on grapher model.

**19.** Answers will vary, depending on grapher model.

**21.** Answers will vary, depending on grapher model.

**23.** Answers will vary, depending on grapher model.

**25.** Window #3

**27.** Window #3

**29.** 5

**31.** $\sqrt{65}$

**33.** $\sqrt{61}$

**35.** $\sqrt{65}$

**37.** $\sqrt{2}|a - b|$

**39.** No

**41.** Yes

**43.** $(10, 10)$; $(-10, 10)$; $(-10, -10)$; $(10, -10)$

**45.** $(5, 5)$; $(-5, 5)$; $(-5, -5)$; $(5, -5)$

**47.** $(x_1 - x_2)^2$

**49.** $(x_1 - x_2)^2 + (y_1 - y_2)^2$

**51.** $d(P, R) = d(Q, R) = \sqrt{17}$; $d(P, Q) = \sqrt{34}$; $(\sqrt{17})^2 + (\sqrt{17})^2 = (\sqrt{34})^2$

**53.** Isosceles

**55.** Isosceles

## SECTION 2.2

**1.** $(0, 6)$, $(3, 0)$

**3.** $(2, 1)$, $(-3, 11)$

**5.** $(3, 6.5)$

**7.** 32

**9.** $-32$

**11.** 9

**13.** 80.625

**15.**

| $x$ | $y$ |
| --- | --- |
| $-3$ | $-18$ |
| $-2$ | $-13$ |
| $-1$ | $-8$ |
| $0$ | $-3$ |
| $1$ | $2$ |
| $2$ | $7$ |
| $3$ | $12$ |

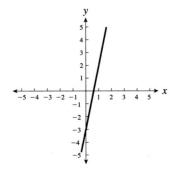

**17.**

| $x$ | $y$ |
| --- | --- |
| $-3$ | $-14$ |
| $-2$ | $-9$ |
| $-1$ | $-4$ |
| $0$ | $1$ |
| $1$ | $6$ |
| $2$ | $11$ |
| $3$ | $16$ |

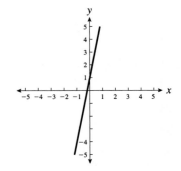

**19.**

| x | y |
|----|----|
| −3 | 14 |
| −2 | 11 |
| −1 | 8 |
| 0 | 5 |
| 1 | 2 |
| 2 | −1 |
| 3 | −4 |

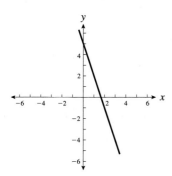

**21.**

| x | y |
|----|----|
| −3 | −3 |
| −2 | 2 |
| −1 | 7 |
| 0 | 12 |
| 1 | 17 |
| 2 | 22 |
| 3 | 27 |

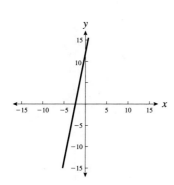

**23.**

| x | y |
|----|--------|
| −3 | −23/3 |
| −2 | −17/3 |
| −1 | −11/3 |
| 0 | −5/3 |
| 1 | 1/3 |
| 2 | 7/3 |
| 3 | 13/3 |

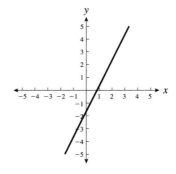

**25.**

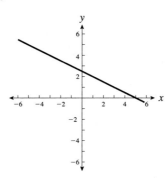

**27.**

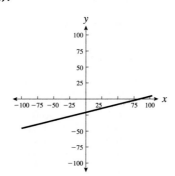

**29.**

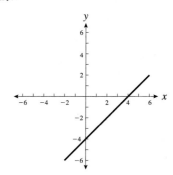

**31.**

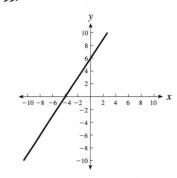

**33.**

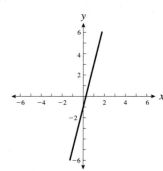

**35.**

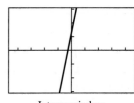

Integer window

**37.**

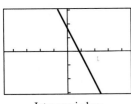

Integer window

**39.**

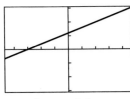

Integer window

**41.**

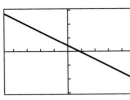

Integer window

**43.**

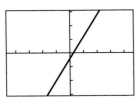

Integer window

**45.** $(-30, -96)$, $(-15, -51)$, $(15, 39)$, $(30, 84)$

Integer window

**47.** $(-30, -16.75)$, $(-15, -9.25)$, $(15, 5.75)$, $(30, 13.25)$

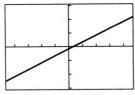

Integer window

**49.** $(-30, 18)$, $(-15, 12)$, $(15, 0)$, $(30, -6)$

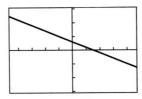

Integer window

**51.** $(-30, 27.75)$, $(-15, 18.375)$, $(15, -0.375)$, $(30, -9.75)$

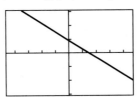

Integer window

**53.** $2x + 9y = 243$

**55.** $y = 3 - x^2$

**57.** Graph (d)

**59.** Graph (a)

**61.**

| $x$ | $y$ |
|---|---|
| $-1.2$ | $2.84$ |
| $-0.9$ | $1.61$ |
| $-0.4$ | $-0.04$ |
| $0.3$ | $-1.51$ |
| $0.8$ | $-1.96$ |
| $1.7$ | $-1.51$ |
| $2.5$ | $0.25$ |

**63.** Written Exercise

**65.** a) $x = 4.5$

b) The graph crosses the $x$-axis at the point $(4.5, 0)$ and this has the same value of $x$ as the solution found in part a).

Integer window

## SECTION 2.3

**1.** $x$-intercept: $(6, 0)$;  $y$-intercept: $(0, -18)$

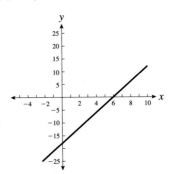

**3.** $x$-intercept: $(5, 0)$;  $y$-intercept: $(0, 4)$

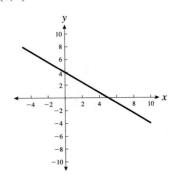

**5.** $x$-intercept: $(3, 0)$; $y$-intercept: $(0, -5)$

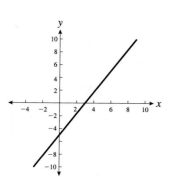

**11.**

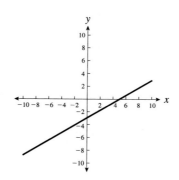

**17.**

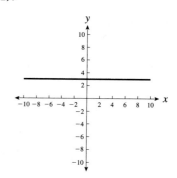

**19.**

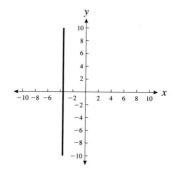

**7.** $x$-intercept: $(-28, 0)$; $y$-intercept: $(0, 12)$

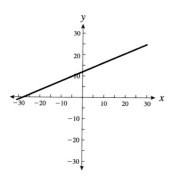

**13.**

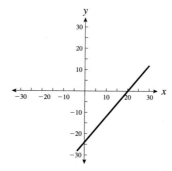

**21.**

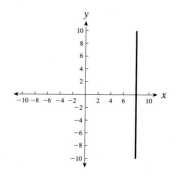

**9.**

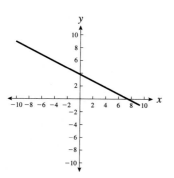

**15.**

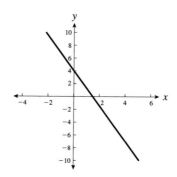

**23.** $x = 3.2$

**25.** $m = 4$

**27.** $m = 3$

**29.** $m = \frac{9}{59} = 0.153$

**31.** $m = \frac{5}{7}$

**33.** $m = \frac{3}{2}$

**35.** $m = -4$

**37.** $m = -\frac{1}{6}$

**39.** $m = \frac{19}{43}$

**41.** $m = 3$

**43.** $m = 0.5$

**45.** $m = -0.5$

**47.** $x = 3$

**49.** $y = 8$

**51.** $x = -0.5$

**53.**

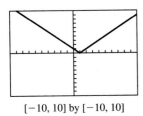

$[-10, 10]$ by $[-10, 10]$

**55.**

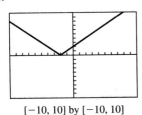

$[-10, 10]$ by $[-10, 10]$

**57.**

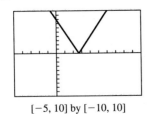

$[-5, 10]$ by $[-10, 10]$

**59.** $y_2 - y_1$

**61.** $\dfrac{y_2 - y_1}{x_2 - x_1}$

**63.** $\dfrac{y_2 - y_1}{x_2 - x_1}$

**65.** a) $I = 0.08x$

b) $y = 1.08x$

c) $m = 1.08$

**67.** Only the points in part (b)

**69.** All three points

**71.** $m = \frac{1}{2}$

**73.** $m = 1$

**75.** Written Exercise

**77.** Written Exercise

**79.**

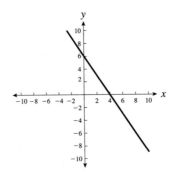

**81.**

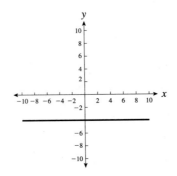

**83.**

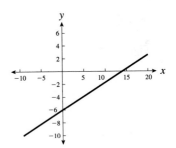

**85.**

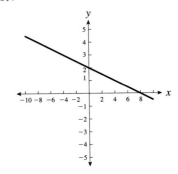

**87.**

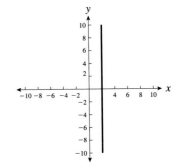

**89.** $x + (-12) = 0$

## SECTION 2.4

**1.** $m = 3$; $y$-intercept: $(0, -4)$

**3.** $m = -\frac{3}{4}$; $y$-intercept: $\left(0, \frac{7}{4}\right)$

**5.** $m = \frac{5}{2}$; $y$-intercept: $\left(0, -\frac{15}{2}\right)$

**7.** $m = 1$; $y$-intercept: $\left(0, \frac{1}{2}\right)$

**9.** $m = \frac{3}{7}$; $y$-intercept: $\left(0, \frac{2}{7}\right)$

**11.** Figure (b)

**13.**

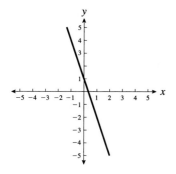

**15.**

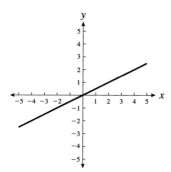

**17.**

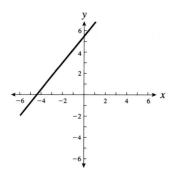

**19.**

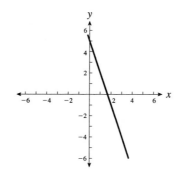

**21.**

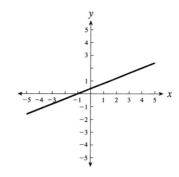

**23.** $y = 5x + 3$

**25.** $y = \frac{3}{4}x + \frac{1}{2}$

**27.** $y = -2.5x - 5.4$

**29.** $m = \frac{3}{2}$; $y$-intercept: $\left(0, -\frac{9}{2}\right)$

**31.** $m = -\frac{5}{2}$; $y$-intercept: $(0, 12)$

**33.** $m = -\frac{5}{3}$; $y$-intercept: $\left(0, \frac{14}{3}\right)$

**35.** $m = -\frac{5}{3}$; $y$-intercept: $(0, 4)$

**37.** $y = 3x - 1$

**39.** $y = -2.5x$

**41.** $y = \frac{1}{2}x + \frac{5}{2}$

**43.** Perpendicular

**45.** Neither

**47.** Neither

**49.** $3x + y = 5$

**51.** $-x + 3y = 7$

**53.** $4x + 3y = 32$

**55.** $x + 2y = -1$

**57.** $-3x + 2y = 3$

**59.** $3x + 4y = 17$

**61.** 20 units

**63.** $b = 6$

**65.** $m = \frac{1}{2}$

**67.** $x + 3y = 7$

**69.** $-(y_2 - y_1)x + (x_2 - x_1)y = x_2 y_1 - x_1 y_2$

**71.** $(-2, 0)$

**73.**

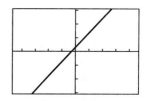

Integer window

**75.** Written Exercise

SECTION 2.5

**1.** $f(0) = 3$; $f(1) = 5$

**3.** $f(3) = 9$; $g(3) = \frac{1}{2}$; $h(3) = 4$

**5.** 304

**7.** $-10$

**9.** $2a + 3$

**11.** $f(a+2) = 2a + 7$; $f(a) + 2 = 2a + 5$

**13.** $f(x) = \frac{3}{5}x - \frac{12}{5}$

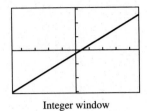

Integer window

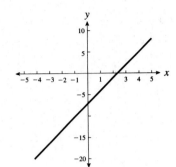

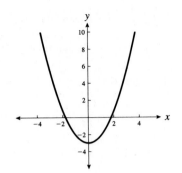

**15.** $f(x) = -\frac{1}{3}x + 12$

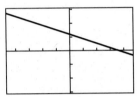

Integer window

**21.**

| $x$ | $g(x)$ |
|-----|--------|
| $-3$ | 23 |
| $-2$ | 20 |
| $-1$ | 17 |
| 0 | 14 |
| 1 | 11 |
| 2 | 8 |
| 3 | 5 |

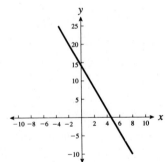

**25.**

| $x$ | $h(x)$ |
|-----|--------|
| $-3$ | 19 |
| $-2$ | 13 |
| $-1$ | 9 |
| 0 | 7 |
| 1 | 7 |
| 2 | 9 |
| 3 | 13 |

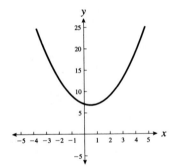

**17.** $f(x) = \frac{1}{3}x + \frac{17}{6}$

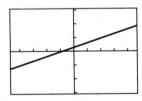

Integer window

**27.** b

**29.** b

**31.** b

**33.** $f(0) = 2$; $f(-1) = -2$; $f(4) = 2$

**35.** $x = 0$ or $x = 4$

**37.** a) Domain: $[-2, 2]$; Range: $[0, 1]$

b) Domain: $(-\infty, \infty)$; Range: $[0, \infty)$

**39.** $(-\infty, \infty)$

**19.**

| $x$ | $f(x)$ |
|-----|--------|
| $-3$ | $-16$ |
| $-2$ | $-13$ |
| $-1$ | $-10$ |
| 0 | $-7$ |
| 1 | $-4$ |
| 2 | $-1$ |
| 3 | 2 |

**23.**

| $x$ | $h(x)$ |
|-----|--------|
| $-3$ | 6 |
| $-2$ | 1 |
| $-1$ | $-2$ |
| 0 | $-3$ |
| 1 | $-2$ |
| 2 | 1 |
| 3 | 6 |

**41.** $(-\infty, \infty)$

**43.** $[0, \infty)$

**45.** $[-5, \infty)$

**47.** $[0, \infty)$

**49.** $(-\infty, \infty)$

**51.** $[0, \infty)$

**53.** Domain: $(-\infty, \infty)$;
Range: $[0, \infty)$

**55.** Domain: $[-3, \infty)$;
Range: $[0, \infty)$

**57.** Domain: $(-\infty, \infty)$;
Range: $[-1, \infty)$

**59.** Domain: $(-\infty, \infty)$;
Range: $[0, \infty)$

**61.** Not a function

| $x$ | $y$ |
|---|---|
| 0 | 0 |
| 1 | $\pm\sqrt{2}$ |
| 2 | $\pm 2$ |
| 3 | $\pm\sqrt{6}$ |
| 4 | $\pm\sqrt{8}$ |

**63.** Function

| $x$ | $y$ |
|---|---|
| 0 | 0 |
| 1 | 0.5 |
| 2 | 2 |
| 3 | 4.5 |
| 4 | 8 |

**65.** Not a function

| $x$ | $y$ |
|---|---|
| 0 | $\pm\sqrt{3}$ |
| 1 | $\pm 2$ |
| 2 | $\pm\sqrt{5}$ |
| 3 | $\pm\sqrt{6}$ |
| 4 | $\pm\sqrt{7}$ |

**67.** a) Function

   b) Not a function

**69.** a) Not a function

   b) Not a function

**71.** $f(x) = 2x - 5$

**73.** $f(x) = x^2 + 5$

**75.** a) $L = x + 20$

   b) $P(x) = 4x + 40$

   c)

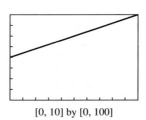

[0, 10] by [0, 100]

   d) $x = 27.5$ ft

**77.** All real numbers except $\pm 1$

**79.** All real numbers except $-2$

**81.** All real numbers except 0, 3

**83.** Range: $[-10.125, \infty)$

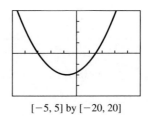

[−5, 5] by [−20, 20]

**85.** Range: $[-18.20, \infty)$

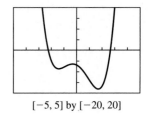

[−5, 5] by [−20, 20]

**87.** Range: $(-\infty, 16.37]$

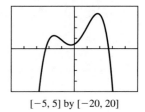

[−5, 5] by [−20, 20]

**89.** a) $x = \dfrac{a + 7}{3}$

   b) Written Exercise

REVIEW EXERCISES

**1.**

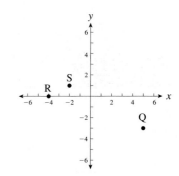

**2.** $(-5, 0), (-5, 150), (15, 150),$ $(15, 0)$

**3.** $\sqrt{116}$

**4.** 5

**5.** $\frac{1}{4}$

**6.** $-3$

**7.**

| $x$ | $y$ |
|-----|-----|
| $-10$ | $16$ |
| $-5$ | $12$ |
| $0$ | $8$ |
| $5$ | $4$ |
| $10$ | $0$ |

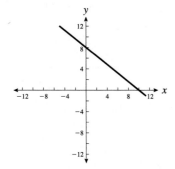

**8.**

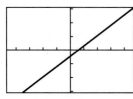

Integer window

**9.**

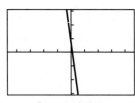

Integer window

**10.** $x$-intercept: $(-28, 0)$,
$y$-intercept: $(0, -21)$

**11.** $x$-intercept: $\left(-\frac{5}{2}, 0\right)$,
$y$-intercept: $\left(0, \frac{9}{2}\right)$

**12.**

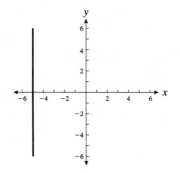

**13.**

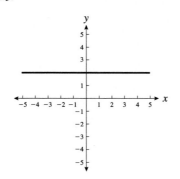

**14.** $m = \frac{16}{11}$

**15.** $m = -\frac{2}{9}$

**16.** $m = -\frac{3}{5}$

**17.** $m = \frac{7}{10}$

**18.** $y = -6$

**19.**

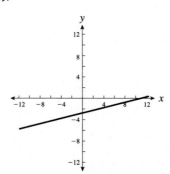

**20.**

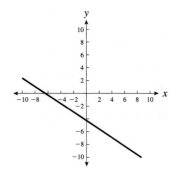

**21.** Perpendicular

**22.** Neither

**23.** $y = \frac{2}{3}x - 2$

**24.** $y = -1.4x + 3$

**25.** $y = -3x - 11$

**26.** $y = \frac{3}{2}x - 14$

**27.** $2x + y = 11$

**28.** $-x + 2y = 0$

**29.** $4x + y = 22$

**30.** $3x + 2y = 11$

**31.** $f(0) = 4$

**32.** $f(2) = 1$

**33.** $x = 3$

**34.** $x = -1$

**35.** $f(-2) = -2; f(a) = a^2 + 3a$

**36.** $g(0) = 1; g(b) = |2b + 1|$

**37.** Window (c)

**38.** Window (c)

**39.** Domain: $[8, \infty)$;
Range: $[0, \infty)$

**40.** Domain: $(-\infty, \infty)$;
Range: $[-16, \infty)$

**41.** This is not a function.

| $x$ | $y$ |
|-----|-----|
| 0 | 0 |
| 1 | $\pm1$ |
| 2 | $\pm2$ |
| 3 | $\pm3$ |
| 4 | $\pm4$ |

**42.** Collinear

**43.** $m = -\frac{3}{5}$

**44.** a) $\frac{3}{5}$; b) $\frac{2}{5}$; slope in Figure (a) is greater

## CHAPTER 2 TEST

**1.** $\sqrt{41}$

**2.** 3

**3.**

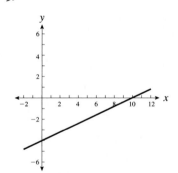

**4.**

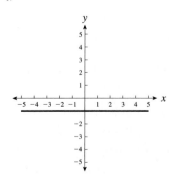

**5.**

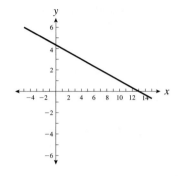

**6.** $\left(\frac{35}{2}, 0\right)$ and $(0, -7)$

**7.** $-\frac{2}{3}$

**8.** $-\frac{6}{7}$

**9.** $x = -2$

**10.** $-x + y = 4$

**11.** $y = 2x + 4$

**12.** $x + 3y = 7$

**13.** $-0.639$

**14.** $3a - 9$

**15.** Domain: $(-\infty, \infty)$; Range: $(-\infty, 5]$

**16.** $\frac{1}{2}$

**17.** Intercepts: $(3, 0)$ and $(0, 5)$

Integer window

**18.** Window (c)

**19.** Window (c)

# Chapter 3

## SECTION 3.1

**1.** $x = 21$

**3.** $x = -4$

**5.** $x = 14$

**7.** $x = -17$

**9.** $x = 7$

**11.** $x = 6$

**13.** $x = 9$

**15.** $x = 7$

**17.** $x = 4$

**19.** $x = -17$

**21.** $x = 16$

**23.** $x = -20$

**25.** $x = 5$

**27.** $x = -3$

**29.** $x = -14$

**31.** $x = -3$

**33.** $2x + 5 = 21$

**35.** $2x = 17 - 5$

**37.** $3x = 17$

**39.** $3x + 2 < 21$

**41.** 14

**43.** $26

**45.** $28,000

**47.** 39 mi

**49.** Answers will vary; $x = -7$

**51.** 2

**53.** $y = -\frac{1}{3}x + \frac{7}{3}$

SECTION 3.2

**1.** $x = 4$

**3.** $x = 1.5$

**5.** $x = 2.5$

**7.** $x = -0.5$

**9.** $x = 8$

**11.** Graphing: $x = 2.35$; Algebraic: $x = \frac{7}{3}$

**13.** Graphing: $x = -2.25$; Algebraic: $x = -\frac{16}{7}$

**15.** Graphing: $x = 1.75$; Algebraic: $x = \frac{95}{54}$

**17.** Graphing: $x = 3.35$; Algebraic: $x = \frac{370}{109}$

**19.** $x = 4.667$

**21.** $x = 6.667$

**23.** $x = 5.238$

**25.** $x = 19.259$

**27.** $5000 - 0.04x$

**29.** $x - 0.07x$

**31.** a) $y = 8x + 125$

b)

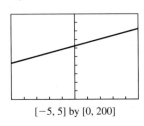

[−5, 5] by [0, 200]

c) March 1995

**33.** a) $y = 65000 - 8125x$

b)

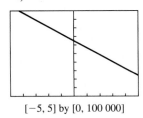

[−5, 5] by [0, 100 000]

c) during 1996

**35.** $55000 - 8000t$; $32000 - 2500t$; yes, during fourth year

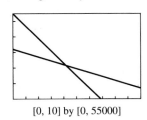

[0, 10] by [0, 55000]

**37.** $x > 2.5$

**39.** $x \geq -3.2$

SECTION 3.3

**1.** Window (b)

**3.** Window (c)

**5.** Window (b)

**7.** $x = -2.640$

**9.** $x = 5.310$

**11.** $x = 1.250$

**13.** $x = 14.000$

**15.** $x = 0.813$

**17.** $x = 11.000$

**19.** No solution

**21.** Two solutions

**23.** No solution

**25.** $x = -2.646, x = 2.646$

**27.** $x = 0.783, x = 10.217$

**29.** $x = 3.184$

**31.** $x = 0.141$

**33.** $23\sqrt{x} = 48$

**35.** $2x^2 = 14 - 6$

**37.** a) 7.071 sec

b) 2304.000 ft

c) 5184.000 ft

d) 2916.000 ft

**39.** $x = 2.727$

**41.**

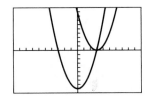

[−10, 10] by [−10, 10]
Standard window

**43.** $x = 400$

**45.** $x = \frac{55}{3}$

SECTION 3.4

**1.** a) The $24 - 2x$ labels change to $30 - 2w$.

b) $A = w(30 - 2w)$

c)

Area

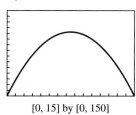

Width

d)

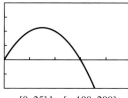

[0, 15] by [0, 150]

e) $x \geq 15$ is not a part of the problem situation since area is no longer positive.

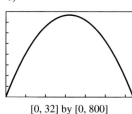

[0, 25] by [−100, 200]

f) 5.2 ft or 9.8 ft

g) 7.5 ft

3. a) Answers will vary.

b) $A = x(96 - 3x)$

c)

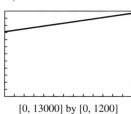

[0, 32] by [0, 800]

d) 10.6 ft or 21.4 ft

e) 768 sq ft

5. a) [0, 3.15] by [0, 40]

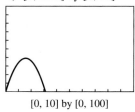

[0, 10] by [0, 100]

b) 0.6 sec or 2.5 sec

c) 39.1 ft after 1.6 sec

d) 80 ft per sec

7. a) $y = 12 + 1.80t$

b)

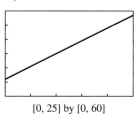

[0, 25] by [0, 60]

c) $55.20

d) 11.75 hr

9. a) $0.09x + 0.07(13,000 - x)$

b) $X$max $= 13,000$; $Y$max $= 1200$

c)

[0, 13000] by [0, 1200]

11. $y = 1.25x$

13. 24 miles; read off $y$-coordinate of point of intersection

15. Written Exercise

17. The temperature with the same number of degrees in Celsius or Fahrenheit

SECTION 3.5

1. Solution

3. Not the solution

5. $(12, -10)$

7. $(-10, 5)$

9. $(-5, 15)$

11. $(25, -15)$

13. $(18, -9)$

15. $(17, 10)$

17. $(10.200, 6.250)$

19. $(-15.500, 20.500)$

21. $(3.500, 18.500)$

23. $(2.250, 3.101)$

25. $(15.510, -26.250)$

27. $(7.143, -1.429)$

29. $(0.800, 0.800)$

31. $(5.000, 2.000)$

33. No solution

35. One solution

37. An infinite number of solutions

39. a) $x + y = 750$

b) $5x + 2.50y = 3250$

c) 550 adults

41. a) Cost: $y = 1.25x + 5.25x + 8000$

b) Revenue: $y = 25x$

c) 433 reproductions

43. $0.25x$

45. $12000 + 125.50x$

47. Answers will vary.

49. No

51. Written Exercise

**53.** $y = \frac{2}{3}x + 2$ and $y = \frac{1}{2}x - 6$; $(-48, -30)$

## SECTION 3.6

**1.** Answers will vary.

**3.** Answers will vary.

**5.** $2(4) + y = 10$ or $8 + y = 10$

**7.** $3(-y + 9) - 2y = 12$ or $27 - 5y = 12$

**9.** $6(y - 5) - y = -5$ or $5y - 30 = -5$

**11.** $(4, 3)$

**13.** $(3, 5)$

**15.** $(29, -14)$

**17.** $(-1, 2)$

**19.** $(8.5, 2.25)$

**21.** $\left(\frac{7}{3}, \frac{2}{3}\right)$

**23.** Answers will vary.

**25.** Answers will vary.

**27.** Answers will vary.

**29.** Answers will vary.

**31.** $(3, 5)$

**33.** $(1, 8)$

**35.** No solution

**37.** $(6, 3)$

**39.** $(225, 700)$

**41.** $(2, -1)$

**43.** $(1, 1)$

**45.** $(-5, 3)$

**47.** $(-2, -6)$

**49.** $(-2, 3.5)$

**51.** $(54, -34)$

**53.** Infinite number of solutions

**55.** $3x + 4y = 144$
$3x - 4y = -48$

**57.** $x + 2y = -12$
$3x - 2y = 36$

**59.** $15x + 4y = 9$
$5x + 12y = 3$

**61.** $x + 45 = 5y$
$6x + 5y = 80$

**63.** $(8, -6)$

**65.** $(2, 8)$

**67.** $(5, -10)$

**69.** a) $x + y = 9500$
$0.05x + 0.075y = 582.50$

b) \$5200 at 5%; \$4300 at 7.5%

**71.** a) $5x + 250y = 175$
$3x + 180y = 108$

b) \$30.00 per day and \$0.10 per mile

**73.** 14 nickels and 9 quarters

**75.** 2 pairs of jeans and 6 pairs of socks

**77.** $25°, 65°, 90°$

**79.** 25, 68

**81.** $\left(\frac{1}{4}, -\frac{1}{2}\right)$

**83.** $A \neq 24$

**85.** $25°, 65°$

**87.** 12 ft by 15 ft

**89.** a) $y = 0.60x$

b) $y = \$53.97$

c) $x = \$113.68$

## REVIEW EXERCISES

**1.** $x = -5$

**2.** $x = 8$

**3.** $x = 6$

**4.** $x = -3$

**5.** $x = 2$

**6.** $x = -2.3$

**7.** $x = -3.692$

**8.** $x = -0.196$

**9.** $x = 6.410$

**10.** $x = 3.840$

**11.** Window (b)

**12.** Window (c)

**13.** One solution

**14.** $x = -5.908, x = 4.908$

**15.** $x = 3.520$

**16.** Solution

**17.** Not a solution

**18.** No solution

**19.** An infinite number of solutions

**20.** $(3.875, 3.625)$

**21.** $(8.333, 0.667)$

**22.** $(20, 12)$

**23.** $(6, 17)$

**24.** $(2, 11)$

**25.** $(-2, 3)$

**26.** $(4, -1)$

**27.** $(5, 5)$

**28.** $(0, 2)$

**29.** $(12, 30)$

**30.** $2\sqrt{x} = 8$

**31.** $0.01x + 0.10y = 8.24$

**32.** $0.65x = \frac{1}{2}x + 3$

**33.** 5

**34.** a) $R = 2(25) + 0.30x$

b)

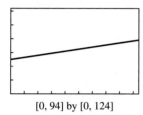

[0, 94] by [0, 124]

c) \$116

d) 300 mi

**35.** a) Cost $= 175 + 12x$, where $x$ is the number of years since 1990

b)

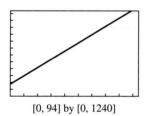

[0, 94] by [0, 1240]

c) \$259

d) The year 2005

**36.** a) $\begin{cases} x + y = 20000 \\ 0.08x + 0.06y = 1431 \end{cases}$

b) \$11,550 at 8%, \$8450 at 6%

CHAPTER 3 TEST

**1.** a) $y_1 = 3x - 8; y_2 = 7x + 5$

b) $y = 3x - 8 - 7x - 5$ or $y = -4x - 13$

**2.** Window (c)

**3.** No solution

**4.** $x = 5.172$

**5.** $x = -10.167, x = -0.333$

**6.** $x = 1.545$

**7.** An infinite number of solutions

**8.** $\left(\frac{3}{5}, -\frac{1}{5}\right)$

**9.** $\left(\frac{125}{17}, -\frac{9}{17}\right)$

**10.** $\left(\frac{7}{8}, -\frac{1}{4}\right)$

**11.** $(-1, 5)$

**12.** $(-3, 8)$

**13.** $2x - 6 = 0.4x$

**14.** a) $\begin{cases} x + y = 29 \\ x - y = 42 \end{cases}$

b) $-\frac{13}{2}, \frac{71}{2}$

**15.** a) Length: $\ell = 40 - 2x$

b) Area: $y = (40 - 2x)(x)$

c) 25 ft by 7.5 ft or 15 ft by 12.5 ft

d) 10 ft

# Chapter 4

SECTION 4.1

**1.** Yes

**3.** No

**5.** Yes

**7.** No

**9.** $[5, \infty)$

**11.** $(4, \infty)$

**13.** $x > 5$

**15.** $x \le 2$

**17.** $x \ge 0$

**19.** $x < 7.25$

**21.** $(-\infty, 6.35)$

**23.** $[2.5, \infty)$

**25.** $(-2.3, \infty)$

**27.** $(-\infty, -3]$

**29.** $[13, \infty); x \ge 13$

**31.** $[-2, \infty); x \ge -2$

**33.** $(-\infty, -55]; x \le -55$

**35.**

**37.**

**39.**

**41.**

**43.** $(5, \infty)$

**45.** $(-\infty, -5]$

**47.** $(-\infty, 2.25)$

**49.** $[-2.03125, \infty)$

**51.** $(-\infty, -2]$

**53.** $(-\infty, -4)$

**55.** $(-\infty, \infty)$

**57.** $[6, \infty)$

**59.** $[5.5, \infty)$

**61.** $(5.25, \infty)$

**63.** $(-\infty, -14]$

**65.** $(-\infty, -38]$

**67.** $[-3.25, \infty)$

**69.** $(-\infty, 5.79)$

**71.** $[3, \infty)$

**73.** $(-\infty, -2.25]$

**75.** a) $\dfrac{267 + x}{4}$

b) $x \geq 93\%$

**77.** 100%

**79.** 520 mi

**81.** After 8 days

**83.** At most 7 coils

**85.** Written Exercise

**87.** $x + 6.5 > 10.35$

**89.** $x > 0$

**91.** $x \geq 3$

**93.** $x \geq 65$

**95.** $\dfrac{x - 5}{2} > 12$

**97.** $a + b + c < 52$

**99.** $x = 1$

**101.** At least 2.5 liters

## SECTION 4.2

**1.** $2x - 11 < 0$

**3.** $-2x - 11 < 0$

**5.** $x + 15 \geq 0$

**7.** $x \geq 0$

**9.** $(-\infty, 2)$

**11.** $(4, \infty)$

**13.** $(-\infty, -2)$

**15.** $[-7, \infty)$

**17.** $(-\infty, -0.667)$

**19.** $[9, \infty)$

**21.** $(-\infty, 7)$

**23.** $(-\infty, 5]$

**25.** $(-\infty, 5)$

**27.** $(26, \infty)$

**29.** $[3, \infty)$

**31.** $(2.167, \infty)$

**33.** $(-\infty, -6.892]$

**35.** a) $y = 4x - 2.75x$

b)

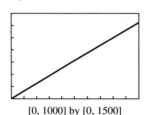

[0, 1000] by [0, 1500]

c) At least 396 boxes

d) More than 800 boxes

**37.** a) $y = 6x - 525$

b)

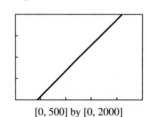

[0, 500] by [0, 2000]

c) At least 338 tickets

**39.** a) $y = 140 + 20x$

b)

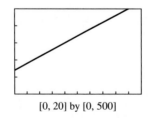

[0, 20] by [0, 500]

c) At least 11 orders

**41.** a) $(0.482, 3.893)$

b) 76.563 ft at 2.188 sec

c) $(2.188, 4.375)$

d) 4.375 sec

**43.** $2(x + 5.6) \geq 2.8$

**45.** Written Exercise

**47.** Written Exercise

**49.**

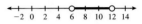

**51.** $\pi r^2 > 17$

**53.** $8x < 96$, where $x$ is the length of one side of the octagon

**55.** Calculator exercise

## SECTION 4.3

**1.**

**3.**

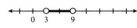

**5.**

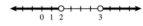

**7.**

**9.** $(-3, 3)$

**11.** $(1, \infty)$

**13.** $(-3, 7)$

**15.** $(-\infty, 2] \cup [4, \infty)$

**17.** No solution

**19.** $x < 1$ or $x > 5$

**21.** $x < 2$ and $x > -5$

**23.** $(4, 10)$

**25.** $[-5.35, 8.9)$

**27.** $(2, 7]$

**29.** $[7.24, 19.5]$

**31.** $5 < x < 8$

**33.** $-3 \le x \le 2$

**35.** $23 \le x < 62.4$

**37.** $7.5 < x < 9.99$

**39.** $(2, 4)$

**41.** $(-\infty, \infty)$

**43.** $(-1, 4)$

**45.** $(5, \infty)$

**47.** Yes

**49.** No

**51.** $(-1, 6)$

**53.** $\left[-\frac{5}{6}, 4\right]$

**55.** $(9.1, 12.8)$

**57.** $[-4, -2)$

**59.** $\left[\frac{22}{3}, 11\right]$

**61.** $(-0.1, 2]$

**63.** $(0.1, 2)$

**65.** $(-5, 6)$

**67.** $[-7.5, -4.5]$

**69.** $(-44, -38)$

**71.** $[-6, -1]$

**73.** $\left(\frac{9}{2}, \frac{17}{3}\right]$

**75.** $(3.5, 17]$

**77.** $56 < x + 32 < 79$

**79.** $92 < \frac{87+x}{2} < 100$

**81.** a) $\frac{279+x}{4}$

b) $85 \le \frac{279+x}{4} \le 92$

c) $61 \le x \le 89$

**83.** $x < 1500$ or $x > 3500$

**85.** Written Exercise

**87.** The distance between 4 and the origin

**89.** Positive

**91.** $|x|$ is nonnegative.


## SECTION 4.4

**1.** $x = -10, x = 10$

**3.** $x = -4, x = 4$

**5.** $x = -14, x = 10$

**7.** a) Yes   b) No

**9.** a) Yes   b) Yes

**11.** a) No   b) Yes

**13.** $x = -2, x = 8$

**15.** $x = -\frac{4}{3}, x = 4$

**17.** $x = \frac{13}{8}, x = \frac{37}{8}$

**19.** $x = -48, x = 24$

**21.** $x = -9, x = 6$

**23.** $x = -\frac{5}{3}, x = 7$

**25.** No solution

**27.** $x = 0, x = 14$

**29.** a) $x = -9, x = 7$

b) $(-9, 7)$

c) $(-\infty, -9) \cup (7, \infty)$

**31.** a) $x = -32, x = 36$

b) $(-32, 36)$

c) $(-\infty, -32) \cup (36, \infty)$

**33.** One interval

**35.** Union of two intervals

**37.** One interval

**39.** $x = -12, x = 10$

**41.** $x = 2, x = 8$

**43.** No solution

**45.** $x = -12, x = 2$

**47.** $x = -3.333, x = 4$

**49.** $x = -10, x = 7$

**51.** No solution

**53.** $x = -17, x = 21$

**55.** $x = -12.100, x = 7.900$

**57.** $x = -7.800, x = 5.250$

**59.** $(-7, 7)$

**61.** $(-2, 6)$

**63.** $(-\infty, -9] \cup [9, \infty)$

**65.** $\left(-\infty, -\frac{15}{2}\right) \cup \left(\frac{21}{2}, \infty\right)$

**67.** $\left[-\frac{103}{33}, \frac{61}{33}\right]$

**69.** $(-12, 10)$

**71.** $(-\infty, -6.500] \cup [11.500, \infty)$

**73.** $(-\infty, \infty)$

**75.** $(-\infty, -8.800) \cup (4.750, \infty)$

**77.** $|x| \le 5$

**79.** $|x| \ge 8$

**81.** $|x - 7| > 2$

**83.** Written Exercise

**85.** $|x| \le 3$

**87.** $|x - 3| > 2$

**89.** $|x - 2| \le 5$

**91.** \$113.75

**93.** 30%

**95.** \$24.95


## SECTION 4.5

**1.** Yes

**3.** Yes

**5.** No

**7.** $m = 0$; solid

**9.** $m = 3$; dashed

**11.** $m = -5$; dashed

**13.** $m = \frac{2}{5}$; dashed

**15.** $m = \frac{1}{2}$; dashed

**17.**

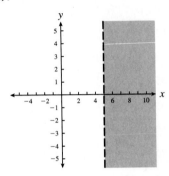

**19.**

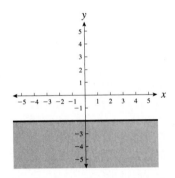

**21.**

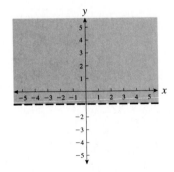

**23.**

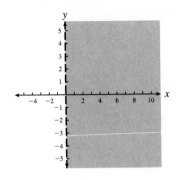

**25.**

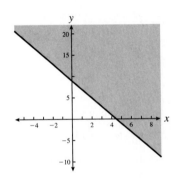

**27.**

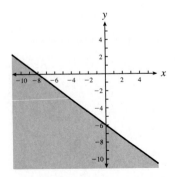

**29.**

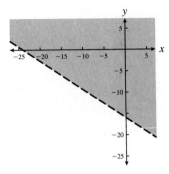

**31.**

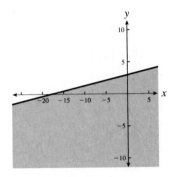

**33.** $y > -\frac{5}{2}x + 7$

**35.** $y \geq -\frac{3}{2}x + 3$

**37.** $y > \frac{2}{3}x - 4$

**39.**

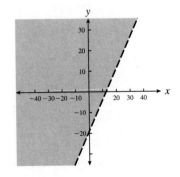

**41.**

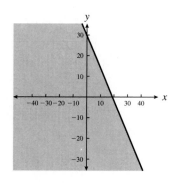

**43.** $2500 \geq 885 + C; C \leq 1615$

**45.** $3x + 5y \geq 2500$

**47.** $x > 0$

**49.** $y > 6$

**51.** $y \leq 325$

**53.** $25 < x < 75$

**55.** Written Exercise

**57.** a)  The amount of money lost at the end of the day

b)  $0.50x + 0.75y < 25$

c)

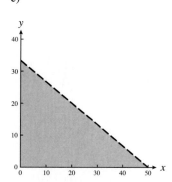

**59.** $y < -3x + 4$

**61.** $y > 5$

**63.** $(-9, 5)$

**65.** $(2, -3)$

**67.** $(-5, 5)$

### SECTION 4.6

**1.** $\begin{cases} 21 > 7 \\ 8 < 10 \end{cases}$

**3.** $\begin{cases} -10 < -5 \\ 1 < 9 \end{cases}$

**5.** a)

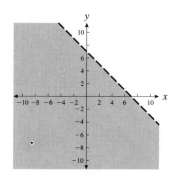

b)

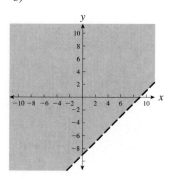

c)

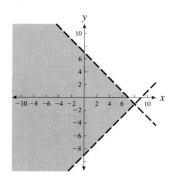

**7.** a)

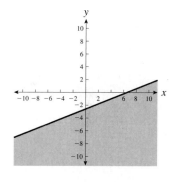

b)

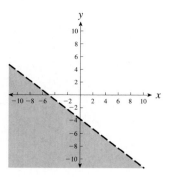

c)

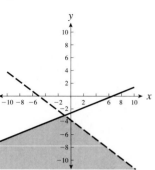

c)

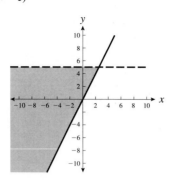

**15.**

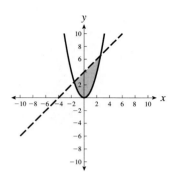

**9.** a)

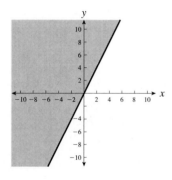

**11.**

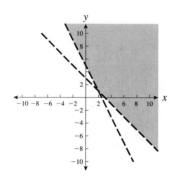

**17.**

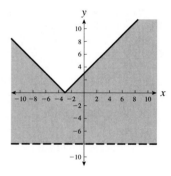

b)

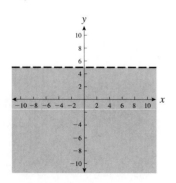

**13.**

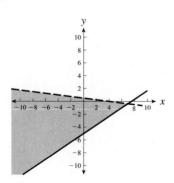

**19.**

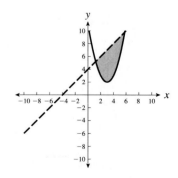

**21.**

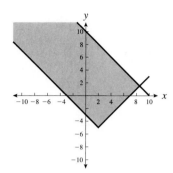

**23.**

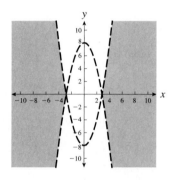

**25.** a) $C(x) = 22,000 + 20.25x$

b) $R(x) = 39.95x$

c) Total cost

d) The solution will indicate when a profit is made.

e) At least 1117 machines

### REVIEW EXERCISES

**1.** $(5.5, \infty)$

**2.** $(-\infty, -2]$

**3.** $(-\infty, 3) \cup [6, \infty)$

**4.** $(4, 10]$

**5.**

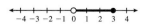

**6.** $(-\infty, 6)$

**7.** $[-1, 5)$

**8.** Yes

**9.** No

**10.** Yes

**11.** $(-\infty, 7)$

**12.** $[1.5, \infty)$

**13.** $[3, 4]$

**14.** $(-8, -5)$

**15.** $(1, 5.5)$

**16.** $[-1.5, 1]$

**17.** $(-\infty, 4) \cup (8, \infty)$

**18.** $[-11, 5]$

**19.** $\left[\frac{2}{3}, 3\right]$

**20.** $(4, \infty)$

**21.** $x = -3, x = \frac{9}{5}$

**22.** $x = -9, x = 16$

**23.** $x = -4, x = 24$

**24.** $x = 20, x = 28$

**25.**

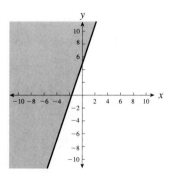

**26.**

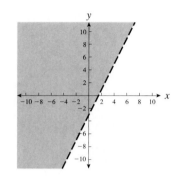

**27.**

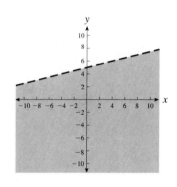

**28.**

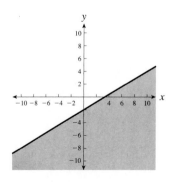

**29.**

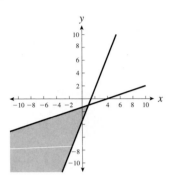

**30.**

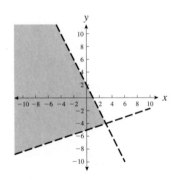

**31.**

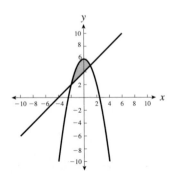

**32.**

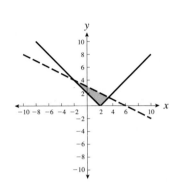

**33.** $|x + 1| < 3$

**34.** $x \leq -5$

**35.** $8.1 < x + 10 < 16.6$

**36.** $|x - 7| \geq 4$

**37.** $2(w + 5) + 2w \geq 40$

**38.** a) $\dfrac{75 + 82 + x}{3}$

b) At least 83%

**39.** a) $x + y \leq 500$

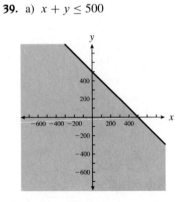

b) $2x + 3y > 1250$

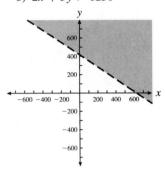

c) Yes, answers will vary.

CHAPTER 4 TEST

**1.** $(-\infty, -2] \cup (5, \infty)$

**2.** $[0, 3]$

**3.** $x = -9, x = \frac{11}{3}$

**4.** $\left[-\frac{4}{5}, \infty\right)$

**5.** $\left(-\infty, \frac{7}{2}\right)$

**6.** $(74, 86)$

**7.** $[-12, 0]$

**8.** $\left(-\infty, -\frac{1}{5}\right)$

**9.** $(-\infty, \infty)$

**10.**

**11.**

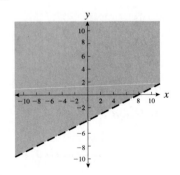

**12.**

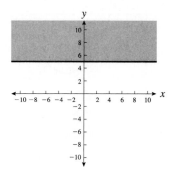

**13.**

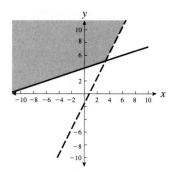

**14.** $|x - 3| < 6$

**15.** a) $2.50 + 0.12x$

    b) $4.66

    c) $2.50 + 0.12x < 4$

    d) 12 checks

## Chapter 5

SECTION 5.1

**1.** 72

**3.** 250

**5.** $\frac{1}{25}$

**7.** $\frac{3}{4}$

**9.** $\frac{29}{4}$

**11.** 1

**13.** $a^6$

**15.** $9x^6$

**17.** $8y^6$

**19.** $\frac{1}{6561}$

**21.** $ab^2$

**23.** $\frac{yz^2}{3x}$

**25.** $\frac{2}{a^6 b^9}$

**27.** $2^{21} a^{24} b^6$

**29.** $\frac{a^2 x}{b}$

**31.** $ax^2$

**33.** $\frac{x^3}{27 y^3}$

**35.** $\frac{2a^4 b^5}{3x^5 y^5}$

**37.** $\frac{3}{8 p^2 q}$

**39.** $1.587 \times 10^9$

**41.** $3.7 \times 10^{-8}$

**43.** $5.79 \times 10^5$

**45.** $1.32 \times 10^{15}$

**47.** $2.3 \times 10^1$

**49.** $5.451 \times 10^4$

**51.** $3.999 \times 10^3$

**53.** 65.45

**55.** $5.892 \times 10^{12}$ in.

**57.** $3.344 \times 10^{15}$ mi

**59.** a) $a^2 \div b$

    b) $(a \div b)^2$

**61.** True; $(xy)^3 = x^3 y^3$

**63.** $x$

**65.** $x = 2$

**67.** $(-4, \infty)$

SECTION 5.2

**1.** Polynomial expression

**3.** Neither

**5.** Polynomial Function

**7.** Binomial; degree: 1; coefficient: 3

**9.** Trinomial; degree: 4; coefficient: 3

**11.** Binomial; degree: 3; coefficient: $-4$

**13.** $a = -3, b = 4, c = -5$

**15.** $a = -1, b = 3, c = 12$

**17.** $2x^2 + x - 3, c$

**19.** $-x^3 - 2x^2 + 3x - 2, d$

**21.** $f(1) = -8, f(13) = -3668$

**23.** $f(0) = -7, f(3) = -28$

**25.**

| Seconds | Distance Traveled |
| --- | --- |
| $-4$ | 763 |
| $-3$ | 169 |
| $-2$ | $-7$ |
| $-1$ | $-23$ |
| 0 | $-17$ |
| 1 | $-7$ |
| 2 | 109 |
| 3 | 553 |
| 4 | 1667 |

**27.** $c$

**29.** $b$

**31.** $5x^2 - 6x + 8$

**33.** $-5y^3 + y^2 - 7y + 5$

**35.** $2x^2y + 4xy + 7y$

**37.** $5p^3 - p^2 + 5p + 1$

**39.** $-7x^2y^2 + 10xy^2 - 2$

**41.** $3x^2 + 4x - 4$

**43.** $-5x^2 + 8x - 3$

**45.** Written Exercise

**47.** $c$

**49.** $\frac{176}{3}\pi$ cc or 184.307 cc

**51.** $192\pi$ sq cm or 603.186 sq cm

**53.** $8.424\pi$ cc or 26.465 cc

**55.** $V = 13.5$ cc, $SA = 40.5$ sq cm

**57.** $2x + x^2$

**59.** $3x - x^2$

**61.** $-4x^{2q} + 10x^q - 7$

**63.** a) Between $x = 3$ and $x = 4$
b) It will cross between $x = 3$ and $x = 4$.
c) Yes

**65.** a) $|-42t + 100|$
b) Answers will vary;

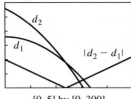

[0, 5] by [0, 300]

c) Yes, at $t = 2.381$ sec

**67.** a)

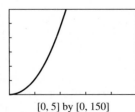

[0, 5] by [0, 150]

b) The volume of the container will be 50 when the radius is $x = 1.338$.

**69.** $A(x) = 5x^2 + 13x + 3$

**SECTION 5.3**

**1.** $2x^3 + 6x$

**3.** $-x^3 - 3x^2 + 2x$

**5.** $10x^4 - 6x^5$

**7.** $15x^5 - 20x^4 + 10x^3 - 15x^2$

**9.** $x^2 + x - 6$

**11.** $6x^2 - x - 2$

**13.** $35x^2 + 11x - 6$

**15.** $3x^3 + 9x^2 + 4x + 12$

**17.** $x^3 - 2x^2 + 3x - 6$

**19.** $6x^5 + 13x^3 + 5x$

**21.** $x^5 - 7x^4 - 6x^2 + 42x$

**23.** Graph (c)

**25.** Graph (a)

**27.** $x^2 + 8x + 15$

**29.** $x^2 + 2x - 8$

**31.** $2x^2 + 7x - 15$

**33.** $3y^2 + 5ay - 2a^2$

**35.** $x$-intercepts: $(-3, 0), (4, 0)$

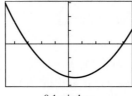

0.1 window

**37.** $x$-intercepts: $(-0.5, 0), (3, 0)$

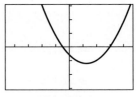

0.1 window

**39.** Written Exercise

**41.** $g(x) = x^2 + 14x + 49$

**43.** $k(x) = x^2 - 18x + 81$

**45.** $g(x) = x^2 - 16x + 64$

**47.** $f(x) = x^2 - \frac{3}{2}x + \frac{9}{16}$

**49.** $h(x) = x^2 - \frac{16}{3}x + \frac{64}{9}$

**51.** $f(x) = x^2 - 0.5x + 0.0625$

**53.** Written Exercise

**55.** $x^5 + 4x^4 - 2x^3 + 5x^2 + 20x - 10$

**57.** $x^5 + 2x^3 + x^2 - 15x - 3$

**59.** $-x^5 + 3x^4 + 5x^3 - 12x^2 - 15$

**61.** $12 = x(x^2 + 4)$

**63.** $V(x) = 6x^2 + 4x$

**65.** $a = 10, \; c = -5$

**67.** $x^3 - 2x^2 - x - 6$

**69.** $9 - x^2 - 2xy - y^2$

**71.** $(x + 2)(2x + 2)$

**73.**

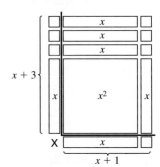

**75.** $x^3 - 2x - 1$
$x^4 - 2x^2 - 2x - 1$
$x^5 - 2x^3 - 2x^2 - 2x - 1$
Pattern:
$(x + 1)(x^n - x^{n-1} - \ldots - 2x - 1) = x^{n+1} - 2x^{n-1} - 2x^{n-2} - \ldots - 2x - 1$

**77.** $(4, 0), (-1, 0), (-5, 0)$

**79.** $(4, 0)$ only

**81.** $(2x + 5)(3x - 1) - \pi x^2$


## SECTION 5.4

**1.** $1, 2, 4, 8, x, 2x, 4x, 8x$

**3.** $1, 2, 3, 6, x, 2x, 3x, 6x, x^2,$
$2x^2, 3x^2, 6x^2$

**5.** $1, x, x^2, (x + 1), x(x + 1),$
$x^2(x + 1)$

**7.** $1, 3, x, y, y^2, y^3, 3x, 3y, 3y^2,$
$3y^3, 3xy, 3xy^2, 3xy^3, xy, xy^2, xy^3$

**9.** $1, 3, 9, x, (2x - 5), (2x - 5)^2,$
$x(2x - 5), x(2x - 5)^2, 3x, 9x,$
$3x(2x - 5), 3x(2x - 5)^2, 3(2x - 5),$
$3(2x - 5)^2, 9(2x - 5), 9(2x - 5)^2,$
$9x(2x - 5), 9x(2x - 5)^2$

**11.** $x = \frac{3}{5}, \ x = -\frac{5}{2}, x = -2$

**13.** $x = \frac{3}{2}, x = \frac{7}{3}, x = 5$

**15.** Answers will vary. One example
is $(x + 3)(x - 1)(x - 5) = 0$.

**17.** Answers will vary. One example
is $(x - 1)(x - 3)(2x - 1) = 0$.

**19.** $d$

**21.** $c$

**23.** $2x$

**25.** $5xy$

**27.** $3$

**29.** $(x + 3)^2$

**31.** $x(3x^3 - 5x + 2)$

**33.** $5x(1 + 5x^2 - 2x)$

**35.** $(x + y)(x + y - 5)$

**37.** $(x + 1)^2(3x - 1)$

**39.** $3xm^2(5xm - 7 + 11x^3)$

**41.** $(x - y)(p - q + r)$

**43.** $(x + y)(x + y)$

**45.** $x = 0, x = -\frac{3}{2}$

**47.** $x = 0, x = \frac{4}{3}$

**49.** $x = 0, x = \pm 2$

**51.** $x = 2, x = -\frac{1}{2}$

**53.** $(y - 3b)(2x - a)$

**55.** $(x - 4)(x - 3)$

**57.** $(2x + 1)(x^2 - 3)$

**59.** $(3x^2 + 1)(x - 2)$

**61.** $0, 4$ or $0, -4$

**63.** $0, 2$ or $-\frac{2}{3}, 0$

**65.** Yes, two linear factors

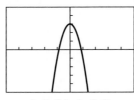

$[-5, 5]$ by $[-5, 5]$

**67.** No linear factors

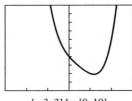

$[-3, 3]$ by $[0, 10]$

**69.** a)

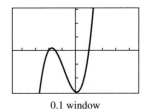

0.1 window

b) $x = 1, x = -1.5, x = -2$
c) $y = (x - 1)\left(x + \frac{3}{2}\right)(x + 2)$

**71.** $-2 < x < 4$

**73.** $6 < x < 7$

**75.** $x < -3$ or $x > 8$


## SECTION 5.5

**1.** Perfect square; $(x - 2)^2$

**3.** Perfect square; $(2y + 3)^2$

**5.** Not a perfect square

**7.** Perfect square; $(x + 2y)^2$

**9.** Perfect square; $(x + \sqrt{2})^2$

**11.** $(x - 3)(x - 7)$

**13.** $(x + 4)(x - 3)$

**15.** $(x - 7)(x + 4)$

**17.** $(x + 7)(x + 2)$

**19.** $(x - 9)(x - 5)$

**21.** $(2x - 1)(x + 3)$

**23.** $(3x - 2)(2x + 3)$

**25.** $(2x + 1)(3x - 5)$

**27.** $(2x - a)(x + 2a)$

**29.** $(3r^2 + s)(r^2 - 2s)$

**31.** $(3q^2 - 4x)(q - 2x)$

**33.** $x = -5, \ x = 4$

**35.** $x = -2, \ x = 5$

**37.** $x = -4$, $x = \frac{1}{2}$

**39.** $x = 5$

**41.** $x = -\frac{3}{2}$, $x = 0$, $x = 1$

**43.** $(x - b)(x^2 + bx + b^2)$

**45.** $(3x + y)(9x^2 - 3xy + y^2)$

**47.** $(4x + 7)(4x - 7)$

**49.** $125(2k + 1)(4k^2 - 2k + 1)$

**51.** Written Exercise

**53.** $126 = (x + 5)x$, where $x$ is width

**55.** $18 = \frac{1}{2}(x + 3)(x)$, where $x$ is altitude

**57.** $(x^2 - 7)(x^2 + 7)$

**59.** $(3x^2 + 2)(x^2 - 4)$

**61.** $(x - 3)(x^2 + 2)$

**63.** $(x + 2)(2x^2 - 1)$

**65.** $(x^t + y^s)(x^t - y^s)$

**67.** Written Exercise

**69.** Written Exercise

## SECTION 5.6

**1.** $x = -3$, $x = 8$

**3.** $x = -3$, $x = 7$

**5.** $x = -4$, $x = \frac{3}{2}$

**7.** $x = -\frac{2}{3}$, $x = 4$

**9.** $x = -\frac{5}{2}$, $x = 1$

**11.** $x = -\frac{2}{5}$, $x = \frac{3}{2}$

**13.** $x = 0$, $x = -1$, $x = 10/3$

**15.** $x = -3$, $x = \frac{1}{2}$

**17.** $x = \frac{5}{2}$, $x = 15$

**19.** $x = 0.382$, $x = 2$, $x = 2.618$

**21.** $x = -3.177$, $x = -0.678$, $x = 1.856$

**23.** $x = -1.162$, $x = 0.321$, $x = 1.341$

**25.** $x(x + 5) = 14$

**27.** $x(x - 9) = 90$

**29.** 2, 7 or $-7$, $-2$

**31.** $-6$, $-15$ or 15, 6

**33.** $L = 40$, $W = 30$

**35.** a) $h(t) = -16t^2 + 128t$

   b) $192 = -16t^2 + 128t$

   c) $t = 2, 6$ sec

**37.** a) $V = 4x^3 - 110x^2 + 750x$

   b) $4x^3 - 110x^2 + 750x = 800$

   c)

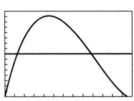

[0, 13] by [0, 1600]

   d) $x = 1.304$, $x = 8.829$

**39.** a) $A(x) = (12 - 2x)^2$, $V(x) = x(12 - 2x)^2$

   b)

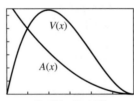

[0, 6] by [0, 130]

   c) $x = 1.817$ in

   d) 127.169 cu in

   e) Yes, the graphs intersect

**41.** Graphical Method

**43.** Graphical Method

## REVIEW EXERCISES

**1.** $\frac{8}{3}$

**2.** 600

**3.** $\dfrac{27x^{12}}{8a^{15}}$

**4.** $\dfrac{y^7}{x}$

**5.** $\dfrac{axy^2}{b^2}$

**6.** $\dfrac{4x^{14}}{y^6}$

**7.** $8.00154 \times 10^{-2}$

**8.** 320, 580

**9.** $7.965 \times 10^{12}$

**10.** a) Trinomial b) Eight c) Four

**11.** $-14$

**12.** $6x^4 - x + 5$

**13.** $6x^2y^3 + 2xy^2 + 3$

**14.** $24x^2 - 41x$

**15.** $x^3 + 5x^2 - 7x - 35$

**16.** $16x^2 - 34x + 15$

**17.** $x^3 + 2x^2 + x - 4$

**18.** $1, 5, x, 5x, y, 5y, xy, 5xy$

**19.** $1, x, x^2, x - 3, x(x - 3),$ $x^2(x - 3)$

**20.** $4x^2y$

**21.** $3x(x - 5)(7x - 10)$

**22.** $x(x + 7)(x + 2)$

**23.** $(3x + 4)(2x - 5)$

**24.** $(x - 6)^2$

**25.** $(2s + t)(s - 2t)$

**26.** $(8x - 1)(x + 2)$

**27.** $(4x + 1)(4x - 1)$

**28.** $(x+1)(x^2-x+1)$

**29.** $(3t+5)^2$

**30.** $y(2y+7)(10y-1)$

**31.** $(3x+5)(x-6)$

**32.** $(x^2-3)(xy+1)$

**33.** $x=-3, x=-\frac{1}{2}$

**34.** $x=-1, x=0$

**35.** $x=\frac{1}{2}, x=4$

**36.** $x=-7, x=-\frac{5}{2}, x=3$

**37.** $x=1.382, x=3.618$

**38.** $x=-1.732, x=-1,$ $x=1.732$

**39.** The cylinder with radius 2 in. and height 2.5 in.

**40.** $V(x)=(36-2x)(24-2x)(x)$

**41.** a) $x+x^2=52$

b) $-7.728$

CHAPTER 5 TEST

**1.** 80

**2.** $\dfrac{x^3}{yz}$

**3.** $\dfrac{a^2}{4b^8}$

**4.** $2.14\times10^4; 0.000627$

**5.** $-4x^3-8x^2-9x+12$

**6.** $x^3+x+10$

**7.** $4x^2-12x+9$

**8.** $rs^2$

**9.** $(2x-5)(x+2)$

**10.** $(2x-3)(4x^2+6x+9)$

**11.** $(x+a)(x-b)$

**12.** $x=-1, x=0, x=1$

**13.** $x=-0.862$

**14.** a) $x(x-6)=27$

b) 3 and 9 or $-3$ and $-9$

**15.** Length: 32 ft; width: 14.5 ft

**16.** $(x-5)(3x-4)$

# Chapter 6

SECTION 6.1

**1.** $\frac{5}{7}$

**3.** $\frac{7}{8}$

**5.** $\frac{5}{8}$

**7.** b

**9.** a

**11.** $x-2$

**13.** $\dfrac{3}{4x^2}$

**15.** $\dfrac{x}{x+7}$

**17.** $\dfrac{3x+7}{x-5}$

**19.** $-1$

**21.** 1

**23.** $\dfrac{xy}{4(x+16)}$

**25.** $\dfrac{4y}{7x}$

**27.** $x>0$

**29.** $\dfrac{x}{2}$

**31.** $\dfrac{x}{y}$

**33.** Yes, $x+3$

**35.** Yes, $x-25$

**37.** Yes, $x-2$

**39.** $-1$

**41.** $\dfrac{1}{2x+3}$

**43.** $-\dfrac{2x}{x+3}$

**45.** $\dfrac{x+7}{2x-7}$

**47.** $5x+2$

**49.** $x+2$

**51.** $\dfrac{x-w}{x+w}$

**53.** $\dfrac{x}{y}$

**55.** $\dfrac{x}{y}+1$

**57.** a) $2.3+0.004x$

b) $\dfrac{2.3+0.004x}{x}$

**59.** a) No

b) No, $\dfrac{1}{x}$

c) No, because there are no common linear factors.

**61.** b

**63.** $x^a-2$

**65.** $\dfrac{x^{2a}+5x^a+25}{x^a+5}$

**67.** $\dfrac{p^x-4}{2p^x-3}$

**69.** Written Exercise

**71.** $\frac{35}{18}$

**73.** $\frac{3}{8}$

**75.** Answers will vary.

**77.** Answers will vary.

**79.** $a^2+2ab+b^2-9$

**81.** $4x^2+20x+25-y^2$

**83.** $-x^2+14x+107.25$

## SECTION 6.2

**1.** $\dfrac{1}{b}$

**3.** $\dfrac{wz}{xy^2}$

**5.** $\dfrac{2a^2b}{5}$

**7.** $a$

**9.** $\dfrac{x^2(x-1)}{x^2+2}$

**11.** $\dfrac{(2x-3)(x+4)}{(x-1)(x+2)}$

**13.** $\frac{1}{2}x(x-2)$

**15.** $\dfrac{x+8}{x-2}$

**17.** $\dfrac{ab(a-b)}{a+b}$

**19.** $\dfrac{1}{(x+7)(x-5)}$

**21.** Graph (a)

**23.** a) $240 \text{ ft}^3$

    b) $8.889 \text{ yd}^3$

**25.** $\dfrac{(x-3)(x+2)}{x+3}$

**27.** $(x-1)(x+2)$

**29.** $-\dfrac{x+4}{x+2}$

**31.** Written Exercise

**33.** $(x+y)^2$

**35.** $\dfrac{a(a-b)}{(a-6)(a+b)}$

**37.** a) $\frac{1}{8}$ min

    b) $\frac{x}{8}$ min

    c) $\$0.04x$

    d) $\frac{x}{24}$ min

**39.** $\dfrac{x+3}{4x(x+4)}$

**41.** $\dfrac{x-1}{x}$

**43.** 1

**45.** $\dfrac{2x-3y}{3x-y}$

**47.** $\dfrac{2x+3}{x-1}$

**49.** $\dfrac{5(x+1)(x-2)(x-4)}{x^2-2}$

**51.** $\dfrac{1}{x(x+1)}$

**53.** $\dfrac{1}{x^{m+n}}$

**55.** $\dfrac{x^m-2}{x^m+3}$

**57.** $\frac{3}{2}$

**59.** $3x^2+x+14$

## SECTION 6.3

**1.** $\dfrac{x+3}{5}$

**3.** $\dfrac{2x+3}{2}$

**5.** $\dfrac{2x+1}{7}$

**7.** $\dfrac{2x}{x+3}$

**9.** $\dfrac{7x+1}{(x+3)(x-2)}$

**11.** 1

**13.** $-1$

**15.** $\dfrac{x+2}{x-3}$

**17.** $\dfrac{x^2+1}{x}$, a

**19.** Written Exercise

**21.** $9x$

**23.** $2x$

**25.** $x-2$

**27.** $\dfrac{35x}{21x^3}$

**29.** $\dfrac{-2(x+3)}{5x(x+3)}$

**31.** $\dfrac{13}{2x}$

**33.** $\dfrac{45-112x}{210x}$

**35.** $\dfrac{73}{80x}$

**37.** $\dfrac{6y^2+9y+27}{y(y+3)}$

**39.** $\dfrac{5x^2-4x}{(x-4)(x+4)}$

**41.** $x(x+1)(x-3)$

**43.** $(x+1)(x-2)(x+3)$

**45.** $\dfrac{8x+4}{(x-1)(x+3)}$

**47.** $\dfrac{3x}{(x+1)(x-2)}$

**49.** $\dfrac{4x^2-8x+5}{(x-1)(x-2)(x+3)}$

**51.** $\dfrac{5x+3}{(x-1)(x+5)}$

**53.** $\dfrac{3000r}{(r+18)(r-18)}$

**55.** $\dfrac{1}{n}+\dfrac{1}{n+1}$ or $\dfrac{2n+1}{n(n+1)}$

**57.** $\dfrac{1}{2n}-\dfrac{1}{2n+2}$ or $\dfrac{2}{(2n)(2n+2)}$

**59.** a) $\dfrac{x}{55}$

    b) $\dfrac{300-x}{65}$

    c) $\dfrac{2x+3300}{715}$

**61.** Answers will vary.

**63.** $\dfrac{1}{R} = \dfrac{R_2 R_3 + R_1 R_3 + R_1 R_2}{R_1 R_2 R_3}$

**65.** $\dfrac{1}{3}$

**67.** $\dfrac{1 + x - x^2}{x^2}$

**69.** $\dfrac{20}{(x-5)(x+5)}$

**71.** $\dfrac{a^2 + a + 1}{a + 1}$

**73.** $\dfrac{7x - 8}{(x+1)(x-2)(2x+3)}$

**75.** $\dfrac{3x^2 - 32x + 86}{(x-5)^3}$

**77.** $\dfrac{x^3 + 3x^2 + 3x + 2}{(x-1)(x^2+x+1)}$

**79.** $\dfrac{x+1}{x^2}$

**81.** $\dfrac{7}{2x}$

**83.** $\dfrac{2x^2 + x + 1}{x^2}$

**85.** $AB = 8,\ A'C' = 5$


**SECTION 6.4**

**1.** $\dfrac{10}{x^2}$

**3.** $\dfrac{y^2}{x}$

**5.** $\dfrac{(x-y)^2}{(x+y)^2}$

**7.** $\dfrac{(x+y)^2}{(x-y)^2}$

**9.** $\dfrac{x+y}{x-y}$

**11.** $\dfrac{(x+3)(x-1)}{(x-2)(x+4)}$

**13.** $\dfrac{5(2y+9)}{81y}$

**15.** $\dfrac{1}{\dfrac{1}{x} + \dfrac{1}{5}}$

**17.** $\dfrac{4x}{15}$

**19.** $y_1 = \dfrac{x^{-2} + 4^{-2}}{x^{-1} + 4^{-1}},\ y_2 = x + 4$

a)

| X | Y₁ | Y₂ |
|---|------|----|
| 1 | .85    | 5  |
| 2 | .41667 | 6  |
| 3 | .29762 | 7  |
| 4 | .25    | 8  |
| 5 | .22778 | 9  |
| 6 | .21667 | 10 |
| 7 | .21104 | 11 |

b)

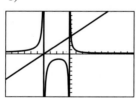

$[-10, 10]$ by $[-10, 10]$
Standard window

**21.** Show that $\dfrac{\text{ft/sec}}{\text{sec}} = \dfrac{\text{ft}}{\text{sec}^2}$.

**23.** $ab^2$

**25.** $x^2 y^2$

**27.** $\dfrac{a^2 b + b^2}{a - b^2}$

**29.** $\dfrac{3xy^2 + 2x^2 y}{2y^2 - x^2}$

**31.** $\dfrac{1 - 2x}{2 + 3x}$

**33.** $-\dfrac{9x + 30}{22x}$

**35.** $\dfrac{a+b}{b}$

**37.** $\dfrac{xy^2 - 5}{xy^2 - 2y}$

**39.** $\dfrac{3x^2 + y^2}{x^2 + 3y^2}$

**41.** $\dfrac{-5a + b}{3a + 2b}$

**43.** $-\dfrac{y}{x}$

**45.** $\dfrac{-(x+3)(x+2)^2}{x(x+1)(8x+19)}$

**47.** Show that the slope of the line through the points $\left(x_1, \dfrac{1}{x_1}\right)$ and $\left(x_2, \dfrac{1}{x_2}\right)$ is $\dfrac{-1}{x_1 x_2}$.

**49.** $\dfrac{y}{x}$

**51.** $\dfrac{1 + xy}{1 - xy}$

**53.** $\dfrac{y + 3x}{y - 2x}$

**55.** $\dfrac{2y + x}{y - 3x}$

**57.** $\dfrac{1}{a} + \dfrac{1}{b}$

**59.** $\dfrac{1}{a + b}$

**61.** Written Exercise

**63.** $\dfrac{a^5 + 4a^3 + 3a}{a^4 + 3a^2 + 1}$

**65.** $2x^2 + 11x - 8$

**67.** $-4x + 9$

**69.** $a^2 + 2a + 1$

**71.** $-8x^2 + 9x - 15$

## SECTION 6.5

**1.** $3x^2 - 5x$

**3.** $3x^4 - 6x^2 + 2$

**5.** $4x + 8 - \dfrac{1}{x}$

**7.** $x^2 - 2x + 3 - \dfrac{4}{x}$

**9.** $x + 3y^2$

**11.** 5

**13.** 0

**15.** 8

**17.** $x^2 - 5x + 1$

**19.** $x^3 + 2x^2 - 5x + 3$

**21.** $x^4 - x^2 + 3$

**23.** $x^2 + 3x + 1 + \dfrac{0}{x^2 - 2x + 3}$

**25.** $x^4 + 2x^2 - 3x + 1 + \dfrac{-3}{2x^2 + 3x + 1}$

**27.** $3x^2 - x + 2 + \dfrac{2x}{x^2 + 1}$

**29.** $3x^2 - x - 1, 0$

**31.** $x^3 + 4x^2 - x + 4, 5$

**33.** $2x^3 - 3x^2 + x - 3, 0$

**35.** $f(-2) = 0$

**37.** $f(3) = 3$

**39.** $f(-1) = -29$

**41.** a) The graph is (a).
   b) The graph is (a).

**43.** False

**45.** False

**47.** a) 180
   b) $48x - 2x^2$
   c) $48 - 2x$

**49.** $\dfrac{x^4 - x^2}{x^2 - 1}$

**51.** Yes

**53.** Yes

**55.** No

**57.** $x^6 - x^5 + x^4 - x^3 + x^2 - x + 1 - \dfrac{2}{x + 1}$

**59.** $(s - 3)(s - 3)$

**61.** $(x - 5)(x - 2)$

**63.** $(2x + 3)(x - 4)$

**65.** $x = -3$

**67.** $x = \frac{1}{2},\ x = 1$

## SECTION 6.6

**1.** Graph (d); two solutions

**3.** Graph (b); two solutions

**5.** $x = 7$

**7.** $x = -2$

**9.** $x = 0.5$

**11.** $x = -2$

**13.** No solution

**15.** $x = -6, x = 2$

**17.** $x = 4$

**19.** $x = 5$

**21.** $x = 5$

**23.** No solution

**25.** $x = 3$

**27.** No solution

**29.** $x = -1$

**31.** No solution

**33.** Written Exercise

**35.** $x = 29$

**37.** $x = -2.25$

**39.** $x = 0$

**41.** $x = -2$

**43.** $\dfrac{1}{x} + \dfrac{1}{x + 1}$

**45.** $\dfrac{1}{2x + 1} + \dfrac{1}{2x + 3}$

**47.** 4 and 6

**49.** $-6$ and $-5$ or 5 and 6

**51.** 4.8 liters

**53.** Sell 46 programs of the 56 purchased

**55.** 1.429 hr

**57.** 9 hr

**59.** a) 66.667 ohms
   b) 142.857 ohms
   c) 125 ohms
   d) 480 ohms

**61.** $x = 5$

**63.** $x = 15$

**65.** a) There is no $y$ value that corresponds to $x = 1$.
   b) Written Exercise

**67.** $b_2 = \dfrac{2A - b_1 h}{h}$

**69.** $x = -5,\ x = \frac{3}{4}$

**71.** $x = 0,\ x = 7$

## REVIEW EXERCISES

**1.** $\frac{4}{7}$

**2.** $\dfrac{8y^2}{5x^3}$

**3.** $\dfrac{3x + 2}{x - 2}$

**4.** $-\dfrac{x}{3}$

**5.** $\dfrac{x+3}{x-3}$

**6.** $\dfrac{1}{6x-5}$

**7.** $\dfrac{y(x-2)}{x(x-1)}$

**8.** $\dfrac{y(x-3)}{2x(x+3)}$

**9.** $\dfrac{2(x+7)}{x+2}$

**10.** $\dfrac{x+2}{(x-2)(x+4)}$

**11.** $\dfrac{x}{36(x-5)}$

**12.** $(x+5)(2x+3)$

**13.** $(2x+1)(6x-5)(3x-4)$

**14.** $\dfrac{x+2}{x-4}$

**15.** $\dfrac{x-1}{x+1}$

**16.** $\dfrac{-4x+15}{x(x-3)}$

**17.** $\dfrac{3x^2+4x-8}{(x-3)(x+2)(x+3)}$

**18.** $\dfrac{(x-3)(x-1)}{5}$

**19.** $\dfrac{x(3x+1)}{(x+2)(3x-1)}$

**20.** $\dfrac{x+3}{x-4}$

**21.** $\dfrac{1}{4}$

**22.** $\dfrac{4x}{x-4}$

**23.** $\dfrac{2(x-1)}{x-2}$

**24.** $\dfrac{b^2(1+a)}{a}$

**25.** $\dfrac{x}{1-xy}$

**26.** $8x+\dfrac{3x^2}{y}$

**27.** $5x^3-9x+\dfrac{2}{x}$

**28.** $x^2-3x+5+\dfrac{1}{2x-3}$

**29.** $3x^2-2x-6+\dfrac{7}{x+2}$

**30.** $4x^2-10x+25-\dfrac{20}{x+2}$

**31.** $x^3+3x^2+6x-3-\dfrac{19}{x-3}$

**32.** $x=3.714$

**33.** $x=0.222$

**34.** $x=\frac{2}{15}$

**35.** $x=2$

**36.** $x=-2$

**37.** $x=-4,\ x=3$

**38.** $\dfrac{1}{x}-\dfrac{1}{x^2}=\dfrac{1}{x+x^2}$

**39.** a) $5x+\dfrac{2}{x}=9.25$

  b) $0.25$ or $1.6$

**40.** $7.5$ hr

CHAPTER 6 TEST

**1.** $\dfrac{3x-1}{2x+3}$

**2.** $(4x-5)(2x+3)(3x+2)$

**3.** $\dfrac{3x}{(x+3)(2x-1)}$

**4.** $\dfrac{-2(x^2+1)}{x^3y}$

**5.** $\dfrac{3x}{2x-3}$

**6.** $\dfrac{6x^2-14x-5}{(x+1)(2x-3)}$

**7.** $\dfrac{xy^2}{2}-\dfrac{3y^2}{x}$

**8.** $\dfrac{x+2}{(x-4)^2}$

**9.** $\dfrac{2-xy}{3}$

**10.** $2x^2-2x-1+\dfrac{2}{2x-1}$

**11.** $2x^2-5x-6-\dfrac{24}{x-3}$

**12.** $x=\frac{11}{5}$

**13.** $x=-2$

**14.** $x=-5.857$

**15.** $\frac{4}{3},\ 4$

# Chapter 7

SECTION 7.1

**1.** $8^{\frac{1}{2}}$

**3.** $12^{\frac{1}{3}}$

**5.** $x^{\frac{1}{2}}$

**7.** $\sqrt{3}$

**9.** $\sqrt[4]{6}$

**11.** $13.856$

**13.** $0.498$

**15.** $0.816$

**17.** $5.278$

**19.** $-0.869$

**21.** $2.774$

**23.** $\frac{5}{4}$

**25.** $5$

**27.** 13

**29.** 0.4

**31.** $-\frac{2}{3}$

**33.** $-6.889$

**35.** 26.851

**37.** 2.378

**39.** 2

**41.** 7

**43.** 2

**45.** $\frac{1}{4}$

**47.** 3

**49.** 7

**51.** 17

**53.** Graph (b)

**55.** Graph (a)

**57.** $\sqrt[3]{8}$

**59.** $4^{\frac{1}{2}}$

**61.** Not a right triangle

**63.** Right triangle

**65.** a) $x = 26$
   b) $x = \sqrt{12}$
   c) $x = \sqrt{18}$
   d) $x = \sqrt{89}$

**67.** 1300 ft

**69.** 8.307 in.

**71.** 6.633 ft

**73.** The square root of a negative number is not a real number.

**75.** Written Exercise

**77.** $8x^3 y^5$

**79.** $375x^5$

**81.** $\dfrac{5y^4}{x}$

**SECTION 7.2**

**1.** $(\sqrt{16})^5$, $\sqrt{16^5}$

**3.** $\sqrt[3]{5^2}$, $(\sqrt[3]{5})^2$

**5.** $\sqrt[5]{3^4}$, $(\sqrt[5]{3})^4$

**7.** $\sqrt[3]{27^2}$, $(\sqrt[3]{27})^2$

**9.** $5^{\frac{1}{2}}$

**11.** $3^{\frac{2}{3}}$

**13.** $x^{\frac{3}{2}}$

**15.** $(2y)^{\frac{3}{4}}$

**17.** $\sqrt{25} = 5$

**19.** $\sqrt[4]{16} = 2$

**21.** $(\sqrt[3]{8})^5 = 32$

**23.** $(\sqrt{9})^3 = 27$

**25.** $x^2$

**27.** $x^{\frac{5}{6}}$

**29.** $x^4$

**31.** $\dfrac{1}{x^{\frac{1}{4}} y}$

**33.** $\dfrac{y}{x^4}$

**35.** $(\sqrt[3]{x})^{19}$

**37.** $(\sqrt{x})^5$

**39.** $(\sqrt[6]{x})^7$

**41.** $\sqrt[6]{x}$

**43.** $\sqrt[4]{x}$

**45.** 25

**47.** $-123$

**49.** $(xy)^{\frac{1}{2}}$

**51.** $7|x|$

**53.** $x^2 y^{\frac{1}{3}}$

**55.** $(x^2 - y^2)^{\frac{1}{2}}$

**57.** 7

**59.** $|y|$

**61.** $x$

**63.** $-x^2$

**65.** a) Answers will vary.
   b) $A = 6\sqrt{11} \approx 19.900 \text{ cm}^2$

**67.** 2

**69.** Answers will vary.

**71.** a) 4
   b) 4
   c) 4

**73.** b

**75.** $\sqrt[3]{x}$

**77.** $\sqrt{x^3}$

**79.** Written Exercise

**81.** $\dfrac{1 + x^2}{x}$

**83.** 0

**SECTION 7.3**

**1.** $\sqrt{15}$

**3.** $\sqrt{21}$

**5.** $5\sqrt{2}$

**7.** $8\sqrt{3}$

**9.** $2\sqrt[3]{5}$

**11.** $5\sqrt{3}$

**13.** $2\sqrt{5}$

**15.** $3\sqrt[3]{2}$

**17.** $2\sqrt{2} \times 10^4$

**19.** $\sqrt{5.2} \times 10^6$

**21.** $\sqrt[3]{7.3} \times 10^5$

**23.** $y^2 \sqrt{y}$

**25.** $xy^2$

**27.** $x^3y^4\sqrt{y}$

**29.** $x^2y^3$

**31.** $2xy^2\sqrt{5}$

**33.** $4a^2b^4\sqrt{2ab}$

**35.** $3\sqrt{5}$

**37.** $8x$

**39.** $x^2y^4\sqrt{10}$

**41.** $72\,ab^2\sqrt{6}$

**43.** $3$

**45.** $2x^3y^3$

**47.** $y_1 = \sqrt{x^2 - 4}, \ y_2 = x - 2$

a)

| X | Y₁ | Y₂ |
|---|------|----|
| 1 | ERROR | −1 |
| 2 | 0 | 0 |
| 3 | 2.2361 | 1 |
| 4 | 3.4641 | 2 |
| 5 | 4.5826 | 3 |
| 6 | 5.6569 | 4 |
| 7 | 6.7082 | 5 |

b)

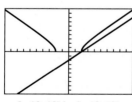

$[-10, 10]$ by $[-10, 10]$
Standard window

**49.** $\sqrt{3}$

**51.** $\frac{4}{5}$

**53.** $\dfrac{2x^3}{y^2}$

**55.** $\dfrac{1}{2xy}$

**57.** $\dfrac{1}{3xy}$

**59.** True

**61.** True

**63.** $\dfrac{x\sqrt{3}}{3}$

**65.** $\dfrac{\sqrt{2x}}{2}$

**67.** $\dfrac{x\sqrt{3}}{3}$

**69.** $\dfrac{2\sqrt[3]{147xy}}{7y}$

**71.** $\dfrac{\sqrt[4]{xy^3z^3}}{z}$

**73.** 63.246 mph

**75.** 3.536 sec

**77.** 3.845 sec

**79.** 38.568 cu units

**81.** $3x^2 + 2x$

**83.** $-2x^2 + 5x + 2$

**85.** $x + 4y + 1$

## SECTION 7.4

**1.** $4\sqrt{3}$

**3.** $\sqrt{11}$

**5.** $7\sqrt{x}$

**7.** $\frac{13}{6}\sqrt{3x}$

**9.** $5\sqrt{5}$

**11.** $28\sqrt{2}$

**13.** $29x\sqrt{5}$

**15.** $12\sqrt[3]{2x}$

**17.** $18x\sqrt{7}$

**19.** $2\sqrt{3} + 6$

**21.** $10\sqrt{3} + 4\sqrt{5}$

**23.** 8

**25.** $3\sqrt{7} - 7$

**27.** $2x\sqrt{3} + 3x$

**29.** $9 + 5\sqrt{6}$

**31.** $y + \sqrt{y} - 20$

**33.** $30y - 41\sqrt{y} - 6$

**35.** $y - 10\sqrt{y} + 25$

**37.** b

**39.** c

**41.** $\frac{4\sqrt{7}}{7}$

**43.** $3\sqrt{3}$

**45.** $12 - \frac{2}{3}\sqrt{3}$

**47.** $\frac{\sqrt{6}}{6}$

**49.** $\dfrac{(3x - 10)\sqrt{3x}}{15x}$

**51.** $\sqrt{2} + 3$

**53.** $7 - \sqrt{31}$

**55.** $\sqrt{x} + 3y$

**57.** $-25\sqrt{3} + 50$

**59.** $3\sqrt{3} - 3\sqrt{2}$

**61.** $\dfrac{5y(\sqrt{y} + 7)}{y - 49}$

**63.** $\dfrac{51 + 14\sqrt{2}}{-47}$

**65.** $\dfrac{x + y - 2\sqrt{xy}}{x - y}$

**67.** a) Answers will vary.

b) 5.179

**69.** $>$

**71.** $<$

**73.** $\dfrac{1 + 2\sqrt{3}}{2}$

**75.** $\dfrac{1 + 5\sqrt[3]{2x}}{2x}$

**77.** $\sqrt{x + y}$

**79.** No solutions

**81.** $6x^4 + 15x^3 + 18x^2$

**83.** $9x^2 - 25$

**85.** $-3 < x < 0$

**SECTION 7.5**

**1.** $x = \pm 2$

**3.** No real solutions

**5.** $x = 8$

**7.** $x = 71$

**9.** No real solutions

**11.** $x = 17$

**13.** $x = 237$

**15.** $x = \frac{17}{2}$

**17.** $x = 13$

**19.** $x = 1,\ x = 13$

**21.** $x = 3 + 2\sqrt{7}$

**23.** $x = 1$

**25.** $x = -2$

**27.** $x = 1$

**29.** $x = 7$

**31.** a) One solution
   b) Now have two solutions

**33.** $15 = \sqrt{x - 4}$

**35.** $25 = \sqrt{2 + x^2}$

**37.** $x = 37$

**39.** $x = 3$

**41.** No solutions

**43.** $x = 7$

**45.** $x = 17$

**47.** $x = 5$

**49.** $d = 256$ ft

**51.** $L = 12.986$ units

**53.** $L = 151.25$ ft

**55.** a) $h(x) = \sqrt{144 - x^2}$
   b) $A = \dfrac{x\sqrt{144 - x^2}}{2}$
   c) $x = 3.106,\ x = 11.591$ in

**57.** $x = -1.827,\ x = 3.381$

**59.** Written Exercise

**61.** $8\sqrt{2}$ ft

**63.** $2\sqrt{10}$ in

**65.** Written Exercise; $x \geq -6.5$

**SECTION 7.6**

**1.** $8 + 2i$

**3.** $-2 - 11i$

**5.** $5 + i$

**7.** $-\sqrt{3} - 12i$

**9.** $1 - 9i$

**11.** $-12$

**13.** $-6$

**15.** $8 + 12i$

**17.** $6 - 4i$

**19.** $4 + 18i$

**21.** $-3 + 3i$

**23.** $23 + 14i$

**25.** $20 + 17i$

**27.** $53$

**29.** $7 - 4i$

**31.** $5$

**33.** $-3 - 4i$

**35.** $-11 - 60i$

**37.** $\frac{16}{41} - \frac{20}{41}i$

**39.** $\frac{6}{13} + \frac{4}{13}i$

**41.** $\frac{3}{13} + \frac{2}{13}i$

**43.** $-\frac{45}{53} - \frac{28}{53}i$

**45.** $\frac{3}{5} + \frac{6}{5}i$

**47.** $-\frac{13}{29} + \frac{11}{29}i$

**49.** $\frac{3}{5} + \frac{4}{5}i$

**51.** $\frac{7}{2} + \frac{3}{2}i$

**53.** $\frac{1}{2} + \frac{3}{2}i$

**55.** $-i$

**57.** $-1$

**59.** $-i$

**61.** $1$

**63.** Written Exercise

**65.** $x = -3,\ x = 8$

**67.** $x = 0,\ x = \frac{7}{2}$

**69.** $x = -\sqrt{17},\ x = \sqrt{17}$

**71.** The sum of $-7 + 3i$ and $2 - 6i$:

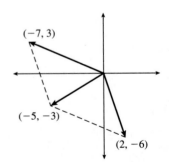

**73.** $-5 + 8i$

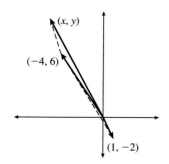

REVIEW EXERCISES

**1.** $(x-1)^{\frac{3}{5}}$

**2.** $\sqrt[3]{(2n)^2}$

**3.** 3.402

**4.** 1.587

**5.** $\frac{1}{7}$

**6.** $\frac{3}{10}$

**7.** 8000

**8.** $xy^2\sqrt[3]{x^2}$

**9.** $2x^2y^4\sqrt{10y}$

**10.** $4\sqrt{2}$

**11.** $2b\sqrt{a}$

**12.** $8\sqrt[3]{x^2y}$

**13.** $\sqrt[12]{x^{11}}$

**14.** $9x^2\sqrt{5}$

**15.** $xz\sqrt[4]{xy^3}$

**16.** $3\sqrt{2} + 2\sqrt{3}$

**17.** $12 - 7\sqrt{3}$

**18.** $x - 16$

**19.** $11 + 2\sqrt{30}$

**20.** 1

**21.** 0

**22.** $3 + 9i$

**23.** $27 + 2i\sqrt{2}$

**24.** $-16 - 30i$

**25.** $34 - 27i$

**26.** $\dfrac{\sqrt{15x}}{3}$

**27.** $9\sqrt{2}$

**28.** $\sqrt{3} + 1$

**29.** $\dfrac{11\sqrt{3}}{3}$

**30.** $\frac{7}{10} + \frac{1}{10}i$

**31.** $\frac{7}{5} + \frac{9}{5}i$

**32.** $\sqrt{161}$

**33.** Right triangle

**34.** $x = -8.426,\ x = 8.426$

**35.** $x = 3.250$

**36.** No solution

**37.** $x = -22$

**38.** $x = 2.8$

**39.** $x = 1,\ x = \frac{5}{4}$

**40.** 320 ft

CHAPTER 7 TEST

**1.** $(4z)^{\frac{1}{3}}$

**2.** 8

**3.** $4x^4y^2\sqrt{5y}$

**4.** $3i\sqrt{7}$

**5.** $6\sqrt{2}$

**6.** $5x^3y^2\sqrt{3x}$

**7.** $3\sqrt{2x} - x\sqrt{6}$

**8.** $\dfrac{x + 2\sqrt{x}}{x - 4}$

**9.** $4 - 15i$

**10.** $8 - 6i$

**11.** $\frac{3}{17} - \frac{5}{17}i$

**12.** 15

**13.** $x = 1.757$

**14.** $x = 2$

**15.** 84.853 ft

# Chapter 8

SECTION 8.1

**1.** $x = \pm 5$

**3.** $x = \pm 3$

**5.** $x = \pm 7i$

**7.** $x = \pm\sqrt{2}$

**9.** $x = \pm 4\sqrt{2}$

**11.** $x = 3 \pm \sqrt{7}$

**13.** $x = -5 \pm 3\sqrt{3}$

**15.** $x = \frac{5 \pm \sqrt{22}}{2}$

**17.** $x = \frac{-7 \pm \sqrt{15}}{3}$

**19.** $x = \frac{-3 \pm 4\sqrt{2}}{4}$

**21.** $x = 2 \pm 2\sqrt{3}$

**23.** $x = \frac{7 \pm \sqrt{41}}{2}$

**25.** $x = 2 \pm \sqrt{11}$

**27.** $x = -3 \pm \sqrt{12}$

**29.** 2 real solutions

**31.** 2 nonreal complex solutions

**33.** 2 nonreal complex solutions

**35.** 2 nonreal complex solutions

**37.** $x = \frac{5 \pm \sqrt{53}}{2}$

**39.** $x = \frac{7 \pm \sqrt{41}}{4}$

**41.** $x = \frac{9 \pm \sqrt{57}}{4}$

**43.** $x = -2 \pm \sqrt{7}$

**45.** $x = \frac{-2 \pm \sqrt{14}}{2}$

**47.** $x = \frac{7 \pm \sqrt{3}i}{2}$

**49.** $x = \frac{15 \pm \sqrt{85}}{10}$

**51.** $x = 1$, $x = \frac{3}{2}$

**53.** $x = \pm 2\sqrt{a}$

**55.** $x = \pm 2\sqrt{7 + a^2}$

**57.** $x^2 + 2x = 4$, $x = -1 \pm \sqrt{5}$

**59.** $x(x + 3) = 15$, $W = \frac{-3 + \sqrt{69}}{2}$ or 2.653 ft, $L = \frac{3 + \sqrt{69}}{2}$ or 5.653 ft

**61.** $8 + 8\sqrt{2}$ cm

**63.** The $x$-coordinates are the same.

**65.** $k = 3$

**67.** Answers will vary.

**69.** $x^3 + 2x^2 - 3x + 4 - (x^3 + x^2 + 2x - 4) = x^2 - 5x + 8$

## SECTION 8.2

**1.** $x_1 = -2 + \sqrt{11}$, $x_2 = -2 - \sqrt{11}$

**3.** $x_1 = \frac{9 + \sqrt{33}}{2}$, $x_2 = \frac{9 - \sqrt{33}}{2}$

**5.** $x_1 = 2$, $x_2 = -\frac{1}{3}$

**7.** $x_1 = \frac{-7 + \sqrt{73}}{4}$, $x_2 = \frac{-7 - \sqrt{73}}{4}$

**9.** $x_1 = \frac{11 + \sqrt{37}}{6}$, $x_2 = \frac{11 - \sqrt{37}}{6}$

**11.** $x_1 = \frac{5 + \sqrt{97}}{6}$, $x_2 = \frac{5 - \sqrt{97}}{6}$

**13.** Nonreal complex numbers

**15.** Real numbers

**17.** Nonreal complex numbers

**19.** Real numbers

**21.** $r_1 = \frac{-1 + \sqrt{29}}{2}$, $r_2 = \frac{-1 - \sqrt{29}}{2}$; real numbers

**23.** $k_1 = \frac{-3 + \sqrt{201}}{24}$, $k_2 = \frac{-3 - \sqrt{201}}{24}$; real numbers

**25.** $v_1 = \frac{-147 + \sqrt{21945}}{24}$, $v_2 = \frac{-147 - \sqrt{21945}}{24}$; real numbers

**27.** $s_1 = \frac{1 + 2i\sqrt{26}}{7}$, $s_2 = \frac{1 - 2i\sqrt{26}}{7}$; nonreal complex numbers

**29.** $z_1 = \frac{-21 + i\sqrt{15}}{6}$, $z_2 = \frac{-21 - i\sqrt{15}}{6}$; nonreal complex numbers

**31.** $t_1 = 3 + \sqrt{10}$, $t_2 = 3 - \sqrt{10}$; real numbers

**33.** Zero

**35.** Two

**37.** Two

**39.** Two

**41.** $x^2 - 2x - 15 = 0$

**43.** $9x^2 + 9x - 10 = 0$

**45.** $x^2 - 4\sqrt{2}x + 6 = 0$

**47.** $x^2 + 4 = 0$

**49.** 29 and 46 or $-46$ and $-29$

**51.** 37 and 71 or $-35.5$ and $-74$

**53.** 1040 ft

**55.** 849 or 6151 computers

**57.** a) $h = -16t^2 + 50t + 300$
b) 244 ft
c) 5 sec

**59.** 1.25 sec

**61.** Show that $\frac{-b + \sqrt{b^2 - 4ac}}{2a} + \frac{-b - \sqrt{b^2 - 4ac}}{2a} = \frac{-b}{a}$.

**63.** c

**65.** $b = 6$, $b = -6$

**67.** $a = -\frac{25}{12}$

**69.** a) $h = 65t - 5t^2$
b) 200 m
c) 3 sec and 10 sec
d) 6.5 sec

## SECTION 8.3

**1.** a)

| Seconds | Distance Traveled |
|---------|-------------------|
| −4 | 48 |
| −3 | 27 |
| −2 | 12 |
| −1 | 3 |
| 0 | 0 |
| 1 | 3 |
| 2 | 12 |
| 3 | 27 |
| 4 | 48 |

b)

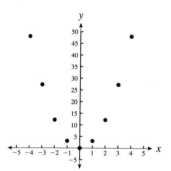

c)

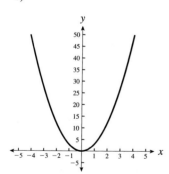

**3.** c

**5.** b

**7.** b

**9.**

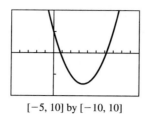

$[-5, 10]$ by $[-10, 10]$

**11.**

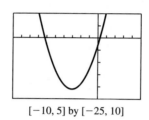

$[-10, 5]$ by $[-25, 10]$

**13.**

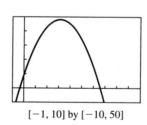

$[-1, 10]$ by $[-10, 50]$

**15.**

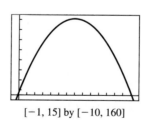

$[-1, 15]$ by $[-10, 160]$

**17.**

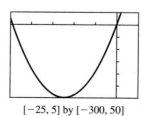

$[-25, 5]$ by $[-300, 50]$

**19.** 2

**21.** 2

**23.** 2

**25.** No real solutions

**27.** $(-5.162, 1.162)$

**29.** $(-8.339, 0.839)$

**31.** $(-\infty, -0.790) \cup (1.124, \infty)$

**33.** $2x^2 < x + 5$, solution exists

**35.** a) $A = x(x + 5)$

b) $x > 0$

c) $2.3 \le x(x + 5) \le 7.8$

d)

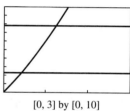

$[0, 3]$ by $[0, 10]$

e) $0.424 \le x \le 1.248$

**37.** a) Nick: $-16t^2 + 60t + 70$

Jane: $-16t^2 + 35t + 105$

b) Nick: 114 ft

Jane: 124 ft

c) $-16t^2 + 60t + 70 < -16t^2 + 35t + 105$

d) $0 \le t \le 1.4$ sec

**39.** One solution, $x = 1$

**41.** $x = -2.200$

**43.** a) 0.31

b) 10%

c) 0.51

## SECTION 8.4

**1.** Vertical slide down 8 units

**3.** Vertical slide up 17 units

**5.** Horizontal slide right 3 units

**7.** Horizontal slide left 5 units

**9.** Vertical slide down 48 units

**11.** Horizontal slide right $\sqrt{2}$ units

**13.** Horizontal slide right 2 units, vertical slide up 4 units

**15.** Horizontal slide left 3 units, vertical slide up 6 units

**17.** Horizontal slide right 0.5 units, vertical slide up $\frac{3}{2}$ units

**19.** Horizontal slide right $\sqrt{5}$ units, vertical slide down $\sqrt{2}$ units

**21.** $c = 16,\ (x + 4)^2$

**23.** $c = 4,\ (x - 2)^2$

**25.** $c = \frac{25}{4},\ \left(x + \frac{5}{2}\right)^2$

**27.** $c = \frac{361}{4},\ \left(x - \frac{19}{2}\right)^2$

**29.** $y = (x - 3)^2 - 4$; horizontal slide right 3 units, vertical slide down 4 units

**31.** $y = (x + 5)^2 - 7$; horizontal slide left 5 units, vertical slide down 7 units

**33.** $y = \left(x + \frac{5}{2}\right)^2 - \frac{73}{4}$; horizontal slide left $\frac{5}{2}$ units, vertical slide down $\frac{73}{4}$ units

**35.** $y = (x + 4)^2 - 19$; horizontal slide left 4 units, vertical slide down 19 units

**37.** $y = (x - 3)^2$

**39.** $y = (x - 2)^2 + 3$

**41.** $y = (x - 4)^2 - 6$

**43.** $y = (x + 1)^2 + 3$

**45.** a) $h_2 = -16(t - 7)^2 + 196$

    b) 5.250 sec

**47.** Crosses

**49.** Does not cross

**51.** Does not cross

**53.** Crosses

**55.** $k < 0$

**57.** $h < 0$

**59.** $y = (x + 4)(x - 6)$

**61.** a) $x^2 + (x + 1)^2 = (x + 3)^2$

    b) 5.464, 6.464, 8.464 units

    c) Answers will vary.

**63.** Written Exercise

**65.** $\left(\frac{\sqrt{10}}{2}, \frac{1}{2}\right), \left(-\frac{\sqrt{10}}{2}, \frac{1}{2}\right)$

## SECTION 8.5

**1.** Stretch

**3.** Stretch

**5.** Shrink

**7.** Stretch

**9.** a) 1

    b) 3

    c) 2

**11.** a) 1

    b) 2

    c) 3

**13.** Answers will vary.

**15.** Answers will vary.

**17.** One is the reflection of the other in the $x$-axis.

**19.** The first is a vertical stretch of 2 of the second.

**21.** The second is a vertical stretch with a stretch factor of 2 of the first.

**23.** One is the reflection of the other in the line $y = 3$.

**25.** i) A vertical stretch with a stretch factor of 3 followed by

    ii) a slide up 4 units.

**27.** i) A slide left 2 units followed by

    ii) a vertical shrink with a shrink factor of 0.3.

**29.** i) A slide right 2 units followed by

    ii) a vertical stretch with a stretch factor of 3 followed by

    iii) a reflection in the $x$-axis.

**31.** i) A slide left 3 units followed by

    ii) a reflection in the $x$-axis followed by

    iii) a slide down 2 units.

**33.**

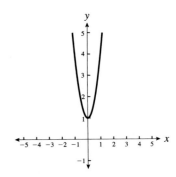

**35.**

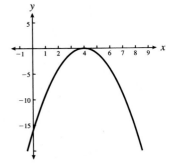

**37.**

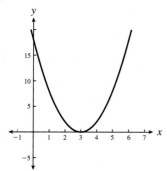

**39.**

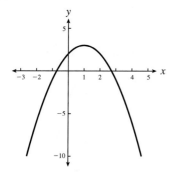

**41.** $y = -x^2 + 4$

**43.** $y = -5(x - 3)^2$

**45.** $y = (x + 2)^2 - 11$

**47.** $y = (x - 4)^2 - 14$

**49.** $y = (x - 2.5)^2 - 9.25$

**51.** $y = 2(x - 1.5)^2 - 11.5$

**53.** $y = 17(x - 1.5)^2 - 57.25$

**55.** $y = 3\left(x + \frac{5}{6}\right)^2 - \frac{61}{12}$

**57.** $y = 3\left(x - \frac{1}{6}\right)^2 - \frac{49}{12}$

**59.** $y = 2(x - 2)^2 - 1$

i) A slide right 2 units followed by

ii) a vertical stretch with a stretch factor of 2 followed by

iii) a slide down 1 unit.

**61.** $y = 5(x + 2.5)^2 - 45.25$

i) A slide left 2.5 units followed by

ii) a vertical stretch with a stretch factor of 5 followed by

iii) a slide down 45.25 units.

**63.** $y = -2(x - 1.5)^2 - 3.5$

i) A slide right 1.5 units followed by

ii) a vertical stretch with a stretch factor of 2 followed by

iii) a reflection in the $x$-axis followed by

iv) a slide down 3.5 units.

**65.** $y = 2(x - 1.75)^2 + 4.875$

i) A slide right 1.75 units followed by

ii) a vertical stretch with a stretch factor of 2 followed by

iii) a slide up 4.875 units.

**67.** $y = \frac{1}{3}(x + 4.5)^2 + 17.25$

i) A slide left 4.5 units followed by

ii) a vertical shrink with a shrink factor of $\frac{1}{3}$ followed by

iii) a slide up 17.25 units.

**69.** $a > 1$

**71.** $k > 0$

**73.** $k = 3$

**75.** Written Exercise

**77.** 4

**79.** c

## SECTION 8.6

**1.** Vertex form

**3.** Vertex form

**5.** Not vertex form

**7.** $(2, 4), x = 2$

**9.** $(-2, -5), x = -2$

**11.** $(-2, 17), x = -2$

**13.** $(-2, -10), x = -2$

**15.** $(-2\pi, -\sqrt{3}), x = -2\pi$

**17.** Answers will vary. One example is $y = -2(x - 1)^2 + 2$.

**19.** $\left(\frac{2}{3}, \frac{11}{3}\right), x = \frac{2}{3}$

**21.** $\left(-\frac{11}{8}, \frac{153}{16}\right), x = -\frac{11}{8}$

**23.** $\left(\frac{5}{14}, \frac{395}{28}\right), x = \frac{5}{14}$

**25.** a)

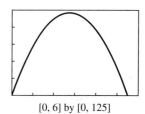

[0, 6] by [0, 125]

b) 112 ft

c) 121 ft at 2.75 sec

d) The ball hits the ground at 5.5 sec; the maximum height is obtained in one-half that time.

**27.** a) Slide down 120 units

b) The maximum height is 120 ft greater

c) $h = -16t^2 + 88t + 120$

**29.** Written Exercise

**31.** Written Exercise

**33.** $g(x) = 2(x - 1)^2 - 8$

**35.** $a = -3$

**37.** $x = \dfrac{a + b}{2}$

**39.** Written Exercise

REVIEW EXERCISES

**1.** $x_1 = 4i, x_2 = -4i$

**2.** $x_1 = -8 + \sqrt{35}$,
$x_2 = -8 - \sqrt{35}$

**3.** $x_1 = 3 + \sqrt{2}, x_2 = 3 - \sqrt{2}$

**4.** $x_1 = \dfrac{-1 + \sqrt{13}}{2}$,
$x_2 = \dfrac{-1 - \sqrt{13}}{2}$

**5.** $x_1 = \dfrac{-5 + \sqrt{29}}{2}$,
$x_2 = \dfrac{-5 - \sqrt{29}}{2}$

**6.** $x_1 = \frac{5}{4}, x_2 = 1$

**7.** $x_1 = \dfrac{7 + \sqrt{177}}{8}$,
$x_2 = \dfrac{7 - \sqrt{177}}{8}$

**8.** $x_1 = \dfrac{7 + \sqrt{69}}{2}$,
$x_2 = \dfrac{7 - \sqrt{69}}{2}$

**9.** $x_1 = \dfrac{1 + \sqrt{129}}{16}$,
$x_2 = \dfrac{1 - \sqrt{129}}{16}$

**10.** $x_1 = \dfrac{-4 + \sqrt{10}}{3}$,
$x_2 = \dfrac{-4 - \sqrt{10}}{3}$

**11.** $(-\infty, 0.382) \cup (2.618, \infty)$

**12.** $(-1.236, 3.236)$

**13.** One

**14.** Two

**15.** $6x^2 + 7x - 3 = 0$

**16.** $x^2 - \sqrt{6}x - 12 = 0$

**17.** Window (c)

**18.** Answers will vary.

**19.** $y = 3(x + 2)^2 + 10$

**20.** $y = -(x - 6)^2 - 3$

**21.** Horizontal slide right 2 units, vertical shrink by a shrink factor of $\frac{1}{5}$, vertical slide up 8 units

**22.** Vertical stretch by a stretch factor of 3, reflection in the $x$-axis, vertical slide up 7 units

**23.** Horizontal slide left 4 units, vertical stretch by a stretch factor of 2, reflection in the $x$-axis

**24.** Horizontal slide left 7 units, vertical slide down 17 units

**25.** Horizontal slide left 1 unit, reflection in the $x$-axis, vertical slide up 5 units

**26.** Horizontal slide right 2 units, vertical stretch by a stretch factor of 3, vertical slide down 8 units

**27.**

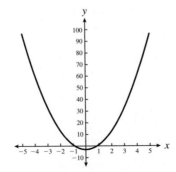

**28.**

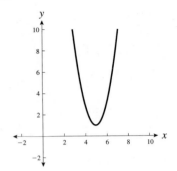

**29.**

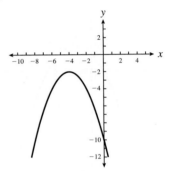

**30.**

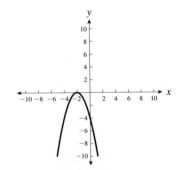

**31.** $(1, -6), x = 1$

**32.** $(-3, 4), x = -3$

**33.** $\left(-\frac{9}{4}, -\frac{121}{8}\right), x = -\frac{9}{4}$

**34.** $(-3, -6), x = -3$

**35.** $x_1 = 2 + \sqrt{4 + 9a}$,
$x_1 = 2 - \sqrt{4 + 9a}$

**36.** $0 < a < 1$

**37.** $a = -\frac{2}{3}$

**38.** 18 and 22.5 or $-22.5$ and $-18$

**39.** 16 ft by 13 ft

**40.** a) 76.563 ft

   b) 2.188 sec

## CHAPTER 8 TEST

**1.** $x_1 = 4 + \sqrt{10}, x_2 = 4 - \sqrt{10}$

**2.** $x_1 = -3, \; x_2 = \frac{5}{2}$

**3.** $x_1 = \frac{1}{2} + \frac{3}{2}i, \; x_2 = \frac{1}{2} - \frac{3}{2}i$

**4.** No real solutions

**5.** $4x^2 - x - 5 = 0$

**6.** Window (b)

**7.** $(-3.589, 2.089)$

**8.** A horizontal slide left 3 units, followed by a vertical stretch by a stretch factor of 4, followed by a vertical slide down 5 units.

**9.** $y = -(x - 1)^2 + 3$; A horizontal slide right 1 unit, followed by a reflection in the $x$-axis, followed by a vertical slide up 3 units.

**10.** $y = \frac{1}{3}(x - 4)^2 - 2$

**11.** $c = -\frac{49}{8}$

**12.** A horizontal slide left 3 units, followed by a reflection in the $x$-axis, followed by a vertical slide down 6 units.

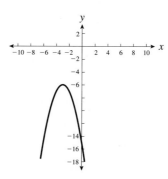

**13.** Vertex: $(2, -5)$, line of symmetry: $x = 2$

**14.** Vertex: $(-0.8, -7.2)$, line of symmetry: $x = -0.8$

**15.** $3.5, -5.5$

## Chapter 9

### SECTION 9.1

**1.** Solution

**3.** Not a solution

**5.** 3

**7.** $-3$

**9.** $(1, 1, -2)$

**11.** $(1, 8, 6)$

**13.** $\left(\frac{76}{9}, \frac{20}{3}, -\frac{43}{9}\right)$

**15.** a)

$$\begin{cases} x + y + z = 50 \\ 3.50x + 2.50y + 1.75z = 120 \\ y = 2x \end{cases}$$

   b) 10 lbs of sour balls, 20 lbs of butterballs and 20 lbs of starlight mints

**17.** One cheeseburger: $0.69
One large drink: $0.99
One order of fries: $0.79

**19.** 5, 8 and 10

**21.** 25 nickels, 12 dimes and 7 quarters

**23.** Written Exercise

**25.** $a = 2, b = 6$ and $c = -12$

**27.** Answers will vary.

### SECTION 9.2

**1.** $2 \times 2$

**3.** $3 \times 3$

**5.** $\begin{bmatrix} 1 & 3 \\ 3 & -2 \end{bmatrix}, \quad \begin{bmatrix} 1 & 3 \\ 3 & -2 \end{bmatrix}$

**7.** No, the matrices are not the same dimension.

**9.** $\begin{bmatrix} 1 & 6 \\ 4 & 1 \end{bmatrix}$

**11.** $\begin{bmatrix} -5 & -9 \\ -13 & 16 \end{bmatrix}$

**13.** $\begin{bmatrix} 180 & 240 \\ 400 & 140 \end{bmatrix}$

**15.** $[M] = \begin{bmatrix} 374 & 320 & 127 \\ 265 & 214 & 154 \end{bmatrix}$

**17.** a) $\begin{bmatrix} 4.7 & 4.5 & 5.4 \\ 4.5 & 5.5 & 6.2 \end{bmatrix}$

   b) Site B for Cashews

$$\begin{bmatrix} -0.3 & -1.5 & 0.2 \\ 0.5 & 0.7 & 0.2 \end{bmatrix}$$

**19.** a) $\begin{bmatrix} 85 & 60 & 85 \\ 75 & 85 & 80 \\ 70 & 85 & 75 \end{bmatrix}$

b) Yes

$\begin{bmatrix} 102 & 72 & 102 \\ 90 & 102 & 96 \\ 84 & 102 & 90 \end{bmatrix}$

**21.** $\begin{bmatrix} 88 & -8 \\ 77 & -70 \end{bmatrix}, \begin{bmatrix} -14 & 36 \\ 47 & -30 \end{bmatrix}$

**23.** $\begin{bmatrix} -181 & 138 \\ 49 & 0 \end{bmatrix},$

$\begin{bmatrix} 201 & 16 \\ 199 & -155 \end{bmatrix}$

**25.** $-11x + 31$

**27.** $x^2 + 6x - 10xy - 30y + 25y^2 + 9$

**29.** $x^2 - y^2 - 10y - 25$

**31.** $x^2 - 6x - 9$

**33.** $-7x - 14$

SECTION 9.3

**1.** $\begin{bmatrix} -5 & 2 \\ -22 & -8 \end{bmatrix}$

**3.** $\begin{bmatrix} 1 & 0 \\ -10 & 9 \end{bmatrix}$

**5.** $\begin{bmatrix} -4 & 2 \\ -32 & 1 \end{bmatrix}$

**7.** $\begin{bmatrix} 1 & 0 \\ 1.667 & -0.333 \end{bmatrix}$

**9.** $\begin{bmatrix} 17 & -6 \\ 13 & -24 \end{bmatrix}$

**11.** $\begin{bmatrix} 3 & 0 \\ -30 & 27 \end{bmatrix}$

**13.** $\begin{bmatrix} 16 & -4 \\ 32 & -30 \end{bmatrix}$

**15.** $-3$

**17.** $\begin{bmatrix} -0.214 & 0.071 \\ -0.036 & 0.179 \end{bmatrix}$

**19.** $\begin{bmatrix} -0.095 & -0.024 \\ 0.262 & -0.060 \end{bmatrix}$

**21.** 2 rows, 3 columns

**23.** $\begin{bmatrix} -7 & 20 & 28 \\ 2 & 13 & -9 \\ -6 & 3 & 22 \end{bmatrix}$

**25.** $\begin{bmatrix} -60 & 96 & 54 \\ 264 & -336 & -432 \end{bmatrix}$

**27.** $\begin{bmatrix} 210 & -264 & -351 \\ -96 & 96 & 216 \end{bmatrix}$

**29.** 36

**31.** $\begin{bmatrix} -0.111 & -0.139 \\ 0.222 & 0.028 \end{bmatrix}$

**33.** $\begin{bmatrix} 1 & 0 \\ 0 & 1 \end{bmatrix}$

**35.** Written Exercise

**37.** Written Exercise

**39.** True

**41.** False

**43.** False

**45.** True, assuming the matrix has an inverse

**47.** False

**49.** $(-2, 1)$

**51.** Infinite number of solutions

**53.** $(-4, -4)$ and $(3, 3)$

**55.** $\begin{cases} x + y = 96 \\ y - x = 6 \end{cases}$, where Mike

studied $x$ hr and Wendy studied $y$ hr; Mike studied 45 hr and Wendy, 51 hr

SECTION 9.4

**1.** $\begin{bmatrix} 2 & -3 \\ 5 & 4 \end{bmatrix}\begin{bmatrix} x \\ y \end{bmatrix} = \begin{bmatrix} 9 \\ 13 \end{bmatrix}$

**3.**
$\begin{bmatrix} 5.1 & -8.6 & 3.1 \\ 3.2 & 2.7 & 6.8 \\ 4.5 & -5.6 & 0.3 \end{bmatrix}\begin{bmatrix} x \\ y \\ z \end{bmatrix} = \begin{bmatrix} 19.7 \\ 18.3 \\ 20.3 \end{bmatrix}$

**5.** $2x - 3y = 12$
$5x - 6y = 15$

**7.** $5.6x - 3.1y \qquad = 15.4$
$-7.2x \qquad + 8.8z = 19.7$
$2.3x + 5.1y \qquad = 18.3$

**9.** $(3, 5)$

**11.** $(5, 9)$

**13.** $(2.5, 8.1)$

**15.** $\left(\frac{75}{23}, -\frac{19}{23}\right)$

**17.** $(1, 2, 6)$

**19.** $(5, -4, 0)$

**21.** $(2, 8, -4)$

**23.** $(2.6, -5.3, 7.8)$

**25.** $(2, -4, -5)$

**27.** Answers will vary.

**29.** $a = 2, b = -1, c = 1$

**31.** $(5.45, 11, 13.7, -9.55)$

**33.** Boat: 5 mph; current: 2 mph

**35.** Rate of plane = 630 mph; rate of wind = 70 mph

**37.** 6 ft, 16 ft, 18 ft

**39.** $(\sqrt{117})^2 + (\sqrt{117})^2$
$= (\sqrt{234})^2$

**41.** $m = 6, b = -13$

## REVIEW EXERCISES

**1.** $-3$

**2.** $-1$

**3.** Answers will vary. One example
is $\begin{cases} x + y + z = 5 \\ 3x - 2y - 2z = 0 \\ x + 3y + 2z = 6 \end{cases}$

**4.** $2 \times 2$

**5.** $2 \times 3$

**6.** $-8$

**7.** $5$

**8.** $\begin{bmatrix} 9 & 3 \\ 4 & 1 \end{bmatrix}$

**9.** $\begin{bmatrix} -3 & -5 \\ 0 & 1 \end{bmatrix}$

**10.** $\begin{bmatrix} 5 & 10 & -5 \\ 0 & 15 & 5 \end{bmatrix}$

**11.** $\begin{bmatrix} 9 & 6 \\ 3 & 0 \end{bmatrix}$

**12.** $\begin{bmatrix} 24 & 10 \\ 10 & 2 \end{bmatrix}$

**13.** $\begin{bmatrix} 9 & 9 \\ 2 & -1 \end{bmatrix}$

**14.** $\begin{bmatrix} 26 & -2 \\ 6 & -2 \end{bmatrix}$

**15.** $\begin{bmatrix} 3 & 3 & -4 \\ 2 & 7 & -1 \end{bmatrix}$

**16.** $\begin{bmatrix} 3 & 5 \\ 4 & 7 \end{bmatrix}$

**17.** $\begin{bmatrix} -4 & 20 \\ 3 & 19 \end{bmatrix}$

**18.** $[\,18 \quad 61\,]$

**19.** $\begin{bmatrix} 9 & -9 \\ 14 & -14 \end{bmatrix}$

**20.** $\begin{bmatrix} 22 & 63 \\ 27 & 85 \end{bmatrix}$

**21.** $0$

**22.** $\begin{bmatrix} 7 & -5 \\ -4 & 3 \end{bmatrix}$

**23.** $\begin{bmatrix} 1 & 0.5 \\ 2 & 1.5 \end{bmatrix}$

**24.** $\begin{bmatrix} 1 & 0 & 0 & 0 \\ 0 & 1 & 0 & 0 \\ 0 & 0 & 1 & 0 \\ 0 & 0 & 0 & 1 \end{bmatrix}$

**25.** Matrix of Coefficients:
$\begin{bmatrix} 4 & 6 \\ -3 & 1 \end{bmatrix}$
Matrix of Constants:
$\begin{bmatrix} 7 \\ 0 \end{bmatrix}$

**26.** Matrix of Coefficients:
$\begin{bmatrix} 6 & 0 \\ 1 & -5 \end{bmatrix}$
Matrix of Constants:
$\begin{bmatrix} 15 \\ 12 \end{bmatrix}$

**27.** $\begin{bmatrix} 2 & -5 \\ 1 & 8 \end{bmatrix} \cdot \begin{bmatrix} x \\ y \end{bmatrix} = \begin{bmatrix} 0 \\ 10 \end{bmatrix}$

**28.** $\begin{bmatrix} 12 & 15 \\ 9 & -3 \end{bmatrix} \cdot \begin{bmatrix} x \\ y \end{bmatrix} = \begin{bmatrix} 7 \\ 8 \end{bmatrix}$

**29.** $\begin{cases} -x + 3y = 2 \\ \quad\quad 2y = -1 \end{cases}$

**30.** $\begin{cases} 3x - 3y + z = 1 \\ 5x \quad\quad + 2z = 3 \\ x + 2y - z = -1 \end{cases}$

**31.** $(5, 8, -1)$

**32.** $(0, 3, -3)$

**33.** $(-2, -1, -4)$

**34.** $(2, -2, 2)$

**35.** $(1, 0.5, -1)$

**36.** $(-3, 8, 11)$

**37.** $(9, 13, 10)$

**38.** $(25, 20, 12)$

**39.** $5, -2$ and $-3$

**40.** Boat rate: 11 mph
Current rate: 3.5 mph

## CHAPTER 9 TEST

**1.** $3 \times 2$

**2.** $17$

**3.** $\begin{bmatrix} 8 & 2 \\ 4 & 7 \end{bmatrix}$

**4.** $\begin{bmatrix} 2 & -6 \\ 2 & -7 \end{bmatrix}$

**5.** $\begin{bmatrix} 21 & 2 \\ 11 & 14 \end{bmatrix}$

**6.** $\begin{bmatrix} 13 & -4 \\ -3 & 0 \\ -9 & 6 \end{bmatrix}$

**7.** $\begin{bmatrix} 1.125 & -1.5 \\ 4.2 & 0.33 \end{bmatrix}$

**8.** $\begin{bmatrix} 0.2 & -0.4 \\ 0.1 & 0.3 \end{bmatrix}$

**9.** $\begin{bmatrix} 5 & -10 \\ 1 & 6 \end{bmatrix} \cdot \begin{bmatrix} x \\ y \end{bmatrix} = \begin{bmatrix} 6 \\ 3 \end{bmatrix}$

**10.** $\begin{cases} x \quad\quad + 4z = 1 \\ 3x - 2y + 2z = 8 \\ \quad\quad 5y - z = 3 \end{cases}$

**11.** $(-3, -1, 2)$

**12.** $(5, 7, -2)$

**13.** $(4, 2, 5)$

**14.** $(0.5, 0.25, 1)$

**15.** a)
$$\begin{cases} x + \quad y = 1200 \\ 0.055x + 0.025y = 56.10 \end{cases}$$

b)  $870 at 5.5%, $330 at 2.5%

---

## Chapter 10

### SECTION 10.1

**1.** $x^2 + y^2 = 16$

**3.** $x^2 + y^2 = 4$

**5.** $(x - 1)^2 + (y - 5)^2 = 144$

**7.** $(x - 4)^2 + (y + 9)^2 = 27$

**9.** $x^2 + (y + 2.1)^2 = 625$

**11.** Center: $(0, 0)$, $r = 9$

**13.** Center: $(0, 0)$, $r = \sqrt{19}$

**15.** Center: $(-7, 5)$, $r = \sqrt{56}$

**17.**

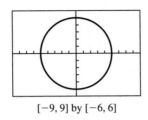

$[-9, 9]$ by $[-6, 6]$

**19.**

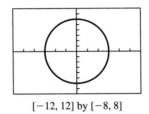

$[-12, 12]$ by $[-8, 8]$

**21.**

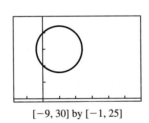

$[-9, 30]$ by $[-1, 25]$

**23.**

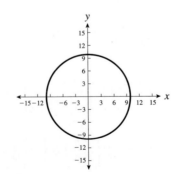

**25.**

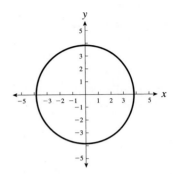

**27.**

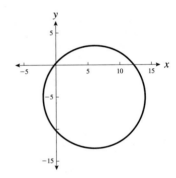

**29.**

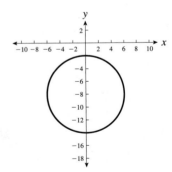

**31.** $x^2 + y^2 = 9$

**33.** $(x - 3)^2 + (y + 2)^2 = 16$

**35.** $(x + 3)^2 + y^2 = 16$

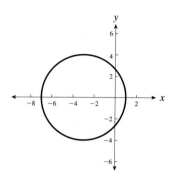

**37.** $(x + 6)^2 + (y - 4)^2 = 36$

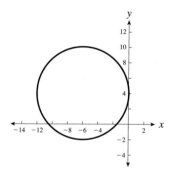

**39.** $(x - 3)^2 + (y + 6)^2 = 64$

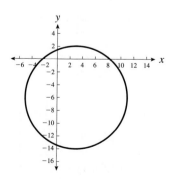

**41.** $(x - 7)^2 + (y + 11)^2 = 100$

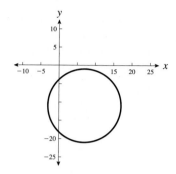

**43.** $x^2 + (y + 4)^2 = 25$

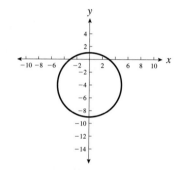

**45.** Written Exercise

**47.** Written Exercise

**49.** 15 ft

**51.** $r = \sqrt{18}$

**53.** $(x - 3)^2 + (y + 2)^2 = 25$

**55.** Answers will vary.

**57.** Yes, it is a right triangle.

**59.** Yes, it is a square.

## SECTION 10.2

**1.** Center $(0, 0)$; vertices $(\pm 6, 0)$, $(0, \pm 2)$

**3.** Center $(0, 0)$; vertices $(\pm 5, 0)$, $(0, \pm 9)$

**5.** Center $(0, 0)$; vertices $(\pm \sqrt{5}, 0)$, $(0, \pm 4)$

**7.**

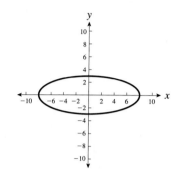

**9.**

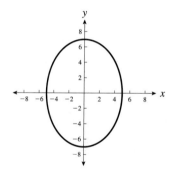

**11.**

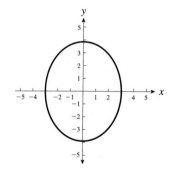

**13.** Center $(1, 5)$; vertices $(1, 8)$, $(1, 2)$, $(-1, 5)$, $(3, 5)$

**15.** Center $(-1, -2)$; vertices $(6, -2)$, $(-1, -7)$, $(-8, -2)$, $(-1, 3)$

**17.** Center $(0, 3)$; vertices $(7, 3)$, $(-7, 3)$, $(0, 1)$, $(0, 5)$

**19.** Center $(5, -3)$; vertices $(5, -3 + \sqrt{12})$, $(5, -3 - \sqrt{12})$, $(5 + \sqrt{7}, -3)$, $(5 - \sqrt{7}, -3)$

**21.**

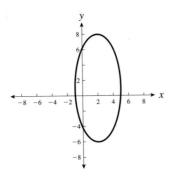

**23.**

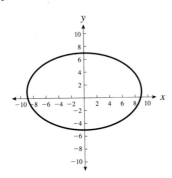

**25.** $\dfrac{x^2}{4} + \dfrac{y^2}{9} = 1$; Center $(0, 0)$; vertices $(\pm 2, 0)$, $(0, \pm 3)$; major axis length 6; minor axis length 4

**27.** $\dfrac{x^2}{36} + \dfrac{y^2}{9} = 1$; Center $(0, 0)$; vertices $(\pm 6, 0)$, $(0, \pm 3)$; major axis length 12; minor axis length 6

**29.** $\dfrac{(x-5)^2}{16} + \dfrac{(y+4)^2}{4} = 1$; Center $(5, -4)$; vertices $(5, -2)$, $(5, -6)$, $(9, -4)$, $(1, -4)$; major axis length 8; minor axis length 4

**31.** $\dfrac{(x-3)^2}{121} + \dfrac{(y+10)^2}{121} = 1$; Center $(3, -10)$; vertices $(3, 1)$, $(3, -21)$, $(-8, -10)$, $(14, -10)$; major axis length 22; minor axis length 22

**33.** $\dfrac{x^2}{4} + \dfrac{y^2}{25} = 1$

**35.** $\dfrac{x^2}{25} + \dfrac{y^2}{16} = 1$

**37.** $\dfrac{x^2}{9} + \dfrac{y^2}{25} = 1$

**39.** $\dfrac{x^2}{16} + \dfrac{y^2}{4} = 1$

**41.** Center $(0, 0)$; vertices $(\pm 6, 0)$, $(0, \pm 3)$

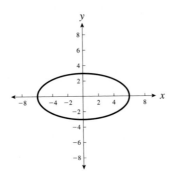

**43.** Center $(0, 0)$; vertices $(\pm 3, 0)$, $(0, \pm 2)$

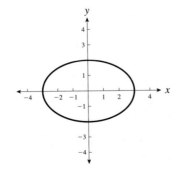

**45.** Center $(-1, 3)$; vertices $(7, 3)$, $(-9, 3)$, $(-1, -1)$, $(-1, 7)$

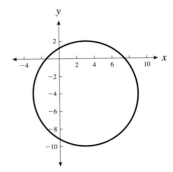

**47.** Center $(3, -4)$; vertices $(9, -4)$, $(-3, -4)$, $(3, 2)$, $(3, -10)$

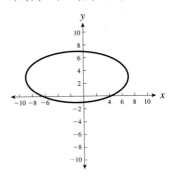

**49.**

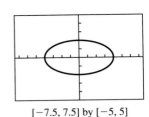

$[-7.5, 7.5]$ by $[-5, 5]$

**51.**

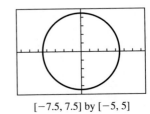

$[-7.5, 7.5]$ by $[-5, 5]$

**53.**

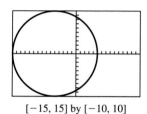

[−15, 15] by [−10, 10]

**55.**

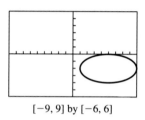

[−9, 9] by [−6, 6]

**57.** 43.301 ft

**59.** a)

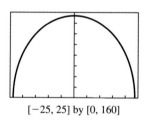

[−25, 25] by [0, 160]

b) 125.084 ft

**61.** $\dfrac{x^2}{25} + \dfrac{y^2}{36} = 1$

**63.** $\dfrac{(x+3)^2}{25} + \dfrac{(y-7)^2}{4} = 1$

**65.** $\dfrac{(x-5)^2}{4} + \dfrac{(y-7)^2}{36} = 1$

**67.** Circle: center $(0, 0)$, $r = 7$

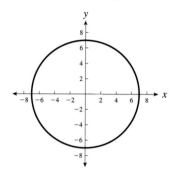

**69.** Parabola: vertex $(0, -5)$, opens up

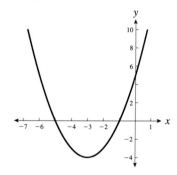

**71.** Parabola: vertex $(-3, -4)$, opens up

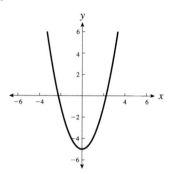

SECTION 10.3

**1.** up

**3.** left

**5.** down

**7.** right

**9.** Vertex: $(5, 0)$, $y = 0$

**11.** Vertex: $(0, 7)$, $y = 7$

**13.** Vertex: $(4, -1)$, $y = -1$

**15.** Vertex: $(8, 5)$, $y = 5$

**17.**

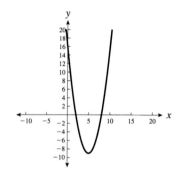

**19.**

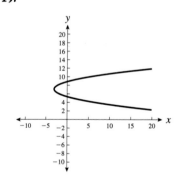

**21.**

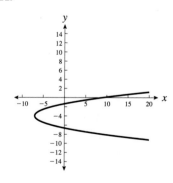

**23.**

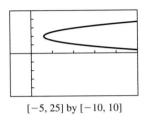

[−5, 25] by [−10, 10]

**25.**

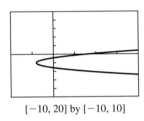

[−10, 20] by [−10, 10]

**27.**

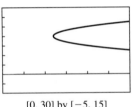

[0, 30] by [−5, 15]

**29.** $y = (x + 1)^2 - 8$

**31.** $x = (y - 4)^2 - 28$

**33.** $y = 3(x + 1)^2 - 17$

**35.** $x = 2(y - 3)^2 - 25$

**37.**

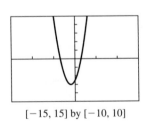

[−15, 15] by [−10, 10]

**39.**

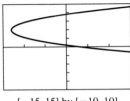

[−15, 15] by [−10, 10]

**41.**

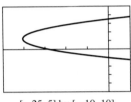

[−25, 5] by [−10, 10]

**43.**

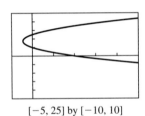

[−5, 25] by [−10, 10]

**45.** Written Exercise

**47.** $k = -9$

**49.** $y = 2 \pm \sqrt{2 + x}$

**51.** $y = \dfrac{-2 \pm \sqrt{2x - 2}}{2}$

**53.** Slide left 3 units

**55.** Slide up 4 units

**57.** A slide down 4 units followed by a slide right 2 units.

**59.** A reflection in the $y$-axis followed by a slide up 6 units followed by a slide right 12 units.

**61.** 95.063 ft

**SECTION 10.4**

**1.** Center $(0, 0)$; vertices $(\pm 3, 0)$

**3.** Center $(0, 0)$; vertices $(0, \pm 4)$

**5.** Center $(0, 0)$; vertices $(0, \pm 9)$

**7.** Center $(0, 0)$; vertices $(\pm 6, 0)$

**9.** Right and left; vertices of f. r. $(4, \pm 6)$, $(-4, \pm 6)$; asymptotes $y = \pm \frac{3}{2}x$

**11.** Up and down; vertices of f. r. $(1, \pm 5)$, $(-1, \pm 5)$; asymptotes $y = \pm 5x$

**13.** Right and left; vertices of f. r. $(2, \pm 3)$, $(-2, \pm 3)$; asymptotes $y = \pm \frac{3}{2}x$

**15.** Up and down; vertices of f. r. $(4\sqrt{2}, \pm 3\sqrt{2})$, $(-4\sqrt{2}, \pm 3\sqrt{2})$; asymptotes $y = \pm \frac{3}{4}x$

**17.** $\dfrac{x^2}{9} - \dfrac{y^2}{81} = 1$; vertices of f. r. $(3, \pm 9)$, $(-3, \pm 9)$

**19.** $\dfrac{y^2}{36} - \dfrac{x^2}{144} = 1$; vertices of f. r. $(12, \pm 6)$, $(-12, \pm 6)$

**21.** $\dfrac{x^2}{\frac{1}{4}} - \dfrac{y^2}{1} = 1$; vertices of f. r. $\left(\frac{1}{2}, \pm 1\right)$, $\left(-\frac{1}{2}, \pm 1\right)$

**23.** $\dfrac{y^2}{1} - \dfrac{x^2}{1} = 1$; vertices of f. r. $(1, \pm 1)$, $(-1, \pm 1)$

**25.** Center $(3, 1)$; vertices $(5, 1)$, $(1, 1)$

**27.** Center $(2, -1)$; vertices $(2, 8)$, $(2, -10)$

**29.**

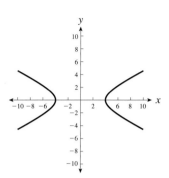

**31.**

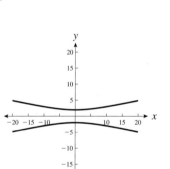

**33.**

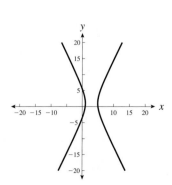

**35.**

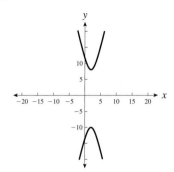

**37.**

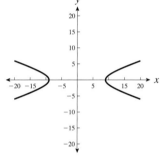

**39.**

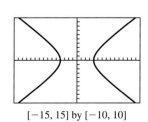

**41.**

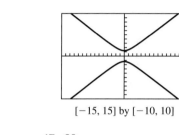

[−15, 15] by [−10, 10]

**43.**

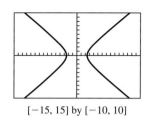

[−15, 15] by [−10, 10]

**45.**

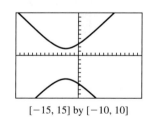

[−15, 15] by [−10, 10]

**47.** 90 m

**49.** Written Exercise

**51.**

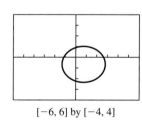

[−15, 15] by [−10, 10]

**53.**

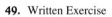

[−6, 6] by [−4, 4]

**55.** Circle

**57.** Hyperbola

**59.** Ellipse

**61.** Parabola

**63.** Parabola

## SECTION 10.5

**1.** 2

**3.** 0

**5.** 4

**7.** $(-1.791, 0.209)$, $(2.791, 4.791)$

**9.** $(3, 4), (-2, -1)$

**11.** $(2.712, 0.356)$, $(-2.212, -2.106)$

**13.** $(3.606, 3.464), (-3.606, 3.464), (3.606, -3.464), (-3.606, -3.464)$

**15.** $(4, 9), (-3, 2)$

**17.** $(3\sqrt{2}, 3\sqrt{2}), (-3\sqrt{2}, -3\sqrt{2})$

**19.** $(-4, -3), (3, 4)$

**21.** $(12, 10), (-12, 10), (12, -10), (-12, -10)$

**23.** $(4.442, 4.729), (-4.442, 4.729), (3.282, -4.229), (-3.282, -4.229)$

**25.** Written Exercise; 0 to 4.

**27.** $-12, -20$ or $20, 12$

**29.** $x = 12, y = 5$

**31.** $L = 3, W = 4$ or $L = 4, W = 3$

**33.** 4

**35.** Written Exercise

**37.** A slide left of 1.5 units followed by a vertical stretch of 3 units followed by a slide down of 20.25 units.

## REVIEW EXERCISES

**1.** $(x - 2)^2 + (y + 3)^2 = 100$

**2.** $(x + 4)^2 + (y - 5)^2 = 36$

**3.** Center $(0, 4)$; $r = 2$

**4.** Center $(5, -2)$; $r = \sqrt{15}$

**5.** Center $(-10, -7)$; vertices $(-10, -2), (-10, -12), (-3, -7), (-17, -7)$

**6.** Center $(-2, 0)$; vertices $(7, 0), (-11, 0), (-2, 10), (-2, -10)$

**7.** Vertex $(-6, -2)$; line of symmetry $y = -2$

**8.** Vertex $(2, 1)$; line of symmetry $y = 1$

**9.** Center $(11, 0)$; vertices $(11, \pm 3)$

**10.** Center $(-7, 9)$; vertices $(-12, 9), (-2, 9)$

**11.** Parabola

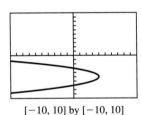
$[-10, 10]$ by $[-10, 10]$

**12.** Circle

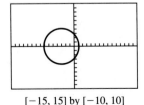
$[-15, 15]$ by $[-10, 10]$

**13.** Hyperbola

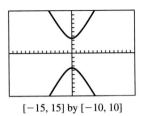
$[-15, 15]$ by $[-10, 10]$

**14.** Ellipse

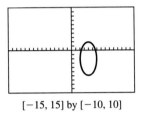
$[-15, 15]$ by $[-10, 10]$

**15.** Ellipse

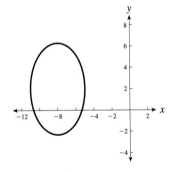

**16.** Hyperbola

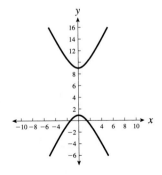

**17.** Circle

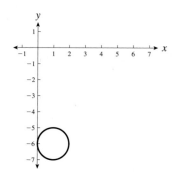

**18.** Parabola

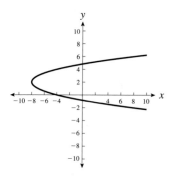

**19.** $(x + 3)^2 + (y + 6)^2 = 6^2$

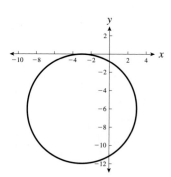

**20.** $\dfrac{(x - 1)^2}{16} + \dfrac{(y + 6)^2}{4} = 1$

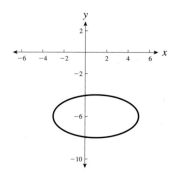

**21.** $\dfrac{x^2}{25} - \dfrac{(y - 2)^2}{4} = 1$

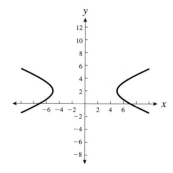

**22.** $y = 2(x - 1)^2 - 9$

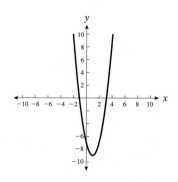

**23.** $\dfrac{x^2}{4} + \dfrac{(y - 3)^2}{25} = 1$

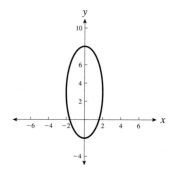

**24.** $(x - 4)^2 + y^2 = 41$

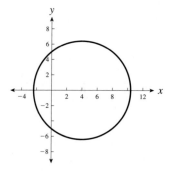

**25.** $x = (y + 8)^2 - 4$

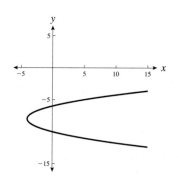

**26.** $\dfrac{x^2}{7} - \dfrac{y^2}{10} = 1$

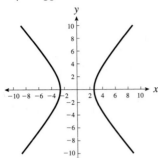

**27.** $h = 8$ or $h = -2$

**28.** $\dfrac{x^2}{36} + \dfrac{y^2}{81} = 1$

**29.** $a = -2$

**30.** Vertices of f. r. $(-2, 5)$, $(6, 5)$, $(6, -7)$, $(-2, -7)$; asymptotes $y = \frac{3}{2}x - 4$, $y = -\frac{3}{2}x + 2$

**31.** Two intersections

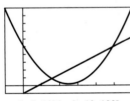

$[-2, 12]$ by $[-10, 100]$

**32.** Four intersections

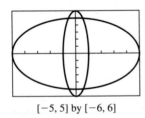

$[-5, 5]$ by $[-6, 6]$

**33.** $(-0.739, -1.784)$, $(-4.061, 8.184)$

**34.** $(1.351, 0.649)$, $(-1.851, 3.851)$

**35.** $(3.000, -4.000)$, $(2.657, -4.236)$

**36.** $(2, \pm\sqrt{20})$, $(-2, \pm\sqrt{20})$

**37.** $(0, 2)$, $\left(-\dfrac{100}{29}, -\dfrac{42}{29}\right)$

**38.** $(3, 0)$, $(4, 1)$

**39.** 15 ft

**40.** a) $\begin{cases} x - y = 23 \\ x^2 - y^2 = 2277 \end{cases}$,
where $x$ and $y$ are integers

    b) 38 and 61

## CHAPTER 10 TEST

**1.** Center $(-4, 5)$; $r = 4$

**2.** Center $(8, -6)$; vertices $(2, -6)$, $(14, -6)$, $(8, 2)$, $(8, -14)$

**3.** Vertex $(-4, 3)$, line of symmetry $y = 3$

**4.** Center $(2, -1)$; vertices $(-5, -1)$, $(9, -1)$

**5.** Ellipse

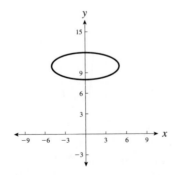

**6.** Circle

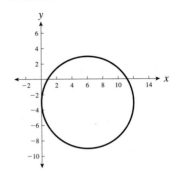

**7.** Parabola

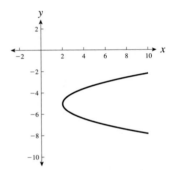

**8.** Hyperbola

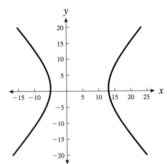

**9.** $(x - 2)^2 + (y + 3)^2 = 16$

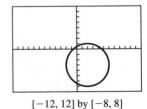

$[-12, 12]$ by $[-8, 8]$

**10.** $\dfrac{x^2}{25} + \dfrac{y^2}{100} = 1$

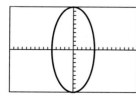

$[-15, 15]$ by $[-10, 10]$

**11.** $x = (y + 6)^2 - 11$

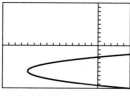

$[-15, 5]$ by $[-10, 10]$

**12.** $\dfrac{(y + 1)^2}{4} - \dfrac{x^2}{4} = 1$

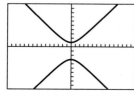

$[-15, 15]$ by $[-10, 10]$

**13.** $(4.270, 4.151), (1, 0)$

**14.** $(0, 1), \left(\frac{8}{5}, -\frac{3}{5}\right)$

**15.** Vertices of f. r. $(5, 1), (-5, 1),$
$(5, -11), (-5, -11)$; asymptotes
$y = \pm\frac{6}{5}x - 5$

---

## Chapter 11

### SECTION 11.1

**1.**

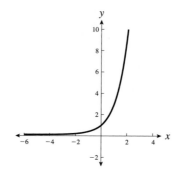

**3.**

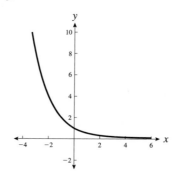

**5.** $x = 2$

**7.** $x = 3$

**9.** $x = -1$

**11.** $x = 3$

**13.** $x = 3$

**15.** 8

**17.** $\frac{1}{3}$

**19.** 101

**21.** 1000.001

**23.** $(0, 1)$

**25.** Functions a and c

---

**27.** Graph (c)

**29.** Graph (d)

**31.** Functions a and c

**33.** Functions a and b

**35.** 52,000 in 1991; 54,080 in 1992;
60,833 in 1995

**37.** a) $P(t) = 325,000(1.025)^t$
b) 45.517 yr

**39.** a) \$6700.48
b) 14.207 yr

**41.** a)

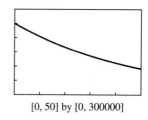

$[0, 50]$ by $[0, 300000]$

b) 45.355 yr

**43.** a) \$16,500
b) \$5220.70
c) $22,000(0.75)^t$ dollars

**45.** $5^x = 15$

**47.** $f(x) = 2x + 5$

**49.** $(2, 4), (4, 16), (-0.767, 0.588)$

**51.** 23,623

**53.** Written Exercise

**55.** \$12,075.00

**57.** A 50% increase; on most copiers
this is a 150% setting

### SECTION 11.2

**1.** One-to-one

**3.** Not one-to-one

**5.** Not one-to-one

**7.** One-to-one

**9.** Not one-to-one

**11.** Not one-to-one

**13.** $f^{-1}(x) = \dfrac{x-3}{2}$

**15.** $f^{-1}(x) = -\dfrac{x-8}{3}$

**17.** $f^{-1}(x) = \dfrac{1}{x}$

**19.** $f^{-1}(x) = \sqrt[3]{x-6}$

**21.**

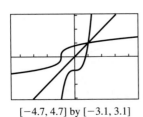

$[-4.7, 4.7]$ by $[-3.1, 3.1]$

**23.**

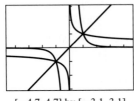

$[-4.7, 4.7]$ by $[-3.1, 3.1]$

**25.** a) Yes

   b) Yes

   c) $(-4, -4)$

**27.**

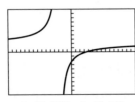

$[-15, 15]$ by $[-10, 10]$

$f^{-1}(x) = \dfrac{3x+7}{2-x}$

**29.**

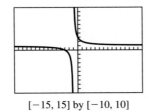

$[-15, 15]$ by $[-10, 10]$

$f^{-1}(x) = \dfrac{x-3}{1-x}$

**31.** The graph of $y = x^3$, the graph of the inverse of $y = x^3$, and the graph of the line $y = x$

**33.** $f(4) = 21$

**35.** $f(5) - g(3) = 27$

**37.** $f(x+h) = x^2 + 2xh + h^2 + 5$

**39.** $f(x+h) - f(x) = 2xh + h^2$

**SECTION 11.3**

**1.** $2^3 = 8$

**3.** $3^2 = 9$

**5.** $10^3 = 1000$

**7.** $\left(\frac{1}{2}\right)^4 = \frac{1}{16}$

**9.** $5^{-3} = \frac{1}{125}$

**11.** $\log_{10} 100 = 2$

**13.** $\log_3 81 = 4$

**15.** $\log_5 125 = 3$

**17.** $\log_9 729 = 3$

**19.** 4

**21.** 3

**23.** $-3$

**25.** $-3$

**27.** $-3$

**29.** 1.544

**31.** 0.512

**33.** 3.219

**35.** 1.447

**37.** $x = 64$

**39.** $x = 3$

**41.** $x = 25$

**43.** $x = 8$

**45.** 244,084 bacteria

**47.** 21st birthday

**49.** $\log_5 x = 3$

**51.** $\log_{10} 100 = 2$

**53.** 3

**55.** 100,000 times

**57.** The graphs are reflections of each other through the line $y = x$.

**59.** $2^{x+5}$

**61.** $3^{2x-10}$ or $9^{x-5}$

**63.** $5^{2x+6}$ or $25^{x+3}$

**SECTION 11.4**

**1.** 16

**3.** 0

**5.** 1

**7.** $\log_2 3 + \log_2 x$

**9.** $\log_3 5 + \log_3(x+3)$

**11.** $\log_2 x + 2\log_2 y$

**13.** $2\log_5 x - 3\log_5 y$

**15.** $\log_4 x - \log_4 9$

**17.** $2\log_8 x$

**19.** $\log_4 5000 + 5\log_4 x$

**21.** $\log_3 5 + 6\log_3 x - \log_3 2 - 3\log_3 y$

**23.** $\log_2 xy^5$

**25.** $\log_3\left(\dfrac{x-4}{x+5}\right)$

**27.** $\log_4 125^2(x+2)^5$

**29.** False

**31.** True

**33.** True

**35.** True

**37.** True

**39.** $\neq$

**41.** $=$

**43.** $=$

**45.** $\neq$

**47.** $x = 9$

**49.** $x = 5$

**51.** $x = 4$

**53.** $x = \pm\frac{1}{3}$

**55.** $\log_5 y = x$

**57.** $\log_3 xy = x$

**59.** $\log_2\left(\dfrac{x}{y}\right) = x$

**61.** $\log_3(x+y) = 4$

**63.** $3^2 = 9$

**65.** $5^{-3} = \dfrac{x}{y}$

**67.** $\log(2x) = \log 2 + \log x$

**69.** $\log x^3 = 3\log x$

**71.** Answers will vary, basically $3 = 3$

**73.** Written Exercise, $10^x > 0$

**75.** Written Exercise

**77.** $x = 315$

**79.** $x = 10, x = 22$

SECTION 11.5

**1.** $x = \frac{3}{2}$

**3.** $x = 14$

**5.** $x = 4$

**7.** $x = 625$

**9.** $x = 4$

**11.** $x = 500$

**13.** $x = 0.111$

**15.** $x = 14.275$

**17.** $x = 3.520$

**19.** $x = 0.831$

**21.** $x = \dfrac{\log 5}{\log 5 - \log 3}$, approx. 3.151

**23.** $x = \dfrac{\log 46}{\log 2}$, approx. 5.524

**25.** $x = \dfrac{\log 7}{\log 7 - 2\log 3}$, approx. $-7.743$

**27.** $x = \dfrac{2\log 6}{\log 6 - \log 4}$, approx. 8.838

**29.** $x = \dfrac{\log 7}{\log 5 + 2}$, approx. 0.313

**31.** $x = 2$

**33.** 14.207 yr

**35.** 29.897 hr

**37.** 13.863 yr

**39.** 2.322

**41.** $-3$

**43.** $-1.771$

**45.** $-0.252$

**47.** $x = 0.250$

**49.** The point in common is $(1, 0)$.

**51.** $x = 19.746$

**53.** $x = -1.465$

REVIEW EXERCISES

**1.** $x = 2$

**2.** $x = -2$

**3.** $x = -2$

**4.** $x = 6$

**5.** Functions a and c

**6.** Functions b and c

**7.** One-to-one

**8.** One-to-one

**9.** $f^{-1}(x) = -\frac{1}{7}x + \frac{5}{7}$

**10.** $f^{-1}(x) = -\frac{1}{4}x - 2$

**11.** $f^{-1}(x) = x^3 + 1$

**12.** $f^{-1}(x) = \sqrt[3]{x} + 3$

**13.** $5^3 = x$

**14.** $x^2 = 4$

**15.** $2^6 = 3x + 4$

**16.** $\log_7 2401 = 4$

**17.** $\log_3 20 = x$

**18.** $\log_2(3x - y) = 4$

**19.** $-2$

**20.** 2

**21.** 2.000

**22.** $-0.301$

**23.** $x = 18$

**24.** $x = 9$

**25.** 4

**26.** 0

**27.** $\log_8 7 + 2\log_8 x - \log_8 3$

**28.** $\log_3(a-1) - \log_3(a+1)$

**29.** $\log_7(a^2 b^3)$

**30.** $\ln\left(\dfrac{x^3}{3}\right)$

**31.** False

**32.** True

**33.** $x = 122$

**34.** $x = 4$

**35.** $x = \dfrac{\log 10 - \log 3}{2\log 3}$,
approx. 0.548

**36.** $x = \dfrac{\log 4}{\log 4 - \log 5}$,
approx. −6.213

**37.** $x = 0.843$

**38.** $x = 1.875$

**39.** a) \$4317.85

   b) 9.006 yr

**40.** 284,632 bacteria

CHAPTER 11 TEST

**1.** $x = \frac{5}{2}$

**2.** Functions b and c

**3.** $f^{-1}(x) = 2x + 12$

**4.** $8^2 = (x-1)^2$

**5.** $\log_3 729 = 6$

**6.** 2

**7.** $x = 4$

**8.** $\log_5 2 + 6\log_5 x$

**9.** $\log(50x)$

**10.** $x = 20$

**11.** $x = \dfrac{\log 3}{2\log 2 - \log 3}$,
approx. 3.819

**12.** $x = \sqrt{30}$

**13.** $x = 1.500$

**14.** 26,356

**15.** 6.439 yr

# Chapter 12

SECTION 12.1

**1.** 5, 7, 9, 11, 13

**3.** 35, 30, 25, 20, 15

**5.** 0.5, 2, 3.5, 5, 6.5

**7.** 17, 21, 25, 29, 33

**9.** 6, 7.5, 9, 10.5, 12

**11.** $x+9$, $x+12$, $x+15$, $x+18$, $x+21$

**13.** $a_7 = 46$

**15.** $a_{12} = -82$

**17.** $a_{20} = 29$

**19.** $a_{25} = 204$

**21.** $a_{43} = -341$

**23.** $a_9 = 38.5$

**25.** $S_{25} = 1200$

**27.** $S_{12} = 108$

**29.** $n = 10$

**31.** $3 - 4\sqrt{3}$

**33.** 36, 42, 48, ...

**35.** When $n = 15$ or $n = 16$,
$S_n = 240$

**37.** When $n = 6$ or $n = 7$,
$S_n = 420$

**39.** 5050

**41.** 2550

**43.** 1001000

**45.** 2550

**47.** $a_n = -15 + 8(n-1)$

**49.** 231

**51.** \$125,000

**53.** \$9600

**55.** Written Exercise

**57.** Written Exercise

**59.** a) −8

   b) 55

   c) When $n = 7$ $S_7 = 70$

   d) The first term

**61.** $\dfrac{1}{x^4}$

**63.** $\dfrac{1}{x^m}$

**65.** $x^3$

**67.** $x^{2m}$

**69.** $x^2$

SECTION 12.2

**1.** −5, −10, −20, −40, −80

**3.** 3, −1.5, 0.75, −0.375, 0.1875

**5.** 0.5, 0.25, 0.125, 0.0625, 0.03125

**7.** 2, 1, $\frac{1}{2}$, $\frac{1}{4}$, $\frac{1}{8}$

**9.** $\frac{1}{16}$, $\frac{1}{32}$, $\frac{1}{64}$, $\frac{1}{128}$, $\frac{1}{256}$

**11.** $x(x+1)^3$, $x(x+1)^4$, $x(x+1)^5$, $x(x+1)^6$, $x(x+1)^7$

**13.** 10,935

**15.** 10, 240

**17.** $-2.621 \times 10^{-14}$

**19.** $2.384 \times 10^{-7}$

**21.** $-1.137 \times 10^{-12}$

**23.** 172.995

**25.** $y_1 = 3 \cdot 4^{x-1}$

**27.** 24.000

**29.** $-6825$

**31.** 8

**33.** $\frac{5}{9}$

**35.** 18, 27, 40.5, 60.75, 91.125

**37.** 0.999

**39.** 5651.369

**41.** a) \$16,777,216

b) \$33,554,431

**43.** 40,203 customers

**45.** 10 g of substance

**47.** a) 12.96 cm

b) 130.56 cm

**49.** Written Exercise

**51.** 3280, 8 terms

**53.** $y = 12$

**55.** $9x^2 - y^2$

**57.** $x^2 + 24x + 144$

**59.** $49x^2 + 14x + 1$

**61.** $x^2 + 4xy + 4y^2 + 8x + 16y + 16$

## SECTION 12.3

**1.** 1, 7, 21, 35, 35, 21, 7, 1

**3.** $x^4 + 4x^3y + 6x^2y^2 + 4xy^3 + y^4$

**5.** $x^3 - 6x^2y + 12xy^2 - 8y^3$

**7.** $32x^5 + 400x^4y + 2000x^3y^2 + 5000x^2y^3 + 6250xy^4 + 3125y^5$

**9.** 120

**11.** 10

**13.** 56

**15.** False

**17.** True

**19.** False

**21.** False

**23.** False

**25.** 10

**27.** 220

**29.** 120

**31.** $286x^{10}y^3$

**33.** $60x^2$

**35.** $-640x^4y$

**37.** $2835x^3y^4$

**39.** $a^7 + 7a^6b + 21a^5b^2 + 35a^4b^3 + 35a^3b^4 + 21a^2b^5 + 7ab^6 + b^7$

**41.** $x^5 + 10x^4y + 40x^3y^2 + 80x^2y^3 + 80xy^4 + 32y^5$

**43.** $16x^4 - 160x^3 + 600x^2 - 1000x + 625$

**45.** $729x^6 + 2916x^5y + 4860x^4y^2 + 4320x^3y^3 + 2160x^2y^4 + 576xy^5 + 64y^6$

**47.** $x^2 + 20\sqrt{x^3} + 150x + 500\sqrt{x} + 625$

**49.** $x^3 - 6\sqrt{x^5y} + 15x^2y - 20\sqrt{x^3y^3} + 15xy^2 - 6\sqrt{xy^5} + y^3$

**51.** $x = -8, 1$

**53.** $x \le -8$ or $x \ge 1$

**55.** $x = \frac{7}{2}, x = -\frac{1}{3}$

**57.** $-\frac{1}{3} < x < \frac{7}{2}$

## SECTION 12.4

**1.** 2520

**3.** $5.449 \times 10^{10}$

**5.** 35

**7.** 1820

**9.** 27 codes

**11.** 10,000 ways

**13.** 40,320 ways

**15.** 160 codes

**17.** 17,576,000 plates

**19.** 175,560 teams, if order matters; 7315, if not

**21.** $6.967 \times 10^{10}$ hands

**23.** $3.917 \times 10^{10}$ committees

**25.** $5.493 \times 10^{26}$ ways

**27.** 10,000 zip codes

**29.** $\dfrac{1 + 2x}{4x - 1}$

**31.** $x = 3 \pm \sqrt{6}$

**33.** $x = \frac{3}{4}, x = -\frac{3}{2}$

## REVIEW EXERCISES

**1.** 9, 10.5, 12

**2.** $-15, -19, -23$

**3.** $-24, 48, -96$

**4.** 363, 121, $\frac{121}{3}$

**5.** 8, 5, 2, $-1, -4$

**6.** 1, 1.6, 2.2, 2.8, 3.4

**7.** 5, 50, 500, 5000, 50000

**8.** $-2, -8, -32, -128, -512$

**9.** 20.4

**10.** 1166

**11.** 6,442,450,944

**12.** $\frac{125}{64}$

**13.** 130

**14.** 1020

**15.** 299,593

**16.** 2.500

**17.** $\frac{2}{3}$

**18.** $n = 8$

**19.** $n = 77$

**20.** 1   6   15   20   15   6   1

**21.** $32x^5 + 80x^4y + 80x^3y^2 + 40x^2y^3 + 10xy^4 + y^5$

**22.** $a^6 - 18a^5 + 135a^4 - 540a^3 + 1215a^2 - 1458a + 729$

**23.** $x^4 - 8ax^3 + 24a^2x^2 - 32a^3x + 16a^4$

**24.** $2187x^7 + 5103x^6 + 5103x^5 + 2835x^4 + 945x^3 + 189x^2 + 21x + 1$

**25.** $96,000x^4$

**26.** $-48,384a^5b^3$

**27.** 210

**28.** $\frac{20}{3}$

**29.** 462

**30.** 56

**31.** 45

**32.** 1,814,400

**33.** True

**34.** True

**35.** $184,467,440,700,000,000

**36.** a) 8.192 cm

   b) 96.482 cm

**37.** 2600 codes

**38.** 60 arrangements

**39.** 28 choices

## CHAPTER 12 TEST

**1.** $\frac{4}{3}, \frac{2}{3}, \frac{1}{3}, \frac{1}{6}, \frac{1}{12}$

**2.** 100, 70, 40, 10, $-20$

**3.** 9, 13.5, 20.25, 30.375, 45.5625

**4.** 552

**5.** 8,388,608

**6.** 3530

**7.** 1,048,575

**8.** 25

**9.** 1300

**10.** $d = 5$

**11.** $x^6 + 18x^5 + 135x^4 + 540x^3 + 1215x^2 + 1458x + 729$

**12.** $1080x^3$

**13.** 5040

**14.** a) $36,120

   b) $285,600

**15.** 24 ways

# INDEX

parabola with horizontal line of symmetry, 635-637, 660
point plotting method, 88
point-plotting method, 124, 136
problem situation representations, 179-193
quadratic functions, 532-554
on rectangular coordinate plane, 76
to solve absolute value inequalities, 290
to solve problem situations, 181-184
to solve systems of equations, 193-206, 225
to support algebraic solutions for inequalities, 236-238
to support algebraic solutions to radical equations, 483-485
to support polynomial operations, 311-312
to support product rule for exponents, 297-298
to support quotient rule for exponents, 301
to support simplification of exponential expressions, 299
systems of inequalities, 281-284, 291
vertical and horizontal lines, 96-97
graphing calculator (*see* grapher)
greater-than inequality, 266
absolute value, 289
greatest common factor (GCF)
reducing numerical fractions, 359-360
guess-and-check strategy
for solving equations, 35-36, 71

**H**

halving function, 669
Heron's formula, 450
higher-order equations, 569-613
horizontal line, 137
graphing, 96-97
horizontal line test, 679, 710
horizontal slide
other transformations combined, 546-549
quadratic function graphing, 537-540
squaring function, 534-535, 565
vertical slide combined, 535-537
hyperbola, 640-651, 660-662
asymptotes, 642, 661
center, 641, 660
centered at origin
graphing, 642-645

defined, 640, 660
foci, 640, 660
fundamental rectangle, 661
centered at origin, 642
illustrated, 615, 641-642
not centered at origin
graphing, 645-649, 662
standard form, centered at origin, 641
standard form, not centered at origin, 646, 661
vertices, 641, 661
hypotenuse
of right triangle, 448

**I**

*i*, 490-491, 502
identity matrices, 591-592, 611
imaginary number, 492, 502
inconsistent system of equations, 203, 225
independent system of equations, 203, 225
index of radical, 446
inequality (*see also* linear inequality; systems of inequalities)
absolute value, 265-267
greater-than, 289
less-than, 289
representations, 266-267
solved algebraically, 267-269
solved graphically, 290
algebraic representations, 240-241
algebraic solutions supported graphically, 236-238, 287
compound, 254, 275
confirming solutions, 234
defined, 275
double, 254, 275
equivalent
to solve linear inequalities, 40-42
greater-than, 266
interval notation, 233, 246
less-than, 265-266
linear, in two variables, 272-280, 290
with non-integer endpoints, solved, 248
nonlinear, systems of, 284-285
problem situations solved, 249-251
real number line intervals, 8-9
solution, 70, 234
solution methods, 34-44
solved algebraically, 71, 287
solved using addition property of

inequality, 235
solved using division property, 235
solved using more than one property, 236
solved with *x*-intercept method, 245-248, 288
special cases, 238-239, 288
symbols, 7, 233
systems of, 280-287
visual representation, 246
infinite geometric series, 730-732, 753
sum, defined, 731
infinite sequence, 715, 751
inflation
algebraic representation, 162
grapher analysis, 162-164
graphical representation, 162
integer, 2, 67
integer window
on grapher, 78-79, 138
intercept
found algebraically, 95-96
intercept method
of graphing linear equation, 95-96, 137
intersection
of compound inequality solutions, 257, 288
interval notation
absolute value inequality solutions, 266
compound inequalities, 254
double inequalities, 255
finite intervals, 8, 68
inequalities, 233, 246
infinite intervals, 8, 68
inverse function, 678-685, 710
defined, 679, 710
finding, 680-683, 710
graphing, 683
inverse matrices, 592-595, 611
defined, 592
inverse of 2x2 matrices, 593
inverse of *nxn* matrix, 594
inverse operations, 16
invertible matrix, 593, 611
irrational numbers, 4, 67
decimal approximations, 448
*e*, 688
isosceles triangle, 81

**L**

LCD (*see* least common denominator